	IB	IIB	IIIA	IVA	VA	VIA	VIIA	O
								2 2 He 4.0026
			2 3 5 B 10.811	2 4 6 C 12.01115	2 5 7 N 14.0067	2 6 8 O 15.9994	2 7 9 F 18.9984	2 8 10 Ne 20.183
			2 8 3 13 Al 26.9815	2 8 4 14 Si 28.086	2 8 5 15 P 30.9738	2 8 6 16 S 32.064	2 8 7 17 Cl 35.453	2 8 8 18 Ar 39.948
28 Ni 58.71	2 8 18 1 29 Cu 63.546	2 8 18 2 30 Zn 65.37	2 8 18 3 31 Ga 69.72	2 8 18 4 32 Ge 72.59	2 8 18 5 33 As 74.9216	2 8 18 6 34 Se 78.96	2 8 18 7 35 Br 79.904	2 8 18 8 36 Kr 83.80
46 Pd 106.4	2 8 18 18 1 47 Ag 107.868	2 8 18 18 2 48 Cd 112.40	2 8 18 18 3 49 In 114.82	2 8 18 18 4 50 Sn 118.69	2 8 18 18 5 51 Sb 121.75	2 8 18 18 6 52 Te 127.60	2 8 18 18 7 53 I 126.9044	2 8 18 18 8 54 Xe 131.30
78 Pt 195.09	2 8 18 32 18 1 79 Au 196.967	2 8 18 32 18 2 80 Hg 200.59	2 8 18 32 18 3 81 Tl 204.37	2 8 18 32 18 4 82 Pb 207.19	2 8 18 32 18 5 83 Bi 208.980	2 8 18 32 18 6 84 Po (210)	2 8 18 32 18 7 85 At (210)	2 8 18 32 18 8 86 Rn (222)

64 Gd 157.25	2 8 18 26 9 2 65 Tb 158.924	2 8 18 28 8 2 66 Dy 162.50	2 8 18 29 8 2 67 Ho 164.930	2 8 18 30 8 2 68 Er 167.26	2 8 18 31 8 2 69 Tm 168.934	2 8 18 32 8 2 70 Yb 173.04	2 8 18 32 9 2 71 Lu 174.97
96 Cm (247)	2 8 18 32 26 9 2 97 Bk (247)	2 8 18 32 27 9 2 98 Cf (249)	2 8 18 32 28 9 2 99 Es (254)	2 8 18 32 29 9 2 100 Fm (253)	2 8 18 32 30 9 2 101 Md (256)	102 No (254)	103 Lr (257)

chemistry for health-related sciences

Curtis T. Sears, Jr. / Conrad L. Stanitski
Georgia State University

Prentice-Hall, Inc., Englewood Cliffs, New Jersey

chemistry for health-related sciences

CONCEPTS AND CORRELATIONS

Library of Congress Cataloging in Publication Data

SEARS, CURTIS T
Chemistry for health-related sciences.

Includes index.
1. Chemistry. I. Stanitski, Conrad L., joint author. II. Title. [DNLM: 1. Chemistry. QD31.2 S439c]
QD31.2.S43 540 75-20283
ISBN 0-13-129429-6

chemistry for health-related sciences concepts and correlations

Curtis T. Sears, Jr./Conrad L. Stanitski

10 9 8 7 6 5 4 3 2

Printed in the United States of America

The cover photograph, courtesy of Dr. Roman Vishniac, is an interference microscopy photomicrograph showing living human skin cells. The original magnification was 460 times; for use on the cover, the subject has been further enlarged photographically to 5700 times.

PRENTICE-HALL INTERNATIONAL, INC., *London*
PRENTICE-HALL OF AUSTRALIA PTY. LIMITED, *Sydney*
PRENTICE-HALL OF CANADA, LTD., *Toronto*
PRENTICE-HALL OF INDIA PRIVATE LIMITED, *New Delhi*
PRENTICE-HALL OF JAPAN, INC., *Tokyo*
PRENTICE-HALL OF SOUTH-EAST ASIA PRIVATE LIMITED, *Singapore*

With love to our children

Amy

Beth

Leslie

Susan

Science is built up of facts as a house is with stones, but a collection of facts is no more a science than a heap of stones a house.

Poincaré (1854–1912)

contents

3
chemical bonding

4
isomerism

5
solutions

6
reaction dynamics

7
introduction to equilibrium

8
acid-base equilibria

9
alcohols and their derivatives

10
acid derivatives

11
amine derivatives

12
energy and biochemical reactions

13
carbohydrate metabolism

14
protein metabolism

15
hormones and vitamins

16
body fluids

17
drugs—chemical agents in disease treatment

appendix one
mathematics review

appendix two
quantum mechanical model of the atom

appendix three
answers to in-chapter exercises

to the reader

Our purpose in writing this text is to provide one specifically designed for a chemistry course for students of the allied health sciences. Topics are introduced and discussed around the unifying theme that chemical structure and biological function are related. The thrust is to use biochemically significant substances and phenomena as a framework to describe chemical principles. Therefore, we have chosen to present only those topics most directly related to understanding biochemical processes, rather than attempt a broad introductory coverage of chemistry. Consequently, some topics usually included in an introductory chemistry course have been shortened in scope or omitted. Among omitted topics are: gas laws, electrolysis, galvanic cells, chemical behavior of individual elements, and industrial processes. For the most part, we have not dwelt upon the historical development of present chemical theories.

The basic principles of atomic structure are established early. Immediately after discussion of atomic structure, all aspects of bonding are considered. The theme of electron transfer focuses the relationships between ion formation, aggregation, and oxidation–reduction. In addition to the usual topics presented under the heading of electron sharing, the ideas and consequences of structural and geometric isomerism are brought forth at this time. The establishment of this working background makes possible the early introduction of biologically important examples. We have attempted to apply chemical principles and show the reasons why health care personnel need to be familiar with them. As a result, the organization and integrated treatment of topics is unusual by comparison with many other texts. For example, in this text: (1) enzymatic catalysis is included within the discussion of reaction rates; (2) oxygen–hemoglobin binding is incorporated into the discussion of acid-base equilibria and buffer action; (3) the studies of aqueous equilibria and factors affecting them are related to diabetes mellitus and respiratory acidosis/alkalosis.

In presenting the nature of biochemically-significant materials, the text continuously uses principles established earlier. It incorporates features of structure and reactivity as they relate to metabolic interactions, drug action, and hereditary diseases. We believe the approach has a sound basis in terms of chemical coherence and avoids the traditional and arbitrary division of chemistry under the headings inorganic, organic, and biological chemistry.

The text is designed to consolidate and build upon principles normally treated in high school chemistry courses, although high school chemistry is not a prerequisite.

We have interspersed exercises throughout each chapter. Each exercise relates to the section under consideration. In this way, the exercises provide continuous points at which understanding of previous material can be checked. End-of-chapter problems of a broader nature are also included. Their solutions require the successful coordination of several concepts and thus serve to intergrate previous material with the chapter under study.

A laboratory program has been developed to accompany this text. This program uses qualitative and quantitative analyses of the body fluids, urine and serum, to amplify the importance of stoichiometry and equilibrium.

We would like to thank the many individuals who have helped bring our ideas to fruition. Special thanks go to the staff at Prentice-Hall, particularly our editor Harry McQuillen for his expertise, coffee, and skyscraper tour. Thanks also to our copy editor Betty Adam for recalling to us principles of freshman English, to Barbara N. Bartlett for her continuous optimism and cheerfulness, and to our typist Marilyn Chapman for deciphering our scratchings. We appreciate the efforts of all our reviewers for their critical comments and suggestions, especially Lawrence Wilkins (Santa Monica College) and Chuck Rose (University of Nevada, Reno). Our families deserve special mention for tolerating missed meals, absence, and occasional displays of maltemper.

Most of all we are indebted to the Georgia State University Chemistry 101 and 102 students who inspired us to undertake this project. These marvelous people are from the allied health programs at Georgia State and from Crawford Long, Grady, and Piedmont Hospitals' Schools of Nursing. If their cooperative spirit shown during this effort translates to their health care practices, getting sick won't be that bad after all.

Throughout the text we have strived to put chemistry into a medical perspective. We solicit your opinions on our efforts.

C.T.S.
C.L.S.
Atlanta, Ga.

1
measurements and descriptions

INTRODUCTION

Alchemy, man's ill-fated pursuit of converting other metals into gold, was the basis from which modern chemistry has evolved through about the last 250 years. During this evolution, certain fundamental descriptions, relationships, and concepts have developed. This chapter is an overview of some of these, many of which are already familiar. You are aware that the stuff you and everything else is made of is called *matter*. Chemistry is the study of the changes that matter undergoes and the energy relationships accompanying these changes.

MEASUREMENT

Everyday experiences relate the need to be able to measure four *basic* quantities: time, temperature, length (distance), and mass (weight).* Other measurements can be derived from these. For example,

$$\text{length} \times \text{length} \times \text{length} = \text{volume} = \text{length}^3$$

$$\frac{\text{length}}{\text{time}} = \text{velocity (speed)}$$

$$\frac{\text{mass}}{\text{volume}} = \frac{\text{mass}}{\text{length}^3} = \text{density}$$

Units are part of a measurement.

The metric system was developed to unify weights and measures used in trade among European nations. It was not specifically developed for scientific work.

Most measurements consist of two parts: a number, which tells the amount of the quantity measured, *and* a unit, which tells what quantity is being measured. For example, you would not say a person is 5 pounds tall or weighs 72 inches. The units would be inappropriate to describe the measurement desired.

Height, weight, body temperature, the distance between Boston and San Francisco, and an 8:00 P.M. date all have meaning because some time ago agreement was reached on a system of units for measurement. The exact value for each of these measurements really depends on an arbitrary, defined reference point. The metric system was developed (1799) by a French committee to create a *uniform* system of weights and measures and has been adopted essentially worldwide.†

THE METRIC SYSTEM

The basic units of the metric system‡ are the *gram* for *mass* measurement and the *meter* for *length* measurement. Each of these is based upon an arbitrarily defined reference point. The gram represents one-thousandth of

*Mass is an essentially fixed quantity for a given amount of material. Weight is the force on an object due to the pull of gravity on that object. Although recognizing this distinction between mass and weight, we will use the terms interchangeably in this book.

†At present the United States is the only major country not using the metric system. Congressional action has already been taken to pave the way for its usage. Studies are now being conducted regarding the feasibility of conversion from our present English system to the metric system. Great Britain began its conversion in 1968.

‡In 1960, an internationally approved modification of the metric system, was adopted. Officially the modified metric system is now called the *Systemé Internationale d'Unites,* the SI system of units. An essential feature of this change is the use of 1000 cc as the basic unit of volume. Also, prior to 1960, the meter was defined as a distance between two marks on a platinum bar.

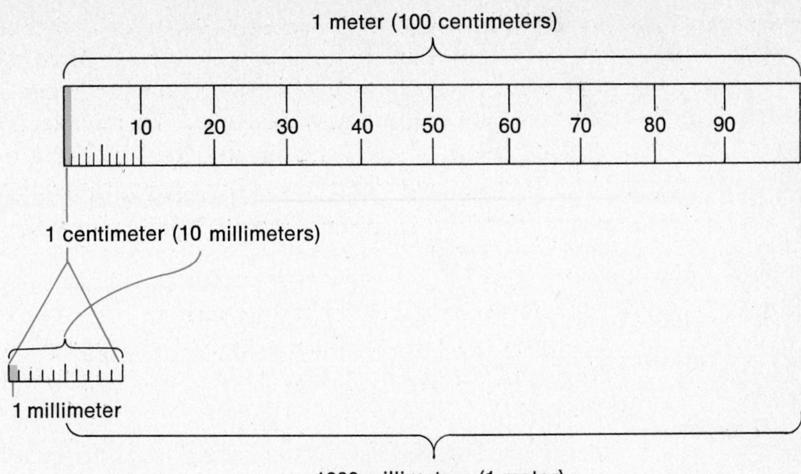

FIGURE 1-1 Relationships between meters, centimeters, and millimeters.

the mass of a block of platinum–iridium alloy kept at the International Bureau of Weights and Measures near Paris, France. The meter is now defined as being a fixed multiple of the wavelength of a certain line in the spectrum of the element krypton. The volume of a cube equals its length times width times height. A cube 10 cm on each side has a volume of 10 cm × 10 cm × 10 cm = 1000 cubic cm = 1000 cc. This volume is also referred to as a *liter*.

For quantities larger or smaller than the basic unit, multiples or fractions of the basic unit are used to create a derived term. The derived term, consisting of a prefix and the basic unit, is always either a multiple or fraction of 10 of the basic unit (see Figure 1-1). The prefix describes the particular multiple or fraction. The most commonly used derived terms are*

PREFIX	RELATION TO BASIC UNIT
kilo	1000 times basic unit
deci	1/10 times basic unit
centi	1/100 times basic unit
milli	1/1000 times basic unit
micro	1/1,000,000 times basic unit
nano	1/1,000,000,000 times basic unit

From Table 1-1, note that the prefix of the derived unit tells the multiple, while the root of the word indicates the basic quantity being measured.

*The prefixes deka (10 times basic unit) and hecto (100 times basic unit) are seldom used.

TABLE 1-1
Basic metric units of mass, length, and volume

	MASS	LENGTH	VOLUME
BASIC UNIT	GRAM (g)	METER (m)	LITER (l)
DERIVED UNITS:	nanogram (ng) (1/1,000,000,000 g)	nanometer (nm) (1/1,000,000,000 m)	nanoliter (nl) (1/1,000,000,000 liter)
	microgram (μg) (1/1,000,000 g)	micrometer (μm) (1/1,000,000 m)	microliter (μl) (1/1,000,000 liter)
	milligram (mg) (1/1000 g)	millimeter (mm) (1/1000 m)	milliliter (ml) (1/1000 liter)
	centigram (cg) (1/100 g)	centimeter (cm) (1/100 m)	centiliter (cl) (1/100 liter)
	decigram (dg) (1/10 g)	decimeter (dm) (1/10 m)	deciliter (dl) (1/10 liter)
	kilogram (kg) (1000 g)	kilometer (km) (1000 m)	kiloliter (kl) (1000 liter)

exercise 1-1

Classify the following data as to the basic quantities measured, i.e., mass, length, or volume: 12.5 km, 10.0 μl, 125 μg, 1.0 mm, 250 mg, 2.54 cm, 7.4 m, 0.46 liter.

Table 1-1 also relates the basic unit to each derived unit. The statement $1\ \text{mg} = \frac{1}{1000}\ \text{g}$ can also be given as $1\ \text{g} = 1000\ \text{mg}$. Some common expressions are

$$1\ \text{g} = 1000\ \text{mg}; \qquad 1 = \frac{1\ \text{g}}{1000\ \text{mg}}; \qquad 1 = \frac{1000\ \text{mg}}{1\ \text{g}}$$

$$1\ \text{liter} = 1000\ \text{ml}; \qquad 1 = \frac{1\ \text{liter}}{1000\ \text{ml}}; \qquad 1 = \frac{1000\ \text{ml}}{1\ \text{liter}}$$

$$1\ \text{m} = 100\ \text{cm}; \qquad 1 = \frac{1\ \text{m}}{100\ \text{cm}}; \qquad 1 = \frac{100\ \text{cm}}{1\ \text{m}}$$

$$1\ \text{kg} = 1000\ \text{g}; \qquad 1 = \frac{1\ \text{kg}}{1000\ \text{g}}; \qquad 1 = \frac{1000\ \text{g}}{1\ \text{kg}}$$

$$1\ \text{g} = 1{,}000{,}000\ \mu\text{g}; \qquad 1 = \frac{1\ \text{g}}{1{,}000{,}000\ \mu\text{g}}; \qquad 1 = \frac{1{,}000{,}000\ \mu\text{g}}{1\ \text{g}}$$

$$1000\ \mu\text{g} = 1\ \text{mg}; \qquad 1 = \frac{1\ \text{mg}}{1000\ \mu\text{g}}; \qquad 1 = \frac{1000\ \mu\text{g}}{1\ \text{mg}}$$

These expressions state an equality between two measurements. Thus the statement *one gram equals one thousand milligrams* can be restated in two ways: (1) There are one thousand milligrams *per* gram (1000 mg/1 g) or (2) one gram *per* thousand milligrams (1 g/1000 mg). These ratios are called *conversion factors*. Their proper use is the key to setting up calculations because units can be handled like numbers in a calculation. Units can be canceled, divided, multiplied, etc.

Conversion factors are ratios between two units.

A general method for problem solving based on the proper use of units is called the *factor-label method.*

EXAMPLE 1-1

Suppose a patient is to receive 250 mg of a particular medication. How many grams does this represent?

The problem is asking, "Is there any relationship between milligrams and grams?" The relationship we know is

$$1000 \text{ mg} = 1 \text{ g}; \qquad \frac{1 \text{ g}}{1000 \text{ mg}}; \qquad \frac{1000 \text{ mg}}{1 \text{ g}}$$

We want to eliminate milligram units and keep grams as our final unit. Setting up the problem so that milligram units cancel will provide the answer:

$$250 \cancel{\text{mg}} \times \frac{1 \text{ g}}{1000 \cancel{\text{mg}}} = \frac{250}{1000} \text{ g} = 0.250 \text{ g}$$

By arranging the units properly, the numerical part of the problem is automatically arranged properly. In Example 1-1 there was no need for guesswork. The units tell you which arrangement to use. You might say why isn't the conversion factor 1000 mg/1 g used? Let's try it and see if grams is our final unit:

$$250 \text{ mg} \times \frac{1000 \text{ mg}}{1 \text{ g}} = 250{,}000 \text{ mg}^2/\text{g}$$

This does not give us the answer to our original question. The units square milligrams per gram are meaningless.

exercise 1-2

Using proper conversion factors and the factor-label method, perform the following conversions: (a) 100 ml to liters, (b) 10 m to km, (c) 10,000 μg to mg, (d) 2.25 liters to ml, (e) 0.08 g to mg.

Several conversion factors can be used together if properly arranged so that appropriate units cancel.

EXAMPLE 1-2

Suppose time would stop exactly as you become 18 years old. How many seconds have you lived?

The question asks about seconds and the problem states time in years. We need conversion factors between years and seconds. Offhand we don't know the single conversion factor to do this but we do know several common ones that we can put together to achieve the same goal.

$$18 \cancel{\text{years}} \times \frac{365 \cancel{\text{days}}}{1 \cancel{\text{year}}} \times \frac{24 \cancel{\text{hr}}}{1 \cancel{\text{day}}} \times \frac{60 \cancel{\text{min}}}{1 \cancel{\text{hr}}} \times \frac{60 \text{ sec}}{1 \cancel{\text{min}}} = 567{,}648{,}000 \text{ sec}$$

exercise 1-3

There are 10 drelbes in one gormp and 12 gormpes in 3 hargens. How many drelbes are there in 24 hargens?

exercise 1-4

Convert the following: (a) 0.042 mg to kg; (b) 0.030 liter to ml, to μl; (c) 5.0 mm to km.

TEMPERATURE SCALES

A nurse records a patient's temperature as 99°. By itself the temperature value has no validity—it is lacking the very important feature of a unit. A temperature of 99°F is not generally a cause for alarm; however, a patient's temperature of 99°C is impossible. Units indeed are significant. In the United States, the Fahrenheit scale (°F) is most commonly used. Elsewhere, the Celsius scale (°C) is the most widely adopted temperature scale.* Pure water has a freezing point of 32°F or 0°C and a boiling point of 212°F or 100°C at 1 atm of pressure.† Figure 1-2 diagrammatically shows this relationship.

From Fig. 1-2, notice that 180 Fahrenheit degrees are equivalent to only 100 Celsius degrees; i.e. $\frac{180}{100} = 1.8$. The correct formula for conversion between temperature scales must also take into account the 32-degree difference in freezing points of water. The final relationship is °F = 1.8°C + 32.

EXAMPLE 1-3

Let's convert normal body temperature of 98.6°F to its equivalent Celsius scale value:

$$°F = 1.8°C + 32$$

*The Celsius scale and the centigrade scale are identical; Celsius is the preferred term.

†The temperature at which a liquid boils under 1 atm of pressure is called the liquid's *normal* boiling point; 1 atm of pressure is a force sufficient to raise a column of mercury to a height of 760 mm.

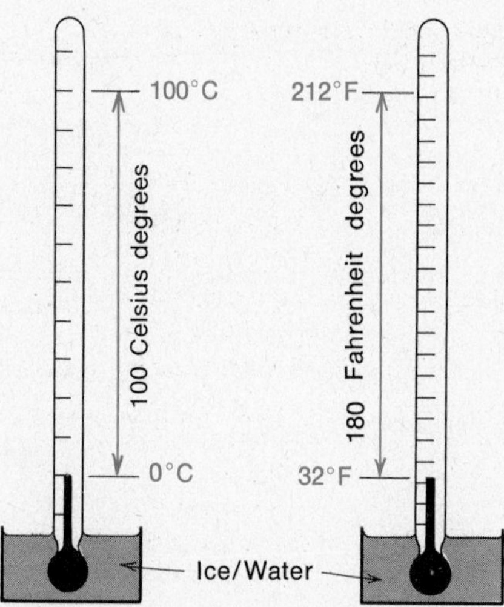

FIGURE 1-2 Comparison of Celsius and Fahrenheit temperature scales.

Rearranging,

$$°F - 32 = 1.8°C$$

Dividing both sides by 1.8,

$$\frac{°F - 32}{1.8} = °C$$

Finally, substituting,

$$\frac{98.6 - 32}{1.8} = °C = \frac{66.6}{1.8} = 37°C$$

exercise 1-5

Diethyl ether is an anesthetic occasionally used during surgery. This liquid has a normal boiling point of 35°C. Diethyl ether has been observed to boil when placed in an open container on a hot day. What is the minimum Fahrenheit temperature at which this will occur?

DENSITY Matter is defined as anything that has mass and occupies space, i.e., has volume. The *ratio* of a *substance's mass to its volume* defines a derived property called *density*.

$$\text{density} = \frac{\text{mass}}{\text{volume}}$$

Although any mass and volume units may be used, densities are usually expressed in grams per cubic centimeter or grams per milliliter. Densities of several common substances are given in Table 1-2.

TABLE 1-2
Densities of common materials measured at specific temperatures (°C)

SUBSTANCE	DENSITY
Pure water	1.00 g/cc at 4°C
Carbon tetrachloride	1.60 g/cc at 20°C
Blood plasma	1.027 g/cc at 25°C
Urine (normal range)	1.003–1.030 g/cc at 25°C
Diethyl ether	0.708 g/cc at 25°C
Chloroform	1.49 g/cc at 20°C
Bone (normal range)	1.7–2.0 g/cc
Diamond	3.01–3.52 g/cc
Ice	0.917 g/cc
Ethyl alcohol	0.785 g/cc at 25°C
Mercury	13.53 g/cc at 25°C
Milk	1.028–1.035 g/cc

exercise 1-6

At 20°C, the volume of 20.5 g of a certain liquid is 25.0 ml. What is the liquid's density at this temperature?

It is important that the temperature at which the density measurement is made be given since substances change volume (expand or contract) with a change in temperature. This volume change causes the density of a substance to vary slightly with even small temperature changes. The change is most drastic for liquids and gases.

Knowing the density of a substance allows the mass or volume of a given amount of the substance to be calculated. Let's use two examples to show this.

EXAMPLE 1-4

Milk from a particular dairy has a density of 1.030 g/ml. How much does 240 ml (approximately an 8-oz portion) of this milk weigh?

Solving the problem then involves cancellation of the volume unit, ml. Thus

$$240\ \cancel{\text{ml}} \times \frac{1.03\ \text{g}}{\cancel{\text{ml}}} = 247.2\ \text{g}$$

EXAMPLE 1-5

We wish to store 500 g of blood plasma at 25°C. What is the minimum-sized vessel that can be used?

The answer sought is in terms of volume and ml is the final unit. We must rearrange terms to cancel the mass unit, grams. The problem has two values in which grams appear: The sample mass is 500 g; the density of the sample is 1.027 g/ml (see Table 1-2). We can express this as an equivalency; 1.027 g = 1 ml and the appropriate conversion factor is 1 ml/1.027 g.

Using the factor-label method we arrange our values such that cancellation of units leads to the desired one.

$$500\ \cancel{\text{g}} \times \frac{1\ \text{ml}}{1.027\ \cancel{\text{g}}} = 486.8\ \text{ml} = \text{sample volume}$$

Therefore a flask having a volume of 486.8 ml is adequate to hold the sample.

exercise 1-7

Using density data from Table 1-2, calculate the following: (a) the mass of three ice cubes, each measuring 4 cm × 4 cm × 3 cm; (b) the mass of a liter of mercury; (c) the volume of a 3.54-g sample of diethyl ether; (d) the number of liters a 15.7-g sample of ethyl alcohol will occupy.

Often, people speak of a substance being *heavier* or *lighter* than another substance, as the case may be. What they actually mean is that the two substances have different densities. One substance may be heavier than another substance but a comparison of their volumes must also be made. One ml of mercury weighs 13.5 g and is 13.5 times heavier than 1 ml of water weighing 1.0 g; however, 13.5 ml of water and 1 ml of mercury *each* weigh 13.5 g.

Objects float because they are less dense than the liquid.

We know ice floats on water. Another way of saying this is that ice, the less dense material, is able to be supported by water, the more dense material. A generalization may be drawn: A liquid or solid will float on a liquid of greater density.

exercise 1-8

The density of lead is 11.3 g/cc. Will a lead block float (a) on mercury? (b) on chloroform? (c) on ethyl alcohol?

exercise 1-9

The density of liquid benzene is 0.89 g/cc at 5.5°C; solid benzene has a density of 1.005 g/cc. Will a benzene "ice cube" float in a glass of liquid benzene?

SPECIFIC GRAVITY

Specific gravity is a comparison of the density of a substance to the density of water measured at the same temperature.

$$\text{specific gravity} = \frac{\text{density of substance}}{\text{density of water}}$$

Both densities are expressed in the *same* units and the units cancel. Thus, specific gravity is a unitless quantity.

A convenient way to measure a liquid's specific gravity is with a hydrometer (see Fig. 1-3). The depth to which the hydrometer sinks in the

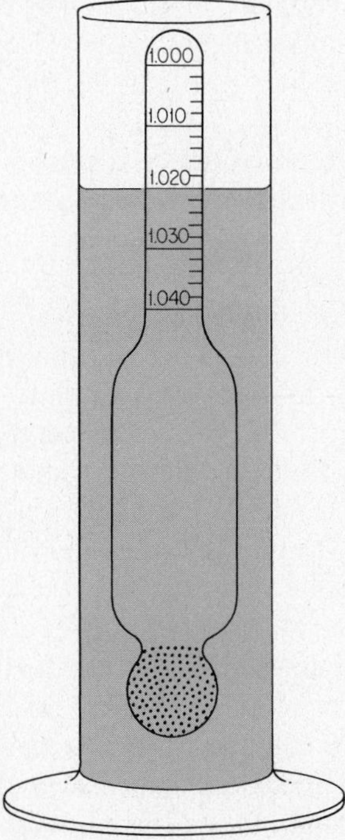

FIGURE 1-3 A hydrometer for measuring specific gravity of liquids having values between 1.000 and 1.040. During measurement, the calibrated stem marking at the liquid surface is the specific gravity of the liquid at a particular temperature. In the figure above, the specific gravity reading is 1.020.

liquid depends on the specific gravity of the liquid; the higher the specific gravity, the higher the hydrometer floats; conversely, it will sink to a lower depth in liquids of lesser specific gravity. The hydrometer stem marking at the liquid level is calibrated to accurately correspond to the actual specific value of the test liquid. In clinical and hospital laboratories, hydrometers are used to determine specific gravities of urine samples. Those data can be helpful in diagnosing certain diseases.

MATTER

The variety of matter we view around us seems to be unlimited. Its obvious diversity leads us to seek a better means of classification. One way to organize a diverse grouping of information is to classify the members of the grouping according to some central theme. The physical state of a sample is such a theme: Three physical states are possible—solid, liquid, and gas. The melting of ice and evaporation of perfume tell us, however, that the physical state of a material depends on conditions such as the temperature and pressure under which the observation is made. We can see that physical state is of limited usefulness as a classification theme.

The states of matter.

Extensive observations over many centuries have led to the concept that a sample of matter may be categorized as being either a mixture or a pure substance. Mixtures contain two or more components that are present in variable amounts; air is an example. Samples of air taken from an urban area, from a rural site, and over the ocean would differ in the type and amount of their constituents.

A pure substance may be further categorized as being either an element or a compound. *An element is a pure substance that cannot be broken down into simpler substances by ordinary means such as heat or light.* To date, the existence of 104 elements has been verified; 89 are naturally occurring on earth, while 15 have been made by physicists and chemists using specialized techniques. Each element is represented by a symbol of one or two letters, generally those of the first letters of the element's name. A list of the name and symbol of each element is given inside the cover of this text.

Each element has its own unique symbol.

A compound is a pure substance which contains two or more elements. The elements in a compound are combined chemically in a fixed relationship by weight so that a compound has a definite, uniform composition. The chemical union of elements to form a compound gives the compound properties distinctly different from those of its component elements. Consider, for example, that the element sodium is a soft metallic lustrous solid that vigorously reacts with water; the element chlorine is a pale green, poisonous gas at normal temperature and pressure. When these two elements react, they produce a white solid that does not react vigorously with water. The white solid is sodium chloride (common table salt), quite different from the elements of which it was formed.

The properties of compounds are very different than the properties of the elements that combined to form them.

A compound is represented by a chemical formula that indicates the elements present in the compound. The formula for water (H_2O) indicates that water consists of the elements hydrogen and oxygen combined in a definite relationship. The formula for hydrogen peroxide (H_2O_2) represents

There are several million compounds known and thousands of new compounds are made each year.

another possible combination. Using appropriate methods, a compound can be decomposed into its component elements. Passing an electric current through water causes its decomposition into the gaseous elements hydrogen and oxygen. Hydrogen peroxide is relatively unstable in the presence of light and decomposes to liberate water and oxygen.

ENERGY All matter has a certain amount of energy associated with it. The amount of energy a particular substance possesses depends on such factors as the composition of the material and its mass. When a substance is transformed into another form of the same substance, e.g., ice into liquid water, or is changed into an entirely different substance, energy changes accompany these processes. Several common experiences attest to this: the burning of a match, melting of ice, metabolism of foodstuffs, light from a flashlight, and the sound of thunder and flashing of lightning during a storm.

There are various forms of energy: heat, light, electrical, and sound are several examples. These forms of energy are interconvertible. For example, energy from the chemical breakdown of foodstuffs (metabolism) provides heat for maintenance of body temperature and energy for muscle action; light striking the eye activates a chemical process that triggers an electrical response (nerve action) culminating in sight.

Energy is commonly described as the ability to do work. Forms of energy are usually classified as potential energy or kinetic energy. Potential energy is stored energy and depends on an object's position or composition. An object possesses kinetic energy when it is moving. In order to move, an object must change position. In doing so, the potential energy is converted into kinetic energy. Consider a situation in which you are standing on the top step of a tall ladder. If you should fall from the ladder, you would rapidly plummet to the ground unless endowed with some rather unusual powers. As you fall, your potential energy is converted into kinetic energy. The energy difference in your potential energy before and after falling manifests itself as kinetic energy as you fall [see Fig. 1-4(a) and (b)].

The various forms of energy are interconverted such that no **net** gain or loss of energy occurs. In the operation of a flashlight, the appearance of light as radiant energy must be accompanied by an equivalent loss of stored energy from the flashlight's battery. In ordinary chemical processes, neither mass nor energy is lost. In nuclear reactions, a small but significant amount of mass is lost, manifesting itself as an equivalent amount of energy. In 1905, Albert Einstein proposed an equation expressing this mass–energy equivalency: $E = mc^2$. This relationship along with the interconversion between various forms of energies leads to the law of conservation of matter and energy: The total energy of the universe is unchanging.

In most chemical reactions, energy changes are manifested as heat (thermal energy). The calorie is a unit used to describe the amount of heat associated with a given change. The *calorie* (cal) is defined as *the amount of heat needed to raise the temperature of one gram of water from 14.5°C to 15.5°C*. A *kilocalorie* (kcal) represents *the amount of heat needed to raise*

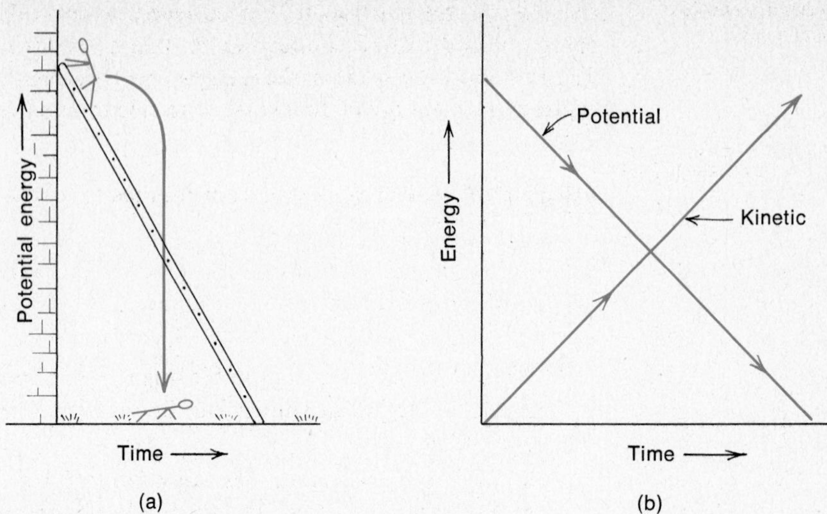

FIGURE 1-4 (a) Change of potential energy during a fall. (b) Conversion of potential and kinetic energy during a fall.

the temperature of one kilogram of water from 14.5°C to 15.5°C. A kilocalorie is equal to the large calorie (Cal) commonly used as a unit in dietary calculations.

Heat may cause a change of state.

When most pure substances are heated, one of two changes occur; either the temperature rises or the substance changes its physical state, i.e., liquifies or vaporizes. The amount of heat required to cause either of these changes depends on the mass and composition of the sample. A gram of water requires the addition of one calorie of heat per 1°C rise in temperature between 0°C and 100°C. Conversely, a calorie of heat is released for each 1°C decrease in temperature as one gram of water cools between 100°C and 0°C. To raise the temperature of a 10-g sample of water from 25°C to 35°C requires $10\ \not{g} \times 1\ \text{cal}/\not{g}°\not{C} \times 10°\not{C} = 100$ cal to be added. *Specific heat* is the term used to describe the amount of heat required to raise the temperature of one gram of a substance by one degree Celsius.

exercise 1-10

Normal body temperature is 98.6°F (37°C). During a high fever, a temperature of 104°F (40°C) may be reached. The water content of an average adult approximates 42 kg. How many calories are required to cause this weight of water to undergo a temperature change from 37°C to 40°C?

Calories are useful for describing energy changes for gram quantities of matter undergoing a change. At certain times, chemists wish to discuss the energy change experienced by a single atom or molecule. Scientists have found it convenient to define a smaller unit of energy, the electron volt (eV). Electrons are negatively charged species that can undergo changes in potential energy. An electron that experiences a one-volt change in potential energy is said to have undergone an energy change of one electron volt. Ten thousand electrons undergoing a one-volt change in potential energy represents an energy of 10,000 eV. One electron volt equals 3.8×10^{-20} cal.

exercise 1-11

Calculate the number of electron volts required to heat 1 g of water from 7°C to 8°C.

CHAPTER SUMMARY

1. There are four basic measurable quantities—time, temperature, length, and mass. Other quantities can be described by combining the basic quantities.
2. Most measurable quantities are expressed by a number and a unit. The number indicates the magnitude of the measurement quantity; the unit indicates the quantity being measured.
3. Units can be handled like numbers during calculations. By setting up conversion factors properly so that only the desired unit remains, calculations are simplified. This is called the factor-label method of problem solving.
4. The metric system uses four basic units—seconds for time, meters for length, liters for volume, and grams for mass. Other units in the metric system are derived as multiple or fractional parts of the basic unit.
5. Density is the ratio of a substance's mass to its volume. Specific gravity relates the density of a substance to the density of water at the same temperature.
6. Matter consists of elements, compounds, and mixtures. Elements are pure substances that cannot be broken down to simpler substances by ordinary means. A compound is a pure substance having a definite composition and containing two or more elements combined chemically. Mixtures contain two or more components present in variable amounts.
7. Energy is described as the ability to do work. Potential energy is stored energy and kinetic energy is the energy of motion. Heat, light, and sound are some examples of forms of energy and are interconvertible.
8. In ordinary chemical processes, neither mass nor energy is lost. The total energy of the universe is unchanging.
9. A calorie is a unit of thermal energy (heat) defined as the amount of heat needed to raise the temperature of 1.0 g of water from 14.5°C to 15.5°C. A kilocalorie equals 1000 cal.
10. An electron volt is another unit of energy. It is used to describe energy changes that individual atoms or molecules undergo.

QUESTIONS

1. Calculate your height in
   (a) Centimeters
   (b) Millimeters
   (1.0 in. = 2.54 cm.)
2. Calculate your weight in
   (a) Kilograms
   (b) Milligrams
   (1 kg = 2.2 lb.)
3. The average adult's body contains about 5 liters of blood.
   (a) How many pints is this (1 liter = 1.06 qt)?
   (b) If 20 drops = 1 ml, how many drops of blood does 5 liters contain?
4. Match the unit in Column A with the correct term in Column B (a choice in column B may be used more than once).

COLUMN A	COLUMN B
liters	mass
millimeters	density
milliliters	energy
grams per cubic centimeter	volume
milligrams	temperature
cubic centimeters	specific gravity
degrees Fahrenheit	specific heat
centimeters	length
kilograms	
$\frac{\text{grams per cubic centimeter}}{\text{grams per cubic centimeter}}$	
micrograms	
grams per milliliter	
kilocalories	
calories per gram per degrees Celsius	
electron volts	
$\frac{\text{grams per milliliter}}{\text{grams per milliliter}}$	

5. Commercially available aspirin tablets for children contain 1.25 grains of aspirin per tablet. Over a period of 12 hr, a child is given six tablets. How many grams of aspirin have been administered (1 g = 15 grains)?
6. The graph at the top of the next page is that of a patient's temperature taken at periodic intervals. Redraw the graph using degrees Fahrenheit as the *Y*-axis.

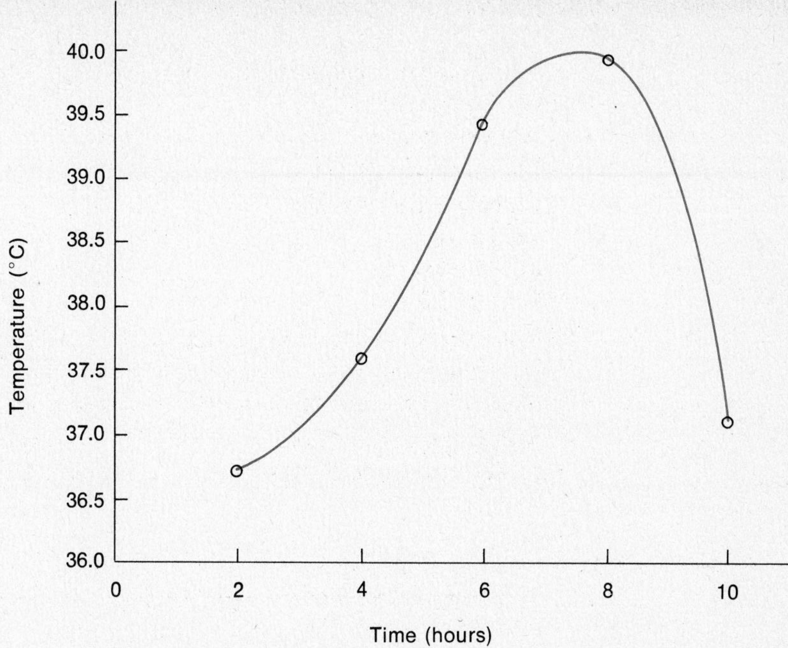

7. A patient's urine sample was found to have a density of 1.028 g/ml. If this individual normally excretes an average of 1500 ml of urine daily, how many grams of urine are eliminated?

8. The following data apply to diethyl ether, a commonly used anesthetic:

$$\text{density at } 20°\text{C} = 0.714 \text{ g/cc}$$

$$\text{specific heat of ether} = 0.54 \text{ cal/g°C}$$

The density of water at 20°C is 0.988 g/cc.

(a) What is the specific gravity of diethyl ether?

(b) Would a 250-ml flask be of sufficient volume to hold $\frac{1}{2}$ lb of diethyl ether (1 lb = 454 g)?

(c) How many calories are required to change the temperature of $\frac{1}{2}$ lb of diethyl ether from 15°C to 25°C?

9. In each of the pairs given below, which quantity is larger?

(a) 1 cm or 1 in.

(b) 1 mm or 1 m

(c) 1 g or 1 lb

(d) 25 g/5 cc or 200 g/50 cc

(e) 60°C or 125°F

(f) 10 cm or 1000 mm

(g) 1 kcal or 100 cal

(h) 10 qt or 10 liters

2
what is an atom?

INTRODUCTION

We put vast amounts and kinds of substances into, on, and through our bodies in an attempt to make us smaller or larger, darker or lighter, younger or older, stronger, healthier, and more attractive. An awareness that these elixirs and potions are composed of atoms, elements, molecules, and compounds is almost taken for granted; however, let us review some of the key events leading to our present belief in and understanding of the atom.

GREEK DEVELOPMENT OF ATOMIC THEORY

The ancient Greek philosophers were the first to begin meaningful speculation, but unfortunately not experimentation, about the nature of their world. One of these men, Democritus (approximately 400 B.C.), is usually credited with using the word *atomos* (Greek for *indivisible*) to describe matter as consisting of indivisible discrete particles. Aristotle rejected this idea and proposed that matter was infinitely divisible. Because of Aristotle's prominence, the Democritian notion of indivisible particles comprising matter was not generally accepted for almost 2000 years.

DALTON'S ATOMIC THEORY

Atoms are too small to be seen even with the most powerful microscope.

In 1804 an English school teacher John Dalton proposed a theory of the nature of matter. The main points of this theory are:

1. Matter consists of small particles called *atoms.*
2. Atoms are not destroyed during chemical reactions but simply become involved in new kinds of chemical combinations (law of conservation of matter).
3. An element's atoms are all alike (a point we shall reexamine later).
4. Elements differ because of the weight and nature of their component atoms.
5. When forming a chemical compound, atoms of the reacting elements combine in a definite ratio (laws of definite and multiple proportions).

These statements are very close to what we accept today as part of our idea of atomic behavior. Unlike the Greek's mere speculation, Dalton's proposals were based upon conclusions drawn from experimental data gathered by Dalton and several of his contemporaries. In these experiments they studied the weight ratios in which different elements combined.

These data formed the basis for a relative scale of atomic weights for the elements known at that time. Thus Dalton's theory proposed not only the existence of atoms but said something about elements, the relative weights of the atoms, and their chemical behavior. It provided the impetus for a flurry of scientific investigation seeking new elements, a refinement of atomic weight values, and an understanding of the forces and energies associated with chemical reactions. Other than describing an atom as a small, indivisible particle, however, Dalton's theory leaves some questions unanswered such as: Why do atoms combine? Why do atoms combine in certain ratios when forming compounds? Why do energy changes accompany chemical changes? There has been a continuing search for answers to these questions since Dalton's time.

EVIDENCE OF SUBATOMIC PARTICLES

During the last half of the 1800's, several significant discoveries were made that showed that the atom was not the ultimate particle but itself was composed of subatomic particles. The initial discovery along these lines came from investigations using evacuated electrical discharge tubes of a design shown in Fig. 2-1.

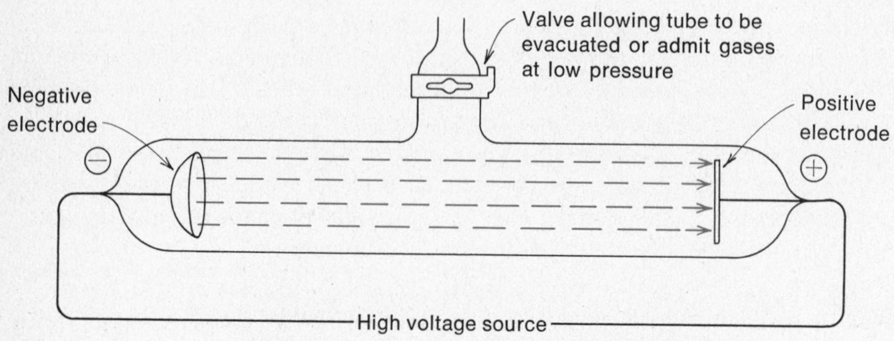

FIGURE 2-1 Diagram of an electrical discharge tube.

Like charges repel; unlike charges attract.

William Crookes and others had demonstrated that a beam of rays traveled between the electrodes of a tube when the tube was evacuated. Furthermore, the beam could be *deflected* by a negative electric charge. Because like charges repel each other, this implied that the beam carried a negative charge.

The negative charge on an electron is the fundamental negative charge.

Using data collected by others and from his own experiments, J. J. Thomson (1897) concluded that the beam consisted of tiny particles each bearing a unit of negative charge. The particles were later named *electrons*. The mass of an electron was determined to be very small—about $\frac{1}{1837}$ that of the lightest atom (hydrogen). (We will return later for a more detailed account of the role electrons play in atomic and chemical phenomena.)

THE NUCLEAR ATOM

Thomson's discovery raised this question: If matter is normally electrically neutral, what and where is the compensating positive charge? Thomson's former student Ernest Rutherford sought an answer. Earlier work by Becquerel, Curie,* and others had demonstrated that atoms of certain elements, radium for example, spontaneously give off radiation of various kinds. Since their exact nature was not immediately known, the radiations were named alpha (α), beta (β), and gamma (γ), the first letters of the Greek alphabet. Rutherford was able to determine certain characteristics of these mysterious radiations.

*Madame Skwoldoska Curie was the first two-time winner of the Nobel Prize in science, sharing one in physics with her husband, Pierre, and the other in chemistry. Pierre Curie met an untimely death by accidentally falling under the wheels of a horse-drawn carriage.

Dr. John Bardeen is the second person to receive two Nobel Prizes in science. His awards, in 1956 and again in 1972, were for his basic research in physics.

NUCLEAR RADIATIONS

TYPE OF RADIATION	CHARGE	RELATIVE ENERGY AND PENETRABILITY
Alpha (α)	+2	Lowest energy: stopped by several centimeters of air, human tissue, or a sheet of paper.
Beta (β)	−1	Higher energy: passes through several meters of air and certain tissues; stopped by several centimeters of aluminum or lead.
Gamma (γ)	0	Highest energy: extremely penetrating

TARGET PRACTICE ON AN ATOMIC LEVEL

Rutherford designed and performed a critical experiment taking advantage of the alpha particles' moderate energy and doubly positive charge. The experiment was designed to observe what happens when alpha radiation is directed toward a very thin metal foil target. To detect what occurred, Rutherford used a screen coated with a substance that would emit flashes of light when struck by alpha radiation. A diagram of the experimental setup is shown in Fig. 2-2.

Possible results from the interaction of the alpha radiation with the foil target might be that the alpha radiation would (1) pass through it, (2) be deflected by it, or (3) be absorbed by it. The experimental results and associated conclusions by Rutherford were

EXPERIMENTAL RESULT	CONCLUSION
1. Most (99.999%) of the alpha particles passed through the foil largely undeflected.	**1.** Most of the foil is *empty space;* the distance between centers of atoms in the foil must be very large compared to the size of the alpha particle.
2. A few, but significant number (1 out of 100,000), of the alpha particles were strongly deflected from their original path.	**2.** Deflection caused by the close approach of an alpha particle (+2 charge) to a dense region of positive charge.

Even as thin foils, metals have definite volume, weight, opaqueness, and other properties normally associated with most solids. Experimental result 1 points out that at the atomic level metals are not the impenetrable materials we ordinarily consider them to be. Since the majority of the alpha particles striking the target passed through undeflected, there must have been nothing present capable of impeding their progress. As an analogy to the experiment, think of tossing small marbles through an open-weave, thin-wire fence; most of the marbles go through the fence without striking the wire.

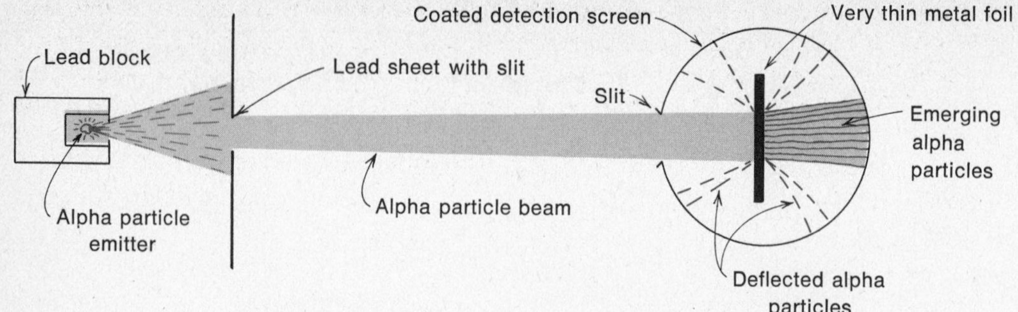

FIGURE 2-2 Rutherford's alpha particle scattering experiment.

The distance between centers of the atoms in the foil must be very large compared with the size of the alpha particle. Thus, Rutherford concluded that the mass of the atom must be concentrated in its central region.

Rutherford still had to explain the small, but significant, number of alpha particles that were strongly scattered from their original path. In fact, some were deflected backward—almost along their original path. Something was in their way! Rutherford was rather surprised at this portion of the experimental data. In his own words he stated "it was almost as incredible as if you fired a 15-inch artillery shell at a piece of tissue paper and it came back and hit you." Rutherford knew, however, that the alpha particle carried a +2 charge. He proposed that the alpha particle's deflection was caused by its close approach to a region of dense positive charge in the atom and its subsequent repulsion by the positively charged region. Since relatively few of the alpha particles were scattered, this indicated that the region of positive charge must be very small compared to the overall size of the atom.

The positively charged nucleus repelled the positively charged alpha *particles.*

Thus, an atom of the foil was characterized by Rutherford as consisting of a very small, dense, positively charged core called a *nucleus* that contains most of the atom's mass. Several years later Rutherford identified the proton as the nuclear particle of unit positive charge. The term *atomic number* is used to describe the number of protons in an atom's nucleus. The atomic number is the unique, identifying feature of an element. All atoms of an element possess this important property—they have the same number of protons. The question of how atoms can be neutral but contain negative electrons can now be answered: In a neutral atom the number of nuclear protons *equals* the number of electrons.

The volume of an atom is about 100 million times the volume of its nucleus.

In 1932, Chadwick, an English physicist, confirmed the existence within the nucleus of a third fundamental subatomic particle, the neutron. This particle has no charge and a mass approximately the same as a proton. Since 1932 scientists have been probing the nucleus with increasingly more energetic atom smashers. This probing has revealed a very large number of other seemingly fundamental subatomic particles. The exact role and chemical significance of these particles remains to be answered by further research. For our purposes, electron, proton, and neutron will suffice for discussion of atomic *and* nuclear behavior.

ATOMIC NUMBER AND MASS NUMBER

The nucleus of each atom contains a certain definite number of protons, depending on the element's atomic number, and a certain, but *not necessarily equal,* number of neutrons. The **mass number** of an atom is the number of protons plus neutrons. Symbolically, an atom of any element may be represented by

Atomic number is the number of protons in an atom.

$$^{\text{mass number (No. of protons + No. of neutrons)}}_{\text{atomic number (No. of protons)}}\text{symbol}$$

ELEMENT	$^{32}_{15}P$	$^{60}_{27}Co$	$^{131}_{53}I$	$^{35}_{17}Cl$	$^{1}_{1}H$
Number of protons	15	27	53	17	1
Number of neutrons (Mass No.–at. No.)	17	33	78	18	0

exercise 2-1

Which of the following have the *same* (a) number of neutrons? (b) Number of protons? (c) Mass number? (d) Number of electrons? $^{65}_{30}Zn$, $^{66}_{30}Zn$, $^{62}_{29}Cu$, $^{66}_{31}Ga$

ISOTOPES

All atoms of an element are alike in the respect that they contain the same number of protons. This does not imply that all atoms of an element have the same number of neutrons. Consider the following cases:

Mass No. **SYMBOL** Atomic No.	$^{1}_{1}H$	$^{2}_{1}H$	$^{3}_{1}H$	$^{86}_{38}Sr$	$^{88}_{38}Sr$	$^{90}_{38}Sr$	$^{127}_{53}I$	$^{131}_{53}I$	$^{135}_{53}I$
Number of protons	1	1	1	38	38	38	53	53	53
Number of neutrons	0	1	2	48	50	52	74	78	82

Isotopes: Greek iso, *same;* top, *place.*

Note that those atoms that have the same number of protons have the same symbol regardless of their number of neutrons. Atoms of the same element (same number of protons) that differ in their number of neutrons are called *isotopes*. Thus Dalton's idea that an element's atoms are all alike is not completely true; they are alike in their number of protons but not necessarily in their number of neutrons. The number of neutrons does not significantly alter their chemical behavior.

exercise 2-2

Following is a list of atoms, some of which are isotopes. Group all isotopes into separate sets. Convince yourself as to your reasoning for classifying the sets of isotopes as you did and why the other remaining atoms were not grouped as isotopes. $^{32}_{15}P$, $^{32}_{16}S$, $^{90}_{38}Sr$, $^{91}_{39}Y$, $^{31}_{14}Si$, $^{33}_{15}P$, $^{131}_{53}I$, $^{60}_{27}Co$, $^{33}_{16}S$, $^{40}_{20}Ca$

NUCLEAR SCHIZOPHRENIA

Previously we mentioned that atoms of certain naturally occurring elements spontaneously emit alpha, beta, and/or gamma radiation. This process is called *radioactivity*. What causes an atom to be radioactive? An isotope's nuclear stability is currently considered to be related to its ratio of neutrons to protons. A given number of protons seems to require an appropriate number of neutrons associated with it to be a stable nucleus. If too few or too many neutrons are present, the nucleus is unstable and releases energy in the form of radiation in an attempt to achieve a less energetic, more stable condition. Thus, certain isotopes of an element may have the proper neutron/proton ratio to be stable, while other isotopes of the same element having unfavorable ratios will be radioactive. All isotopes of an element will have essentially the same chemical behavior regardless of their nuclear stability or instability.

By giving off radiation, a radioactive atom loses energy.

When a radioactive nucleus emits radiation, a nuclear rearrangement occurs during this radioactive decay process. Sensitive electronic instruments such as Geiger–Müller counters and scintillation counters are used to measure the emitted radiation during these nuclear processes. The radioactive decay may be symbolized by means of a balanced equation using the following ground rules:

1. To balance the equation, the *total* mass numbers on each side of the equation must be the same.
2. The total charge on both sides must be equal.
3. Alpha particles are nuclei of helium atoms containing two protons and two neutrons, symbolized $^{4}_{2}He$.
4. Beta radiation resembles high-speed electrons with negligible mass (compared to protons and neutrons), symbolized $^{0}_{-1}e$.
5. Although gamma radiation usually accompanies either alpha or beta emission, it is normally not represented in the nuclear equation.
6. Atomic numbers and the corresponding symbols of the elements can be found in the periodic chart inside the cover of this book.

You are not expected to know whether a particular radioisotope is an alpha or beta emitter. Either this information is given or sufficient other information is present that will allow you to deduce this fact.

Examples of such radioactive decay equations are

ALPHA EMISSION

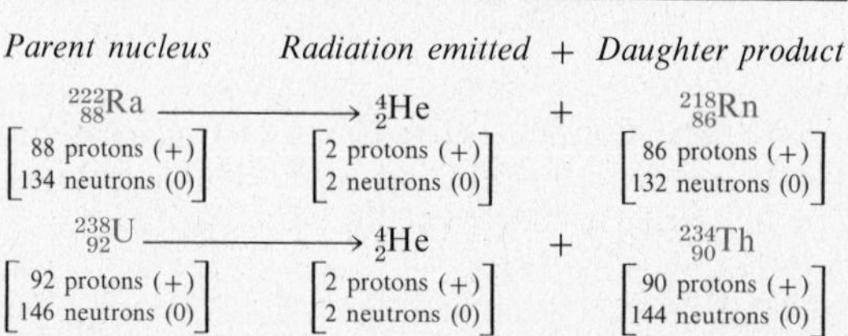

Note that in releasing an alpha particle the parent nucleus has lost two protons and two neutrons. By comparison, the daughter product will have an atomic number two less and a mass number four less than the parent nucleus.

BETA EMISSION

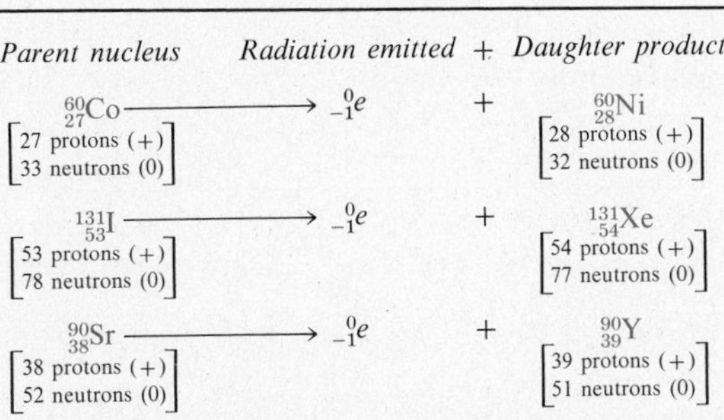

Parent nucleus
Radiation emitted + Daughter product
$^{60}_{27}Co \longrightarrow {}^{0}_{-1}e + {}^{60}_{28}Ni$
[27 protons (+), 33 neutrons (0)]
[28 protons (+), 32 neutrons (0)]
$^{131}_{53}I \longrightarrow {}^{0}_{-1}e + {}^{131}_{54}Xe$
[53 protons (+), 78 neutrons (0)]
[54 protons (+), 77 neutrons (0)]
$^{90}_{38}Sr \longrightarrow {}^{0}_{-1}e + {}^{90}_{39}Y$
[38 protons (+), 52 neutrons (0)]
[39 protons (+), 51 neutrons (0)]

Note that in beta emission the mass number of the daughter product is unchanged but its atomic number increases by one. It might be useful to think of this as arising from the conversion of a neutron into a proton plus a beta particle: ${}^{1}_{0}n \longrightarrow {}^{1}_{1}p + {}^{0}_{-1}e$. The proton is retained by the nucleus and the beta particle is emitted.

exercise 2-3

For each of the following equations, correctly identify the missing species indicated by the blank.

(a) $^{226}_{88}\mathbf{Ra} \longrightarrow$ ______ $+ {}^{222}_{86}\mathbf{Rn}$
(b) $^{234}_{90}\mathbf{Th} \longrightarrow$ ______ $+ {}^{0}_{-1}e$
(c) $^{210}_{83}\mathbf{Bi} \longrightarrow {}^{0}_{-1}e +$ ______

The rate at which a certain radioactive isotope decays is a fixed, constant quantity. The half-life is the period of time required for half of a sample of a radioactive element to decay. The half-life is independent of the amount of sample. For example, ^{131}I is a beta-emitting radioisotope of iodine.* The thyroid gland concentrates iodine and radioactive iodine is often administered to detect thyroid disorders. After a 100-mg dose of ^{131}I is administered as a salt solution of sodium iodide, the following decay scheme occurs:

^{131}I is read as iodine 131

Start		After 8 days		After 16 days		After 24 days		After 32 days	
100 mg (^{131}I)	$\xrightarrow{8\text{ days}}$	50 mg	$\xrightarrow{8\text{ days}}$	25 mg	$\xrightarrow{8\text{ days}}$	12.5 mg	$\xrightarrow{8\text{ days}}$	6.25 mg	$\xrightarrow{\text{etc.}}$

*Commonly the atomic number is not given with an element's symbol and mass number. This is because, by knowing an element's symbol, its atomic number can be obtained from a periodic chart.

Note that the amount of time (8 days) required for half the preceding amount to decay does **not** change even though the amount of iodine-131 remaining is decreasing. While the half-life of a particular radioisotope does not vary, all radioactive isotopes do not decay at the same rate. Table 2-1 shows the wide variation in half-life of some selected radioisotopes.

TABLE 2-1
Radioisotopes and half-life

RADIOISOTOPE	RADIATION EMITTED	HALF-LIFE
^{14}C	beta	5.7×10^3 yr
^{24}Na	beta, gamma	15.0 hr
^{32}P	beta	14.3 days
^{238}U	alpha, gamma	4.5×10^9 yr
^{68}Cu	beta, gamma	32 sec

exercise 2-4

How long will it take a 200-mg sample of ^{24}Na to decay until only 12.5 mg of ^{24}Na remains?

exercise 2-5

^{35}S, a beta emitter, has a half-life of 87 days. Starting with a 400-mg sample of ^{35}S, how much of the sample remains as ^{35}S after 435 days?

RADIOACTIVITY, TRACERS, AND MEDICAL APPLICATIONS

We have discussed the fact that unstable nuclei emit radiation of various types. Atoms with unstable nuclei also undergo virtually the same types of chemical actions in the body as nonradioactive nuclei. If the level of radioactivity could be controlled by limiting the amount of radioactive sample present, could the controlled use of radiation serve a useful biochemical purpose? The answer is yes. Low-level radiation from a wide variety of radioisotopes has many applications in biochemistry and medicine for diagnostic and curative purposes. Radioisotopes are widely used when there is a need to follow the rate or uptake of the metabolic path a substance has in the body. Because the radioisotopes emit radiation that can be detected by scanning with a suitable instrument, their path can be monitored (traced). The types of radioisotopes that permit this sort of activity are commonly called *tracers*. The amount of radiation emitted is governed by the amount of radioisotope present.

Let's briefly consider several of the many medical applications using radioactive tracers in diagnosis and treatment of human disorders.

Iodine is chemically stored (taken up) by the thyroid gland, one of the sites controlling body metabolism. A normal-functioning thyroid gland takes up about 12 percent of a dose of iodine within a few hours. A patient suspected of having a thyroid disorder is given a drink of water containing a small amount of iodine-131 in the form of sodium iodide. Iodine-131 is a beta emitter. By scanning the neck region with an appropriate detector, the level of radioactivity due to the uptake of iodine-131 by the thyroid gland can be monitored. An uptake greater than normal may indicate a hyperthyrod condition; an uptake less than normal may be related to a hypothyroid condition. Larger concentrations of iodine-131 are given for the treatment of thyroid gland cancer. The internally emitted beta radiation can irradiate and destroy cancerous tissue in a very localized area.

Cancer cells metabolize faster than normal cells. All cells need phosphorus for metabolism and an accelerated uptake of phosphorus-32 can be used as a tracer to locate the site of brain tumors. It is also used to study phosphorus uptake during bone and teeth formation.

As a tracer, radioactive iron-59 can be used to evaluate iron concentration in blood, an important consideration in the diagnosis of anemia and other blood diseases.

The body's state of health is closely associated with its ability to take up and transport the element sodium in the form of sodium ions. Radioactive sodium-24 is a beta emitter. A solution of sodium chloride can be injected into a patient's bloodstream and its rate of transport across blood vessel membranes measured. This transport rate can then be compared with that of healthy individuals. Sodium-24 can also be used to trace the site of a blood clot. The patient suspected of having a blood clot in a limb has a sodium chloride solution containing sodium-24 injected into his bloodstream. The path of the injected fluid is carefully followed by a radiation detector. If a clot is present, it will restrict circulation and the radiation level from the sodium-24 will be high on one side and low on the opposite side of the clotting site.

ELECTRONIC STRUCTURE OF THE ATOM

The preceding sections described the nucleus of the atom. Rutherford's scattering experiment led to the concept of a small, dense, positive nucleus with negative electrons about it. We now turn our attention to those electrons and ask about them: What keeps them outside the nucleus? As negative charges, why aren't they drawn into the positive nucleus because of attraction between unlike charges? The answers to these questions come from studying a seemingly unrelated topic—the study of light.

Light is a form of energy exhibiting wave-like characteristics and wavelengths can be assigned to it.

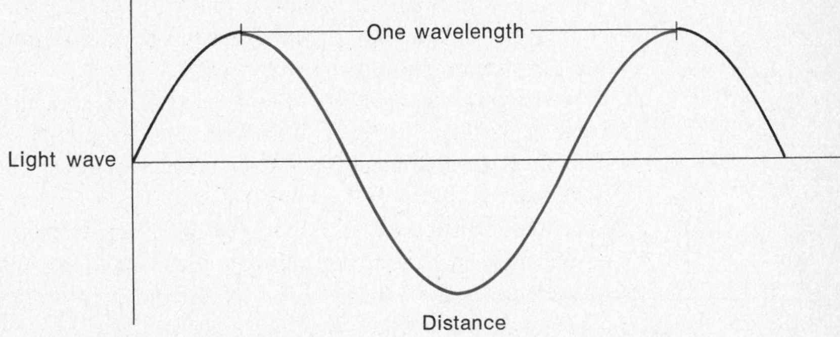

The distance between adjacent wave crests is called a *wavelength*. Light doesn't appear as a series of squiggles to us, however, because visible light consists of extremely short wavelengths, far too short for us to detect visually. *White light,* like that produced by a glowing object such as the sun, is a mixture of many different wavelengths. This mixture of visible light ranges from wavelengths of 400 nm (violet light) to 700 nm (red light). You have probably already observed this when watching sunlight strike a crystal chandelier or light from an incandescent bulb pass through a prism. The dispersion of the light beam into a *continuous* spectrum of component colors occurs because in passing through the prism the components with longer wavelengths (the red end of the spectrum) are bent less than those with shorter wavelengths (the violet end of the spectrum). Figure 2-3 shows a diagram of such a simplified spectroscope.

Recall that $1\ nm = 1 \times 10^{-9}\ m$.

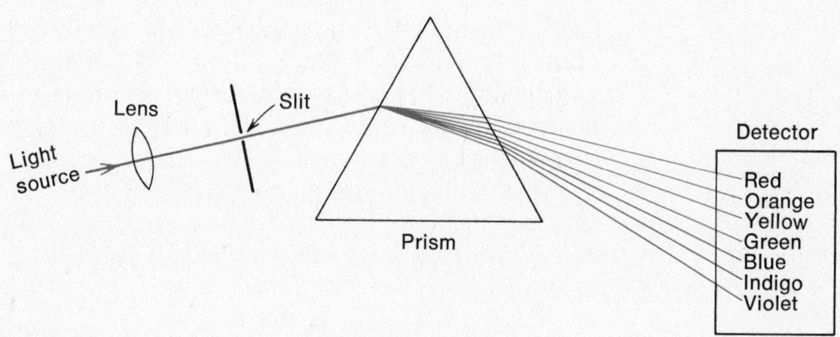

FIGURE 2-3 A simplified spectroscope.

The German physicist Max Planck demonstrated that a relationship existed between energy and wavelength. This relationship indicates that shorter wavelengths have more energy than longer wavelengths; i.e., green light is more energetic than orange light. By experimentally measuring a wavelength, the energy corresponding to that wavelength can be calculated.

The longer the wavelength of light, the lower its energy.

Even before Thomson's and Rutherford's discoveries, spectroscopists carefully measured and tabulated wavelength and energy values. Besides white light, they also obtained values using a different type of light source.

If the light source is changed from a glowing solid to a heated gaseous sample of an element, hydrogen for example, a continuous spectrum of colors is not observed. Instead, the resulting spectrum consists of a series of narrow lines of color called a *line spectrum* (Fig. 2-4).
The lines correspond to discrete, not continuous, energies (wavelengths). Spectroscopists derived equations that could accurately account for the observed wavelengths and energies but they were unable to explain why a spectrum of a gaseous element exhibited only certain discrete energy (wavelength) values.

FIGURE 2-4 Line spectrum of atomic hydrogen.

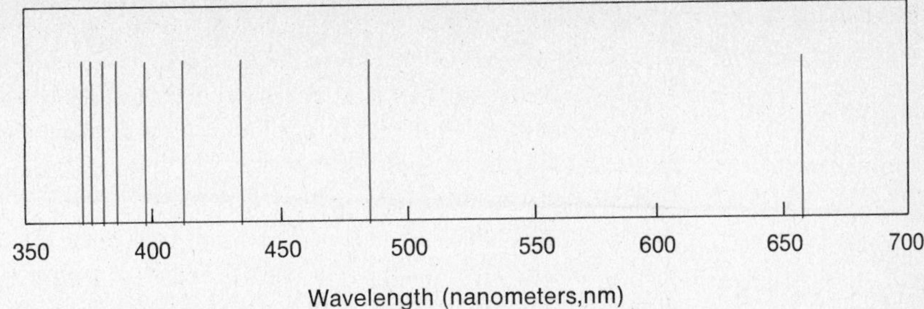

THE BOHR ATOMIC THEORY

The laws of physics change.

Niels Bohr, a young Danish physicist who had worked with both Thomson and Rutherford, set out to resolve the line spectra problem. Bohr knew that the laws of classical physics predicted that an electron traveling about the nucleus should radiate (lose) energy and ultimately spiral into the nucleus. At this point Bohr made a bold departure from classical physics. He said that electrons may not have a continuous range of energies associated with them as they orbit the nucleus. Instead, he proposed that an orbiting electron could have only certain discrete levels of energy; i.e., the electron's energy is quantized. The allowed energy levels are called shells and designated by the letters *K*, *L*, *M*, *N*, . . . as their energy increases (*K* shell is lowest in energy). This is shown in Fig. 2-5.

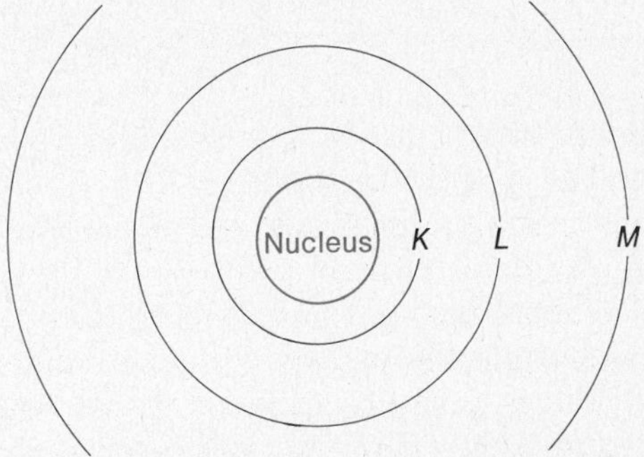

FIGURE 2-5 Model of Bohr atom.

Another way of expressing this diagrammatically is shown in Fig. 2-6. The integers 1, 2, 3, . . . are frequently used for *K*, *L*, *M*, . . . , respectively. This diagram may remind you of a ladder. Like those of a ladder, the steps are quantized. There are no half steps allowed.

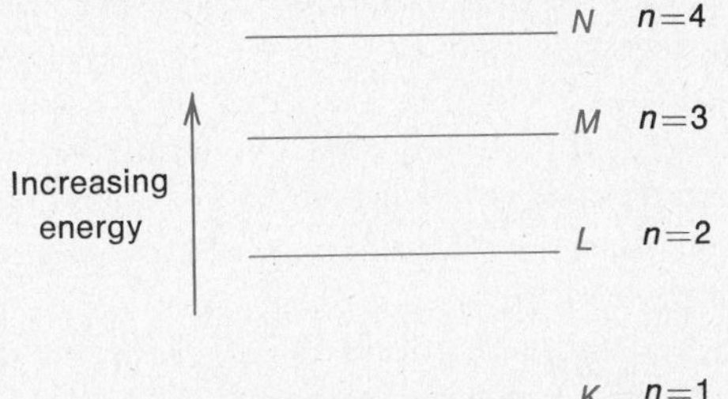

FIGURE 2-6 Atomic electron energy levels.

Bohr explained the line spectrum of hydrogen by saying that in order to change energy levels an electron had to absorb or emit energy. Through thermal or electrical excitation an electron can absorb a specific amount of energy and shift from a lower to a higher energy level. When the electron shifts from a higher to a lower energy level, it emits light of an energy exactly that of the amount it previously absorbed.

Some electron transitions between energy levels are shown in Fig. 2-7. The transition between the *K* and *N* shells represents a greater amount of energy than a transition between the *K* and *L* or *K* and *M* shells. Think of the ladder analogy again. You can go directly from the ground to the first, second, third, or even higher step. Falling from the eighth step to the ground will hurt you more severely (lose more energy) than falling from the first step.

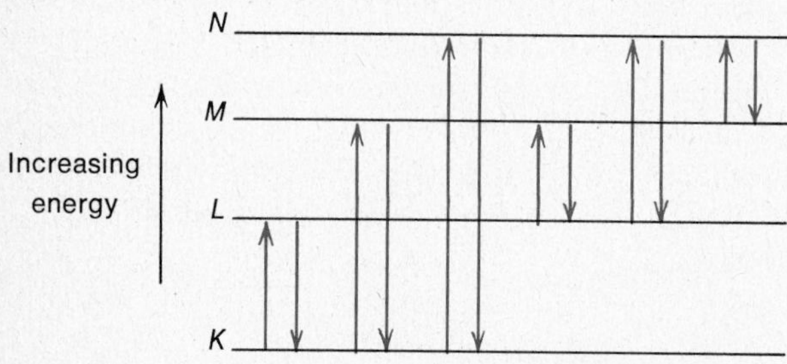

FIGURE 2-7 Possible electron transitions.

Only certain transitions are allowed, and only those energies of radiation corresponding to these transitions will be emitted. Because there are no partial steps on the energy ladder, a continuous range of transitions will not occur and the observed light will not contain a continuous range of energies (wavelengths) but will exhibit a line spectrum. Bohr derived a series of equations with which the energies of these transitions could be *predicted.* The calculated values using the equations agreed very well with the experimentally obtained data. This is the true proof of any theory—agreement between theory-based prediction and observed facts.

Electrons are not considered to be localized in one region.

Bohr's original theory has since been modified to describe electron behavior in atoms more complex than hydrogen. This description uses complex mathematical relationships that ascribe wave-like properties to electrons whose distance from the nucleus is treated in probability terms. This area of study is called *quantum mechanics* (a brief discussion of quantum theory is in Appendix 2).

ELECTRON DISTRIBUTION

The Bohr model of the atom is that of a miniature solar system in which electrons in energy levels (shells) circle the dense, positive nucleus. The moving electrons may be visualized as generating a sphere of negative charge around the nucleus.

Electron distribution around the nucleus follows certain rules. The *maximum* number of electrons in the first (*K*) shell is two; in the second

SHELL	K	L	M	N	O	P	Q
Shell number, *n*	1	2	3	4	5	6	7
Maximum number	2	8	18	32	50	72	98
Observed number	2	8	18	32	32	8	2

FIGURE 2-8 Maximum and observed electron distribution by shells

(*L*) shell it is eight. For any shell, its *maximum* number of electrons is $2\mathbf{n}^2$ where **n** is the shell number (see Fig. 2-8). The first shell (*K* shell) must have two electrons in it before electrons can begin filling the second shell; the second shell must have an octet of electrons before the third shell filling starts. Beginning with the third (*M*) shell, a shell must have at least an octet of electrons before filling of the next higher shell occurs.

The electron distribution of some selected elements is shown below in diagrammatic form. The nucleus of the element's most abundant isotope is represented with p = proton, n = neutron, and curved lines = energy levels (shells).

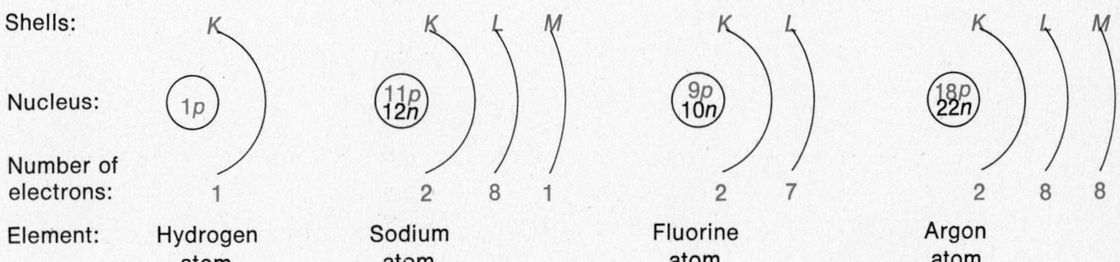

Table 2-2 lists the electron distribution of the first 20 elements. Note the sequence of filling successive shells.

The sequence of filling the shells is

$$\begin{array}{ccccc} K & L & M & N & O \\ 2 \longrightarrow & 8 \longrightarrow & 8 \longrightarrow & 2 & \\ & & 18 \longrightarrow & 8 \longrightarrow & 2 \end{array}$$

In Table 2-2, note particularly the electron distribution for potassium and calcium. These elements are the beginning of the trend where an octet of electrons marks the use of the next higher shell for subsequent filling. This occurs even though the shell containing an octet of electrons does not have its maximum number of electrons. We shall see shortly that there is a significance associated with an octet of electrons. Also, note that certain elements have the same number of electrons in the outer shell. This observation will soon be of value to you.

Two electrons are in the N *shell before the* M *shell is filled.*

TABLE 2-2
Electron arrangement for elements 1 to 20

ELEMENT	SYMBOL	ATOMIC NUMBER	ELECTRON SHELLS *K*	*L*	*M*	*N*
Hydrogen	H	1	1			
Helium	He	2	2 (major shell filled)			
Lithium	Li	3	2	1		
Beryllium	Be	4	2	2		
Boron	B	5	2	3		
Carbon	C	6	2	4		
Nitrogen	N	7	2	5		
Oxygen	O	8	2	6		
Fluorine	F	9	2	7		
Neon	Ne	10	2	8 (major shell filled)		
Sodium	Na	11	2	8	1	
Magnesium	Mg	12	2	8	2	
Aluminum	Al	13	2	8	3	
Silicon	Si	14	2	8	4	
Phosphorus	P	15	2	8	5	
Sulfur	S	16	2	8	6	
Chlorine	Cl	17	2	8	7	
Argon	Ar	18	2	8	8	
Potassium	K	19	2	8	8	1
Calcium	Ca	20	2	8	8	2

exercise 2-6

Atoms of which of the following elements have (a) the same number of outer shell electrons and (b) an octet of electrons in their outer shell? Phosphorus (P), sodium (Na), potassium (K), neon (Ne), helium (He), nitrogen (N), argon (Ar)

THE PERIODIC CHART

One of the techniques scientists use to interpret large amounts of data is to seek similarities in the data that would allow classification. A listing of the presently known elements represents such a collection of data. Are there similarities among elements? In 1860 about 70 elements were known. The entire family of the noble gas elements—helium, neon, argon, krypton, xenon, and radon—were among those still to be discovered. During this time the German Lothar Meyer and the Russian Dimitri Mendeleev independently devised a classification scheme of the elements based on reoccurring (periodic) similarities in their properties and reactions. Elements similar in behavior were grouped together as a family. Using his classification scheme based upon similarities among known elements, Mendeleev recognized there were elements missing. He then predicted the existence and properties of these undiscovered elements. Shortly thereafter new elements were discovered that exhibited properties in agreement with Mendeleev's predictions.

In the modern periodic chart, elements are classified consecutively according to their atomic number (see inside cover of book). Each box in the chart contains the following information.

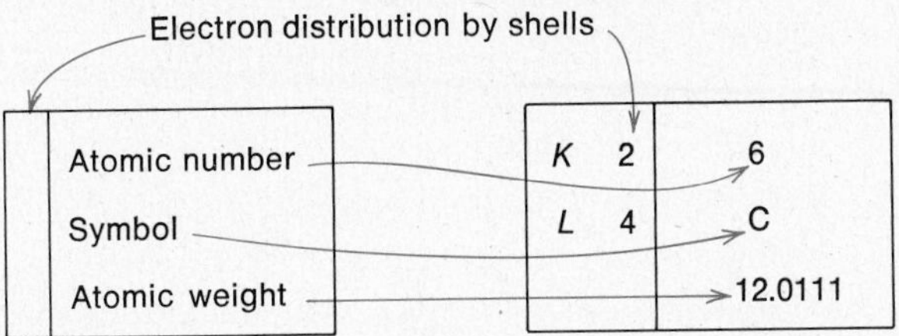

A pattern appears in which elements with similar properties are placed beneath each other. Such vertical columns are called *groups* or *families. Note that atoms of all elements in a group have the same number of outer shell electrons.* Each horizontal row is termed a *period.* The large block of elements between Groups IIA and IIIA are called the *transition metals.* Elements 57 to 71 are the lanthanide series (rare earths) and correctly belong to Period 6. Elements 89 to 104, the actinide series, are in period 7. Both of the series are separated from the main body of the chart for the convenience of keeping the chart a reasonable width.

Let's look at the block diagram of the periodic chart in Fig. 2-9 and the periodic chart on the inside cover. This examination leads to several generalizations:

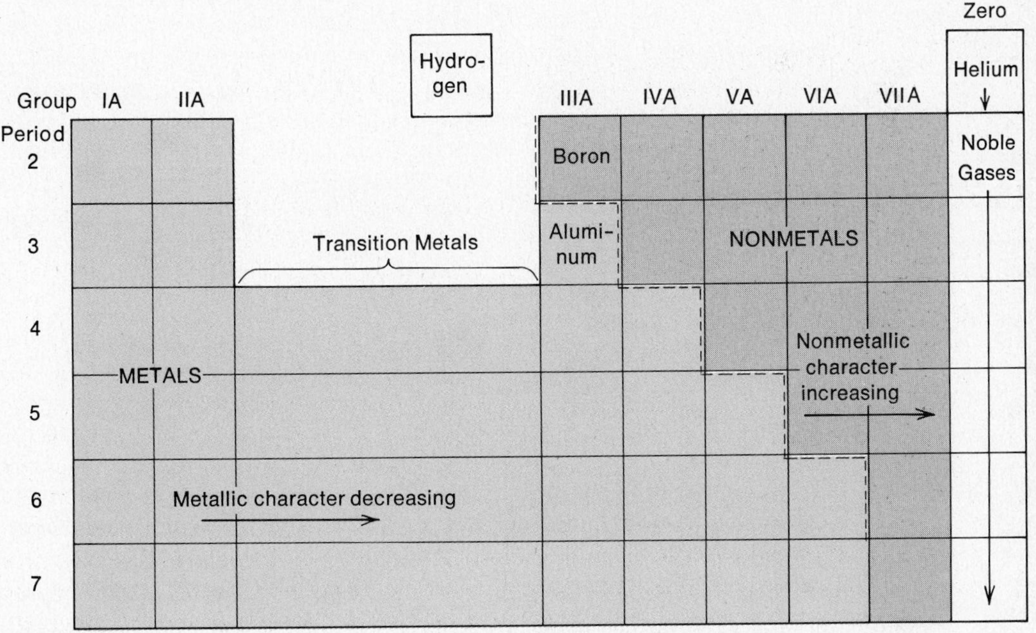

FIGURE 2-9 General periodic chart locations of metals and nonmetals.

1. Most elements are metals.
2. Metals are distributed from the extreme left on through the middle of the chart with metallic character decreasing from left to right.
3. Most metal atoms are characterized by having one, two, or three electrons in their outer (valence) shell.
4. An uneven division exists between the locations of metallic and nonmetallic elements in the chart. Elements bordering it exhibit both metallic and nonmetallic properties.
5. Nonmetals are grouped largely toward the right side of the chart. Atoms of these elements are characterized by having four, five, six, or seven electrons in their highest occupied energy level.
6. Noble gas elements complete the right side of the chart and their atoms, except helium with only two, have an octet of electrons in their outer shells. This octet of electrons is sometimes called a *noble gas arrangement* and represents a condition of chemical stability.
7. Hydrogen exhibits both metallic and nonmetallic properties.

exercise 2-7

Do this exercise without a periodic chart.

Give the electron configuration (distribution by shells) for atoms with the number of electrons given below. Also indicate whether the atom's element is metallic or nonmetallic. 11, 7, 3, 19, 16

ATOMIC SIZE AND PERIODICITY

An important consideration during chemical reactions is the size of the interacting atoms. The fact that atoms of all the elements are *not* the same size may surprise you. Atomic diameters are often expressed in units called *angstroms* where 1 angstrom (Å) equals 10 nanometers. Atomic diameters range from 0.74 Å for hydrogen to 5.24 Å for cesium. The number and distribution of an atom's electrons account for its size. A change in the number and distribution of electrons alters an atom's size, making atomic size a periodic property. Consider the sequence of lithium, sodium, and potassium:

		K	*L*			*K*	*L*	*M*		*K*	*L*	*M*	*N*
	Li	)	)		Na	)	)	)	K	)	)	)	)
		2	1			2	8	1		2	8	8	1
Atomic diameters:		3.10 Å				3.80 Å				4.70 Å			

Notice that as the number of occupied shells increases, the size of the atom increases; therefore, atomic size increases down a group in the periodic table. This trend applies to both metals and nonmetals.

exercise 2-8

For each of the following pairs of atoms, indicate which atom would have the *smaller* atomic diameter. Li and K; O and S; Mg and Ca; Br and Cl

IONIZATION POTENTIAL AND ELECTRON DISTRIBUTION

So far we have dealt with electrically neutral, uncharged atoms containing equal numbers of protons and electrons. Metals have been described as elements whose atoms have one, two, or three electrons in their outer shell.

Ionization potential is the term for the energy needed to remove electrons from atoms. Since electron energies are quantized, the ionization potential will be a certain discrete amount of energy, in units of electron volts, for a specific gaseous atom. For any metal atom, this may be represented by

$$\text{M} + \text{ionization potential} \longrightarrow n \text{ electrons} + \text{M}^{n+} \qquad n = 1, 2, \text{ or } 3$$

Group IA

$$\text{Li} + 5.4\ \text{eV} \longrightarrow e^- + \text{Li}^+$$
$$\text{Na} + 5.2\ \text{eV} \longrightarrow e^- + \text{Na}^+$$
$$\text{K} + 4.35\ \text{eV} \longrightarrow e^- + \text{K}^+$$

Group IIA

$$\text{Mg} + 22.7\ \text{eV} \longrightarrow 2e^- + \text{Mg}^{2+}$$
$$\text{Ca} + 18.0\ \text{eV} \longrightarrow 2e^- + \text{Ca}^{2+}$$

When more than one electron is lost, they are removed one at a time in a stepwise fashion. For calcium, this occurs as

$$\text{Ca} + 6.1\ \text{eV} \longrightarrow e^- + \text{Ca}^+$$
$$\text{Ca}^+ + 11.9\ \text{eV} \longrightarrow e^- + \text{Ca}^{2+}$$
$$\text{Net result: Ca} + 18\ \text{eV} \longrightarrow 2e^- + \text{Ca}^{2+}$$

A trend is present in the data above: Ionization potentials decrease going down a group. As the size of atoms increases down a group, the outer shell becomes farther away from the influence of the atom's positive nucleus. Therefore, it becomes easier (requires less energy) to remove electrons from the outer shell.

Let us examine more carefully the action of removing outer shell electrons from metal atoms. We have come to associate stability with an octet of electrons in the outer shell of an atom. How can a metal atom attain eight electrons in its outer shell? Let's use sodium and magnesium atoms as examples:

	Na^0 + energy $\longrightarrow e^-$ +	Na^+	Compared to neon (Ne)
Number of protons	11(+)	11(+)	10(+)
Number of electrons	11(−)	10(−)	10(−)
Net charge	0	1+	0
Electron distribution by shells			
	K 2, L 8, M 1	K 2, L 8	K 2, L 8

	Mg^0 + energy ⟶ $2e^-$ +	Mg^{2+}	Compared to neon (Ne)
Number of protons	12(+)	12(+)	10(+)
Number of electrons	12(−)	10(−)	10(−)
Net charge	0	2+	0
Electron distribution by shells	K L M: 2 8 2	K L: 2 8	K L: 2 8

When an atom loses (an) electron(s), it becomes positively charged.

The metal atom loses the appropriate number of electrons to achieve an octet of electrons in its remaining outer shell. It then has the same electron distribution as an atom of the noble gas element preceding it in the periodic table (neon in this case). In the electron-releasing process the electrically neutral atom becomes positively charged because of an insufficient number of electrons remaining to offset the unchanged positive charge of the nucleus.

A charged atom is called an *ion* and we can speak of metal atoms losing electrons to become positive metal ions. The positive charge acquired by the ion equals the number of electrons lost. A metal atom will lose electrons from its outer shell until it reaches an electron distribution like that of the preceding noble gas. Atoms of elements in Groups IA, IIA, and IIIA have one, two, and three electrons in their outer shells and lose electrons to form 1+, 2+, and 3+ ions, respectively. The key in determining what charge a metal ion will have is to relate this charge to the outer shell electron distribution in the neutral atom.

exercise 2-9

Write the electron distribution for a neutral atom of each of the following elements: Li, Ca, Al. Write the electron distribution and charge of their corresponding ions.

ELECTRON AFFINITY AND ELECTRON DISTRIBUTION

Ionization potential increases across a period; thus the ionization potential of nonmetals is so large that electrons are not removed under ordinary circumstances. Instead, nonmetal atoms gain sufficient electrons to have an outer shell octet of electrons. The number of electrons gained by a nonmetal atom depends on the number of outer shell electrons needed to complete an octet. This may be generalized by

For nonmetals, usually n = *1 or 2.*

$$\underset{\text{Nonmetal atom}}{\text{NM}} + ne^- \longrightarrow \underset{\text{Nonmetal ion}}{\text{NM}^{n-}} + \text{energy}$$

Some examples of this are

In general, the number of electrons a nonmetal atom gains equals 8 minus its group number, e.g., fluorine gains one electron: 8 − 7 = 1.

Element	Fluorine			Fluoride ion	Compared to neon (Ne)
Symbol	F	+	e^- ⟶	F^- + energy	
Number of protons	9(+)			9(+)	10(+)
Number of electrons	9(−)			10(−)	10(−)
Net charge	0			1−	0
Electron distribution by shells	K 2, L 7			K 2, L 8	K 2, L 8

Element	Chlorine			Chloride ion	Compared to argon (Ar)
Symbol	Cl	+	e^- ⟶	Cl^- + energy	
Number of protons	17(+)			17(+)	18(+)
Number of electrons	17(−)			18(−)	18(−)
Net charge	0			1−	0
Electron distribution by shells	K 2, L 8, M 7			K 2, L 8, M 8	K 2, L 8, M 8

When a neutral atom gains electrons, it becomes negatively charged.

The gain of electrons by a nonmetal atom causes the electrically neutral atom to become negatively charged because the number of electrons now exceeds the unchanged number of nuclear protons. Nonmetal atoms gain electrons and become negative ions. This is in contrast to metal atoms losing electrons to become positive metal ions.

Generally, atoms in Groups IVA and VA do not form ions.

Neutral atoms and their ions differ in numbers of electrons.

We have seen that an atom's outer shell electron distribution enables us to predict what kind of ion that atom will form. If the atom has one, two, or three outer shell electrons, it will lose these electrons to achieve electron distribution of the preceding noble gas and become a 1+, 2+, or 3+ metal ion, respectively. An atom that has six or seven outer shell electrons gains two electrons or one electron, respectively, to achieve the electron distribution of the next noble gas. This gain of electrons produces 2− and 1− nonmetal ions. Whether applied to a metal or a nonmetal, the *most* significant difference between an ion and its corresponding neutral atom is the number of electrons.

exercise 2-10

For atoms of each of the following nonmetallic elements, write (a) the electron distribution for the neutral atom and (b) electron distribution and charge for the corresponding nonmetal ion: oxygen, sulfur.

exercise 2-11

How many electrons and protons are present in (a) a neutral bromine atom and (b) a bromide ion (Br^{-1})?

The change in energy when a neutral atom acquires electrons is called *electron affinity*. The electron affinity is a measure of an atom's electron-accepting ability. The electron affinity of nonmetal atoms is greater than that of metal atoms.

RELATIVE WEIGHTS OF ATOMS AND ATOMIC WEIGHT

We have discussed at some length the structure of an atom. It was mentioned previously that most of an atom's mass is in its nucleus due to the presence of protons and neutrons there. All atoms of an element have the same number of protons but not necessarily the same number of neutrons. Isotopes are atoms with differing numbers of neutrons and the same number of protons. Recall atoms of the same element can differ in mass (weight). What about the *relative* weights of atoms of different elements? Are atoms of one element heavier or lighter than atoms of another element? Since atoms of other elements differ in their numbers of protons and neutrons, it is almost intuitive that we feel elements should have differing weights. The question arises, How much heavier are atoms of one element than those of another element? This is like asking the question How much longer is one piece of string compared to another piece of string? An arbitrary reference point is needed to answer. In the case of the string we might use inches or feet as our arbitrary reference point since the length of an inch or a foot has already been arbitrarily defined as 1.0 foot = 12 inches. What is needed in the case of atomic weights is a reference point of weight with which all other atomic weights can be compared.

Why can't individual atoms be weighed in grams?

In 1961 an internationally used atomic weight (mass) scale based on a particular isotope of carbon, ^{12}C (carbon-12) was adopted. One atomic mass unit (amu) is defined as one twelfth the mass of a carbon-12 atom. Thus, carbon-12 has an atomic weight of 12 amu. The experimentally determined atomic weights of helium, titanium, and lead are 4.00, 48.00, and 207 amu, respectively. These atomic weights indicate that atoms of these respective elements are close to $\frac{1}{3}$, 4, and 17.3 times as heavy as carbon-12 atoms. Notice that we are considering relative weights—how much an atom of an element weighs compared with the weight of an atom of an arbitrary reference standard. An instrument called a *mass spectrograph* can be used to determine experimentally the relative weights of atoms of various elements.

Let us consider some examples of how the data from several typical mass spectrographic analyses can be used to calculate the atomic weights of helium, titanium, and lead as described above.

It is experimentally determined that helium atoms are only $\frac{1}{3}$ the mass of carbon-12 atoms. Therefore,

$$\text{atomic weight of helium} = \text{experimental ratio of } \frac{\text{weight of element}}{\text{weight of carbon-12}} \times 12 \text{ amu}$$

$$= \frac{1}{3} \times 12.00 \text{ amu} = 4.00 \text{ amu}$$

Similarly for titanium (Ti) and lead (Pb),

$$\text{atomic weight of titanium} = \frac{4}{1} \times 12.00 \text{ amu} = 48.0 \text{ amu}$$

$$\text{atomic weight of lead} = \frac{17.3}{1} \times 12.00 \text{ amu} = 207 \text{ amu}$$

exercise 2-12

Atoms of elements X, Y, and Z are, respectively, $\frac{1}{2}$, 12, and 15 times as heavy as carbon-12 atoms. What are the atomic weights of elements X, Y, and Z?

WHY AREN'T ATOMIC WEIGHTS OF THE ELEMENTS USUALLY WHOLE NUMBERS?

The relative abundance of isotopes can be determined in a mass spectrometer.

A quick scan of the periodic chart reveals that most atomic weight values are not whole numbers. Why is this so? It is because the atomic weight of an element is really an average. This average takes into account both the kind and number of atoms of the various naturally occurring isotopes of the element. Let's turn to the relative abundance data for chlorine to see how this is calculated. There are three times as many ^{35}Cl atoms as ^{37}Cl atoms or three out of every four chlorine atoms are ^{35}Cl atoms. Thus we can say that if you could reach into a container of naturally occurring chlorine atoms and draw out a handful (figuratively speaking), three-fourths of the atoms would be ^{35}Cl and one-fourth would be ^{37}Cl. What would the average weight of atoms in your sample be? This can be determined by relating the mass and relative abundances (fraction) of the atoms in your sample:

	mass (weight)	× *fraction*	=	weight contribution
chlorine-35	35 amu	× $\frac{3}{4}$ or 0.75	=	26.25 amu
chlorine-37	37 amu	× $\frac{1}{4}$ or 0.25	=	9.25 amu
	atomic weight of chlorine		=	35.5 amu

Another way of looking at this would be to say that if *all* (100 percent) chlorine atoms were only ^{35}Cl, the atomic weight of chlorine would be 35.00 amu; if all the atoms were ^{37}Cl, the atomic weight would correspondingly be 37.00 amu. If half the atoms were ^{35}Cl and the other half ^{37}Cl atoms, chlorine's atomic weight would be 36.00 amu. Since chlorine's atomic weight is 35.5 amu, this indicates that there are more ^{35}Cl than ^{36}Cl atoms and more than half of the atoms are ^{35}Cl; consequently less than half must be ^{37}Cl atoms.

Atomic weight is the average weight of atoms of an element; mass number is given for individual atoms.

Please note that the atomic weight value for chlorine or any other element is a *proportional average* of the kind and number of isotopes making up that element. It *does not* indicate the weight (mass) of any particular atom of the element. The mass number of an atom furnishes this information. To use an analogy, think about the weight of each individual class member (that student's *mass number*). Multiplying each of the individual weights by the appropriate fraction and summing up these values would give you a proportional average weight for the class that may not coincide with any individual's real weight.

exercise 2-13

Four-fifths of naturally occurring boron atoms are boron-11. The remainder is boron-10. Calculate the atomic weight of boron.

CHAPTER SUMMARY

1. Ancient Greek philosophers speculated about the nature of matter. This speculation led to the proposal that matter consists of discrete, indivisible particles called *atoms*.
2. John Dalton (1804) proposed that atoms make up matter and combinations of atoms form new substances, an element's atoms are alike and elements differ because of the weight and nature of their atoms, and reacting elements combine in a definite ratio.
3. Present atomic theory proposes that atoms consist of subatomic particles. The most important fundamental particles are the electron (-1 charge), the proton ($+1$ charge), and the neutron (0 charge). In a neutral atom, the number of protons equals the number of electrons. Protons and neutrons are of approximately equal mass and both are much heavier (about 2000 times) than electrons.
4. From experimental data, Lord Rutherford formulated a model of an atom. The model depicts an atom as having a dense, positively charged nucleus containing essentially all the atom's mass. Later work by Rutherford and Chadwick indicated that an atom's protons and neutrons are located in its nucleus. The number of protons in an atom is called the *atomic number*. The sum of protons and neutrons in an atom's nucleus is called the *mass number*.
5. Atoms of the same element have the same atomic number. Atoms of the same element having differing numbers of neutrons are called *isotopes*. The stability of a nucleus seems to be related to its number of protons and neutrons. Unstable nuclei emit energy in the form of alpha (${}^{4}_{2}He$), beta (${}^{0}_{-1}e$) and/or gamma (${}^{0}_{0}\gamma$) radiation. The amount of time required for a radioactive element to decay to half its previous amount is called its *half-life*. Various radioactive isotopes can be used clinically as tracers.
6. Niels Bohr proposed that electrons occupy discrete energy levels (shells) about an atom's nucleus; i.e., an electron's energy is quantized. In order to occupy other energy levels, an electron has to lose energy or gain energy. As long as electrons remain in the same energy level, energy is neither gained nor emitted.
7. In order of increasing energy, the energy levels are termed *K*, *L*, *M*, *N*, *O*, ... Each energy level can hold a certain maximum number of electrons. The distribution of electrons in energy levels about a

nucleus can be determined by using certain guidelines. The *K* shell contains a maximum of two electrons. For other energy levels the presence of an octet of electrons results in stability.

8. The outer shell electrons are the ones primarily responsible for an atom's chemical characteristics. Based on the number of outer shell electrons in its atoms, an element can be generally classified as a metal or nonmetal. Metal atoms have one, two, or three electrons in their outer shell; nonmetal atoms have five, six, or seven electrons in their outer shell. Noble gases have an octet of electrons in their outer shell (helium only has two).
9. The modern periodic chart arranges elements in the order of their atomic numbers. Vertical columns called *groups* contain elements whose atoms have the same number of electrons in their outer shells. Therefore, these elements have similar chemical behavior.
10. The number and distribution of electrons in an atom largely govern the size of the atom. As the number of occupied shells increases, the atomic size increases. Thus, atomic size increases down a group in the periodic table.
11. Ionization potential is the energy required to remove electron(s) from atoms. Metal atoms lose the appropriate number of electrons to achieve an octet of electrons in their outer shells. When a neutral atom loses electrons, a positively charged ion results.
12. Nonmetal atoms gain electrons in order to achieve an octet of outer shell electrons. By gaining electrons, nonmetal atoms form negative ions. Electron affinity is the change in energy when a neutral atom acquires electrons.
13. Atoms have weight. Currently, an atomic weight scale based on ^{12}C is used. By definition, one atomic mass unit is defined as one-twelfth the mass of a carbon-12 atom. Because atoms of an element may contain different numbers of neutrons, atomic weights are a kind of average that is usually not a whole-numbered value. Atomic weight and mass number are not the same.

QUESTIONS

1. The passage of chloride ions (Cl^-) is vital to the proper functioning of our cells. The diameter of a chloride ion is 3.62 Å (1 Å = 1×10^{-8} cm). Calculate the diameter of a Cl^- in centimeters.
2. Chloride ions (Cl^-) are small enough to pass through (across) cell membranes. This is vital to the proper functioning of our cells. The diameter of a chloride ion is 3.62 Å (1 Å = 1×10^{-8} cm).
   (a) Calculate the diameter of a chloride ion in both millimeters and inches.
   (b) How many chloride ions placed side by side so that they were just touching would be required to make a 1-in. chain of chloride ions?

(c) Would the diameter of a chlorine atom be larger, smaller, or the same as that of a chloride ion?

3. The symbolism for various elements is listed below. Identify what various ones have in common (not all necessarily have the same common characteristics).

$^{32}_{15}P$, $^{40}_{20}Ca$, $^{37}_{17}Cl$, $^{79}_{35}Br$, $^{24}_{11}Na$, $^{39}_{19}K$, $^{7}_{3}Li$, $^{19}_{9}F$, $^{32}_{16}S$, $^{41}_{19}K$, $^{24}_{12}Mg$

4. Discuss any differences and similarities between a potassium atom and a potassium ion (K^+).

5. What would be the charge, if any, on the following particles?
   (a) 20 protons, 20 neutrons and 18 electrons
   (b) 3 protons, 4 neutrons, and 2 electrons
   (c) 15 protons, 17 neutrons, and 15 electrons
   (d) 15 protons, 17 neutrons, and 18 electrons
   (e) 16 protons, 16 neutrons, and 18 electrons

6. Without consulting a periodic table but by using electron distributions,
   (a) Predict in which group of the periodic table each of the following elements would occur: nitrogen (element 7), calcium (20), beryllium (4), aluminum (13), chlorine (17), fluorine (9)
   (b) Categorize each of the previous elements as being either a metal or a nonmetal.

7. Discuss the differences between mass number and atomic weight.

8. Complete the following crossword puzzle:

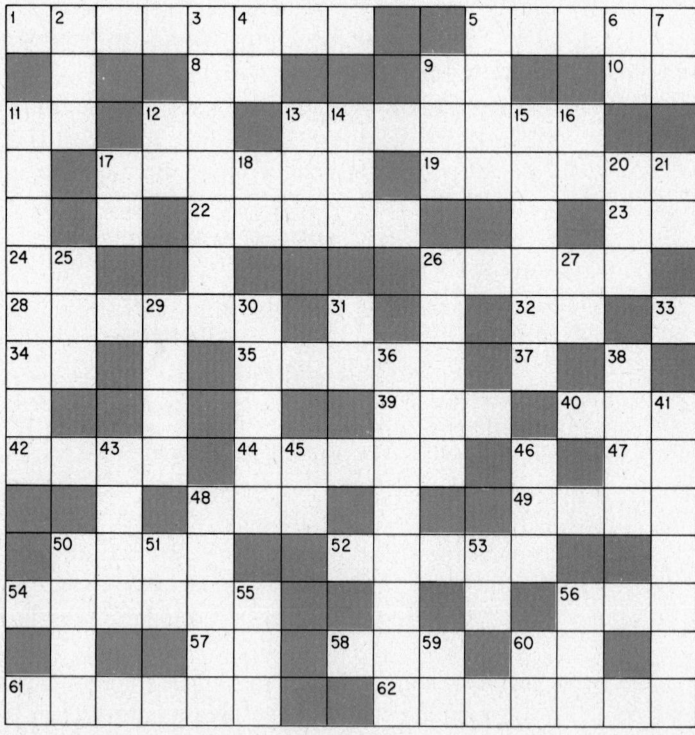

ACROSS

1. An element that can form 3+ ions
5. Refers to *K*, *L*, . . .
8. A Group IVA element (symbol)
9. An element that forms 2+ ions
10. A Group VIIA element (symbol)
11. A noble gas element (symbol)
12. Atomic number 42 (symbol)
13. One of the states of matter (pl.)
17. A widely used system of measurement
19. First proposed quantization of energy
22. Chemical properties are largely determined by an atom's ______ shell electron(s)
23. A noble gas
24. A radioactive, Group IIA element (symbol)
26. Elements in Groups IA or IIA are classified as a ______
28. An element that can form 2− ions
32. Symbol for element 76
33. A Group IA element (symbol)
34. A Group IVA element (symbol)
35. Eight electrons in an outer shell
37. The most abundant element in air (symbol)
39. Allow, permit
40. Within a group, the element with the smallest atoms is at the ______
42. A noble gas
44. A unit of volume
47. A Group IIA element (symbol)
48. Number of electrons in a neon atom

DOWN

2. Robert E. ______
3. $^{2}_{1}H$ is an ______ of hydrogen
4. Both an element and a coin
5. Attached to a mast
6. Pound (abb.)
7. Has 103 protons
9. A drinking utensil
11. The lightest element
12. Self
13. Was lighted
14. A form of water
15. Proposed an atomic theory in the 19th century
16. Element 50 (symbol)
17. Milligrams (abb.)
18. A transition element (symbol)
20. Unit of energy (abb.)
21. Superman's weakness (symbol)
25. A cutting tool
26. A unit of length
27. A toxic, Group VA element (symbol)
29. To form a F^{-} ion, a fluorine atom must ______ an electron
30. The gases helium through radon
31. Atomic number 85 (symbol)
36. Negatively charged particle
38. Electron ______ by a potassium atom forms a potassium ion
41. Positively charged subatomic particles
43. Many chemicals have a distinctive ______
45. Opposite of out
46. A state of matter
48. A Scandinavian elf
50. Radiation bearing a negative charge
51. A greeting
53. Milliliter (abb.)
55. Unit of time
56. An appendage

49. Too
50. He proposed electron energy must be quantized
52. Basic units of matter
54. Horizontal row of periodic table
56. The negative charge of a sulfur ion
57. Element 57 (symbol)
58. A charged atom
60. Has 27 protons (symbol)
61. Another term for group in the periodic table
62. Site of an atom's mass
59. A Greek letter
60. A Group VIIA element (symbol)

3
chemical bonding

INTRODUCTION

In Chapter 2 we studied the fundamental aspects of atomic structure. Modern ideas about the nature of the atom were drawn from the results of hundreds of experiments performed over a period of approximately half a century. Yet, even while the ideas of atomic structure were being developed, other groups of scientists were already asking such questions as Why do atoms combine with one another? What holds atoms together in compounds? These groups of research workers recorded many observations about reactions between the elements and between compounds in the hope that their observations would be useful in answering these questions after an understanding of atomic structure was gained.

We are more fortunate in our study of chemical bonding since we already possess an insight into the nature of atoms. Armed with this knowledge we can more readily enter the realm of chemical bonding.

WHY DO ATOMS COMBINE WITH EACH OTHER?

exo *means* out of; endo *means* into.

Many chemical reactions release energy, usually in the form of heat, although other forms of energy may also be released (light, electrical, sound, etc.). Heat-producing reactions are termed *exothermic*. The group of chemical changes that absorb heat upon reaction are called *endothermic*. The terms exothermic and endothermic apply specifically to reactions involving changes in *heat* energy. More general descriptions would be *exergonic* and *endergonic,* which mean energy-releasing and energy-absorbing reactions, respectively.

The fact that energy is released during reaction implies that energy must be stored in the reacting species. Indeed, the law of conservation of energy* tells us that the energy must have been present in the reacting species since energy can be neither created nor destroyed. A change in the form of energy is permissible, e.g., from chemical energy to electrical energy as in nerve impulse reactions. Stored energy is called *potential energy* and during an exothermic reaction we witness the change of potential energy into heat that is released to the surroundings. A decrease in potential energy leads to a more stable system.

A basic tendency of nature is to reduce the potential energy of a system whenever possible. Examples of this tendency are water running downhill, a battery losing its charge, and a spring coming unwound. Recognition of this tendency of nature to reduce potential energy coupled with the observation that many chemical changes yield heat leads us to conclude tentatively that *atoms combine with one another because potential energy is decreased and a more stable system is formed in the process.*

exercise 3-1

(a) What must be happening during an endothermic reaction?
(b) Photosynthesis, the process by which plants convert carbon dioxide and water

*Recall that mass and energy are interconvertible and, strictly speaking, there is a single law of conservation of mass and energy. The conversion of mass to energy (and vice versa) is governed by the equation first formulated by Einstein: $E = mc^2$. During chemical reactions, however, there is no detectable change in mass; thus, we may refer separately to a law of conservation of energy and a law of conservation of mass.

to starch, is an endergonic reaction. How does the formation of starch by plants and the consumption of starch by animals fit into the topic of this section?

WHAT HOLDS ATOMS TOGETHER IN COMPOUNDS?

In attempts to answer this question, nineteenth-century chemists compiled many facts about known chemicals. These facts were gathered by observation of both physical and chemical properties of compounds. Gradually they recognized that chemical compounds could be divided into two broad categories. Some of the properties of these two categories are given below.

CATEGORY I (ionic)	CATEGORY II (covalent)
High-melting solids (melt above 200°C)	Gases, liquids, or low-melting solids (melt below 200°C)
Many are soluble in water	Most are insoluble or only slightly soluble in water
When heated sufficiently to cause melting, the molten compound conducts an electrical current	Will not conduct an electric current in gas or molten form
A direct electrical current (dc) will decompose the compound to the elements from which it was made	An electrical current will usually not decompose the compound to its elements

Today we call those compounds in Category I *ionic* compounds and those in Category II *covalent* or *molecular* compounds. These classifications, in themselves, contribute nothing to our understanding of what holds the compounds together; however, the observations leading to the categorization of the compounds do provide a basis for understanding chemical bonding. The fact that an electric current can reverse the combination (uncombine?) of the atoms in the Category I compounds suggests the chemical bonding force in these compounds is of an electrical nature. Additionally, the fact molten ionic compounds will conduct a current suggests that the atoms have in some manner acquired electric charges. Furthermore, since the atoms in these compounds are attracted to each other, some must have a positive charge and some, a negative charge. Thus, the attraction between opposite charges bonds (holds) the atoms together in ionic compounds.

An ionic bond is an attraction between oppositely charged ions.

An example of an ionic compound is sodium chloride, which has a high melting temperature (801°C), dissolves in water, can be decomposed to elemental sodium and chlorine by an electric current, and will conduct a current when melted. Therefore, sodium chloride belongs in Category I. Let us investigate its behavior a little more fully.

Let's melt a sample of sodium chloride in a suitable container and attach a source of direct electric current as shown in Fig. 3-1. We find sodium metal is produced at the negatively charged electrode and chlorine gas is produced at the positively charged electrode. This means that in sodium

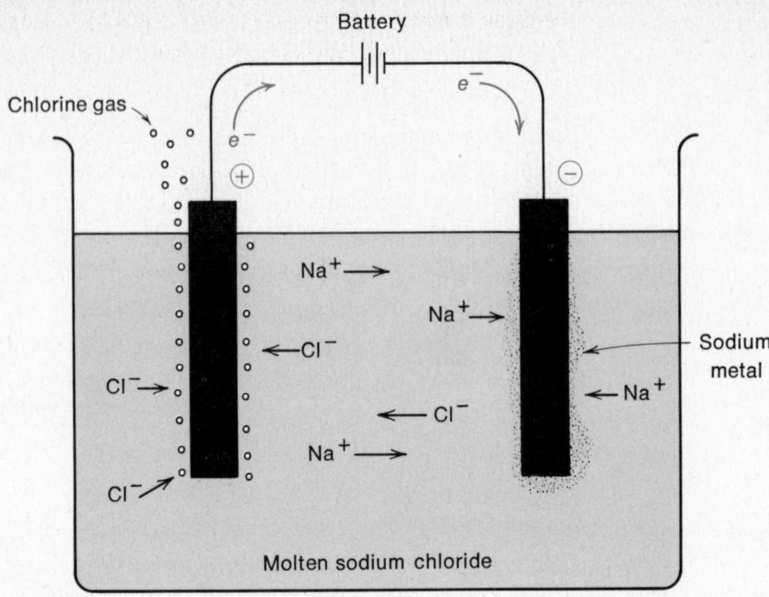

FIGURE 3-1 Decomposition of molten NaCl by an electric current.

chloride the sodium atom must have had a positive charge. (Why else would it be attracted toward a negative charge?) Similarly the chlorine atom must have a negative charge in sodium chloride. (Why?) When an electric current is passed through molten sodium chloride, the positively charged sodium ions move to the negative electrode and pick up electrons to form sodium metal. The negatively charged chloride ions migrate to the positive electrode and give up electrons to produce elemental chlorine gas. In this manner the electrical circuit is completed and an electric current flows.

In what ways do ions differ from atoms?

Positively charged ions are *cations* (kat′ī′ans) and negatively charged ions are *anions* (an′ī′ans). Let us again note that the properties of ions are vastly different than are the properties of the atoms from which they arise. Sodium chloride, containing sodium ions and chloride ions, is a white crystalline substance, dissolves readily in water and is nontoxic. Elemental sodium is a silverly white soft metal that quickly reacts with the oxygen in air and undergoes violent reaction with water. Elemental chlorine is a yellowish-green, highly toxic gas. When elemental sodium and chlorine are mixed, they undergo a very vigorous reaction, releasing a great deal of energy in the form of heat and light. From these observations we can see that the properties of sodium and chloride ions differ considerably from those of elemental sodium and chlorine atoms.

We might now inquire how and why ions are formed.

ION FORMATION

In our discussion of chemical periodicity in Chapter 2 we discussed ionization potential, electron affinity, and the stability of the electron arrangement of the noble gases. We found that relatively little energy is required to remove

Atoms of the noble gases Ne, Ar, Kr, Xe, and Rn have an octet of electrons in their outer shell.

an electron from an atom that has one electron in its outermost shell. When this electron is lost, the resulting ion is positively charged and has an electron arrangement identical to the neutral atoms of the preceding element—a noble gas. Thus, we can explain how a sodium ion could be formed—by the loss of an electron.

What about the chloride ion in sodium chloride? How does it acquire a negative charge? Recall from our discussion of electron affinity that a chlorine atom will give off energy when it gains an electron and acquires the electron arrangement of argon, the next noble gas.

This line of reasoning suggests that when elemental sodium is combined with elemental chlorine to produce sodium chloride, electrons are transferred from sodium atoms to chlorine atoms to produce sodium ions and chloride ions. The transfer of electrons *to produce an ionic compound* is accompanied by a decrease in potential energy and the release of this energy as heat and light.

WHERE DOES THE ENERGY RELEASED COME FROM WHEN SODIUM CHLORIDE IS FORMED?

We might be tempted at this point to conclude that energy is given off during the electron transfer process. A look at some facts concerning the electron transfer reveals that this cannot be true. The energy required to remove an electron from a sodium atom is 5.2 eV.

$$\mathrm{Na} + 5.2\ \mathrm{eV} \longrightarrow \mathrm{Na}^+ + e^-$$

The energy given off when chlorine gains an electron is only 3.8 eV.

This energy is measured experimentally.

$$\mathrm{Cl} + e^- \longrightarrow \mathrm{Cl}^- + 3.8\ \mathrm{eV}$$

A quick calculation reveals that we *are actually short* an amount of energy equivalent to 1.4 eV to achieve the electron transfer.

$\mathrm{Na} \longrightarrow \mathrm{Na}^+ + e^-$	5.2 eV required
$\mathrm{Cl} + e^- \longrightarrow \mathrm{Cl}^-$	3.8 eV released
$\mathrm{Na} + \mathrm{Cl} \longrightarrow \mathrm{Na}^+ + \mathrm{Cl}^-$	1.4 eV net required for electron transfer

When opposite charges come together, energy is given off.

Obviously, the energy released when the *compound* sodium chloride is formed must come from a source we have not yet considered. We have overlooked the attraction of the positive sodium ion for the negative chloride ion. The attraction between these oppositely charged ions releases an additional 7.0 eV.

$$\underset{\text{separated ions}}{\mathrm{Na}^+ + \mathrm{Cl}^-} \longrightarrow \underset{\text{bonded ions}}{\mathrm{Na}^+\mathrm{Cl}^-} + 7.0\ \mathrm{eV}$$

The additional energy more than accounts for the 1.4 eV lacking for the electron transfer process. (Remember this extra 7.0 eV is unavailable unless the ions are formed.) The extra 7.0 eV of energy released when the newly formed ions come together yields an extra 5.2 eV of potential energy that can be lost as heat and light when sodium combines with chlorine to give sodium chloride.

SODIUM CHLORIDE MOLECULES?

We cannot mix a single sodium atom with one chlorine atom when sodium and chlorine are combined and form sodium chloride. Instead, billions of sodium atoms mingle with billions of chlorine atoms to produce billions of sodium and chloride ions. To understand what happens when all these ions are produced, we must consider what the environment around an ion is like.

A charge creates an electric field around it. This field extends in all directions around each ion much like light radiates in all directions from a light bulb.

Opposite charges attract; like charges repel.

A chloride ion is surrounded by six Na^+ ions: one above, one below; one in front, another in back, and one on each side.

When ions are formed near each other, their electric fields interact. Since these fields extend in all directions, each individual ion interacts with all nearby ions. Each positively charged ion repels all the other positively charged ions around it and attracts all the negatively charged ions around it. Similarly, the negative ions attract positive ions and repel negative ions. Thus, the ions will clump together in an arrangement of alternating cations and ions. When the arrangement of sodium and chloride ions in sodium chloride is determined with the aid of X-ray crystallography, we find the ions are arranged in a regular array of alternating Na^+ and Cl^- ions. The attraction between opposite charges and repulsions between like charges cause the ions to arrange in a pattern of alternating cations and anions.

If we look at Figs. 3-2 and 3-3, we see that the attraction of a sodium ion for a chloride ion on one side of it should be no greater than the attraction for a chloride ion on the other side. If each chloride ion surrounding a sodium ion is equally attracted to it, we cannot say a particular chloride ion belongs to a particular sodium ion. In fact, looking at Fig. 3-3 showing the actual arrangement of the ions, we conclude each sodium ion is equally attracted to the six chloride ions that surround it. Likewise, each chloride ion is surrounded by and equally attracted to six sodium ions. Consequently, we cannot single out a particular Na^+Cl^- pair and call the pair a molecule. Chemists regard *ionic solids merely as orderly collections of oppositely charged ions.*

Ionic compounds do not exist as molecules at room temperature.

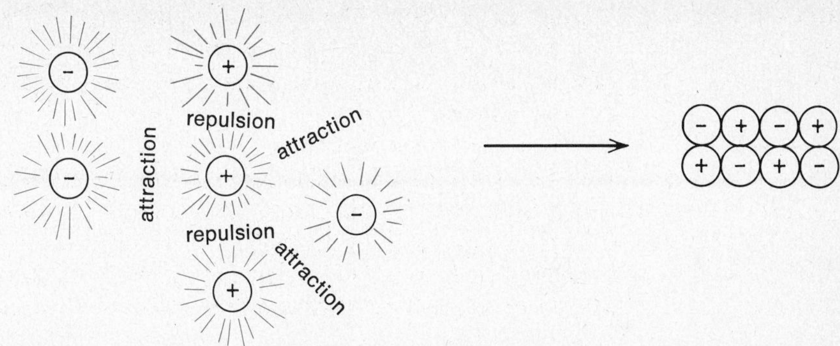

FIGURE 3-2 The attractions between opposite charges and repulsions between like charges cause the ions to arrange in a pattern of alternating cations and anions.

FIGURE 3-3 Arrangement of sodium and chloride ions in sodium chloride.

OTHER IONS AND IONIC COMPOUNDS

Experimentation has shown in general that the energy considerations (ionization potential, electron affinity, and attraction between opposite charges) are such that when elements of Groups IA, IIA, or IIIA (except boron) are combined with elements of either VIA or VIIA, ionic compounds result. Furthermore, we find that each of the ions produced has the electron arrangement of a noble gas. The metals (Groups IA, IIA, and IIIA) form cations and the nonmetals (Groups VIA and VIIA) form anions.

In general, the number of electrons a nonmetal atom will gain equals 8 − its group number, e.g., oxygen, gains two electrons: 8 − 6 = 2.

For example, magnesium is found to lose two electrons and attain the electron arrangement of neon. Oxygen in Group VIA can gain two electrons and attain the electron arrangement of neon. Thus, magnesium can react with oxygen in an electron transfer process to produce the ionic compound magnesium oxide (MgO) containing Mg^{2+} and O^{2-} ions.

Since we are concerned only with the outermost electrons of an atom in an electron transfer process, we can use a shorthand way of representing the outermost electron arrangement. In this shorthand notation we shall represent the outermost electrons by dots and use the elemental symbol to represent the nucleus *plus* the inner electrons. Under this system the first 20 elements would be depicted as illustrated below.

It doesn't matter which side of the symbol the dots are placed; ·H, H·, $\dot{H}$, and $\underset{\cdot}{H}$ all mean the same thing.

H·

Li·   ·Be·   ·$\dot{B}$·   ·$\dot{\underset{\cdot}{C}}$·   ·$\ddot{\underset{\cdot}{N}}$·   :$\ddot{\underset{\cdot}{O}}$·   :$\ddot{\underset{\cdot\cdot}{F}}$·   :$\ddot{\underset{\cdot\cdot}{Ne}}$:

Na· ·Mg· ·Al· ·Si· ·P· :S· :Cl· :Ar:

K· ·Ca·

exercise 3-2

(a) Explain in detail what the symbol C, Ne, and Cl represent in the context in whicl they are used above.
(b) Explain why the dot representation of oxygen and sulfur are the same (the symbo is surrounded by six dots in each).

These dot representations are usually referred to as Lewis dot diagrams. We can use these diagrams to predict the ions each element will form. Remember that metals form cations and that nonmetals form anions. Sodium has but one outer electron and if this electron is lost, it will acquire the electron arrangement of neon.

$$\mathrm{Na}\cdot \longrightarrow \mathrm{Na}^{+} + e^{-}$$

Beryllium is predicted to lose two electrons and form the Be^{2+} ion. Sulfur, on the other side of the periodic chart, will gain two electrons to form S^{2-}.

$$2e^{-} + :\ddot{\underset{\cdot}{\mathrm{S}}}\cdot \longrightarrow :\ddot{\underset{\cdot\cdot}{\mathrm{S}}}:^{2-}$$

exercise 3-3

Predict the ions expected to be formed by Al, F, Ca, Cs, I, and Se.

PREDICTING AND WRITING FORMULAS FOR IONIC COMPOUNDS

Formulas for ionic compounds are written so that total number of plus (+) charges equals total number of minus (−) charges.

Formulas are chemical shorthand for showing the composition of compounds. We have already used the formulas NaCl and MgO to represent sodium chloride and magnesium oxide, respectively. Formulas tell us more than just what elements are present. They also tell us the ratio in which the elements combine to produce the compound. The formula NaCl tells us, for instance, that one chloride ion is required for each sodium ion. This is the simplest ratio in which sodium ions and chloride ions can combine to produce an electrically neutral compound. Since, as we have seen, ionic compounds do not contain molecules, *the formulas written for ionic compounds merely represent the smallest whole-number ratio of cations and anions that yield an electrically neutral formula.* Such formulas are called *empirical* formulas.

If the ratio between cations and anions is something other than 1:1, numerical subscripts are used. In the compound magnesium chloride, there are two chloride ions for each magnesium ion. The formula is written $MgCl_2$. When no subscript appears in the formula, the subscript is understood to be 1. If there weren't at least one ion of that particular element in the ratio, we wouldn't bother to write the symbol.

Our ability to predict the charges on ions allows us to predict and to write formulas for most ionic compounds produced by reaction between Group IA, IIA, and IIIA elements and Group VIA and VIIA elements. By convention, the positive ion is written first.

EXAMPLE 3-1

Predict and write the formula for the ionic compound produced by reaction between calcium and fluorine.

Calcium has the dot diagram $\cdot \mathrm{Ca} \cdot$ and should form a Ca^{2+} ion by loss of two electrons.

$$\cdot \mathrm{Ca} \cdot \longrightarrow \mathrm{Ca}^{2+} + 2e^-$$

Fluorine has the dot diagram $:\ddot{\underset{\cdot\cdot}{\mathrm{F}}}\cdot$ and should form a F^- ion by gain of one electron.

$$e^- + :\ddot{\underset{\cdot\cdot}{\mathrm{F}}}\cdot \longrightarrow :\ddot{\underset{\cdot\cdot}{\mathrm{F}}}:^-$$

Since two negative charges are required to neutralize two positive charges, the simplest ratio between Ca^{2+} and F^- is CaF_2.

EXAMPLE 3-2

Predict and write the formula for the ionic compound produced by reaction between aluminum and oxygen.

Aluminum forms a 3+ ion:

$$\cdot\dot{\mathrm{Al}}\cdot \longrightarrow \mathrm{Al}^{3+} + 3e^-$$

and oxygen forms a 2− ion:

$$:\ddot{\underset{\cdot}{\mathrm{O}}}\cdot + 2e^- \longrightarrow :\ddot{\underset{\cdot\cdot}{\mathrm{O}}}:^{2-}$$

When total plus *charge equals total* minus *charge, the* net *charge is zero*

The combination of one aluminum ion with one oxygen ion would give a formula with a +1 charge. This would clearly be unsatisfactory. The simplest satisfactory formula would be one containing two aluminum ions and three oxygen ions: Al_2O_3. In this way we have a total of six positive charges and six negative charges that cancel each other.

$$\underset{\text{positive charge}}{[2(+3)} + \underset{\text{negative charge}}{3(-2)]} = [(+6) + (-6)] = \underset{\text{net charge}}{0}$$

exercise 3-4

Write formulas for the compounds expected to be formed between Na and Br, Li and O, Mg and S, Mg and F, K and S.

Up to this point, there are a number of elements we have ignored that will form ionic compounds with the Group VIA and VIIA elements. These elements are the transition metals, the lanthanides and the actinides. All of these elements form positive ions.

It is difficult to predict the charge of the ions that these elements will form. In fact, many of these elements form more than one cation! In

Note that some transition metal atoms can form more than one cation, e.g., Fe^{2+} and Fe^{3+}.

biological systems we will be concerned with only a very small number of these elements and if we will spend just a few minutes learning the charges on several transition metal ions, we can then continue. These important ions are Mn^{2+}, Fe^{2+}, Fe^{3+}, Co^{2+}, Co^{3+}, Ni^{2+}, Cu^{+}, Cu^{2+}, and Zn^{2+}.

exercise 3-5

Write formulas for the ionic compounds formed between Mn and Cl, Zn and O, Cu^{+} and O, Cu^{2+} and O, Fe^{3+} and S.

NAMING IONIC COMPOUNDS

When we meet people, we generally find out their names. This helps us to remember and identify them later. Similarly, we should learn the names of the compounds with which we have become acquainted. People's names have an orderly arrangement: a first or given name followed by a last or surname. Zoraster Knoppenheimerschmidt for example. Simple (containing only two elements) ionic compounds also have a first name and last name. The first name is simply the elemental name of the cation. The last name is generally the first syllable of the anion's elemental name followed by *ide*. Thus, NaCl is sodium chlor*ide;* Al_2O_3 is aluminum ox*ide*. In those cases where more than one positive ion is possible for the metal atom, the charge on the ion is given in Roman numerals in parentheses. $FeCl_2$ and $FeCl_3$ are iron(II) chloride and iron(III) chloride, respectively. Likewise, FeO and Fe_2O_3 are, respectively, iron(II) oxide and iron(III) oxide.

The use of Roman numerals indicates the charge on cations.

exercise 3-6

Name all the compounds in Exercises 3-4 and 3-5.

CHEMICAL EQUATIONS

Chemical formulas are a shorthand way of conveying information about the composition of a compound. Similarly, a chemical equation is a shorthand way of describing the changes that take place during a chemical reaction. Since mass or atoms are neither created nor destroyed during a chemical change, chemical reactions are always balanced to show the same numbers of atoms of each element on both sides of the equation. This is useful because the resulting equation tells us the relative numbers of the different atoms that are undergoing reaction.

Generally, in a chemical equation, substances written to the left of the arrow are called reactants; *substances to the right of the arrow are called* products.

Experimentally, we observe that orange-red copper metal reacts with yellow sulfur powder to produce black copper(II) sulfide. The reaction may be represented by the balanced equation

$$Cu + S \longrightarrow CuS$$

Returning to our old friend sodium chloride, we can write the equation

$$Na + Cl \longrightarrow NaCl$$

The arrow is read as the word yields.

An arrow is used instead of an equals sign in chemical equations because a mixture of the reactants is different than the products. In this case, a mixture of sodium atoms and chlorine atoms are not the same as sodium chloride, which is an aggregation of sodium and chloride ions. The equation would be read *sodium plus chlorine yields sodium chloride.*

Chemists would not be satisfied with the way we have written the equation for sodium chloride formation. Certain elements or compounds occur as molecules and that fact must be shown in the equation. It turns out, as we shall see shortly, that elemental chlorine is normally found in pairs. These pairs of chlorine atoms are called *chlorine molecules* and are written Cl_2.

Thus, we need to rewrite our equation to convey this information. Starting over, we first write

$$Na + Cl_2 \longrightarrow NaCl$$

Now the equation is unbalanced; the left-hand side has two atoms of chlorine and the right-hand side has only one. To correct this we must also have two chlorides on the right-hand side. We can't change the formula of sodium chloride so the only choice is to have two NaCl formed. Thus

We should not write $NaCl_2$. This formula would indicate that two Cl^- ions combine with one Na^+ ion; this would be electrically unbalanced.

$$Na + Cl_2 \longrightarrow 2NaCl$$

But now the sodium is unbalanced. This can be corrected by beginning with two sodium atoms:

$$2Na + Cl_2 \longrightarrow 2NaCl$$

and finally we have achieved a correct balanced equation.

Determining the coefficients in a chemical equation is called *balancing* the equation and is essentially a trial and error process. In most cases, there are not any general sets of procedures that will allow us to write balanced equations. When attempting to write equations, we need to know which elements exist as diatomic (two atoms) molecules. These are H_2, N_2, O_2, F_2, Cl_2, Br_2, and I_2.

Memorize these diatomic elements.

To write balanced equations, first write the formulas of the reactants correctly; then write the formulas of the products correctly; next, use only coefficients to balance the equation. Additional changes in subscripts are not permitted.

Let's try writing and balancing another equation, this time for the reaction between aluminum and oxygen. Attempts would probably look something like the following:

$$Al + O_2 \longrightarrow Al_2O_3$$
$$2Al + O_2 \longrightarrow Al_2O_3$$
$$2Al + 3O_2 \longrightarrow Al_2O_3$$
$$2Al + 3O_2 \longrightarrow 2Al_2O_3$$
$$4Al + 3O_2 \longrightarrow 2Al_2O_3 \quad \text{Correct balanced equation, Whew!}$$

exercise 3-7

Write balanced equations from the elements for the formation of the following compounds: MgI_2, Fe_2O_3, Li_2O, BeO, and KBr.

Not all reactions involve the combination of elements. Two compounds or a compound and an element could react. Balancing equations for such reactions follows the same general procedure. Consider the reaction between magnesium oxide and hydrochloric acid:

$$MgO + HCl \longrightarrow MgCl_2 + H_2O \qquad \text{unbalanced}$$
$$MgO + 2HCl \longrightarrow MgCl_2 + H_2O \qquad \text{balanced}$$

Consider the reaction between aluminum and hydrochloric acid:

$$Al + HCl \longrightarrow AlCl_3 + H_2 \qquad \text{unbalanced}$$
$$2Al + 6HCl \longrightarrow 2AlCl_3 + 3H_2 \qquad \text{balanced}$$

Don't worry for now how to predict the products of reactions between compounds. We will consider this in due time. You should practice how to balance equations now though. A few are provided in Exercise 3-8 for your convenience.

exercise 3-8

Balance the following equations.

(a) $Na + H_2O \longrightarrow NaOH + H_2$

(b) $CuS + O_2 \longrightarrow CuO + SO_2$

(c) $N_2 + H_2 \longrightarrow NH_3$ **(certain bacteria do this)**

(d) $CO_2 + H_2O \longrightarrow C_6H_{12}O_6 + O_2$ **(green plants do this when the sun shines)**

(e) $C_{16}H_{32}O_2 + O_2 \longrightarrow CO_2 + H_2O$ **(you do this every day, even when the sun doesn't shine)**

OXIDATION AND REDUCTION (YOU CAN'T HAVE ONE WITHOUT THE OTHER)

Metals have been shown to be electron donors and nonmetals can act as electron acceptors. The transfer of electrons from metal atoms to nonmetal atoms involves two simultaneous processes: (1) loss of an electron (or electrons) by an atom (or ion) and (2) gain of an electron (or electrons) by another atom (or ion). Such electron transfer reactions are collectively known as oxidation-reduction reactions or more simply, *redox* reactions. Many redox reactions are energy-releasing. We take advantage of this fact in constructing batteries. Nature also takes advantage of this exergonic feature of redox reactions to provide energy to run our bodies. For this reason we need to investigate redox reactions in a little more detail.

Oxidation refers to the loss of one or more electrons by an atom or molecule. Conversely, reduction is the gain of one or more electrons by an atom or molecule.

EVENT	**DESCRIPTION**
$Li \longrightarrow Li^+ + e^-$	Oxidation (loss of electron from lithium atom)
$Fe^{2+} \longrightarrow Fe^{3+} + e^-$	Oxidation (oxidation of Fe^{2+} ion to Fe^{3+} ion)
$Br^0 + e^- \longrightarrow Br^-$	Reduction (gain of electron by bromine atom to form bromide ion)
$S^0 + 2e^- \longrightarrow S^{2-}$	Reduction of sulfur to sulfide ion
$2H^+ + 2e^- \longrightarrow H_2$	Reduction of hydrogen ion

exercise 3-9

Classify each of the following events as either an oxidation or reduction process:
(a) Na^0 + energy $\longrightarrow e^- + Na^+$
(b) $Ca^{2+} + 2e^- \longrightarrow Ca^0$ + energy
(c) Conversion of an oxygen atom to an oxide ion.
(d) Formation of elemental sulfur from a sulfide ion.

The substance that loses electrons during reaction is said to be *oxidized.* The substance gaining electrons is *reduced.* Returning once again to reaction between sodium and chlorine to produce sodium chloride, we may divide the overall reaction into two parts:

$$Na\cdot \longrightarrow Na^+ + e^-$$

and

$$:\underset{\cdot\cdot}{\overset{\cdot\cdot}{Cl}}\cdot + e^- \longrightarrow :\underset{\cdot\cdot}{\overset{\cdot\cdot}{Cl}}:^-$$

In the first equation shown, sodium loses an electron. This equation represents the oxidation half reaction. Sodium is oxidized. We call it a *half reaction* because it cannot take place unless something is present to gain the electron, in this case chlorine. The second equation shows chlorine gaining an electron; therefore, chlorine is reduced. It depicts the reduction half reaction. When the half reactions are combined to show the overall reaction, it is termed a *redox reaction.*

$$2Na + Cl_2 \longrightarrow 2NaCl$$

Since the chlorine causes the sodium atom to lose an electron, chlorine is called an *oxidizing agent*. Similarly, because sodium causes chlorine to gain an electron, sodium is a *reducing agent*. Terms relating to oxidation and reduction processes are summarized in Table 3-1.

Because redox reactions are electron transfer processes, the charge on atoms or ions change during reaction. To symbolize this and to show more clearly which atoms gain or lose electrons, the term *oxidation state,* also called *oxidation number* has been devised. The oxidation state (or oxidation number) of an atom is given by a number and a sign. An algebraic increase in this number indicates the atom has been oxidized during reaction. A

TABLE 3-1
Oxidation–Reduction relationships

Oxidation	A loss of electrons
Reduction	A gain of electrons
Substance oxidized	Loses electrons; is the reducing agent
Substance reduced	Gains electrons; is the oxidizing agent
Oxidizing agent	Causes another substance to lose electrons; it gains electrons
Reducing agent	Causes another substance to gain electrons; it loses electrons

decrease indicates the atom has been reduced. The oxidation state of an atom or ion is determined according to the following conventions:

Elemental form *means that the element is* not combined with any other element.

1. The oxidation state of an element in its elemental form is defined as zero.
2. The oxidation state of a simple cation or anion is equal to the charge on the ion.*
3. Hydrogen in compounds normally has an oxidation state of +1. We can ignore the few exceptions.
4. Oxygen in compounds normally has an oxidation state of −2. An exception will be considered later.
5. The algebraic sum of the oxidation states of ions in an ionic compound is 0 when the empirical formula is written correctly.

Again consider the reaction

$$2Na + Cl_2 \longrightarrow 2NaCl$$

During reaction the oxidation state of sodium goes from 0 to +1, an increase in oxidation state; thus sodium is oxidized. Chlorine also undergoes an oxidation state change, 0 to −1; this is a decrease in oxidation number or reduction. During a redox reaction, not all atoms or ions undergo oxidation or reduction. For example, in the reaction between zinc and hydrogen chloride

$$Zn + 2HCl \longrightarrow ZnCl_2 + H_2$$

Chlorine has a −1 oxidation state in both the reactant and product.

*Sometimes the oxidation state of a simple ion is called its *valence* or *electrovalence*. The term *valence* is incorrect when used *in this context*.

exercise 3-10

Below are several balanced equations. Determine the oxidation state of each of the elements in both the reactants and the products. Decide whether oxidation or reduction has occurred. What is the oxidizing agent? What is the reducing agent? Write the oxidation half reaction and the reduction half reaction.

(a) $2Mg + O_2 \longrightarrow 2MgO$
(b) $MgO + 2HCl \longrightarrow MgCl_2 + H_2O$
(c) $2Al + Fe_2O_3 \longrightarrow Al_2O_3 + 2Fe$
(d) $2Al + 6HCl \longrightarrow 2AlCl_3 + 3H_2$
(e) $CuCl_2 + Zn \longrightarrow ZnCl_2 + Cu$

Metabolites are molecules from food that undergo reactions in the body.

Extremely important electron transfer processes take place in living cells. A number of molecules known as *cytochromes* are in the mitochondria. These cytochrome molecules transfer electrons from metabolites to oxygen. These redox reactions supply the majority of the energy required by the cell to go about its life processes. At the heart of these complex cytochrome molecules are iron ions. As the electrons move from the metabolites to the cytochromes and ultimately to oxygen, the iron changes oxidation states between +2 and +3. This change in oxidation states of the iron releases energy to the cell. To understand the process more clearly, let us look at the sequence of reactions taking place in Fig. 3-4.

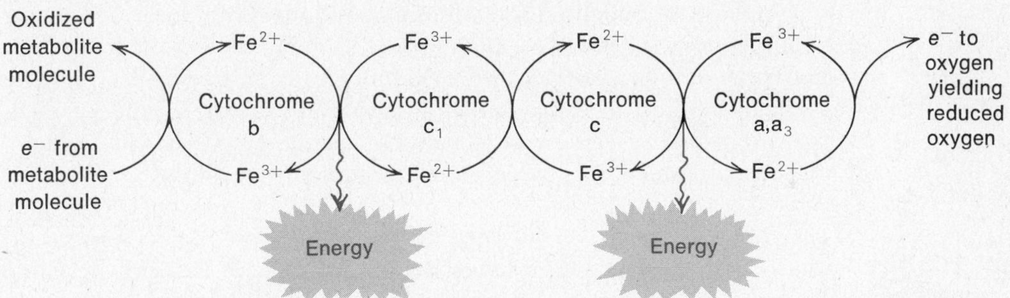

FIGURE 3-4 The cytochrome chain.

$Fe^{2+} \rightarrow e^- + Fe^{3+}$ (oxidation)
$e^- + Fe^{3+} \rightarrow Fe^{2+}$ (reduction)
This electron transfer process can be represented

Fe^{3+} Fe^{2+}
Fe^{2+} Fe^{3+}

We see an electron from a metabolite molecule (how the electron comes to this point from food molecules is a topic for later chapters) is transferred to cytochrome b, reducing the Fe^{3+} to Fe^{2+}. The Fe^{3+} in cytochrome c oxidizes the now Fe^{2+} of cytochrome b and itself is reduced to Fe^{2+}. During this redox reaction, energy is released to the cell. The electron passes next to cytochrome a in another redox reaction and then on to cytochrome a_3. The transfer from cytochrome a to cytochrome a_3 is another point at which energy is released to the cell. In the last step the electron passes to an oxygen atom reoxidizing the cytochrome a_3. The reduced oxygen ion picks up hydrogen ions given off by the metabolite molecule producing a molecule of water. Examination of Fig. 3-4 representing the electron transport system reveals it functions much like a bucket brigade. There is a steady stream of electrons moving through the system with oxygen as the only atom permanently reduced and used up. The metabolite molecule is permanently oxidized and consumed.

COMPOUNDS IN CATEGORY II

Earlier we saw that compounds are divided into two categories: ionic compounds and covalent compounds. Now that we have acquired an understanding of ionic compounds, let us turn our attention to the covalent ones. Review of the properties of covalent compounds leads us to conclude that they do not consist of ions since they do not conduct an electric current and are low melting solids or liquids or gases. Yet we find large amounts of energy are given off when atoms combine to produce these compounds. For example, when hydrogen combines with chlorine to produce hydrogen chloride gas, the equivalent of 4.5 eV of energy is liberated. This amount of energy is close to that given off when sodium combines with chlorine to produce sodium chloride. Nevertheless, pure hydrogen chloride has all the classic properties of a covalent compound.

From this type of experimental data, it is concluded that atoms may combine in small groups called *molecules. Within the molecule the forces of attraction between atoms is very large but the forces of attraction between neighboring molecules is generally small.* Another way of stating this belief is that a group of atoms forming a molecule behaves relatively independently of the groups of atoms in nearby molecules. We might liken this situation to an apartment building filled with families, in which a family represents a molecule with a close association among its members. Yet the interaction of a family in one apartment with a family in the next apartment may be minimal. Contrast this situation with that of an ionic compound in which each individual ion interacts with all the surrounding ions. How can we explain the attraction between atoms in molecules without the formation of ions?

LEWIS' THEORY OF COVALENT BONDING

Shortly after Bohr proposed his model of the atom, G. N. Lewis provided the first suitable explanation of covalent bonding. He noticed that most covalent compounds contain an even number of electrons. For example, water (H_2O) contains 10 electrons, 1 from each hydrogen plus 8 from the oxygen. Similarly, carbon dioxide (CO_2) contains 22 electrons, 8 from each oxygen atom plus 6 from carbon.

Atoms may share electrons.

This observation suggested to Lewis that covalent bonding is a result of a tendency of electrons to pair during the bonding process. From this reasoning, he concluded *a covalent bond is the result of the sharing of a pair of electrons between two bonded atoms.* Initially he envisioned that each of the combining atoms contributed one electron to the shared pair.

The simplest covalent compound is hydrogen gas (H_2). Recalling the dot diagrams, we may picture the formation of an H_2 molecule from hydrogen in the following manner:

$$\mathrm{H\cdot + H\cdot \longrightarrow H:H}$$

In this way, Lewis has given us a basic understanding of the distinction between covalent and ionic bonding. Both types of bonding are of an electrical nature. The distinction between bond types is in the way the electrical forces operate. *Ionic bonding is the result of attraction between ions of unlike charge; covalent bonding is the result of attraction between positively charged nuclei and negatively charged pairs of electrons.*

The reaction between two hydrogen atoms to form H_2 lowers the potential energy of hydrogen atoms by the equivalent of 4.5 eV. How can this be? How is it possible to lower potential energy by bringing negatively charged electrons from different atoms near one another. Lewis was able to answer these questions in the following way. Because the electron pair shared by the atoms is negatively charged and the nuclei of the atoms are positively charged, both hydrogen nuclei are simultaneously attracted to the electron pair, with a net result of an increase in attraction. Consider the situation as two hydrogen atoms approach one another, as shown in Fig. 3-5 (solid double-headed arrows represent attraction; dotted ones, repulsion).

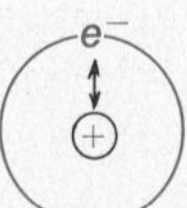

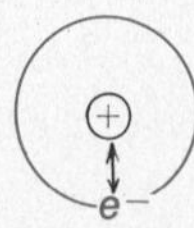

Isolated H atoms-no interaction

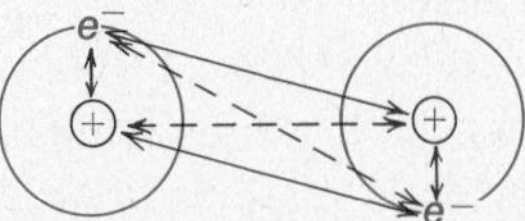

Approaching atoms

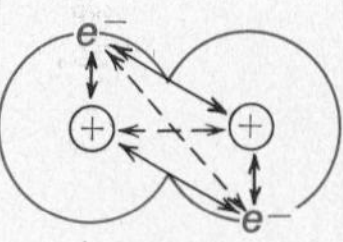

Bonded atoms

FIGURE 3-5 Attractive and repulsive forces between covalently bonded atoms.

As the atoms approach one another, two additional attractive forces begin to develop as do two repulsive forces; however, the interaction between electric charges increases as the distance decreases. As the atoms bond, the distance between the attracting nuclei and electrons is shorter than the distance between the repelling electrons and the repelling nuclei. Therefore, the two new attractive forces are larger than the two repulsive forces. Comparing the unbonded atoms and the bonded atoms we find the net attractive forces are larger in the H_2 molecule than twice the attractive force in a single atom. The increased attraction lowers the potential energy and the formation of an H_2 molecule from two hydrogen atoms is exergonic.

UNBONDED ATOMS	BONDED ATOMS
Two strong attractive forces. (The attraction of each nucleus for its own electron.)	Four strong attractive forces and two weaker repulsive forces.

PREDICTING AND WRITING FORMULAS OF COVALENT COMPOUNDS

Experience has shown that covalent compounds are formed when nonmetals react with other nonmetals. In general these elements will form a sufficient number of bonds to attain a noble gas electron arrangement. With the exception of hydrogen, this means they will try to involve eight electrons in the outer shell by electron sharing. As we will see later, this is not a universal rule. With rare exception, the noble gas electron arrangement rule is true for hydrogen and the second period elements carbon through neon. For nonmetals with higher atomic numbers, this holds true in a large number of compounds but many exceptions exist. One might ask why rules are made if there are many exceptions. It is better to have rules allowing us to be correct most of the time than to have no guidance and the possibility of never being correct. The situation is similar to when we first learned to conjugate verbs. First we learned a set of rules for the conjugation of regular verbs. These rules were followed shortly thereafter by a long list of irregular verbs that do not obey the rules and must be learned separately. This is the way with some reactions between elements. Let us first look at those reactions that follow the rules.

Hydrogen will combine with all the nonmetals (except boron) to form compounds that follow the rules. Consider the hydrogen fluoride molecule HF. The dot diagrams of hydrogen and fluorine are $H\cdot$ and $:\!\ddot{\underset{\cdot}{F}}\!:$. If the hydrogen and fluorine atoms each contribute an unpaired electron to forming a shared electron pair both atoms *appear* to achieve a noble gas electron arrangement

$$H\cdot + \cdot\ddot{\underset{\cdot\cdot}{F}}\!: \longrightarrow H\!:\!\ddot{\underset{\cdot\cdot}{F}}\!:$$

This way of representing the HF molecule is known as a *Lewis dot structure.*

Oxygen has only six outer electrons. To attain the desired eight electrons, it must form two covalent bonds. Since hydrogen forms only one bond, two hydrogen atoms are necessary to satisfy each oxygen atom.

$$H\cdot + \cdot\ddot{\underset{\cdot\cdot}{O}}\cdot + \cdot H \longrightarrow H\!:\!\ddot{\underset{\cdot\cdot}{O}}\!:\!H$$

Carbon, with only four outer electrons, needs to form four bonds with hydrogen to acquire a share in eight outer electrons.

$$\begin{array}{ccccc} & H & & & \\ & \cdot & & & \\ H\cdot & \cdot C \cdot & \cdot H & \longrightarrow & \begin{array}{c} H \\ H\!:\!\overset{\cdot\cdot}{\underset{\cdot\cdot}{C}}\!:\!H \\ H \end{array} \\ & \cdot & & & \\ & H & & & \end{array}$$

Similarly, carbon will combine with four fluorine atoms.

$$\begin{array}{ccccc} & :\!\ddot{F}\!: & & & \\ & \cdot & & & \\ :\!\ddot{\underset{\cdot\cdot}{F}}\cdot & \cdot C \cdot & \cdot\ddot{\underset{\cdot\cdot}{F}}\!: & \longrightarrow & \begin{array}{c} :\!\ddot{F}\!: \\ :\!\ddot{\underset{\cdot\cdot}{F}}\!:\!\overset{\cdot\cdot}{\underset{\cdot\cdot}{C}}\!:\!\ddot{\underset{\cdot\cdot}{F}}\!: \\ :\!\underset{\cdot\cdot}{F}\!: \end{array} \\ & \cdot & & & \\ & :\!\underset{\cdot\cdot}{F}\!: & & & \end{array}$$

With the exception of hydrogen and boron, we can determine the minimum number of covalent bonds a nonmetal will form by subtracting the number of outermost electrons from eight. (Why is hydrogen an exception?)

exercise 3-11

Predict the molecular formulas for the following combinations of elements and write their Lewis dot structures.
(a) F and F
(b) N and H
(c) P and Cl
(d) C and Cl
(e) Si and F

MULTIPLE BONDS

Carbon and oxygen will combine to produce carbon dioxide (CO_2). According to our rule for predicting numbers of bonds, carbon will form four bonds and oxygen, two bonds. On this basis the formula CO_2 seems entirely reasonable. When we try to write a Lewis dot structure, however, a difficulty arises. How can a carbon atom form four bonds to only two oxygen atoms? The answer lies in allowing two bonds to form between each oxygen and the carbon atom. If one bond is a shared pair of electrons, then two bonds must be two shared pairs of electrons or four electrons. The Lewis dot structure of CO_2 must then be

$$\ddot{\underset{\cdot\cdot}{O}}::C::\ddot{\underset{\cdot\cdot}{O}}$$

This is the only way in which one carbon atom and two oxygen atoms can combine to satisfy all the rules of writing Lewis dot structures.

A double bond is two pairs of electrons shared between two atoms.

A two-electron bond is known as a *single* bond. A four-electron bond is called a *double* bond. Another compound containing a double bond is formaldehyde (CH_2O). The Lewis dot structure of formaldehyde is

```
H     ..
 :C::O
H     ..
```

A line is used to represent a pair of electrons.

Getting tired of writing all those dots? Let's simplify the procedure a little by using a line to represent a *pair* of electrons. Under this system the Lewis structure of formaldehyde is written

```
H      _
 \
  C=O
 /     ‾
H
```

Electrons shared by atoms are called bonding *electrons. Outer shell electrons that are not shared are called* nonbonding *electrons.*

Sometimes a double bond isn't sufficient to satisfy all the rules for writing Lewis structures and we need to consider triple bonds. Acetylene with the molecular formula C_2H_2 is such a case. The Lewis structure of acetylene is

$$H{-}C{\equiv}C{-}H$$

exercise 3-12

All the following contain multiple bonds. Write the Lewis structure of each. The underlined atom is the one to which all others are bonded.
(a) H$\underline{C}$N (c) $\underline{C}S_2$
(b) $\underline{C}OCl_2$ (d) N_2

Before leaving this section it should be pointed out that in general only carbon, nitrogen, oxygen, and sometimes sulfur and phosphorus form multiple bonds.

COORDINATE COVALENT BONDS

To this point in our discussion of covalent bonds, we have only considered a contribution of one electron by each of the atoms sharing the electron pair. Chemists have found a number of molecules that cannot be explained in this manner. Carbon monoxide is such a molecule. If we only allow each atom to contribute a single electron toward the formation of each bond, we find it impossible to obey all the requirements for writing Lewis structures. For example, two attempts are illustrated below:

$$1.\ \cdot\dot{\underset{\cdot}{C}}\cdot + \cdot\ddot{\underset{\cdot}{O}}\colon \longrightarrow |C{=}\overline{\underline{O}}$$

$$2.\ \cdot\dot{\underset{\cdot}{C}}\cdot + \cdot\ddot{\underset{\cdot}{O}}\colon \longrightarrow C{\equiv}O|$$

Both attempts are unsatisfactory; in reaction 1 the carbon is surrounded by only 6 electrons; in reaction 2 the oxygen is surrounded by 10 electrons. If we envision one atom as donating both electrons of the shared pair, however, we can quickly arrive at a suitable electron arrangement. In order to visualize this more readily, the electrons on carbon are represented by dots and the electrons on oxygen by crosses.

$$\cdot\dot{\underset{\cdot}{C}}\cdot + {}^{\times}\overset{\times\times}{\underset{\times}{O}}{}^{\times}_{\times} \longrightarrow \dot{\underset{\cdot}{C}}{:}{}^{\times}_{\times}\overset{\times\times}{O}{}^{\times}_{\times} \longrightarrow {:}C{:}{}^{\times}_{\times}\overset{\times\times}{O}{}^{\times}_{\times} \text{ or } |C{\equiv}O|$$

A bond formed by the contribution of two electrons from a single atom is called a *coordinate covalent bond.*

This view of bonding should be accepted with a great deal of caution. All electrons are identical. There is no way we can determine which atom the electrons came from or how the bond is actually formed. We have merely constructed a convenient way of keeping track of electrons and developed some minimal understanding of why certain molecules exist. A covalent bond is the sharing of an electron pair between two atoms. Whether we envision the formation of the bond as resulting from the contribution of one electron by each atom or both electrons by a single atom, the result is the same: shared electron pairs.

With the above disclaimer in mind, let us see how we can use the idea of a coordinate covalent bond to explain an important biological process—the transport of oxygen from the lungs to the cells of animals, including man. Hemoglobin in the red blood cells is responsible for the movement of oxygen molecules in these organisms. We are in a position to investigate in more detail how this occurs.

Each hemoglobin molecule contains four subunits called *heme*. The heme unit itself is rather complex with an Fe^{2+} ion at its center. It is this iron ion that actually does the work of carrying the oxygen molecule. By the use of some sophisticated techniques, chemists have found heme to have the following structure:

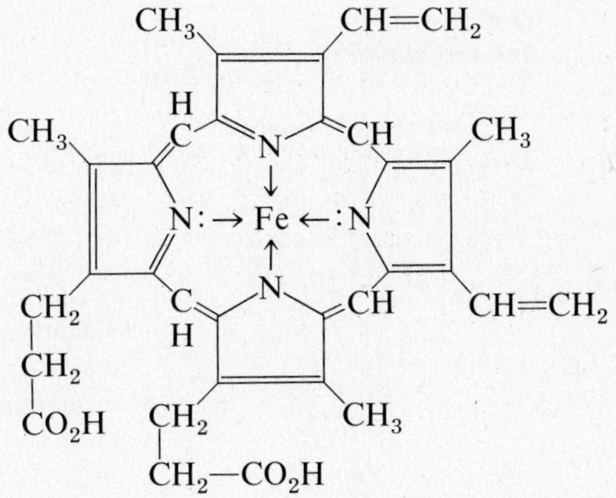

CH_3 $CH{=}CH_2$
H
C N CH CH_3
CH_3
N: → Fe ← :N
C N CH $CH{=}CH_2$
CH_2 H
CH_2
CO_2H CH_2 CH_3
$CH_2{-}CO_2H$

The iron ion is bonded to four nitrogen atoms contained in a complex molecule known as *protoporphyrin IX*. The iron ion is capable of accepting a pair of electrons from an oxygen molecule to form a coordinate covalent bond. The exact arrangement of the oxygen molecule around the iron is yet unknown. One arrangement that has been suggested is

Coordinate covalent bonds are necessary for life!

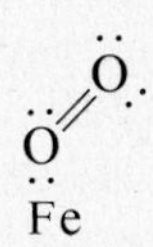

:Ö:
:Ö
Fe

In the lungs, molecular oxygen bonds to the iron ions of the heme units in hemoglobin to produce oxyhemoglobin. The oxyhemoglobin is carried to other regions of the body where it releases the oxygen molecule for use by the cells. The manner in which the body regulates the bonding and release of oxygen will be discussed in Chapter 11.

Unfortunately, the iron in hemoglobin will also form a coordinate covalent bond with certain other smaller molecules if they are present in the lungs. Carbon monoxide is one such molecule and, even worse, the bond

If hemoglobin is bonded to CO, it cannot bond to O_2.

formed between hemoglobin and carbon monoxide is more stable than the bond between hemoglobin and oxygen. This means that hemoglobin will preferentially bond with carbon monoxide rather than oxygen if both are present in the lungs. So great is this preference, that when only one out of every thousand gas molecules in the lungs is carbon monoxide, a sufficient amount of hemoglobin is bonded to carbon monoxide to cause paralysis or death because of a lack of oxygen movement to the cells.

POLYATOMIC IONS

Just as there are charged atoms called ions, there exists a number of charged particles containing two or more atoms. Because the atoms within these particles are bonded covalently, we might consider them to be charged molecules. These species are referred to as polyatomic ions*. A list of important polyatomic ions is given in Table 3-2. You should memorize the

TABLE 3-2
Polyatomic ions

FORMULA AND CHARGE	NAME
$(NH_4)^+$	ammonium
$(OH)^-$	hydroxide
$(CO_3)^{2-}$	carbonate
$(HCO_3)^-$	monohydrogen carbonate (bicarbonate)
$(PO_4)^{3-}$	phosphate
$(HPO_4)^{2-}$	monohydrogen phosphate
$(H_2PO_4)^-$	dihydrogen phosphate
$(SO_4)^{2-}$	sulfate
$(Cr_2O_7)^{2-}$	dichromate
$(MnO_4)^-$	permanganate
$(NO_3)^-$	nitrate
$(C_2O_4)^{2-}$	oxalate
$(C_2H_3O_2)^-$	acetate

formulas, names, and charges of these ions. Polyatomic ions often remain as a unit during many chemical processes.

Memorize Table 3-2. Index cards with the name of a polyatomic ion on one side and its formula and charge on the other side are useful aids in learning the polyatomic ions.

The procedure for writing formulas of compounds containing polyatomic ions is the same as for simple ionic compounds. For example, sodium hydroxide is written NaOH, calcium nitrate is written $Ca(NO_3)_2$, ammonium sulfate is written $(NH_4)_2SO_4$. Total positive charge must equal total negative charge in a correctly written formula. Note also the use of parentheses around polyatomic ions. Parentheses are used around a polyatomic ion if more than one of that ion is needed to write the formula for the compound correctly e.g., $Ca(NO_3)_2$.

*An older term for polyatomic ions is radicals. This term has been dropped because polyatomic ions is more descriptive of the nature of the particles.

exercise 3-13

Write formulas for the following compounds:

(a) sodium carbonate
(b) aluminum sulfate
(c) ammonium nitrate
(d) calcium dihydrogenphosphate
(e) magnesium hydroxide

The Lewis structures for polyatomic ions are a little more difficult to write than those of neutral molecules. The reason for this is in many cases the *central* atom forms a different number of bonds than that predicted by subtracting the group number from eight. For example, in a neutral compound, nitrogen forms three bonds; however, in the ammonium ion, nitrogen forms four bonds. The idea of surrounding each atom with a noble gas electron arrangement remains valid.

Some modified rules are helpful for writing Lewis structures of polyatomic ions.

1. Determine the number of electrons to be distributed by summing the total number of outer shell electrons of each atom. Adjust this total to compensate for the charge on the ion. For instance, subtract one electron from the total for a 1+ ion or add two electrons to the total for a 2− ion.
2. Arrange the atoms and join them by single bonds.
3. Deduct two electrons for each bond written in step 2 from the total number of electrons found in step 1. Distribute the remaining electrons around the atoms so as to give each atom a noble gas electron arrangement. (It may be necessary to form multiple bonds in some instances.)

You should work through these examples by writing out each of the steps.

Two examples are

EXAMPLE 3-3

Hydroxide ion, $(OH)^-$

(a) Oxygen has six outer electrons; hydrogen, one outer electron, this plus one electron for the negative charge yields a total of eight electrons.
(b) Write the basic structure: O—H.
(c) The remaining number of electrons to be distributed is six (eight minus two used for the bond). Hydrogen already has a noble gas electron arrangement; therefore, we place the remaining electrons around oxygen: $(|\overline{\underline{O}}—H)^-$.

EXAMPLE 3-4

Carbonate ion, $(CO_3)^{2-}$

(a) A total of 24 electrons (4 from carbon, 6 from each oxygen, plus 2 to account for the charge).
(b) The basic structure is

```
   O
   |
   C
  / \
 O   O
```

(c) There are 18 electrons left to be distributed (24 − 6 = 18). If these 18 electrons are distributed evenly between the three oxygen atoms, we find the carbon atom is unsatisfied:

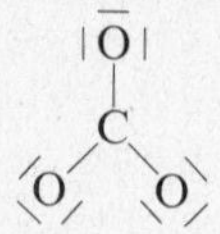

$$\begin{array}{c} |\overline{O}| \\ | \\ C \\ \diagup \quad \diagdown \\ \langle\underline{O}| \quad |\underline{O}\rangle \end{array}$$

It doesn't matter which oxygen atom is written with the double bond.

The carbon atom can be satisfied if a double bond is formed between one of the oxygen atoms and carbon.

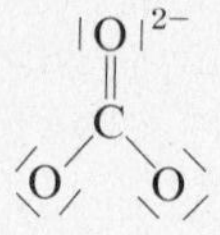

$$\left[\begin{array}{c} |O| \\ \| \\ C \\ \diagup \quad \diagdown \\ \langle\underline{O}| \quad |\underline{O}\rangle \end{array}\right]^{2-}$$

exercise 3-14

Write Lewis structures for each of the following ions. (The central atom is underlined and, with the exception of ammonium ion, the hydrogen atoms are bonded to oxygen.)

(a) $(\underline{N}H_4)^+$ **(d)** $(H\underline{C}O_3)^-$
(b) $(\underline{S}O_4)^{2-}$ **(e)** $(H_2\underline{P}O_4)^-$
(c) $(\underline{P}O_4)^{3-}$

NONCONFORMING MOLECULES

Earlier the existence of molecules that do not conform to the rules for writing Lewis structures were mentioned. There are several ways some molecules may violate the rule of attaining a noble gas arrangement of electrons around each atom in the molecule. In some molecules, an atom may be surrounded by less than eight outer shell electrons; in other molecules, an atom may have more than eight electrons involved in outer shell electron sharing.

Boron doesn't have enough electrons to share to form four bonds.

For example, boron is surrounded by less than eight electrons in many of its compounds. Reference to the dot diagram for boron ($\cdot\dot{B}\cdot$) explains why boron is an exception to the rules. Boron atoms have only three electrons in their outer shell. Thus, boron atoms can form a maximum of three covalent bonds by sharing their outer shell electrons. Boron trifluoride is an example:

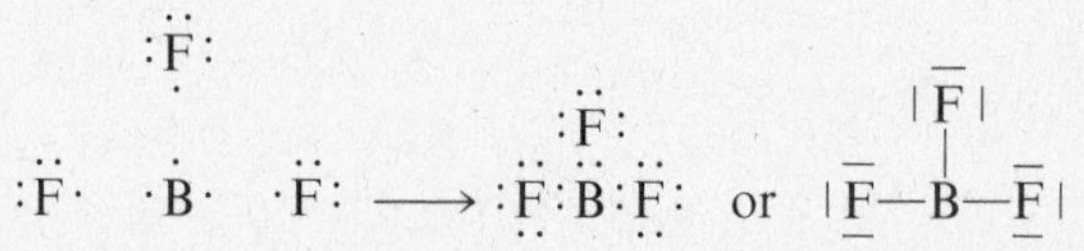

$$:\ddot{F}\cdot \quad \cdot\dot{B}\cdot \quad \cdot\ddot{F}: \;(\text{with } :\ddot{F}\cdot \text{ above B}) \longrightarrow :\ddot{F}:\underset{\displaystyle :\ddot{F}:}{}\!\!\!\!\!\!\overset{\displaystyle :\ddot{F}:}{B}:\ddot{F}: \quad \text{or} \quad |\underline{F}-\overset{\displaystyle |\overline{F}|}{\underset{}{B}}-\underline{F}|$$

Nonmetals of the third, fourth, and higher periods form compounds in which their atoms are surrounded by more than eight electrons in the outer shell. Examples of such compounds are PCl_5, SF_6, and IF_7. More than eight electrons in the outer shell is possible because their outer shells have a capacity for more than eight electrons. Atoms of the third period element phosphorus have their outer electrons in the *M* shell. The *M* shell has a maximum capacity of 18 electrons. You are not expected to predict the formulas of compounds in which an atom has more than a noble gas electron arrangement.

A very interesting and important example of a molecule that doesn't have the expected electron arrangement is elemental oxygen (O_2).

We might expect oxygen to have the Lewis structure

$$\overline{\underline{O}}=\overline{\underline{O}}$$

Experimental evidence indicates that this structure is incorrect, however. The oxygen molecule actually has two unpaired electrons. The following structure shows two unpaired electrons:

$$|\overline{\underset{\cdot}{O}}-\overline{\underset{\cdot}{O}}|$$

Unfortunately, this structure is also unsatisfactory. Besides not obeying the rules for writing Lewis structures, it does not agree with other experimental evidence indicating that the oxygen molecule actually has a double bond.

This example serves to point out the inadequacies of the Bohr model of the atom and of Lewis structures. While these ideas are useful, they are not infallible. There is nothing wrong with the oxygen molecule; what is wrong is the inability of the theories we have been using to explain the nature of the oxygen molecule. Lest you think that scientists are incapable of explaining the oxygen molecule, let us note that a different theory of atomic and molecular structure can explain the nature of oxygen. Quantum mechanics yields a satisfactory answer to the question of electron arrangement in oxygen molecules. Unfortunately, it is a much more difficult theory to understand and we shall not attempt an explanation in this text.

MILKSHAKES, POLAR BONDS, AND ECTRONEGATIVITY

So far in this chapter we have discussed two kinds of bonding: ionic and covalent. In the first case there is a complete transfer of electrons; in the other electrons are evenly shared between two atoms. Bridging these two extremes is a large number of bonds in which there is an unequal sharing of electrons. Such bonds are called *polar covalent bonds*. To appreciate better what is meant by an unequal sharing of electrons, consider the following situation.

Suppose while you and I are walking downtown we pass an ice cream parlor and decide to have a milkshake. Unfortunately, milkshakes cost 50¢ and we each have only 25¢. Since half a shake is better than none, we order one milkshake with two straws. After the milkshake is served and the waitress paid, there are three things which could happen:

1. We each drink half the milkshake;
2. One of us drinks faster than the other and consequently gets more than a fair share;

3. One of us pushes the other off the chair onto the floor and drinks the whole milkshake.

In the first two cases the milkshake is shared; while in the third, one person has the whole thing.

Now, equate the milkshake to a pair of electrons formed by the contribution of one electron by each of two atoms. Situation 1 in the analogy above corresponds to a covalent bond with equal sharing of electrons. Situation 2 corresponds to a covalent bond with unequal sharing of electrons—one atom is thirstier for electrons than the other. The third situation corresponds to an ionic bond. One atom loses out completely and the other takes both electrons.

How can we tell how "thirsty" an atom is for electrons? Linus Pauling introduced an electronegativity scale in 1932 that enables us to predict when an electron pair will be unequally shared. Electronegativity is defined as the ability of an atom in a molecule to attract electrons to itself. It is difficult to understand the manner Pauling used to calculate values for electronegativity. A simpler description of electronegativity was offered by Mulliken, who suggested that electronegativity is the average of the ionization potential and the electron affinity of an atom:

$$\text{electronegativity} = \frac{\text{ionization potential} + \text{electron affinity}}{2}$$

The greater the electronegativity value of an element, the greater is its ability to attract electrons. Examination of a table of electronegativity values (Table 3-3) reveals that electronegativity increases left to right across the periodic chart and from the bottom to the top of a group.

TABLE 3-3
Electronegativity values

H 2.1																	He —
Li 1.0	Be 1.5											B 2.0	C 2.5	N 3.0	O 3.5	F 4.0	Ne —
Na 0.9	Mg 1.2											Al 1.5	Si 1.8	P 2.1	S 2.5	Cl 3.0	Ar —
K 0.8	Ca 1.0	Sc 1.3	Ti 1.5	V 1.6	Cr 1.6	Mn 1.5	Fe 1.8	Co 1.8	Ni 1.8	Cu 1.8	Zn 1.6	Ga 1.6	Ge 1.8	As 2.0	Se 2.4	Br 2.8	Kr —
Rb 0.8	Sr 1.0	Y 1.2	Zr 1.4	Cb 1.6	Mo 1.8	Tc 1.9	Ru 2.2	Rh 2.2	Pd 2.2	Ag 1.9	Cd 1.7	In 1.7	Sn 1.8	Sb 1.9	Te 2.1	I 2.5	Xe —
Cs 0.7	Ba 0.9	57-71 1.1-1.2	Hf 1.3	Ta 1.5	W 1.7	Re 1.9	Os 2.2	Ir 2.2	Pt 2.2	Au 2.4	Hg 1.9	Tl 1.8	Pb 1.8	Bi 1.9	Po 2.0	At 2.2	Rn —
Fr 0.7	Ra 0.9																

The intensity of the background color in this table roughly parallels the value of the electronegativity.

Bonded atoms having the same electronegativity form nonpolar covalent bonds.

When two atoms with the same electronegativity bond together, both have equal ability to attract electrons and an equal sharing of electrons results. If the bonded atoms have unequal electronegativities, the electrons are unequally shared. The bonding pair of electrons will, on the average, be closer to the atom with the greater electronegativity. This unequal sharing creates a negative end of the bond and a positive end of the bond. Remember this is *not* a complete transfer of an electron; it is merely a movement of shared electrons toward the more electronegative element.

Comments

Difference in electronegativity between bonded atoms	Bond type
0.0	Nonpolar covalent
Greater than 0.0 to 1.7	Polar covalent
Greater than 1.7	Ionic

Consider the molecule HCl. The electronegativities of hydrogen and chlorine are 2.1 and 3.0, respectively. Thus, the bonding pair of electrons are, on the average, closer to the chlorine atom than to the hydrogen atom. The result is a *polar covalent bond*. The direction of the electron shift is indicated by an arrow. The head of the arrow points toward the more negative end of the bond: $\overset{\delta+ \longrightarrow \delta-}{\text{H—Cl}}$. The lowercase Greek letter delta (δ) is used to represent a partial (incomplete) unit of charge.

exercise 3-15

Decide which of the following *bonds* are polar and indicate the direction of polarity by an arrow.

(a) H—H (d) N—N
(b) C—Cl (e) Br—Br
(c) H—O (f) C—O

MOLECULAR SHAPES

We shall very frequently find that not only do we need to know the formula and electron arrangement of a compound, but we also need to know the shape of the molecule. The spatial arrangement of atoms in a molecule can be calculated via some complex mathematical equations. Fortunately, there is also a much simpler, although naive, approach to predicting molecular shapes from the Lewis structure of molecules. This simpler approach makes the assumption that although electrons will pair together to form bonds, the electron *pairs* will arrange themselves around an atom to be as far apart as possible. We shall consider only those cases where there are two, three, or four pairs of electrons around an atom. Secondly, we shall agree that the shape of a molecule is described by the spatial location of the nuclei with respect to one another.

Beryllium fluoride molecules have the Lewis structure

$$|\overline{\underline{F}}\text{—Be—}\overline{\underline{F}}|$$

There are two pairs of electrons around the beryllium atom. If they are placed on opposite sides of the Be atom, they will be as far apart as possible. Such an arrangement will cause the nuclei of the three atoms in BeF_2 to be in a straight line. Thus, BeF_2 is said to be a linear molecule with 180° bond angles.

$$\text{F—Be—F} \qquad \theta = 180°$$

An example of a molecule with three electron pairs around the central atom is BCl_3. The three electron pairs will be farthest apart when placed at the corners of an equilateral triangle:

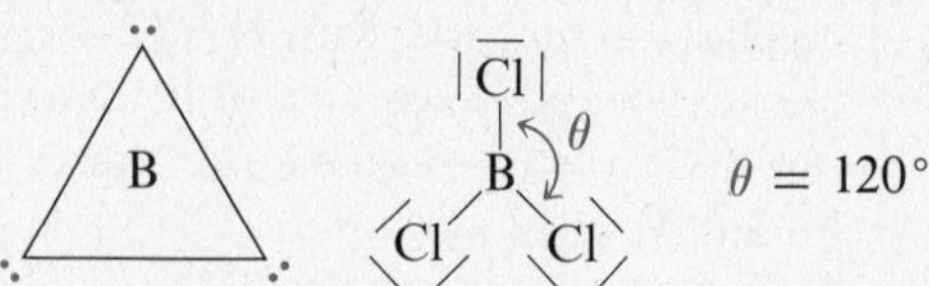

The resultant molecule is described as a planar triangular molecule and has bond angles of 120°.

There are numerous examples of molecules in which the central atom is surrounded by four electron pairs. The hydrogen compounds of the elements carbon through fluorine, CH_4, NH_3, H_2O, HF, each have four pairs of electrons around the central atom. Four distinctly different situations are represented by these molecules:

1. CH_4—four bonding pairs of electrons surround the central atom.
2. NH_3—three bonding pairs and one nonbonding pair of electrons surround the central atom.
3. H_2O—two bonding pairs and two nonbonding electron pairs around the central atom.
4. H—F—one bonding pair and three nonbonding electron pairs surround the central atom.

The last situation, that of HF, is a trivial one. Two atoms can always be connected by a straight line with a linear molecule as the only possible result.

Placement of the four electron pairs at the corners of a tetrahedron yields the maximum separation. The tetrahedral methane molecule has 109°28′ H—C—H bond angles [see Fig. 3-6(a) and (b)].

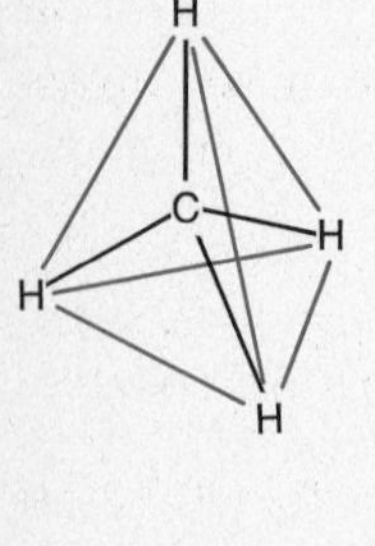

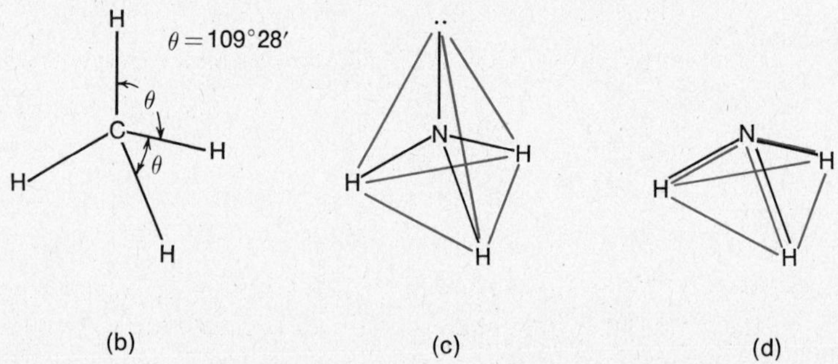

FIGURE 3-6 Molecular shapes. (a) Methane, tetrahedral. (b) Tetrahedral angles equal 109°28′. (c) Bonding and nonbonding electron pairs in ammonia. (d) Ammonia, trigonal pyramid.

The nitrogen atom of ammonia also has four pairs of electrons at the corners of a tetrahedron. Only three pairs are bonding pairs, however; the remaining corner of the tetrahedron is occupied by a nonbonding electron pair. Since the position of the nuclei, and not that of the electrons, determines the molecular shape, the shape of the molecule is described as a triangular pyramid. Carefully examine Fig. 3-6(c) and (d) until you are certain you understand the difference between a tetrahedron and a triangular pyramid.

Similarly, the four electron pairs surrounding the oxygen atom of water are placed at the corners of a tetrahedron. Since only two of these electron pairs are used in bonding, the shape of the water molecule is bent or V-shaped.

TOTAL NUMBER OF ELECTRON PAIRS AROUND CENTRAL ATOM	NUMBER OF BONDING PAIRS	NUMBER OF NON-BONDING PAIRS	MOLECULAR SHAPE	EXAMPLE
2	2	0	linear	BeF_2
3	3	0	triangular planar	BCl_3
4	4	0	tetrahedral	CH_4
4	3	1	triangular pyramid	NH_3
4	2	2	V-shaped	H_2O

The molecular shapes of molecules containing multiple bonds can also be predicted by treating the electrons in a multiple bond as if they were a single pair of electrons. (Why?) Carbon dioxide with two double bonds on the carbon atoms is a linear molecule:

$$\overline{\underline{O}}{=}C{=}\overline{\underline{O}}$$

exercise 3-16

Predict the shapes of the following molecules and ions (the central atom is underlined).

(a) $H_2\underline{S}$ (d) $\underline{P}O_4^{3-}$
(b) $\underline{C}OH_2$ (e) $\underline{C}S_2$
(c) $\underline{N}H_4^+$

DIPOLE MOMENTS

The presence of polar bonds within a molecule may give rise to a dipolar molecule, depending on the molecular shape. A dipolar molecule is one that will take a preferred orientation in the presence of an electrical field.

Consider the H—Cl molecule. The hydrogen–chlorine bond is polar, giving the chlorine a partial negative charge and the hydrogen a partial

Dipole moments are caused by an unequal sharing of electrons by atoms in molecules.

positive charge. In the presence of an electrical field, the hydrogen chloride molecule would be expected to orient itself such that the chlorine is pointed toward the positive pole of the field and the hydrogen toward the negative pole.

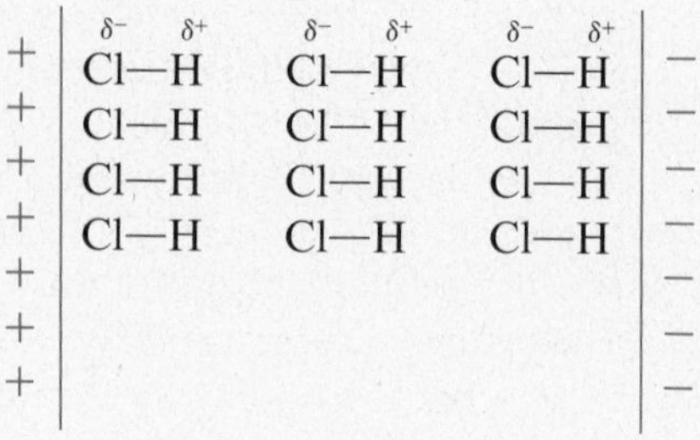

```
+ | δ−  δ+    δ−  δ+    δ−  δ+  | −
+ | Cl—H     Cl—H     Cl—H     | −
+ | Cl—H     Cl—H     Cl—H     | −
+ | Cl—H     Cl—H     Cl—H     | −
+ | Cl—H     Cl—H     Cl—H     | −
+ |                            | −
+ |                            | −
```

A molecule that will orient itself in an electrical field is said to possess a *permanent dipole moment* and is called a *polar molecule.*

Shape is an important factor in determining whether a molecule has a permanent dipole moment. Beryllium fluoride, for example, has *polar bonds* but because the molecule is linear, it does *not* possess a permanent dipole moment.

$$\overleftarrow{\text{F—Be}}\overrightarrow{\text{—F}}$$

Both ends of the molecule are negatively charged and the molecule can find no preferential orientation in an electric field.

In contrast to BeF_2, water molecules are V-shaped and as such find a preferred alignment in the field.

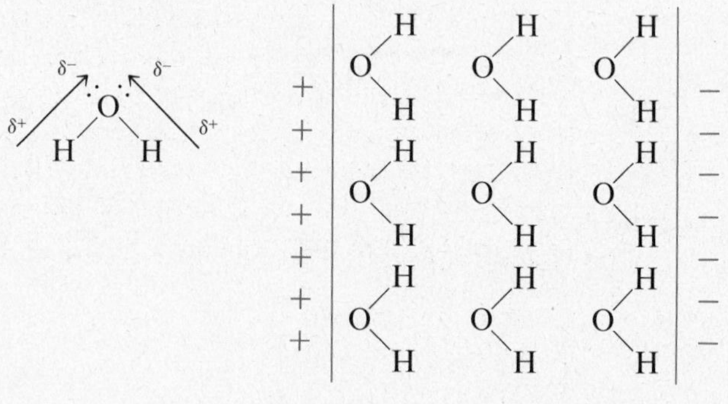

```
δ−      δ−
   ↖ :O: ↗
δ+  H    H  δ+

+ |  H      H      H   | −
+ | O      O      O    | −
+ |  H      H      H   | −
+ |  H      H      H   | −
+ | O      O      O    | −
+ |  H      H      H   | −
+ |  H      H      H   | −
  | O      O      O    |
  |  H      H      H   |
```

We can say, then, that polar bonds are a necessary, but not sufficient, condition to give a molecule a permanent dipole moment.

exercise 3-17

Decide which of the following molecules will have a permanent dipole moment. For those molecules having no permanent dipole, give the reason why they do not.

(a) CO_2 (d) $H-\overset{\overset{H}{|}}{\underset{\underset{H}{|}}{C}}-\overline{O}H$

(b) CH_4 (e) $CHCl_3$

(c) CH_2O (f) $H-\overset{\overset{H}{|}}{\underset{\underset{H}{|}}{C}}-\overline{N}H_2$

INTERMOLECULAR DIPOLE FORCES AND HYDROGEN BONDING

Oxygen and hydrogen chloride are both diatomic molecules of similar size and weight. The boiling and melting points of these substances should be similar if the forces holding the molecules together in the liquid and solid states are the same. The data in Table 3-4 reveal that the melting and boiling points of these compounds are considerably different.

TABLE 3-4
Normal melting and boiling points of HCl and O_2

SUBSTANCE	MP (°C)	BP (°C)
O_2	−218	−183
HCl	−112	− 84

Consideration of the bonding within these molecules allows us to give an explanation for the differences in their melting and boiling points. The oxygen–oxygen bond in O_2 is nonpolar, whereas the hydrogen–chlorine bond in hydrogen chloride is polar. In the liquid and solid states of these substances the molecules are relatively close to one another. We know ions arrange themselves in the solid so as to place oppositely charged ions as close together and like charged ions as far apart as possible. Thus, it is reasonable to expect polar molecules to arrange themselves so as to place the positive end of one molecule as close as possible to the negative end of a nearby molecule, as shown in Fig. 3-7.

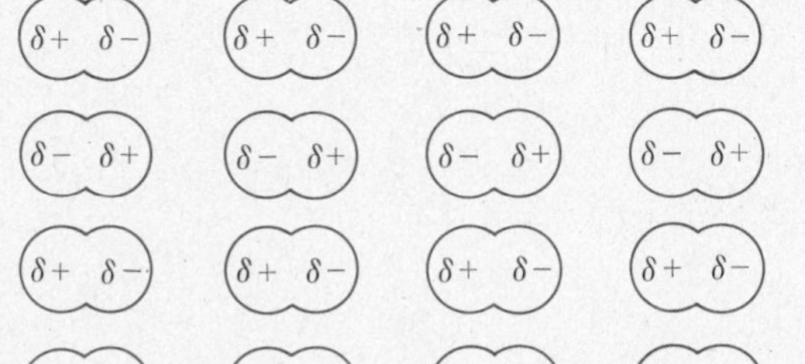

FIGURE 3-7 Arrangement of polar molecules in a solid.

Melting and boiling points are measures of intermolecular forces.

This attraction between adjacent molecules means that more energy is required to separate them. Consequently, the melting and boiling points should be higher for polar molecules than for nonpolar molecules of approximately the same size and weight.

exercise 3-18

Predict which member of the following pairs of compounds should have the higher melting and boiling points.
(a) N_2 or CO **(d) O_2 or H_2S**
(b) NH_3 or CH_4 **(e) HCN or H_2C_2**
(c) CH_2O or O_2

The melting and boiling points of ammonia, water, and hydrogen fluoride are considerably higher than would be expected even from consideration of their polar character. The relatively high boiling points of these three substances implies the presence of an abnormally strong intermolecular force. X-ray examination of the arrangement of water molecules in an ice crystal suggests the origin of this intermolecular force. The observed arrangement (see Fig. 3-8) indicates an attraction between the hydrogen atoms of one molecule and the oxygen atom of a neighboring water molecule. Experimentation has shown this attraction is much stronger than normal dipole attractions. So strong is this attraction that chemists have termed it a *hydrogen bond.* Hydrogen bonds are frequently represented by a dotted line to indicate the attraction is weaker than a true covalent bond but stronger than normal dipolar attractions.

Hydrogen bonds occur between a hydrogen atom of one molecule and a very electronegative atom of another molecule.

Hydrogen bonding seems to be a special case of dipolar attraction in which the hydrogen atom of one molecule is attracted toward a nonbonding pair of electrons on an atom of an adjacent molecule. It results from the very polar character of bonds between hydrogen and nitrogen, oxygen, or fluorine.

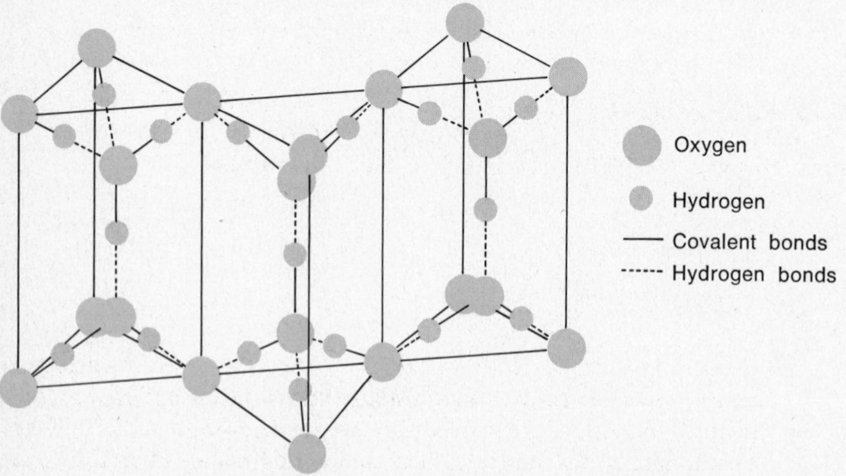

FIGURE 3-8 Arrangement of water molecules in ice.

In addition to explaining the relatively high melting and boiling points of NH_3, H_2O, and HF, the extreme polarity of the hydrogen oxygen bond in water explains the solubility of ionic substances in water. We shall also see in later chapters that hydrogen bonding plays an extremely important role in chemistry of life process. Hydrogen bonding is, for example, responsible for the unique properties of proteins.

FORMULA WEIGHTS

Since atoms possess mass (weight), common sense tells us that molecules must also weigh something. The weight of a molecule is simply the sum of the atomic weights of the individual atoms making up the molecule. This weight is referred to as the *formula weight.*

The formula weight of hydrogen chloride (HCl) is the sum of the atomic weights of hydrogen and chlorine:

$$1.008 \text{ amu} + 35.453 \text{ amu} = 36.461 \text{ amu}$$

The formula weight of a carbon dioxide molecule is 44 amu. This is found by adding the weights of one carbon atom and two oxygen atoms:

$$12.011 \text{ amu} + 2(15.999 \text{ amu}) = 44.009 \text{ amu}$$

Although ionic compounds do not occur as molecules, it is frequently useful to know the formula weights of these compounds. The procedure is the same as for molecular compounds: Sum the weights of the individual atoms in the formula. Calcium phosphate $[Ca_3(PO_4)_2]$ has a formula weight of 310.17 amu.

Weight of calcium	3×40.08 amu	= 120.24 amu
Weight of phosphorus	2×30.97 amu	= 61.94 amu
Weight of oxygen	8×15.999 amu	= 127.99 amu
	Formula weight of $Ca_3(PO_4)_2$	= 310.17 amu

exercise 3-19 **Determine the formula weights of**

(a) NaCl
(b) Na_2SO_4
(c) $(NH_4)_2CO_3$
(d) $C_6H_{12}O_6$(glucose)
(e) $C_{738}H_{1166}O_{208}N_{203}S_2Fe_4$ (hemoglobin)

CHAPTER SUMMARY

1. Chemical reactions are accompanied by energy changes. An exergonic reaction releases energy; an endergonic reaction absorbs energy. Exothermic and endothermic apply specifically to changes in heat energy. Exothermic is the release of heat and endothermic is the absorption of heat.

2. Energy can be stored in chemical bonds. Stored energy is potential energy. The stored energy is released during exergonic reactions; whereas energy is stored in bonds during endergonic reactions.
3. Bonds between atoms in compounds may be classified as ionic bonds or as covalent bonds. Ionic bonding is the attraction between oppositely charged atoms (called *ions*); covalent bonding is the sharing of (an) electron pair(s) between two atoms.
4. Positively charged atoms or groups of atoms are called *cations*. Simple cations are formed from atoms by the loss of (an) electron(s).
5. Negatively charged atoms or groups of atoms are called *anions*. Simple anions are formed by atoms gaining electrons.
6. Ionic compounds are aggregates of oppositely charged ions arranged in an orderly fashion. Covalent compounds exist as small discrete particles called *molecules*.
7. Formulas of ionic compounds state the simplest whole-number ratio of ions that can combine to produce an electrically neutral compound.
8. Formulas of covalent compounds give the number and kind of atoms in molecules of the compound.
9. Balanced chemical equations are written to show the ratio in which substances react and the formulas of the products and the ratios in which they are produced. Balancing equations is essentially a trial and error method.
10. Oxidation is the loss of electrons by an atom or ion; reduction is the gain of electrons by an atom or ion; both processes must occur simultaneously.
11. Multiple bonds may be formed between atoms in covalent compounds. A double bond is two shared pairs of electrons; a triple bond, three shared pairs of electrons. A triple bond is the maximum number of bonds that can be formed between atoms.
12. Coordinate covalent bonds may be formed if one atom contributes both electrons to be shared by the bonded atoms.
13. Unequal sharing of electrons leads to polar covalent bonds. The amount of bond polarity can be predicted from the electronegativity of the bonded atoms. Polar bonds within a molecule may cause the molecule to be polar.
14. Molecular shapes can be predicted from the number of electron pairs around the atoms of the molecule.
15. Intermolecular hydrogen bonding causes a strong attraction between molecules of some compounds resulting in high melting and boiling points.
16. The formula weight of a compound is the sum of the weights of the individual atoms in the compound's formula.

QUESTIONS

1. The Lewis dot diagrams for the hypothetical elements A and Z are given below. For each pair, predict the formula of the compound that would be expected to result from their combination. Predict whether the compound would be ionic or covalent.

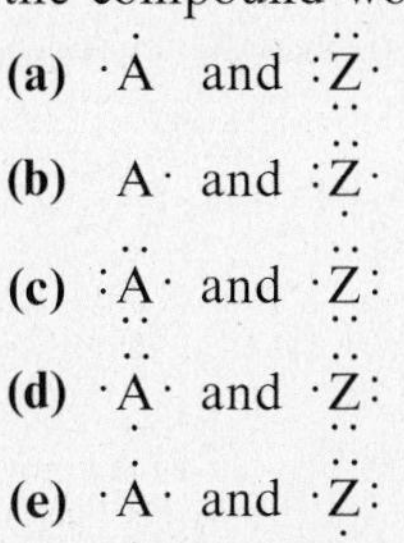

   (a) $\cdot\overset{\cdot}{\text{A}}$ and $:\overset{\cdot\cdot}{\underset{\cdot\cdot}{\text{Z}}}\cdot$

   (b) $\text{A}\cdot$ and $:\overset{\cdot\cdot}{\underset{\cdot}{\text{Z}}}\cdot$

   (c) $:\overset{\cdot\cdot}{\underset{\cdot\cdot}{\text{A}}}\cdot$ and $\cdot\overset{\cdot\cdot}{\underset{\cdot\cdot}{\text{Z}}}:$

   (d) $\cdot\overset{\cdot\cdot}{\underset{\cdot}{\text{A}}}\cdot$ and $\cdot\overset{\cdot\cdot}{\underset{\cdot\cdot}{\text{Z}}}:$

   (e) $\cdot\overset{\cdot}{\text{A}}\cdot$ and $\cdot\overset{\cdot\cdot}{\underset{\cdot}{\text{Z}}}:$

2. The Group II element calcium forms the compound $CaCl_2$. The Group VI element sulfur forms the compound SCl_2. Explain why an atom of both a Group II and a Group VI element will combine with two atoms of chlorine to form a compound.

3. Predict the product and write balanced equations for
   (a) $Si + Cl_2 \longrightarrow$
   (b) $N_2 + H_2 \longrightarrow$
   (c) $Li + I_2 \longrightarrow$
   (d) $Al + S \longrightarrow$
   (e) $Mg + O_2 \longrightarrow$

4. Write formulas for
   (a) potassium bromide **(c)** potassium phosphate **(e)** calcium sulfide **(b)** ammonium chloride **(d)** ammonium dichromate **(f)** iron(III) sulfate

5. Write Lewis structures for (the central atom is underlined)
   (a) $\underline{P}H_3$ **(d)** $\underline{B}F_3$
   (b) $\underline{B}F_4^-$ **(e)** $\underline{C}H_2Cl_2$
   (c) $\underline{C}_2H_4$ **(f)** $\underline{N}O_3^-$

6. The two reactions below are known to occur.
   Reaction (a)

   $NH_3 + HCl \longrightarrow NH_4Cl$ (contains NH_4^+ and Cl^- ions)

   Reaction (b)

   $BF_3 + HF \longrightarrow HBF_4$ (contains H^+ and BF_4^- ions)

   In reaction (a), explain why the neutral molecule combines with an H^+ ion (from HCl) to produce a cation but in reaction (b) the neutral molecule combined with F^- (from HF) to produce an anion.

7. Using the electronegativity values in Table 3-3, arrange the following bonds in order of decreasing polarity.
   (a) C—H **(b)** C—F **(c)** N—N **(d)** N—H **(e)** B—Cl

8. The molecular shape of NH_3 is said to be a triangular pyramid, while NH_4^+ is said to be tetrahedral. Explain.

9. Explain why the boiling point of HCl is higher than that of N_2.

4
isomerism

WHAT IS ISOMERISM?

Perhaps it occurred to you while studying Chapter 3 that there may be more than one way to bond the atoms of covalent compounds. For example, there are two ways to arrange the atoms in a compound with the formula C_2H_6O. We may begin in one case by bonding the two carbon atoms together and then attach the oxygen atom:

Why would it be incorrect to start by writing C—H—C or C—H—O?

```
  |  |
 —C—C—O—
  |  |
```

This arrangement leaves six bonding sites, one for each hydrogen atom:

```
   H H
   | |
 H—C—C—O—H
   | |
   H H
```

On the other hand we might have started by attaching both carbon atoms to the oxygen:

```
  |     |
 —C—O—C—
  |     |
```

This also leaves six bonding sites for the hydrogen atoms:

```
   H     H
   |     |
 H—C—O—C—H
   |     |
   H     H
```

Both of the structures we have drawn obey all the bonding rules we have learned.

Which of the two structures is correct? As it turns out, both structures are! The first structure drawn is that of ethyl alcohol. The second is that of dimethyl ether. The chemical and physiological properties of ethyl alcohol and dimethyl ether are vastly different, although the chemical formula for both is the same: C_2H_6O. We now find outselves in the position of not only needing to know chemical formulas but also needing to know the arrangement of the atoms in a molecule. *Compounds possessing the same molecular formula but different structural arrangements of atoms are called structural isomers.* The structural arrangement of atoms in a molecule can only be determined from a knowledge of chemical and physical properties of the compound or from the way they are formed in a chemical reaction. There are several different kinds of isomerism that will be studied in this chapter. The greatest number of compounds displaying isomerism are those contain-

Isomers always *have the same molecular formula.*

ing carbon. The compounds most intimately associated with life contain carbon. For the latter reason, our discussion of isomerism will be limited to carbon-containing compounds.

HYDROCARBONS

The simplest carbon-containing molecules exhibiting isomerism are those of carbon and hydrogen collectively called *hydrocarbons*. The hydrocarbons are generally divided into four subclasses depending on particular structural features. These subclasses, alkanes, alkenes, alkynes, and aromatics, are shown in Fig. 4-1.

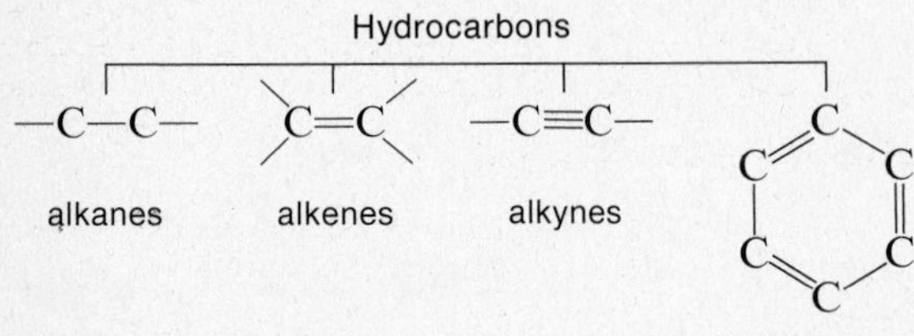

FIGURE 4-1 Characteristic grouping of the four subclasses of hydrocarbons.

Alkanes are characterized by having only single bonds between the carbon atoms. You are already familiar with the simplest alkane, methane (CH_4). The alkanes are also known as saturated hydrocarbons. The word *saturated* means the molecule contains only single bonds. A general formula for most alkanes is $C_nH_{(2n+2)}$.

A molecule of an alkane has six C atoms. How many H atoms does it have?

Alkenes are the group of hydrocarbons containing a carbon–carbon double bond in the molecule. Occasionally, these compounds are also referred to as olefins. Most alkenes have the general formula C_nH_{2n}.

Alkynes, or acetylenes as they are also known, contain carbon–carbon triple bonds. The general formula of alkynes is $C_nH_{(2n-2)}$.

The aromatic hydrocarbons have a six-membered carbon ring containing three double bonds.

WRITING STRUCTURES AND NAMING THE ALKANES

Because of the great number of carbon-containing compounds, an elaborate naming system has been developed to distinquish between compounds. The names of alkanes always end in *ane*. This ending signifies the compound is a hydrocarbon containing only single bonds.

The remaining portion of the name tells the number and the arrangement of the carbon atoms in the molecule. Table 4-1 gives the relationship between the name and the number of carbon atoms contained in a continuous chain in an alkane. This table should be memorized now.

Table 4-1 contains molecular and condensed structural formulas. Why are both important and what does the condensed structural formula mean? To answer these questions we shall examine the first few members of the alkane family more closely.

Since we are already familiar with the structure of methane, we will

TABLE 4-1
Names and formulas of unbranched alkanes

NUMBER OF CARBON ATOMS IN A CONTINUOUS CHAIN	NAME*	MOLECULAR FORMULA $[C_nH_{(2n+2)}]$	CONDENSED STRUCTURAL FORMULA
1	*meth*ane	CH_4	CH_4
2	*eth*ane	C_2H_6	CH_3CH_3
3	*prop*ane	C_3H_8	$CH_3CH_2CH_3$
4	*but*ane	C_4H_{10}	$CH_3CH_2CH_2CH_3$
5	*pent*ane	C_5H_{12}	$CH_3CH_2CH_2CH_2CH_3$
6	*hex*ane	C_6H_{14}	$CH_3CH_2CH_2CH_2CH_2CH_3$
7	*hept*ane	C_7H_{16}	$CH_3CH_2CH_2CH_2CH_2CH_2CH_3$
8	*oct*ane	C_8H_{18}	$CH_3CH_2CH_2CH_2CH_2CH_2CH_2CH_3$

*The italicized portion of the name indicates the number of carbon atoms in a continuous chain.

begin our discussion with ethane (C_2H_6). Application of the bonding rules learned in Chapter 3 allows us to write the structure of ethane:

```
   H  H
   |  |
H—C—C—H
   |  |
   H  H
```

Why does hydrogen form only one bond?

The carbon atoms must be bonded to one another since each hydrogen can form but one bond. The condensed structural formula (CH_3CH_3, given in Table 4-1) is a shorthand way of indicating the Lewis structure. It is understood that each of the hydrogen atoms is bonded to the preceding carbon atom and the molecule contains a carbon–carbon single bond.

The Lewis structure of propane (C_3H_8) is written

```
   H    H    H
   |    |    |
H—C————C————C—H
   |    |    |
   H    H    H
```

There are several important features about this structural representation for us to notice. One of these features is that the bonding of all the hydrogen atoms is not identical. Two of the hydrogen atoms in propane are bonded to a carbon atom that is bonded to two other carbon atoms. The other two carbon atoms each bear three hydrogen atoms. The propane molecule is said to have two different kinds of hydrogen, meaning there are two different bonding environments for the hydrogen atoms in propane.

Another feature of the propane structure is that the three carbon atoms are *not* in a straight line. The four bonds around each carbon atom are directed toward the corners of a tetrahedron. This means the actual shape of the molecule is

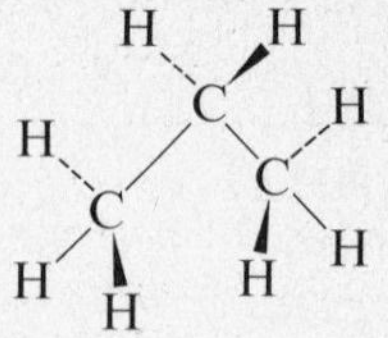

A line indicates a bond in the plane of the paper. A wedge (—) indicates the bond coming out of the plane of the paper toward you. A dashed line indicates a bond going behind the plane of the paper.

with bond angles of 109°28′. (If you have access to a set of molecular models, build a model of propane, paying particular attention to the shape of the molecule and the presence of two different types of hydrogen atoms.)

Yet another bonding feature we need to be aware of is that atoms (or groups of atoms) connected by a *single* bond can rotate about the bond without breaking or affecting the strength of the bond. Furthermore, the basic structure of the molecule is unchanged by rotation about the single bond. Although there is rotation about each single bond, the three carbon atoms in propane remain in a continuous chain.

The four carbon atoms in butane form a continuous chain with the Lewis structure:

These four carbons are not actually in a straight line.

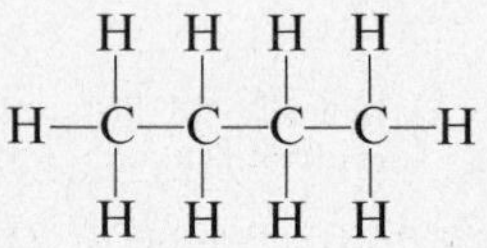

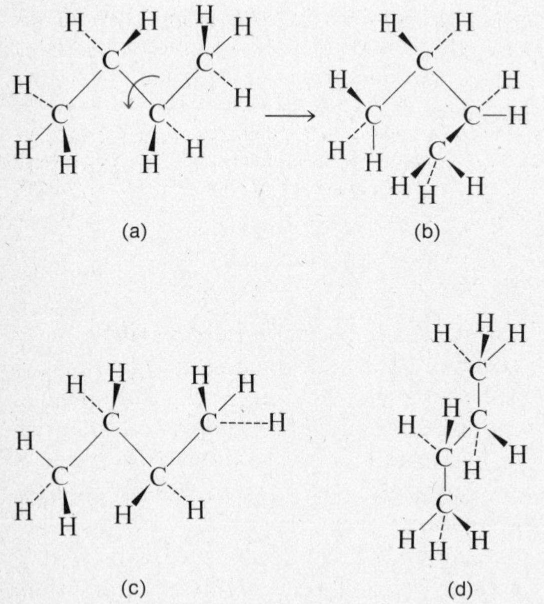

(a) (b)

(c) (d)

FIGURE 4-2 Possible shapes of a butane molecule. Notice how rotation about a C-C single bond in (a) leads to (b).

The condensed structural formula is written $CH_3CH_2CH_2CH_3$. Because of rotation around single bonds, the actual shape of a butane molecule at any given instance could be one of those shown in Fig. 4-2. (A set of molecular models is useful for seeing the different shapes a butane molecule may have.)

exercise 4-1

Write Lewis and condensed structures for natural gas (methane), bottled gas (propane), and the fluid in butane cigarette lighters. Speculate on the term *octane* as associated with gasolines.

BRANCHED ALKANES

The atoms in C_4H_{10} can be arranged in more than one way. The four carbon atoms do not have to form a continuous chain as they do in butane. Another possible arrangement is

```
     H  H  H
     |  |  |
  H—C—C—C—H
     |  |  |
     H  |  H
     H—C—H
        |
        H
```

Although this compound has the same molecular formula as butane (C_4H_{10}), the structure is different. Since the longest continuous chain of carbon atoms is three, the compound is named as a propane (*prop* means three carbon atoms in a continuous chain). Clearly, the compound is not propane though. Propane has the formula C_3H_8. *Alkanes in which all the carbon atoms are not included in a continuous chain are called branched alkanes.* Branched alkanes are named according to the number of carbon atoms forming the *longest continuous* chain.

exercise 4-2

For each of the compounds below, find the longest continuous chain of carbon atoms. On the basis of the chain length, decide what each compound will be named (e.g., as a ______ane). Also decide the continuous chain alkane of which each is an isomer.

```
               H            H
               |            |
        H  H—C—H   H—C—H   H
        |      |            |      |
(a) H—C——————C——————————C——————C—H
        |      |            |      |
        H      H            H      H
```

```
          H  H  H  H  H
          |  |  |  |  |
(b)    H—C—C—C—C—C—H
          |  |  |  |  |
          H  |  H  H  H
          H—C—H
             |
             H
```

```
                   H
                   |
                H—C—H
                   |
                H—C—H
                   |
             H  H  |  H
             |  |  |  |
    (c) H—C—C—C—C—H
             |  |  |  |
             H  |  H  H
                H—C—H
                   |
                H—C—H
                   |
                   H
```

```
            H                 H                              H
            |                 |                              |
      H  H—C—H   H  H    H—C—H                     H  H—C—H   H   H
      |     |     |  |      |                       |     |     |   |
(d) H—C—————C—————C—C———————C—H            (e) H—C—————C—————C—C—H
      |     |     |  |      |                       |     |     |   |
      H     H     H  |      H                       H     H     |   H
                  H—C—H                                       H—C—H
                     |                                          |
                     H                                          H
```

After we decide upon the longest continuous chain of carbon atoms and the corresponding name, the branches (also called side chains) are named. If these side chains contain only carbon and hydrogen atoms joined by single bonds, they are called *alkyl groups*. The names of alkyl groups are based on the number of carbon atoms contained in them. *An alkyl group has one less hydrogen atom than the corresponding alkane.* The simplest alkyl group is the methyl group (—CH_3). The name of an alkyl group is often obtained by changing the *ane* ending of the alkane to *yl* (e.g., CH_4, meth*ane;* —CH_3, methyl). Thus,

```
     H      H      H
     |      |      |
   H—C——————C——————C—H
     |      |      |
     H   H—C—H     H
            |
            H
```

is named 2-methylpropane. *Methyl* and *propane* are written as one word. The 2 indicates the methyl group is joined to the second carbon of the propane chain. A condensed structural formula for 2-methylpropane can be written

```
CH3CHCH3
    |
   CH3
```

(Why shouldn't we simply write C_4H_{10} for 2-methylpropane?)

exercise 4-3

Draw Lewis structures for each alkyl group in Table 4-2.

Notice that the name of the last entry in Table 4-2 is not obtained directly from the name of the alkane. *The names and structures of the alkyl groups in Table 4-2 should be memorized now.*

Let us emphasize at this time that the alkyl groups are *not stable molecules.* One carbon atom of each alkyl group has one less than the desired number of bonds and does not possess a noble gas electron arrangement.

TABLE 4-2
Alkyl groups

CONDENSED STRUCTURAL FORMULA	NAME	PARENT ALKANE
$CH_3—$	methyl	methane
$CH_3CH_2—$	ethyl	ethane
$CH_3CH_2CH_2—$	propyl	propane
$(CH_3)_2CH—$	isopropyl	propane

This remaining bonding site on the alkyl group is used to attach it to the longer carbon chain of the branched alkane.

We can *imagine* that an alkyl group is formed by removal of one hydrogen atom from an alkane, leaving an unused bonding site. For example, removal of a hydrogen from methane leaves a methyl group.

$$\underset{\text{methane molecule}}{CH_4} \xrightarrow[\text{of hydrogen atom}]{\text{imagined removal}} \underset{\text{methyl group}}{—CH_3}$$

Now we can see the importance of recognizing the different types of hydrogen atoms in an alkane. The imagined removal of a hydrogen atom from an end carbon of propane yields a propyl group; whereas, the imagined removal from the middle hydrogen of propane gives an isopropyl group.

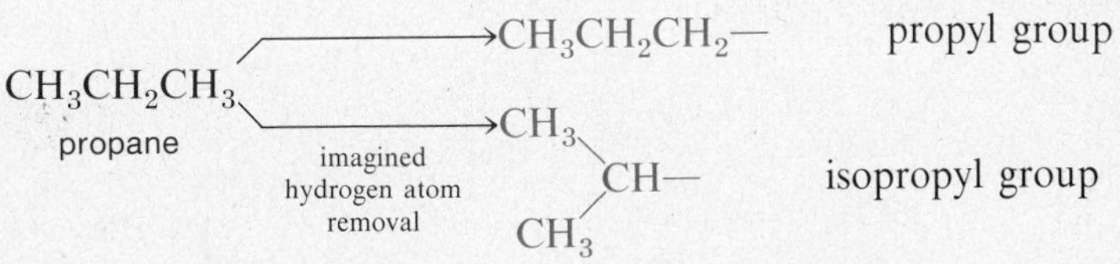

$$\underset{\text{propane}}{CH_3CH_2CH_3} \rightarrow CH_3CH_2CH_2— \quad \text{propyl group}$$

$$\underset{\text{propane}}{CH_3CH_2CH_3} \xrightarrow[\text{removal}]{\text{imagined hydrogen atom}} (CH_3)_2CH— \quad \text{isopropyl group}$$

exercise 4-4

Show the structures of the two alkyl groups that may be derived from 2-methylpropane (do not try to name them).

exercise 4-5

Identify the longest continuous chain in the following alkanes. Name each of the side chain alkyl groups. Also write the Lewis structure of each.

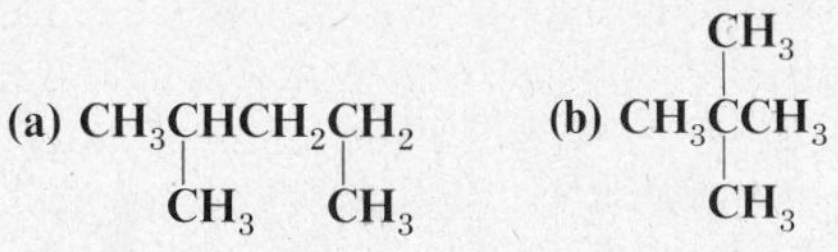

(a) $CH_3CH(CH_3)CH_2CH_2(CH_3)$ — i.e., $CH_3\underset{|}{C}HCH_2\underset{|}{C}H_2$ with CH_3 on the second and fourth carbons

(b) $CH_3C(CH_3)_2CH_3$ — i.e., $CH_3\underset{|}{\overset{|}{C}}CH_3$ with CH_3 above and below the central carbon

(c)
$$\begin{array}{c} CH_3 \\ | \\ CH_3CCH_3 \\ | \\ CH_2 \\ | \\ CH_2 \\ | \\ CH_3 \end{array}$$

(d)
$$\begin{array}{l} \qquad\qquad\quad CH_3 \\ \qquad\qquad\quad | \\ \qquad\quad CH_3{-}CH \\ \qquad\qquad\quad | \\ CH_3CH_2CH_2CHCHCH_3 \\ \qquad\qquad\qquad\quad | \\ \qquad\qquad\qquad CH_2CHCH_2CH_3 \\ \qquad\qquad\qquad\qquad | \\ \qquad\qquad\qquad\quad CH_2CH_2CH_3 \end{array}$$

NAMING BRANCHED ALKANES

At this juncture we are able to name unbranched alkanes, recognize the longest continuous chain, and identify the side chains for branched alkanes. We next need to learn the rules enabling us to name branched alkanes. These rules are international ones, agreed upon by chemical societies throughout the world. The rules for naming compounds are referred to as the IUPAC System of Nomenclature* and are summarized in Table 4-3.

TABLE 4-3
Nomenclature of alkanes

1. Saturated hydrocarbons are classed as alkanes and always have the ending ane.
2. Branched alkanes are named on the basis of the longest continuous chain of carbon atoms in the molecule. (Unbranched alkanes are named as given in Table 4-1.)
3. To indicate the location of the branches, the carbon atoms of the longest continuous chain are numbered consecutively. The chain is numbered such that the first branch has the lowest possible number.
4. The side chains (branches) are named as indicated in Table 4-2.
5. When two or more identical side chains are contained in the same molecule, the number of identical groups is given by *di* for two, *tri* for three, etc. The location of *each* side chain is given by the number assigned to the carbon to which the branch alkyl group is attached.
6. In the name, hyphens are used to separate numbers from letters; commas, to separate numbers. Letters are not separated by spaces or punctuation marks.

EXAMPLE 4-1

$$\begin{array}{l} \quad\;\; CH_3 \\ \quad\;\;\, | \\ CH_3CCH_2CH_2CH_3 \\ \quad\;\;\, | \\ \quad\;\; CH_3 \end{array}$$

The longest continuous chain is five carbons. Therefore, the compound is named as a pentane. Since there are two methyl groups present as side chains, the compound is a dimethylpentane. The location of each methyl group is carbon number 2. Hence, the compound's name is 2,2-dimethylpentane. (Why not 4,4-dimethylpentane?)

*IUPAC is the abbreviation for International Union of Pure and Applied Chemistry.

EXAMPLE 4-2

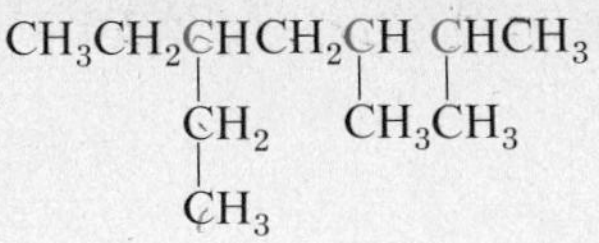

$$\begin{array}{l}
CH_3CH_2CHCH_2CH\ CHCH_3 \\
\qquad\quad\ \ |\qquad\quad\ |\quad\ \ | \\
\qquad\quad CH_2\quad CH_3CH_3 \\
\qquad\quad\ \ | \\
\qquad\quad CH_3
\end{array}$$

The longest continuous chain of carbon atoms is seven. Two methyl and one ethyl side chains are present. Accordingly, the compound is named 5-ethyl-2,3-dimethylheptane. (Remember rule 3 dealing with the numbering of carbon chains.)

exercise 4-6

Name the compounds in Exercises 4-2 and 4-5 [omit 4-5(d)].

ALKENES

Alkenes, commonly called olefins, are the group of hydrocarbons containing carbon–carbon double bonds. The names of alkenes have the characteristic ending ene.

The rules governing the naming of alkenes are relatively simple and are given in Table 4-4.

TABLE 4-4
Alkene nomenclature

1. Characteristic ending of the name is ene.
2. Select the longest continuous chain containing the double bond. Name the chain by changing the *ane* ending of the corresponding alkane to *ene.*
3. Number the chain so as to give the first carbon atom of the double bond the lowest possible number.
4. Identify side chains in the same manner as for alkanes. Specify the location of side chains according to the numbering assigned in rule 3.

Sometimes more than one name is acceptable for a compound as in this example. The first one is preferred.

EXAMPLE 4-3

$CH_3CH_2CH_2CH{=}CH_2$

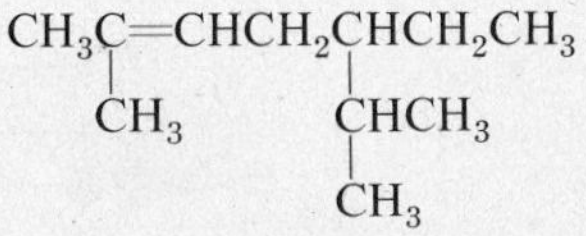

1-pentene

$$\begin{array}{l}
CH_3C{=}CHCH_2CHCH_2CH_3 \\
\qquad\ |\qquad\qquad\ | \\
\quad\ CH_3\qquad\ CHCH_3 \\
\qquad\qquad\qquad\ | \\
\qquad\qquad\quad\ CH_3
\end{array}$$

5-*iso*propyl-2-methyl-2-heptene
or
5-ethyl-2,6-dimethyl-2-heptene

exercise 4-7

Name the following alkenes:

(a) $CH_3CH{=}CH_2$

(b) $(CH_3)_2C{=}CH_2$

(c) $CH_3CH_2CH{=}CH_2$

(d)
$$\begin{array}{l}
\qquad\ CH_3\qquad\qquad\qquad\quad CH_3 \\
\qquad\ \ |\qquad\qquad\qquad\quad\ / \\
CH_3CHCH_2CH_2CH{=}C \\
\qquad\qquad\qquad\qquad\quad\ \backslash \\
\qquad\qquad\qquad\qquad\qquad CH_3
\end{array}$$

(e) 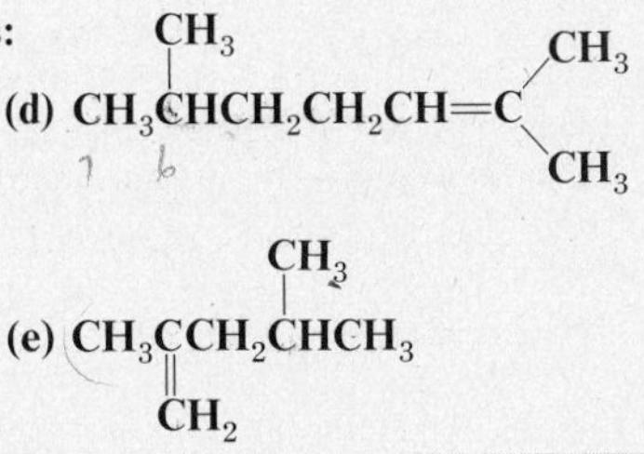
$$\begin{array}{l}
\qquad\qquad\qquad CH_3 \\
\qquad\qquad\qquad\ | \\
CH_3CCH_2CHCH_3 \\
\qquad\ \| \\
\qquad CH_2
\end{array}$$

The simplest alkene ($CH_2{=}CH_2$) is correctly named ethene; however, it is generally referred to by the older name ethylene. We shall often find examples of carbon compounds that are generally referred to by older or trivial names. Although inconvenient, it will be necessary to learn the trivial names for several compounds to be able to converse meaningfully with chemists, pharmacists, physicians, and nutritionists.

CIS AND *TRANS* ISOMERISM (ARE YOU ON MY SIDE?)

The presence of a double bond in alkenes makes possible another type of isomerism. This new kind of isomerism is geometric or *cis–trans* isomerism. Geometric isomerism is the result of the fact that rotation around a double bond is not possible.

Consider 2-butene ($CH_3CH{=}CHCH_3$). Because there is no rotation around the double bond, there are two ways of arranging the carbon atoms (structures 1 and 2).

$$\begin{array}{ccc} CH_3 & & CH_3 \\ & C{=}C & \\ H & & H \end{array} \qquad \begin{array}{ccc} CH_3 & & H \\ & C{=}C & \\ H & & CH_3 \end{array}$$

structure 1 structure 2

The methyl groups in structure 1 are on the same side of the double bond; whereas in structure 2 the methyl groups are on opposite sides of the double bond. Structure 1 cannot be changed to structure 2 without breaking a bond, an event that does not happen under normal circumstances. Compounds corresponding to both structures have been isolated and studied.

When two identical groups (or atoms) are on the same side of a double bond, they are said to be cis *to each other. If two identical groups are on opposite sides of the double bond, they are* trans *to each other.* The compound whose molecules have structure 1 is *cis*-2-butene; structure 2 is that of *trans*-2-butene.

Notice that if there are two identical groups (or atoms) on the *same* carbon atom involved in a double bond, there is no possibility for geometric isomerism. Thus, there is only one possible structure for 1-butene:

$$\begin{array}{ccc} CH_3CH_2 & & H \\ & C{=}C & \\ H & & H \end{array}$$

exercise 4-8

With the aid of a molecular model set, convince yourself there are two isomers of 2-butene and only one for 1-butene.

If you don't have a model set, make models out of toothpicks and jellied candy (gumdrops or marshmallows). (You can eat the candy later.)

We can state two requirements for geometric isomerism:

1. Restricted rotation about a bond must be present in a molecule. This

restricted rotation may be caused by the presence of either a double bond or ring formation.

2. No carbon atom whose rotation is restricted may bear two identical groups or atoms.

exercise 4-9

Decide whether geometric isomerism is possible for each of the following. Draw Lewis structures and name each isomer.

(a) $CH_2{=}C(CH_3)_2$

(b) $CH_3CH{=}CHCH_2CH_3$

(c) 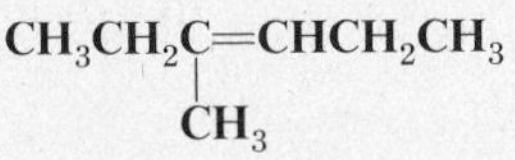 $CH_3CH_2\underset{\displaystyle |\atop \displaystyle CH_3}{C}{=}CHCH_2CH_3$

(d) 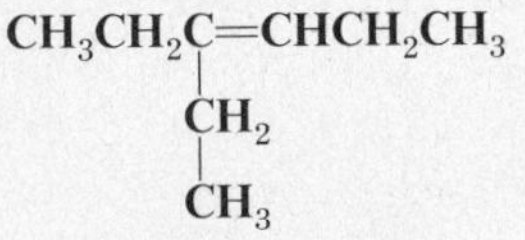 $CH_3CH_2\underset{\displaystyle |\atop \displaystyle CH_2 - CH_3}{C}{=}CHCH_2CH_3$ (the side chain CH_2CH_3 is attached vertically to C)

(e) 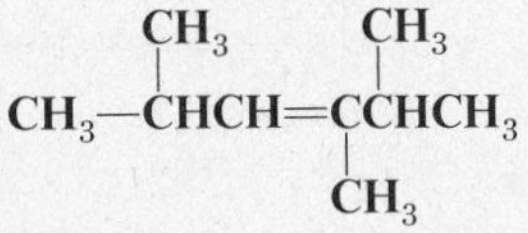 $CH_3{-}\overset{\displaystyle CH_3\atop \displaystyle |}{C}HCH{=}\overset{\displaystyle CH_3\atop \displaystyle |}{\underset{\displaystyle |\atop \displaystyle CH_3}{C}}CHCH_3$

ALKYNES Hydrocarbons containing a carbon–carbon triple bond are classified as alkynes. The characteristic suffix indicating the presence of a carbon–carbon triple bond is *yne*. See Table 4-5 for alkyne nomenclature. The first member

TABLE 4-5
Alkyne nomenclature

1. Characteristic ending of the name is yne.
2. Select the longest continuous carbon chain containing the triple bond. Name the chain by changing the *ane* ending of the corresponding alkane to *yne*.
3. Number the chain so as to give the first carbon atom of the triple bond the lowest possible number.
4. Identify side chains in the same manner as alkenes.

Compounds containing multiple (double or triple) bonds are said to be unsaturated. *What do you think the term* polyunsaturated *means?*

of this series of compounds is ethyne (H—C≡C—H). Ethyne is more generally known by its trivial name, acetylene. In fact, the whole class of compounds is more frequently referred to as acetylenes rather than alkynes. There are few alkynes of biological significance in mammals, although they are prevalent in plants and molds. A few alkynes are important drugs. For example, norethindrone, the principal component of the birth control pill Enovid contains a carbon–carbon triple bond. Several antibiotics (mycomycin, isomycomycin, and nemotin) also contain triple bonds between carbon atoms.

CYCLIC COMPOUNDS All the hydrocarbons we have considered so far are open-chain compounds. The molecule has two ends, like a piece of string has two ends. The ends of the string can be tied together to form a loop or "string ring." We might ask if the ends of a carbon atom chain can be bonded (tied) together to produce hydrocarbon rings. To join the terminal carbon atoms of an open chain into a ring requires one additional bonding site on each of these carbon atoms. For example, a four-carbon open-chain alkane has a combined total of 10 additional bonding sites on the carbon atoms; the corresponding carbon ring has only 8 bonding sites left after joining the carbon atoms together. Compare

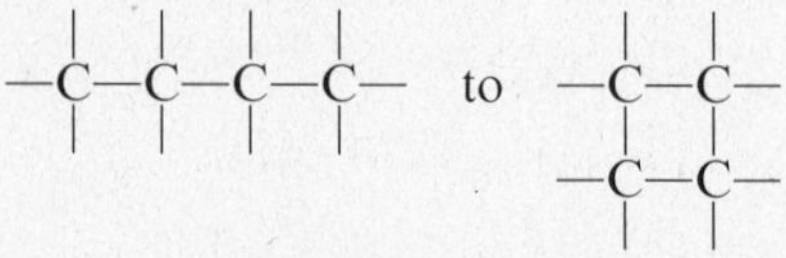

$$-\overset{|}{\underset{|}{C}}-\overset{|}{\underset{|}{C}}-\overset{|}{\underset{|}{C}}-\overset{|}{\underset{|}{C}}- \quad \text{to} \quad \begin{array}{c} | \quad | \\ -C-C- \\ | \quad | \\ -C-C- \\ | \quad | \end{array}$$

Compounds in which the carbon atoms form a ring are called *cyclic compounds*. Cyclic hydrocarbons containing only single bonds are cycloalkanes. They are named by using the prefix *cyclo-* before the name of the corresponding open-chain alkane. Thus,

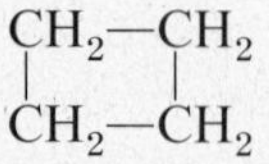

$$\begin{array}{c} CH_2-CH_2 \\ | \qquad | \\ CH_2-CH_2 \end{array} \qquad \begin{array}{c} CH_2 \\ CH_2 \quad CH_2 \\ | \qquad | \\ CH_2 \quad CH_2 \\ CH_2 \end{array} \qquad \begin{array}{c} CH_2 \\ CH_2 \quad CH-CH_3 \\ | \qquad | \\ CH_2-CH_2 \end{array}$$

cyclobutane 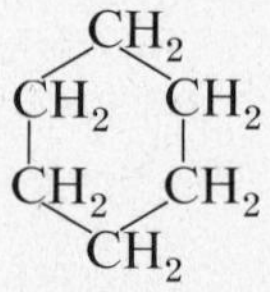 cyclohexane 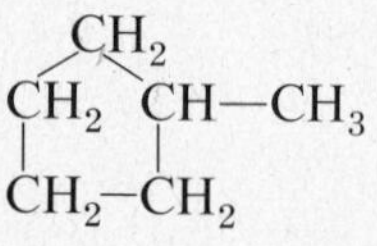 methylcyclopentane

When two or more groups are attached to carbon atoms in the ring, their relative positions must be given. Position of the groups are specified by numbering the atoms comprising the ring. Numbers are assigned the ring atoms so as to give the attached groups the lowest possible numbers. Two examples are

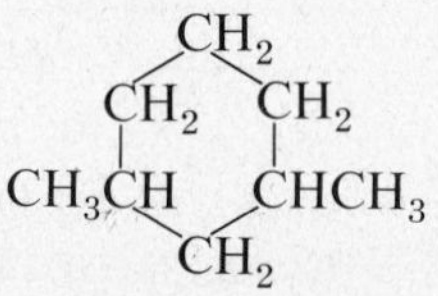

$$\begin{array}{c} CH_2 \\ CH_2 \quad CH_2 \\ | \qquad | \\ CH_3CH \quad CHCH_3 \\ CH_2 \end{array} \qquad \begin{array}{c} CH_2 \quad CH_3 \\ CH_2 \quad C \\ | \qquad | \quad CH_3 \\ CH_2-CHCH_2CH_3 \end{array}$$

1,3-dimethylcyclohexane (not 1,5-dimethylcyclohexane)

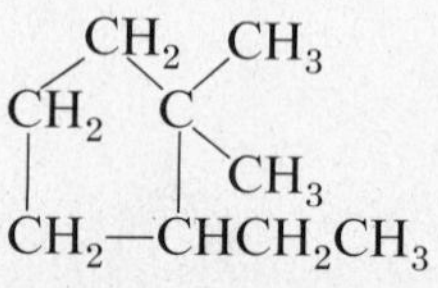

1,1-dimethyl-2-ethylcyclopentane (not 2,2-dimethyl-1-ethylcyclopentane)

Because carbon atoms always form four bonds, a shorthand notation for writing cyclic compounds is used. This is illustrated below:

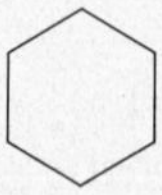

CH_3 CH_3 CH_3 CH_3

cyclohexane 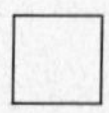 cyclobutane 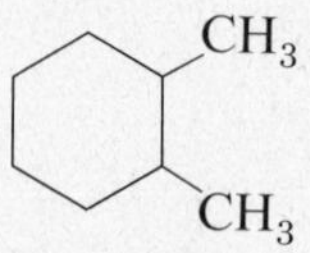 1,2-dimethylcyclohexane 1,1-dimethylcyclohexane

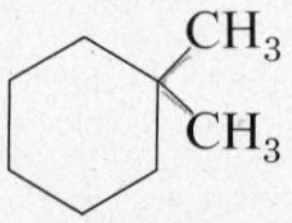

Each point of the figure represents the position of a carbon atom, while each line represents a bond between two carbon atoms. It is *understood* that the remaining bonding sites on each carbon atom are occupied by hydrogen. Thus, carbons 1 and 2 of 1,2-dimethylcyclohexane each have one hydrogen atom bonded to them; the remaining carbon atoms in the ring are each bonded to two hydrogen atoms.

exercise 4-10

Write the molecular formulas for each of the following (e.g., the molecular formula of △ is C_3H_6).

(a) CH_3 CH_3

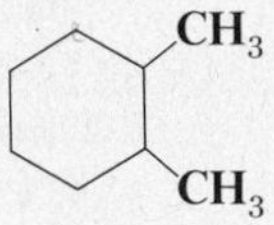

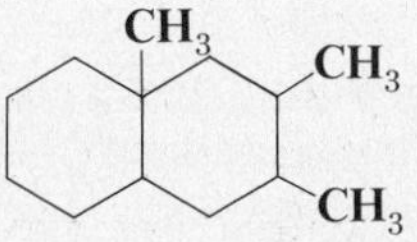

(b) CH_3 CH_3 CH_3CH_2

(c) CH_3 CH_3 CH_3

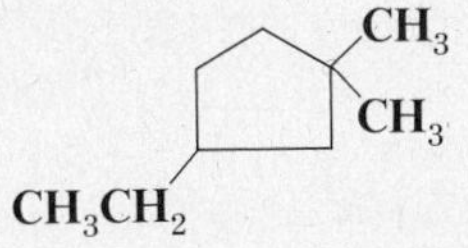

(d)

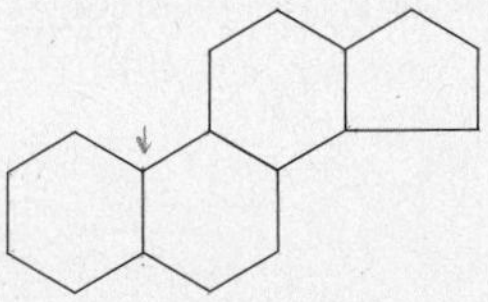

(This is the basic carbon arrangement of steroids. Cholesterol, testosterone, progesterone, and cortisone are all steroids.)

Many cyclic hydrocarbons contain carbon–carbon double bonds in the ring. These compounds are cycloalkenes. (Cycloalkynes are rare and so we may ignore their existence here.)

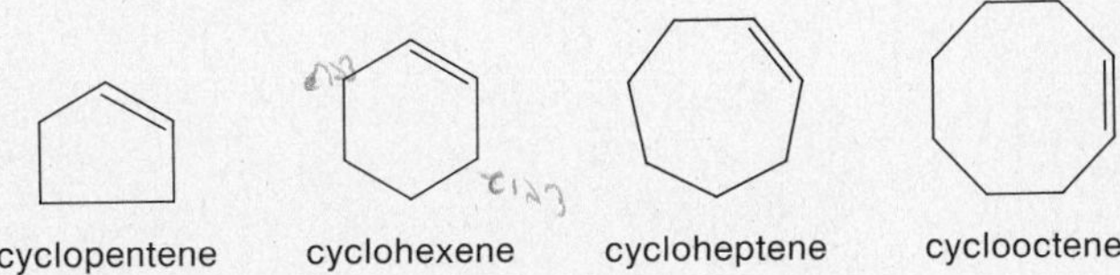

exercise 4-11

Write the molecular formulas for each of the following:

(a)

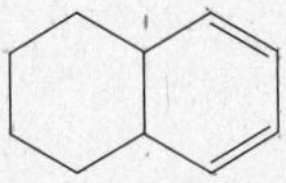

(b) cyclobutene

(c)

(d) CH_2 $CH{=}CH_2$

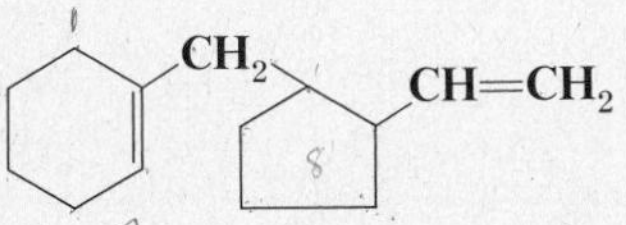

(e)

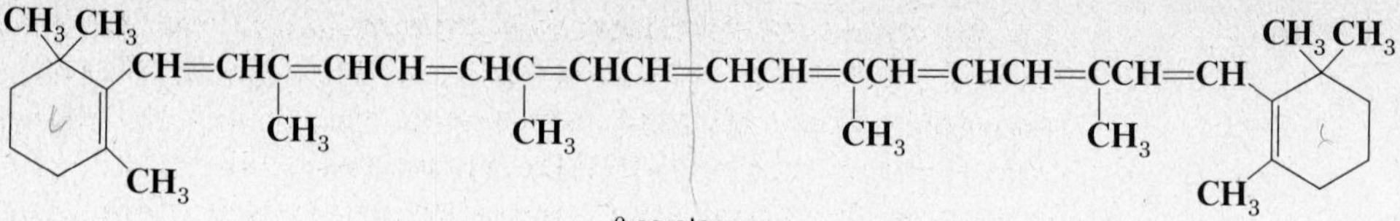

β-carotene

(β-Carotene is partly responsible for the orange color of carrots and of leaves in the autumn. β-Carotene is converted to vitamin A in the liver.)

AROMATIC HYDROCARBONS All the hydrocarbons we have studied so far are collectively known as *aliphatic hydrocarbons*. Aromatic hydrocarbons are a special class of hydrocarbons whose molecules have a six-membered carbon ring containing what appears to be alternating single and double carbon–carbon bonds. The term *aromatic* (a derivative of the word *aroma*) was applied to these compounds long before their structure became known. Thus, aromatic was used to denote their strong odor. While many aromatic compounds are very pleasant smelling compounds used in the perfume and flavorings industries, others have distinctly unpleasant odors.

The simplest aromatic compound is benzene, discovered by Faraday in 1824. Benzene has the molecular formula C_6H_6 and its structure can be written

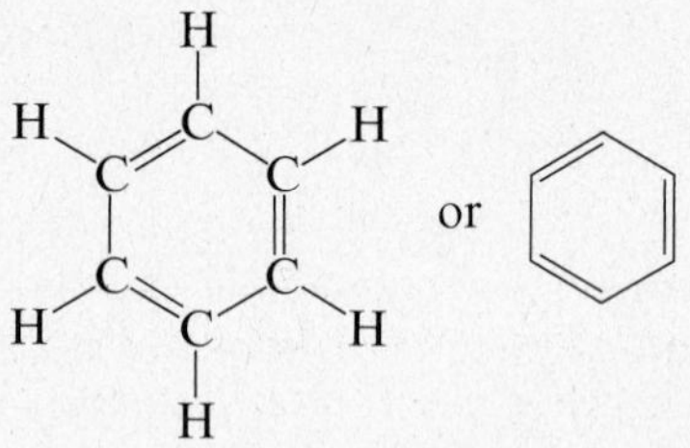

or

When chemists began to investigate the properties and reactions of aromatic substances, they discovered some rather puzzling facts about these compounds. For instance, it would seem that if the ring structure contains alternating single and double bonds, there should be two isomers of 1,2-dimethylbenzene—structures 1 and 2.

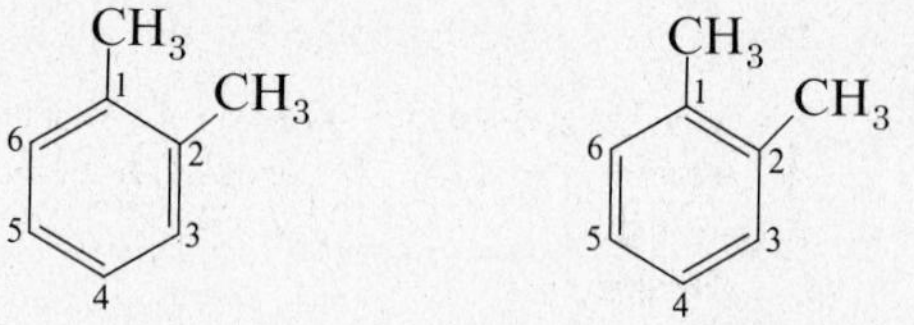

structure 1

structure 2

Only one 1,2-dimethylbenzene has ever been isolated. Kekulé, in 1865, attempted to resolve the discrepancy between the expected number of isomers of substituted benzene molecules and the number actually observed. Kekulé suggested that the double bonds change positions very rapidly. If the double bonds change positions fast enough, the two isomers of 1,2-dimethylbenzene would be interconverted so quickly that they could not be separated. Thus, it would appear that there is only one isomer.

Since Kekulé's time, chemists have experimentally determined that all the bonds in the ring are identical. (See Appendix 2.)

Our modern view of aromatic compounds describes the six electrons used for double-bond formation in structures 1 and 2 as being in a molecular electron shell. Thus, these six electrons are shared by all the carbon atoms in the ring. Chemists, in an effort to show the distribution of these six electrons in a molecular orbit, frequently depict the aromatic ring as

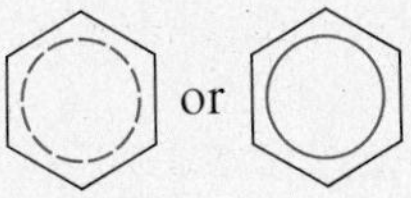

or

The (dotted) circle in the ring represents the six electrons in the molecular electron shell. Accordingly, the structure of 1,2-dimethylbenzene is represented as

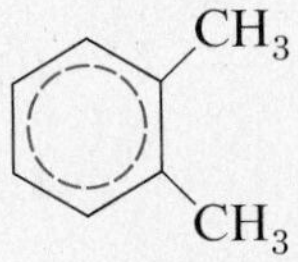

CH_3

CH_3

REACTIVITY OF HYDROCARBONS

All hydrocarbons can be burned to produce carbon dioxide and water. Combustion of hydrocarbons is a primary source of energy in our lives. We burn gasoline, a mixture of hydrocarbons, to power our cars. Methane (natural gas) is burned in many homes for heat.

Balanced equations for the combustion of methane and for one component of gasoline are

$$CH_4 + 2O_2 \longrightarrow CO_2 + 2H_2O$$
$$2C_8H_{18} + 25O_2 \longrightarrow 16CO_2 + 18H_2O$$

exercise 4-12

Write balanced equations for the combustion of

(a) butane **(d) *trans*-2-hexene**
(b) 1-hexene **(e) benzene**
(c) *cis*-2-hexene

Notice from Exercise 4-12 that regardless of the class of hydrocarbon (alkane, alkene, or aromatic) or the isomer (structural or geometric), all hydrocarbons burn to ultimately produce carbon dioxide and water. In contrast to the combustion reaction, the vast majority of organic reactions are dependent on classification of the reactants.

The characteristic reactions of alkenes are addition reactions.

Alkanes undergo very few reactions. In fact, the only reaction of alkanes we will discuss is combustion. Alkenes, however, undergo a large number of reactions. Typically, reactions of alkenes involve the double bond. During these reactions the carbon–carbon double bond is converted to a carbon–carbon single bond. Simultaneously, a new bond is formed by each of the carbon atoms of the original double bond. Such reactions are called *addition reactions*.

An example of an addition reaction is that between 1-butene and hydrogen in the presence of platinum.

$$CH_3CH_2CH{=}CH_2 + H_2 \xrightarrow{Pt} CH_3CH_2\underset{\displaystyle H}{\underset{|}{C}}H-\underset{\displaystyle H}{\underset{|}{C}}H_2$$

Notice that in this reaction only the double-bond portion of the alkene molecule reacts. The rest of the molecule remains unchanged. This observation suggests that other alkenes should react with hydrogen in the presence of platinum. Experimentation shows this to be true. For example, *cis*-2-butene and cyclohexene will both react with hydrogen.

$$\text{(CH}_3\text{)(H)C=C(H)(CH}_3\text{)} + H_2 \xrightarrow{Pt} CH_3CH_2CH_2CH_3$$

and

$$\text{cyclohexene} + H_2 \xrightarrow{Pt} \text{cyclohexane (H, H added)}$$

Under appropriate conditions, water will react with compounds containing double bonds. In the presence of sulfuric acid (H_2SO_4) both *cis*-2-butene and cyclohexene react with water:

$$\text{(CH}_3\text{)(H)C=C(H)(CH}_3\text{)} + H_2O \xrightarrow{H_2SO_4} CH_3\underset{\displaystyle H}{\underset{|}{C}}H-\underset{\displaystyle OH}{\underset{|}{C}}H-CH_3$$

$$\text{cyclohexene} + H_2O \xrightarrow{H_2SO_4} \text{cyclohexane (H, OH added)}$$

Again we should notice that only the double-bond portion of the alkene takes part in the reaction. The rest of the molecule remains unchanged. We will find throughout our studies of organic compounds in later chapters the hydrocarbon portions of the molecules containing *only single bonds* are unreactive. For this reason, chemists look for the reactive atom or group of atoms in organic molecules. This reactive atom or group of atoms is called a *functional group*.

Memorize the functional groups and their names from Table 4-6.

Alkenes contain the functional group $>C{=}C<$. The presence of a carbon–carbon double bond is also the basis for classification of a particular hydrocarbon as an alkene. Thus, *a functional group both classifies a compound and tells us something about the chemical properties and reactivity of the compound.*

TABLE 4-6
Selected functional groups of organic compounds

FUNCTIONAL GROUP	NAME OF COMPOUND CLASS	EXAMPLE
—	alkane	$CH_3CH_2CH_2CH_3$
$>C{=}C<$	alkene	CH_3[$CH{=}CH_2$]
$-C{\equiv}C-$	alkyne	CH_3-[$C{\equiv}C$]$-CH_2CH_3$
benzene ring	aromatic	[benzene ring] with CH_2CH_3 and CH_3 substituents
$-\overset{\vert}{\underset{\vert}{C}}-OH$	alcohol	CH_3[CH_2OH]
$-\overset{\vert}{\underset{\vert}{C}}-O-\overset{\vert}{\underset{\vert}{C}}-$	ether	CH_3[CH_2-O-CH_2]CH_3
$-C(=O)-O-H$	acid	CH_3[$C(=O)-O-H$]
$-\overset{\vert}{\underset{\vert}{C}}-NH_2$	amine	cyclohexyl[$-NH_2$]
$-\overset{\vert}{\underset{\vert}{C}}-C(=O)-N<$	amide	CH_3-[$C(=O)-N(H)$]$-CH_2CH_3$
$-\overset{\vert}{\underset{\vert}{C}}-C(=O)-O-\overset{\vert}{\underset{\vert}{C}}-$	ester	CH_3-[$C(=O)-O-CH_2$]CH_3

It will be very useful for us to learn to recognize a few functional groups at this time. Additional functional groups will be introduced in later chapters.

FUNCTIONAL GROUPS OF ORGANIC MOLECULES

We are already familiar with three functional groups: (1) carbon–carbon double bonds, (2) carbon–carbon triple bonds, and (3) aromatic rings. Some functional groups and illustrative compounds are given in Table 4-6.

exercise 4-13

Classify each of the following compounds according to their functional group. (Some compounds may contain more than one functional group and thus be placed in more than one classification.)

(a) 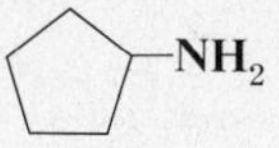 cyclopentyl–NH_2

(b) 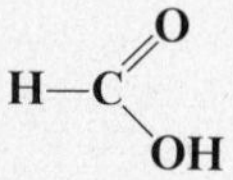 $H—C(=O)OH$
(first discovered in ants)

(c) 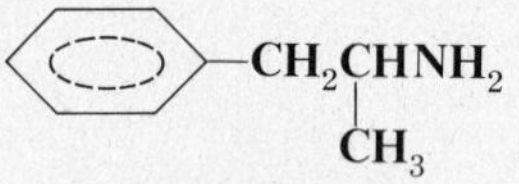 C_6H_5–CH_2CHNH_2 with CH_3 on CH
benzedrine
(a stimulant)

(d) 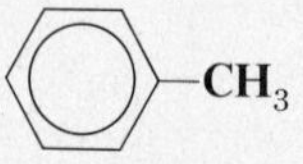 C_6H_5–CH_3
toluene
(a paint solvent)

(e) 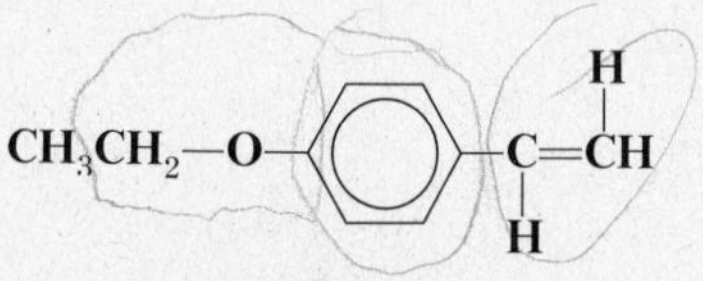 $CH_3CH_2—O—C_6H_4—C(H)=CH(H)$
paravinylanisole
(the active component of anise flavor)

(f) 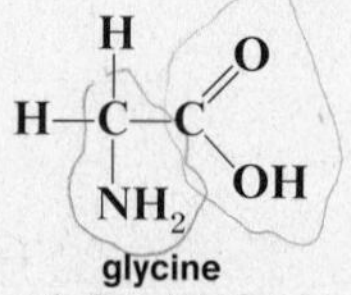 $H—C(H)(NH_2)—C(=O)OH$
glycine
(a breakdown product of many proteins)

(g) 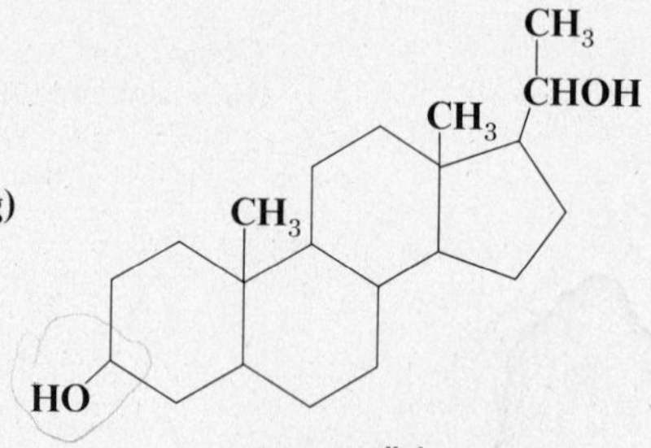 steroid ring system with HO, CH_3, CH_3, and CH_3–CHOH side chain
pregnanediol
(a female sex hormone)

(h) 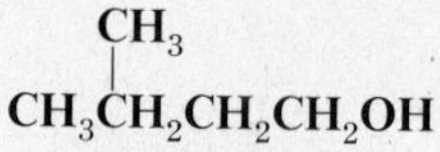 $CH_3CH_2CH_2CH_2OH$ with CH_3 on the second carbon
3-methyl-1-butanol
(frequently found in bourbon and other distilled whiskies; a cause of hangovers)

(i) 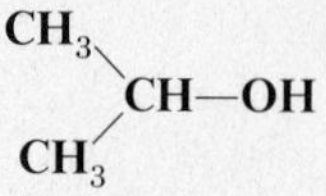 $(CH_3)_2CH—OH$
isopropyl alcohol
(rubbing alcohol)

(j) 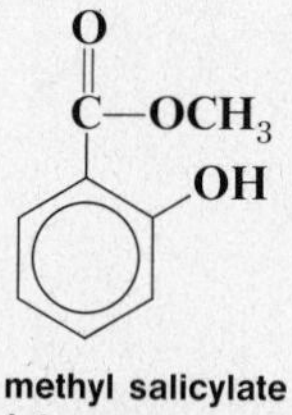 benzene ring bearing $C(=O)—OCH_3$ and OH
methyl salicylate
(oil of wintergreen, a component of rubbing compounds for sore muscles)

POLARITY OF ORGANIC COMPOUNDS

The electronegativity of carbon and hydrogen are very similar; consequently, there is very little polar character associated with carbon–hydrogen bonds. Accordingly, hydrocarbons are essentially nonpolar compounds.

For our purposes, we may regard C—H bonds as nonpolar.

Nitrogen and oxygen are more electronegative than carbon. Therefore C—N, C—O, N—H, and O—H bonds have a substantial polar character. The polar character of these bonds manifests itself in the behavior of alcohols, acids, and amines. All of these compounds are polar.

Let us consider alcohols first. The alcohols have the characteristic group

```
 |
—C—O—H
 |
```

Both the C—O bond and the O—H bond are polar. The Lewis structure of ethyl alcohol is

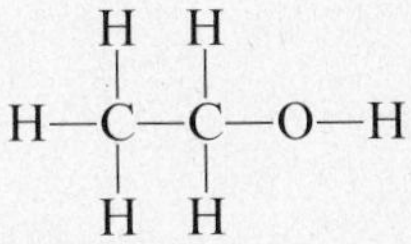

```
   H  H
   |  |
H—C—C—O—H
   |  |
   H  H
```

The oxygen atom in ethyl alcohol is surrounded by four pairs of electrons; two of these electron pairs are bonding pairs of electrons. Thus, the shape of the molecule around the oxygen atom is bent or V-shaped.

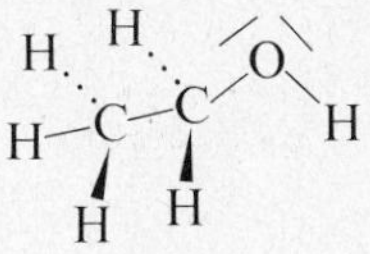

A result of the bond angles around oxygen and the polar character of O—H and C—O bonds is that the alcohol molecule is polar.

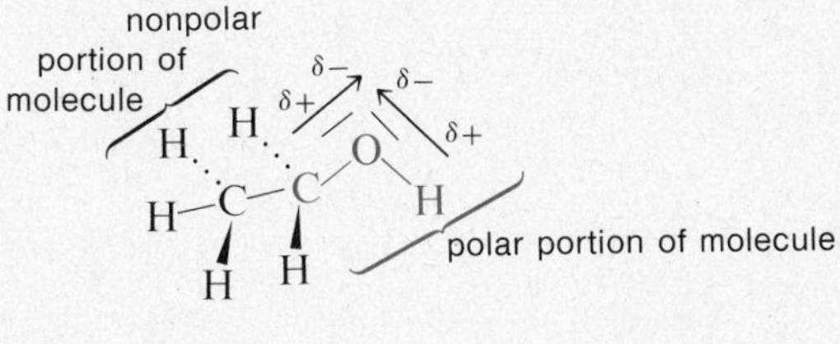

nonpolar portion of molecule

$\delta-$ $\delta+$ $\delta-$ $\delta+$

polar portion of molecule

Referring to Fig. 3-8, compare hydrogen bonding in water to that in ethyl alcohol.

In this respect the alcohol molecule is similar to water. We might then expect to find some similarities in the behavior of alcohols and water. One such similarity is the observance of hydrogen bonding. Figure 4-3 shows hydrogen bonding in liquid water and Fig. 4-4 shows hydrogen bonding in liquid ethyl alcohol.

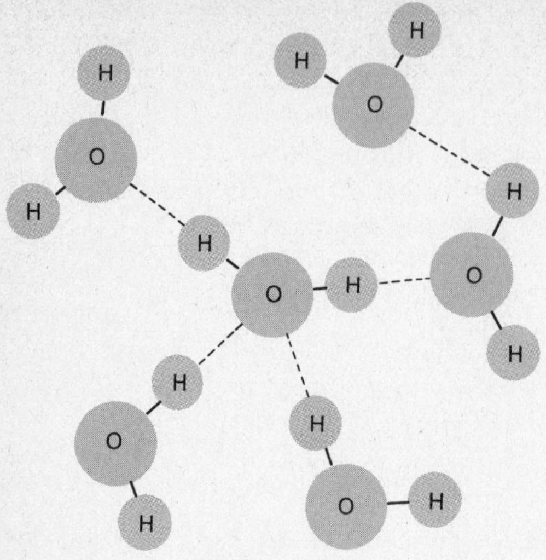

FIGURE 4-3 Hydrogen bonding in liquid water.

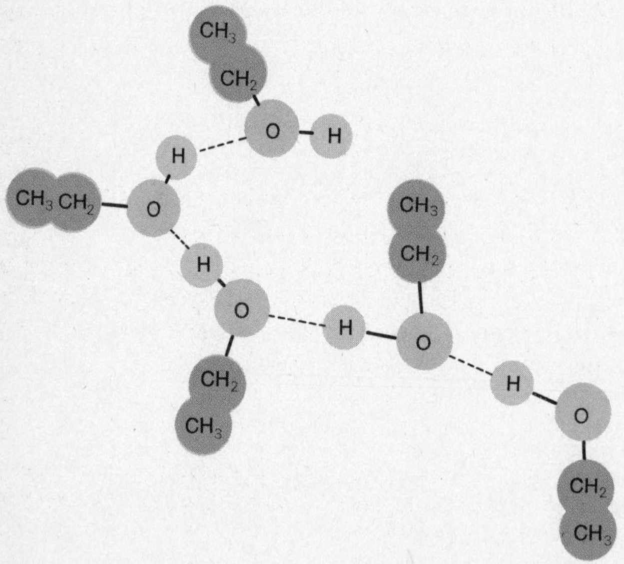

FIGURE 4-4 Hydrogen bonding in liquid ethyl alcohol.

exercise 4-14

Describe, on a molecular level, what might occur when ethyl alcohol is poured into water or vice versa.

The electronegativity of nitrogen is less than that of oxygen. Thus, N—C and N—H bonds are less polar than C—O and O—H bonds. Nevertheless, amines are polar molecules and hydrogen bonding can be experimentally observed.

exercise 4-15

Apply your knowledge from Chapter 3 to show that CH_3NH_2 (methylamine) is a polar molecule. Make a rough sketch showing how hydrogen bonding can occur in liquid methylamine.

Organic acids are even more polar than alcohols or amines. The Lewis structure depicting the shape of acetic acid molecules shows why the molecule is so polar.

$$CH_3-\overset{\delta+}{C}(=O^{\delta-})-\overset{\delta-}{O}-\overset{\delta+}{H}$$

The partial positive charge on the carbon atom created by the double-bonded oxygen actually causes the bonding pair of electrons between the oxygen and hydrogen atoms of the O—H portion to move farther away from the hydrogen than might be expected. We will see the consequences of this in Chapter 8, which deals with the behavior of acids and bases dissolved in water.

CHAPTER SUMMARY

1. Isomerism is possible for covalent compounds. Two kinds of isomerism are structural isomerism and geometric isomerism.
2. Hydrocarbons are divided into the subclasses alkanes, alkenes, alkynes, and aromatics. Alkanes contain only single bonds; alkenes, double bonds; and alkynes, triple bonds. Aromatic compounds contain the  group.
3. Hydrocarbons are named on the basis of the longest continuous chain of carbon atoms. Although often written in a straight line, the carbon atoms are not in a straight line.
4. Geometric isomerism is the result of restricted rotation around carbon–carbon bonds. Restricted rotation can arise from double bonds, triple bonds, or ring formation.
5. All hydrocarbons can be burned in oxygen to form carbon dioxide and water.
6. Except for the combustion reaction, alkanes are normally unreactive.
7. The reactive site of organic molecules is the functional group. Two functional groups are carbon–carbon double bonds and carbon–carbon triple bonds. These two functional groups undergo addition reactions.

8. Organic compounds are grouped in classes according to the functional groups present in the molecule.
9. Organic compounds containing oxygen or nitrogen atoms are usually polar compounds.

QUESTIONS

1. Given below are pairs of structural formulas for organic compounds. In each case, decide whether the members of each pair are structural isomers, geometric isomers, identical, or unrelated.

(a) $CH_3CH_2CH_2CH_2OH$ and $CH_3CH_2CH(OH)CH_3$ (OH on the third carbon)

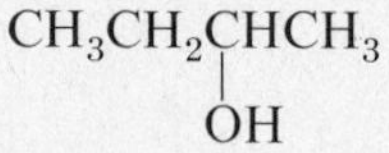

(b) 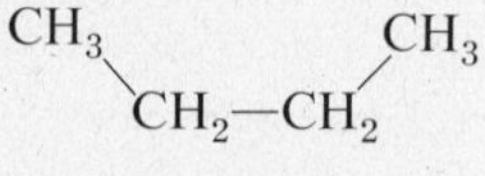 $CH_3-CH_2-CH_2-CH_3$ (drawn with both CH_3 groups pointing up) and $CH_3-CH_2-CH_2-CH_3$ (drawn as a zigzag chain)

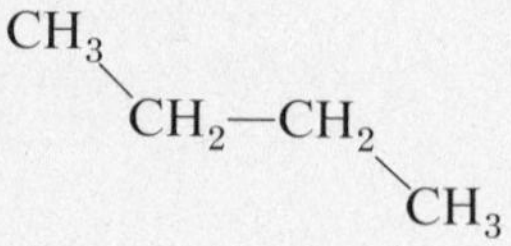

(c) 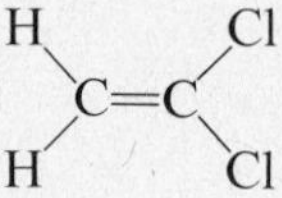 $(H)(H)C=C(Cl)(Cl)$ and $(Cl)(Cl)C=C(H)(H)$

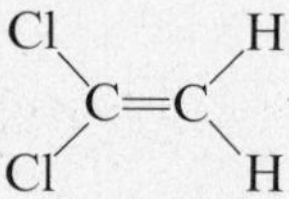

(d) 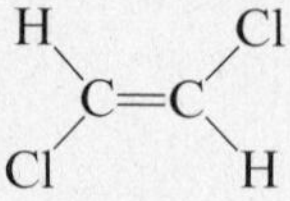 $(H)(Cl)C=C(Cl)(H)$ [H top left, Cl bottom left; Cl top right, H bottom right] and $(H)(H)C=C(Cl)(Cl)$ [H top left, H bottom left; Cl top right, Cl bottom right]

(e) $(H)(Cl)C=C(Cl)(H)$ [H top left, Cl bottom left; Cl top right, H bottom right] and $(Cl)(H)C=C(Cl)(H)$ [Cl top left, H bottom left; Cl top right, H bottom right]

(f) $(H)(Cl)C=C(Cl)(H)$ [H top left, Cl bottom left; Cl top right, H bottom right] and $(Cl)(H)C=C(H)(Cl)$ [Cl top left, H bottom left; H top right, Cl bottom right]

(g) $CH_3-C(=O)-CH_3$ and $CH_3-C(=O)-O-CH_3$

(h) $CH_3-C(=O)-OH$ and $H-C(=O)-O-CH_3$

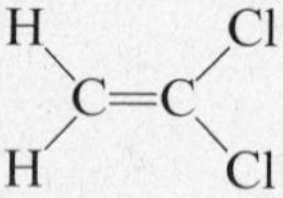

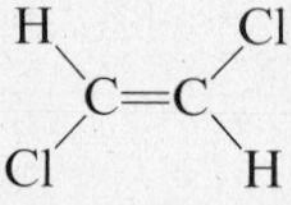

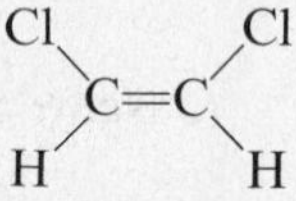

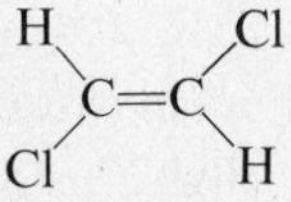

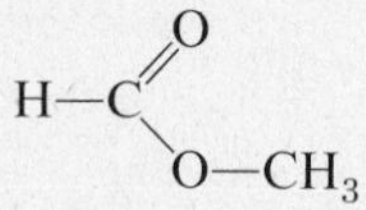

(i) $CH_3CH=CH_2$ and cyclo-$(CH_2)_3$ (a three-membered ring of CH_2 groups)

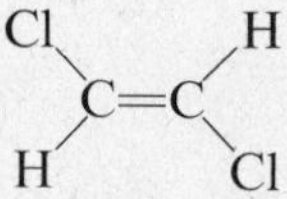

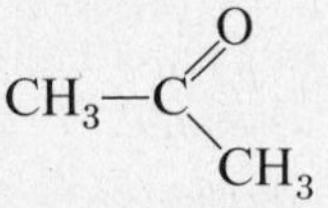

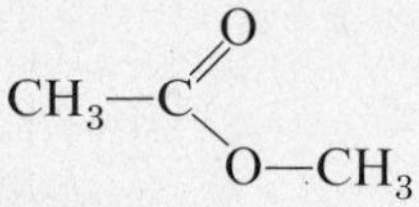

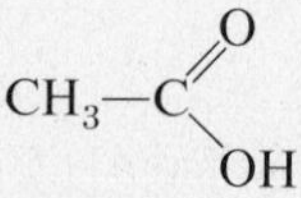

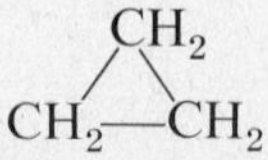

(j) $CH_3CH_2C{\equiv}CCH_2CH_3$ and $CH_3CH{=}CH{-}CH{=}CHCH_3$

(k) $CH_3CH_2C{\equiv}CCH_3$ and $CH_3C{\equiv}CCH_2CH_3$

2. Write the structural formula for the product expected from each of the following reactions and balance the equation. If no product is expected, write N.R.

(a) 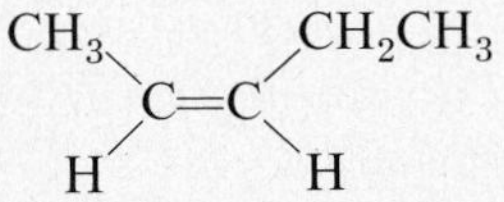 $\underset{H}{\overset{CH_3}{}}\!\!>C{=}C<\!\!\underset{H}{\overset{CH_2CH_3}{}} + H_2 \xrightarrow{Pt}$

(b) 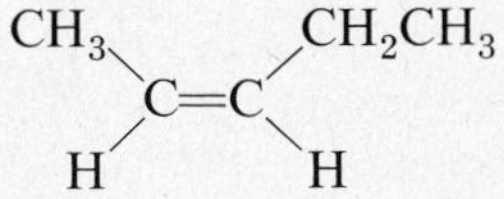 $\underset{H}{\overset{CH_3}{}}\!\!>C{=}C<\!\!\underset{H}{\overset{CH_2CH_3}{}} + O_2 \longrightarrow$

(c) $CH_3CH_2CH_3 + H_2 \xrightarrow{Pt}$

(d) $CH_3CH_2CH_3 + H_2O \xrightarrow{H_2SO_4}$

(e) $CH_3CH_2CH_3 + O_2 \longrightarrow$

(f) 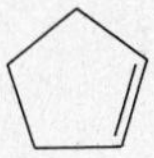 cyclopentene $+ H_2O \xrightarrow{H_2SO_4}$

(g) 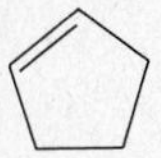 cyclopentene $+ H_2 \xrightarrow{Pt}$

3. Name the following:

(a) 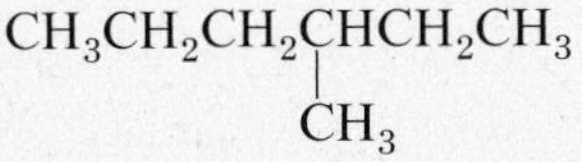 $CH_3CH_2CH_2\underset{\displaystyle CH_3}{\underset{|}{C}}HCH_2CH_3$

(b) $CH_3CH_2CH_2CH_2CH{=}CH_2$

(c) 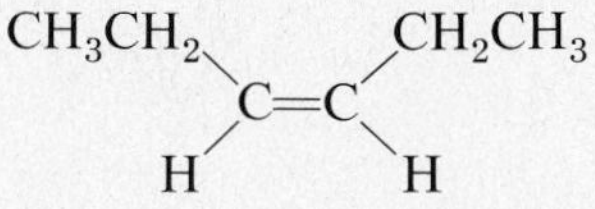 $\underset{H}{\overset{CH_3CH_2}{}}\!\!>C{=}C<\!\!\underset{H}{\overset{CH_2CH_3}{}}$

(d) $CH_3CH_2C{\equiv}CCH_2CH_3$

(e) 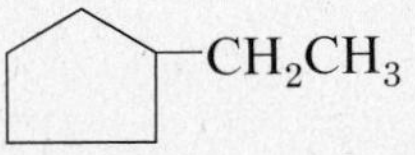 cyclopentane$-CH_2CH_3$

(f) 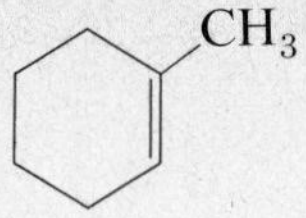 1-methylcyclohexene ring structure with CH_3 substituent

(g) 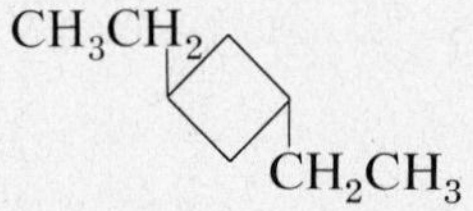 cyclobutane ring with CH_3CH_2 and CH_2CH_3 substituents

4. Draw structures for each of the following:
   (a) methylcycloheptane
   (b) *cis*-2-pentene
   (c) 3-heptyne
   (d) 3-ethylhexane
   (e) 3,3-dimethyloctane
   (f) 3,4-dimethyloctane
   (g) 3,3,4-trimethylheptane
5. Explain why geometric isomerism is not possible for alkynes.
6. Using only methyl groups and the appropriate functional group, give an example of a structural formula for each of the following classes of compounds.
   (a) An acid
   (b) An amide
   (c) An amine
   (d) An alcohol
   (e) An ether

5
solutions

INTRODUCTION

In the preceding chapters we have primarily dealt with pure compounds. This chapter is concerned with mixtures of substances. Mixtures can be divided into two general categories: heterogeneous mixtures and homogeneous mixtures. Mixtures in the latter category are called *solutions*. Here, homogeneous means that visually the material appears uniform throughout. Thus, by visual observation it is not possible to distinguish between solutions and pure substances.

Although we normally think of solutions as liquids, there are gaseous solutions and solid solutions. Air is an example of a gaseous solution consisting primarily of a mixture of elemental nitrogen and oxygen. Certain alloys (such as 14-karat gold, a mixture of gold, silver, and copper) are examples of solid solutions. We will mainly be concerned with liquid solutions. Just because the solution is a liquid does not mean, however, that all the individual components from which the solution was prepared were liquids. For example, mixing sugar and water will produce a liquid solution, as would mixing the gas carbon dioxide with water. When a solid is mixed with a liquid to produce a solution, the solid is the *solute* and the liquid, the *solvent*. Similarly, the gas is the *solute* and the liquid, the *solvent* in a solution prepared from mixing a gas with a liquid. When a solution is prepared from two liquids, the distinction between solute and solvent is less clear. Normally, the liquid present in the larger amount is called the solvent and the one in lesser quantity, the solute.

WHY DO SUBSTANCES DISSOLVE?

Why does the grease float instead of sinking to the bottom?

Some materials dissolve in certain solvents and others do not dissolve. For example, when a spoonful of sugar is put into coffee or tea, the sugar dissolves but the spoon does not! Common experience also shows us that grease does not dissolve in water; instead, it floats on the surface of the water. Grease dissolves in cleaning fluid, and it appears to dissolve in water when soap or detergent is present. Why do these substances behave in this fashion? What is different about cleaning fluids that enables them to dissolve grease while water will not? In order to answer these questions we need to consider the molecular structure, bonding characteristics, and the forces of attraction between the molecules of both the potential solutes and solvents.

INTERMOLECULAR FORCES

In Chapter 3 we saw that the strong electrical attractions between oppositely charged ions caused them to gather in an orderly arrangement. These strong forces allow little movement of the ions, with the result that ionic compounds are high melting solids.

During our studies of covalent compounds, we found that some were composed of polar molecules and others of nonpolar molecules. We saw that polar molecules are attracted to one another. The attraction between polar molecules can be accounted for by attraction between dipoles or by hydrogen bonding. These forces are weaker than the forces between ions and the molecules may move past one another more easily than ions may move. Consequently, polar molecules are generally liquids or low melting solids.

Without intermolecular forces, all covalent compounds would be gases.

Intermolecular attractive forces between nonpolar molecules have not been considered up to this point. Yet it is evident that attractive forces must exist between nonpolar molecules. Otherwise, why would nonpolar molecules come together to form liquids and solids? In the absence of any intermolecular forces, the molecules would not be expected to come together to form a liquid or a solid. The attractive forces between nonpolar molecules are called *London forces* or *dispersion forces*. In nonpolar compounds, the London forces account for all the attraction between molecules. Detailed studies of these forces reveal that they gradually increase as the size and molecular weight of the molecules increase.

It is found that London forces also exist between molecules of polar compounds and slowly increase with increasing molecular weight. (Discussion of the origin of these forces is beyond the scope of this book but may be found in many general college chemistry texts.)

Generally speaking, except when hydrogen bonding is involved, the London forces account for the bulk of the attractive forces between polar molecules. Thus, in the absence of hydrogen bonding, the London forces are mainly responsible for intermolecular attraction in both nonpolar and polar compounds. When hydrogen bonding is involved, it is primarily responsible for intermolecular attraction. Now let's see how this discussion helps answer the questions raised in the preceding section.

WHY SUBSTANCES DISSOLVE

Experimentally it is found that carbon tetrachloride dissolves in hexane but is insoluble in water. The molecular shape of carbon tetrachloride is tetrahedral, consequently CCl_4 is a nonpolar substance. Hexane is also nonpolar. (Why?) Thus, London forces account for the attraction between CCl_4 molecules. Similarly, London forces hold the hexane molecules together in the liquid state. Since London forces exist between all molecules and only slowly increase with molecular size, we might expect that the attraction between a CCl_4 molecule and a hexane molecule would be about the same as the attraction between two CCl_4 molecules or between two hexane molecules.

Following this reasoning, the sum of the attractive forces between the molecules in a mixture of carbon tetrachloride and hexane would be approximately the same as the sum of the attractive forces in pure carbon tetrachloride and those in pure hexane. In agreement with the experimental results we predict that carbon tetrachloride will dissolve in hexane since there is little or no change in the attractive forces upon mixing. These arguments hold true for mixtures of most nonpolar compounds and we may formulate a general rule: *Nonpolar compounds dissolve in nonpolar solvents.*

This generalization is sometimes stated as like dissolves like.

Since carbon tetrachloride is insoluble in water, we must conclude by the arguments in the paragraphs above that the sum of the attractive forces in a mixture of carbon tetrachloride and water would be less than the sum of those in pure carbon tetrachloride plus those in pure water. Recall from Chapter 3 that water possesses a high molecular polarity and that hydrogen bonding occurs between water molecules. Both of these factors lead to large attractive forces between water molecules. Carbon tetrachloride is nonpolar; consequently, there should be little attraction between CCl_4 and H_2O molecules. Under these circumstances, water molecules will associate with each other in preference to associating with CCl_4 molecules. Thus, water and carbon tetrachloride are mutually insoluble because of the lack of association between the molecules of these two compounds. On the basis of many experiments we can say that *nonpolar compounds are insoluble in water because the attractive forces between nonpolar molecules and water molecules are less than the attractive forces between water molecules.*

In Exercise 4-24 you were asked to describe on a molecular level what might happen when ethyl alcohol is poured into water. Ethyl alcohol contains an O—H group and hydrogen bonding is known to occur between ethyl alcohol molecules. It seems reasonable then to expect that ethyl alcohol molecules could hydrogen bond to water molecules. The possibility of hydrogen bonding between ethyl alcohol and water molecules should mean that ethyl alcohol will dissolve in water. In agreement with this prediction, we experimentally find ethyl alcohol and water are soluble in one another in all proportions.

exercise 5-1

Predict which pairs of compounds should form a solution upon mixing.

(a) CH_3CO_2H and water yes
(b) benzene and hexane yes
(c) hexane and water no
(d) CH_3CO_2H and CH_3CH_2OH yes
(e) methylamine (CH_3NH_2) and water yes
(f) carbon tetrachloride and benzene yes

exercise 5-2

Glucose has the molecular formula $C_6H_{12}O_6$ and is very soluble in water. Which functional group is glucose more likely to contain: ether or alcohol groups? Explain.

more polar

THE SOLUBILITY OF IONIC COMPOUNDS

In Chapter 3 we saw that crystalline ionic compounds are collections of ions held together by the attractive forces between the oppositely charged cations and anions. When an ionic compound is dissolved, a breakdown of this orderly arrangement of ions occurs. Interionic forces must be overcome during the dissolving process. Thus, if an ionic compound is dissolved, it seems reasonable to assume that the attractive forces between the ions and the solvent are greater than the interionic forces in the crystal. Chemists have found that, in general, ionic compounds are insoluble in most liquids except water (and a few other very polar liquids). This observation suggests that an electrical attraction between the charged ions and partially charged atoms on polar molecules is responsible for the dissolving of ionic compounds.*

Consider a crystal of sodium chloride made up of alternating sodium and chloride ions. When the salt crystal is placed into water, the water molecules will turn so as to place the partially positively charged hydrogen atoms of nearby water molecules near negatively charged chloride ions on the surface of the salt crystal. Other water molecules will be oriented so

*Some ionic compounds dissolve in solvents other than water because they undergo a definite chemical reaction with the solvent. Such cases are excluded from consideration here.

as to place the partially negatively charged oxygen atom near the positive sodium ions on the crystal surface. This process is called *solvation;* when water is the solvent, it may be called *hydration*. The hydration process reduces the attraction between the oppositely charged ions in the crystal. The result of this reduced attraction between ions is a breaking away of ions from the crystal surface. As each ion is hydrated and breaks away, another ion in the crystal is exposed to the water and the hydration–breaking away process may be repeated. Thus, the crystal is literally picked apart and the ions dispersed among the water molecules as the compound dissolves (see Fig. 5-1).

exercise 5-3

It is found that a given amount of sodium chloride will dissolve faster if the crystals are ground to fine powder before adding them to the water. Explain why this should be true in terms of the dissolving process.

From your chemistry laboratory experience, you probably realize that not all ionic compounds are water soluble. This should not be surprising if it is remembered that there are two opposing forces that influence the solubility of ionic compounds: (1) the attraction between oppositely charged ions, tending to keep the ions in the crystal, and (2) the attraction of the polar solvent molecules for the ions, tending to pull the ions away from the crystal. Since both forces originate from the electrical attraction between opposite charges, any change in the ions that serves to increase the attraction between the ions in the crystal (such as a larger ionic charge) will also

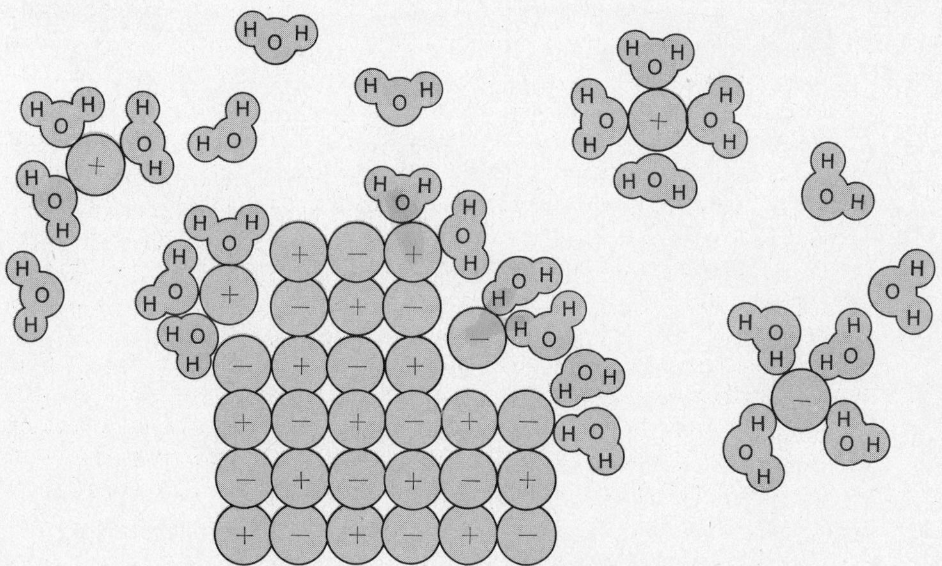

FIGURE 5-1 The dissolving of an ionic compound in water.

TABLE 5-1
Solubility guidelines for ionic compounds

1. Ionic compounds of the alkali metals and ammonium ion are soluble.
2. All ionic compounds containing acetate and nitrate ions are soluble.
3. All metal chlorides are soluble except those of Ag^+, Hg_2^{2+}, and Pb^{2+}.
4. All carbonates and phosphates, except those of the alkali metals and ammonium ion, are insoluble.
5. Most metal hydroxides are insoluble except those of the alkali metals and alkaline earths. Hydroxides of the latter are only slightly soluble.

increase the attraction between the ions and the water molecules. Consequently, we find no simple rules enabling us to predict the solubility of ionic compounds. It has been possible, however, to compile some crude but useful guidelines to the solubility of ionic compounds. These guidelines are given in Table 5-1.

ACTION OF SOAPS AND DETERGENTS

From everyday experience we know that a soap or detergent will apparently cause grease to dissolve in water. Soap and detergent molecules consist of a relatively large portion that is nonpolar or only slightly polar and an ionic group as the other portion. Several such molecules are shown in Fig. 5-2.

The reason grease or fats and oils don't dissolve in water is because they are relatively nonpolar molecules and as we saw earlier, the very polar water molecules prefer to associate with one another rather than with the nonpolar molecules. If we examine the structure of soap and detergent molecules, we see that one portion of the molecules would prefer to associate with the nonpolar fat or oil molecules and the other portion with the water molecules. The result of this is to break up grease globs and disperse them throughout the water as shown in Fig. 5-3. Notice in this figure that the grease is not dispersed throughout the water as individual molecules; microscopic droplets of a mixture of grease and soap are dispersed in the water. These droplets are called *micelles* (pronounced "my-cells"). The micelles disperse in the water because of the attraction between the ionic portion of the soap molecules and the polar water molecules. Also, notice that the micelles are negatively charged, causing them to repel one another, which prevents them from clumping together again. Because of the presence of these microscopic micelles, chemically a true solution has not been formed; instead an *emulsion* or *suspension* has been produced. For this reason, soaps and detergents are classified, along with other molecules possessing similar types of structural features, as *emulsifying agents* or *emulsifiers*.

Emulsifiers are added to shortening so the shortening will mix better with water when making baked goods.

You may be wondering what the difference is between a soap and a detergent. In a soap the ionic portion of the molecule is always a

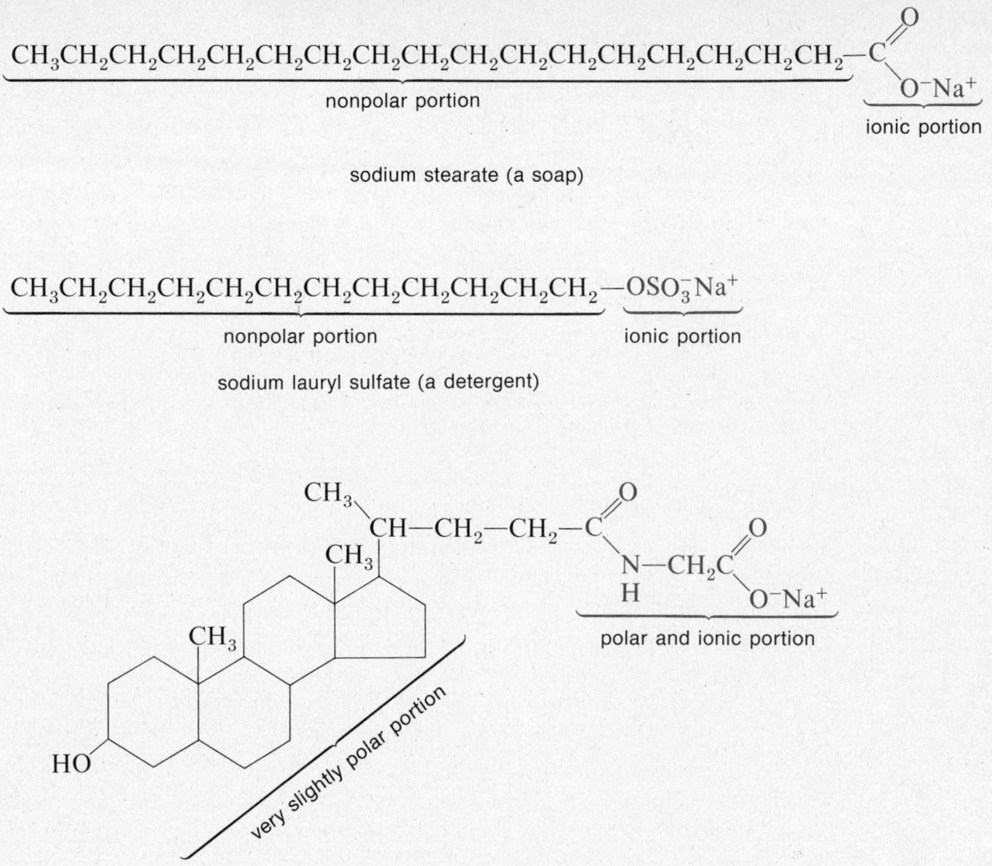

$$\underbrace{CH_3CH_2CH_2CH_2CH_2CH_2CH_2CH_2CH_2CH_2CH_2CH_2CH_2CH_2CH_2CH_2CH_2}_{\text{nonpolar portion}}-\underbrace{C(=O)O^-Na^+}_{\text{ionic portion}}$$

sodium stearate (a soap)

$$\underbrace{CH_3CH_2CH_2CH_2CH_2CH_2CH_2CH_2CH_2CH_2CH_2CH_2}_{\text{nonpolar portion}}-\underbrace{OSO_3^-Na^+}_{\text{ionic portion}}$$

sodium lauryl sulfate (a detergent)

HO– (steroid ring system with two CH_3 groups) –$CH(CH_3)-CH_2-CH_2-C(=O)-\underbrace{N(H)-CH_2C(=O)O^-Na^+}_{\text{polar and ionic portion}}$

very slightly polar portion

sodium glycolithocholate (in human bile)

FIGURE 5-2 Structure of a typical soap, detergent, and bile salt.

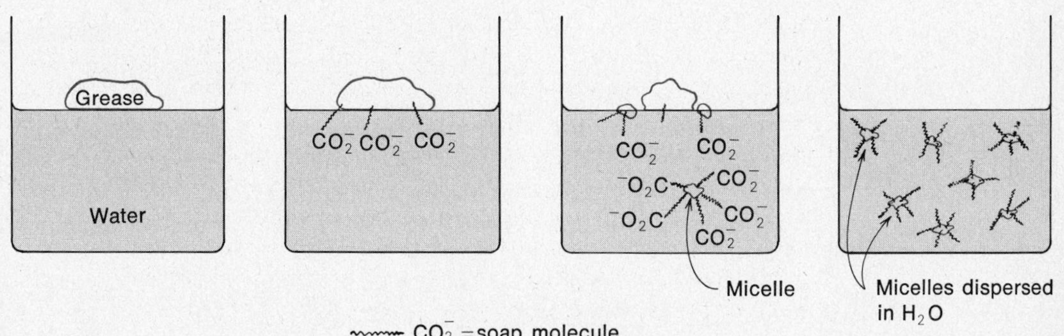

FIGURE 5-3 The process by which a soap disperses grease in water.

$$-C\begin{matrix}{}^{\displaystyle /\!\!/ O}\\ {}_{\displaystyle \backslash O^-Na^+}\end{matrix} \quad \text{or} \quad -C\begin{matrix}{}^{\displaystyle /\!\!/ O}\\ {}_{\displaystyle \backslash O^-K^+}\end{matrix}$$

group;* whereas in a detergent the ionic group is —O—SO_3^- Na^+ or possibly some other ionic grouping. Detergents are more frequently used than soaps because they are soluble in hard water, which contains dissolved calcium and magnesium compounds. The calcium ions in the hard water will combine with the soap molecules to form an insoluble precipitate. This precipitate forms curds that adhere to clothes, giving them a grayish cast. Detergents do not form an insoluble compound with calcium ions.

exercise 5-4

Write the formula of the insoluble precipitate that is formed by reaction between sodium stearate and calcium ions.

Figure 5-2 includes the steriod compound sodium glycolithocholate, which is one of a group of steroid compounds known as bile salts. The other members of this group of compounds have closely related structures. The bile salts exhibit detergent or soap action and behave as emulsifying agents in the duodenum (upper portion of the small intestines). They are stored in the gallbladder and released into the duodenum during digestion. Their purpose is to disperse dietary fats in the watery environment of our digestive tract. Fat that has been dispersed in this fashion is more easily attacked by the digestive juices than large fat particles. Persons who have had their gallbladder removed are often placed on restricted fat diets because of the lesser amounts of bile salts present to aid in fat digestion. Although the structures are very different from the bile salts, both blood and lymph contain compounds exhibiting a detergent action that aids in the transport of fat and other nonpolar molecules.

CONCENTRATION OF SOLUTIONS

Up to this point, we have only considered whether one substance will dissolve in another. We have not yet considered how much of a solute will dissolve in a given quantity of a solvent. The concentration of a solution is a measure of the amount of solute in a quantity of solvent. Solution concentration can be expressed qualitatively or quantitatively. *Qualitative* expressions of solution concentration are given by the terms saturated, concentrated, or dilute.

A *saturated* solution is one which contains the maximum amount of solute which can be dissolved in a given quantity of solvent at a given temperature. For example, the maximum amount of sodium chloride that will dissolve in 100 ml of water at 25 °C is 36.2 g. Thus, a solution containing

*Soaps containing Na^+ ions are known as hard soaps and those containing K^+ ions are soft soaps.

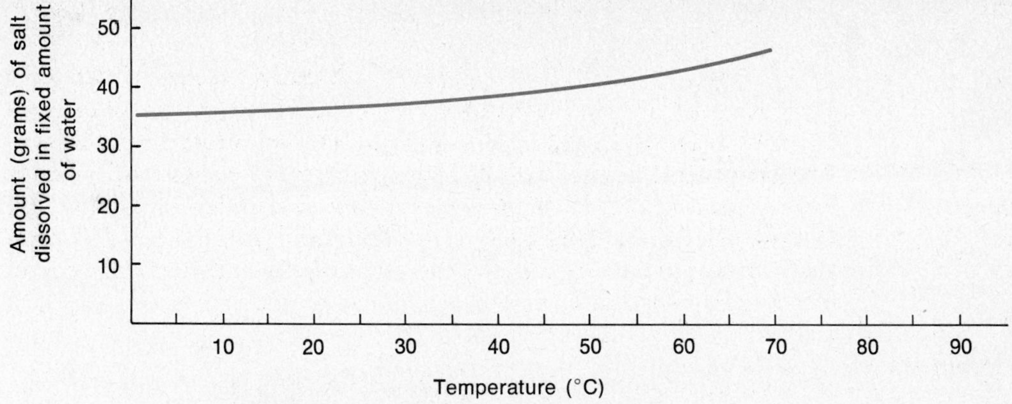

FIGURE 5-4 Solubility of sodium chloride.

36.2 g of NaCl dissolved in 100 ml of water is said to be a *saturated* solution of sodium chloride.

The *solubility* of a compound is the amount that will dissolve in a given quantity of a solvent to produce a saturated solution and is usually expressed as the number of grams of solute that will dissolve in 100 g of solvent at a given temperature. Thus, the solubility of a compound is an experimentally determined quantity and the solubility values for a large number of common substances have been tabulated in chemical reference books.

The solubility of solids in liquids generally increases with increasing temperature; however, the solubility of gases in liquids decreases with increasing temperature.

For most aqueous solutions, an increase in temperature of the solution increases the amount of solid that can dissolve in a given amount of water (i.e., the solubility of solids increases with an increase in temperature). By raising the temperature of the salt solution from 25°C to 35°C, additional sodium chloride will dissolve. Adding additional sodium chloride would eventually lead to a saturated solution at 35°C. By raising the temperature still higher, say to 65°C, more salt could be added before the solution would once again become saturated at that temperature. Graphically, this can be represented as shown in Fig. 5-4. Each point on the graph represents a condition of a saturated solution at that temperature.

The greater the amount of solute present in a given amount of solution, the greater the concentration of the solution.

The terms *concentrated* and *dilute* are used to describe solutions containing less solute than a saturated solution of the substance would contain. These descriptions are imprecise and the only definitive statement that can be made is that a dilute solution of a substance contains less of that substance in a given quantity of solvent than does a concentrated solution in the same solvent. Both concentrated and dilute solutions of a substance are said to be *unsaturated* solutions since they contain less solute than is required to produce a saturated solution.

QUANTITATIVE ASPECTS OF SOLUTIONS

Qualitative descriptions of solutions may at times be useful but in general are unsatisfactory. In chemistry, medicine, and other fields it is often necessary to know exactly how much of a substance is contained in a given amount of a solution. Medicinal dosages generally require the administration of prescribed amount of a therapeutic agent of known concentration. A large number of ways for quantitatively expressing concentrations of solutions have been devised. We shall limit our study to several of the most common methods of expressing concentration.

PERCENT CONCENTRATION

% concentration = $\frac{\textit{grams of solute}}{\textit{100 ml of solution}}$

The definition of percent concentration we shall use is the so-called weight–volume percent, defined as the *grams of solute per 100 ml of solution.* A 5 percent glucose solution is one containing 5 g of glucose per 100 ml of solution. To prepare 100 ml of 5 percent glucose solution, weigh 5 g of glucose and add to it enough water to give a *total volume of solution* of 100 ml. Notice that it is the volume of the solution and *not* the volume of water added which must be 100 ml.

EXAMPLE 5-1

How would one prepare 6 liters of 5 percent glucose solution?

$$\text{amount of glucose} = (6\ \cancel{\text{liters}})(1000\ \cancel{\text{ml}}/1\ \cancel{\text{liter}})(5\text{ g glucose}/100\ \cancel{\text{ml}}\text{ solution})$$
$$= 300\text{ g of glucose}$$

Weigh out 300 g of glucose and add enough water to make a total volume of 6 liters.

exercise 5-5 **Describe how the following solutions would be prepared.**

(a) 500 ml of 5 percent boric acid (sometimes used for eye wash)
(b) 1000 ml of 0.9 percent sodium chloride (physiological salt solution)
(c) 2 liters of 0.1 percent silver nitrate
(d) 30 ml of 0.25 percent Benadryl® elixir

Frequently, we find that a solution of known concentration has already been prepared and the problem is to determine the amount of solution to be used in order to have a specified amount of solute.

EXAMPLE 5-2

How many milliliters of 5 percent glucose solution should be used to have 10 g of glucose?

$$(10\ \cancel{\text{g}}\text{ glucose})(100\text{ ml solution}/5\ \cancel{\text{g}}\text{ glucose}) = 200\text{ ml}$$

exercise 5-6

Determine the amount (in milliliters) of solution that will contain the indicated amount of solute.

(a) 12 g of glucose from 2 percent glucose
(b) 3.6 g of NaCl from 0.9 percent NaCl
(c) 150 g of boric acid from 5 percent boric acid
(d) 0.04 g of $AgNO_3$ from 0.1 percent $AgNO_3$

exercise 5-7

Determine the number of grams of solute contained in the following solutions.

(a) 750 ml of 5 percent glucose

(b) 350 ml of 0.9 percent NaCl
(c) 15 ml (1 tablespoon) of 0.25 percent Benadryl®
(d) 0.5 liter of 0.1 percent $AgNO_3$

INTRODUCTION TO MOLARITY

In order for a chemical reaction to occur it is necessary to mix the reactants. The formation of a solution containing the reactants is an ideal way of achieving the desired mixing. For this reason the majority of chemical reactions, including biological reactions, takes place in solution. This ordinarily means, except for reactions between gases, that in the chemistry laboratory each of the reactants are first dissolved in a suitable liquid solvent and then mixed together.*

Suppose we wish to study the reaction between Br_2 and $FeBr_2$ to produce $FeBr_3$. One approach we might try is to prepare in water 100 ml of 1 percent Br_2 solution and 100 ml of 1 percent $FeBr_2$ solution. To mix the reactants we begin to pour the reddish-brown bromine solution into the light green solution of $FeBr_2$. As we pour the bromine solution into the $FeBr_2$ solution, we notice the characteristic color of the bromine solution disappears, indicating that reaction is occurring to produce $FeBr_3$. When all the bromine solution has been added, the reaction mixture has a red-brown color. This indicates that all the bromine didn't react. Why should there be any unreacted bromine? After all, we did mix 1 g of Br_2 with 1 g of $FeBr_2$.

To help us answer this question, let's look at the balanced equation for the reaction:

$$2FeBr_2 + Br_2 \longrightarrow 2FeBr_3$$

Reactants Products

The balanced equation tells us that two formula units of iron(II) bromide are required for each bromine molecule. This seems to indicate that we should use twice as much $FeBr_2$ as the amount of Br_2. With this in mind, let's see what happens if we add 100 ml of 1 percent Br_2 solution to 200 ml of 1 percent $FeBr_2$ solution. During the initial addition of the Br_2 solution we again observe the disappearance of the characteristic bromine color; but after 100 ml of the bromine solution has been added, the final reaction mixture has a red-brown color, again indicating the presence of unreacted bromine. Since even in this second experiment there is some unreacted bromine left, we must conclude that *there is something fundamentally wrong with the approach we have taken.*

The trouble lies in our means of measuring the amounts of reactants used. The balanced equation told us that two formula units of $FeBr_2$ react with one formula unit of Br_2. We mixed 2 g of $FeBr_2$ with 1 g of Br_2. Therefore, we should *not* expect 2 g of $FeBr_2$ to react with 1 g of Br_2 unless two formula units of $FeBr_2$ weigh exactly the same as one formula unit of Br_2. Obviously, two formula units of $FeBr_2$ do not weigh the same as one

*Occasionally a liquid may function both as a reactant and as the solvent.

formula unit of Br_2! A formula unit of Br_2 weighs 160 amu and a formula unit of $FeBr_2$ weighs 215.6 amu. Thus, the problem confronting us is that although percent concentration is a way of expressing the amount of solute in a given quantity of solution, it is not very useful in determining the amount of reactants that should be mixed together for a particular chemical reaction. What we really need is a method of expressing concentration that directly gives us information about the number of formula units of a solute in a given quantity of solution. The problem we have just uncovered is one that chemists faced over 150 years ago. A young Italian chemist—Amedeo Avogadro—contributed greatly to the resolution of this problem.

THE MOLE CONCEPT

To aid in understanding how chemists express concentration in terms of formula units of solute in a given quantity of solution, we shall continue our study of the reaction between $FeBr_2$ and bromine. This time we shall conduct the experiment in a way that will allow us to measure the amount of bromine solution that reacts with 100 ml of 1 percent $FeBr_2$ solution. This is accomplished by use of the apparatus shown in Fig. 5-5. First, 100 ml of 1 percent $FeBr_2$ solution is placed in the flask. The 1 percent bromine solution is put in the buret and the initial volume reading recorded. The stopcock is opened and the bromine solution allowed to slowly run into the

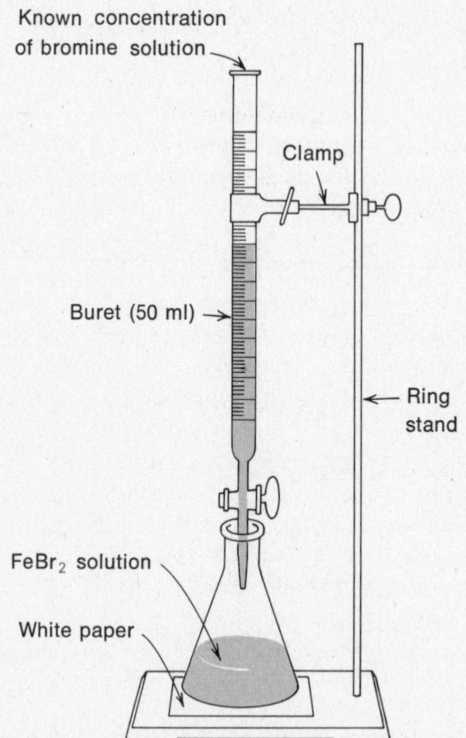

FIGURE 5-5 Apparatus for titration of $FeBr_2$ solution with Br_2 solution.

flask containing the $FeBr_2$ solution. The bromine reacts with the $FeBr_2$, evidenced by the disappearance of the characteristic bromine color. During the bromine addition the concentration of $FeBr_2$ is continually diminishing because of reaction with bromine. Eventually, as we continue to add bromine solution, the point is reached where the bromine color no longer disappears. This point is called the *end point* of the reaction and signifies that all the $FeBr_2$ has reacted. When the end point is reached, we close the stopcock and record the buret reading. In our experiment, the amount of 1 percent bromine solution added to reach the end point was 37.1 ml.

The data gathered in the experiment are not in the most useful form. To make the experiment more meaningful, we need to convert to the actual number of grams of reactant used. The number of grams of $FeBr_2$ reacting is

$$(1.0 \text{ g}/100 \cancel{\text{ml}})(100 \cancel{\text{ml}}) = 1.000 \text{ g of FeBr}_2$$

The number of grams of Br_2 reacting is

$$(1.0 \text{ g}/100 \cancel{\text{ml}})(37.1 \cancel{\text{ml}}) = 0.371 \text{ g of Br}_2$$

Thus, we have determined that 1.00 g of $FeBr_2$ reacts with 0.371 g of Br_2. If we use less bromine than this, there will be some unreacted $FeBr_2$; if we use more Br_2, there will be unreacted bromine. These quantities can be expressed as a ratio: $\dfrac{0.371 \text{ g Br}_2}{1.000 \text{ g FeBr}_2}$.

If we repeat the experiment using 200 ml of 1 percent $FeBr_2$ solution, we find that the end point of the reaction is reached with the addition of 74.20 ml of 1 percent Br_2 solution. In this case the number of grams of $FeBr_2$ equals

$$(1.0 \text{ g}/100 \cancel{\text{ml}})(200 \cancel{\text{ml}}) = 2.00 \text{ g of FeBr}_2$$

and the number of grams of Br_2 reacting is

$$(1.0 \text{ g}/100 \cancel{\text{ml}})(74.20 \cancel{\text{ml}}) = 0.7420 \text{ g Br}_2$$

The ratio of reacting quantities is

$$\frac{0.7420 \text{ g Br}_2}{2.00 \text{ g FeBr}_2} = \frac{0.3710 \text{ g Br}_2}{1.00 \text{ g FeBr}_2}$$

Notice that the ratio of the grams of Br_2 to the grams of $FeBr_2$ is the same in both experiments and that the use of twice as much $FeBr_2$ requires twice as much Br_2 to react with it.

These last two experiments tend to confirm our supposition that chemicals react with one another in a fixed ratio of quantities and that we need to develop a different method of expressing the reacting quantities.

If we compare the ratios of reacting quantities as determined from our experiments to the ratio predicted by the chemical equation, we get a very interesting result. Our experiments tell us the ratio is 0.3710 g Br_2/1.00 g $FeBr_2$. The balanced equation tells us $FeBr_2$ should react with Br_2 in the ratio 160 amu Br_2/431.2 amu $FeBr_2$. (Remember that two formula units of $FeBr_2$ weigh 431.2 amu.) Reduction of the latter ratio to its simplest form yields the ratio

$$\frac{0.3710 \text{ amu } Br_2}{1.000 \text{ amu } FeBr_2}$$

The numerical values of the ratios are identical! Indeed they should be identical since mathematically these weight units factor out and the ratios are unitless numbers. The weight units we actually use to measure the quantities of Br_2 and $FeBr_2$ are immaterial as long as the unitless $Br_2/FeBr_2$ weight ratio is 0.3710/1.00.

EXAMPLE 5-3

How much bromine is required to convert 3.0 g of $FeBr_2$ to $FeBr_3$?

$$(3.0 \text{ g } FeBr_2)\left(\frac{0.3710 \text{ g } Br_2}{1.000 \text{ g } FeBr_2}\right) = 1.113 \text{ g } Br_2$$

exercise 5-8

Determine the quantity (in appropriate units) of bromine required to react with the indicated quantity of $FeBr_2$.

(a) 1 g $FeBr_2$
(b) 20 g $FeBr_2$
(c) 431.2 g $FeBr_2$
(d) 2.15 lb $FeBr_2$
(e) 4.31 tons $FeBr_2$

The reaction between $FeBr_2$ and Br_2 serves to illustrate a very important principle of chemistry: *The reacting weight ratios are fixed by the formula weights of the reacting substances and by the ratio in which the formula units combine.*

To be certain that you understand this principle, let's consider a few examples.

EXAMPLE 5-4

Carbon and oxygen react to produce carbon dioxide. The balanced equation for this reaction is

$$C + O_2 \longrightarrow CO_2$$

The equation tells us one formula unit of carbon combines with one formula unit of oxygen. One formula unit of carbon weighs 12 amu and one formula unit of oxygen weighs 32 amu. The reacting weight ratio is

$$\left(\frac{1 \text{ \sout{formula unit} C}}{1 \text{ \sout{formula unit} O}}\right)\left(\frac{12 \text{ \sout{amu} carbon}}{32 \text{ \sout{amu} oxygen}}\right) = 0.375 \text{ carbon}/1.00 \text{ oxygen}$$

EXAMPLE 5-5

The overall reaction for the metabolism of glucose by the body is

$$C_6H_{12}O_6 + 6O_2 \longrightarrow 6CO_2 + 6H_2O$$

The equation tells us that one formula unit of glucose reacts with six formula units of oxygen. One formula unit of glucose weighs 180 amu. One formula unit of oxygen weighs 32 amu. Since six formula units of oxygen are required for each formula unit of glucose, six times 32 amu or 192 amu of oxygen are required for 180 amu of glucose. The reacting weight ratios of oxygen and glucose are

$$\left(\frac{6 \text{ \sout{formula units} } O_2}{1 \text{ \sout{formula unit} glucose}}\right)\left(\frac{32 \text{ \sout{amu} oxygen}}{180 \text{ \sout{amu} glucose}}\right) = \frac{192 \text{ oxygen}}{180 \text{ glucose}}$$

$$= \frac{1.066 \text{ oxygen}}{1 \text{ glucose}} = \text{reacting weight ratio}$$

EXAMPLE 5-6

Aluminum sulfate [$Al_2(SO_4)_3$] is the active ingredient in styptic pencils and can be prepared in the laboratory by the reaction

$$2Al(OH)_3 + 3H_2SO_4 \longrightarrow Al_2(SO_4)_3 + 3H_2O$$

Find the number of grams of H_2SO_4 required to react with 78 g of $Al(OH)_3$.

The balanced equation tells us that two formula units of $Al(OH)_3$ react with three formula units of H_2SO_4. Therefore the reacting weight ratios are

$$\left(\frac{3 \text{ \sout{formula units} } H_2SO_4}{2 \text{ \sout{formula units} } Al(OH)_3}\right)\left(\frac{98 \text{ \sout{amu} } H_2SO_4}{78 \text{ \sout{amu} } Al(OH)_3}\right) = \frac{294\ H_2SO_4}{156\ Al(OH)_3}$$

$$= \frac{1.885\ H_2SO_4}{1.00\ Al(OH)_3} = \text{reacting weight ratio}$$

If we wish to react 78 g of $Al(OH_3)$, then the weight of H_2SO_4 required is

$$78 \text{ g } Al(OH)_3 \left(\frac{1.885\ H_2SO_4}{1\ Al(OH)_3}\right) = 147 \text{ g } H_2SO_4 \text{ required}$$

Example 5-6 demonstrates that because the reacting weight ratios are unitless numbers, we can use any weight unit we find convenient and be assured that we will mix the reactants in the proper proportions as long as we take into account the coefficients of the reactants in the balanced equation. Since the gram is a convenient weight unit and the formula weight of a substance is important in determining the reacting weight ratio, chemists find it useful to define a quantity called the *gram formula weight. The gram formula weight is that quantity of a substance that when measured in grams is numerically equal to the formula weight of that substance.* Let us carefully note that the gram formula weight is *not* the weight of one formula unit of a substance. The weight of one formula unit of a substance is expressed in atomic mass units. For example, the formula weight of a formula unit (or molecule) of bromine is 160 amu; whereas, the gram formula weight of bromine is 160 g.

Let's return for a moment to the reaction between $FeBr_2$ and Br_2. The balanced equation indicates 160 amu of Br_2 should react with 2(215.6) or 431.2 amu of $FeBr_2$. Since the reacting weight ratio is independent of the weight unit used, we expect that 160 g of Br_2 should react with 2(215.2 g) or 431.2 g of $FeBr_2$. Similarly, we could calculate the number of grams of oxygen required to react with 180 g of glucose according to the equation in Example 5-5. The reacting weight ratio $\frac{6(32 \text{ amu}) \text{ oxygen}}{180 \text{ amu glucose}}$ can also be given as $\frac{6(32 \text{ g}) \text{ oxygen}}{180 \text{ g glucose}}$; thus six times the gram formula weight of oxygen or 192 g of oxygen is required.

Because the gram formula weight of a compound is so important in determining the amounts of compounds to be measured in the laboratory, the name *mole* has been given to this quantity. For our purposes, *one mole of a compound is a quantity of that compound equal to its gram formula weight.** For example, 1 mole of glucose weighs 180 g since its gram formula weight is 180 g.

exercise 5-9

How much does 1 mole of each of the following weigh?

(a) $FeBr_2$

(b) $Al_2(SO_4)_3$

(c) CH_3CH_2OH **(ethyl alcohol)**

(d) CO_2H / $-O-\overset{O}{\overset{\|}{C}}-CH_3$ **(aspirin)** (benzene ring bearing both groups)

(e) $C_6H_8O_6$ **(vitamin C)**

exercise 5-10

Calculate the number of moles represented by the given quantity of each of the following:

(a) 215.6 g of $FeBr_2$

(b) 23 g of ethyl alcohol

(c) 3.6 g of aspirin

(d) 240 g of water (1 cup of water)

(e) the moles of glucose in 1 liter of 5 percent glucose solution

*The word *mole* is an abbreviation for gram molecular weight. The definition of a mole recommended by the Commission on Symbols, Terminology and Units of the International Union of Pure and Applied Chemistry is "A mole is an amount of substance, of specified chemical formula, containing the same number of formula units (atoms, molecules, ions, electrons, quanta, or other entities) as there are atoms in 12 grams (exactly) of the pure nuclide ^{12}C.

Some thought about gram formula weights and moles will reveal that equal numbers of moles always contain equal numbers of formula units.* Thus there are as many Br_2 molecules in 1 mole of Br_2 as there are glucose molecules in 1 mole of glucose.

Let us return again to the reaction between $FeBr_2$ and Br_2. The balanced equation

$$2FeBr_2 + Br_2 \longrightarrow 2FeBr_3$$

Substances react on a mole-to-mole basis.

tells us that two formula units of $FeBr_2$ will react with one formula unit of Br_2. Since equal numbers of moles always contain equal numbers of formula units, the equation also tells us that 2 moles of $FeBr_2$ will react with 1 mole of Br_2. Translated into weight units, the reacting weight ratios are identical. The ratio

$$\frac{160 \text{ amu } Br_2}{2(215.6 \text{ amu}) \; FeBr_2}$$

equals the ratio

$$\frac{160 \text{ g } Br_2}{2(215.6 \text{ g}) \; FeBr_2}$$

Suppose we wish to know how much $FeBr_2$ will react with 1.6 g of Br_2. Correctly set up, the mathematical equation to solve this problem is

$$\text{No. of g } FeBr_2 = 1.6 \cancel{\text{g } Br_2} \left(\frac{1 \cancel{\text{mole } Br_2}}{160 \cancel{\text{g } Br_2}}\right)\left(\frac{2 \cancel{\text{moles}} \; FeBr_2}{1 \cancel{\text{mole}} \; Br_2}\right)\left(\frac{215.6 \text{ g } FeBr_2}{1 \cancel{\text{mole}} \; FeBr_2}\right)$$

$$= 4.3 \text{ g } FeBr_2$$

exercise 5-11

Aspirin can be synthesized by the reaction

CO_2H, OH (benzene ring) + CH_3—C(=O)—O—C(=O)—CH_3 $\longrightarrow$ CO_2H, O—C(=O)—CH_3 (benzene ring) + CH_3CO_2H

salicylic acid | acetic anhydride | aspirin | acetic acid

How many grams of acetic anhydride is required to react with 27.6 g of salicylic acid?

27.6g Sa × 102.04g aa / 138.07 Sa = 20.4g AA

*A natural question is How many formula units are in a mole? Actually the answer to this question is immaterial for our purposes. In fact, chemists used the mole concept for over 75 years before J. B. Perrin determined in 1908 that there are 6.02×10^{23} formula units in a mole. The number 6.02×10^{23} is known as Avogadro's number in honor of Avogadro's contribution to the development of the mole concept.

MOLARITY Recall from p. 114 the need for expressing concentration in terms of formula units of solute in a given quantity of solution. Since the mole is a measure of the number of formula units, the number of moles of solute in a given quantity of solution should be a useful way of expressing concentration.

The molarity of a solution is defined as the number of moles of solute per liter of solution. A solution containing 1 mole of solute per liter of solution is said to be *1 molar,* abbreviated 1 *M.*

A 1 *M* solution of glucose contains 1 mole of glucose per liter. Since the gram formula weight of glucose is 180 g, 1 liter of 1 *M* glucose solution could be prepared by weighing 180 g of glucose and adding enough water to yield 1 liter of solution. Greater or lesser amounts of solution can be prepared by using different quantities of solute and solvent.

1M = 1 mole/l

EXAMPLE 5-7

100 ml of 1 *M* glucose can be prepared by using 18 g of glucose and enough water to give a solution of 100 ml. The amount of glucose required is calculated in the following manner.

grams glucose required = (0.1 ~~liter~~)(1 ~~mole~~/~~liter~~)(180 g glucose/1 ~~mole~~ glucos
= 18 g glucose

EXAMPLE 5-8

Solution concentrations may be greater or less than 1 Molar. 500 ml of 0.3 *M* glucose can be prepared by dissolving 27 g of glucose in enough water to yield 500 ml of solution. The grams of glucose needed are calculated:

grams glucose required = (0.5 ~~liter~~)(0.3 ~~mole~~/1 ~~liter~~)(180 g glucose/~~mole~~ gluco
= 27 g glucose required

A general equation to calculate the number of grams of solute required is

An important equation! grams solute required = (amount, in liters, of solution desired)(molarity of desired solution)(gram formula weight of solute)

exercise 5-12

Calculate the grams of solute required to prepare each of the following solutions.

(a) 250 ml of 0.5 *M* NaCl
(b) 3 liters of 0.02 *M* I_2
(c) 250 cc of 0.2 *M* $NaHCO_3$
(d) 0.60 liter of 4 *M* NH_4Cl
(e) 0.5 liter of 0.15 *M* sucrose (molecular formula: $C_{12}H_{22}O_{11}$)

Frequently we will find that a solution has already been prepared and we need to calculate the volume of solution to use in order to obtain a specified number of moles of solute. This calculation is easily performed via the equation

Another important equation!

$$\text{volume of solution needed} = \frac{\text{moles solute desired}}{\text{solution molarity}}$$

Another useful calculation involves determining the number of moles of solute contained in a given quantity of a solution of known concentration. The equation

Still another important equation. You should be able to apply these three equations.

$$\text{moles solute} = (\text{solution molarity})(\text{volume, in liters})$$

allows this calculation to be made.

exercise 5-13

(a) How much 0.1 *M* NaCl solution contains 0.03 mole of NaCl?
(b) How many moles of glucose are in 3 liters of 0.5 *M* glucose?
(c) How many grams of NaCl are in 1500 ml of 0.4 *M* NaCl?
(d) 300 cc of 0.2 *M* sucrose contains ______ moles of sucrose.
(e) 0.25 mole of $NaHCO_3$ are contained in ______ ml of 0.1 *M* $NaHCO_3$.

EQUIVALENTS

$$1 \text{ mole } FeBr_3 \times \frac{1 \text{ mole } Fe^{3+}}{1 \text{ mole } FeBr_3} = 1 \text{ mole } Fe^{3+}$$

$$1 \text{ mole } FeBr_3 \times \frac{3 \text{ moles } Br^-}{1 \text{ mole } FeBr_3} = 3 \text{ moles } Br^-$$

The equivalent is a unit frequently used in reporting clinical data. This unit is closely related to the mole and is most often used to express the quantity of a particular ion present in a solution. *The number of equivalents* (abbreviated eq) *of an ion is equal to the number of moles of that ion times the charge on the ion.* Thus 1 mole of sodium ions equals 1 equivalent of sodium ions; *1 mole of calcium ions equals 2 equivalents of calcium ions.* One mole of $FeBr_3$ contains 1 mole of Fe^{3+} ions and 3 moles of Br^- moles; hence 1 mole of $FeBr_3$ contains 3 equivalents of Fe^{3+} and 3 equivalents of Br^-.

exercise 5-14

Calculate the number of equivalents of the underlined ion contained in each of the following quantities:
(a) 2 moles of Na<u>OH</u>
(b) 0.5 mole of <u>Al</u>Cl_3
(c) 20 g of <u>Na</u>OH
(d) 156 g of $Ca_3(\underline{PO_4})_2$
(e) 0.25 mole of $Ca\underline{Cl}_2$

Just as the concentration of solutions can be expressed in moles per liter, the concentration of an ion in a solution can be expressed in equivalents per liter. To determine the number of equivalents of an ion in a given quantity of a solution of known concentration, we multiply the concentration expressed in equivalents per liter by the volume in liters.

EXAMPLE 5-9

Calculate the number of equivalents of chloride ion in 500 ml of a solution containing 0.3 eq/liter of chloride ion.

$$\text{eq of } Cl^- = (0.3 \text{ eq}/\cancel{\text{liter}})(0.5 \ \cancel{\text{liter}}) = 0.15 \text{ eq of } Cl^-$$

EXAMPLE 5-10

Calculate the number of equivalents of aluminum in 500 ml of 0.05 *M* $Al_2(SO_4)_3$.

$$500\,\cancel{\text{ml}}\,(1\,\cancel{\text{liter}}/1000\,\cancel{\text{ml}})(0.05\,\cancel{\text{moles } Al_2(SO_4)_3}/\cancel{\text{liter}})$$
$$(2\,\cancel{\text{moles } Al^{3+}}/\cancel{\text{mole } Al_2(SO_4)_3})(3\text{ eq } Al^{3+}/\cancel{\text{mole } Al^{3+}}) = 0.15\text{ eq } Al^{3+}.$$

exercise 5-15

Calculate the number of equivalents of the underlined ion in the given quantity of the following solutions.
(a) 2 liters of a solution containing 0.05 eq/liter of $\underline{K}^+$
(b) 250 ml of a solution containing 0.12 eq/liter of $\underline{Ca}^{2+}$
(c) 10 ml of 1.2 *M* $\underline{Ca}Cl_2$
(d) 10 ml of 1.2 *M* $Ca\underline{Cl_2}$
(e) 5 liters of 0.001 *M* $Na_2\underline{SO_4}$

Since the concentration of ions in body fluids is relatively low, it is often given in milliequivalents per liter. There are 1000 milliequivalents (meq) in 1 eq. A solution containing 0.01 eq/liter of Cl^- contains

$$(0.01\,\cancel{\text{eq}}/\text{liter})(1000\text{ meq}/\cancel{\text{eq}}) = 10\text{ meq/liter}$$

The number of milliequivalents of an ion contained in a given quantity of a solution is calculated by multiplying the *concentration of the ion expressed in equivalents per liter by the volume in milliliters.*

exercise 5-16

Calculate the milliequivalents of HCO_3^- ion in each of the following:
(a) 0.5 eq of $NaHCO_3$
(b) 0.72 moles of $Ca(HCO_3)_2$
(c) 100 ml of a solution containing 0.1 eq/liter of HCO_3^-
(d) 250 ml of 0.2 *M* $NaHCO_3$
(e) 5 liters of 0.001 *M* $LiHCO_3$

DILUTIONS

Laboratory supply rooms commonly stock concentrated solutions that can be diluted to yield solutions of lesser concentration. When a solution is diluted, the solute/solution ratio is decreased. To accomplish this, all we have to do is mix more solvent with the concentrated solution. After addition of more solvent we end up with a larger volume of solution of a lesser concentration. For example, if we have 1 liter of 1 *M* glucose and we add enough water to give a volume of 2 liters, the resulting solution must be 0.5 *M*. Remember that 1 liter of 1 *M* glucose contains 1 mole of glucose. After we add more water, the same mole of glucose is dissolved in 2 liters of solution. The molarity must now be

$$1\text{ mole glucose}/2\text{ liters} = 0.5\text{ mole glucose}/1\text{ liter} = 0.5\,M\text{ glucose}$$

The total number of moles of glucose in the concentrated solution equals the number of moles in the diluted solution since no glucose was

added or removed. *In the general case, the moles of solute in the concentrated solution equals the moles of solute in the diluted solution,* expressed mathematically by Eq. 5-1:
Equation 5-1:

$$\text{moles of solute}_{(\text{solution A})} = \text{moles of solute}_{(\text{solution B})}$$

Remember for any solution concentration expressed in molarity that Eq. 5-2 must be true.

Equation 5-2:

$$\text{moles of solute} = (\text{molarity})(\text{volume in liters})$$

Substitution of Eq. 5-2 into Eq. 5-1 yields Eq. 5-3.
Equation 5-3:

$$[\text{molarity}_{(\text{solution A})}][\text{volume in liters}_{(\text{solution A})}] = [\text{molarity}_{(\text{solution B})}][\text{volume in liters}_{(\text{solution B})}]$$

Suppose that 50 ml of 0.5 *M* glucose has been diluted to 125 ml and we wish to know the concentration of the diluted solution. Application of Eq. 5-3 yields

$$(0.5\ M)(0.050\ \text{liter}) = (0.125\ \text{liter})(M)$$

$$M = \frac{(0.5\ M)(0.050\ \cancel{\text{liter}})}{(0.125\ \cancel{\text{liter}})} = 0.2\ M$$

Seldom do we find that someone has diluted the solution and left behind the procedure they used but did not bother to calculate the new concentration. Occasionally someone has diluted a solution and left it unlabeled. In this case, the unlabeled solution should be discarded to prevent the possibility of a serious accident. The usual situation is that we must carry out the dilution and need to know how much of the concentrated solution to use. For instance, suppose we need 500 ml of 0.5 *M* glucose and all that is available is a 2.0 *M* glucose solution. Again we apply Eq. 5-3:

$$(2.0\ M)(\text{volume}) = (0.5000\ \text{liter})(0.5\ M)$$

$$\text{volume} = \frac{(0.500\ \text{liter})(0.5\ \cancel{M})}{(2.0\ \cancel{M})} = 0.125\ \text{liter} = 125\ \text{ml}$$

Thus, 125 ml of the 2.0 *M* glucose solution should be diluted to 500 ml to have a 0.5 *M* glucose solution. *Caution: Do not add 500 ml of water.*

exercise 5-17

(a) How much 0.5 *M* NaCl should be diluted to give 2 liters of 0.01 *M* NaCl?
(b) What is the concentration of the solution resulting from the dilution of 30 ml of 0.1 *M* $NaHCO_3$ to 100 ml?
(c) How would 500 cc of 0.2 *M* glucose be prepared from 2 *M* glucose?
(d) How would you prepare 4 liters of 0.3 *M* NaCl from 5 *M* NaCl?

In the clinical situation, percent concentration is used much more frequently than molarity is; yet there remains the need to carry out dilutions of solutions. Since percent concentration is the grams of solute per 100 ml of solution, it is possible to find an equation similar to Eq. 5-3 that can be used in dilution problems involving percent concentration. Equation 5-4 is the needed relationship.
Equation 5-4:

$$[\%\ \text{concentration}_{(\text{solution A})}][\text{volume in milliliters}_{(\text{solution A})}]$$
$$= [\%\ \text{concentration}_{(\text{solution B})}][\text{volume in milliliters}_{(\text{solution B})}]$$

Suppose the doctor prescribes injection of 1 cc of 1.25 percent chlorpromazine and the only solution available is 2.5 percent chlorpromazine. How should one proceed to administer the drug?

$$(2.5\%)(\text{volume in milliliters}) = (1.25\%)(1\ \text{ml})$$

$$\text{volume} = \frac{(1.25\%)(1\ \text{ml})}{(2.5\%)} = 0.5\ \text{ml}$$

Thus, the correct procedure would be to use 0.5 ml of 2.5 percent chlorpromazine solution that has been diluted to 1.0 ml for the injection.

exercise 5-18

(a) How much 5 percent glucose solution should be diluted to give 750 ml of 2 percent glucose solution?
(b) 25 ml of a NaCl solution was diluted to 150 ml. The resulting solution had a concentration of 0.9 percent. What was the concentration of the original solution?
(c) How would 400 cc of 5 percent glucose be prepared from 20 percent glucose?

TITRATIONS

The procedure described on p. 114 for determining the amount of bromine in a solution of known concentration that will react with a given quantity of a $FeBr_2$ solution of known concentration is an application of a general technique called *titration*. Titrations are useful for experimentally determining the coefficients of the reactants in chemical equations. Since these coefficients represent the relative numbers of moles that react, it is convenient to use molarity to express solution concentration. Example 5-11 illustrates the use of a titration to determine the coefficients of a chemical equation.

EXAMPLE 5-11

Ascaridole was at one time used as an anthelmintic (*anti,* against; *helminthos,* worms) and chemists found it can be synthesized from α-terpinene, a constituent of coriander oil. α-Terpinene ($C_{10}H_{16}$) reacts with bromine. In an experiment it was found that 10 ml of a 0.01 *M* solution of α-terpinene required the addition of 13.33 ml of 0.015 *M* Br_2 solution to reach the end point.

From these data, determine the number of moles of bromine that react with 1 mole of α-terpinene and write a balanced equation for the reaction assuming that a bromoterpinene is the only product.

$$\text{moles } \alpha\text{-terpinene reacting} = (0.01\ \cancel{\text{liter}})(0.01\ \text{mole}/\cancel{\text{liter}}) = 0.0001\ \text{mole}$$

$$\text{moles } Br_2 \text{ reacting} = (0.01333\ \cancel{\text{liter}})(0.015\ \text{mole}/\cancel{\text{liter}}) = 0.0002\ \text{mole}$$

The ratio of the moles of reactants is

$$\frac{0.0002\ \text{mole } Br_2}{0.0001\ \text{mole } \alpha\text{-terpinene}} = \frac{2\ \text{moles } Br_2}{1\ \text{mole } \alpha\text{-terpinene}}$$

Therefore the balanced equation must be

$$C_{10}H_{16} + 2Br_2 \longrightarrow \text{bromoterpinene}$$

Since bromoterpinene is the only product, its formula must be $C_{10}H_{16}Br_4$ and a tetrabromoterpinene has been produced; thus,

$$C_{10}H_{16} + 2Br_2 \longrightarrow C_{10}H_{16}Br_4$$

Although titrations are sometimes used to determine the mole ratio in which reactants combine, they are most often used to analyze a solution to determine the concentration of a particular solute. If a titration is used to analyze a solution, it is necessary to know the balanced equation for the reaction that takes place.

Suppose that we have a solution of sodium chloride of unknown concentration and it is necessary to determine its concentration. It is known that a solution of sodium chloride will react with a solution of silver nitrate to produce insoluble silver chloride according to the equation

$$NaCl + AgNO_3 \longrightarrow AgCl + NaNO_3$$

In order to find the concentration of the sodium chloride solution we would measure out a known volume of the NaCl solution and titrate it with (add to it) a solution of known concentration of $AgNO_3$ until the formation of the insoluble AgCl ceases. When the formation of AgCl ceases, all the NaCl must have reacted and the end point of the reaction has been reached.* From the data collected the molarity of NaCl can be determined.

*It is necessary to use an indicator to tell when the end point of the reaction has been reached. General discussion of indicators is in Chapter 8.

EXAMPLE 5-12

In an experiment, 37.2 ml of 0.115 *M* $AgNO_3$ was required to titrate 25.0 ml of a NaCl solution to the end point. What is the molarity of the NaCl solution?

The balanced equation tells us that 1 mole of $AgNO_3$ reacts with 1 mole of NaCl. Therefore, at the end point, the moles of NaCl present in the NaCl solution equals the moles of $AgNO_3$ added. The moles of $AgNO_3$ added equals (Molarity of $AgNO_3$)(volume of $AgNO_3$ added). Thus at the end point,

$$\text{moles NaCl} = \text{moles } AgNO_3 = (\text{molarity } AgNO_3)(\text{volume } AgNO_3 \text{ added})$$
$$\text{moles NaCl} = (0.115 \text{ mole}/\cancel{\text{liter}})(0.0372 \cancel{\text{liter}}) = 0.00428 \text{ mole}$$

Since this amount of NaCl was originally present in 25 ml of NaCl solution, the molarity of the original NaCl solution must be

$$\frac{0.00428 \text{ mole}}{0.025 \text{ liter}} = 0.171\ M$$

Many tests performed in hospital laboratories employ titrations. One example is the determination of the concentration of oxalate ions in urine. In one procedure, calcium chloride is added to a urine sample to precipitate the insoluble calcium oxalate. The calcium oxalate is redissolved in sulfuric acid and titrated with a potassium permanganate solution. The balanced equation for the reaction between calcium oxalate and potassium permanganate is

$$8H_2SO_4 + 5CaC_2O_4 + 2KMnO_4 \longrightarrow 10CO_2 + 8H_2O + 5CaSO_4 + 2MnSO_4 + K_2SO_4$$

EXAMPLE 5-13

In a typical oxalate determination, 10 ml of urine was treated with $CaCl_2$ and the calcium oxalate formed redissolved in 1 *M* sulfuric acid solution. To reach the titration end point, 8.7 ml of 0.00102 *M* $KMnO_4$ solution was required. What was the concentration of oxalate ions in the urine sample?

The chemical equation tells us that 5 moles of CaC_2O_4 reacts with 2 moles of $KMnO_4$. Therefore at the end point the moles of CaC_2O_4 in the sample must equal $\frac{5}{2}$ the moles of $KMnO_4$ added. Thus;

$$\text{moles } CaC_2O_4 = \tfrac{5}{2}\,(0.0087 \cancel{\text{liter}}\ KMnO_4 \text{ added})\left(0.00102\,\frac{\text{mole } KMnO_4}{\cancel{\text{liter}}}\right)$$
$$\text{moles } CaC_2O_4 = 0.000008974 = 8.974 \times 10^{-6} \text{ mole } CaC_2O_4$$

To calculate the concentration in moles per liter, we must divide the moles of solute in the sample by the volume (in liters) of the sample.

Since 1 mole of oxalate ions ($C_2O_4^{2-}$) are contained in 1 mole of CaC_2O_4, the urine sample contained 8.974×10^{-6} mole of $C_2O_4^{2-}$.

Therefore the concentration of $C_2O_4^{2-}$ in the urine sample is

$$\frac{8.974 \times 10^{-6} \text{ mole}}{0.010 \text{ liter}} = 8.974 \times 10^{-4} \text{ mole/liter}$$

exercise 5-19

(a) A 50-ml sample of a NaCl solution was titrated to the end point with 37.5 ml of 0.02 *M* $AgNO_3$. What is the concentration of the NaCl solution expressed in moles per liter?

(b) A 25-ml sample of $FeBr_2$ solution was titrated to the end point with 40 ml of 0.05 *M* Br_2 solution. What is the molarity of the $FeBr_2$ solution?

CHAPTER SUMMARY

1. Solutions are homogenous mixtures. Intermolecular attractive forces play an important part in the dissolving process. In general, nonpolar substances dissolve in nonpolar substances and polar or ionic compounds dissolve in polar solvents.
2. Soaps and detergents are molecules with both a nonpolar portion and an ionic portion. This dual nature of the soap and detergent molecules allows them to behave as emulsifying agents.
3. Concentration of a solution describes the amount of solute dissolved in a given amount of solvent. Concentration can be described qualitatively or quantitatively. Qualitative descriptions include saturated or unsaturated. Percent concentration and molarity are two quantitative ways of describing a solution.
4. Percent concentration is defined as the grams of solute per 100 ml of solution.
5. Molarity is defined as the number of moles of solute per liter of solution. A mole of a substance is an amount of that substance, in grams, equal to its formula weight.
6. A balanced equation is useful in calculating the amounts of reactants that will react with another. It is also possible to calculate the amount of a product that can be formed from a given amount of reactants with the aid of a balanced equation.
7. Concentration of an ion in a solution can be expressed in equivalents per liter as well as moles per liter. The number of equivalents is calculated by multiplying the charge on the ion by the number of moles of the ion.
8. Solutions can be diluted by the addition of more solvent.
9. Titrations are a useful experimental technique for determining the concentration of a solution.

QUESTIONS

1. Discuss why iodine (I_2) is relatively insoluble in water but very soluble in CCl_4.
2. How would the following solutions be prepared?
   (a) 2 liters of 1 *M* NaCl
   (b) 2 liters of 5 percent NaCl
   (c) 100 ml of 0.3 *M* Na_2SO_4

(d) 300 ml of 1.2 *M* NH_4NO_3
(e) 250 cc of 1.2 percent LiOH
(f) 750 ml of 0.9 *M* NaCl
(g) 5000 ml of 2.5 percent Na_3PO_4

3. Which member of the following pairs of solutions is more concentrated?
(a) 0.1 percent or 0.1 *M* NaCl
(b) 10 percent or 1 *M* $C_6H_{12}O_6$
(c) 10 percent or 1 *M* KCl
(d) 2 percent or 2 *M* NH_3
(e) 5 percent or 0.5 *M* LiF

4. Hydrogen and oxygen react to form water according to the equation

$$2H_2 + O_2 \longrightarrow 2H_2O$$

If 8 g of hydrogen and 1 g of oxygen are mixed and allowed to react,
(a) How many grams of water are produced?
(b) How many grams of hydrogen are left?
(c) How many grams of oxygen are left?

5. When 20 ml of 0.5 *M* NaCl is diluted to 100 ml with water, what is the molarity of the resultant solution?

6. When 10 ml of 1 percent sucrose ($C_{12}H_{22}O_{11}$) is added to 40 ml of water, what is the percent concentration of the resultant solution?

7. How many milliliters of 0.05 *M* $AgNO_3$ would be required to react with 50 ml of 0.1 *M* NaCl?

8. 5 ml of 5 *M* NaCl was added to 50 ml of 0.01 *M* $AgNO_3$. Perform appropriate calculations to determine if all the $AgNO_3$ reacted.

9. With the aid of Table 5-1, predict whether a precipitate would be formed if solutions of the compounds in each of the following pairs were mixed. Write a balanced equation for the formation of all precipitates.
(a) NH_4Cl and $AgNO_3$
(b) NH_4Cl and KNO_3
(c) $NaC_2H_3O_2$ and KNO_3
(d) NaCl and $CaBr_2$
(e) $CaCl_2$ and $(NH_4)_3PO_4$
(f) Li_2CO_3 and $Ca(NO_3)_2$
(g) $(NH_4)_2CO_3$ and NaCl

6

reaction dynamics

INTRODUCTION

A reaction rate tells how much material reacts in a given amount of time.

Although we have studied many reactions and the properties of the compounds produced during chemical reactions, we have considered neither how chemical reactions occur nor why some reactions take place faster than others. Information about reaction rates and how reactions take place is sought through a subspecialty of chemistry called *reaction kinetics* or *reaction dynamics*.

ENERGETICS

In Chapter 3, reactions were described as either exergonic or endergonic. *Exergonic reactions are energy yielding reactions and are termed exothermic if the energy is evolved as heat. Endergonic reactions absorb energy and if the energy source is heat, they are called endothermic reactions.* Since nature tends to move toward lower energy, we would expect the energetics of a reaction to determine the tendency of a reaction to occur. In accordance with our expectations, chemists have found that most exothermic reactions have a tendency to occur.* One might also be tempted to suggest that exothermic reactions should take place easily and rapidly upon mixing the reactants, while endothermic reactions should occur very slowly if at all.

Before succumbing to this temptation, let's first reflect on a few common experiences. Wood, paper, and gasoline, among other things, will all react with oxygen to evolve heat in a process we call *burning*. The combustion of these materials is an exothermic reaction and has a great tendency to occur. On the other hand, the reaction of these materials with oxygen at normal temperatures does not take place very rapidly. This is very fortunate since most of us would find it inconvenient to sit in a burning chair reading a book that is going up in flames! Thus, although there may be a great tendency for a reaction to occur, this does not necessarily lead to a rapid reaction. There seems to be a barrier preventing these reactions from occurring at a rapid rate at ordinary temperatures.

A clue to the nature of this barrier to reaction comes from considering what we do to make combustible materials burn. To ignite combustible materials we heat them in the presence of oxygen. Before you protest too loudly that you use a burning match to start a fire, let's point out that a match is merely a convenient way of heating materials. There are many observations we can make to show it is not necessary to have an open flame to start a fire. For example, if we place a piece of paper on a hot electric burner, it will burst into flame. Electric charcoal starters for use in barbecue grills ignite the charcoal by heating it. Electric wiring or a fireplace chimney with inadequate insulation may start house fires by heating combustible materials. Similarly, fires can be started by the heat generated from rubbing two sticks together or focusing sunlight with a magnifying glass on combustible materials. Hopefully by now you are convinced that a flame is not

*For some very complex reasons, whose discussion is beyond the scope of this text, there exist some exothermic reactions that have little tendency to occur. The student who wishes to pursue this point is advised to read the chapters on thermodynamics found in many general chemistry texts.

required to start a fire. Heat is a form of energy. Apparently the barrier to exergonic reactions can be overcome by adding energy to the reactants. In recognition of this fact, the barrier is called an *energy barrier* and the energy required to overcome the barrier is known as the *activation energy*, abbreviated E_a.

THE ORIGIN OF THE ENERGY BARRIER

During a chemical change, reactant molecules are transformed into product molecules. Bonds in the reactant molecules are broken and new bonds made in forming product molecules during a reaction. Wood and paper are primarily cellulose, which can be represented by the formula $(C_6H_{10}O_5)_n$, where n is a very large number. The reaction for the combustion of cellulose can be written

$$(C_6H_{10}O_5)_n + 6nO_2 \longrightarrow 6nCO_2 + 5nH_2O$$

exercise 6-1

Write the balanced equation for the combustion of cellulose if n in the equation above equals 500, 10,000, or 100,000.

From examination of this equation it is obvious that an oxygen–oxygen bond in the oxygen molecule must be broken so that the oxygen atoms can form new bonds with either carbon or hydrogen to produce the products. Additionally, carbon–hydrogen bonds, carbon–carbon bonds, and perhaps carbon–oxygen or oxygen–hydrogen bonds in cellulose must be broken in order for new bonds to be formed to produce CO_2 and H_2O.

Recall our discussions about bonding in Chapter 3. We concluded that chemical bonds are formed because bond formation is accompanied by a decrease in potential energy. This energy may be released as heat and light. The lowering of potential energy results in a more stable system. This means that to break a bond, energy must be put back into the system in order to increase the potential energy. This is shown diagrammatically in Fig. 6-1.

The necessity to break bonds during the course of a reaction means that energy is required to start a reaction even if that reaction is exothermic. This energy requirement constitutes a barrier to the reaction taking place. It turns out that there are also other factors that influence the size of the energy barrier for a given reaction. Consequently, it is not possible for chemists to calculate the activation energy for a reaction from knowledge of the energy required to break certain bonds. Activation energies for reactions must be determined experimentally for each reaction.

The activation energy is the difference in energy between the top of the energy barrier and the reactants.

Chemists have determined the activation energies for a great many reactions and found that the activation energy for some reactions is relatively small, while for others it is quite large. Further, *for exergonic reactions the activation energy is unrelated to the overall energy change for the reaction.* Additionally, *the activation energy and not the overall energy change determines the rate of reaction.*

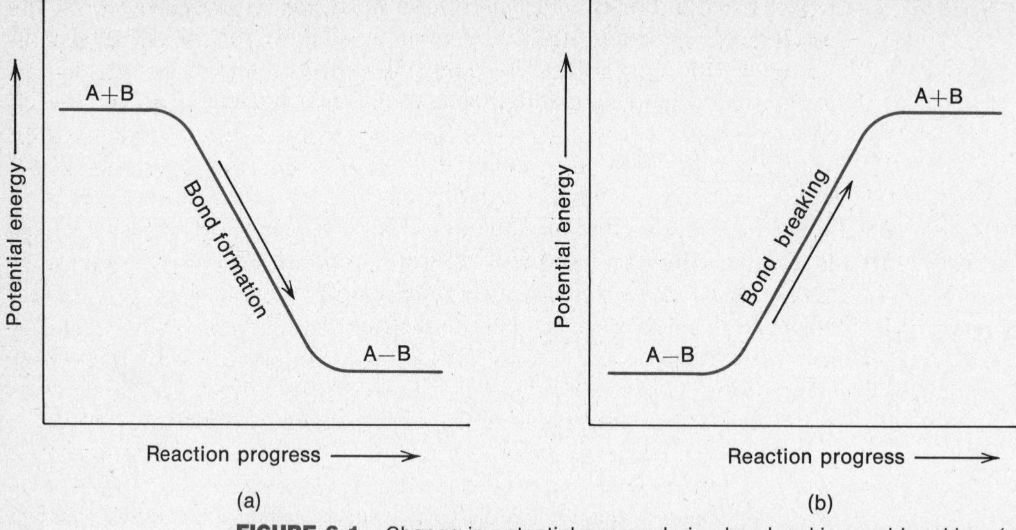

FIGURE 6-1 Change in potential energy during bond making and breaking. (a) Potential energy decreases during bond formation between atoms A and B. (b) Potential energy increases during breaking of the A-B bond.

The relationships among the activation energy, the overall energy change, and the energy of the reactants and products can be represented diagrammatically as shown in Fig. 6-2. Notice in Fig. 6-2(a) for an exergonic reaction that the energy of the products is less than that of the reactants. Consequently, the products of the reaction are more stable than the reactants

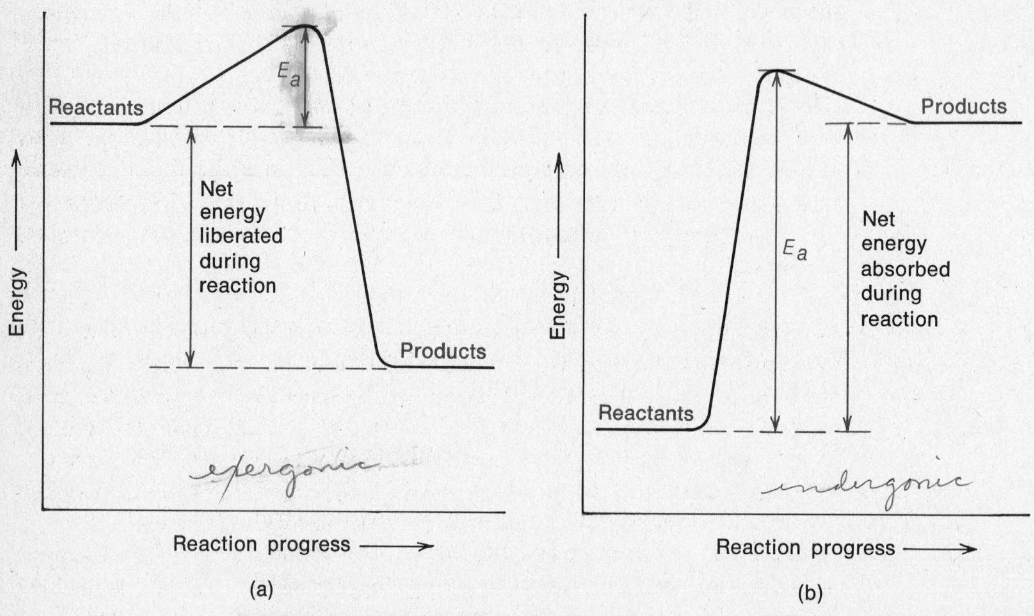

FIGURE 6-2 (a) Reaction energy diagram for an exergonic reaction. (b) Reaction energy diagram for an endergonic reaction.

The difference in the energies of the products and reactants equals the energy change during reaction.

since there is an overall decrease in energy during reaction. Conversely, the products of the endergonic reaction shown in Fig. 6-2(b) are less stable than reactants as a result of the increase in energy during reaction. Also notice *the activation energy of the endergonic reaction is larger than the overall energy change for the reaction.* This relationship between the activation energy and the total energy change is true for all endergonic reactions.

HOW DOES THE ACTIVATION ENERGY DETERMINE REACTION RATES?

The key to understanding how the activation energy determines the rate of reaction lies in remembering that while we tend to think of large amounts of substances reacting in a given time span, the reaction occurs on a molecular basis. For this reason, it is not necessary that all the reactant molecules acquire sufficient energy to overcome the energy barrier at the same time. Instead, the reaction rate depends on the fraction of the total number of molecules that at any given instant possesses enough energy to react.

Consider burning a piece of paper. It isn't necessary to heat the entire piece of paper to give all the molecules enough energy to react with oxygen. We only need to heat a corner of the paper to give just a small fraction of the total number of molecules in the paper enough energy to react. As the paper begins to react, the rate of burning is relatively slow. Since this is an exothermic reaction, more energy is given off in going from the peak of the energy barrier to the products than was absorbed to get the reactant molecules to the peak of the energy barrier. As this energy is given off, it can be used to supply the activation energy to other reactant molecules, causing the reaction to become self-sustaining. Not only is the reaction self-sustaining, but the reaction rate will continually increase (as long as there is a sufficient supply of unreacted molecules) since with the passage of time more and more energy is available to give an increasingly larger fraction of the molecules the required activation energy.

The self-sustaining nature of combustion reactions is a property of all exergonic reactions. As a result, it is necessary to control exergonic reactions either by limiting the amount of reactant present or by providing a means to dissipate energy as it released during reaction. Endergonic reactions on the other hand are more easily controlled since there is a net absorption of energy.

KINETIC MOLECULAR THEORY

We have just seen that the more heat available to the reactants, the faster the reaction rate will be since a larger fraction of the molecules will possess the required activation energy for reaction. From everyday experience we know that as heat is applied to a substance, it becomes hotter and the temperature of the substance rises. Therefore, *increasing the temperature at which a reaction is occurring must increase the rate of reaction.* Conversely, a decrease in temperature results in a decreased reaction rate.

Results from studies on the behavior of gases led to the development of the kinetic molecular theory during the mid-1800's. This theory relates temperature to the kinetic energy of molecules and aids in developing a fuller understanding of how heat affects reaction rates. Kinetic molecular theory states that *the average kinetic energy of a collection of molecules is directly proportional to temperature.* Kinetic energy is the energy of motion and the kinetic energy of a moving body is equal to one-half times its mass multiplied by the square of its velocity (kinetic energy $= \frac{1}{2}mv^2$). This means that molecules are in constant motion and the higher the temperature, the faster the molecules are moving.* Also, according to the theory, the motion of the molecules is completely random.

Notice that while kinetic molecular theory tells us something about the *average* kinetic energy of a collection of molecules, it is silent about the kinetic energy of any particular molecule. This means that some molecules may have large kinetic energies, while others possess little kinetic energy. The distribution of kinetic energies among a large number of molecules can be accurately described mathematically. The easiest way of understanding this description is by a graph obtained from plotting the number of molecules versus their kinetic energy. The kinetic energy distribution for a large collection of molecules at two different temperatures is shown in Fig. 6-3. Notice the considerable range of kinetic energies of the molecules at each temperature and that the average kinetic energy is higher at 225 °C than at 25 °C.

Suppose we place a line on the energy distribution graph at the energy equal to the activation energy of a reaction. The result is shown in Fig. 6-4.

Obviously, the fraction of the total number of molecules possessing the activation energy for reaction is smaller at 25 °C than at 225 °C. Hence the rate of reaction should be less at 25 °C than at 225 °C, in agreement with our earlier discussions.

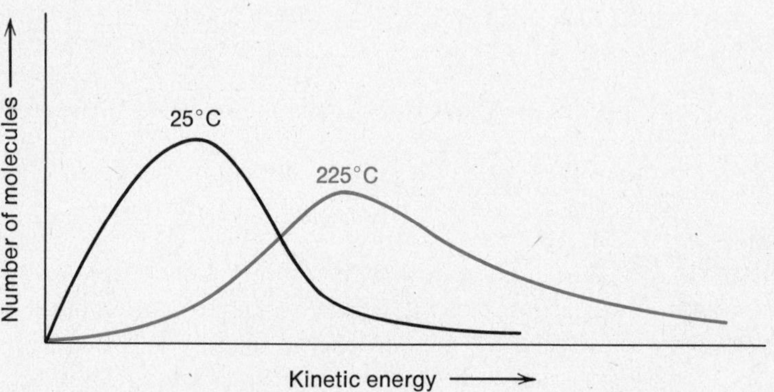

FIGURE 6-3 The distribution of molecular kinetic energies at room temperature (25°C) and 225°C.

*Since molecular motion increases with increasing temperatures and the heat content of an object increases with temperature, scientists have concluded that heat is a result of molecular motion.

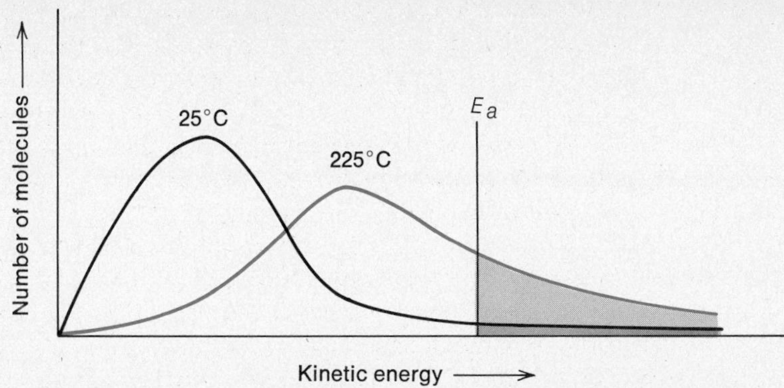

FIGURE 6-4 The number of molecules possessing kinetic energy equal to or greater than the activation energy of reaction at each temperature is proportional to the shaded areas.

exercise 6-2

Explain in terms of the kinetic energies of molecules why the rate of an exothermic reaction increases with the passage of time. Explain why it is possible to control the rate of many exothermic reactions by placing the reaction vessel in a container of ice water.

KINETIC AND POTENTIAL ENERGY—HIT 'EM AGAIN, HARDER

At this point you may be wondering how the kinetic energy of the molecules supplies the required activation energy since the barrier to reaction is a potential energy barrier. The answer becomes clear if we consider what must happen in order for reaction to occur. If two molecules are to react, they must come in contact with one another. Kinetic molecular theory tells us that molecules are in constant random motion. The random motion of the molecules means that they are constantly bumping into one another. You have probably learned that during the collision of two objects, their kinetic energy is momentarily converted to potential energy; this potential energy is reconverted to kinetic energy as the objects bounce away from each other.

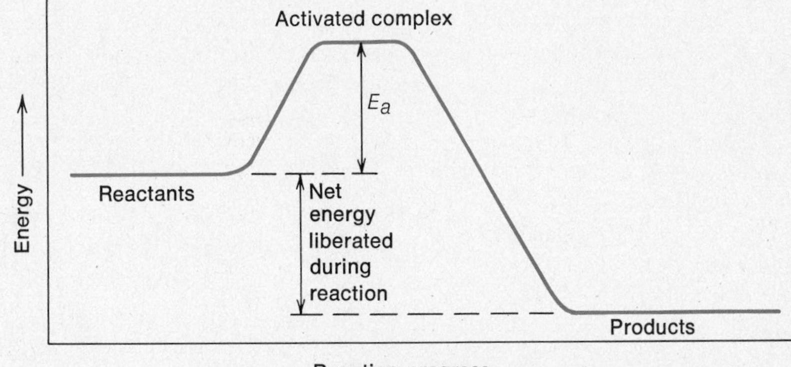

FIGURE 6-5 A reaction energy diagram showing the energy of the activated complex.

When molecules collide with one another, their kinetic energy is momentarily converted to potential energy. If the collision is between molecules with high kinetic energies, the potential energy at impact will equal or exceed the activation energy for reaction. In this case the reactant molecules may stick together and react instead of immediately bouncing apart. The species formed during collision of these energetic molecules is called an *activated complex*. Figure 6-5 is a reaction energy diagram for an exergonic reaction, showing the position of the activated complex.

If the collision between reactant molecules is not energetic enough, the activated complex will not be formed and the molecules will simply bounce away from each other without reaction.

CONCENTRATION AND THE RATE LAW

We have just seen that the reaction rate is dependent on the collision energy of the reactant molecules. The energy of the collisions depends solely on the temperature. Consequently, one way of controlling reaction rates is by controlling the temperature at which the reaction occurs. However, there are many instances in which it is not possible to control the temperature at which reaction occurs or else it is undesirable to change the reaction temperature. Such is the case with the reactions occurring in our bodies. Body temperature is normally 37.0°C and it is undesirable to change body temperature from this value. If control is to be maintained over body reactions, there must exist other means for controlling reaction rates.

In a situation in which it is not possible to control the energy of collisions, we might try to control the number of collisions per time interval. If we can increase the number of collisions occurring between energetic molecules per time interval, the reaction rate should increase. Correspondingly, a decrease in the number of collisions should decrease the rate.

The likelihood of collisions between molecules can be controlled by changing the concentrations of the reactants. An increase in concentration crowds more molecules into a given volume, resulting in an increase in the number of collisions during a given time period. As the collision rate increases, the reaction rate increases. Thus, the reaction rate is proportional to the concentration of the reactants.

In 1864, Guldberg and Waage first expressed the relationship between reaction rates and reactant concentration. They found that, in general, for any single step reaction of the type

$$a\text{A} + b\text{B} + \cdots \longrightarrow z\text{Z} + y\text{Y} + \cdots$$

that the reaction rate at any given temperature will obey the equation

$$\text{reaction rate} = k[\text{A}]^a[\text{B}]^b$$

This equation is known as the *rate law* for the reaction and k is a proportionality constant called the *specific rate constant*.* The brackets [ ] in the rate equation indicate that the concentration of the reactants is expressed as molarity. Since the reaction rate is dependent on the number of reactant molecules in a given volume, molarity is the most convenient method for expressing concentration.

Sucrose (cane sugar) will react with water to produce glucose plus fructose according to the equation

$$\underset{\text{sucrose}}{C_{12}H_{22}O_{11}} + H_2O \longrightarrow \underset{\text{glucose}}{C_6H_{12}O_6} + \underset{\text{fructose}}{C_6H_{12}O_6}$$

Accordingly, the rate equation for this reaction is written

$$\text{rate} = k[\text{sucrose}][H_2O]$$

Suppose, for example, that the reaction rate is r when the sucrose concentration is 0.1 M and the water concentration† is 0.1 M. If, in another experiment, the concentration of sucrose is doubled while the water concentration remains 0.1 M, the reaction rate must double; it equals $2r$.

exercise 6-3

The reaction rate for the reaction of sucrose with water at 25°C can be represented as *r* when the concentration of sucrose is 0.1 *M* and that of water is 0.1 *M*.
(a) What would be the reaction rate at 25°C if [sucrose] = 0.1 *M* and $[H_2O]$ = 0.05 *M*?
(b) What would be the reaction rate at 25°C if [sucrose] = $[H_2O]$ = 0.2 *M*?
(c) What would be the reaction rate at 25°C if [sucrose] = 0.2 *M* and $[H_2O]$ = 0.05 *M*?

During reaction the reactants are used up. As a result, their concentration is decreasing with the passage of time, causing the rate of reaction to decrease with time. This feature of reaction rates has significant biological implications. The body must be able to conduct a wide variety of reactions and control the rate at which these reactions take place. The control mechanisms are very complex; yet they must be able to respond very rapidly to changes in a person's activities. Secondly, regardless of activities, the body must conduct these reactions at some minimal constant rate necessary to sustain life. One means by which the reactions can be controlled is by controlling the concentrations of reactants.

Because we do not ingest food continuously, the body must have the means to store nutrients when they are in abundant supply and release them from storage when required. If the body did not have this capability, the

*The specific rate constant varies with temperature changes since the reaction rate changes with temperature and the rate equation as written does not include a temperature. Consequently, it is necessary to measure reaction rates at a constant temperature.

†This reaction is being conducted in a solvent other than water.

reactions might occur too rapidly shortly after a meal. Several hours later, with depletion of the nutrient concentration, the reaction rates would become too low for good health. Two examples of substances that are stored and released as needed are glucose, which can be stored as glycogen and fat, which can be stored in fat depots.

The normal range of values for glucose concentration in the blood of healthy individuals is 0.065 to 0.095 percent. If the glucose concentration exceeds this value, the person is said to be *hyperglycemic;* if below this level, *hypoglycemic*. Glucose concentration within normal range provides a smooth flow of glucose into body tissue for use in cells for energy production. The movement of glucose from the bloodstream into the cells maintains cell concentration of glucose at a level that will ensure a glucose reaction rate necessary to meet the energy demands for the cell.* If after ingestion of a meal the blood glucose levels tend to exceed the normal concentration range, glucose is removed from the bloodstream and stored as glycogen in the liver. Later as the blood glucose level is diminished by cell utilization, the liver glycogen is broken down to glucose and released into the bloodstream to maintain normal concentration.

exercise 6-4

Express normal value range for blood glucose concentrations in milligrams of glucose per 100 ml of blood.

CATALYSIS

Chemists have found that the addition of small amounts of certain substances to a reaction mixture can have a dramatic effect on the rate of reaction. These substances are called *catalysts*. A catalyst is defined as *a substance which changes the rate of a reaction and can be recovered essentially unaltered.* Through general usage, catalyst has come to mean a substance which increases the reaction rate; substances decreasing reaction rates are called *inhibitors*.

The catalyst (or inhibitor) is not shown in a balanced chemical equation as a reactant or product since it is recovered unchanged after reaction. Nevertheless, it is necessary to convey to the reader that a catalyst or inhibitor has been used in the reaction. This information is usually given by writing the catalyst (or inhibitor) above or below the arrow in the balanced equation.

Catalysts and inhibitors are found in, or used in, the manufacture of many materials affecting our daily lives. Catalysts are used in the manufacture of hydrogenated vegetable shortenings. Catalysts are used in devices to control exhaust emissions from automobiles. Inhibitors are put in antifreeze to slow down the corrosion rate in automobile cooling systems. Paints also contain inhibitors to slow the attack of oxygen and make repainting

*Cell concentrations of glucose are controlled by both the blood glucose concentration and by insulin concentration. The action of insulin is discussed in Chapter 13.

less frequent. Biologically, catalysts and inhibitors are of great importance. Nature employs catalysts in almost all biological reactions. These biological catalysts are given the special name *enzymes*. Many drugs used in medical treatment of disease are inhibitors.

Catalysts lower the activation energy of a reaction.

Catalysts take part in the reaction in such a way that they are regenerated during reaction. Their role in a reaction is to lower the activation energy for a reaction by providing an alternate way for reaction to occur. In many cases the lowering of the activation energy is accomplished by combination of the catalyst with one of the reactants to form an intermediate or activated complex.

Consider the general reaction represented by the equation

$$\mathrm{A + B \xrightarrow{catalyst} Z + Y}$$

The catalyst combines with A to give an activated complex:

$$\mathrm{A + catalyst \longrightarrow catalyst - A}$$

This activated complex then interacts with the other reactant to give the products and regenerate the catalyst:

$$\mathrm{catalyst - A + B \longrightarrow Z + Y + catalyst}$$

The activation energy for the formation of the activated complex between catalyst and reactant is lower than that for the activated complex formed between the reactants. A reaction energy diagram for an uncatalyzed and catalyzed reaction is given in Fig. 6-6.

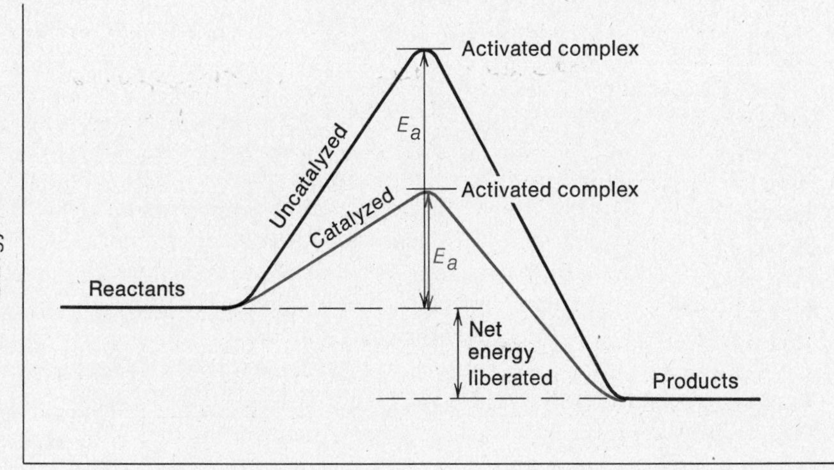

FIGURE 6-6 Comparison of reaction energies for an uncatalyzed and catalyzed reaction.

ENZYMATIC CATALYSIS

Enzymes are the catalysts for the chemical reactions of life. In many respects these are the most remarkable compounds known. An enzyme is a very complex molecule consisting of a high molecular weight protein molecule. The chemical nature of proteins will be discussed in Chapter 11. It will suffice for our purposes here to know that proteins are made up of covalently bonded carbon, hydrogen, nitrogen, oxygen, and sulfur with molecular weights in the hundreds of thousands.

In addition to the protein part, the enzyme may also consist of another smaller portion. This smaller portion may be either a covalent molecule or a metal ion. The protein portion, called the *apoenzyme*, must combine with the smaller portion known as a *cofactor* to produce the active enzyme. To distinguish between the two types of cofactors, the cofactor is called a *coenzyme* if it is an organic molecule and an *activator* if it is a metal ion. Minerals are the body's source of metal ions to act as enzyme activators. Among the required metal ions are: K^+, Mg^{2+}, Ca^{2+}, Zn^{2+}, Mn^{2+}, Co^{2+}, Fe^{2+}, and Cu^{2+}. Several of the vitamins serve as coenzymes in the chemical machinery of the cell. Thus, the necessity of including minerals and vitamins in the daily diet can be understood. Without these substances, the body's enzymes would be inactive and reaction rates would be too slow to maintain proper growth and maintainance.

Basically, an enzyme functions as a catalyst in the same way that laboratory catalysts function. The enzyme combines with one of the reactants to form an activated complex. This activated complex interacts with the other reactant to produce the products and regenerate the enzyme. In the case of enzymatic catalysis, the reactants are generally referred to as *substrates*. The sequence can be represented by the equation

$$\underset{\text{enzyme}}{E} + \underset{\text{substrate}}{S} \longrightarrow \underset{\substack{\text{enzyme—substrate}\\ \text{activated complex}}}{E{-}S} \longrightarrow \underset{\text{enzyme}}{E} + \underset{\text{products}}{P}$$

reactants

Although enzymes are very large and complex molecules, not all the enzyme molecule is directly involved in the catalytic activity. Apparently there exists several relatively small areas of the enzyme that are important to its catalytic activity. One of these areas is the *active site* that actually catalyzes the reaction. Generally, the cofactor forms a portion of this active site. The relationship of the cofactor to the active site is shown in Fig. 6-7. Other areas of the enzyme molecule serve as binding sites. These binding sites serve to hold the substrate on the enzyme in the most favorable position for reaction to occur on the *active site*. The binding of the substrate to the enzyme occurs primarily via dipolar forces and hydrogen bonding. These features of enzymes are represented schematically in Fig. 6-8 for a reaction involving breakage of a bond.

The presence of binding sites on the enzymes conveys the property of specificity to enzymes. *Specificity* means the enzyme will catalyze some reactions but will not catalyze other very closely related reactions. For

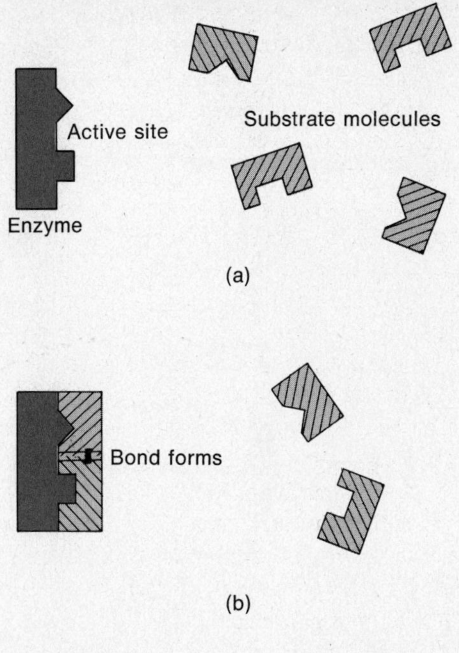

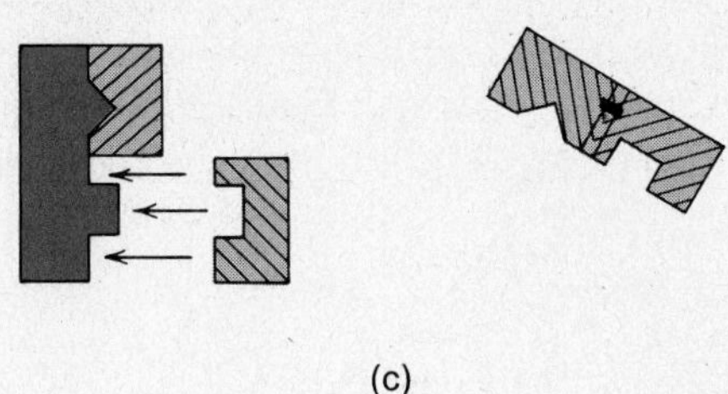

FIGURE 6-7 (a) The active site of the enzyme molecule catalyzes the reaction. (b) Molecules react while attached to the surface of the enzyme. (c) The same enzyme may be used again.

example, the enzyme urease will catalyze the hydrolysis of urea to carbon dioxide and ammonia.

$$\underset{\text{urea}}{H_2N{-}\overset{\overset{\displaystyle O}{\|}}{C}{-}NH_2} + H_2O \xrightarrow{\text{urease}} 2NH_3 + CO_2$$

The compound biuret can also be chemically hydrolyzed to ammonia and carbon dioxide.

$$\underset{\text{biuret}}{H_2N{-}\overset{\overset{\displaystyle O}{\|}}{C}{-}\overset{\overset{\displaystyle H}{|}}{N}{-}\overset{\overset{\displaystyle O}{\|}}{C}{-}NH_2} + 2H_2O \longrightarrow 2CO_2 + 3NH_3$$

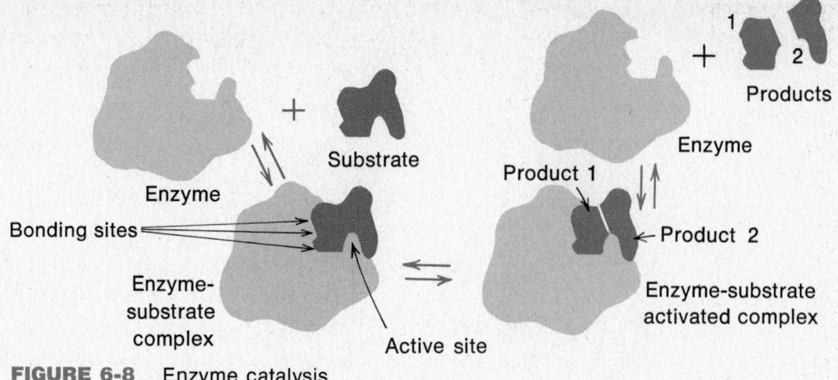

FIGURE 6-8 Enzyme catalysis.

Both urea and biuret belong to the general class of organic compounds known as *amides*. (Recall the general formula of amides is $R{-}\overset{\overset{\displaystyle O}{\|}}{C}{-}NH_2$). Although urease will catalyze the hydrolysis of urea, it will not catalyze the hydrolysis of biuret or any other amide. For this reason, urease is said to possess *absolute specificity*.

Other enzymes are less specific. They may possess *reaction specificity* or *group specificity*. Dehydrogenases and esterases are examples of reaction specific enzymes. Dehydrogenases are enzymes catalyzing the removal of hydrogen atoms from the substrate and are important in the metabolism of organic acids resulting from the breakdown of fats. Esterases catalyze the hydrolysis of esters. The enzymatic hydrolysis of an ester is shown in Fig. 6-9.

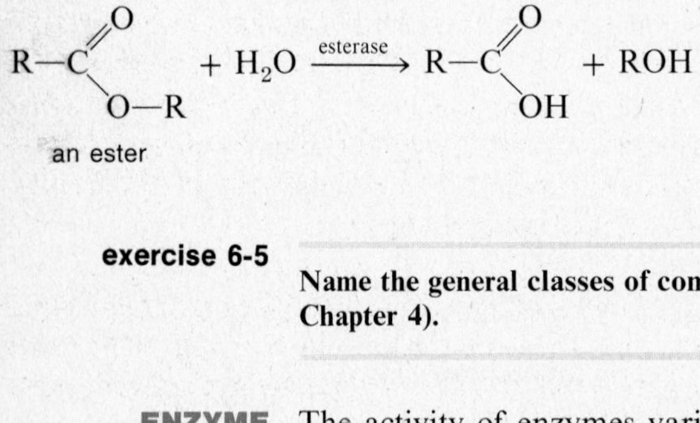

$$\underset{\text{an ester}}{R{-}C({=}O){-}O{-}R} + H_2O \xrightarrow{\text{esterase}} R{-}C({=}O){-}OH + ROH$$

FIGURE 6-9 Enzymatic hydrolysis of an ester.

exercise 6-5

Name the general classes of compounds resulting from the hydrolysis of esters (see Chapter 4).

ENZYME ACTIVITY The activity of enzymes varies. The turnover number is used to compare the activity of one enzyme to another. The *turnover number* of an enzyme may be defined as the number of substrate molecules whose reaction is catalyzed per enzyme molecule per minute. Typical turnover numbers range from 10,000 to 5,000,000. The enzyme catalase, which catalyzes the breakdown of hydrogen peroxide to water and oxygen, exhibits the highest turnover number. Thus, in one minute the breakdown of 5 million molecules of hydrogen peroxide is catalyzed by a *single* catalase molecule. Considering this phenomenal reaction rate and the fact that blood contains catalase, it is not surprising that a solution of hydrogen peroxide fizzes when it is poured on an open cut or abrasion.

$$\underset{\text{hydrogen peroxide}}{2H_2O_2} \xrightarrow{\text{catalase}} 2H_2O + O_2$$

Obviously from our discussion of reaction rates and mode of enzyme action, the overall reaction rate is dependent on both the concentration of the substrate and the enzyme.

In the physiological situation, the substrate concentration is held nearly constant (recall our earlier discussion of blood glucose concentration) and the reaction rates controlled by varying the concentration of the enzyme. It is quite easy to understand why nature prefers to control reaction rates by controlling active enzyme concentration rather than by changing substrate concentration. Many substances are the starting materials for more than one reaction. Therefore, if the concentration of the substance is changed, the reaction rate of several reactions will be influenced. If the concentration of a specific enzyme is changed, however, the rate of only one reaction is affected. The rates of all other reactions for which that substance is a starting material remain unaffected.

Inhibitors decrease the active enzyme concentration.

The active enzyme concentration can be controlled in several ways. One of the most important is the use of enzyme inhibitors. Enzyme inhibitors are substances that will bind to the enzyme in such a way as to decrease the ability of the enzyme to bind the substrate. When this occurs, the enzyme is rendered inactive and the *effective concentration* of the enzyme is decreased. In some cases a product of the enzyme-catalyzed reaction behaves as an inhibitor. If the concentration of the reaction product is high, a substantial amount of the enzyme is bound to the product. This decreases the amount of enzyme available to bind the substrate and thereby to catalyze the reaction. On the other hand, if the product concentration is low, very little of the enzyme is rendered inactive. In other cases, hormones control the active enzyme concentration. Some hormones behave as enzyme inhibitors, while others stimulate enzyme production.

POISON AND DRUG ACTION

Many poisons exert their influence by behaving as enzyme inhibitors. Since enzymes are present in the body in exceedingly small amounts, it becomes clear why small doses of certain poisons can cause death. Hydrogen cyanide ($H{-}C{\equiv}N$) and cyanide ion (CN^-) in small concentrations can cause death because cyanide has the ability to form coordinate covalent bonds with metal ions. If the metal ion is an enzyme activator, coordination of the metal ion by cyanide renders the enzyme inactive.

Copper ions (Cu^{2+}) function as the activator for the enzyme cyctochrome oxidase necessary to catalyze the reactions of the cyctochrome chain (Chapter 3). If the reactions of the cyctochrome chain do not take place rapidly enough, energy production of the cell is so diminished that the cell dies.

Phosphonate ester–based nerve gases and insectides behave by inactivating the enzyme acetylcholine esterase. Nerve impulses are transmitted from one nerve cell to another by acetylcholine. Upon the arrival of a nerve impulse at the end of a nerve cell, acetylcholine is released. Acetylcholine moves across the cell junction transmitting the signal to the second nerve cell. Once the signal has been transferred, the acetylcholine must be removed from the nerve cell junction before further transmission across the junction can occur. The situation is somewhat analogous to making several telephone calls. After one telephone call is completed, it is necessary to break the circuit by hanging up the phone before the next call can be made. In the case of nerve cells the circuit is broken by destruction of acetylcholine.

Acetylcholine is destroyed by the hydrolysis reaction

$$\underset{\text{acetylcholine}}{(CH_3)_3\text{—}\overset{+}{N}\text{—}CH_2\text{—}CH_2\text{—}O\text{—}\overset{\overset{\displaystyle O}{\|}}{C}\text{—}CH_3} + H_2O \longrightarrow$$

$$\underset{\text{choline}}{(CH_3)_3\text{—}\overset{+}{N}\text{—}CH_2\text{—}CH_2OH} + \underset{\text{acetic acid}}{CH_3CO_2H}$$

This reaction is catalyzed by the enzyme acetylcholine esterase. Acetylcholine esterase has a turnover number of 300,000, thus ordinarily allowing a very rapid transmission of signals across the nerve cell junction. Certain phosphonate esters can bond to acetylcholine esterase. After *tricking* the enzyme into accepting them as if they were acetylcholine, no reaction occurs and the enzyme is inactivated. Inactivation of the enzyme allows acetylcholine to build up in the nerve junctions causing cessation of nerve transmission. The resultant loss of communication between the brain and other organs of the body quickly leads to unconsciousness and death. As little as one drop (approximately 0.05 ml) of some phosphonates is sufficient to kill an adult human. Consequently, extreme care must be used in handling insecticides such as parathion. Compounding the hazards of handling parathion is the fact that it can be absorbed directly through skin. The structural formula of parathion is shown in Fig. 6-10.

Parathion is preferable to chlorinated hydrocarbons such as DDT because parathion is decomposed fairly rapidly by water and it does not accumulate in the environment.

Biochemists, in conjunction with microbiologists, employed by the pharmaceutical industry search for differences in the metabolic pathways between man and bacteria. Any difference in these metabolic paths holds

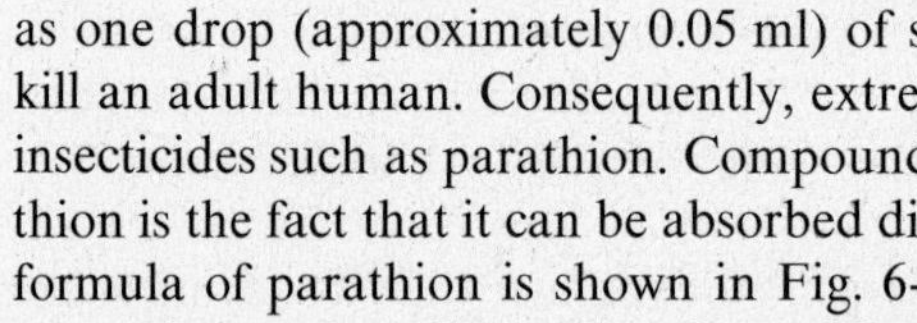

$$\begin{array}{l} CH_3\text{—}CH_2\text{—}O\text{—}\overset{\overset{\displaystyle S}{\|}}{\underset{|}{P}}\text{—}O\text{—}C_6H_4\text{—}NO_2 \\ \qquad\quad CH_3\text{—}CH_2\text{—}O \end{array}$$

FIGURE 6-10 Parathion—a phosphonate insecticide.

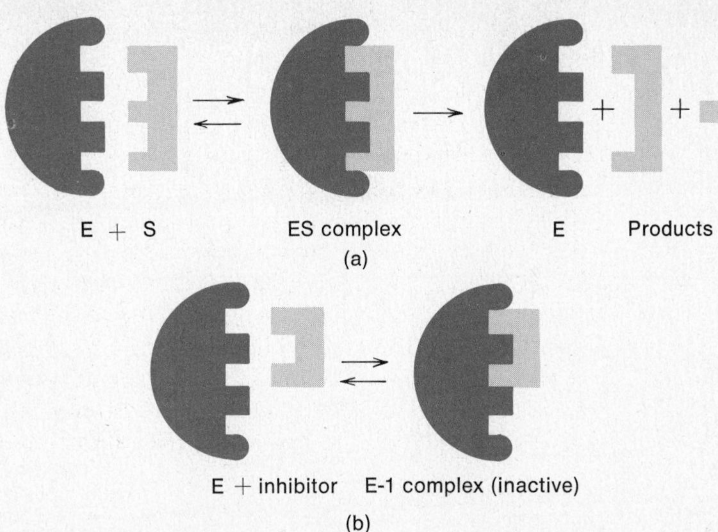

FIGURE 6-11 (a) Action of an enzyme on its substrate. (b) Deactivation of an enzyme by an inhibitor having a shape similar to that of the substrate.

the potential for discovery of a substance (a drug) that will inhibit a vital bacterical enzyme but not affect any human enzymes. Such a search is rarely completely successful and instead we settle for compounds that are more effective at inhibiting bacterial enzymes than at inhibiting human enzymes (see Fig. 6-11). This means that the administering of drugs is always accompanied by some danger to the patient. Thus, the physician is constantly put in the position of trying to prescribe the drug and drug dosage that will be most effective against a bacterial infection and with the least risk to the patient.

Two examples of drugs whose actions are relatively well understood are penicillin and the sulfa drugs. Penicillin inhibits the enzymes involved in cell wall construction of certain bacteria. Although animals do not construct cell walls,* the fact that some people exhibit allergies to penicillin tells us that penicillin is capable of interfering with human metabolism.

The effectiveness of sulfa drugs is based on the fact that folic acid is a vital coenzyme for both bacteria and humans. The difference arises in that folic acid is a vitamin for humans and must be included in the human diet but bacteria manufacture their own folic acid. The structure of folic acid is shown in Fig. 6-12(a).

Bacteria manufacture folic acid by combining pteridine, *para*-aminobenzoic acid, and glutamic acid. This reaction, like all biological reactions, is catalyzed by an enzyme. The bacterial enzyme involved in folic acid synthesis exhibits reaction specificity rather than absolute specificity.

*A cell wall is a fairly rigid entity surrounding bacteria and plant cells. Animal cells are surrounded by a more flexible cell membrane.

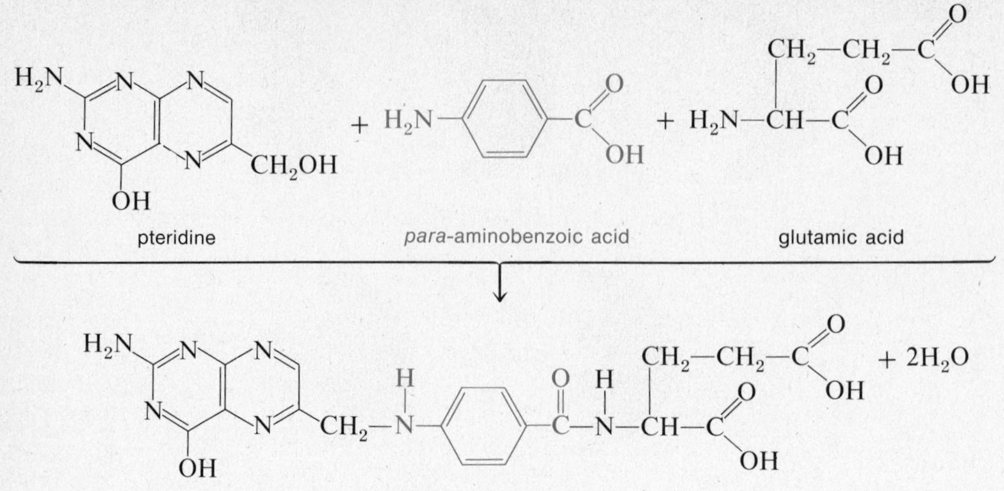

pteridine + *para*-aminobenzoic acid + glutamic acid → folic acid + $2H_2O$

(a)

pteridine + sulfanilamide + glutamic acid → altered folic acid + NH_3 + H_2O

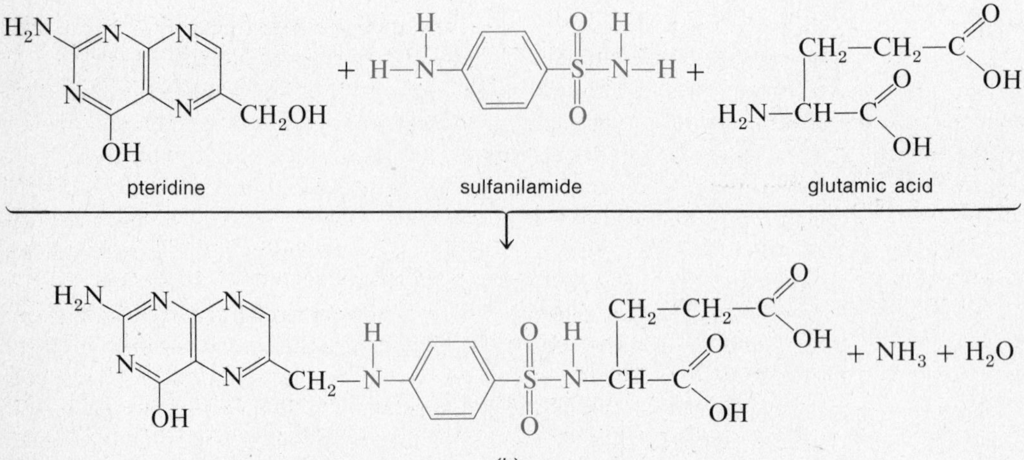

(b)

FIGURE 6-12 (a) Folic acid. (b) Altered folic acid.

Apparently the structure of sulfa drugs resembles the structure of *para*-aminobenzoic acid closely enough to trick the enzyme into using the sulfa drug in the place of *para*-aminobenzoic acid. Incorporation of the sulfa drug leads to the formation of the altered folic acid molecule, shown in Fig. 6-12(b). Unfortunately for the bacteria, the altered folic acid will not serve as an effective coenzyme and ultimately leads to death of the bacteria. Since the patient continues to receive folic acid via his diet, the patient's metabolic processes continue unhindered.

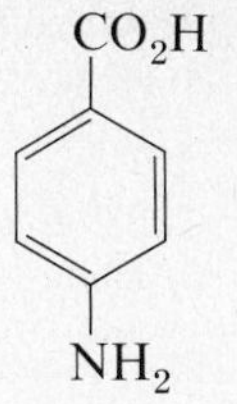

CO_2H / NH_2
para-aminobenzoic acid

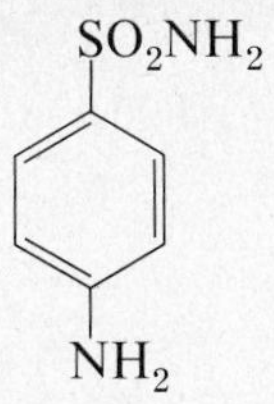

SO_2NH_2 / NH_2
sulfanilamide (a sulfa drug)

COUPLED REACTIONS

Metabolic reactions can be divided into two classes: *anabolism,* the synthesis of cell constituents; and *catabolism,* the degradation of cell constituents. In general, catabolic reactions are exergonic and anabolic reactions are endergonic. The net energy required for endergonic reactions to occur is supplied by the net energy evolved during exergonic reactions. This process requires that the reactions be coupled to one another. The process is not completely efficient and some of the energy from exergonic reactions is always lost as heat.

In order for two reactions to be successfully coupled, the net energy evolved by an exergonic reaction must be greater than the net energy required for the endergonic reaction. Since the exergonic reaction supplies the energy required for the endergonic one, it follows that the reaction rate for the endergonic reaction is dependent on the rate of the exergonic reaction.

ANOTHER APPLICATION OF KINETIC MOLECULAR THEORY

When water is left undisturbed in an open container, it disappears and we say the water has evaporated. Similarly, if a bottle containing isopropyl alcohol (rubbing alcohol) or ether is left uncapped at room temperature, the liquid soon evaporates. We can observe that diethyl ether evaporates more quickly than isopropyl alcohol, which in turn evaporates more quickly than water at the same temperature.

Kinetic molecular theory states that the average kinetic energy of a collection of molecules is proportional to the temperature. Some molecules may be moving very rapidly, a few very slowly, and the majority at intermediate speeds. Recall that attractive forces exist between the molecules of a liquid. In order for a molecule to escape from its companions in the liquid state, it must overcome these attractive forces. Only molecules having high kinetic energies possess sufficient energy to overcome these attractive forces and escape from the liquid surface. Upon escaping, these molecules enter the gaseous state and we say the liquid is *evaporating*. If the most energetic molecules escape from the liquid, the *average kinetic energy* of the remaining liquid decreases. Consequently, the temperature of the liquid remaining also decreases. Thus, evaporation is a cooling process.

exercise 6-6

Given the fact that the evaporation rate of diethyl ether is greater than that of isopropyl alcohol, which is greater than that of water, what can be said about the

relative strength of the attractive forces in these liquids? What will happen to the rate of evaporation as the temperature is increased?

exercise 6-7

When a child runs a fever of 102°F or above, a common technique to lower the child's body temperature is to administer an alcohol rub or sponge bath. Explain the molecular basis for this action.

CHAPTER SUMMARY

1. An energy barrier exists for all reactions. The need to break bonds in the reactants is partially responsible for the barrier. The activation energy for a reaction is the amount of energy required to overcome the reaction barrier.
2. The rate of reaction is dependent on the activation energy for the reaction.
3. The average kinetic energy of a collection of molecules is directly proportional to temperature. An increase in temperature increases reaction rates because the average kinetic energy of the molecules is increased.
4. Reaction rates are increased by an increase in concentration of the reactants.
5. Catalysts are substances that alter reaction rates but can be recovered from the reaction mixture unchanged. Biological catalysts are called *enzymes*. Catalysts increase reaction rates by lowering the activation energy of the reaction. Compounds which decrease reaction rates are commonly called *inhibitors*.
6. Enzymes usually consist of a protein portion called the *apoenzyme* and a smaller nonprotein portion called a *cofactor*. Metal ions and vitamins are common cofactors.
7. Enzymes have a small region called the *active site* that actually catalyzes the reaction.
8. Enzymes exhibit specificity for a particular compound or reaction or type of functional group.
9. The activity of an enzyme is measured by the turnover number. The rate of enzymatic reactions is dependent on both substrate and enzyme concentration.
10. Many poisons and drugs exert their effect by behaving as enzyme inhibitors.
11. Kinetic molecular theory is useful in explaining evaporation rates.

QUESTIONS

1. The conversion of glucose to carbon dioxide and water might be represented by the following reaction coordinate diagram:

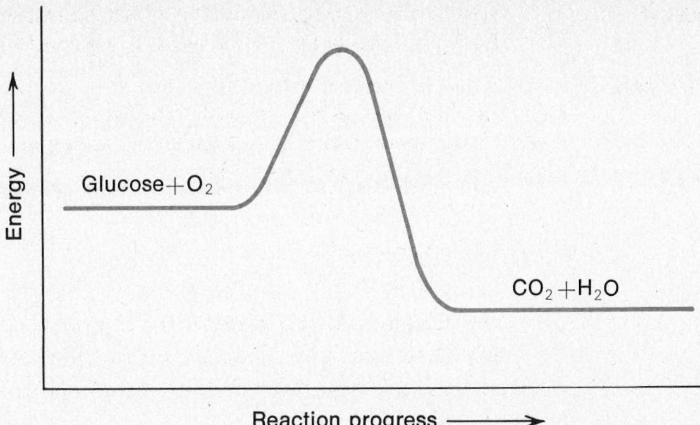

(a) Is the reaction exergonic or endergonic? Explain.
(b) What effect would a decrease in temperature have on the rate of this reaction?
(c) What effect would the addition of a catalyst have on the rate of this reaction? Explain how the catalyst causes the effect.
(d) What effect would increasing the concentration of oxygen have on the reaction rate?

2. Given below are values of the activation energy for three hypothetical reactions. Decide which reaction will occur most rapidly at 25°C. Explain why you choose your answer.

Reaction	**E_a (kcal)**
A	7
B	10
C	13

3. Suppose the reaction $2NO + O_2 \longrightarrow 2NO_2$ is a single-step reaction.
(a) Write the rate law equation for this reaction.
(b) What effect would doubling the oxygen concentration have on the reaction rate?
(c) What effect would doubling the NO concentration have on the reaction rate? Halving the NO concentration?

4. The activation energy for the reaction $2H_2O_2 \longrightarrow 2H_2O + O_2$ is found to be 18 kcal. The addition of finely divided iron particles decreases the activation energy to 13 kcal; platinum decreases it to 12 kcal; catalase decreases it to 5 kcal. Under which of the four possible conditions would the reaction proceed at the greatest rate?

5. A solution of a particular enzyme is brown. Experimenters found that when a colorless solution of the substrate was added to the enzyme

solution, the mixture turned green permanently. Offer an explanation of these observations.

6. The compound ethylenediaminetetracetate reacts with metal ions by the formation of coordinate covalent bonds. This compound is extremely effective in removing metal ions and decreasing metal ion concentration in solution. Explain why ethylenediaminetetracetate inactivates many enzymes.
7. Explain the following observations:
   (a) Warm water evaporates faster than cold water.
   (b) Evaporation of water from the skin cools it.
   (c) On a hot, humid summer day, water condenses on the outside of a glass containing a cold beverage.

7
introduction to equilibrium

INTRODUCTION

The previous two chapters have dealt with solution-making processes and the rates at which chemical reactions occur. The nature of solute/solvent mixtures will be examined in this chapter. We shall also seek to gain an understanding of equilibrium by initially considering a solution's behavior with respect to rates of dissolving and crystallization. The equilibrium concept will be extended to processes such as osmosis and chemical reactions. Eventually, a relationship will be formulated allowing us to describe quantitatively a chemical system at equilibrium.

TYPES OF SOLUTE/SOLVENT MIXTURES: TRUE SOLUTIONS, SUSPENSIONS, AND COLLOIDAL DISPERSIONS

Glucose, urea, and ethanol are examples of molecules small enough to form true solutions.

In Chapter 5 we discussed solutions as uniform solute/solvent mixtures and have assumed that the solute would remain uniformly dispersed throughout the solvent. A *true solution* may be described as one in which particles (ions and/or molecules) with diameters of 0.5 to 2.5 Å are dispersed in a solvent. These solutions are homogeneous and transparent to light. The component particles are too small to be removed by filtration or to be viewed with electron microscopes. To appreciate the size of these particles, consider that approximately 100 million particles, each 2.5 Å in diameter, would form a row only *1 in.* long when placed side by side.

Opaque or translucent dispersions of solute particles in a solvent are called *suspensions*. Particles in a suspension have diameters greater than 1000 Å, are filterable, and have masses great enough to settle under the force of gravity.

Muddy water is a typical example of a suspension. We shall not be concerned with properties of suspensions in future discussions.

Is it possible to have a dispersion with particles of average diameters less than 1000 Å but larger than those in a true solution? Yes, it's possible and particles in this size range are called colloids (Greek, *glue-like*). Their dispersions in a solvent are called *colloidal dispersions*. Some common materials present in colloidal form are milk, fruit jellies, gelatin desserts, whipped cream, clouds, and many components of body fluids, including protoplasm. Despite containing larger-sized particles than those in true solutions, colloidal dispersions do **not** settle out upon standing.

SOLUBILITY AND EQUILIBRIUM

A reversible process is one that can occur in either direction.

Consider the following set of events. It's breakfast time and a quick cup of coffee helps get you started. You add a spoonful, or more, of sugar to the cup of coffee, absentmindedly stirring it. The sweetened taste of the coffee assures you that the sugar dissolved. It's almost classtime and you rush off before finishing your coffee. Later you return and realize you would like another cup of coffee. Your only cup has the remains of your previous coffee in it and you quickly make two observations regarding the cup's contents: (1) Some of the coffee (solvent) has evaporated and (2) tiny sugar crystals form a ring inside the cup where the coffee level was before evaporation occurred. The initial dissolving of the sugar represents one process; the crystallization of sugar from the saturated solution is the reverse process. This is evidence that a process may be reversible. Chemists have long accepted the concept that reversible processes occur constantly in a saturated solution. Experimental data show that a solution is the net result of two processes: (1) the rate of solute dissolving in a solvent and (2) the rate of solute coming out of solution (crystallization).

Before any solute is added, only pure solvent is present. When a small amount of solute is added, the rate of dissolving is far greater than rate of crystallization. As more solute dissolves in the fixed amount of solvent, the difference between the rates of dissolving and crystallization continuously decreases. Eventually, the amount of dissolved solute is such that the rate of solute going into solution (rate 1) equals the rate of solute coming out of solution (rate 2). When this occurs, a saturated solution exists and under these conditions the solute is said to be at equilibrium with the solution.

A system at equilibrium is one in which the rate of a process in one direction is equal to the rate of a process in the reverse direction; i.e.,

***rate** of forward reaction = **rate** of reverse reaction*

The symbolism $\rightleftharpoons$ is often used to denote equilibrium. For example, the equation representing the hypothetical reaction $A + B \rightleftharpoons Z + Y$ is interpreted to mean that **at equilibrium** (*when equilibrium is reached*) the rate of conversion of A and B to Z and Y (forward reaction) is the same as the rate of conversion of Z and Y to A and B (reverse reaction). Both reactions are occurring but there is no **net** change in the system since the rates of these opposing reactions are equal.

OSMOSIS

The phenomenon of osmosis, or at least the term, is probably familiar to you from a biology course. Osmosis is a mechanism by which water flows in an attempt to equalize solution concentrations between a cell and its environment. Let's begin our discussion with a definition: Osmosis is the flow of water through an osmotic membrane from a region of pure water or a region of lower solute concentration (greater water concentration) to a region of higher solute concentration (lesser water concentration). An osmotic membrane is one that allows the passage of water molecules but not solute particles.

Consider a U-tube separated into two compartments C and S by an osmotic membrane, o.m. (see Fig. 7.1). Initially, compartment C has a higher solute concentration than compartment S. Osmosis will occur and *water* will flow through the osmotic membrane from the region of lower solute concentration (compartment S) to the region of higher solute concentration (compartment C). The flow continues until the water level has risen to a height in tube C great enough so that its weight exerts a sufficient downward pressure on the membrane so that no **net** transfer of water occurs across the membrane. This downward pressure is called the *osmotic pressure* of the solution. Theoretically the osmotic pressure is controlled solely by the concentration of solute particles present. As the solute concentration changes by water transfer, a point is reached where the rate of water flow in one direction equals the rate of flow in the reverse direction and equilibrium

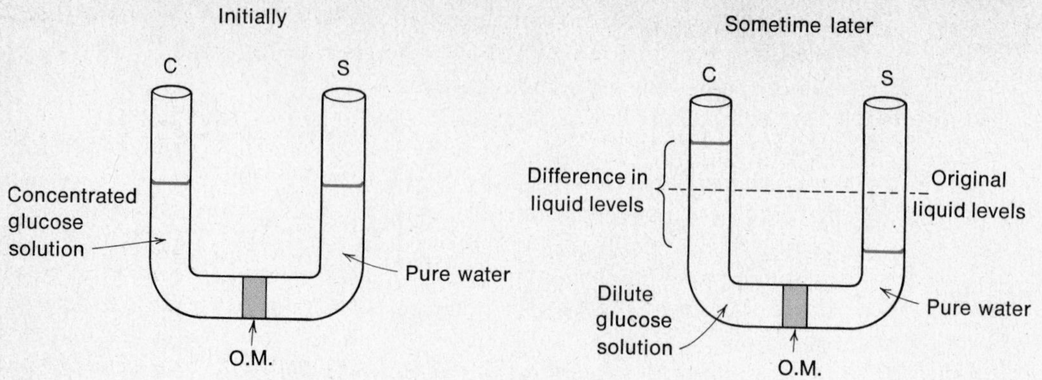

FIGURE 7-1 Solution concentrations and osmosis.

is established. At that point, because the rates of opposing water passage are equal, there is no **net** transfer of water across the membrane.

The osmotic pressure of two solutions, A and B, with respect to each other may be described by one of the following conditions:

Condition 1:

$$\text{osmotic pressure}_A = \text{osmotic pressure}_B$$

Solutions are *isotonic.*

Condition 2:

$$\text{osmotic pressure}_A > \text{osmotic pressure}_B$$

Solution A is *hypertonic* to solution B.

Condition 3:

$$\text{osmotic pressure}_A < \text{osmotic pressure}_B$$

Solution A is *hypotonic* to solution B.

exercise 7-1

What would be an alternative way of describing the relationships between solutions A and B in Condition 2? Condition 3?

Normal saline solution is a 0.9 percent sodium chloride solution, which is isotonic with red blood cells. Water is hypotonic to normal saline and red blood cells placed into pure water swell and rupture, an event called *hemolysis.* Osmosis will occur with a flow of water from the region of lower solute concentration (pure water) across the red blood cell membrane into the region of higher solute concentration (red blood cell). This massive influx of water distends and then ruptures the cell membrane.

exercise 7-2

Predict and explain what will occur if red blood cells are placed into a 9.0 percent solution of sodium chloride.

exercise 7-3

Why is it essential that the concentration of normal saline solution used during intravenous maintenance of body fluid ion levels not vary drastically from 0.9 percent?

OSMOLARITY

A 1.0 *M* aqueous glucose solution is experimentally determined to have a certain osmotic pressure. Another experiment shows the osmotic pressure of a 1.0 *M* sodium chloride solution to be *twice* that of the glucose solution. How can this be if the solutions are both 1.0 *M*?

Electrolytes exist as ions in solution; nonelectrolytes are undissociated in solution.

The answer lies in the behavior of each solute upon dissolving. Two possible behaviors are (1) a solute dissolves to yield a solution that does **not** conduct an electric current (such solutes are *undissociated in solution* and are termed *nonelectrolytes*) and (2) a solute dissolves giving a solution that **does** conduct an electric current (such solutes *dissociate* to produce ions *in solution* and are termed *electrolytes*).

The seeming anomaly of differences in osmotic pressure between glucose and sodium chloride solutions of equal concentration can be explained on this basis. In solution, glucose molecules remain undissociated. When solid sodium chloride is placed into water, it dissociates into sodium ions and chloride ions. Due to solute dissociation, a 1.0 *M* sodium chloride solution contains 1.0 mole of sodium ions and 1.0 mole of chloride ions per liter of solution; 1.0 *M* glucose contains only 1.0 mole of undissociated glucose molecule per liter of solution—only half the number of particles than are present in the sodium chloride solution.

From observations such as this, chemists conclude that osmotic pressure depends on the concentration of solute particles. Therefore, the osmotic pressure of a solution depends on whether the solute is an electrolyte or a nonelectrolyte. The osmotic pressure of 1.0 *M* sodium chloride is twice that of 1.0 *M* glucose since 1.0 *M* sodium chloride contains twice as many particles as an equal volume of 1.0 *M* glucose.

Table 7-1 lists several examples of electrolytes and nonelectrolytes and their behavior as solutes.

exercise 7-4

What is the number of moles of particles in 1 liter of a 1.0 *M* solution of *each* of the following solutes? (a) potassium chloride, (b) ribose, a nonelectrolyte; (c) calcium nitrate; (d) ammonium chloride.

Water balance between cells and the interstitial fluid in which the cells are bathed depends on the osmotic pressure of the cellular and interstitial

TABLE 7-1
Solute dissociation behavior

SOLUTE (1 mole)	BEHAVIOR IN AQUEOUS (aq) SOLUTION	MOLES OF PARTICLES IN SOLUTION PER MOLE OF SOLUTE
Sodium chloride (NaCl)	$NaCl \xrightarrow{H_2O} Na^+ (aq) + Cl^- (aq)$	2: 1 Na^+ and 1 Cl^-
Sodium carbonate (Na_2CO_3)	$Na_2CO_3 \xrightarrow{H_2O} 2Na^+ (aq) + CO_3^{2-} (aq)$	3: 2 Na^+ and 1 CO_3^{2-}
Glucose (dextrose)($C_6H_{12}O_6$)	$C_6H_{12}O_6 \xrightarrow{H_2O} C_6H_{12}O_6 (aq)$	1: undissociated
Sodium phosphate (Na_3PO_4)	$Na_3PO_4 \xrightarrow{H_2O} 3Na^+ (aq) + PO_4^{3-} (aq)$	4: 3 Na^+ and 1 PO_4^{3-}
Sucrose ($C_{12}H_{22}O_{11}$)	$C_{12}H_{22}O_{11} \xrightarrow{H_2O} C_{12}H_{22}O_{11} (aq)$	1: undissociated
Glycerol ($C_3H_8O_3$)	$C_3H_8O_3 \xrightarrow{H_2O} C_3H_8O_3 (aq)$	1: undissociated
Potassium nitrate (KNO_3)	$KNO_3 \xrightarrow{H_2O} K^+ (aq) + NO_3^- (aq)$	2: 1 K^+ and 1 NO_3^-

Osmolarity is a measure of the number of particles in a solution.

fluids. Therefore, this balance is related to solute particle concentration. Physiologists often use **osmolarity** in discussing osmotic behavior of various solutes.

An operational definition of osmolarity (osmol) is

$$\text{osmolarity} = \text{molarity} \times (\text{moles of particles/mole solute})$$

Table 7-2 shows how this relationship can be applied to individual solutions of several solutes:

TABLE 7-2
Osmolarity of four solutions

SOLUTION	DISSOCIATION BEHAVIOR	OSMOLARITY, (osmol) = (molarity) × (moles of particle/mole of solute)
1.0 *M* sodium chloride	$NaCl \xrightarrow{H_2O} Na^+ (aq) + Cl^- (aq)$	$1.0\ M \times 2 = 2.0$ osmol
1.0 *M* glucose	$C_6H_{12}O_6 \xrightarrow{H_2O} C_6H_{12}O_6 (aq)$	$1.0\ M \times 1 = 1.0$ osmol
2.0 *M* sucrose	$C_{12}H_{22}O_{11} \xrightarrow{H_2O} C_{12}H_{22}O_{11} (aq)$	$2.0\ M \times 1 = 2.0$ osmol
1.0 *M* sodium phosphate	$Na_3PO_4 \xrightarrow{H_2O} 3Na^+ (aq) + PO_4^{3-} (aq)$	$1.0\ M \times 4 = 4.0$ osmol

The greater the osmolarity of a solution, the greater the osmotic pressure the solution exhibits.

Thus, for solutions of nonelectrolytes, osmolarity equals molarity. The osmolarity of solutions of electrolytes equals the molarity multiplied by the moles of particles of solute dissociating in solution. The term *milliosmol,* which equals $\frac{1}{1000}$ osmol, is also commonly used.

exercise 7-5

Using Tables 7-1 and 7-2, calculate the osmolarity of the following solutions: (a) 1.0 *M* potassium nitrate, (b) 1.5 *M* glucose, (c) 0.05 *M* sodium carbonate, (d) 2.0 *M* glycerol, and (e) 0.05 *M* sodium phosphate.

exercise 7-6

The osmolarity of a potassium chloride solution is 4.0 osmol. What would be the molarity of each of the following solutions so that the resulting solution would be equal to the osmolarity of the potassium chloride solution? (a) sucrose, (b) sodium chloride, (c) sodium carbonate

DIALYSIS

What are the differences and similarities between osmotic membranes and dialyzing membranes?

An osmotic membrane is permeable to water molecules but not solute particles. A cell restricted to operating under these conditions could neither take in nutrients nor liberate waste materials. Living systems must obviously be capable of more than just the transfer of water molecules. Membranes that are permeable to flow in *either* direction of true solution-sized particles, including solvent molecules, are called *dialyzing membranes*. When functioning properly, dialyzing membranes are *not* permeable to particles of colloid or suspension size such as starch or high molecular weight proteins.

If an aqueous mixture of colloidal, suspension, and true solution-sized particles is placed on one side of a dialyzing membrane and pure water on the other side, the true solution-sized particles will pass through the dialyzing membrane into the pure water. This process is called *dialysis*. Schematically, this is shown in Fig. 7-2.

A practical use of this phenomenon is the purification of proteins and vaccines. The impure material is placed into a dialysis bag in a setup similar to that in Fig. 7-2. The dialyzable impurities pass through the bag leaving the purified protein or vaccine inside the bag. Milk used in low salt (sodium) diets can also be obtained by this method.

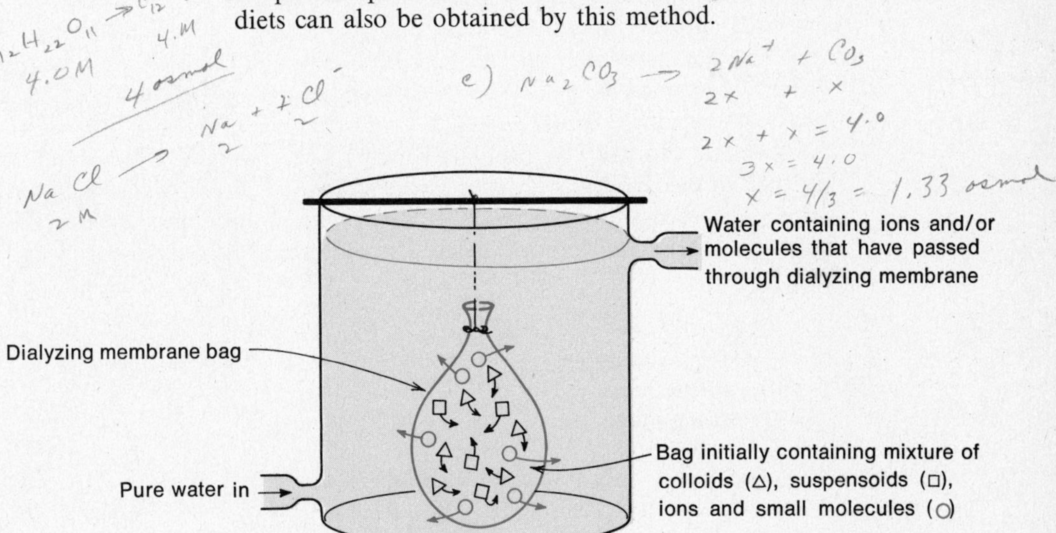

FIGURE 7-2 Dialysis apparatus.

KIDNEY FUNCTION AND DIALYSIS

Physiologically, nearly all body membranes function as dialyzing membranes. The most dramatic physiological example of dialysis is the maintainence of the blood's solute and electrolyte balance through kidney functions. The following data may give you an appreciation of the magnitude of the task handled routinely by the kidneys during normal *daily* operation. The total body water of a 70 kg adult (about 150 lb) is approximately 40 liters (57 percent of the total body mass). Normal fluid intake including fluid synthesized by the body is about *2400 ml/day*. Of this amount, an average *1500 ml* (63 percent) is lost in the urine; perspiration, feces, diffusion through the skin, and evaporation through the lungs account for the remainder.

In generating the 1.5-liter average daily volume of urine, the kidneys filter blood at a rate averaging 125 ml/min, equivalent to about 180 liters daily. The blood is brought to the kidneys by the large renal arteries and after filtration it leaves via the large renal veins. Over 99 percent of the fluid is reabsorbed by kidney action. Less than 1 percent of the fluid is not reabsorbed and is eliminated as urine—an aqueous solution of dissolved electrolytes and organic materials. Thus the kidneys serve *both* to control the concentration of body fluid constituents and to provide a means for excretion of metabolic waste products.

Figures 7-3 and 7-4 show the main structural features of the kidneys. To meet the necessarily high filtration rate efficiently, an extensive system of nephrons provides the large surface area required for filtration. Filtration occurs at each nephron according to the ability of a substance to pass through the membrane. Blood plasma proteins, red and white blood cells, and fats normally do not pass through the dialyzing glomerulus membranes. They are returned to the bloodstream through the pertibular capillaries.

Why don't proteins, blood cells, and fats normally pass through the glomerulus membranes?

Plasma containing dissolved electrolyte ions such as sodium, potassium, and chloride ions and small organic molecules such as glucose, creatinine, and urea passes through the dialyzing membranes into the Bowman's capsules. The glomerular filtrate then passes through the proximal tubules where reabsorption of most of the water, glucose, and most of the electrolytes occurs as the tubule winds among a dense network of fine capillaries. By reabsorption through the capillaries, the reabsorbed materials are thus returned to the circulatory system. Creatinine, urea, and uric acid tend not to be reabsorbed and these materials, along with nonreabsorbed waste and electrolytes, eventually become urine. The continuing filtrate passes through

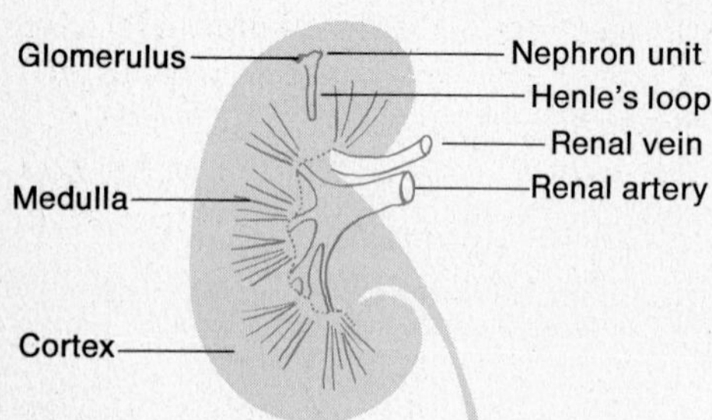

FIGURE 7-3 Section through a kidney.

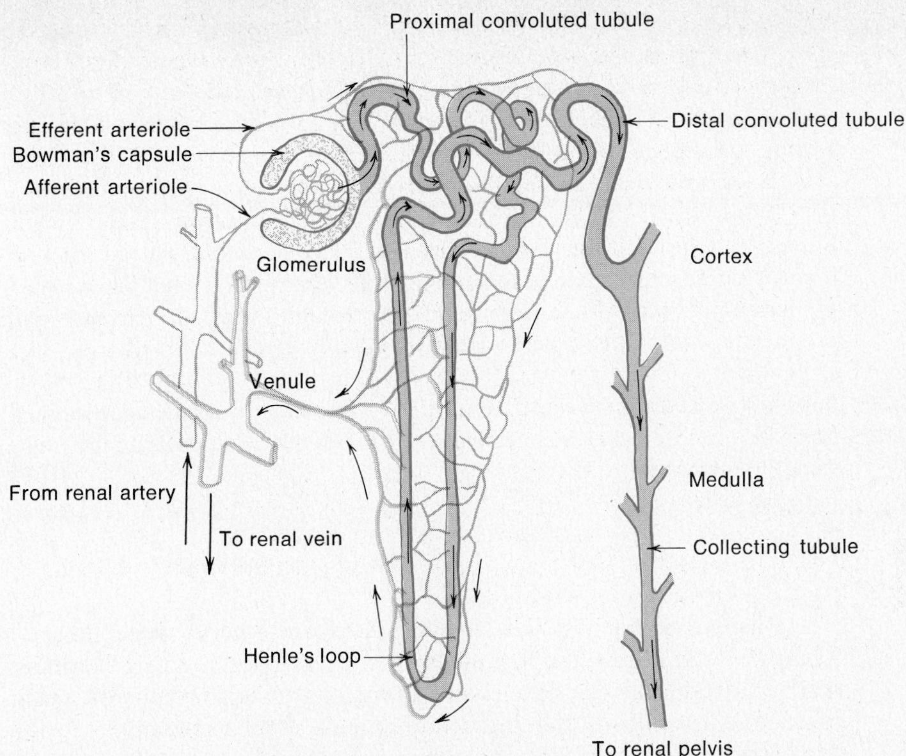

FIGURE 7-4 A kidney tubule and its blood supply.

Normal urine components are materials the body does not need at that time.

the loop of Henle, then through the distal tubules, and finally into the collecting tubules that empty into the pelvis of the kidneys. The net result is formation of urine by the action of the tubules separating the desired components of the tubular fluid from the undesired components in order to maintain proper electrolyte levels.

REACTION REVERSIBILITY

The solution-making process, osmosis, and dialysis are all examples of a dynamic situation. Each involves a reaction in one direction in competition with a reverse reaction. In such cases, the forward *and* reverse reactions occur *simultaneously*. There are some chemical reactions, however, that proceed only in the forward direction with reactants being converted to products. Such reactions are called *irreversible reactions*. In effect, the products are not able to chemically *regroup* and to be reconverted to reactants by the reverse reaction. Because of this irreversibility, it is not possible for an irreversible reaction to establish an equilibrium between reactants and products. The thermal decomposition of common table sugar (sucrose is another name) exemplifies an irreversible reaction:

$$C_{12}H_{22}O_{11}\,(s) \xrightarrow{\triangle\triangle} 12C\,(s) + 11H_2O\,(g)$$

△ signifies heat
△△ signifies even more heat

No conditions are known under which elemental carbon can directly react with water to form sucrose. The reaction above cannot be an equilibrium reaction. (If it could equilibrate, rain could convert piles of coal to sugar and change Appalachia to Candyland.) Baking a cake, frying meat, and boiling an egg are examples of irreversible reactions.

Reversible reactions are those in which the reverse reaction can compete significantly with the forward reaction. Equilibrium is reached when the *rates* of forward and reverse reactions become equal. At equilibrium there is no *apparent* change in the reaction mixture as no **net** loss or gain of material occurs. At equilibrium the concentrations of reactants and products do **not** have to be equal; they usually are not.

CHEMICAL EQUILIBRIUM

Let's first look at the establishment of an equilibrium by using a generalized reversible reaction in which substance *R* is converted chemically into substance P according to the equation

$$R \rightleftharpoons P \qquad \begin{array}{l}\text{R represents reactant}\\ \text{P represents product}\end{array}$$

Before reaction, no product P has formed and its concentration is zero. As the reaction begins, the rate of the forward reaction is greater than the reverse reaction rate since the concentration of R is large compared to the concentration of P. As the reaction continues, R is converted to P, R's concentration decreases, and the rate of the *forward* reaction slows (recall that a change in concentration affects reaction rates). As P is produced, its concentration increases and the reverse reaction's rate increases. When the forward and reverse *rates* become equal, equilibrium is established and no net change in the concentrations of R and P occurs as long as neither R nor P is lost from the reaction mixture. This is shown in Fig. 7-5.

At equilibrium, how do the concentrations of R and P compare? We find there are only three possible situations for all equilibria that can be represented by $R \rightleftharpoons P$.

Although in situations 2 and 3, the concentrations of the products and reactants are unequal, the rates of the forward and reverse reactions are equal.

SITUATION 1. The concentration of P is approximately equal to the concentration of R at equilibrium: $[P] = [R]$.

SITUATION 2. At equilibrium, the concentration of P is *greater than* the concentration of R: $[P] > [R]$. This situation is described by saying the equilibrium favors product formation, i.e., the forward reaction.

SITUATION 3. The concentration of P is *less than* the concentration of R: $[P] < [R]$ at equilibrium. In such a condition, the equilibrium favors the reactants, i.e., the reverse reaction.

In most chemical reactions that reach equilibrium, situation 2 or situation 3 apply: $[P] \neq [R]$ at equilibrium. Either product formation (forward reaction) or reactants are favored depending on reaction conditions.

Many physiologically and biochemically significant processes involve

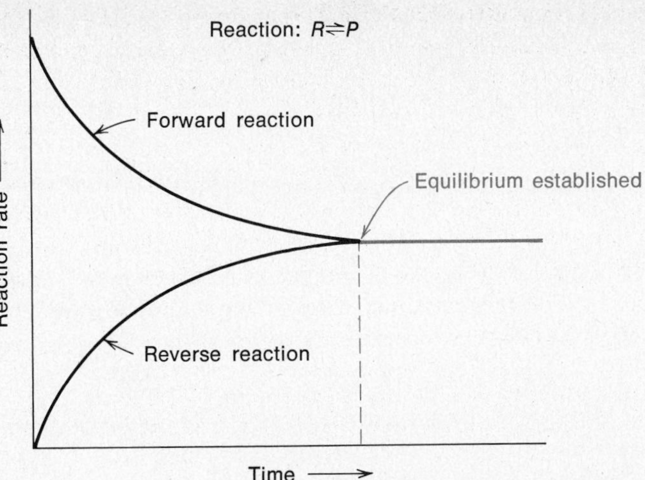

FIGURE 7-5 Relationship of reaction rates for a reversible reaction and establishment of equilibrium.

reversible reactions that may establish equilibrium. In the blood, the normal range of glucose concentration is 65 to 100 mg of glucose per 100 ml of blood. This level is adjusted in several ways, three of which are by (1) dietary intake (eating), (2) exercise, and (3) the use of stored glucose called *glycogen.* Under normal conditions, dietary intake provides sufficient glucose to maintain this level and any excess glucose ingested can be chemically converted to glycogen and stored in the muscles and liver. This is to say that an equilibrium exists between glucose formation and glycogen formation:

$$x\text{glucose} \underset{\text{R}}{\overset{\text{F}}{\rightleftharpoons}} x\text{H}_2\text{O} + \text{glycogen}$$

At equilibrium,

rate of forward reaction (glycogen formation) =
rate of reverse reaction (glucose formation)

If the rates of these two opposing reactions are not equal, the system is **not** at equilibrium.

There are two ways in which the glucose/glycogen equilibrium can be established.

METHOD 1: React glycogen with water and, under appropriate conditions, equilibrium will be established with the formation of glucose from glycogen decomposition.

METHOD 2: Start with glucose and, under appropriate conditions equilibrium will be reached with the conversion of glucose into some glycogen and water.

The equilibrium concentrations of glucose and glycogen from method 1 are experimentally determined to be the *same* as those from method 2. *The direction from which equilibrium is approached does not alter* (*determine*) *the concentrations of products and reactants at equilibrium.*

LE CHATELIER'S PRINCIPLE— REVERSIBLE REACTIONS, LIKE PEOPLE, RESPOND TO STRESS

Let's now think about three situations in which the position of the glycogen/glucose equilibrium is altered by placing a chemical demand (stress) on it.

SITUATION 1. *Cause:* Not enough glucose is taken in by eating. *Result:* Insufficient glucose is present for the reverse rate to equal the forward rate. The forward rate is less than the reverse rate. This imbalance causes a net conversion of glycogen to glucose that continues until equilibrium is reestablished.

SITUATION 2. *Cause:* An excess of glucose is ingested by a person. *Result:* The glucose concentration is increased by the addition of glucose. The forward reaction rate increases converting the excess glucose to glycogen.

SITUATION 3. *Cause:* A person participates in strenuous exercise. *Result:* Glucose is metabolized to furnish energy required during exercise. This lowers the concentration of glucose. The forward reaction rate no longer equals the reverse reaction rate and glycogen is converted into glucose until equilibrium is reestablished.

The applied stress comes from outside the chemical system.

Each situation represents a condition in which the status quo of the glycogen/glucose system is changed by the applied stress. The stress causes the system to shift temporarily from an equilibrium to a nonequilibrium condition and the system responds until equilibrium is reestablished. After studying a broad range of chemical equilibria, the French chemist Henri Le Chatelier (1888) pointed out the effect a stress would have on a chemical system at equilibrium: *When a change (stress) is placed on a chemical system which is at equilibrium, the system responds so as to counteract the imposed change and to reestablish equilibrium.*

exercise 7-7

Urea formation is an important method by which the body purges itself of the waste materials ammonia and carbon dioxide via the urine. Urea formation can be represented in a simplified way by the net equation

$$2NH_3 + CO_2 \rightleftharpoons H_2O + \underset{\text{urea}}{H_2N{-}\overset{\overset{\displaystyle O}{\|}}{C}{-}NH_2}$$

Using Le Chatelier's principle, predict what effect the following modifications would have on the equilibrium above: (a) addition of more ammonia; (b) removal of urea; (c) decreasing the concentration of carbon dioxide.

For a reversible reaction to achieve equilibrium, both reactants *and* products must be present. Le Chatelier's principle allows us to predict what effect a disruption will have on this equilibrium. Is it possible to have conditions that prevent a reversible reaction from achieving equilibrium?

Consider carbonated beverages containing water and carbon dioxide gas under pressure in equilibrium with H_2CO_3. By equation, we can represent this as

$$H_2CO_3 \underset{\text{high pressure}}{\rightleftharpoons} H_2O\ (l) + CO_2\ (g)$$

When the bottle cap is removed, a fizzing occurs caused by the escape of carbon dioxide gas from the beverage. The condition that allowed equilibrium to initially be present—the capped bottle—has been removed, allowing the gas to escape. Since CO_2 is lost from the system, the reverse reaction can no longer take place and equilibrium is destroyed. It can be said that the reaction has been driven to completion. *In a reversible reaction involving gaseous reactants and/or products, equilibrium can only be established in a closed container that prevents loss of the gaseous species.*

exercise 7-8

If a flip-top can of carbonated beverage is opened, the beverage "goes flat". Explain.

Let's continue our disruptive activities with another experiment. We have on hand a saturated aqueous solution of table salt (NaCl). The solid and solution are in equilibrium:

$$NaCl(s) + H_2O \rightleftharpoons Na^+(aq) + Cl^-(aq)$$

To this solution we add several drops of a 0.1 *M* silver nitrate solution that contains $Ag^+(aq)$ and $NO_3^-(aq)$ ions. Immediately, an insoluble white solid begins to precipitate from solution. Addition of silver ions causes formation of *insoluble* silver chloride:

$$Ag^+(aq) + Cl^-(aq) \longrightarrow AgCl(s)$$

The stress on the equilibrium is the removal of Cl^- ions from solution caused by the addition of Ag^+ ions.

By chemical combination with silver ions, chloride ions are removed from solution. This disrupts the sodium chloride solubility equilibrium. In an effort to restore equilibrium, some of the solid sodium chloride dissolves. Further addition of silver ions will result in more silver chloride formation, again disrupting the sodium chloride equilibrium. Continued addition of silver ions eventually causes all the solid sodium chloride to dissolve. *In a reaction involving aqueous ions, the formation of an insoluble product drives the reaction to completion.*

Aspirin (acetylsalicylic acid) can be synthesized by the reaction between salicylic acid and acetic acid in alcohol; concentrated sulfuric acid acts as a catalyst in this reaction.

$$\text{salicylic acid} + CH_3{-}C(=O)OH \overset{H_2SO_4}{\rightleftharpoons} \text{aspirin (acetylsalicylic acid)} + H_2O$$

salicylic acid　　acetic acid　　aspirin (acetyl-salicylic acid)

How could we adjust the experimental conditions so that a maximum amount of aspirin is formed, i.e., drive the reaction to completion? One way would be to extract the aspirin as it forms in the reaction mixture. An easier way is to put a drying agent into the mixture to absorb the water as it is formed by the forward reaction.

In the last two examples, *reactions were driven to completion by chemically tying up product materials.* As in the case of CO_2 escape, *the reversibility of reaction is impossible because reagents have become unavailable for further reaction in the system.*

Reactions are driven to completion because a continuously applied stress prevents equilibrium from being restored.

It is also possible to drive a reaction to completion by using an excess of one of the reactants. Excess acetic acid could be used to cause complete conversion of the salicylic acid to aspirin. With respect to the salicylic acid, the reaction has been driven to completion; there is excess unreacted acetic acid.

Recall that a catalyst speeds up a reaction by lowering the activation energy. Since the activation energy is lowered for both the forward and reverse reactions, the presence of the catalyst does not effect the equilibrium position.

In Chapter 3 it was noted that, in the lungs, oxygen combines with hemoglobin (HHb) to form oxyhemoglobin:

$$HHb + O_2 \rightleftharpoons HHbO_2$$

In the lungs, the relatively high concentration of oxygen causes the forward reaction to be favored. The oxyhemoglobin is transported by the blood to the cells. Within cells that require oxygen, the oxygen concentration is low. As oxyhemoglobin is carried in a capillary adjacent to the cell, the low oxygen concentration creates a stress on the equilibrium above and oxygen is liberated. It diffuses across the cell membrane into the cell.

exercise 7-9

Explain what effect the presence of carbon monoxide would have on the oxyhemoglobin/hemoglobin equilibrium.

QUANTITATIVE ASPECTS OF EQUILIBRIUM

We have been qualitatively describing and predicting behavior of equilibria. Now let's extend our understanding by applying these principles in a *quan-*

titative way. As before, the equilibrium R $\rightleftharpoons$ P will serve as our initial example.

Let's put a concept from Chapter 6 to use. The rate of the forward reaction may be expressed $\overrightarrow{\text{rate}} = k_1[\text{R}]$ where k_1 is the rate constant for the forward reaction. The reverse reaction rate is given by $\overleftarrow{\text{rate}} = k_2[\text{P}]$ where k_2 is the rate constant for the reverse reaction. Equilibrium is reached when the forward and reverse rates are equal and thus no **net** change in concentrations occurs. This means that

$$k_1[\text{R}] = k_2[\text{P}] \quad \text{or} \quad \frac{[\text{P}]}{[\text{R}]} = \frac{k_1}{k_2} = K$$

Dividing one constant by a second constant mathematically yields a new constant, K. Since this new constant K expresses a relationship of a chemical system at equilibrium, it is called an *equilibrium constant* and is frequently denoted as K_{eq}. *Notice that the numerical value of* K, *depends on the equilibrium concentrations of P and R.*

For this hypothetical reaction, let us assume $K = 10$. We can write

$$K = \frac{[\text{P}]}{[\text{R}]} = 10$$

What must be the ratio of the equilibrium concentrations of P and R to satisfy this relationship? The value of K tells us that at equilibrium

$$K = \frac{[\text{P}]}{[\text{R}]} = 10 = \frac{10}{1}$$

Suppose that we *started* with *11* moles of R in a liter container and let the reaction proceed to equilibrium. In this case, equilibrium will be established when the reaction mixture contains 10 moles of P and 1 mole of R in the liter container. If initially 22 moles of R are present, equilibrium is established when 20 moles of P and 2 moles of R are present in the reaction mixture.

exercise 7-10

With reference to the equilibrium system above, if the equilibrium concentration of R is 0.5 mole/liter, what is the equilibrium concentration of P?

exercise 7-11

Describe three ratios of concentrations of P and R at equilibrium that satisfy the equilibrium constant value of 10. Use values other than given in the examples above.

exercise 7-12

A sample taken from a reaction mixture of R and P is analyzed. The equilibrium constant for this reaction has been determined to be 10. The ratio of [P]/[R] determined by analysis equals 2. How would you interpret these data?

THE EQUILIBRIUM CONSTANT EXPRESSION

So far we've used a two-component equilibrium system of R and P. How can we write the equilibrium expression for a reaction system involving more than two components? Over 100 years ago, the Norwegian chemists C. M. Guldberg and P. Waage proposed a concept that became a cornerstone for describing chemical equilibrium. Extended from its original form, their idea applies to equilibria that can be symbolically generalized by the *balanced* chemical equation

$$aA + bB + \cdots \rightleftharpoons zZ + yY + \cdots$$

A and B are reactants; Z and Y are products. The notation $\cdots$ implies that neither reactants nor products are necessarily limited to one or two species. The lowercase letters are the coefficients in the *balanced* equation.

You have already seen how the relationships involving R and P were developed. For the generalized equilibria above, the equilibrium constant expression is written

$$K = \frac{[Z]^z \times [Y]^y \times \cdots}{[A]^a \times [B]^b \times \cdots}$$

Points to be noted about the expression above are

1. The equilibrium constant expression is a *ratio relating concentrations of product and reactant species present at equilibrium.*
2. Start by writing the properly balanced chemical equation representing the equilibrium. In this *generalized* case, it is the one given above.
3. *By convention,* the concentration of each product species is placed in the numerator; the concentration of each product species is multiplied by the other product concentration terms; the concentration of each reactant species is placed in the denominator; the concentration of each reactant species is multiplied by the other reactant concentration terms.
4. The coefficients in the chemical equation are exponents for the appropriate concentration term in the ratio.

Let's try our generalized relationship on a specific chemical reaction. Maltose, a carbohydrate, can be converted to the less complex carbohydrate, glucose. The reaction is catalyzed by the enzyme maltase.

First, the unbalanced word equation

$$\text{maltose} + \text{water} \xrightleftharpoons{\text{maltase}} \text{glucose}$$

Next, the unbalanced chemical equation

$$C_{12}H_{22}O_{11} + H_2O \rightleftharpoons C_6H_{12}O_6$$

Finally, the *balanced* chemical equation

$$C_{12}H_{22}O_{11} + H_2O \rightleftharpoons 2C_6H_{12}O_6$$

The equilibrium constant expression for this reaction is

$$K = \frac{[\text{glucose}]^2}{[\text{maltose}] \times [\text{water}]} = \frac{[C_6H_{12}O_6]^2}{[C_{12}H_{22}O_{11}] \times [H_2O]}$$

Notice the coefficient 2 for glucose in the balanced chemical equation appears as an exponent for the glucose concentration term in the equilibrium constant expression.

Consider the conversion of ammonia and carbon dioxide to water and urea ($H_2N\text{—}\overset{\overset{\displaystyle O}{\|}}{C}\text{—}NH_2$), an important process by which the body incorporates two metabolic waste materials into one that can be excreted. Let's try writing an equilibrium constant expression for urea formation (also see Exercise 7-8). First we need the properly *balanced* equation:

$$2NH_3 + CO_2 \rightleftharpoons H_2O + \underset{\text{(urea)}}{H_2N\text{—}\overset{\overset{\displaystyle O}{\|}}{C}\text{—}NH_2}$$

Next, follow instructions given for the generalized case:

$$K = \frac{[H_2O] \times [\text{urea}]}{[NH_3]^2 \times [CO_2]}$$

Note the squaring of the ammonia term. The other species all had coefficients of 1 in the balanced equation and an exponent of 1 is understood for their terms in the K expression.

Now try your skill on two other equilibria given in Exercises 7-13 and 7-14. *Hint:* Be sure the chemical equations are balanced.

exercise 7-13 **Methyl salicylate (oil of wintergreen) is a common component in several commercial analgesics used to alleviate muscle soreness. Its synthesis involves the reaction between salicylic acid and methyl alcohol:**

$$\underset{\text{salicylic acid}}{C_6H_4(OH)\text{—}C(=O)\text{—}OH} + \underset{\text{methyl alcohol (methanol)}}{CH_3\text{—}OH} \rightleftharpoons \underset{\text{methyl salicylate}}{C_6H_4(OH)\text{—}C(=O)\text{—}O\text{—}CH_3} + H_2O$$

Write the equilibrium constant expression for the reaction above.

exercise 7-14

In a sturdy, sealed container, hydrogen peroxide can establish an equilibrium with its decomposition products, oxygen and water:

$$H_2O_2\,(l) \rightleftharpoons H_2O\,(l) + O_2\,(g)$$

(a) Why is a sealed container necessary for establishment of this equilibrium?
(b) Write the equilibrium constant expression for this reaction.

THE EQUILIBRIUM CONSTANT— ITS CHEMICAL MEANING

So far we have found that the equilibrium constant can be expressed as a *ratio* relating the concentrations of product and reactant species present at equilibrium. The magnitude of the numerical value of this ratio quantitatively tells us something about whether reactants or products are favored in the reversible reaction.

In a system where the equilibrium constant is very large, say 10^7, there would be a negligible amount of reactants remaining at equilibrium. The reaction essentially would go to completion. Thus, *the larger the equilibrium constant is, the more highly favored the product formation will be.*

For a reaction whose equilibrium constant equals 1×10^{-6}, only a very small amount of reactants have been converted to products. Therefore, *the smaller the equilibrium constant is, the less favored the product formation will be.*

An important reaction in carbohydrate metabolism is the conversion of glucose-1-phosphate to glucose-6-phosphate. The forward reaction is catalyzed by the enzyme phospho-glucomutase:

$$\text{glucose-1-phosphate} \rightleftharpoons \text{glucose-6-phosphate}$$

By measuring the concentrations of product and reactant at equilibrium, the equilibrium constant for this reaction has been experimentally determined to equal 20. Symbolically we could write

$$K = \frac{[\text{glucose-6-phosphate}]}{[\text{glucose-1-phosphate}]} = 20$$

exercise 7-15

Refering to the equilibrium above, (a) which substance is in greater concentration at equilibrium? (b) If the equilibrium concentration of glucose-6-phosphate is 0.6 mole/liter, what is the equilibrium concentration of glucose-1-phosphate?

exercise 7-16

The air we breathe is mainly composed of nitrogen (78 percent by volume) and oxygen (21 percent by volume). The following reaction between those two elements to form nitric oxide is known to occur under appropriate conditions:

$$N_2\,(g) + O_2\,(g) \rightleftharpoons 2NO\,(g)$$

Nitric oxide is a colorless, poisonous gas. At normal body temperature, the equilibrium constant for this reaction is about 1×10^{-12}. Comment on the likelihood of nitric oxide formation in the air at normal body temperature (37°C).

Another example of an equilibrium reaction involves the common substance water. In Chapter 3, water molecules were described as being highly polar because of electronegativity differences between bonded hydrogen and oxygen atoms. Hydrogen bonding results from weak attractions between partially positively charged hydrogen atoms on one water molecule and unshared electron pairs on oxygen of neighboring water molecules. If the $\overset{\delta+}{H}—\overset{\delta-}{O}$ bond in water is so polar, is it possible for the H to be pulled away by attraction toward another water molecule? This could be done if the H—O bond ruptures such that both bonding pair electrons remained on oxygen. This would form a hydroxide ion (OH^-) and a hydronium ion (H_3O^+). Experimental evidence indicates this occurs. Figure 7-6 illustrates this process.

Net result: $$2H_2O \rightleftharpoons H_3O^+ + OH^-$$

This equation represents the *ionization* of water into hydronium and hydroxide ions.

The term *ionization* is used to mean the formation of ions due to the breaking of a polar covalent bond. The electron pair remains on the more electronegative atom. This action leads to a negative ion formed by the species retaining the electron pair and a positive ion resulting from the species that has lost its share of the electron pair. In a 1-liter sample of pure water at 24°C, the hydronium ion concentration is experimentally determined to be 1.0×10^{-7} *M* and hydroxide ion concentration equals 1.0×10^{-7} *M*. The smallness of these concentrations is indication that the ionization of water molecules occurs only very slightly. Even though slight, this ionization is still significant. The concentration of unionized water in a liter of water can be calculated knowing the density and gram formula weight of water:

$$\left(\frac{1000\ \text{ml}}{1\ \text{liter}}\right) \times \left(\frac{1\ \text{g}}{1\ \text{ml}}\right) \times \left(\frac{1\ \text{mole water}}{18\ \text{g water}}\right) = \frac{55.5\ \text{moles water}}{1\ \text{liter water}}$$

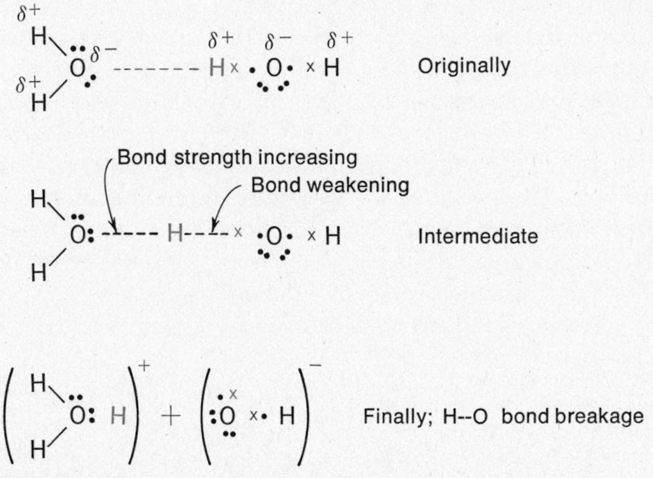

FIGURE 7-6 Ionization of water.

Thus the concentration of unionized water is far greater than either the concentration of hydronium or hydroxide ions.

Let's write the equilibrium constant expression for the ionization of water.

$$2H_2O \rightleftharpoons H_3O^+ + OH^-$$

$$K = \frac{[H_3O^+] \times [OH^-]}{[H_2O]^2}$$

We have just seen that at 55.5 *M* the concentration of unionized water is far greater than the ionized species. There are proportionally so many unionized water molecules compared to ionized water molecules that the concentration of unionized water molecules is essentially constant at 55.5 *M*.

Rearranging the equation above,

$$\underbrace{K \times [H_2O]^2 = \overset{}{\underset{\uparrow}{K}} \times \underset{\uparrow}{(55.5\ M)^2}}_{\text{two constants}} = \underset{\substack{\uparrow \\ \text{a new constant}}}{K_w} = [H_3O^+] \times [OH^-]$$

Mathematically, whenever two constants are multiplied, a new constant results. K_w above is a very important equilibrium constant representing the ionization of water. Now let's calculate the value for K_w using our experimental data:

$$K_w = [H_3O^+] \times [OH^-] = (1 \times 10^{-7}) \times (1 \times 10^{-7}) = 1 \times 10^{-14}$$

Since water is a material common to all cells, the relationship above is extremely significant in its application to physiological and biochemical processes. It will be discussed in detail in Chapter 8.

HOMEOSTASIS

In living cells, very few individual reactions ever reach equilibrium because these reactions are often coupled with each other. This sequencing of reactions places a continuous stress preventing equilibrium from being reached in any previous reactions. You would be dead if all cells in your body achieved equilibrium.

Consider that your body temperature serves as a useful guide to your physiological condition. The normal 98.6°F value and its relative constancy results from the *net* metabolic changes in your body even though your physical activities are not constant. You eat, sleep, work, walk, sit, sing—all different activities—and somehow your body accommodates these diverse functions without a drastic change in body temperature even in summer's heat and winter's cold.

A prominent American physiologist Walter B. Cannon commented on this in 1933:

> The constant conditions which are maintained in the body might be termed *equilibria*. That word, however, has come to have fairly exact meaning as applied to relatively simple physicochemical states, in closed systems where known forces are balanced. The coordinated physiological processes which maintain most of the steady states in the organism are so complex and so peculiar to living beings—involving, as they may, the brain and nerves, the heart, lung, kidneys and spleen, all working cooperatively—that I have suggested a special designation for these states: *homeostasis*. The word does not imply something set and immobile, a stagnation. It means a condition—a condition which may vary, but which is *relatively* constant.

Thus, homeostasis involves a total system whose internal environment is constantly compromising by shifting materials from site to site in response to stresses placed upon it so that no **net** change in the environment occurs. Water enters or leaves cells depending on osmotic pressure changes; chloride ions enter cells and bicarbonate ions leave; oxygen and carbon dioxide mutually exchange cellular occupancy: All these processes occur because a stimulus causes an organism to initiate a sequence of metabolic actions attempting to restore normalcy to the system.

CHAPTER SUMMARY

1. True solutions contain solute particles having diameters less than those of colloidal dispersions, which in turn are smaller than particles of suspensions. True solutions and colloidal dispersions do not settle out upon standing.
2. Solubility involves two processes: (a) solute dissolving and (b) solute coming out of solution. When the rate of process (a) equals the rate of process (b), the solution is at equilibrium.
3. Osmosis is the flow of water through an osmotic membrane from a region of lesser solute concentration to a region of greater solute concentration. The terms *isotonic, hypotonic,* and *hypertonic* relate osmotic pressures of two solutions.
4. Solutes that do not dissociate in solution are called *nonelectrolytes*. Their solutions do not conduct an electric current. Solutes that dissociate in solution are termed *electrolytes*. They form solutions that do conduct an electric current. Osmolarity is a quantity that reflects this behavior:

$$\text{osmolarity} = \text{molarity} \times \text{moles of particles per mole of solute}$$

5. Dialysis occurs when particles of true solution size (ions and small molecules) pass through a dialyzing membrane. Kidney tubules dialyze blood, thus controlling body fluid constituents. Substances that are not reabsorbed by kidney action form urine.

6. Irreversible reactions are ones in which only the forward reaction occurs. Equilibrium cannot be established because the reverse reaction does not take place. Reversible reactions are those in which the reverse reaction competes with the forward reaction.
7. In order to reach equilibrium, a system must have a forward and a reverse reaction (process). Equilibrium is reached when the rate of the forward reaction equals the rate of the reverse reaction. The symbolism $\rightleftharpoons$ is used to denote equilibrium.
8. Le Chatelier's principle points out that when a change is placed on a chemical system at equilibrium, the system shifts in the direction that offsets the change. Changes in concentration are stresses that can cause an equilibrium to shift.
9. A reversible reaction may be driven to completion by eliminating the availability of a product in the reaction mixture. Gas formation and precipitation both cause removal of products from the mixture, thus preventing the reverse reaction from occurring.
10. An equilibrium constant is a ratio expressing a relationship between concentrations of product and reactant species in a system at equilibrium. The larger the equilibrium constant, the more favored the forward reaction will be.
11. Water dissociates slightly, liberating equal concentrations of hydronium and hydroxide ions:

$$2H_2O \rightleftharpoons H_3O^+ + OH^-$$

The equilibrium constant K_w for this reaction is

$$[H_3O^+] \times [OH^-] = 1 \times 10^{-14}$$

12. Homeostasis relates to a dynamic situation in which varying factors interact to keep a system relatively constant.

QUESTIONS

1. During periods of fever, a patient is instructed to increase fluid intake. In terms of osmosis, why is this done?
2. Soaking cucumbers in a saturated salt solution causes the cucumbers to shrivel and become pickles. Explain in terms of osmosis.
3. Hot dogs can be cooked by placing them in boiling water. After the hot dogs cook, the water tastes salty.
   (a) Assuming the hot dogs did not split open during cooking, explain the saltiness of the water.
   (b) Upon cooking, the hot dogs swell. Explain.

4. A 0.9 percent (weight/volume) solution of NaCl is isotonic with body fluids. What is the osmolarity of such a solution?
5. How many osmoles of glucose would be required to make 500 ml of solution having the same osmolarity as 0.9 percent NaCl?
6. A listed first aid procedure as an antidote for silver nitrate poisoning is to administer sodium chloride (common salt) and water. Can you give a chemical basis for this suggested procedure?
7. The equilibrium constant for the following reaction is 0.3:
   $(\text{glycogen})_x + x\text{ATP} \longrightarrow x\text{ADP} + x\text{glucose-1-phosphate}$
   (a) Write the equilibrium constant expression for this reaction.
   (b) At equilibrium, which reaction is favored—the forward or reverse reaction?
   (c) In which direction will the equilibrium shift if an increase in glucose-1-phosphate concentration occurs?

8
acid-base equilibria

INTRODUCTION

In Chapter 7, the formation of ions from water is described as ionization. Strictly speaking, ionization and dissociation describe two different processes. For our purposes, the terms will be used interchangeably.

The water making up about 57 percent of an adult's body weight performs a variety of functions, some of which are a solvent to dissolve and transport substances necessary for metabolism so that new cells may be formed, a solvent to dissolve and transport metabolic waste products from the body, a heat-exchange liquid that can help to maintain a stable body temperature, an osmotic fluid whose release and intake maintains the concentration of solute materials within normal limits, and a reactant in many of the body's chemical processes. In this chapter, water and its dissociation equilibrium are used to extend the earlier concepts of chemical bonding and equilibria to the ideas of acids and bases. Following this, acid–base equilibria are studied in relation to physiologically important processes such as oxygen–hemoglobin binding, buffers and blood pH control, and respiratory acidosis–alkalosis.

K_w REVISITED

Earlier the idea was presented that unionized water is in equilibrium with hydronium ions (H_3O^+) and hydroxide ions (OH^-).

$$2H_2O \rightleftharpoons H_3O^+ + OH^-$$

The balanced equation says that in pure water, for each hydronium ion formed, a hydroxide ion will also be formed. This is verified experimentally and the concentration of hydronium and hydroxide ions are found to be

$$[H_3O^+] = [OH^-] = 1 \times 10^{-7}\,M$$

From these data we may write the K_w expression

$$K_w = [H_3O^+] \times [OH^-] = 1 \times 10^{-14}$$

In *pure* water,

$$[H_3O^+] = [OH^-] = 1 \times 10^{-7}\,M$$

and we say water is *neutral*. Actually, *any aqueous solution* that has the property of

$$[H_3O^+] = [OH^-] = 1 \times 10^{-7}\,M$$

is also a *neutral* solution.

ACIDIC SOLUTIONS

Is it possible to have an aqueous solution where the hydronium ion concentration does *not* equal the hydroxide ion concentration? Recall how the hydronium ion was formed in the water dissociation. A partially negative oxygen in a water molecule picked off the partially positive hydrogen from another polar water molecule. What happens if water is not the only material present? Suppose we add a substance composed of polar molecules contain-

HCl is called hydrogen chloride. *An aqueous solution of HCl is called* hydrochloric acid.

ing partially positive hydrogen atoms to a sample of pure water. In this case, hydrogen ions from the molecules of the added substance could be transferred to the water molecules to form hydronium ions. Recall from Chapter 3 that HCl is an example of a highly polar molecule. Let's add some HCl to a liter of water and measure the hydronium ion concentration in relation to the amount of HCl added. Table 8-1 lists the results of this experiment.

TABLE 8-1
Relationship between $[H_3O^+]$ and $[OH^-]$ in aqueous HCl solution

TOTAL AMOUNT HCl ADDED	$[H_3O^+]$	$[OH^-] = \frac{K_w}{[H_3O^+]} = \frac{1 \times 10^{-14}}{[H_3O^+]}$
0.000 mole/liter	$1 \times 10^{-7}\ M$	$\frac{1 \times 10^{-14}}{1 \times 10^{-7}} = 1 \times 10^{-7}\ M$
0.001 mole/liter	$1 \times 10^{-3}\ M$	$\frac{1 \times 10^{-14}}{1 \times 10^{-3}} = 1 \times 10^{-11}\ M$
0.010 mole/liter	$1 \times 10^{-2}\ M$	$\frac{1 \times 10^{-14}}{1 \times 10^{-2}} = 1 \times 10^{-12}\ M$
0.100 mole/liter	$1 \times 10^{-1}\ M$	$\frac{1 \times 10^{-14}}{1 \times 10^{-1}} = 1 \times 10^{-13}\ M$

Two conclusions may be drawn from these data:

1. Because the $[H_3O^+]$ increased upon addition of HCl, the HCl must have donated hydrogen ions to the water. *Substances that donate hydrogen ions in aqueous solution are defined as acids*. Table 8-2 is a list of several common laboratory acids.

TABLE 8-2
Common laboratory acids

NAME	FORMULA	DISSOCIATION
Hydrochloric	HCl	$HCl \xrightarrow{H_2O} H^+(aq) + Cl^-(aq)$
Hydrobromic	HBr	$HBr \xrightarrow{H_2O} H^+(aq) + Br^-(aq)$
Hydroiodic	HI	$HI \xrightarrow{H_2O} H^+(aq) + I^-(aq)$
Perchloric	$HClO_4$	$HClO_4 \xrightarrow{H_2O} H^+(aq) + ClO_4^-(aq)$
Nitric	HNO_3	$HNO_3 \xrightarrow{H_2O} H^+(aq) + NO_3^-(aq)$
Sulfuric	H_2SO_4	$H_2SO_4 \xrightarrow{H_2O} 2H^+(aq) + SO_4^{2-}(aq)$

Acids that liberate 1 mole of hydrogen ion per mole of acid are called *monoprotic;* those that liberate 2 moles of hydrogen ions per mole acid are *diprotic*.

2. The hydronium and hydroxide ion concentrations are no longer equal. The $[H_3O^+]$ is greater than the $[OH^-]$. *Any aqueous solution in which the hydronium ion concentration is greater than the hydroxide ion concentration is called an acidic solution.*

Recall that the equilibrium constant K_w is the product of the hydronium and hydroxide ion concentrations. Although the product of these two terms must equal 1×10^{-14}, *the H_3O^+ and OH^- concentrations are not necessarily equal.* Notice that the addition of HCl results in an increase in $[H_3O^+]$, causing a stress on the water dissociation equilibrium. In response to this stress, a decrease in $[OH^-]$ occurs. Le Chatelier's principle allows you to predict this result.

By now we are familiar with the idea that a hydronium ion is the combination of a hydrogen ion (H^+) with a water molecule. Recognizing this, let's make a modification and abbreviate H_3O^+ as H^+. *These two terms will henceforth be used interchangeably.*

So far then, if

$[H^+] = [OH^-]$ solution is *neutral*

if

$[H^+] > [OH^-]$ solution is *acidic*

exercise 8-1

Complete the table below by filling in the blanks.

SOLUTION	$[H^+]$	$[OH^-]$	ACIDIC, NEUTRAL, OR NEITHER
A	$1 \times 10^{-4}\ M$	______	______
B	______	$1 \times 10^{-9}\ M$	______
C	$1 \times 10^{0}\ M$	______	______
D	______	$1 \times 10^{-6}\ M$	______

BASIC SOLUTIONS

We just saw that $[H^+]$ does not have to equal $[OH^-]$. In acidic solution, $[H^+] > [OH^-]$. Solution D in Exercise 8-1 is neither acidic nor neutral. It appears that we need another category of aqueous solutions.

Sodium hydroxide is soluble in water (Chapter 5). It will dissolve, dissociating into Na^+ (aq) and OH^- (aq). Let's add solid NaOH to a liter of water and measure the $[H^+]$ that results as NaOH is added. Table 8-3 contains the experimental data.

The terms basic *and* alkaline *may be used interchangeably.*

In each case above, upon addition of NaOH, the $[H^+]$ is less than the $[OH^-]$. *Aqueous solutions in which the hydronium ion concentration is less than the hydroxide concentration are called basic (alkaline) solutions.* Sub-

TABLE 8-3
Relationship between $[H^+]$ and $[OH^-]$ in aqueous NaOH solution

TOTAL AMOUNT NaOH (S) ADDED	$[H^+]$	$[OH^-] = \frac{K_w}{[H^+]} = \frac{1 \times 10^{-14}}{[H^+]}$
0.000 mole/liter	1×10^{-7}	$\frac{1 \times 10^{-14}}{1 \times 10^{-7}} = 1 \times 10^{-7}$
0.001 mole/liter	1×10^{-11}	$\frac{1 \times 10^{-14}}{1 \times 10^{-11}} = 1 \times 10^{-3}$
0.010 mole/liter	1×10^{-12}	$\frac{1 \times 10^{-14}}{1 \times 10^{-12}} = 1 \times 10^{-2}$
0.100 mole/liter	1×10^{-13}	$\frac{1 \times 10^{-14}}{1 \times 10^{-13}} = 1 \times 10^{-1}$

TABLE 8-4
Common laboratory bases

An alkali is a base.

NAME	FORMULA	AQUEOUS DISSOCIATION
Sodium hydroxide	NaOH	$NaOH \xrightarrow{H_2O} Na^+ (aq) + OH^- (aq)$
Potassium hydroxide	KOH	$KOH \xrightarrow{H_2O} K^+ (aq) + OH^- (aq)$
Barium hydroxide	$Ba(OH)_2$	$Ba(OH)_2 \xrightarrow{H_2O} Ba^{2+} (aq) + 2OH^- (aq)$

stances that *liberate hydroxide ions* in aqueous solution are called *bases*. Table 8-4 lists several common laboratory bases.

exercise 8-2

In terms of Le Chatelier's principle, explain why $[H^+]$ decreases when a base is added to water.

AN ALTERNATE METHOD OF EXPRESSING $[H^+]$—WHOLE NUMBER pH CONCEPT

Table 8-5 is a summary of data we now have regarding various aqueous hydrogen and hydroxide ion concentrations.

The first two rows of Table 8-5 can be used to see that a relationship exists between hydrogen ion concentrations and hydroxide concentrations. This relationship can be likened to a seesaw. As the hydrogen ion concentration goes up, the hydroxide ion concentration goes down; as the hydroxide ion concentration goes up, the hydrogen ion concentration goes down. In spite of these changes, their *product* remains 1×10^{-14} (see Fig. 8-1).

Going from a hydrogen ion concentration of 1×10^0 *M* to 1×10^{-14} *M* represents a concentration change of 100 trillion! In 1909 Sörensen, a bacteriologist, was studying bacterial growth under varying conditions from highly alkaline to highly acidic media. Because of the broad range of $[H^+]$

TABLE 8-5
Relationships among [H⁺], [OH⁻], and pH

← $[H^+]$ INCREASING —— $[H^+]$ DECREASING →

	HIGHLY ACIDIC ↓			$[H^+] > [OH^-]$ ← ACIDIC →			SLIGHTLY ACIDIC ↓	$[H^+] = [OH^-]$ NEUTRAL	SLIGHTLY ALKALINE ↓			$[H^+] < [OH^-]$ ← ALKALINE (basic) →			HIGHLY ALKALINE ↓
$[H^+]$	1×10^{0}	1×10^{-1}	1×10^{-2}	1×10^{-3}	1×10^{-4}	1×10^{-5}	1×10^{-6}	1×10^{-7}	1×10^{-8}	1×10^{-9}	1×10^{-10}	1×10^{-11}	1×10^{-12}	1×10^{-13}	1×10^{-14}
$[OH^-]$	1×10^{-14}	1×10^{-13}	1×10^{-12}	1×10^{-11}	1×10^{-10}	1×10^{-9}	1×10^{-8}	1×10^{-7}	1×10^{-6}	1×10^{-5}	1×10^{-4}	1×10^{-3}	1×10^{-2}	1×10^{-1}	1×10^{0}
$[H^+] \times [OH^-]$	1×10^{-14}	1×10^{-14}	1×10^{-14}	1×10^{-14}	1×10^{-14}	1×10^{-14}	1×10^{-14}	1×10^{-14}	1×10^{-14}	1×10^{-14}	1×10^{-14}	1×10^{-14}	1×10^{-14}	1×10^{-14}	1×10^{-14}
pH	0	1	2	3	4	5	6	7	8	9	10	11	12	13	14

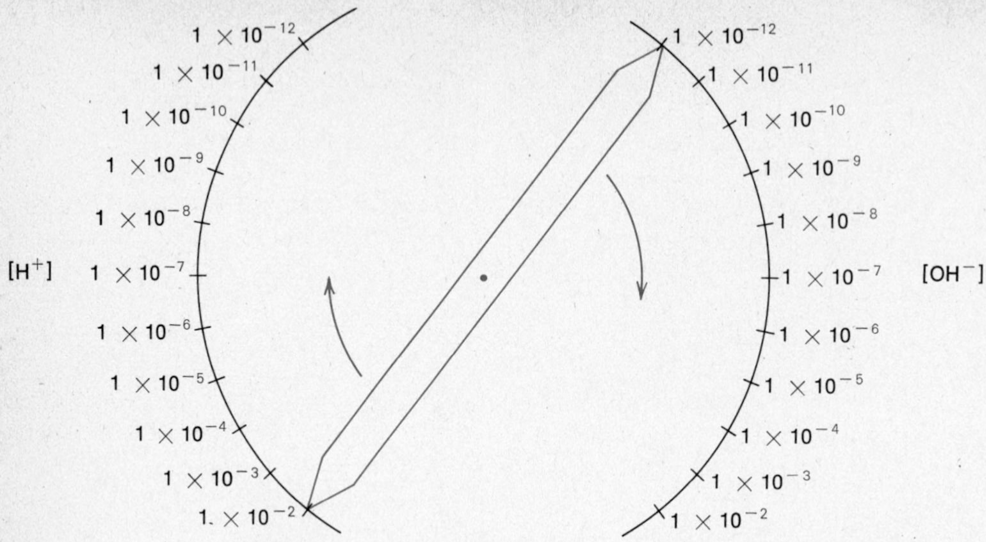

FIGURE 8-1 $[H^+]$ versus $[OH^-]$. As $[H^+]$ increases, $[OH^-]$ decreases; as $[H^+]$ decreases, $[OH^-]$ increases.

to be studied, Sörenson* proposed a more convenient alternate method for expressing hydrogen ion concentration. A working definition of this alternate method is

If n *does not equal 1, the pH does not equal the numerical value of the exponent.*

$$[H^+] = n \times 10^{-m}$$

where n is some number. If $n = 1$, the pH exactly equals the numerical value of the exponent, m.

Let's try out this new relationship.

IF $[H^+]$ IS	THEN pH IS
1×10^{-2}	2
1×10^{-7}	7
1×10^{-9}	9
1×10^{0}	0

*Sörenson's proposal was to define pH as the negative logarithm of the $[H^+]$, i.e., $pH = -\log [H^+]$. A logarithm is the power to which some number is raised. For example, 100 may be written as 10^2; the logarithm (log) of 100 = 2. For a solution in which $[H^+] = 0.01$ or 10^{-2}, the $pH = -\log [H^+] = -(-2) = 2$. Thus above we make the simplification of $[H^+] = 10^{-pH}$.

Notice that as the $[H^+]$ decreases, the pH increases.

Referring to Table 8-5 we see, if

$$[H^+] = 1 \times 10^{-7} \qquad pH = 7 \qquad \text{Solution is } \textit{neutral.}$$

If

$$[H^+] > 1 \times 10^{-7} \qquad pH < 7 \qquad \text{Solution is } \textit{acidic.}$$

If

$$[H^+] < 1 \times 10^{-7} \qquad pH > 7 \qquad \text{Solution is } \textit{alkaline.}$$

A change in pH of one unit changes the $[H^+]$ by a factor of 10! Fluctuations of even less than one pH unit represent considerable changes in hydrogen ion concentration (see Fig. 8-2).

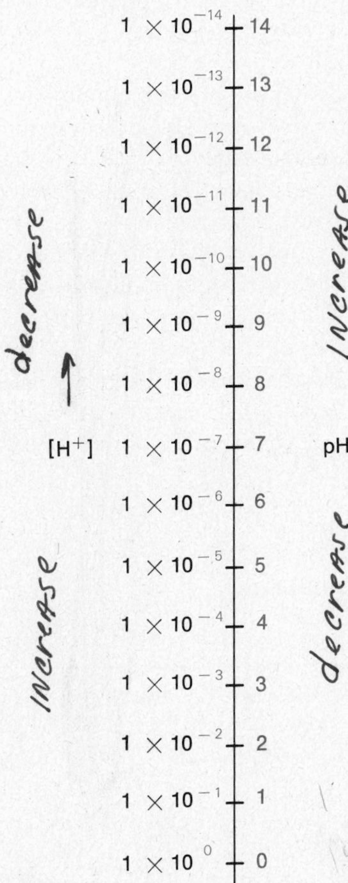

FIGURE 8-2 Relationship between $[H^+]$ and pH. The greater the hydrogen ion concentration, the lower the pH.

exercise 8-3

Complete the table:

$[H^+]$	pH	ACIDIC, BASIC, OR NEUTRAL SOLUTION
1×10^{-4}	4.0	acidic
1×10^{-8}	8.0	Basic
1×10^{-3}	3.0	acidic
1×10^{-13}	13.0	Basic
1×10^{-1}	1.0	ACIDIC

Note the broad normal range of pH for urine in contrast to the very limited normal range of blood plasma. In certain respects, the wide fluctuation of pH of urine is in response to maintainence of blood plasma's pH within a narrow range.

NONWHOLE-NUMBER pH—AN APPROXIMATION APPROACH

Table 8-6 contains pH values that are not whole numbers. Does this mean our working definition is wrong? Not really; it simply means that *n* has a value other than 1 in the equation

$$[H^+] = n \times 10^{-m}$$

In this text, we shall be satisfied with determining an approximate pH for a solution whose hydrogen ion concentration contains a value for *n* other than 1.

Citric acid is common to all citrus fruits and is one of the substances responsible for their particularly tart taste. Let's calculate an approximate pH for a citric acid solution having a hydrogen ion concentration of 9.0×10^{-3} *M*.

TABLE 8-6
pH values of some common substances

ACIDIC ($pH < 7$)	NEUTRAL ($pH = 7$)	ALKALINE ($pH > 7$)
Limes, 2.0	Pure water, 7.0	Blood plasma, 7.3–7.5
Urine, 4.8–6.9	Urine, 7.0	Urine, 7.1–8.4
Soft drinks, 2.0–4.0		Milk of magnesia [$Mg(OH)_2$], 10
Stomach acid, 1.0–3.0		Fresh egg white, 7.6–8.0
Tomatoes, 4.0		
Beer, 4.0–5.0		
Milk, 6.3–6.6		
Grapefruit, 3.0		

From our working definition of $[H^+] = n \times 10^{-m}$, we can substitute 9.0 for n and 3 for m. *If* the $[H^+] = 1 \times 10^{-3}$ *M*, the pH would be 3.0; *if* the $[H^+] = 10 \times 10^{-3}$ *M*, this can be rewritten as 1×10^{-2} *M* and the pH would be 2.0 (10^{-2} is 10 times larger than 10^{-3}). The hydrogen ion concentration of our citric acid sample is 9.0×10^{-3} *M*, which is between 1×10^{-3} *M* and 10×10^{-3} *M* (1×10^{-2} *M*). Therefore its pH lies between 2 and 3 and we have accomplished our goal of determining an approximate pH. The pH associated with a given hydrogen ion concentration lies within the pH range calculated by this approximation technique. The exact pH value may be calculated using logarithms. For the example above, your authors calculate a value of 2.05.

exercise 8-4

Several solutions are tested and found to have the following $[H^+]$ or $[OH^-]$ concentrations: A, $[H^+] = 2.0 \times 10^{-2}$ *M*; *B*, $[H^+] = 6.2 \times 10^{-7}$ *M*; C, $[H^+] = 4.5 \times 10^{-1}$ *M*; $[OH^-] = 5 \times 10^{-2}$ *M*. For each of these solutions, (a) determine its pH range, and (b) describe the solution as highly acidic, slightly acidic, neutral, slightly alkaline, or highly alkaline.

exercise 8-5

Urine's pH can range from 4.5 to 8.4. What is the approximate hydrogen ion concentration range corresponding to these limits?

NEUTRALIZATION —QUALITATIVE ASPECTS

A salt is an ionic compound containing ions other than H^+ or OH^-.

Acids (hydrogen ion donors) and bases (hydroxide ion donors) react to form water and a class of compounds called *salts*. This reaction is called a *neutralization* reaction. Let's see what happens when equal volumes of 0.1 *M* hydrochloric acid and sodium hydroxide solutions react:

In the HCl solution,

$$HCl + H_2O \longrightarrow H_3O^+(aq) + Cl^-(aq)$$

or simply

$$HCl \longrightarrow H^+(aq) + Cl^-(aq)$$

In the NaOH solution,

$$NaOH(s) \xrightarrow{H_2O} Na^+(aq) + OH^-(aq)$$

Upon mixing the two solutions, we don't anticipate reaction between the species of like charge, i.e., Na^+ with H_3O^+. We have already described the reaction occurring between hydrogen ions and hydroxide ions, however;

$$H^+(aq) + OH^-(aq) \longrightarrow H_2O$$

Thus, the acidic component of the solution (H^+) has been neutralized by the presence of the basic component (OH^-) to produce the neutral species (H_2O). The salt sodium chloride, being soluble, remains dissolved.

The net reaction for neutralization may be generalized as follows:

$$\underbrace{H^+(aq) + X^-(aq)}_{\text{an acid}} + \underbrace{M^+(aq) + OH^-(aq)}_{\substack{\text{a base}\\\text{(metal hydroxide)}}} \longrightarrow \underbrace{HOH}_{\text{water}} + \underbrace{M^+(aq) + X^-(aq)}_{\text{salt}}$$

Note that the particular salt formed depends on the cation of the base and the anion of the acid used. The salt's solubility depends on its component ions. The solid crystalline salt is obtained upon evaporation of the water.

exercise 8-6

(a) What salt is produced by the reaction between potassium hydroxide and hydrochloric acid?
(b) What base and acid react to form barium chloride?

exercise 8-7

A slurry of magnesium hydroxide and water is called *milk of magnesia.* It is commercially available and when taken internally acts as an antacid to offset excess stomach acidity. Assuming the stomach acid is essentially HCl, write a balanced equation to represent this reaction.

NEUTRALIZATION—QUANTITATIVE ASPECTS

The salt sodium chloride is made by the neutralization reaction between sodium hydroxide and hydrochloric acid. Let's set up an experiment that allows us to determine when all the acid has been neutralized by the controlled addition of base. The most common way to carry out this experiment is to perform a titration. We know that 1 mole of hydroxide ions neutralizes 1 mole of hydrogen ions. By using a sodium hydroxide solution whose concentration is accurately known (a standardized solution), the experiment will also enable us to calculate the concentration of acid in the sample.

A carefully measured volume of hydrochloric acid of unknown concentration is placed into a flask. A few drops of an indicator, a substance that changes color when all the acid is neutralized, is also added. A standardized sodium hydroxide solution is placed in a buret. The base is slowly added to the flask.* Upon initial addition of base, hydroxide ions react with H^+ ions in the acid solution to produce water. Sodium and chloride ions remain dissolved. As additional base is added, more hydrogen ions are neutralized. Thus, as the titration proceeds, the concentration of H^+ is continuously decreasing from its original value. Eventually, when sufficient base is added to neutralize all the hydrogen ions present from the dissociation of HCl in the acid sample, the indicator changes color. The point at which the number of moles of hydroxide ion added equals the number of moles of H^+ liberated by HCl in the original acid sample is called the *end point* of the titration.

*Acid–base titrations are also performed by adding standardized acid to a solution of base whose concentration is to be determined.

Example 8-1

If 25 ml of 0.100 M NaOH was required to titrate 50.0 ml of an HCl solution to the end point, calculate the concentration of the HCl solution. At the end point we know that

No. of moles OH^- added = No. of moles H^+ originally present in HCl

No. of moles OH^- = volume of base × molarity of base =
0.025 ~~liter~~ × 0.100 mole/~~liter~~ = 0.0025 mole OH^-

No. of moles H^+ = No. of moles OH^- = 0.0025 mole H^+ in original sample

Therefore, 0.0025 mole of H^+ were present in 0.050 liter of HCl and the original HCl concentration was 0.0025 mole/0.050 liter or 0.050 mole/liter (0.050 M).

A generalized equation can be written for the neutralization of **monoprotic** acids:

(volume of acid, liters) × (molarity of acid) =
(volume of base, liters) × (molarity of base)

By using an experimentally determined volume of standardized base to neutralize a known volume of acid, the concentration of the acid can be calculated.

exercise 8-8

It required 37.0 ml of 0.050 *M* NaOH to neutralize 100 ml of a patient's gastric contents. Assuming HCl is the only acid present in the gastric contents, what is the molarity of HCl present?

ANOTHER DESCRIPTION OF ACIDS AND BASES—A GIVE AND TAKE PROPOSITION

In describing the addition of HCl to water, we found that an acidic solution resulting from HCl furnished hydrogen ions more readily than water.

$$HCl + H_2O \longrightarrow H_3O^+ + Cl^-$$

In this reaction, the function of HCl is that of an acid because it donates hydrogen ions. What role does water play? It accepts hydrogen ions from HCl. If an acid is a hydrogen ion donor, what should we call a hydrogen ion acceptor? Brønsted and Lowry independently suggested that the term *base* be used for *any species that accepts hydrogen ions from some other substance.* According to the Brønsted–Lowry description, water functions as a base in the reaction above by accepting hydrogen ions.

Acids are H^+ donors; bases are H^+ acceptors.

You might now wonder if our previous description of a base as a hydroxide ion donor is incorrect. Let's reexamine the neutralization reaction between sodium hydroxide and hydrochloric acid from the Brønsted–Lowry viewpoint.

$$Na^+(aq) + OH^-(aq) + H^+(aq) + Cl^-(aq) \longrightarrow HOH + Na^+(aq) + Cl^-(aq)$$

What role does each reactant play? HCl dissociates and functions as an acid by donating hydrogen ions. The hydroxide ions from sodium hydroxide accept (combine with) these hydrogen ions to form water. Thus, hydroxide ions function as a base—a hydrogen ion acceptor. The Brønsted-Lowry description is a generalized one in which hydroxide ion is one of many possible bases.

exercise 8-9

Using Brønsted-Lowry terminology, what role does water play in the equilibrium:

$$2H_2O \rightleftharpoons H_3O^+ + OH^-$$

RELUCTANT HYDROGEN ION DONORS—WEAK ACIDS

Acetylsalicylic acid, commonly called aspirin, contains a carboxyl group ($-\overset{\overset{\displaystyle O}{\|}}{C}-OH$). Acids containing this group are called *carboxylic* acids. As noted in Chapter 4, the carboxyl group is very polar. In aqueous solution, a carboxylic acid can act as a hydrogen ion donor:

$$R-C(=O)OH + H_2O \longrightarrow H_3O^+ + R-C(=O)O^{(-)}$$

When a 0.10 *M* solution of hydrochloric acid and a 0.10 *M* solution of acetylsalicylic acid are prepared, their acidities differ. Experimentally, the *hydrogen ion* concentration of the hydrochloric acid is measured as 1.0×10^{-1} *M*; that of the acetylsalicylic acid solution is 5.6×10^{-3} *M*. Both solutions are definitely acidic, indicating that both acids release hydrogen ions into solution.

$$\underset{0.10\,M}{HCl} \longrightarrow \underset{0.10\,M}{H^+(aq)} + \underset{0.10\,M}{Cl^-(aq)}$$

$$\underset{0.10\,M}{C_6H_4(COOH)-O-\overset{\overset{\displaystyle O}{\|}}{C}-CH_3} \longrightarrow \underset{0.0056\,M}{H^+(aq)} + \underset{0.0056\,M}{C_6H_4(C(=O)O^{(-)})-O-\overset{\overset{\displaystyle O}{\|}}{C}-CH_3(aq)}$$

The fact that the hydrogen ion concentration in the acetylsalicylic acid is considerably less than that in the hydrochloric acid solution indicates acetylsalicylic acid is a much poorer hydrogen ion donor than hydrochloric acid.

The hydrogen ion concentration in the hydrochloric acid solution results from the essentially *complete* dissociation of 0.1 *M* HCl into 0.1 mole of hydrogen ions and 0.1 mole of chloride ions per liter. The much lower concentration of hydrogen ions in the acetylsalicylic acid solution is due to the *incomplete* dissociation of acetylsalicylic acid into ions. This means that a significant amount of *undissociated* acetylsalicylic acid molecules are present in solution. The term *weak acid* is used to describe *an acid that incompletely dissociates in aqueous solution.* Most acids are of this type including those of physiological significance. *An acid that is essentially completely dissociated is termed a strong acid.* A strong acid has a *strong tendency* to liberate hydrogen ions. The acids listed in Table 8-2 are all strong acids. Although both solutions in the example above are 0.10 *M*, they contain differing hydrogen ion concentrations because HCl is a strong acid, whereas acetylsalicylic acid is a weak acid. Be careful not to confuse the terms strong and concentrated; they are **not** synonymous.

Because undissociated acetylsalicylic acid molecules along with hydrogen and acetylsalicylate ions are present in solution, an equilibrium is established.

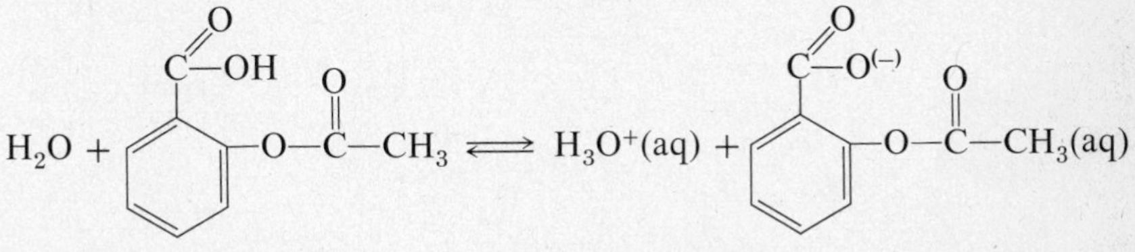

$$H_2O + C_6H_4(COOH)\text{—O—}\overset{\overset{O}{\|}}{C}\text{—}CH_3 \rightleftharpoons H_3O^+(aq) + C_6H_4(COO^{(-)})\text{—O—}\overset{\overset{O}{\|}}{C}\text{—}CH_3(aq)$$

In the forward reaction, *acetylsalicylic acid* functions as an acid by donating hydrogen ions. In the reverse reaction, *acetylsalicylate ions* function as a base by combining with hydrogen ions. An acid–base pair, such as acetylsalicylic acid–acetylsalicylate ion, is called a *conjugate acid–base pair*. In this example, acetylsalicylate ion is the conjugate base of the acetylsalicylic acid. This can be generalized symbolically:

$$H_2O + \underset{\text{acid}}{HB} \rightleftharpoons \underset{\text{conjugate base}}{B^-} + H_3O^+$$

In the dissociation of a *weak* acid (HB) the conjugate base (B^-) is such a powerful hydrogen ion acceptor that the *reverse* reaction is highly favored. Thus the acid remains largely undissociated.

Let's look at another example of a conjugate acid–base pair relationship. Carbonic acid

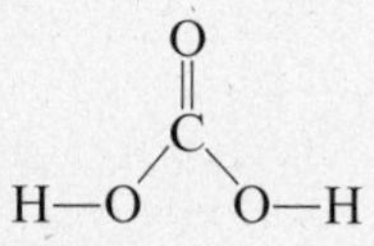

$$\text{H—O—}\overset{\overset{O}{\|}}{C}\text{—O—H}$$

dissociates into hydrogen and bicarbonate ions:

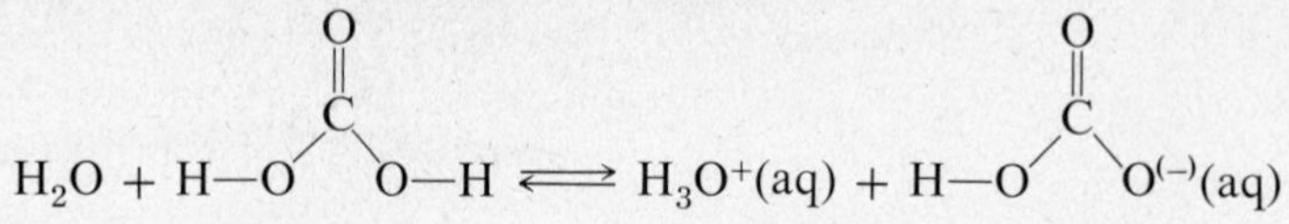

$$H_2O + H{-}O{-}\overset{\overset{\displaystyle O}{\|}}{C}{-}O{-}H \rightleftharpoons H_3O^+(aq) + H{-}O{-}\overset{\overset{\displaystyle O}{\|}}{C}{-}O^{(-)}(aq)$$

carbonic acid (H_2CO_3) — bicarbonate ion (HCO_3^-)

Thus, bicarbonate ion is the conjugate base of carbonic acid. This relationship will be taken up again as it applies to the control of acidity and alkalinity.

Notice in the examples above that water behaves as a base and hydronium ion is its conjugate acid.

exercise 8-10

Lactic acid $\left(CH_3{-}\overset{\overset{\displaystyle OH}{|}}{\underset{\underset{\displaystyle H}{|}}{C}}{-}C(=O){-}OH\right)$ **is a waste product of anaerobic glucose metabolism.**

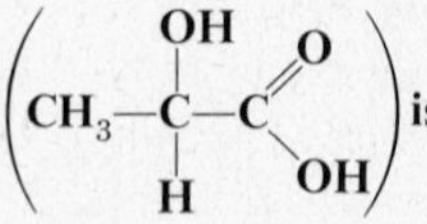

(a) Write an equilibrium equation representing its dissociation.
(b) Which of the equilibrium species is the conjugate base of lactic acid?

exercise 8-11

Given the equation

$$H_2O + CO_3^{2-} \rightleftharpoons HCO_3^- + OH^-$$

(a) What roles do water and carbonate ion play in the reaction?
(b) What are the relationships between water and hydroxide ion and between carbonate and bicarbonate ion?

K_a—THE EQUILIBRIUM CONSTANT EXPRESSION FOR ACID DISSOCIATION

In Chapter 7, we discussed the equilibrium constant expression for a system at equilibrium. The expression tells the ratio of product and reactant concentrations at equilibrium. In the preceding section, the incomplete dissociation of a weak acid was described as an equilibrium situation. Since it is an equilibrium involving dissociation of an acid, we'll call the dissociation equilibrium constant K_a. Let's begin by considering the acetylsalicylic acid dissociation using the abbreviations HASA and ASA^-, as shown below.

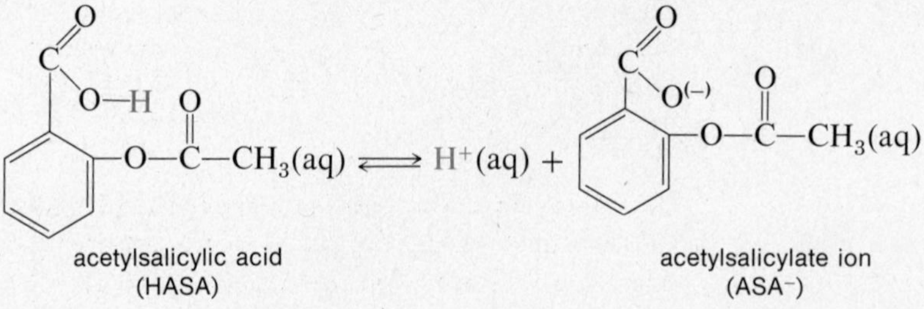

$$C_6H_4(COOH){-}O{-}\overset{\overset{\displaystyle O}{\|}}{C}{-}CH_3(aq) \rightleftharpoons H^+(aq) + C_6H_4(COO^{(-)}){-}O{-}\overset{\overset{\displaystyle O}{\|}}{C}{-}CH_3(aq)$$

acetylsalicylic acid (HASA) — acetylsalicylate ion (ASA^-)

Since all coefficients in the balanced equation are 1, all exponents in the equilibrium constant expression are also 1.

$$K_a = \frac{[H^+] \times [ASA^-]}{[HASA]}$$

Note that the numerical value of K_a *is determined by the amount of acid dissociated into hydrogen and conjugate base ions compared to the amount of undissociated acid.* To better appreciate this, let's look at some experimental data. When 0.100 mole of solid acetylsalicylic acid is added to enough water to make a liter of solution, some of the acetylsalicylic acid dissociates into hydrogen and acetylsalicylate ions. Upon reaching equilibrium, the hydrogen ion concentration is measured experimentally and found to be 5.6×10^{-3} *M*. Since in the dissociation of HASA an ASA^- ion is released each time a H^+ is formed, the concentration of ASA^- is also 5.6×10^{-3} *M*. Because the dissociation of acetylsalicylic acid is the source of H^+ and ASA^- ions, the hydrogen and acetylsalicylate ion concentrations tell us how much of the acid dissociated. This allows the amount of undissociated acid at equilibrium to be calculated:

original amount HASA – amount dissociated
= amount undissociated HASA at equilibrium

$$0.100 \text{ mole/liter} - 0.0056 \text{ mole/liter} = 0.0944 \text{ mole/liter}$$

These data are summarized below.

AT START	**AT EQUILIBRIUM**
$[H^+]$ from HASA = 0.00 *M*	$[H^+]$ = 0.0056 *M*
$[ASA^-]$ from HASA = 0.00 *M*	$[ASA^-]$ = 0.0056 *M*
[HASA] = 0.100 *M*	[HASA] = [HASA] originally – amount dissociated = 0.100 *M* – 0.0056 *M* = 0.0944 *M*

Using these experimental data, we can calculate the dissociation constant K_a for acetylsalicylic acid by substituting the experimental values into the K_a expression.

$$K_a = \frac{[H^+] \times [ASA^-]}{[HASA]} = \frac{(0.0056\ M)(0.0056\ M)}{(0.0944\ M)} = 3.3 \times 10^{-4} \text{ mole/liter}$$

K_a values for other acids can be obtained similarly. Table 8-7 lists a series of carboxylic acids. The dissociation equation and dissociation constant is given for each acid. In the discussions to follow we shall be using these data.

TABLE 8-7
Dissociation behavior of selected carboxylic acids

NAME	COMMENTS	STRUCTURAL FORMULA	ABBREVIATED FORMULA	DISSOCIATION [(aq) implied]	K_a
Lactic	Waste product of carbohydrate metabolism	$CH_3{-}C(OH)(H){-}C(=O)OH$	HLac	$HLac \rightleftharpoons H^+ + Lac^-$	1.4×10^{-4}
Pyruvic	Compound vital in metabolism	$CH_3{-}C(=O){-}C(=O)OH$	HPy	$HPy \rightleftharpoons H^+ + Py^-$	3.2×10^{-3}
Carbonic	Found in carbonated beverages	$H{-}O{-}C(=O){-}O{-}H$	H_2Car	$H_2Car \rightleftharpoons H^+ + HCar^-$	4.5×10^{-7}
Acetylsalicylic	Active ingredient in asprin	$C_6H_4(C(=O)OH)(O{-}C(=O){-}CH_3)$	HASA	$HASA \rightleftharpoons H^+ + ASA^-$	3.3×10^{-4}
Salicylic	Formerly used as an analgesic	$C_6H_4(C(=O)OH)(OH)$	HSal	$HSal \rightleftharpoons H^+ + Sal^-$	1.0×10^{-3}

Acetic	Found in vinegar and poor wines	$CH_3-C(=O)-OH$	HOAc	$HOAc \rightleftharpoons H^+ + OAc^-$	1.8×10^{-5}
Citric	Found in citrus fruits; important intermediate in carbohydrate metabolism	$HO-C(=O)-CH_2-C(OH)(COOH)-CH_2-C(=O)-OH$	HCit	$HCit \rightleftharpoons H^+ + Cit^-$	8.4×10^{-4}
Ascorbic	Vitamin C	$C_6H_8O_6$	HAsc	$HAsc \rightleftharpoons H^+ + Asc^-$	8.1×10^{-5}
Cholic	Most abundant bile acid	$C_{22}H_{40}O_5$	HCol	$HCol \rightleftharpoons H^+ + Col^-$	—
Acetoacetic	Metabolic breakdown product of fats	$CH_3-C(=O)-CH_2-C(=O)-OH$	HAcAc	$HAcAc \rightleftharpoons H^+ + AcAc^-$	—
β-Hydroxybutyric	Metabolic breakdown product of fats	$CH_3-CH(OH)-CH_2-C(=O)-OH$	HBut	$HBut \rightleftharpoons H^+ + But^-$	2.10^{-5}
Hippuric	Formed in the liver	$C_9H_9O_3N$	HIp	$HIp \rightleftharpoons H^+ + Ip^-$	2.3×10^{-4}
Formic	Found in bees and ants	$H-C(=O)-O-H$	HFor	$HFor \rightleftharpoons H^+ + For^-$	1.8×10^{-4}

For a monoprotic acid we can write a *generalized* dissociation equation and dissociation constant expression:

$$\underset{\text{conjugate acid}}{HB} \rightleftharpoons H^+ + \underset{\text{conjugate base}}{B^-} \qquad K_a = \frac{[H^+] \times [B^-]}{[HB]}$$

exercise 8-12

With the aid of Table 8-7, write dissociation constant expressions for each of the following acids: carbonic, salicylic, and pyruvic.

The ratio of dissociation product concentrations to undissociated acid concentration determines the numerical value of K_a. The less the tendency an acid has to dissociate, *the weaker the acid* is by comparison with other acids and *the smaller its* K_a *value*. Referring to Table 8-7 we see that lactic acid is a weaker acid than salicylic acid; acetic acid is a stronger acid than carbonic acid.

exercise 8-13

Using Table 8-7, arrange the following acids in order of *increasing* acid strength: lactic, pyruvic, acetic, acetylsalicylic, salicylic, and hippuric.

exercise 8-14

Arrange the hydrogen ion accepting ability of the conjugate bases for the acids in Exercise 8-13 in increasing order of their hydrogen ion accepting ability.

USING THE K_a VALUE TO DETERMINE $[H^+]$ OF A WEAK ACID

From experimental data we have calculated equilibrium constants for weak acids. Suppose we know K_a and wish to calculate the hydrogen ion concentration in a solution of a weak acid.

EXAMPLE 8-2:

Calculate the hydrogen ion concentration of a 1.00 *M* solution of citric acid.

From Table 8-7;

$$HCit \rightleftharpoons H^+ + Cit^- \qquad K_a = 8.4 \times 10^{-4} = \frac{[H^+] \times [Cit^-]}{[HCit]}$$

We have used a certain amount of citric acid (HCit) in preparing the solution. When equilibrium is established, some of the citric acid will have dissociated. Since we don't know how much citric acid has dissociated, let x equal this quantity. The chemical equation tells us that for each citric acid molecule that dissociates, one hydrogen ion and one citrate ion are produced. Therefore, at equilibrium, $[H^+] = [Cit^-] = x$. These data are readily organized in tabular form.

For a change in concentration, a plus (+) sign indicates an increase in concentration; a minus (−) sign indicates a decrease.

MOLARITY/ SPECIES	[H^+]	[Cit^-]	[HCit]
Initially	0.000	0.000	1.00
Change	$+x$	$+x$	$-x$
At equilibrium	x	x	$1.00 - x$

Notice that at equilibrium the [HCit] is less than its original value. If the concentration of a species is reduced, a minus sign is used; if the concentration increases, a plus sign is used. Substituting these values into the K_a expression,

$$K_a = 8.4 \times 10^{-4} = \frac{[H^+] \times [Cit^-]}{[HCit]} = \frac{(x)(x)}{1.00 - x} = \frac{x^2}{1.00 - x}$$

To solve this expression for the exact value of x would involve the use of the quadratic equation. To avoid this we can make a simplifying assumption based on our chemical intuition. The value for K_a is small, indicating citric acid is only slightly dissociated. Therefore x, the amount dissociated, will be small compared to the original amount of undissociated acid, 1.00 *M*. We shall assume the value $1.00 - x$ is approximately 1.00. Now we solve the equation.

$$8.4 \times 10^{-4} = \frac{x^2}{1.00}$$

$$(8.4 \times 10^{-4})(1.00) = x^2$$

$$x^2 = 8.4 \times 10^{-4}$$

$$x = \sqrt{8.4 \times 10^{-4}} \cong 3 \times 10^{-2}\,M = [H^+] = [Cit^-]$$

EXAMPLE 8-3:

What is the pH of a 0.010 *M* ascorbic acid solution?

In order to find the pH, we must first know the hydrogen ion concentration. Table 8-7 furnishes us with pertinent data.

$$HAsc \rightleftharpoons H^+ + Asc^- \qquad K_a = 8.1 \times 10^{-5} = \frac{[H^+] \times [Asc^-]}{[HAsc]}$$

MOLARITY/ SPECIES	[H^+]	[Asc^-]	[HAsc]
Initially	0.000	0.000	0.010
Change	$+x$	$+x$	$-x$
At equilibrium	x	x	$0.010 - x$

Again we note that K_a is small and we approximate $0.010 - x$ as 0.010.

$$8.1 \times 10^{-5} = \frac{[H^+] \times [Asc^-]}{[HAsc]} = \frac{(x)(x)}{0.010} = \frac{x^2}{1 \times 10^{-2}}$$

$$(8.1 \times 10^{-5})(1 \times 10^{-2}) = x^2$$

$$x^2 = 8.1 \times 10^{-7} = 81 \times 10^{-8}$$

$$x = \sqrt{81 \times 10^{-8}} = \sqrt{81} \times \sqrt{10^{-8}}$$

$$x = 9 \times 10^{-4} = [H^+] = [Asc^-]$$

Since the $[H^+]$ is 9×10^{-4} *M*, the pH is approximately 3, a moderately acidic solution.

exercise 8-15

Ever been stung by a bee? Formic acid is the main component of the bee's deposit. Swelling and itchiness attest to its irritability. What is the approximate pH of a 0.10 *M* formic acid solution? Make any necessary approximations.

BUFFERS—pH CONTROL AGENTS

Aqueous solutions in the body generally have a very narrow range of hydrogen ion concentration within which the body functions best and a limited hydrogen ion concentration range in which the organism can function at all. One reason for this is that enzymes have an optimum pH for maximum efficiency.

Enzyme activity and its relation to pH is discussed in Chapter 11.

The extracellular fluid, including blood plasma, is normally regulated at a hydrogen ion concentration of approximately 4×10^{-8} *M*, corresponding to a pH of 7.4. Cellular metabolic waste products that function as acids (e.g., lactic acid) or bases (e.g., bicarbonate ion), upon entering the extracellular fluid, can alter its pH. The H^+ value can vary from a lower limit of 1.6×10^{-8} (pH = 7.8) to an upper limit of 1.2×10^{-7} (pH = 7.0) but only for brief periods or else death occurs. The normal pH of arterial blood is 7.4. *Whenever the pH falls within the range of slightly less than 7.4–7.0,* a person is said to have *acidosis* caused by an *increase* in hydrogen ion concentration. *Alkalosis* occurs *within the range 7.4–7.8* due to a *decrease* in hydrogen ion concentration. How does the body manage to prevent continued acidosis or alkalosis under normal conditions? What we need is something to neutralize the acidic and/or alkaline waste products of metabolism.

The body uses three primary methods to accomplish this: (1) If the blood hydrogen ion concentration changes from its normal range, the kidneys either reabsorb fewer hydrogen ions causing an acidic urine or reabsorb additional hydrogen ions resulting in an alkaline urine. Thus, the fluctuation in urine pH reflects its role in maintaining blood pH within the very narrow normal range. (2) Pulmonary ventilation is altered, changing the rate at which carbon dioxide is removed from body fluids. The change in CO_2 concentration helps return hydrogen ion concentration to normal limits. (3) Buffers, a special system of hydrogen ion donors and hydrogen ion acceptors, are present in body fluids.

The regulating of hydrogen ion concentration in body fluids is sometimes referred to as acid–base balance. The utility of the K_a expression in calculating hydrogen ion concentration has been shown. The idea that an acid and its anion comprise a conjugate acid–base pair has also been established. Let's put these two concepts together to understand method 3 above. The generalized K_a expression

$$\underset{\text{acid}}{HB} \rightleftharpoons H^+ + \underset{\text{conjugate base}}{B^-} \qquad K_a = \frac{[H^+] \times [B^-]}{[HB]}$$

can be rewritten

$$[H^+] = \frac{K_a \times [HB]}{[B^-]} = K_a \times \frac{[HB]}{[B^-]} = K_a \times \frac{[\text{undissociated acid}]}{[\text{conjugate base}]}$$

Since the dissociation constant K_a has a constant value, the solution's hydrogen ion concentration depends on the concentration *ratio* of undissociated acid HB to its conjugate base B^-. To better understand the role of each of these in hydrogen ion control, let's conduct an experiment. In Example 8-2 we saw that a 1.00 *M* citric acid solution had a hydrogen ion concentration of 3×10^{-2} *M* due to the hydrogen ions released by partial dissociation of citric acid (HCit).

$$\text{HCit} \rightleftharpoons H^+(aq) + \text{Cit}^-(aq)$$

How do we know sodium citrate will dissociate completely?

To this solution, let's add a small amount of solid sodium citrate. Sodium citrate is an ionic compound that dissociates *completely,* releasing sodium and citrate ions:

$$\text{NaCit (s)} \xrightarrow{H_2O} Na^+ \text{(aq)} + \text{Cit}^- \text{(aq)}$$

What effect does addition of sodium citrate have on the citric acid dissociation equilibrium? Adding citrate ions to the solution has placed a stress on this equilibrium. Le Chatelier's principle tells us the equilibrium will shift to offset this stress. In attempting to offset the addition of citrate ions, hydrogen ions combine with citrate ions to produce citric acid. In doing so they cause the concentration of hydrogen ions to decrease; hence, the solution should be less acidic and have a hydrogen ion concentration less than 3×10^{-2} *M.* Experimentally the $[H^+]$ is determined to be 1×10^{-5} *M.* Le Chatelier was right.

What is happening to the pH as sodium citrate is added?

Adding more sodium citrate further reduces the hydrogen ion concentration. After addition, the $[H^+]$ is 1×10^{-6} *M.* This means that citrate ions, regardless of their source, can function as a base by combining with hydrogen ions. A solution of citric acid and sodium citrate contains a hydrogen ion donor, HCit, and a hydrogen ion acceptor, Cit^-. *By adjusting how much of each is present, i.e., the HCit/Cit⁻ ratio, the solution's hydrogen ion concentration can be regulated.* This is just the kind of system needed for acid–base balance in body fluids.

Such a *weak acid/conjugate base pair* is called a *buffer pair. A buffer solution contains a weak acid/conjugate base pair.* Presence of the weak acid and its conjugate base allows the pH of a solution to remain relatively constant despite addition of limited quantities of acids or bases from other sources.

Generalizing we could say

$$B^- + H^+ \text{ added to solution} \longrightarrow HB$$

$$HB \rightleftharpoons H^+ + B^-$$

$$H^+ + \text{Base added to solution} \longrightarrow \text{HBase}$$

A note of caution: A buffer system's ability to offset addition of acid or base without a significant change in pH is limited. There is a point at which addition of a huge excess of either acid or base overwhelms the buffer's capacity for offsetting this addition. At that point the pH changes sharply.

BODY FLUID BUFFERS

Body fluids contain three major buffer systems: the carbonic acid/bicarbonate buffer; the $H_2PO_4^-$–HPO_4^{2-} buffer, and the protein buffer. The body uses each of these to carry out necessary buffering action.

Dissolved carbon dioxide gas from metabolism combines with water in blood plasma to form unstable carbonic acid (H_2CO_3).

$$CO_2\,(g) + H_2O\,(l) \rightleftharpoons H_2CO_3\,(aq)$$

This equation indicates the amount of H_2CO_3 present is related to the amount of CO_2 present. Recall that adjustment of pulmonary ventilation adjusts CO_2 concentration.

Carbonic acid is a weak acid that dissociates:

$$\underset{\text{(acid)}}{H_2CO_3} \rightleftharpoons H^+\,(aq) + \underset{\substack{\text{bicarbonate ion} \\ \text{(conjugate base)}}}{HCO_3^-\,(aq)}$$

Since the amount of CO_2 present governs the amount of H_2CO_3, it also affects the concentration of hydrogen and bicarbonate ions.

$$K_a = \frac{[H^+] \times [HCO_3^-]}{[H_2CO_3]}$$

$$[H^+] = K_a \times \frac{[H_2CO_3]}{[HCO_3^-]} = K_a \times \frac{[\text{conj. acid}]}{[\text{conj. base}]}$$

From the above we see that the $[H_2CO_3]/[HCO_3^-]$ governs the $[H^+]$ of a solution in which they are present. Normally 20 times as much bicarbonate ion is present as carbonic acid. Carbonic acid is capable of neutralizing base added to the solution; bicarbonate ion can neutralize hydrogen ions from an acid added to the solution, e.g., lactic acid.

$$H_2CO_3 + \text{Base} \longrightarrow \text{HBase} + HCO_3^-$$

$$HCO_3^- + H^+ \longrightarrow H_2CO_3 \xrightarrow{\text{carbonic anhydrase}} CO_2 + H_2O$$

The enzyme carbonic anhydrase catalyzes the breakdown of the carbonic acid formed into water and carbon dioxide. The latter is expelled by the lungs. Because of their ability to neutralize hydrogen ions, bicarbonate ions are occasionally called the *alkaline reserve* of the body fluids. Bicarbonate is present in higher concentration than other buffer conjugate bases.

exercise 8-16

Think of the equilibria among CO_2, H_2CO_3, H^+, and HCO_3^- in water. What effect would an increase in CO_2 concentration have on the concentration of hydrogen ion?

increase in H^+

exercise 8-17

What would the $[H^+]$ be if carbonic acid/bicarbonate was the only buffer system present in the blood and $[HCO_3^-] = [H_2CO_3]$?

The dihydrogen phosphate ($H_2PO_4^-$)/monohydrogen phosphate ($HPO_4{}^{2-}$) buffer pair is unusual in that an anion functions as an acid:

$$\mathrm{HO{-}P({=}O)(O^{(-)}){-}O{-}H} \rightleftharpoons H^+ + \mathrm{H{-}O{-}P({=}O)(O^{(-)}){-}O^{(-)}} \qquad K_a = 6.2 \times 10^{-8}$$

dihydrogen phosphate ion (acid) — monohydrogen phosphate ion (conjugate base)

4X greater in plasma

Functioning as a weak acid, $H_2PO_4^-$ releases hydrogen ion to neutralize base added from an external source. The conjugate base HPO_4^{2-} accepts H^+ added from some external source. The plasma concentration of monohydrogen phosphate is normally four times greater than that of dihydrogen phosphate.

weak acid

INDICATORS—A TYPE OF WEAK ACID

The ripening of many fruits is accompanied by a color change due to a change in hydrogen ion concentration within the fruit. The color change is due to a compound present in the fruit whose color differs depending on pH of the medium. Such a compound is called an acid–base indicator. Most acid–base indicators are themselves weak acids whose degree of dissociation is pH dependent. For a generalized case, the dissociation of an indicator, HIn, can be expressed

$$\underset{\text{weak acid, color 1}}{\mathrm{HIn}} \rightleftharpoons H^+ + \underset{\text{conjugate base, color 2}}{\mathrm{In^-}} \qquad K_{\mathrm{HIn}} = \frac{[H^+] \times [\mathrm{In}^-]}{[\mathrm{HIn}]} = [H^+] \times \frac{[\mathrm{In}^-]}{[\mathrm{HIn}]}$$

From the equilibrium constant expression above, we see that the $[In^-]/[HIn]$ ratio depends on the $[H^+]$. Applying Le Chatelier's principle in relation to the chemical equation, in a highly acidic solution, In^- combines with H^+ shifting the equilibrium so that the color of the HIn form predominates; in highly basic solution, the HIn concentration is reduced by combination with base and the color of the In^- form predominates. Litmus is an indicator dye extracted from lichens; when it is impregnated on absorbent paper, it forms the common laboratory indicator, litmus paper. It turns *blue* in *basic* solution and *red* in *acidic* solution.

$$\underset{\text{red}}{\text{HLit}} \rightleftharpoons H^+ + \underset{\text{blue}}{\text{Lit}^-}$$

Other acid–base indicators commonly used are phenolphthalein, methyl orange, and bromthymol blue.

NITROGEN COMPOUNDS AS H+ ACCEPTORS

When HCl is added to water, hydronium ions are formed. An unshared electron pair on an oxygen atom of a water molecule functions as a base by forming a coordinate covalent bond with a hydrogen ion from the HCl.

$$\underset{\text{base}}{H_2\ddot{O}:} + H^+ \longrightarrow \underset{\text{conjugate acid}}{[H_3\ddot{O}]^+}$$

An ammonia molecule ($:NH_3$) is analogous to a water molecule by virtue of having an unshared electron pair on the nitrogen atom. Consider the reaction between ammonia and hydrochloric acid.

$$\underset{\text{base}}{H_3N:}\,(g) + H^+\,(aq) + Cl^-\,(aq) \longrightarrow \underset{\text{conjugate acid}}{[NH_4]^+}\,(aq) + Cl^-\,(aq)$$

By using its unshared electron pair to form a coordinate covalent bond with a hydrogen ion, ammonia functions as a base. The result is an ammonium ion (NH_4^+), a species distinctly different than ammonia.

exercise 8-18

Describe why the reaction between ammonia and hydrochloric acid can be considered a neutralization reaction.

Amines are a group of compounds related to ammonia by having one or more alkyl groups attached directly to the nitrogen atom. The alkyl groups need not all be alike. The simplest open-chain amines are named by indicating the alkyl group(s) name(s) and adding the suffix *amine.*

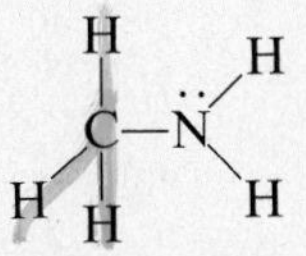

H_3C-NH_2

methylamine

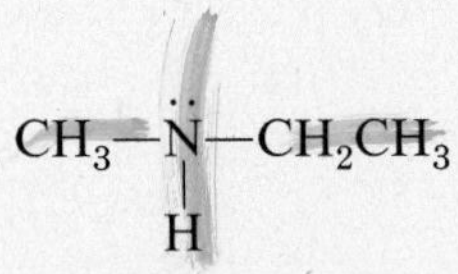

$CH_3-\ddot{N}H-CH_2CH_3$

methylethylamine

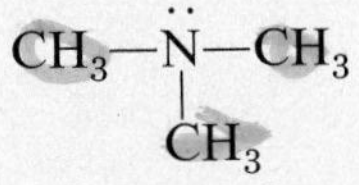

$CH_3-\ddot{N}(CH_3)-CH_3$

trimethylamine

$H_2\ddot{N}-(CH_2)_6-\ddot{N}H_2$

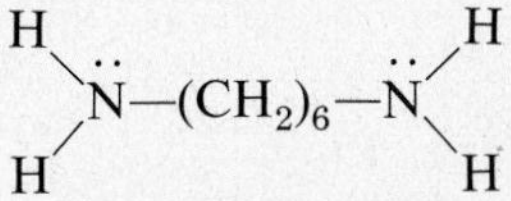

hexamethylenediamine

The amine nitrogen can also be part of a ring system or a side chain attached to a ring system (see Fig. 8-3). Note that one, two, or all three hydrogens on an ammonia molecule may be replaced by alkyl groups.

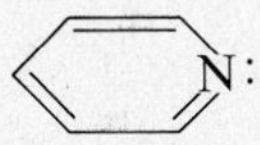

pyridine

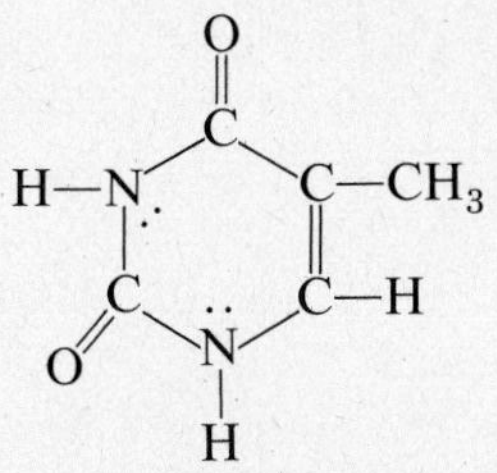

thymine

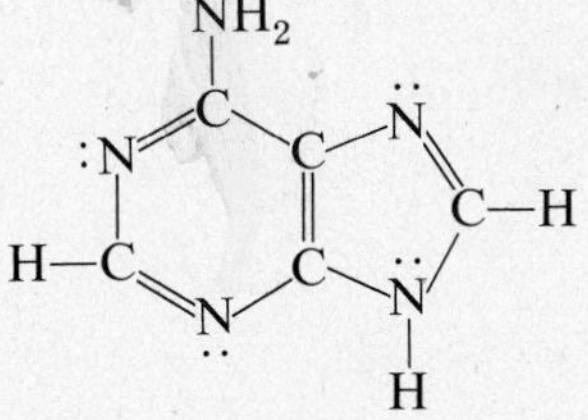

adenine

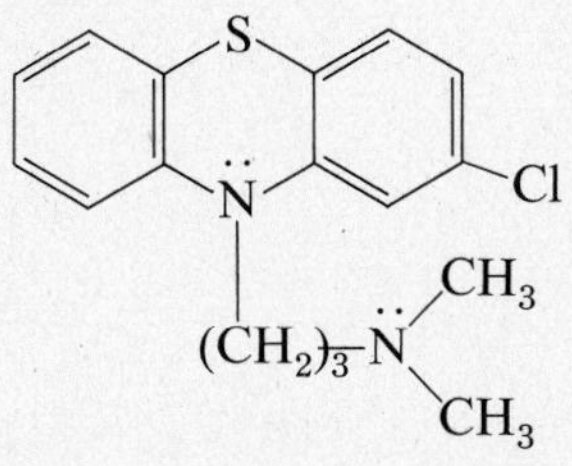

chlorpromazine (a tranquilizer)

FIGURE 8-3 Structures of several cyclic amines.

Many of the low molecular weight amines are volatile, water soluble, and odorous. Ever smell a fish? The odor is primarily due to the presence of amines in the fish's body fluid, e.g., dimethylamine and trimethylamine are common constituents. Putrescine [$H_2N(CH_2)_4NH_2$] and cadaverine [$H_2N(CH_2)_5NH_2$] are decomposition products of flesh and both names are rather descriptive.

Because of the presence of unshared electrons on nitrogen, amine molecules can also act as hydrogen ion acceptors. Methyl amine reacts with hydrochloric acid to produce methylammonium chloride.

$$CH_3NH_2\text{:}\ (g) + HCl\ (aq) \longrightarrow [CH_3NH_3]^+ + Cl^-\ (aq)$$

base — conjugate acid methylammonium — chloride

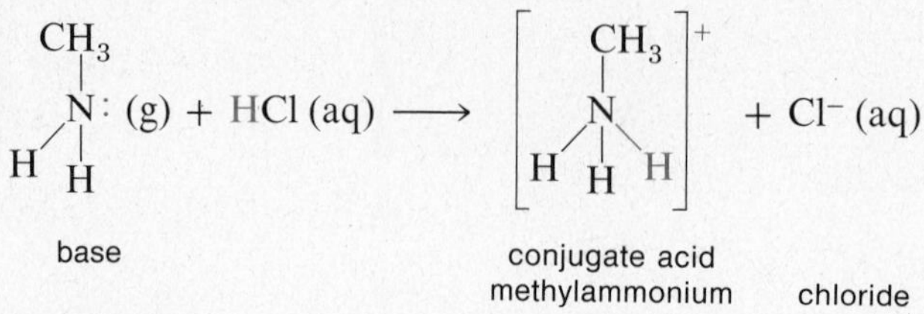

This reaction is analogous to the one between ammonia and HCl. Methylammonium chloride is water soluble, which isn't surprising in view of the fact that it is an ionic compound.

exercise 8-19

What type of reagent could react with methylammonium chloride to regenerate methylamine directly?

Higher molecular weight amines are usually not water soluble. Included in this group are various pharmaceuticals, some of which may require administration by injection. The question becomes how to convert the amine to a water-soluble form suitable for injection. By using the hydrogen ion accepting property of the amine molecule's nitrogen, a water-soluble salt is formed. Hydrochloric acid is commonly used for this purpose and the resulting salt is named using the suffix *hydrochloride*. Chlorpromazine is an anxiety-reducing drug. Its administration can lessen patient irritability and activity without diminishing mental capability. Although the drug is not water soluble, it can be converted to its soluble hydrochloride by reaction with hydrochloric acid.

$$\text{chlorpromazine (ring N–}(CH_2)_3\text{–}N(CH_3)_2) + HCl\ (aq) \longrightarrow \text{(ring N–}(CH_2)_3\text{–}N^{\oplus}H(CH_3)_2) + Cl^-\ (aq)$$

chlorpromazine — chlorpromazine hydrochloride

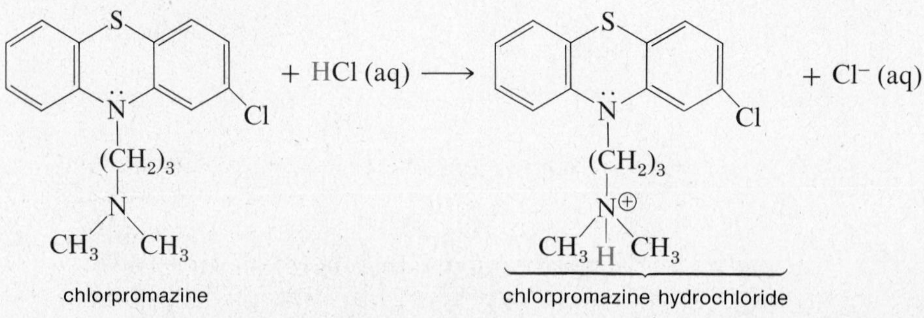

exercise 8-20

In addition to intramuscular injection of the hydrochloride, chlorpromazine can also be taken orally. How is it possible for the body to absorb this insoluble material?

exercise 8-21

Novocaine is the common name used for the local anesthetic procaine hydrochloride. The structure of procaine is

$$\text{H}_2\text{N}-\text{C}_6\text{H}_4-\text{C}(=\text{O})-\text{O}-(\text{CH}_2)_2-\text{N}(\text{CH}_2\text{CH}_3)_2$$

Write the structural formula for the hydrochloride formed by reacting hydrochloric acid and procaine.

AMINO ACIDS Since the carboxyl group $\left(-\overset{\displaystyle O}{\overset{\|}{C}}-OH\right)$ can release a hydrogen ion and an amine group $\left(-N\begin{matrix} R \\ R \end{matrix}\right)$ (R represents hydrogen, an alkyl group, or a ring structure) acts as a base, any molecule in which both of these functional groups are present should have some rather unusual but useful properties. In nature, approximately 20 compounds are found that fit this description. They are called *amino acids* and their structural formulas and names are given in Table 8-8. Take time to examine several features of these compounds carefully. Note a similarity of their molecular structure—the amino group is on the carbon adjacent to the carboxyl group carbon. This adjacent carbon is called the *alpha* (α) carbon; hence, the amino acids are called *alpha amino acids.* In Chapter 11 considerable attention is given to reactions between amino acids that lead to the formation of extended sequences of amino acids called *proteins*.

All the naturally occurring amino acids can be represented as

$$\text{R}-\text{C}(\text{H})(\text{NH}_2)-\text{C}(=\text{O})-\text{OH}$$

R is a substituent group bonded to the alpha carbon. The variation in the nature of R gives amino acids their diversity. Another feature to note is that *most* amino acids contain *an equal number of carboxyl and amine groups;* these amino acids are called *neutral amino acids,* e.g., alanine, serine, and proline. Those containing *more carboxyl than amine groups* are called *acidic*

Which amino acids are acidic amino acids? Basic ones?

TABLE 8-8
Common amino acids (R grouping underlined)

$HO-C(=O)-C(H)(NH_2)-\underline{R}$	General formula
$\underline{-H}$	Glycine
$\underline{-CH_3}$	Alanine
$\underline{-C(H)(CH_3)-CH_3}$	Valine*
$\underline{-CH_2-C(H)(CH_3)-CH_3}$	Leucine*
$\underline{-C(H)(CH_3)-CH_2-CH_3}$	Isoleucine*
$\underline{-CH_2-OH}$	Serine
$\underline{-C(H)(OH)-CH_3}$	Threonine*
$\underline{-CH_2-SH}$	Cysteine
$HO-C(=O)-C(H)(NH_2)-CH_2-S-S-\underline{CH_2}-C(H)(NH_2)-C(=O)-OH$	Cystine
$\underline{-CH_2CH_2-S-CH_3}$	Methionine*
$\underline{-CH_2-C(=O)-OH}$	Aspartic acid
$\underline{-CH_2CH_2-C(=O)-OH}$	Glutamic acid
$\underline{-(CH_2)_4-NH_2}$	Lysine*
$\underline{-CH_2CH_2CH(OH)CH_2NH_2}$	Hydroxylysine
$HO-C(=O)-C(H)(NH_2)-\underline{R}$	General formula
$\underline{-CH_2-C_6H_5}$ (benzene ring)	Phenylalanine*
$\underline{-CH_2-C_6H_4-OH}$ (benzene ring)	Tyrosine
$\underline{-(CH_2)_3-N(H)-C(=NH)-NH_2}$	Arginine
$\underline{-CH_2-C{=}CH,\ H-N,\ N,\ C-H}$ (imidazole ring)	Histidine
$\underline{-CH_2-}$ (indole ring, $N-H$)	Tryptophan*
$HO-C(=O)-C(H)-\underline{CH_2-CH_2-CH_2}-N(H)$ (ring)	Proline
$HO-C(=O)-C(H)-\underline{CH_2-C(H)(OH)-CH_2}-N(H)$ (ring)	Hydroxyproline

*These are the *essential* amino acids that cannot be synthesized by the body at a rate sufficient for normal growth and body metabolism. Therefore, they *must* be components among proteins in the normal diet. Nutritional studies have shown these amino acids to be indispensable for proper body well-being. The term *essential* is an unfortunate choice since all amino acids are essential for normal body metabolism.

amino acids; those whose molecules have *more amine than carboxyl groups* are termed *basic amino acids.* Also note the formation of cystine by coupling two cysteine molecules and ring formation in proline and hydroxyproline by using the amine nitrogen.

Amino acids are crystalline solids having high melting temperatures in relation to their low molecular weight. In addition, most are readily water soluble. It appears that they meet the criteria expected of ionic compounds; but how can a carbon compound achieve such status? An amino acid contains a carboxyl group capable of partially dissociating with the liberation of hydrogen ions like any normal carboxylic acid.

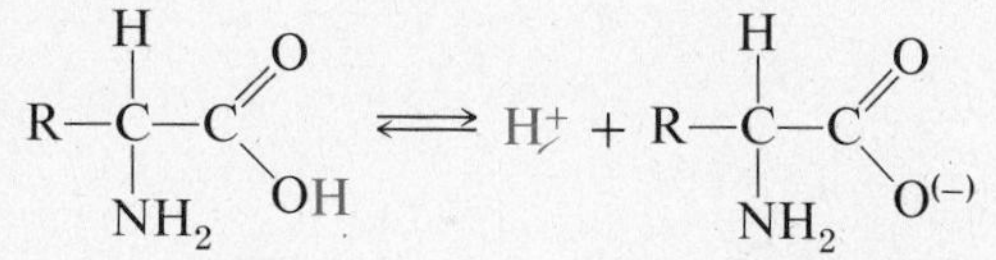

$$\underset{\text{acid}}{R-\overset{\displaystyle H}{\underset{\displaystyle NH_2}{\overset{|}{\underset{|}{C}}}}-C\begin{matrix}\nearrow O\\ \searrow OH\end{matrix}} \rightleftharpoons H^+ + \underset{\text{conjugate base}}{R-\overset{\displaystyle H}{\underset{\displaystyle NH_2}{\overset{|}{\underset{|}{C}}}}-C\begin{matrix}\nearrow O\\ \searrow O^{(-)}\end{matrix}}$$

In general then, equilibrium can be established. The equilibrium can be disrupted if hydrogen ions are removed. The molecule has a built-in hydrogen ion acceptor, the amine group. Net result:

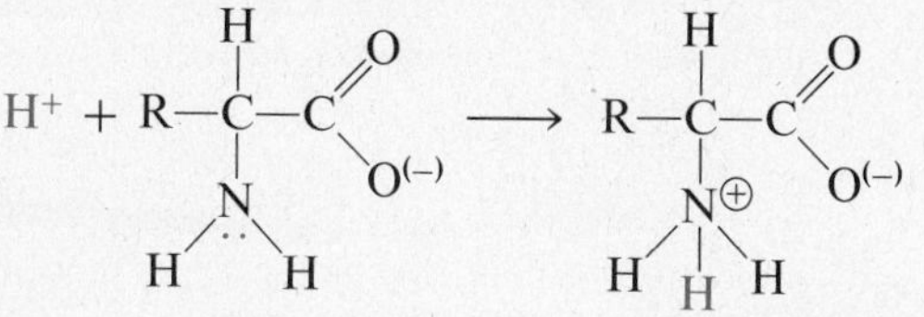

$$H^+ + \underset{\text{base}}{R-\overset{\displaystyle H}{\underset{\displaystyle \ddot{N}H_2}{\overset{|}{\underset{|}{C}}}}-C\begin{matrix}\nearrow O\\ \searrow O^{(-)}\end{matrix}} \longrightarrow \underset{\text{conjugate acid (net charge is zero, zwitterion form)}}{R-\overset{\displaystyle H}{\underset{\displaystyle N^{\oplus}H_3}{\overset{|}{\underset{|}{C}}}}-C\begin{matrix}\nearrow O\\ \searrow O^{(-)}\end{matrix}}$$

Zwitterion: a covalent compound containing equal numbers of positive and negative charges.

This act of internal, self-neutralization is so highly favored that amino acids essentially exist in the zwitterion form. Thus, they resemble ionic compounds.

exercise 8-22

Using Table 8-8, write structural formulas for the zwitterion form of each of the following amino acids: glycine, valine, tyrosine, proline, arginine, and glutamic acid.

Why do basic and acidic amino acids have different isoelectric points?

Note that in the zwitterion form the molecule has no **net** charge; it is said to be *isoelectric*. Zwitterion formation is dependent on the hydrogen ion concentration of the solution. Appropriate adjustment of the pH of the solution can ensure that the amino acid molecules present will be in the isoelectric form. *The pH at which an amino acid exists in its isoelectric (zwitterion) form is called the isoelectric point* for that amino acid. Basic and acidic amino acids would have differing isoelectric points. The tendency for H^+ dissociation and acceptance among neutral amino acids is not quite identical and thus each possesses a slightly different isoelectric point.

AMINO ACIDS AS BUFFERS In its zwitterion form, an amino acid can function as a buffer A solution of alanine at its isoelectric point can act accordingly:

$$H_3C-CH(NH_3^{\oplus})-COO^{(-)}$$

no ***net*** *charge*
(alanine-zwitterion form)

added H^+ → $H_3C-CH(NH_3^{\oplus})-COOH$ ***net*** *charge,* +

added base → $HBase + H_3C-CH(\ddot{N}H_2)-COO^{(-)}$ ***net*** *charge,* –

exercise 8-23

Starting with the zwitterion form, write equations using structural formulas to show the buffering action of each of the following amino acids: leucine, valine, and serine.

Amino acids are the *building blocks* of proteins. Because not all amino acids are neutral, the presence of acidic or basic amino acids provide *extra* carboxyl and amine groups in the protein. Thus, proteins can function as buffers and their presence in blood helps to maintain blood pH at or near 7.4. Proteins are the major buffers in the blood.

pH VARIATION AND AMINO ACID STRUCTURE It was mentioned previously that a buffer has a limited capacity for offsetting acid or base addition. Because its isoelectric point is pH dependent, what might be expected if a solution of amino acid is swamped by addition of *excess* base or acid? Under these conditions neither the undissociated amino acid nor the zwitterion form is present. If *excess acid* is present, the *positive species* is mainly present (see page 205); *excess base* causes the net *negatively charged species* to *predominate*.

Does the idea of net charges and pH adjustment have any practical application? Mixtures of amino acids can be separated using these principles. Because glutamic acid contains an additional carboxyl group and lysine an additional amine group, the isoelectric point of these two amino acids will differ and also be different than that of the neutral amino acid alanine. The solution pH will govern whether the *net* charge on each of these amino acids is positive, negative, or no net charge (isoelectric). The pH of a solution containing these three amino acids can be adjusted such that each amino acid in the mixture will have a different charge. Placing a positive and

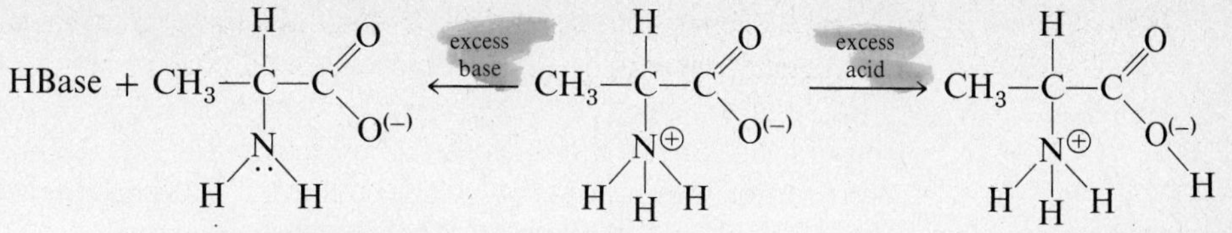

$$\text{HBase} + CH_3\text{—}CH(NH_2)\text{—}COO^{(-)} \xleftarrow{\text{excess base}} CH_3\text{—}CH(NH_3^{\oplus})\text{—}COO^{(-)} \xrightarrow{\text{excess acid}} CH_3\text{—}CH(NH_3^{\oplus})\text{—}COOH$$

basic form
(net charge, −1)

alanine
(zwitterion form,
no net charge)

acidic form
(net charge, +1)

$$HOOC\text{—}CH_2CH_2\text{—}CH(NH_3^{\oplus})\text{—}COO^{(-)}$$

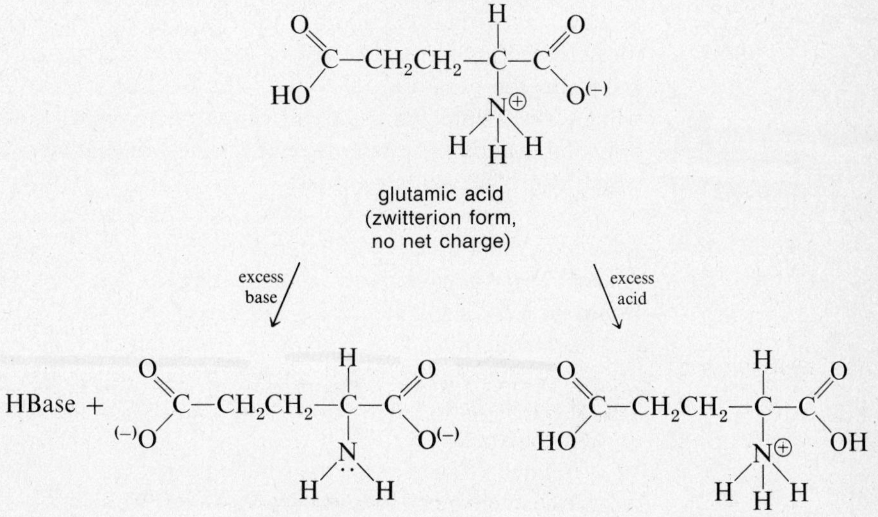

glutamic acid
(zwitterion form,
no net charge)

excess base ↙ excess acid ↘

$$\text{HBase} + {}^{(-)}OOC\text{—}CH_2CH_2\text{—}CH(NH_2)\text{—}COO^{(-)} \qquad HOOC\text{—}CH_2CH_2\text{—}CH(NH_3^{\oplus})\text{—}COOH$$

basic form
(net charge, −2)

acidic form
(net charge, +1)

$$H_2N\text{—}(CH_2)_4\text{—}CH(NH_3^{\oplus})\text{—}COO^{(-)}$$

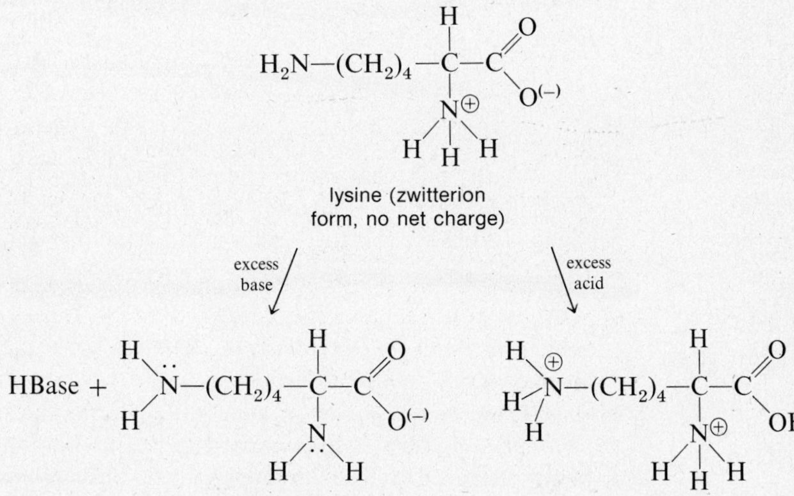

lysine (zwitterion
form, no net charge)

excess base ↙ excess acid ↘

$$\text{HBase} + H_2N\text{—}(CH_2)_4\text{—}CH(NH_2)\text{—}COO^{(-)} \qquad H_3N^{\oplus}\text{—}(CH_2)_4\text{—}CH(NH_3^{\oplus})\text{—}COOH$$

basic form
(net charge, −1)

acidic form
net charge, +2)

negative electrode into the solution creates an electric field causing migration of the negative species to the positive electrode; positive species are attracted to the negative electrode. Because they lack a *net* charge, zwitterion isoelectric species are immobile in the electric field. Thus separation of the mixture into its components can be achieved. Such a technique using an electrical field and pH control is called *electrophoresis*.

An interesting chemical situation relating pH to amino acid solubility is found in the condition cystinuria. Individuals suffering from this condition have a genetic defect in their kidney tubules that prevents cystine from being retained in normal circulation. Instead, cystine is released into the urine in increased amounts. The pH of urine may vary from 4.8 to 8.4, usually falling in the range of 5.0 to 7.0. The isoelectric point of cystine is 5.02, which falls within the usual pH range of urine. At a pH of 5.02, cystine is insoluble and precipitates from the urine as insoluble masses called *calculi*, which can block kidney tubules.

Amino acids are least soluble at their isoelectric point.

pH AND OXYGEN TRANSPORT

To meet normal metabolic demands, the human body requires an enormous amount of oxygen on a continuous basis. The solubility of oxygen in blood would be very slight if not for the presence of the oxygen carrier, hemoglobin. Hemoglobin functions as a reversible oxygen carrier, picking up oxygen in the lungs and releasing it in tissues for use in cellular metabolism. The maintenance of a relative constancy of 7.4 of blood pH is vital for oxygen transport.

Inhaled oxygen from the lungs diffuses into the capillaries due to a difference in oxygen concentrations between the lungs and capillaries. Upon entering a capillary, oxygen diffuses into the red blood cells where it combines with hemoglobin (HHb) to form oxyhemoglobin ($HHbO_2$). Combination with oxygen promotes the loss of a hydrogen ion from histidine in hemoglobin. The oxygenation reactions may be summarized

$$O_2 + HHb \rightleftharpoons HHbO_2 \rightleftharpoons H^+ + HbO_2^- \qquad (8\text{-}1)$$

Hemoglobin Oxyhemoglobin

From our very experience of being alive we know that oxygen uptake and release occurs; but what causes the equilibria to be shifted in our favor as needed? Recalling Le Chatelier's principle, high oxygen concentration in the lungs favors the forward reaction. Additionally, bicarbonate ions in the red blood cells combine with hydrogen ions forming carbonic acid (H_2CO_3);

$$HCO_3^- + H^+ \rightleftharpoons H_2CO_3$$

bicarbonate ions

Both of these factors favor oxyhemoglobin formation.

Oxyhemoglobin is then carried to the tissues where carbon dioxide is being produced by cellular metabolism. The CO_2 diffuses into the red blood cells where it forms carbonic acid by combining with water. Carbonic acid dissociation provides bicarbonate and hydrogen ions:

$$H_2O + CO_2 \rightleftharpoons H_2CO_3 \rightleftharpoons H^+ + HCO_3^- \qquad (8\text{-}2)$$

Why do HCO_3^- ions diffuse out of the red blood cells as the HCO_3^- concentration increases?

What does release of hydrogen ions do to the equilibria in Eq. 8-1? It places a stress on the equilibria; oxyhemoglobin combines with hydrogen ion forming $HHbO_2$, which splits into oxygen and hemoglobin. The oxygen diffuses into the cells for use in cellular metabolism. Thus oxygen transport is a pH controlled process.

According to Eq. 8-2, as hydrogen ions combine with oxyhemoglobin, bicarbonate ion concentration in the red blood cells increases. When this occurs, bicarbonate ions diffuse out of the red blood cell into the plasma. To balance the loss of negative ions by the red blood cell, chloride ions migrate from the plasma into the cell, a process called the *chloride* shift.

RESPIRATORY ACIDOSIS AND ALKALOSIS

From the preceding discussion, carbonic acid and bicarbonate levels in the blood, if uncompensated, can change the blood's pH. By decreasing the amount of breathing (hypoventilation), insufficient carbon dioxide is released. Holding your breath will do this. *The increased level of carbon dioxide in the blood leads to increased carbonic acid and hydrogen ion levels and respiratory acidosis results.* Fainting can occur from mild acidosis; coma can be the result of severe acidosis. Alternatively, by hyperventilating (overbreathing), an excessive amount of carbon dioxide is expelled. Equation 8-2 indicates that excessive loss of carbon dioxide shifts the equilibrium to compensate for CO_2 loss. Hydrogen ions and bicarbonate ions combine to form carbonic acid, which decomposes to water and carbon dioxide. *The reduction in hydrogen ion concentration leads to respiratory alkalosis.*

Acidosis results from an increase in H^+ concentration; alkalosis, from a decrease in H^+ concentration.

exercise 8-24

A first aid measure for a patient in alkalosis due to hyperventilating is to have the patient breathe into a paper bag. Explain the basis of this procedure in view of the ideas presented above.

METABOLIC ACIDOSIS*— DIABETES MELLITUS

Normal carbohydrate metabolism results in the formation of acidic byproducts, which can be termed *metabolic acids*. Lactic, citric, and pyruvic acids are examples of metabolic acids. Their concentration is ordinarily such that normal body buffer action prevents continuing acidosis. In diabetes mellitus, *abnormal* carbohydrate metabolism leads to acidosis. Lack of insulin secretion by the pancreas prevents glucose from being metabolized. In its continuing need for energy, the body instead metabolizes stored fats. Two of the products of fat metabolism are

*Metabolic alkalosis is not nearly as common as metabolic acidosis. Ingestion of excessive alkaline drugs such as sodium bicarbonate or excessive loss of stomach hydrochloric acid through excessive vomiting of gastric contents are two causes of occasional metabolic alkalosis.

acetoacetic acid $\left(CH_3-\overset{\overset{\displaystyle O}{\|}}{C}-CH_2-C\begin{matrix}\nearrow O\\ \searrow OH\end{matrix}\right)$ and β-hydroxybutyric acid $\left(CH_3-\underset{\underset{\displaystyle H}{|}}{\overset{\overset{\displaystyle OH}{|}}{C}}-CH_2-C\begin{matrix}\nearrow O\\ \searrow OH\end{matrix}\right)$. Because these acids are formed continuously by a person suffering from untreated diabetes mellitus, their dissociation increases the hydrogen ion concentration in extracellular fluids causing metabolic acidosis. If untreated, severe acidosis leads to coma and even death.

STRONG AND WEAK ACIDS—A REVIEW

In this chapter, acids have been classified as strong or weak according to their dissociation behavior; the former essentially dissociating completely and the latter only slightly. Weak acids differ in their tendency to dissociate and we used dissociation constant values to differentiate among them; the smaller the value of K_a, the weaker the acid. The incomplete dissociation of weak acids allows establishment of the equilibrium

$$\underset{\text{acid}}{HB} \rightleftharpoons H^+ + \underset{\text{conjugate base}}{B^-}$$

The conjugate base plays an important role in the extent to which an acid dissociates. A conjugate base B^- that avidly accepts hydrogen ions causes the acid HB to be largely undissociated, i.e., a weak acid. Conversely, a conjugate base B^- that is a poor hydrogen ion acceptor essentially allows the complete dissociation of the acid HB, i.e., a strong acid. We can summarize by stating a rule of thumb: *Strong acids have weak conjugate bases; weak acids have strong conjugate bases. The stronger the conjugate base, the weaker its acid will be.* Table 8-9 is a compilation of relative acid strengths based on extensive experimental data. In aqueous solution, the strength of any acid is relative to the $B^- + H_3O^+ \rightleftharpoons H_2O + HB$ equilibrium.

CHAPTER SUMMARY

1. The dissociation constant for water K_w is

$$[H_3O^+] \times [OH^-] = 1 \times 10^{-14}$$

The product of the two terms is constant; as $[H_3O^+]$ changes, $[OH^-]$ changes in the opposite way. Aqueous solutions can be neutral, acidic, or basic (alkaline). Aqueous solutions in which $[H_3O^+] = [OH^-]$ are

TABLE 8-9
Relative acid strengths in aqueous solution

ACID STRENGTH DECREASING ↓ — CONJUGATE BASE STRENGTH INCREASING ↓

ACID		K_a	CONJUGATE BASE
$HClO_4$	Perchloric	Very large	ClO_4^- (perchlorate)
H_2SO_4	Sulfuric	Very large	HSO_4^- (monohydrogen sulfate)
HCl	Hydrochloric	Very large	Cl^- (chloride)
HNO_3	Nitric	Very large	NO_3^- (nitrate)
H_3O^+	Hydronium ion		H_2O
HSO_4^-	Monohydrogen sulfate	2×10^{-2}	SO_4^{2-} (sulfate)
H_3PO_4	Phosphoric	8×10^{-3}	$H_2PO_4^-$ (dihydrogen phosphate)
$CH_3—C(=O)—C(=O)OH$	Pyruvic	3×10^{-3}	$CH_3—C(=O)—C(=O)O^{(-)}$ (pyruvate)
$C_6H_4(OH)—C(=O)OH$	Salicylic	1×10^{-3}	$C_6H_4(OH)—C(=O)O^{(-)}$ (salicylate)
$C_6H_8O_6$	Citric	8×10^{-4}	citrate
$CH_3—CH(OH)—C(=O)OH$	Lactic	1.4×10^{-4}	$CH_3—CH(OH)—C(=O)O^{(-)}$ (lactate)
$C_6H_8O_6$	Ascorbic	8×10^{-5}	ascorbate
$CH_3—C(=O)OH$	Acetic	1.8×10^{-5}	$CH_3—C(=O)O^{(-)}$ (acetate)
$H—O—C(=O)—OH$	Carbonic	4.5×10^{-7}	$H—O—C(=O)—O^{(-)}$ (bicarbonate)
NH_4^+	Ammonium ion	5.6×10^{-10}	NH_3 (ammonia)
HCO_3^-	Bicarbonate ion	1×10^{-11}	CO_3^{2-} carbonate

termed neutral; where $[H_3O^+] > [OH^-]$, these solutions are acidic; basic solutions are those in which $[H_3O^+] < [OH^-]$. The terms H_3O^+ and H^+ are used interchangeably.

2. $[H^+]$ and pH are related by the expression

$$[H^+] = n \times 10^{-pH}$$

As $[H^+]$ decreases, pH increases; as $[H^+]$ increases, pH decreases. A solution having a pH less than 7.0 is acidic; a pH of 7.0 indicates a neutral solution; a basic solution's pH is greater than 7.0.

3. Salts are formed by the reaction between an acid and a base. Acids are hydrogen ion donors; bases are hydrogen ion acceptors. Neutralization occurs when hydrogen ions from an acid react with hydroxide ions or other bases. Titrations can be used to determine the concentration of either a base or an acid.
4. Acids that only partially dissociate are called *weak acids; strong acids* are essentially completely dissociated. An acid forms its conjugate base by losing hydrogen ions. Weak acids have strong conjugate bases; strong acids have weak conjugate bases.
5. K_a is the dissociation constant for an acid. Generalizing,

$$\mathrm{HB} \rightleftharpoons \mathrm{H}^+ + \mathrm{B}^- \qquad K_a = \frac{[\mathrm{H}^+] \times [\mathrm{B}^-]}{[\mathrm{HB}]}$$

   Weak acids have K_a values less than 1; the weaker the acid, the smaller its K_a value.
6. The K_a expression can be used to calculate [H$^+$], [conjugate base], or the concentration of undissociated acid at equilibrium.
7. A buffer maintains the pH of a solution relatively constant when limited quantities of acids or bases are added from external sources. Buffers generally contain a weak acid and a soluble salt containing the conjugate base. The weak acid neutralizes added base; the conjugate base neutralizes added acid.
8. Body fluids contain three major buffer systems: (a) HCO_3^-/H_2CO_3; (b) $HPO_4^{2-}/H_2PO_4^-$; and (c) proteins, the major buffer. Blood is normally buffered to maintain a pH of 7.4.
9. Nitrogen compounds accept hydrogen ions by using unshared electron pairs on the nitrogen atoms for bonding with hydrogen ions.
10. Amino acid molecules contain at least one carboxylic acid group ($-CO_2H$) and an amine group [$-NH_2(R_2)$] on the carbon atom adjacent to the acid group (the alpha carbon). Amino acids have the general formula

$$\mathrm{R}-\underset{\displaystyle \mathrm{NH_2}}{\overset{\displaystyle \mathrm{H}}{\underset{|}{\overset{|}{\mathrm{C}}}}}-\mathrm{CO_2H}$$

   where R is a side chain that may also contain either acid or amine groups. Amino acids whose molecules contain equal numbers of amine and acid groups are termed *neutral* amino acids; those with more acid groups than amine groups are called *acidic* amino acids; *basic* amino acids have molecules that contain more amine groups than acid groups.
11. Acid groups in an amino acid can donate hydrogen ions; amine groups in amino acids can accept hydrogen ions. Thus, the net charge of an

amino acid molecule is determined by the pH of the solution. The isoelectric point of an amino acid is the pH at which its molecules have no net charge.

12. Oxygen uptake by hemoglobin to form oxyhemoglobin is reversible and pH dependent. Hemoglobin releases hydrogen ions when combining with oxygen to form oxyhemoglobin. An increase in hydrogen ion concentration causes the reverse reaction to occur, releasing oxygen.
13. Carbonic acid, formed by the combination of carbon dioxide and water, dissociates into hydrogen ions and bicarbonate ions:

$$CO_2 + H_2O \rightleftharpoons H_2CO_3 \rightleftharpoons H^+ + HCO_3^-$$

Hypoventilation leads to respiratory acidosis due to CO_2 buildup; hyperventilation can cause respiratory alkalosis. Both conditions result from stress on the carbonic acid equilibria.

14. Diabetes mellitus causes the production of acidic waste products. If untreated, this overtaxes the blood's buffer systems and metabolic acidosis results.

QUESTIONS

1. Vinegar, lemon juice, or orange juice are commonly listed as antidotes in the home first aid treatment of someone who has swallowed lye (sodium hydroxide). What chemical principle is the basis for this procedure?
2. A home remedy to relieve *acid indigestion* is to drink a baking soda ($NaHCO_3$) solution. Explain its action in relieving excess stomach acid distress.
3. What will be the charge, if any, on the species produced when
   (a) Bicarbonate ion loses a hydrogen ion?
   (b) Ammonium ion loses a hydrogen ion?
   (c) Carbonate ion gains a hydrogen ion?
   (d) Acetic acid loses a hydrogen ion?
   (e) Sulfate ion gains two hydrogen ions?
   (f) Dihydrogen phosphate ion gains a hydrogen ion?
4. Select the conjugate acid-base *pairs* for each of the following:
   (a) $HPO_4^{2-} + H_2O \rightleftharpoons H_2PO_4^- + OH^-$
   (b) $\underset{\text{oxyhemoglobin}}{HHbO_2} + HCO_3^- \rightleftharpoons H_2CO_3 + HbO_2^-$
   (c) $CH_3{-}NH_2 + HCl \rightleftharpoons (CH_3{-}NH_3)^+ + Cl^-$
5. Usually it requires 37.0 ml of 0.05 *M* NaOH to neutralize 100.0 ml of gastric juice. Assuming HCl to be the only acid present in the gastric juice,

(a) Calculate the molarity of the HCl present.
(b) Calculate the number of grams of HCl per 100 ml of gastric juice.

6. Vinegar and lemon juices are often used on salads. Could an equal volume of 0.1 *M* HCl be used as well?
7. When lemon juice is added to tea, the color of the tea changes. Explain.
8. Under normal metabolic conditions, the body's daily lactic acid production is about 0.100 mole. Using data from Table 8-9, answer the following:
   (a) How many grams of lactic acid are produced daily?
   (b) What is the molarity of a solution resulting from dissolving 0.100 mole of lactic acid in sufficient water to make 35 ml of solution?
   (c) Calculate the hydrogen ion concentration of solution (b) above. Approximate square root values that may occur during calculations.
   (d) What is the approximate pH of solution (c)?
9. Asparagus contains aspartic acid (H_2Asp). If the dissociation of aspartic acid is

$$H_2Asp \rightleftharpoons 2H^+ + Asp^{2-}$$

write the equilibrium constant expression for aspartic acid.

10. Blood is normally buffered to a pH of 7.4 (hydrogen ion concentration of 4×10^{-8} *M*). *If* bicarbonate and carbonic acid were the only buffers present, calculate the ratio of H_2CO_3/HCO_3^- required for this buffering action.
11. Using data regarding the monohydrogen and dihydrogen phosphate buffer system given in the chapter, calculate the hydrogen ion concentration in blood (assuming no other buffers are present).
12. Urea ($H_2N{-}\overset{\overset{\displaystyle O}{\|}}{C}{-}NH_2$) can function as a base when reacting with dilute hydrochloric acid. Write an equation to represent this reaction.
13. Is the chloride ion concentration of venous red blood cells greater than, equal to, or less than that of arterial red blood cells? Explain your answer.

9
alcohols and their derivatives

INTRODUCTION

The structures and reactions of most biochemically important molecules are quite complex. Fortunately, their structures and reactions can be readily learned and understood in terms of the behavior of simpler organic compounds. To gain the background necessary to understand biological reactions we shall study several classes of the simpler organic compounds here and in Chapter 10.

Our study of organic compounds will be limited to those classes of compounds and reactions having the most numerous applications to biochemical reactions. You should, however, be aware that we will only be taking a very brief glimpse at organic chemistry. This area of study is the largest of the traditional fields of chemistry (general, inorganic, analytical, organic, and physical), involving over 40 percent of the world's chemists. Our knowledge of organic chemistry makes possible the manufacture of an almost endless list of products including such commodities as gasoline, paints, plastics, synthetic fibers, inks, dyes, drugs, and perfumes. The field of organic chemistry is so enormous that many introductory organic textbooks are well over a thousand pages long.

Thousands of new organic compounds are synthesized yearly. Many have pharmaceutical value.

CLASSIFICATION AND NOMENCLATURE

In Chapter 4 we saw that an alcohol is characterized by a *hydroxyl group* (—O—H) bonded to a saturated carbon atom. Since the hydrocarbon group in a simple alcohol is an alkyl group, we may write ROH as a general symbol for simple alcohols. Alcohols vary in their water solubility. Low molecular weight alcohols, those containing up to five or six carbon atoms, are water soluble by virtue of their ability to hydrogen bond with water. Monohydroxy alcohols containing more than six carbon atoms are water insoluble as the behavior of the hydrocarbon portion of the molecule predominates.

The possibility of structural isomerism exists if the alcohol molecules contain three or more carbon atoms. Alcohols are divided into subclasses depending on the number of carbon atoms bonded to the carbon atom attached to the hydroxyl group. The four possible structures for alcohols with the molecular formula $C_4H_{10}O$ are shown in Fig. 9-1. A *primary* alcohol is one in which the hydroxyl group is attached to a carbon atom forming only *one* carbon–carbon bond—structures (a) and (b) in Fig. 9-1. *Secondary* alcohols—structure (c)—have *two* carbon–carbon bonds on the carbon with the hydroxyl group, while *tertiary* alcohols—structure (d)—have *three.* General symbols for each of the alcohol subclasses are

Alcohols are classified as being primary, secondary, or tertiary.

$$\underset{\text{primary alcohol}}{\mathrm{R{-}\overset{\displaystyle H}{\underset{\displaystyle H}{\overset{|}{\underset{|}{C}}}}{-}OH}} \qquad \underset{\text{secondary alcohol}}{\mathrm{R{-}\overset{\displaystyle R}{\underset{\displaystyle H}{\overset{|}{\underset{|}{C}}}}{-}OH}} \qquad \underset{\text{tertiary alcohol}}{\mathrm{R{-}\overset{\displaystyle R}{\underset{\displaystyle R}{\overset{|}{\underset{|}{C}}}}{-}OH}}$$

The simplest alcohol, CH_3OH, is also classed as a primary alcohol although it does not have a carbon–carbon bond.

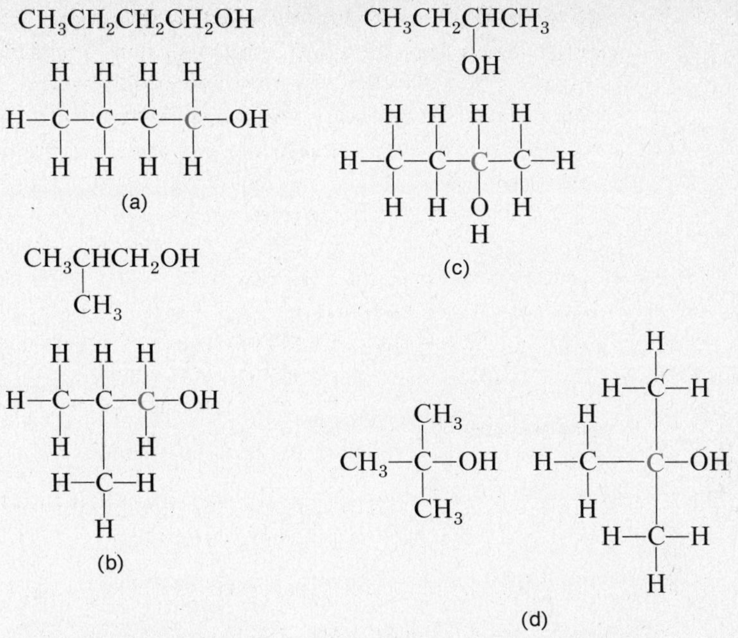

$CH_3CH_2CH_2CH_2OH$

$$\begin{array}{ccccccccc} & H & & H & & H & & H & \\ & | & & | & & | & & | & \\ H- & C & - & C & - & C & - & C & -OH \\ & | & & | & & | & & | & \\ & H & & H & & H & & H & \end{array}$$

(a)

$CH_3CH_2CHCH_3$ (with OH on the third carbon)

$$\begin{array}{ccccccccc} & H & & H & & H & & H & \\ & | & & | & & | & & | & \\ H- & C & - & C & - & C & - & C & -H \\ & | & & | & & | & & | & \\ & H & & H & & O & & H & \\ & & & & & | & & & \\ & & & & & H & & & \end{array}$$

(c)

CH_3CHCH_2OH (with CH_3 on the second carbon)

$$\begin{array}{ccccccc} & H & & H & & H & \\ & | & & | & & | & \\ H- & C & - & C & - & C & -OH \\ & | & & | & & | & \\ & H & & | & & H & \\ & & H- & C & -H & & \\ & & & | & & & \\ & & & H & & & \end{array}$$

(b)

$$\begin{array}{ccc} & CH_3 & \\ & | & \\ CH_3- & C & -OH \\ & | & \\ & CH_3 & \end{array}
\qquad
\begin{array}{ccccc} & & & H & \\ & & & | & \\ & & H- & C & -H \\ & H & & | & \\ & | & & | & \\ H- & C & - & C & -OH \\ & | & & | & \\ & H & & | & \\ & & H- & C & -H \\ & & & | & \\ & & & H & \end{array}$$

(d)

FIGURE 9-1 Structures of alcohols with the formula $C_4H_{10}O$.

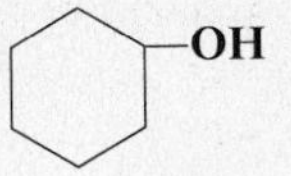

exercise 9-1

Classify each of the following as primary, secondary, or tertiary alcohols:

(a) CH_3CH_2OH

(b) $CH_3CH_2CH_2OH$

(c) CH_3CHCH_3 (with OH on the middle carbon)

(d) cyclohexane ring–OH

(e) cyclohexane ring bearing CH_3 and OH on the same carbon

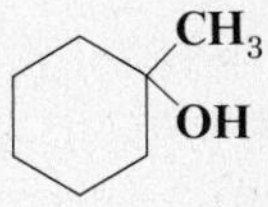

(f) cyclohexane ring–CH_2OH

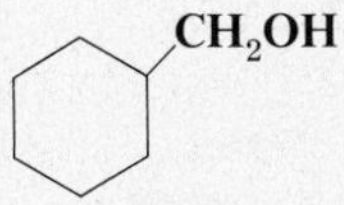

exercise 9-2

Write all the isomers for alcohols with the molecular formula $C_5H_{12}O$, and classify each as a primary, secondary, or tertiary alcohol.

The common and most widely used method of naming alcohols simply gives the name of the alkyl group followed by the word *alcohol.* Thus, CH_3OH, CH_3CH_2OH, and CH_3CHCH_3 (with HO on the middle carbon) are methyl alcohol, ethyl alcohol, and isopropyl alcohol, respectively. In the IUPAC system of naming, the *e* ending of the alkane name is replaced by an *ol*. The position of the OH group on the carbon chain is given by a numeral. The longest continous carbon chain containing the OH group is numbered to give the hydroxyl group the lowest value. Appropriate numbers and prefixes are used to indicate the presence and position of other substituents. For example,

$$\underset{\displaystyle CH_3}{CH_3\underset{|}{C}H}CH_2\underset{\displaystyle OH}{\underset{|}{C}H}CH_3$$

4-methyl-2-pentanol

Phenol is commonly called carbolic acid.

Common and IUPAC names for several simple alcohols are given in Table 9-1. Phenol, the last entry in the table, has a hydroxyl group attached to an aromatic ring instead of an alkyl group. Aromatic alcohols are collectively known as *phenols*, with the simplest member called *phenol*. In addition to their properties as alcohols, phenols are also very weakly acidic

TABLE 9-1
Common and IUPAC names of several simple alcohols

STRUCTURE	SUBCLASS	COMMON NAME	IUPAC NAME
CH_3OH	Primary	Methyl alcohol	Methanol
CH_3CH_2OH	Primary	Ethyl alcohol	Ethanol
$CH_3CH_2CH_2OH$	Primary	*n*-Propyl alcohol	1-Propanol
$CH_3\underset{\displaystyle OH}{\underset{\mid}{C}H}CH_3$	Secondary	Isopropyl alcohol	2-Propanol
$CH_3CH_2CH_2CH_2OH$	Primary	*n*-Butyl alcohol	1-Butanol
$CH_3\underset{\displaystyle CH_3}{\underset{\mid}{C}H}\ CH_2OH$	Primary	Isobutyl alcohol	2-Methyl-l-propanol
$CH_3CH_2\underset{\displaystyle OH}{\underset{\mid}{C}H}CH_3$	Secondary	Secondary butyl alcohol	2-Butanol
$CH_3—\overset{\displaystyle CH_3}{\underset{\displaystyle CH_3}{\overset{\mid}{\underset{\mid}{C}}}}—OH$	Tertiary	Tertiary butyl alcohol	2-Methyl-2-propanol
$\underset{\displaystyle OH}{\underset{\mid}{CH_2}}—\underset{\displaystyle OH}{\underset{\mid}{CH_2}}$	Primary diol	Ethylene glycol	1,2-Ethanediol
$\underset{\displaystyle OH}{\underset{\mid}{CH_2}}—\underset{\displaystyle OH}{\underset{\mid}{CH}}—\underset{\displaystyle OH}{\underset{\mid}{CH_2}}$	Triol	Glycerol or glycerine	1,2,3-Propanetriol
C_6H_5—OH (benzene ring with —OH)	Phenol (aromatic alcohol)	Phenol (carbolic acid)	Phenol

and quite irritating to tissue. Our primary interest in phenols will be in their behavior as alcohols.

exercise 9-3

Name the following alcohols by the IUPAC system.

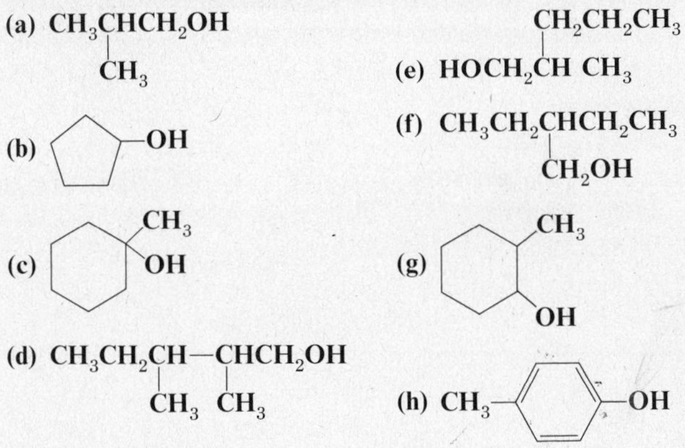

(a) $CH_3\underset{|}{C}HCH_2OH$ with CH_3 on the second carbon

(b) cyclopentane ring —OH

(c) cyclohexane ring with CH_3 and OH on the same carbon

(d) $CH_3CH_2CH—CHCH_2OH$ with CH_3 on the third and fourth carbons

(e) $HOCH_2CH\ CH_3$ with $CH_2CH_2CH_3$ on the CH

(f) $CH_3CH_2CHCH_2CH_3$ with CH_2OH on the middle carbon

(g) cyclohexane ring with CH_3 on one carbon and OH on the adjacent carbon

(h) CH_3—(benzene ring)—OH

exercise 9-4

Write structures for the following alcohols.
(a) isopropyl alcohol
(b) 2-methyl-2-heptanol
(c) cyclobutyl alcohol
(d) 3-methyl-1-cyclopentanol
(e) 1,3-propanediol

PREPARATION OF ALCOHOLS

Alcohols can be prepared in both the living cell and in the chemistry laboratory by the addition of water to a carbon–carbon double bond. In the laboratory, the reaction requires a strong acid catalyst—usually sulfuric acid.

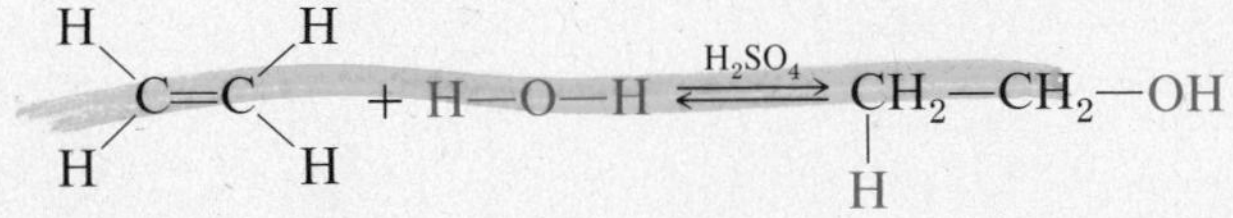

$$\begin{matrix} H & & H \\ & C{=}C & \\ H & & H \end{matrix} + H{-}O{-}H \xrightleftharpoons{H_2SO_4} \underset{\displaystyle H}{\underset{|}{CH_2}}{-}CH_2{-}OH$$

The reaction is a reversible one, and maximum yields of alcohol are obtained by using a large excess of water to drive the reaction to completion.

It is possible to write two different structures for the addition of water to most olefins. For example, the addition of water to propene could give rise to either 1-propanol or 2-propanol.

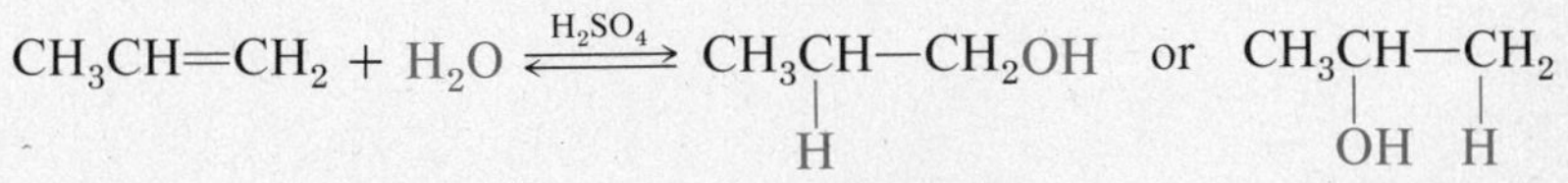

$$CH_3CH{=}CH_2 + H_2O \xrightleftharpoons{H_2SO_4} \underset{\text{1-propanol}}{CH_3\underset{\displaystyle H}{\underset{|}{C}}H{-}CH_2OH} \quad \text{or} \quad \underset{\text{2-propanol}}{CH_3\underset{\displaystyle OH}{\underset{|}{C}}H{-}\underset{\displaystyle H}{\underset{|}{CH_2}}}$$

Actually 2-propanol is the *predominant product* from the reaction, with only *trace* amounts or *no* 1-propanol produced. Studies of many addition reactions reveal whenever two products are possible, one is usually formed in large amounts and relatively little of the other is produced. These studies led to the formulation of Markovnikov's rule, enabling one to predict which isomer will predominate. Markovnikov's rule states *whenever an unsymmetrical reagent adds to a carbon–carbon double bond, the positive portion of the adding reagent bonds to the least substituted carbon atom of the double bond.*

This formidable-sounding rule can be restated in more down-to-earth language as "he who has, gets"; meaning that the hydrogen atom of the adding reagent will go on the carbon atom of the double bond already having the greater number of hydrogen atoms. The remaining portion of the adding reagent bonds to the carbon atom of the original double bond having the fewer hydrogen atoms. The addition of water to a double bond involves the addition of a hydroxyl group to the other carbon atom. In the case of the addition of water to propene, the hydrogen from water bonds to the No. 1 carbon atom and the hydroxyl group to the No. 2 carbon, yielding 2-propanol.

$$\overset{3}{C}H_3\overset{2}{C}H{=}\overset{1}{C}H_2 + H_2O \xrightleftharpoons{H_2SO_4} \overset{3}{C}H_3\underset{\displaystyle OH}{\overset{2}{C}H}{-}\underset{\displaystyle H}{\overset{1}{C}H_2}$$

propene — 2-propanol

exercise 9-5

Write the structure of the major product expected from the acid-catalyzed addition of water to each of the following alkenes:

(a) $CH_3CH_2CH{=}CH_2$

(b) $(CH_3)_2C{=}CH_2$

(c) $(CH_3)_2C{=}C(CH_3)H$

(d) $(CH_3CH_2)(CH_3)C{=}C(H)CHCH_3$ with CH_3 on the CH (i.e., $CH_3CH_2(CH_3)C{=}CH{-}CH(CH_3)_2$)

(e) 1-methylcyclohexene (cyclohexene ring with CH_3 on the double-bond carbon)

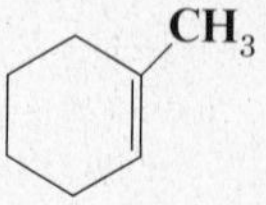

(f)  methylenecyclopentane (cyclopentane ring with ${=}CH_2$)

(g) $CH_3CH{=}CHCH_3$

(h) $CH{-}CO_2H \; \| \; CH{-}CO_2H$ ($HO_2C{-}CH{=}CH{-}CO_2H$)

exercise 9-6

The acid-catalyzed addition of water to $CH_3CH_2CH{=}CHCH_3$ leads to the formation of approximately equal amounts of 2-pentanol and 3-pentanol. Explain.

DEHYDRATION OF ALCOHOLS

In the previous section it was pointed out that the addition of water to an alkene is a reversible reaction. Thus, for example, ethylene can be prepared by the dehydration of ethanol:

$$CH_3CH_2OH \xrightleftharpoons{H_2SO_4} CH_2{=}CH_2 + H_2O$$

To maximize the yield of olefin, an excess of sulfuric acid is used. The sulfuric acid acts as a desiccant, absorbing the water as it is formed. Removal of the water in this manner drives the reaction to completion as predicted by Le Chatelier's principle.

It is necessary, however, to exercise some caution in interpreting the word *reversible* as applied to the hydration of alkenes/dehydration of alcohols reactions. Just as the addition of water to an alkene could, in theory, lead to the formation of more than one isomer of the alcohol, the elimination of water from an alcohol can also lead to the formation of more than one alkene. For instance, the elimination of water from 2-butanol could conceivably lead to both 1-butene and 2-butene.

$$\underset{\text{2-butanol}}{CH_3CH_2\underset{\displaystyle OH}{\underset{|}{C}}HCH_3} + H_2SO_4 \rightleftharpoons \underset{\text{2-butene}}{CH_3CH{=}CHCH_3} + \underset{\text{1-butene}}{CH_3CH_2CH{=}CH_2}$$

Under the usual reaction conditions, 2-butene is the predominate product. In general we find *the elimination of water from an alcohol leads to the most highly substituted double bond possible*. In other words, the carbon atoms of the double bond will have the fewest *total* number of hydrogen atoms bonded to them as possible. (Of course the hydrogen and hydroxyl groups must be removed from adjacent carbon atoms). Notice there are only two hydrogen atoms bonded to the double-bonded carbon atoms in 2-butene; whereas, there are three in 1-butene.

Let's look at another example of this aspect of the dehydration of alcohols. The dehydration of 2-methyl-2-butanol yields principally 2-methyl-2-butene, *not* 2-methyl-1-butene.

$$\underset{\text{2-methyl-2-butanol}}{CH_3\overset{\displaystyle CH_3}{\overset{|}{\underset{\underset{\displaystyle H}{|}}{\underset{\displaystyle O}{\underset{|}{C}}}}}{-}CH_2CH_3} + H_2SO_4 \rightleftharpoons \underset{\text{2-methyl-2-butene}}{(CH_3)_2C{=}CH{-}CH_3} \quad \textbf{not} \quad \underset{\text{2-methyl-1-butene}}{CH_2{=}C(CH_3)CH_2CH_3}$$

Recall the earlier remark about interpretation of the word *reversible* as it applies to these reactions. If we add water to 1-butene, the product is 2-butanol:

$$CH_3CH_2CH{=}CH_2 + H_2O \xrightleftharpoons{H_2SO_4} CH_3CH_2\underset{\displaystyle OH}{\underset{|}{C}}HCH_3$$

1-butene — 2-butanol

The elimination of water from 2-butanol, however, yields 2-butene and *not* 1-butene.

$$CH_3CH_2\underset{\displaystyle OH}{\underset{|}{C}}HCH_3 + H_2SO_4 \rightleftharpoons CH_3CH{=}CHCH_3$$

Thus, the reaction is reversible with respect to starting with an alkene, producing an alcohol followed by regeneration of an alkene. The reaction is not necessarily reversible, however, with respect to starting with a particular alkene, producing an alcohol, and regenerating the starting alkene. In Chapter 12 some examples are given of the importance of this feature of the reversible addition of water to alkenes in cellular reactions. Nature utilizes these addition/elimination reactions of water to alkenes to synthesize compounds required by living organisms.

exercise 9-7

Write the structure of the alkene expected by dehydration of each of the following alcohols:

(a) $CH_3CH_2CH_2OH$

(b) $CH_3\overset{\displaystyle CH_3}{\underset{\displaystyle CH_3}{C}}{-}OH$

(c)  cyclohexane ring with —OH

(d) 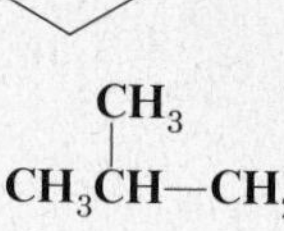 cyclohexane ring with CH_3 and OH on adjacent carbons

(e) 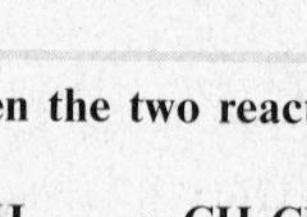 cyclohexane ring with CH_3 and OH on the same carbon

(f) $CH_3\overset{\displaystyle CH_3}{\overset{|}{C}}H{-}CH_2{-}\underset{\displaystyle OH}{\underset{|}{C}}H{-}\overset{\displaystyle CH_3}{\overset{|}{C}}H{-}CH_3$

exercise 9-8

Explain the difference between the two reaction sequences below:

Sequence 1 $\quad CH_3CH_2CH{=}CH_2 \longrightarrow CH_3CH_2\underset{\displaystyle OH}{\underset{|}{C}}H{-}CH_3 \longrightarrow CH_3CH{=}CHCH_3$

Sequence 2 $\quad CH_3CH{=}CHCH_3 \longrightarrow CH_3CH_2\underset{\displaystyle OH}{\underset{|}{C}}H{-}CH_3 \longrightarrow CH_3CH{=}CHCH_3$

So far we have discussed only *intramolecular* elimination of water from alcohols. *Intra*molecular means within the molecule; the molecule of water eliminated comes from within each molecule of alcohol. Alcohols may also undergo an *intermolecular* elimination of water. *Inter*molecular means between two molecules. Thus, two molecules of alcohol come together to eliminate a molecule of water. The other product formed during this reaction is an *ether*. In the general case the reaction can be written

$$\text{R—O—H} + \text{H—O—R} \xrightarrow{H_2SO_4} H_2O + \underset{\text{an ether}}{\text{R—O—R}}$$

or simply

$$2\text{ROH} \xrightarrow{H_2SO_4} H_2O + \text{R—O—R}$$

A specific example is the formation of diethyl ether from ethyl alcohol.

$$\underset{\text{ethyl alcohol}}{2CH_3CH_2OH} \xrightarrow{H_2SO_4} \underset{\text{diethyl ether}}{CH_3CH_2\text{—O—}CH_2CH_3} + H_2O$$

Notice that the catalyst for the formation of ethers from alcohols is the same as for the formation of alkenes from alcohols. A consequence of this is that in the chemistry laboratory a mixture of alkene and ether is usually obtained from the dehydration of alcohols. In most cases chemists can adjust the reaction temperature to obtain primarily alkene or primarily ether. You will not be asked to learn the conditions under which either olefin or ether is primarily produced.

Whenever ethers are mentioned, most people think of diethyl ether, used as an anesthetic. In fact, even chemists often refer to diethyl ether simply as ether. Chemically, diethyl ether (a liquid at room temperature; bp, 35 °C,) is an important solvent, dissolving fats and other slightly polar compounds. Medically, diethyl ether has long been employed as a general anesthetic, its use having first been reported for this purpose in 1842 by Dr. Crawford W. Long of Atlanta, Georgia. Drawbacks to ether are its highly flammable nature, irritation of respiratory tissues, and the tendency to cause nausea. Other ethers, principally methylpropyl ether, are also used as general anesthetics.

Dimethyl ether, a gas at room temperature, is used as an aerosol spray propellant, although its use in this capacity is declining because of its high flammability. Butylated hydroxy anisole (BHA), another ether, is used as a preservative for grains and cereals. Usually the packaging materials are impregnated with BHA instead of direct addition of the ether to the product. Ethylene oxide is a cyclic ether used as a fumigant and is one of few ethers that will readily undergo reaction with water. It reacts with water to produce ethylene glycol, the principal ingredient of permanent antifreeze.

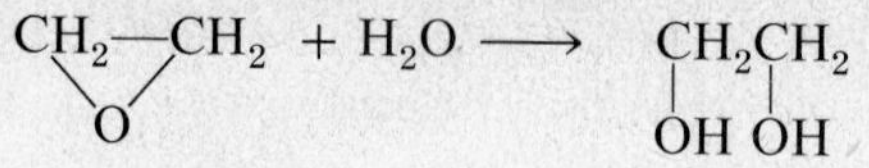

$$\underset{\text{ethylene oxide}}{\overset{\;\;\;\text{O}}{CH_2—CH_2}} + H_2O \longrightarrow \underset{\text{ethylene glycol (1,2-ethanediol)}}{\underset{|\quad\;\;|}{\overset{}{CH_2CH_2}}\atop{OH\;OH}}$$

exercise 9-9

Write the structure of the ether that can be produced from each of the following alcohols:

(a) CH_3OH

(b) $(CH_3)_2CH—OH$

(c) 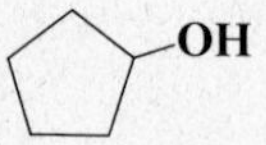 cyclopentyl—OH

(d) 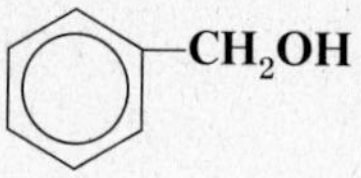 $C_6H_5—CH_2OH$

exercise 9-10

Tetrahydrofuran is an ether that can be prepared by *intramolecular* elimination of water from $HOCH_2CH_2CH_2CH_2OH$ (1,4-butanediol). Write the structure of tetrahydrofuran.

exercise 9-11

Heating benzyl alcohol ($C_6H_5—CH_2OH$) in the presence of sulfuric acid does not lead to olefin formation but will yield an ether. Explain.

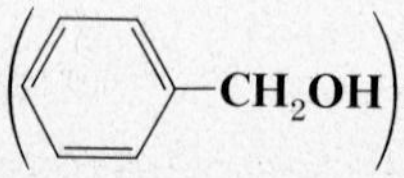

DEHYDROGENATION OF ALCOHOLS—OXIDATION

Under appropriate conditions hydrogen can be removed from primary and secondary alcohols. Two hydrogen atoms (the equivalent of H_2) are removed from each alcohol molecule. *One of the hydrogen atoms removed is the hydroxyl group hydrogen atom and the other is from the carbon atom bonded to the hydroxyl group.* Removal of the hydrogen atoms is accompanied by formation of a double bond between the carbon and oxygen atoms. An *aldehyde* results from the removal of the hydrogen atoms from a *primary alcohol:*

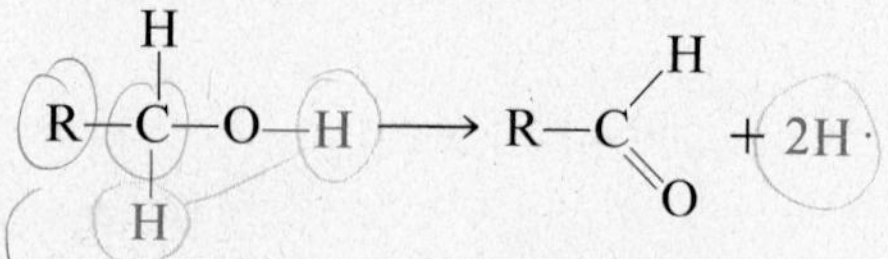

$$\underset{\text{primary alcohol}}{R—CH_2—O—H} \longrightarrow \underset{\text{aldehyde}}{R—C(H)=O} + 2H\cdot$$

Dehydrogenation of a *secondary alcohol* yields a *ketone:*

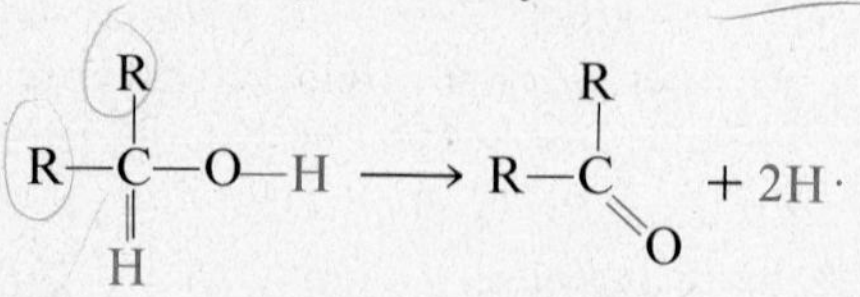

$$\underset{\text{secondary alcohol}}{R—CH(R)—O—H} \longrightarrow \underset{\text{ketone}}{R—C(R)=O} + 2H\cdot$$

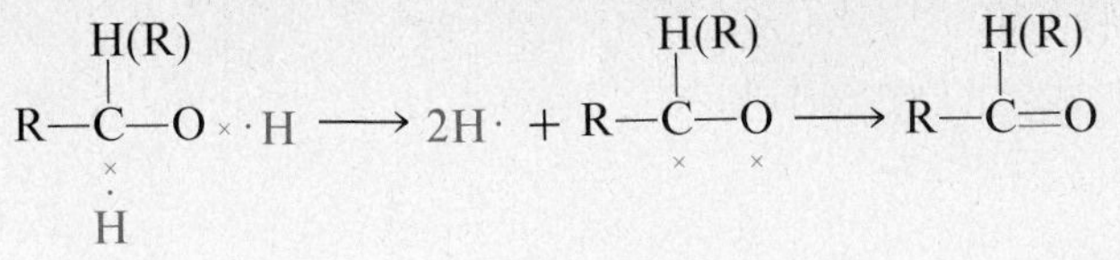

$$\underset{\substack{\times \\ \cdot \\ H}}{\overset{\substack{H(R) \\ |}}{R{-}C{-}O}} \times \cdot H \longrightarrow 2H\cdot + \overset{\substack{H(R) \\ |}}{R{-}C{-}O} \longrightarrow \overset{\substack{H(R) \\ |}}{R{-}C{=}O}$$

FIGURE 9-2 Dehydrogenation of an alcohol is oxidation. — is a two-electron covalent bond; × and · represent individual electrons.

Loss of hydrogen atoms is oxidation.

Removal of hydrogen means electrons are removed from the alcohol molecule. Loss of electrons is oxidation; thus, alcohols may be oxidized to produce aldehydes and ketones. The loss of electrons by the alcohol is depicted in Fig. 9-2. Oxidation of a *primary alcohol* yields an *aldehyde;* oxidation of a *secondary* alcohol, a ketone.

Recall from Chapter 3 that whenever oxidation occurs, reduction must simultaneously take place. Furthermore, an oxidizing agent is needed to cause oxidation. Suitable oxidizing agents to oxidize alcohols are potassium permanganate ($KMnO_4$) and potassium dichromate ($K_2Cr_2O_7$). During reaction, manganese or chromium atoms gain electrons and thereby are reduced. The hydrogen atoms from the alcohol combine with oxygen from the oxidizing agent to produce water. Balanced equations for the general reactions in acidic reaction media are

$$5RCH_2OH + 2KMnO_4 + 3H_2SO_4 \longrightarrow 5RCHO + 2MnSO_4 + 8H_2O + K_2SO_4$$

$$3RCH_2OH + K_2Cr_2O_7 + 4H_2SO_4 \longrightarrow 3RCHO + Cr_2(SO_4)_3 + K_2SO_4 + 7H_2O$$

Balanced equations are written to convey quantitative information to the reader. Since organic reactions are frequently accompanied by side reactions, conversion of one organic compound to another often is not quantitative. For this reason, coupled with primary interest in the changes the organic compound undergoes, balanced equations are not usually written for reactions involving organic compounds. For instance, the oxidation of an alcohol by potassium dichromate is often simply written

$$R{-}CH_2OH + K_2Cr_2O_7 \xrightarrow{H^+} RCHO$$

or even as

$$\underset{\text{primary alcohol}}{R{-}CH_2OH} \xrightarrow[H^+]{K_2Cr_2O_7} \underset{\text{aldehyde}}{RCHO}$$

exercise 9-12

Write the structural formula of the organic product expected from treating each of the alcohols below with $K_2Cr_2O_7$. If no reaction is expected, write N.R. and explain why.

(a) CH_3CH_2OH

(b) $\underset{\substack{| \\ OH}}{CH_3CHCH_3}$

CH₃-C=O-H

CH

(c) 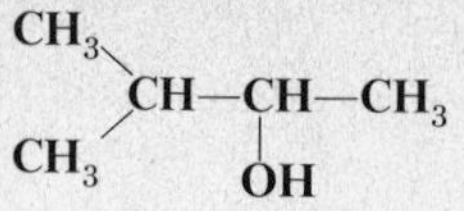 $\begin{matrix} CH_3 \\ \quad \diagdown \\ \quad CH{-}CH{-}CH_3 \\ \quad \diagup \quad\;\; | \\ CH_3 \quad\;\; OH \end{matrix}$ **(d)** $HOCH_2CH_2CH_2OH$

(e) 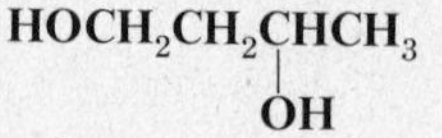 $HOCH_2CH_2\underset{\displaystyle OH}{\underset{|}{C}}HCH_3$ **(f)** 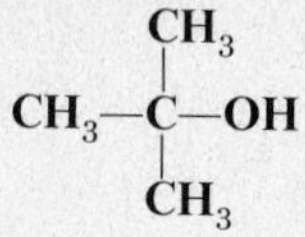 $CH_3{-}\overset{\displaystyle CH_3}{\overset{|}{\underset{\displaystyle CH_3}{\underset{|}{C}}}}{-}OH$

Exercise 9-12(f) is a tertiary alcohol. The correct answer to this problem is N.R. Tertiary alcohols cannot be oxidized because the carbon atom bonded to the hydroxyl group has no hydrogen atom bonded to it (see Fig. 9-3).

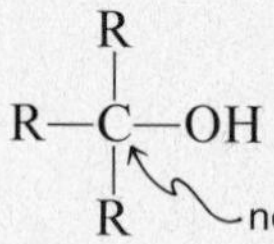

$$R{-}\overset{\displaystyle R}{\overset{|}{\underset{\displaystyle R}{\underset{|}{C}}}}{-}OH$$

no hydrogen on this carbon atom

FIGURE 9-3 A tertiary alcohol cannot be oxidized.

OXIDATION OF ALDEHYDES

Aldehydes can be further oxidized to acids. This oxidation involves the incorporation of oxygen into the aldehyde molecule.

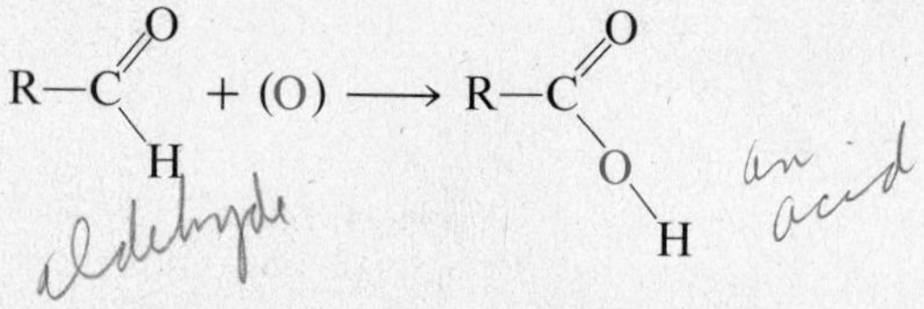

$$R{-}C\overset{\displaystyle O}{\underset{\displaystyle H}{\lessgtr}} + (O) \longrightarrow R{-}C\overset{\displaystyle O}{\underset{\displaystyle O{-}H}{\lessgtr}}$$

aldehyde   an acid

A gain of oxygen atoms is oxidation.

In order to oxidize a carbon-to-hydrogen bond, it is necessary that an activating group be bonded to the carbon atom. An oxygen atom may serve as an activating group. Carbon–carbon single bonds cannot normally be oxidized. Therefore, ketones are usually resistant to further chemical oxidation.

The aldehydes are removed by distillation.

Both potassium permanganate and potassium dichromate can oxidize aldehydes to acids. In fact, aldehydes are more easily oxidized than are alcohols. This means that aldehydes are difficult to prepare. As some of the primary alcohol is oxidized to the aldehyde, the aldehyde produced will be further oxidized in preference to oxidizing the remaining alcohol. Fortunately, many aldehydes have lower boiling points than the alcohols from which they are prepared. Thus, low molecular weight aldehydes can be removed from the reaction mixture as they are formed. Removal of the aldehyde from the reaction mixture prevents further oxidation to the acid.

exercise 9-13

Decide which of the following alcohols can be oxidized to acids. Write the structure of the acid produced.

(a) CH_3CH_2OH

(b) CH_3CHCH_3 (with OH on the central carbon)

$$\begin{array}{c} CH_3\underset{\displaystyle OH}{\underset{|}{C}}HCH_3 \end{array}$$

(c) cyclohexane ring bearing OH

(d) $C_6H_5{-}CH_2OH$

(e) $HOCH_2CH_2CH_2CHO$

(f) $CH_3CH_2\underset{\displaystyle OH}{\underset{|}{C}}HCH_2CH_2OH$

With one exception, acids are resistant to further oxidation. The

exception is formic acid $\left(H{-}C\begin{matrix}{}^{\nearrow O}\\ {}_{\searrow OH}\end{matrix}\right)$ which has a carbon atom bonded to both hydrogen and oxygen atoms.

$$H{-}C\begin{matrix}{}^{\nearrow O}\\ {}_{\searrow OH}\end{matrix} + (O) \longrightarrow H{-}O{-}\overset{\overset{\displaystyle O}{\|}}{C}{-}OH$$

The product of formic acid oxidation is our old friend carbonic acid (H_2CO_3), which decomposes to carbon dioxide and water. Thus, the oxidation of formic acid can be written

$$H{-}C\begin{matrix}{}^{\nearrow O}\\ {}_{\searrow OH}\end{matrix} + (O) \longrightarrow H_2O + CO_2$$

Simple tests involving a visual change during reaction have been developed to identify the presence of aldehydes. These tests are based on the fact that aldehydes are more easily oxidized than alcohols. Strong oxidizing agents like $KMnO_4$ and $K_2Cr_2O_7$ are capable of oxidizing both aldehydes and alcohols. Weak oxidizing agents such as silver(I) and copper(II) ions are able to oxidize aldehydes but not alcohols.

The complex ion $[Ag(NH_3)_2{}^+]$ results from formation of coordinate covalent bonds between ammonia and silver ions.

An ammoniacal solution of silver ions known as *Tollen's reagent* oxidizes aldehydes to acids. The purpose of the ammonia is to form the complex ions $Ag(NH_3)_2^+$ preventing the precipitation of insoluble silver hydroxide. The silver ions are reduced to metallic silver, forming a silver mirror on the inside of the glass reaction vessel. The formation of the silver mirror gives visual confirmation of the presence of an aldehyde in the sample tested.

$$RCHO + \underset{\text{Tollen's reagent}}{2Ag(NH_3)_2^+} + 2OH^- \longrightarrow RCO_2^-NH_4^+ + \underset{\text{metallic silver}}{2Ag\downarrow} + 3NH_3 + H_2O$$

Notice that the ammonium salt of the organic acid is produced in the basic solution of Tollen's reagent.

exercise 9-14

Glass mirrors are produced commercially by treating a clean glass surface with an aqueous solution of formaldehyde by HCHO and Tollen's reagent. Write an equation for the reaction that takes place.

Complexing agents form coordinate covalent bonds to metal ions.

The bright blue *Fehling's* and *Benedict's solutions* are basic solutions of copper(II) ions. Copper(II) sulfate is the usual source of copper ions. Fehling's solution contains sodium potassium tartrate and Benedict's solution contains sodium citrate as complexing agents. These complexing agents prevent the formation of insoluble copper(II) hydroxide. Both Fehling's and Benedict's solutions yield an orange to red precipitate of copper(I) oxide upon reaction with aldehydes. The overall reaction is

$$\underset{}{RCHO} + \underset{\text{blue}}{2Cu^{2+}(aq)} \xrightarrow{NaOH} RCO_2^-Na^+ + \underset{\text{red}}{Cu_2O\downarrow}$$

α-Hydroxy ketones have an —OH group on the carbon next to the ketone group.

Other easily oxidized compounds also react with Tollen's reagent and Fehling's or Benedict's solution. One such group of compounds are α-hydroxy ketones $\left(\begin{array}{c} R-\underset{\displaystyle OH}{\underset{|}{C}H}-\underset{\displaystyle O}{\underset{\|}{C}}-R \end{array}\right)$. The alcohol group in these compounds is activated toward oxidation by the ketone group next to it. The reactions of aldehydes and α-hydroxy ketones with these weak oxidizing agents are important in the study of sugars and other carbohydrates.

exercise 9-15

Write an equation for the reaction of $CH_3\underset{\displaystyle OH}{\underset{|}{C}H}-\underset{\displaystyle O}{\underset{\|}{C}}-CH_3$ with Benedict's solution.

ADDITION REACTIONS OF ALDEHYDES AND KETONES

Just as carbon–carbon double bonds undergo addition reactions, carbon–oxygen double bonds also undergo addition reactions.

Refer back to Chapter 4 for the products of hydrogen addition to C=C bonds.

The addition of hydrogen to the carbon–oxygen double bonds converts aldehydes and ketones to alcohols. In the laboratory, a catalyst such as platinum metal is required for the hydrogenation of aldehydes or ketones:

$$R-\overset{\displaystyle O}{\overset{\|}{C}}-H + H_2 \xrightarrow{Pt} R-\overset{\displaystyle H}{\overset{|}{\underset{\displaystyle H}{\underset{|}{C}}}}-O-H$$

$$R-\overset{\displaystyle O}{\overset{\|}{C}}-R + H_2 \xrightarrow{Pt} R-\overset{\displaystyle H}{\overset{|}{\underset{\displaystyle OH}{\underset{|}{C}}}}-R$$

exercise 9-16

Hydrogenation of aldehydes and ketones is often called *reduction.* Explain why the term is appropriate.

exercise 9-17

Write the structural formula of the product expected from each of the indicated reactions:

(a) $CH_3CH_2CHO + H_2 \xrightarrow{Pt}$

(b) $CH_3\underset{\underset{O}{\|}}{C}CH_3 + H_2 \xrightarrow{Pt}$

(c) $CH_3\underset{\underset{O}{\|}}{C}CH_2CH_2CHO + H_2 \xrightarrow{Pt}$

(d) $CH_2{=}CHCH_2CHO + 2H_2 \xrightarrow{Pt}$

(e) 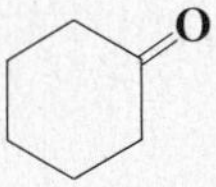 cyclohexanone $+ H_2 \xrightarrow{Pt}$

Aldehydes and ketones also undergo addition reactions with alcohols. *Addition of an alcohol to an aldehyde yields a hemiacetal:*

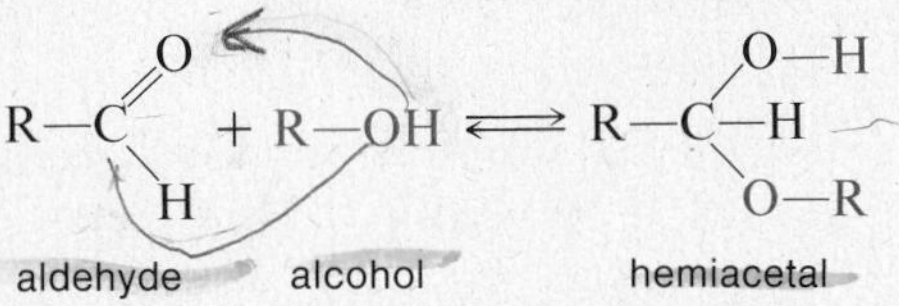

$$\underset{\text{aldehyde}}{R{-}\overset{O}{\overset{\|}{C}}{-}H} + \underset{\text{alcohol}}{R{-}OH} \rightleftharpoons \underset{\text{hemiacetal}}{R{-}\underset{O-R}{\overset{O-H}{C}}{-}H}$$

The hydroxyl group hydrogen of the alcohol bonds to the oxygen atom of the aldehyde, and the alkoxy portion (R—O—) of the alcohol bonds to the carbon atom of the original aldehyde group. The reaction does not require a catalyst and an equilibrium mixture of unreacted aldehyde, unreacted alcohol, and hemiacetal is formed spontaneously upon mixing of the aldehyde and alcohol. The position of the equilibrium depends on which aldehyde and alcohol are mixed.

Hemiacetals are said to react with Tollen's reagent and Fehling's or Benedict's solution since some free aldehyde is always present. As the aldehyde reacts, the equilibrium is upset, and some of the hemiacetal must decompose to free aldehyde and alcohol in order to restore equilibrium. Consequently, if sufficient oxidizing agent is present, all the hemiacetal will eventually decompose.

Remember Le Chatelier's principle!

Reaction also occurs between *alcohols and ketones to produce hemiketals:*

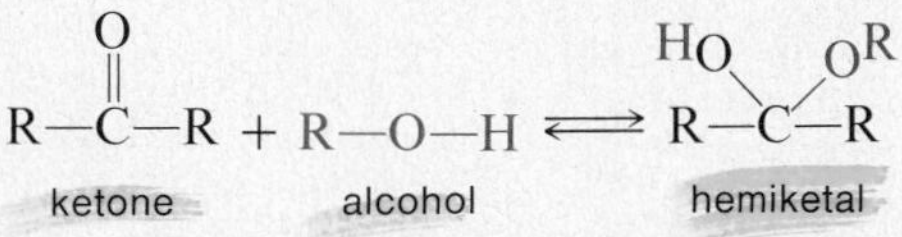

$$\underset{\text{ketone}}{R{-}\overset{O}{\overset{\|}{C}}{-}R} + \underset{\text{alcohol}}{R{-}O{-}H} \rightleftharpoons \underset{\text{hemiketal}}{R{-}\overset{HO\quad OR}{C}{-}R}$$

Hemiketal formation is not as favorable as hemiacetal formation. Since simple ketones cannot be oxidized further, their hemiketals do not react with Tollen's or Benedict's and Fehling's solutions.

exercise 9-18

Write the structural formula of the product expected to be produced during the indicated reaction:

(a) $CH_3CHO + CH_3OH$

(b) $CH_3CH_2CHO + CH_3OH$

(c) $CH_3CHO + CH_3CH_2OH$

(d) $CH_3\overset{\displaystyle O}{\overset{\|}{C}}CH_3 + CH_3OH$

(e) $2CH_3CHO + HOCH_2CH_2OH$

Some molecules may contain both an alcohol group and an aldehyde group. In this case an intramolecular reaction may occur to form a *cyclic hemiacetal.* For example,

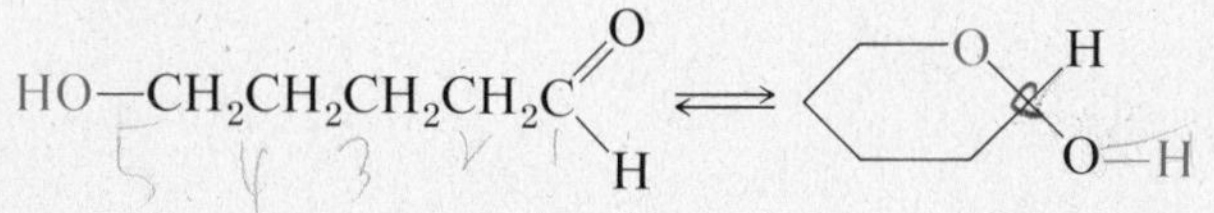

$$HO{-}CH_2CH_2CH_2CH_2C(=O)H \rightleftharpoons \text{(six-membered ring: O, C(H)(O{-}H))}$$

Cyclic hemiacetal (or cyclic hemiketal) formation is particularly favorable when a five- or six-membered ring can be formed. Under these conditions, almost all the material is present in the hemiacetal or hemiketal form.

exercise 9-19

Write the structural formula of the cyclic hemiacetal or hemiketal formed:

(a) $CH_3CH_2{-}\underset{\displaystyle O}{\underset{\|}{C}}{-}CH_2CH_2\underset{\displaystyle OH}{\underset{|}{C}H}CH_3$

(b) $CH_3\underset{\displaystyle OH}{\underset{|}{C}H}CH_2CH_2CH_2CHO$

(c) $HOCH_2CH_2CH_2CH_2\underset{\displaystyle O}{\underset{\|}{C}}CH_3$

(d) $HOCH_2CH_2CH_2CHO$

(e) $HOCH_2CH_2CH_2\underset{\displaystyle OH}{\underset{|}{C}H}CHO$

Hemiacetals and hemiketals will react with additional alcohol in the presence of an acid catalyst to produce acetals and ketals, respectively. Water is also produced during reaction.

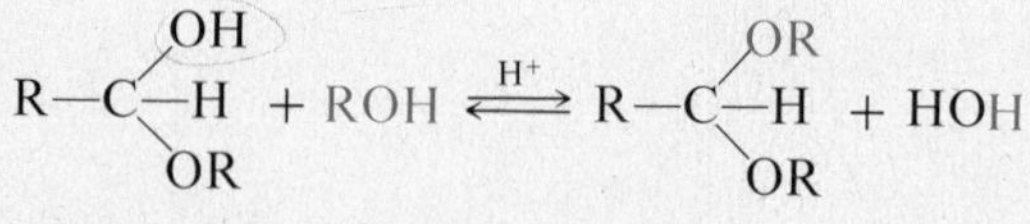

$$R{-}\overset{\displaystyle OH}{\underset{\displaystyle OR}{C}}{-}H + ROH \underset{}{\overset{H^+}{\rightleftharpoons}} R{-}\overset{\displaystyle OR}{\underset{\displaystyle OR}{C}}{-}H + HOH$$

hemiacetal   acetal

$$\underset{\text{hemiketal}}{\mathrm{R{-}\overset{\overset{\large OH}{/}}{\underset{\underset{\large OR}{\backslash}}{C}}{-}R}} + \mathrm{ROH} \overset{\mathrm{H^+}}{\rightleftharpoons} \underset{\text{ketal}}{\mathrm{R{-}\overset{\overset{\large OR}{/}}{\underset{\underset{\large OR}{\backslash}}{C}}{-}R}} + \mathrm{HOH}$$

Although the reaction is an equilibrium reaction, acetals and ketals can be isolated in the pure state if the catalyst is removed. After the catalyst is removed, neither the forward nor the reverse reactions occur at an appreciable rate. The easiest way to remove the catalyst is to add base to neutralize it. Thus, acetals, in contrast to hemiacetals, will *not* react with Tollen's reagent and Fehling's or Benedict's solutions.

CARBOHYDRATES

Poly means many; *thus, polyhydroxy aldehydes are aldehyde molecules containing many —OH groups.*

One of the three basic foodstuffs, carbohydrates or saccharides are composed of carbon, hydrogen, and oxygen. Their name was originally derived from the general formula $C_x(H_2O)_y$. Structurally, these compounds are *polyhydroxyaldehydes* or *polyhydroxyketones* and their derivatives. Those saccharides containing an aldehyde group are called *aldoses,* while those with a ketone group are *ketoses.*

Lower molecular weight carbohydrates are known as sugars and their names generally have the characteristic ending *-ose.* Some examples are sucrose (cane sugar), lactose (milk sugar), glucose (also known as dextrose), and fructose (levulose). *Monosaccharides* (simple sugars) are further classified according to the number of carbon atoms in a continuous chain. A three-carbon monosaccharide is a triose; four carbons, a tetrose; five carbons, a pentose; and six carbons, a hexose. A disaccharide results from the joining of two monosaccharides. Carbohydrates containing from three up to about eight monosaccharide units are *oligosaccharides*. Those containing more than about eight monosaccharide units are polysaccharides. The classifications of carbohydrates are given in Table 9-2.

STEREOISOMERISM

Before embarking on our study of carbohydrate chemistry, we need to understand a third kind of isomerism. This type of isomerism, called *stereoisomerism*, is more subtle than either structural or geometric isomerism and has its origin in the tetrahedral bonding arrangement around carbon atoms. To start our study of stereoisomerism, consider placing four different groups around a carbon atom in a tetrahedral arrangement. As we attempt to do this, we find there are two different structures possible [Fig. 9-4(a) and (b)].

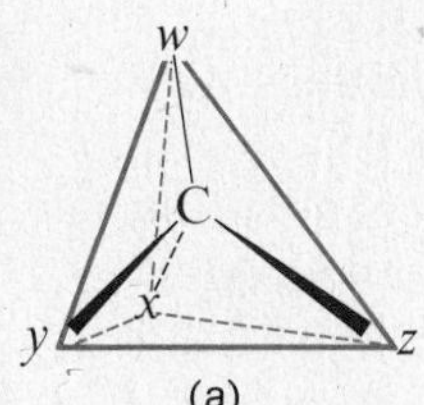

(a)

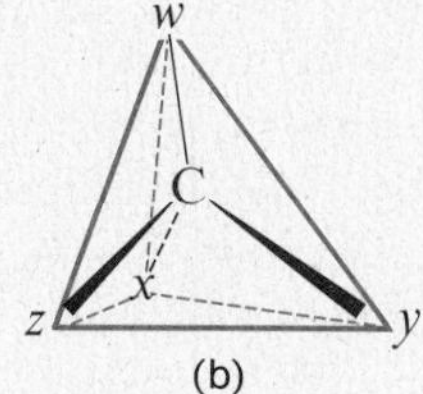

(b)

FIGURE 9-4 Tetrahedral carbon atoms with four different groups attached to the central carbon.

TABLE 9-2
The classification of carbohydrates

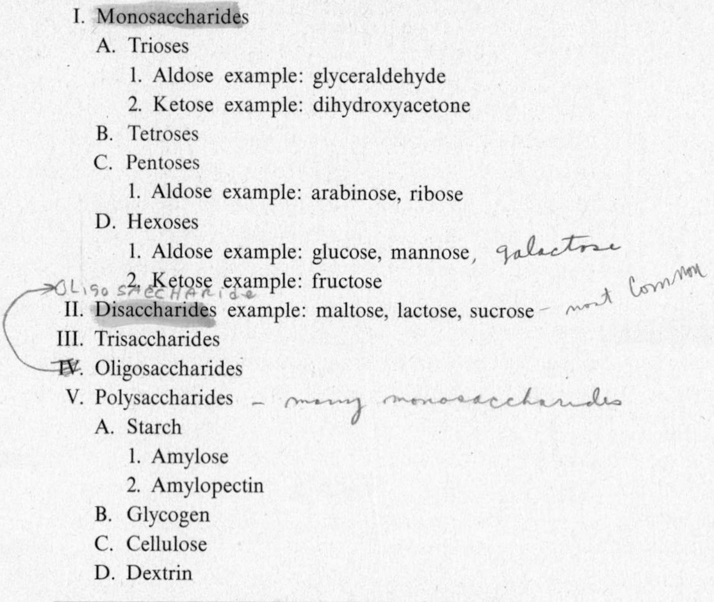

I. Monosaccharides
- A. Trioses
  - 1. Aldose example: glyceraldehyde
  - 2. Ketose example: dihydroxyacetone
- B. Tetroses
- C. Pentoses
  - 1. Aldose example: arabinose, ribose
- D. Hexoses
  - 1. Aldose example: glucose, mannose
  - 2. Ketose example: fructose

II. Disaccharides example: maltose, lactose, sucrose

III. Trisaccharides

IV. Oligosaccharides

V. Polysaccharides
- A. Starch
  - 1. Amylose
  - 2. Amylopectin
- B. Glycogen
- C. Cellulose
- D. Dextrin

In these figures the black wedges mean groups *y* and *z* are pointing forward out of the plane of the paper at you, while the black dotted line indicates group *x* is behind the plane of the page. Group *w* and the carbon atom lie in the plane of the paper. Structural formulas written in this way are called *perspective* formulas. A tetrahedron is shown in color to aid in visualizing the shape of the molecule.

Notice the difference in the positions of groups *y* and *z* in the two figures. Although the molecules are identical in all other respects, the two molecules are nonsuperimposable. Nonsuperimposable means that one cannot be set down on the other with identical groups touching. Since the two molecules shown in 9-4(a) and (b) have the same formula but are not identical, they must be isomers of one another. This type of isomerism, neither structural nor geometric, is known as *stereoisomerism.*

To aid in understanding stereoisomerism, let's turn away from molecules for the moment and instead consider our hands. Both of our hands are identical in having a thumb and four fingers arranged in a particular sequence such that the index finger is next to the thumb, the middle finger next to index finger, and followed by the ring finger, and the little finger in that order. Place your hands palms down on the desk. Your hands look the same except, as you know, one is a right hand and the other the left one. Now without moving your left hand try to superimpose your right hand on the left one by placing it palm down on your left one. Your hands are nonsuperimposable; your right thumb lies over the left little finger.

You might argue that if you put your hands back-to-back, that thumb would touch thumb, index finger touch index finger, etc. This is true. However, you still would not have superimposed your hands as one palm would be facing downward and the other upward. Similarily if you try to turn structure 9-4(b) around to put group y on the left and z on the right, they would no longer be pointing out of the paper at you but would then be behind the plane of the paper.

Build models of these molecules.

Because holding your hands back-to-back is a little uncomfortable, hold them palm-to-palm. Although you still have not superimposed your hands, an interesting relationship between your hands reveals itself. Your hands are mirror images of one another. Stand in front of a mirror and prove this relationship to yourself. Hold your hands up with the palm of the right hand facing the mirror and the palm of the left hand facing you. The reflection in the mirror will look identical to your left hand (excluding scars and jewelry). Similarily, the molecules in Fig. 9-4 are mirror images and their perspective formulas are shown below in Fig. 9-5.

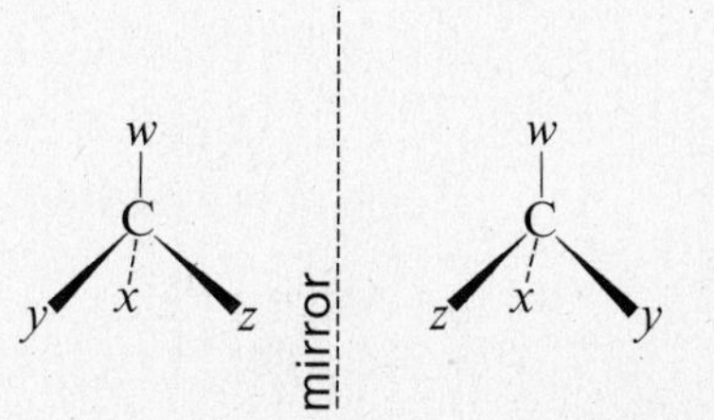

w, C, y, x, z | mirror | w, C, z, x, y

FIGURE 9-5 Mirror images.

The term chiral *is increasingly used in preference to the term* asymmetric.

Molecules that are nonsuperimposable mirror images of one another are called *enantiomers. The presence of an asymmetric carbon atom in a molecule may lead to stereoisomerism. An asymmetric carbon atom is defined as one with four different groups attached to it.*

In order to be asymmetric, a carbon must have four groups attached to it and the groups must be different.

Lactic acid is a compound containing an asymmetric carbon atom. The two enantiomers of lactic acid are shown in Fig. 9-6. These mirror images

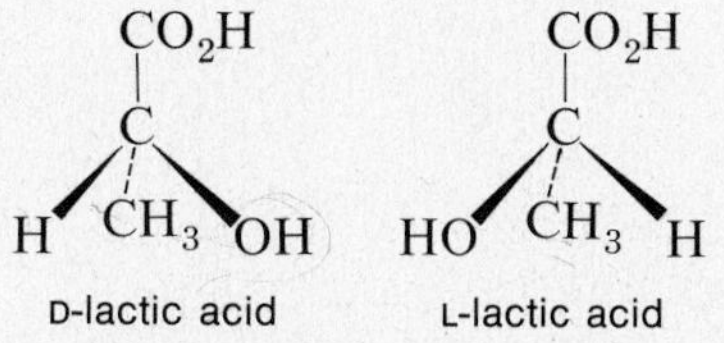

CO_2H, C, H, CH_3, OH — D-lactic acid

CO_2H, C, HO, CH_3, H — L-lactic acid

FIGURE 9-6 Enantiomers of lactic acid.

or enantiomers are called an *enantiomeric pair*. One of the isomers is referred to as the D-enantiomer and the other as the L-enantiomer (D, dextro; L, levo).

A substantial portion of the physical and chemical properties of the two enantiomers are identical. For example, the melting points and solubility of D- and L-lactic acid are the same. Both isomers are weak acids with identical dissociation constants. There are, however, some important differences in their behavior. The most notable differences are their interactions with living systems. In the 1840's Louis Pasteur found that molds could only

A racemic mixture is one that contains equal amounts of the two mirror images.

metabolize D-lactic acid. When the mold was placed in a solution containing mixture of D- and L-lactic acid, it grew until the D-lactic acid was used up; then, although the solution contained L-actic acid, growth ceased. Other investigators have also shown that most living organisms have a preference for one enantiomer over the other. In human muscle tissue only D-lactic acid is produced. It should be noted that not all organisms favor the same isomer. For example, some microorganisms produce L-lactic acid as the product of sucrose metabolism.

exercise 9-20

Draw Lewis structures for each of the following molecules. Decide which of the compounds contain an asymmetric carbon atom(s). Put an asterisk next to the asymmetric carbon atoms in the Lewis structures. Example:

$$\mathrm{H{-}\overset{*}{C}(CH_3)(CO_2H){-}OH}$$

(a) $CH_3CH(OH)CH_2CH_3$

(b) $CH_3CH(OH)CH_3$

(c) $CH_3{-}C({=}O){-}CH_2CH_3$

(d) $CH_3C({=}O){-}CH(OH)CH_3$

(e) $CH_3CH{=}CHCH_2CH_3$

(f) $HO_2CCH_2C(OH)(CO_2H)CH_2CO_2H$

(g) $HO_2CCH_2CH(OH)CO_2H$

(h) $HO_2CCH(OH){-}CH(OH)CHO$

(i) $CH_3CH(OH){-}CH(OH)CH_3$

(j) $CH_3CH(OH){-}CH(OH){-}CH(OH)CH_3$

HOW TO WRITE STEREOISOMERS—FISCHER PROJECTION FORMULAS

Drawing structures for stereoisomers could be a time-consuming process if always written in the perspective form shown in Figs. 9-4, 9-5, and 9-6. In an effort to simplify the drawing of stereoisomers, chemists use *Fischer projection formulas*. Certain rules are followed so that everyone understands what is meant by the formula.

1. If the compound contains an aldehyde, ketone, or carboxylic acid group, the *Fischer projection formulas are always written with this group at the top of* the structure.
2. The groups or atoms that project out of the plane of the paper toward the viewer in the perspective formula are always written in horizontal positions.
3. Groups or atoms that lie in or behind the plane of the paper in the perspective formula are always written in vertical positions.

These rules are illustrated in Fig. 9-7 for D- and L-lactic acid.

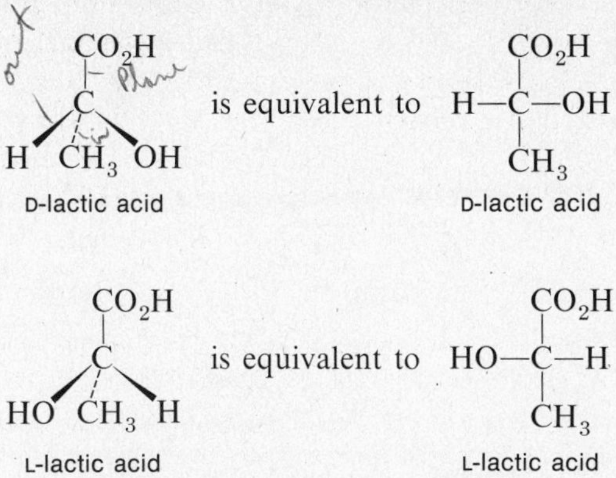

CO_2H / C / H, CH_3, OH (perspective) is equivalent to $H{-}\overset{\displaystyle CO_2H}{\underset{\displaystyle CH_3}{C}}{-}OH$

D-lactic acid   D-lactic acid

CO_2H / C / HO, CH_3, H (perspective) is equivalent to $HO{-}\overset{\displaystyle CO_2H}{\underset{\displaystyle CH_3}{C}}{-}H$

L-lactic acid   L-lactic acid

FIGURE 9-7 Fischer projection formulas for D-and L-lactic acid.

exercise 9-21

Below are written perspective formulas for several compounds. Write the Fischer projection formulas for each and for their mirror images (label the mirror images as such).

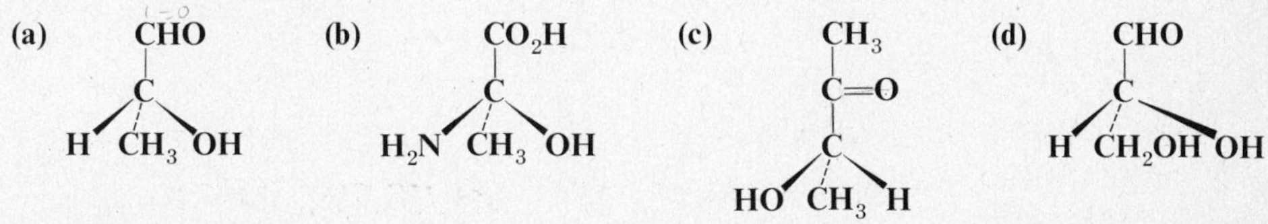

(a) CHO / C / H, CH_3, OH

(b) CO_2H / C / H_2N, CH_3, OH

(c) CH_3 / C=O / C / HO, CH_3, H

(d) CHO / C / H, CH_2OH, OH

exercise 9-22

Given below is a list of several perspective formulas and a list of Fischer projection formulas. Match the Fischer projection formula to its corresponding perspective formula.

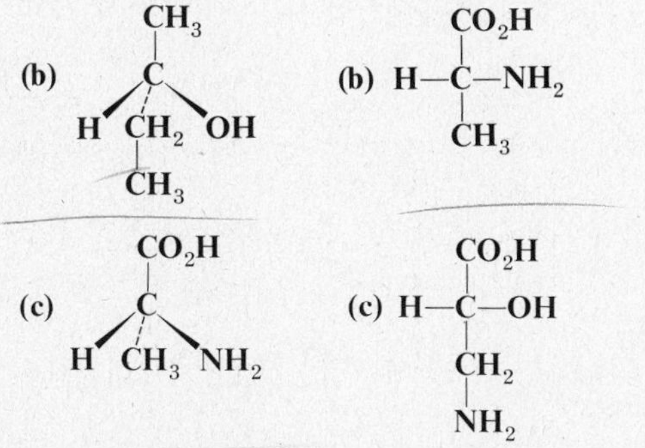

Column A	Column B
(a) CH_3 / C / HO, CH_2, H / CH_3	(a) $H{-}\overset{\displaystyle CH_3}{C}{-}OH$ / CH_2 / CH_3
(b) CH_3 / C / H, CH_2, OH / CH_3	(b) $H{-}\overset{\displaystyle CO_2H}{\underset{\displaystyle CH_3}{C}}{-}NH_2$
(c) CO_2H / C / H, CH_3, NH_2	(c) $H{-}\overset{\displaystyle CO_2H}{C}{-}OH$ / CH_2 / NH_2

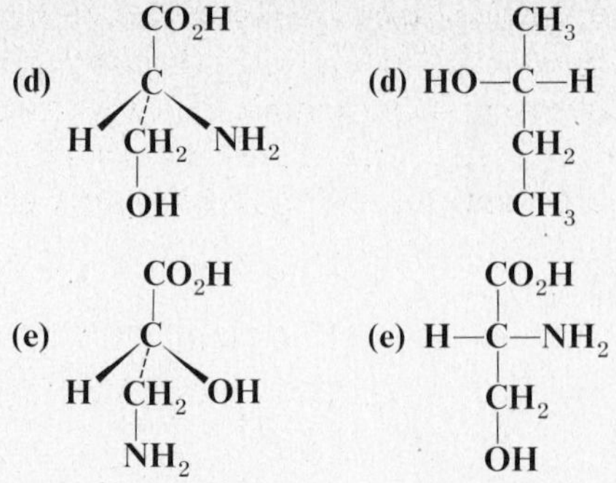

(d) [perspective formula: C bearing CO_2H (up), H, CH_2OH, NH_2] (d) [Fischer projection: CH_3 / HO—C—H / CH_2 / CH_3]

(e) [perspective formula: C bearing CO_2H (up), H, CH_2NH_2, OH] (e) [Fischer projection: CO_2H / H—C—NH_2 / CH_2 / OH]

exercise 9-23

Given below are Fischer projection formulas for several compounds. Draw the perspective formulas for each and for their mirror images (label the mirror images as such).

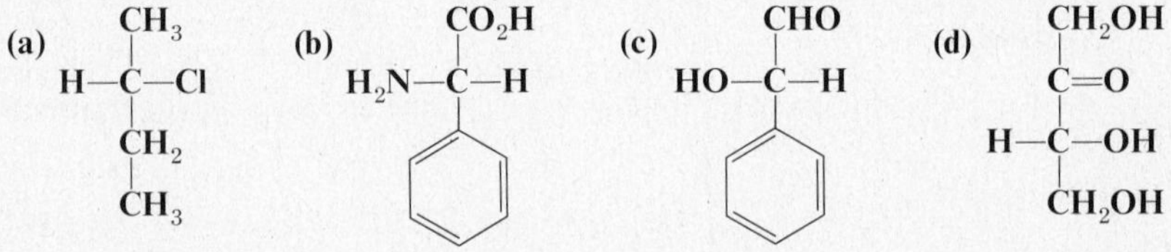

(a) CH_3 / H—C—Cl / CH_2 / CH_3

(b) CO_2H / H_2N—C—H / C_6H_5 (phenyl)

(c) CHO / HO—C—H / C_6H_5 (phenyl)

(d) CH_2OH / C=O / H—C—OH / CH_2OH

COMPOUNDS CONTAINING MORE THAN ONE ASYMMETRIC CARBON ATOM

Exercise 9-20 contained several compounds with more than one asymmetric carbon atom. The compound

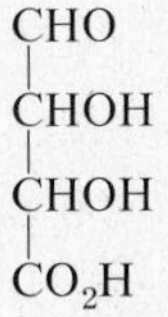

CHO
|
CHOH
|
CHOH
|
CO_2H

is one of these. Let's try to draw Fischer projection formulas for the stereoisomers of this compound.

One isomer would be

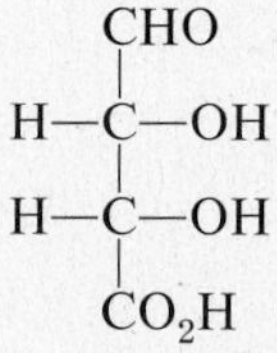

CHO
|
H—C—OH
|
H—C—OH
|
CO_2H

This isomer would have an enantiomer whose Fischer projection formula is

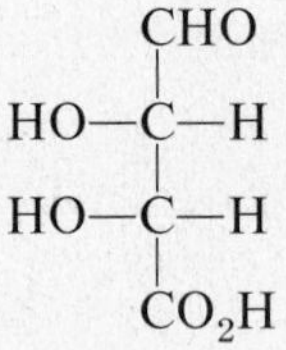

CHO
|
HO—C—H
|
HO—C—H
|
CO_2H

(*Note:* Since the aldehyde and carboxylic groups do not contain asymmetric carbon atoms, they do not have nonsuperimposable mirror images. Thus, using the aldehyde group as an example, it doesn't matter whether we write —CHO or OHC—.) These two enantiomers are not the only possible stereoisomers of the compound. Another possible stereoisomer is

```
      CHO
       |
   H—C—OH
       |
  HO—C—H
       |
      CO2H
```

This stereoisomer also has a mirror image

```
      CHO
       |
  HO—C—H
       |
   H—C—OH
       |
      CO2H
```

Consequently, there are a total of four stereoisomers of

```
  CHO
   |
  CHOH
   |
  CHOH
   |
  CO2H
```

consisting of two pairs of enantiomers. Stereoisomers that are not enantiomers are called ***diastereomers***. These relationships are shown in Fig. 9-8.

```
                              CHO
                               |
                              CHOH
                               |
                              CHOH
                               |
                              CO2H
                               |
        ----------------------------------------------
       |              |                |              |
      CHO            CHO              CHO            CHO
       |              |                |              |
   H—C—OH        HO—C—H           H—C—OH        HO—C—H
       |              |                |              |
   H—C—OH        HO—C—H          HO—C—H          H—C—OH
       |              |                |              |
      CO2H           CO2H             CO2H           CO2H
      (a)            (b)              (c)            (d)
```

FIGURE 9-8 (a) and (b) are enantiomers (mirror images); (c) and (d) are enantiomers; (a) and (c) are diastereomers; (b) and (c) are diastereomers; (b) and (d) are diastereomers; (a) and (d) are diastereomers.

Be sure you know the difference between enantiomers and diastereomers.

The maximum possible number of stereoisomers can be determined from the number of asymmetric carbon atoms in a compound.

The maximum number of stereoisomers is 2^n, where n is the number of asymmetric carbon atoms. The number of enantiomeric pairs is given by $(2^n)/2$, which equals $2^{(n-1)}$. In the case we have been considering, there are two asymmetric carbon atoms. Thus the total number of stereoisomers is $2^2 = 4$ and the number of enantiomeric pairs is $2^{2-1} = 2^1 = 2$. The structure

$$\begin{array}{l} CHO \\ | \\ CHOH \\ | \\ CHOH \\ | \\ CHOH \\ | \\ CH_2OH \end{array}$$

containing three asymmetric carbon atoms would be expected to have $2^3 = 8$ stereoisomers, comprised of $2^{3-1} = 2^2 = 4$ enantiomeric pairs.

exercise 9-24

Predict the number of stereoisomers and enantiomeric pairs for each of the following.

(a) $\begin{array}{l} CHO \\ | \\ CHOH \\ | \\ CH_2OH \end{array}$

(b) $\begin{array}{l} CO_2H \\ | \\ CHOH \\ | \\ CHOH \\ | \\ CH_3 \end{array}$

(c) $\begin{array}{l} CH_2OH \\ | \\ C{=}O \\ | \\ CHOH \\ | \\ CH_2OH \end{array}$

(d) $\begin{array}{l} CH_3 \\ | \\ CHOH \\ | \\ C{=}O \\ | \\ CHOH \\ | \\ CH_2OH \end{array}$

(e) $\begin{array}{l} CHO \\ | \\ CHOH \\ | \\ CHOH \\ | \\ CHOH \\ | \\ CHOH \\ | \\ CH_2OH \end{array}$

CARBOHYDRATE FAMILIES

All aldoses can be prepared by a complex series of reactions from one of the enantiomeric forms of glyceraldehyde. The two enantiomers of glyceraldehyde are

$$\begin{array}{c} CHO \\ | \\ HO{-}C{-}H \\ | \\ CH_2OH \end{array} \qquad \begin{array}{c} CHO \\ | \\ H{-}C{-}OH \\ | \\ CH_2OH \end{array}$$

L-glyceraldehyde D-glyceraldehyde

The glyceraldehyde molecule whose Fischer projection formula shows the —OH group of the asymmetric carbon on the left is defined as L-glyceraldehyde. The glyceraldehyde with the —OH group of the asymmetric carbon on the right is D-glyceraldehyde. Most naturally occurring monosaccharides are related to D-glyceraldehyde.

The relationship of the four possible aldotetroses to D- and L-glyceraldehyde is shown in Fig. 9-9.

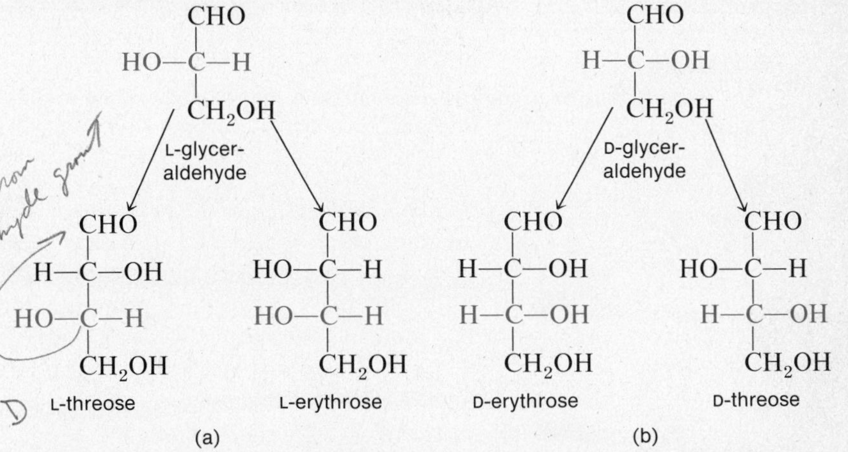

L-glyceraldehyde	D-glyceraldehyde
CHO	CHO
HO—C—H	H—C—OH
CH_2OH	CH_2OH

L-threose	L-erythrose	D-erythrose	D-threose
CHO	CHO	CHO	CHO
H—C—OH	HO—C—H	H—C—OH	HO—C—H
HO—C—H	HO—C—H	H—C—OH	H—C—OH
CH_2OH	CH_2OH	CH_2OH	CH_2OH
(a)	(a)	(b)	(b)

FIGURE 9-9 The four aldotetroses and their relationship to D- and L-glyceraldehyde. Structures in part (a) are the aldotetroses derived from L-glyceraldehyde. Structures in part (b) are the aldotetroses derived from D-glyceraldehyde.

We need not worry about the reactions by which glyceraldehyde is converted to aldotetroses other than to note that when glyceraldehyde is converted to an aldotetrose, the carbon atom of the aldehyde group of glyceraldehyde is reduced, becoming the new asymmetric carbon atom. The added carbon atom becomes the aldehyde group of the aldotetrose produced. The important thing for us to notice is that the asymmetric carbon of glyceraldehyde becomes the asymmetric carbon atom farthest from the aldehyde group in the new aldose. This carbon atom is shown in color in Fig. 9-9. *All aldoses in which the —OH group of the asymmetric carbon farthest from the aldehyde group is written on the right of the Fischer projection formula are said to be related to D-glyceraldehyde. Whenever the name of a compound is preceded by D, it means the compound is related to D-glyceraldehyde. Those compounds whose names are preceded by L are related to L-glyceraldehyde.*

exercise 9-25

Classify each of the following as related to D- or L-glyceraldehyde.

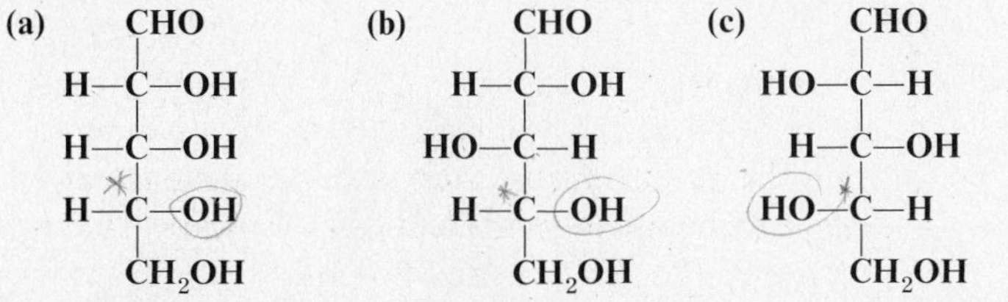

(a)	(b)	(c)
CHO	CHO	CHO
H—C—OH	H—C—OH	HO—C—H
H—C—OH	HO—C—H	H—C—OH
H—C—OH	H—C—OH	HO—C—H
CH_2OH	CH_2OH	CH_2OH

(d)

```
       CHO
        |
  H—C—OH
        |
  H—C—OH
        |
HO—C—H
        |
       CH₂OH
```

(e)

```
       CHO
        |
HO—C—H
        |
HO—C—H
        |
HO—C—H
        |
       CH₂OH
```

(f)

```
       CHO
        |
HO—C—H
        |
  H—C—OH
        |
  H—C—OH
        |
       CH₂OH
```

exercise 9-26

Refer to the structures given in Fig. 9-7 and explain what is meant by D-lactic acid and L-lactic acid.

The D-family of aldoses up to the hexoses are shown in Fig. 9-10. Only a few of these sugars are widespread in nature. The pentose D-ribose is common to all living organisms as a constituent of the ribose nucleic acids (RNA). D-Arabinose and D-xylose are primarily found in plants. Gum arabic, which is used as a thickening agent in foods, is primarily a polymer of arabinose. D-xylose can be obtained by hydrolysis of corncobs, straw, or wood. Overwhelmingly the hexoses are the most common monosaccharides. Almost all the carbohydrates in our diet are hexoses or their derivatives. Among the eight aldohexoses of the D-family, only three are accepted as naturally occurring: glucose, mannose, and galactose. Glucose is by far the most prevalent of the three.

Glucose (also called dextrose) is of such great importance to human metabolism that you should memorize its structural formula at this time. Glucose is the normal sugar occurring in blood and is one of the primary sources of energy to the cells. Galactose is found combined with glucose to form the disaccharide lactose or milk sugar. Additionally, galactose is found in nervous and brain tissues. Relatively little is known about the importance of mannose in human metabolism.

In addition to the aldoses, there is one ketose of importance in human metabolism, D-fructose, whose structure is

```
       CH₂OH
        |
        C=O
        |
HO—C—H
        |
  H—C—OH
        |
  H—C—OH
        |
       CH₂OH
```

Fructose is a member of the D-family of carbohydrates as shown by the position of the —OH group on the asymmetric carbon atom farthest from the ketone group. Actually, the arrangement about the three asymmetric carbon atoms in fructose is identical to the arrangement of the three

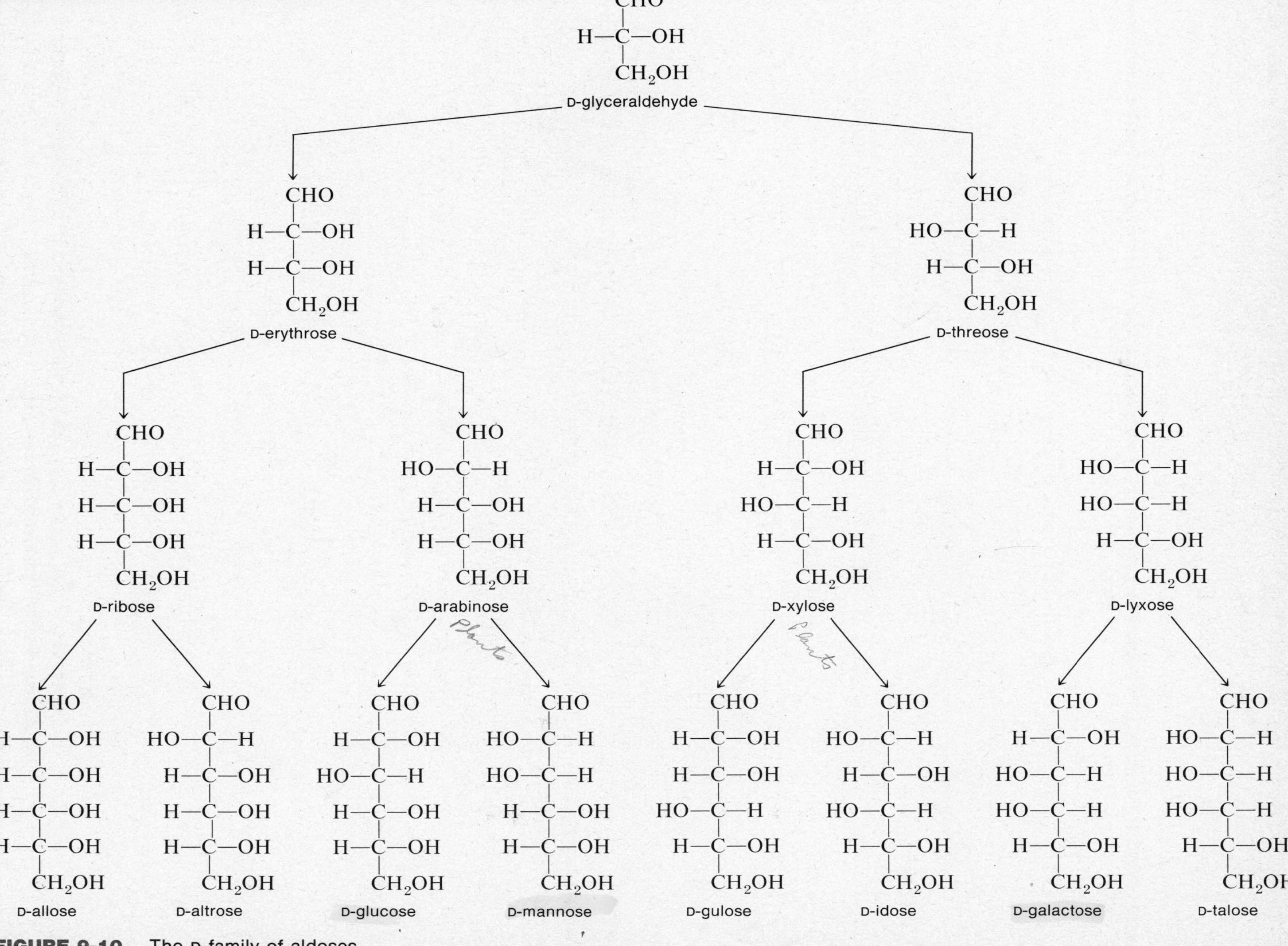

FIGURE 9-10 The D family of aldoses.

asymmetric carbon atoms farthest from the aldehyde group in D-glucose. The importance of this in metabolism is seen in Chapter 13.

```
      CHO                    CH2OH
       |                       |
    H—C—OH                    C=O
       |                       |
   HO—C—H                  HO—C—H
       |                       |
    H—C—OH                  H—C—OH
       |                       |
    H—C—OH                  H—C—OH
       |                       |
      CH2OH                  CH2OH
   D-glucose               D-fructose
```

Fructose is the sweetest of all sugars. A standard sweetness scale assigns sucrose (cane sugar) a value of 100. On this scale fructose has a sweetness value of 135; glucose, 75. Greater than 50 percent of the sugar in honey is fructose, giving honey its extremely sweet taste. It is very difficult to make fructose crystallize from solution. Furthermore, the presence of fructose prevents glucose and sucrose from crystallizing. This peculiar characteristic of fructose keeps honey from crystallizing over long periods of time. Candy and jelly makers take advantage of this fact by adding invert sugar containing approximately 30 percent fructose to their products to prevent an undesirable grainy texture.

REACTIONS OF MONOSACCHARIDES

The reactions of monosaccharides are those to be expected from the functional groups present. For example, the aldehyde or ketone group can be reduced to produce a polyhydroxy alcohol. Reduction of D-glucose with H_2 gas in the presence of a catalyst produces the polyhydroxy alcohol sorbitol.

```
    H   O
     \ //
      C                         CH2OH
      |                           |
   H—C—OH                      H—C—OH
      |                           |
  HO—C—H        H2/Pt         HO—C—H
      |       ———————>            |
   H—C—OH                      H—C—OH
      |                           |
   H—C—OH                      H—C—OH
      |                           |
     CH2OH                       CH2OH
   D-glucose                   sorbitol
```

In the presence of the appropriate enzymes, glucose can be reduced to sorbitol by living organisms. Biological reduction of glucose is important to reproduction of mammals. In the seminal vesicles, glucose is reduced to sorbitol. This reduction is followed by oxidation of sorbitol to fructose.

Fructose is abundant in seminal fluid and is the primary energy source for sperm cells. This sequence of reactions can be represented:

$$\begin{array}{c} \text{H}\diagdown\;\;/\!\!/\text{O} \\ \text{C} \\ | \\ \text{H—C—OH} \\ | \\ \text{HO—C—H} \\ | \\ \text{H—C—OH} \\ | \\ \text{H—C—OH} \\ | \\ \text{CH}_2\text{OH} \\ \text{D-glucose} \end{array} \xrightarrow{\text{NADH}_2} \begin{array}{c} \text{CH}_2\text{OH} \\ | \\ \text{H—C—OH} \\ | \\ \text{HO—C—H} \\ | \\ \text{H—C—OH} \\ | \\ \text{H—C—OH} \\ | \\ \text{CH}_2\text{OH} \\ \text{sorbitol} \end{array} \xrightarrow{\text{NAD}} \begin{array}{c} \text{CH}_2\text{OH} \\ | \\ \text{C=O} \\ | \\ \text{HO—C—H} \\ | \\ \text{H—C—OH} \\ | \\ \text{H—C—OH} \\ | \\ \text{CH}_2\text{OH} \\ \text{D-fructose} \end{array}$$

The reducing and oxidizing agents are $NADH_2$ and NAD, respectively. The structures and functions of these agents are discussed in Chapter 12.

Just as simple aldehydes and ketones undergo reaction with alcohols to produce hemiacetals and hemiketals, so will monosaccharides. A major difference is, however, that in monosaccharides the alcohol group and the aldehyde or ketone group are contained in the same molecule. The result of this is that monosaccharides undergo *intramolecular* reaction to produce hemiacetals or hemiketals.

Recall from p. 228 that formation of five- and six-membered ring hemiacetals or hemiketals is particularly favorable. Aldohexoses form six-membered ring hemiacetals by reaction of the aldehyde group with the alcohol group of the No. 5 carbon atom.

$$\begin{array}{c} \text{H}\diagdown\;\;/\!\!/\text{O} \\ {}^{1}\text{C} \\ | \\ \text{H—}{}^{2}\text{C—OH} \\ | \\ \text{HO—}{}^{3}\text{C—H} \\ | \\ \text{H—}{}^{4}\text{C—OH} \\ | \\ \text{H—}{}^{5}\text{C—OH} \\ | \\ {}^{6}\text{CH}_2\text{OH} \\ \text{D-glucose} \\ \text{(open-chain form)} \end{array} \rightleftharpoons \begin{array}{c} \text{H}\diagdown\;\;/\text{OH} \\ {}^{1}\text{C}\text{———}\rceil \\ | \qquad\quad | \\ \text{H—}{}^{2}\text{C—OH} \;\; | \\ | \qquad\quad | \\ \text{HO—}{}^{3}\text{C—H} \;\;\; | \\ | \qquad\quad | \\ \text{H—}{}^{4}\text{C—OH} \;\; | \\ | \qquad\quad | \\ \text{H—}{}^{5}\text{C———O} \\ | \\ {}^{6}\text{CH}_2\text{OH} \\ \alpha\text{-D-glucose} \\ \text{(hemiacetal form)} \end{array}$$

Notice that when the hemiacetal is formed, a new asymmetric carbon atom results. Whenever a new asymmetric carbon atom is formed in this way, both stereoisomers about that carbon atom can be produced. Since reaction involves many molecules rather than a single molecule, the result is a mixture of both isomers. For glucose the formation of both isomers can be represented as follows:

```
HO   H               H    O                H    OH
  \ /                 \  //                 \  /
   C ─────┐            C                     C ─────┐
   |      |            |                     |      |
H─C─OH    |         H─C─OH                H─C─OH    |
   |      |  ⇌         |          ⇌          |      |
HO─C─H    |         HO─C─H                HO─C─H    |
   |      |            |                     |      |
H─C─OH    |         H─C─OH                H─C─OH    |
   |      |            |                     |      |
H─C──────O          H─C─OH                H─C──────O
   |                   |                     |
  CH2OH               CH2OH                 CH2OH
```

β-D-glucose (one hemiacetal form) | D-glucose (open-chain form) | α-D-glucose (one hemiacetal form)

Very little glucose exists in the open-chain form.

The hemiacetal form of glucose in which the —OH group is written on the right hand side of the Fischer projection formula is the alpha form: α-D-glucose. The beta form (β-D-glucose) has the —OH group on the left-hand side. Notice that α-D-glucose and β-D-glucose are diastereomers and *not* mirror images. An equilibrium mixture of glucose consists of 0.02 percent of the open-chain form, 37 percent of the α-isomer, and 63 percent of the β-isomer.

exercise 9-27

Refer to Fig. 9-10 for the structures of galactose and mannose. Write the Fischer projection formulas for the α and β hemiacetal forms of these sugars.

HAWORTH FORMULAS FOR SUGARS

Following this procedure will allow you to write Haworth formulas for all aldohexoses in their hemiacetal form.

Fischer projection formulas for the cyclic hemiacetal forms of sugars are awkward looking and do not portray the structure very well. Remember that while we might write straight chains of carbon atoms, the bond angles between singly bonded carbon atoms are really 109°28′. A better, although still not entirely accurate, way of depicting the hemiacetal forms of sugars are *Haworth formulas*. The Haworth formula emphasizes the ring structure of the hemiacetal form of monosaccharides. The conventional means of writing the Haworth formula for a six-membered ring is to draw a hexagon. The oxygen atom of the ring is shown in the back right-hand corner of the hexagon; the remaining corners of the hexagon are understood to be occupied by carbon atoms. Thus, to write the Haworth formula for any six-membered ring hemiacetal of a monosaccharide, you initially write

```
  ____O
 /     \
 \_____/
```

. Since the hemiacetal is formed between the aldehyde group of the No. 1 carbon and the —OH group on the No. 5 carbon, the carbon atoms of the ring are numbered

```
     ⑤____O
     /      \
  ④ <        > ①
     \______/
    ③        ②
```

If the hexose is a member of the L family, the No. 6 C atom is written downward.

For all members of the D-family of hexoses, the No. 6 carbon is written attached to the No. 5 carbon in an upward position.

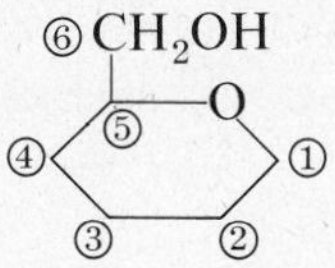

To complete the writing of the Haworth formula, all groups written to right of the Fischer projection formula are written downward; all groups on the left of the Fischer projection are written upward in the Haworth formula. The Fischer projection and the Haworth formulas for β-D-glucose and α-D-glucose are

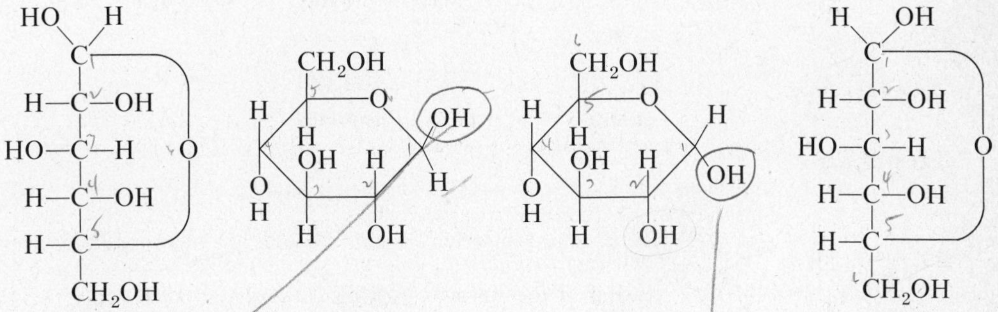

Fischer projection — Haworth formula

β-D-glucose

Haworth formula — Fischer projection

α-D-glucose

In keeping with the writing of other ring compounds, the hydrogen atoms are not shown but are understood to be present. Thus, the Haworth formulas for β-D-glucose and α-D-glucose become

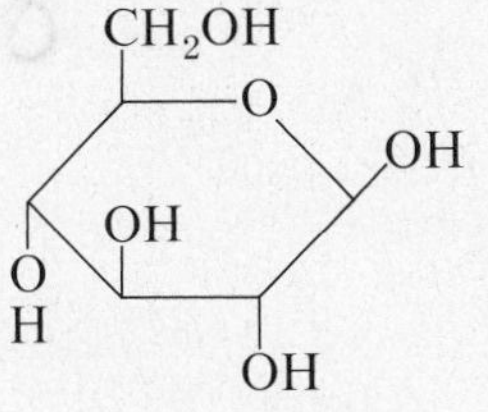

β-D-glucose

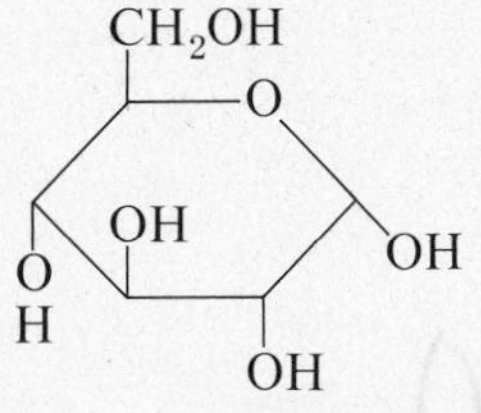

α-D-glucose

exercise 9-28

Write the Haworth formulas for the following:

(a)

```
  H   OH
   \ /
    C ─────┐
    |      |
 H──C──OH  |
    |      |
HO──C──H   O
    |      |
HO──C──H   |
    |      |
 H──C──────┘
    |
    CH2OH
```

α-D-galactose

(c)

```
  H   OH
   \ /
    C ─────┐
    |      |
HO──C──H   |
    |      |
HO──C──H   O
    |      |
 H──C──OH  |
    |      |
 H──C──────┘
    |
    CH2OH
```

α-D-mannose

(e)

```
  H   OH
   \ /
    C ─────┐
    |      |
 H──C──OH  |
    |      |
 H──C──OH  O
    |      |
 H──C──OH  |
    |      |
 H──C──────┘
    |
    CH2OH
```

α-D-allose

(b)

```
 HO   H
   \ /
    C ─────┐
    |      |
 H──C──OH  |
    |      |
HO──C──H   O
    |      |
HO──C──H   |
    |      |
 H──C──────┘
    |
    CH2OH
```

β-D-galactose

(d)

```
 HO   H
   \ /
    C ─────┐
    |      |
HO──C──H   |
    |      |
 H──C──OH  O
    |      |
HO──C──H   |
    |      |
 H──C──────┘
    |
    CH2OH
```

β-D-idose

exercise 9-29

D-Ribose forms a five-membered ring hemiacetal by reaction of the —OH group on the No. 4 carbon with the aldehyde.

(a) Draw the Fischer projection formula for one of the two possible hemiacetal forms.

(b) By extension of the rules for writing Haworth formulas, write the Haworth formula for the hemiacetal form drawn in part (a).

The ketohexose called *fructose* forms a cyclic hemiketal in equilibrium with the open-chain ketone form. The cyclic form of fructose is a five-membered ring. The formation of the hemiketal generates a new asymmetric carbon atom giving rise to a mixture of α-D-fructose and β-D-fructose. The Fischer projection of the open-chain and Haworth formulas for the cyclic forms are

```
 1CH2OH      O                                 1CH2OH                  HO       O
     \     /   \                                  |                       \   /   \
      C2         C5                            2C═O                        C2       C5
    /   \       /  \                              |                       /  \     /  \
  HO     3───4      6CH2OH   ⇌     HO──3C──H    ⇌    1CH2OH   3───4      6CH2OH
         |   |                                    |                           |   |
         OH  OH (H above)                   H──4C──OH                        OH  OH
                                                  |
                                            H──5C──OH
                                                  |
                                                 6CH2OH
```

β-D-fructose

fructose (open chain)

α-D-fructose

REDUCING PROPERTIES OF SUGARS

Aldehydes and α-hydroxy ketones are easily oxidized compounds, reacting readily with Tollen's reagent and Fehling's or Benedict's solutions. Since monosaccharides are polyhydroxy aldehydes and polyhydroxy ketones, we would expect them to react readily with mild oxidizing agents. Although monosaccharides exist primarily in the cyclic hemiacetal or hemiketal form at equilibrium, there is always present a small amount of the open-chain free aldehyde or α-hydroxy ketone form. As the open-chain form of the monosaccharide reacts with the oxidizing agent, some of the hemiacetal or hemiketal form reverts to the open-chain species in an attempt to maintain equilibrium. Eventually, if enough oxidizing agent is present, all the monosaccharide is converted to the open-chain form and is oxidized. Recall that oxidation cannot occur without reduction and since the monosaccharides are oxidized, they must be reducing agents. *All sugars that will react with Tollen's reagent or either Fehling's or Benedict's solutions are known as reducing sugars. All monosaccharides are reducing sugars.*

The suffix uria *indicates the compound is found in urine.*

Monosaccharides may appear as an abnormal constituent of urine. The most common cause of a monosaccharide in urine is diabetes mellitus resulting in the presence of glucose in the urine. This condition is known as glucosuria, and the glucose concentration in the urine of a diabetic individual may range from 0.5 to 12.0 percent.

Benedict's solution has been widely used to test urine for the presence of glucose.* The amount and color of the precipitate formed presumably gives an indication of the amount of glucose in the urine sample. Unfortunately, the Benedict's test is not specific for glucose. Although glucose is the most likely cause of reaction when urine is treated with Benedict's solution, it is not the only possible cause. A reaction will occur due to presence of many different possible reducing substances.

The suffix semia *indicates the compound is found in blood.*

Other possible substances in urine capable of reducing the copper(II) ions are the five-carbon sugars, arabinose and xylulose; the seven-carbon sugar, mannoheptulose; galactose; the disaccharide, lactose; creatinine; and glucuronic acid or its derivatives. Creatinine and glucuronic acid are normal components of urine. Arabinose in urine may result from ingestion of large amounts of fruit or fruit juices. Similarly, mannoheptulose is found in the urine of individuals who have eaten large amounts of avocado. Lactose is frequently found in the urine of lactating women. The presence of either xylulose or galactose in urine results from hereditary defects. The excretion of xylulose is apparently harmless. Galactose in urine is a symptom of galactosemia. Fortunately, galactosemia is relatively rare, ordinarily easily detectable in early infancy and can be controlled by excluding galactose from the diet. Galactosemia results from the congenital absence of an enzyme that catalyzes the conversion of galactose to glucose. The absence

*The Benedict's test is available as a tablet formed from the dry reagents used in preparing Benedict's solution. Although slightly less sensitive than the solution, it is suitable when used properly. The tablets are sold as Clinitest Tablets®, Ames Company, Elkhart, Indiana.

of the enzyme allows galactose to accumulate in the body, eventually causing damage to the brain, liver, and eyes.

To lessen the possibility of mistaking other reducing substances for the presence of glucose in urine, more selective tests have recently been developed. One of these tests involves the use of the enzyme glucose oxidase. The enzyme is specific for reaction with glucose; consequently, no reaction is observed in the absence of glucose.

DISACCHARIDES

Hemiacetals will react with alcohols in the presence of an acid catalyst. Since aldohexoses exist primarily in the hemiacetal form, we would expect them to react with alcohols. Glucose, for example, would be expected to react with methyl alcohol to produce a methyl acetal.

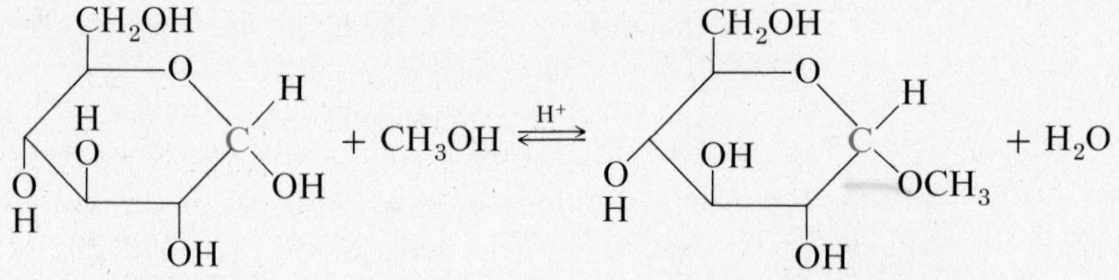

$$\alpha\text{-D-glucose} + CH_3OH \overset{H^+}{\rightleftharpoons} \text{methyl-}\alpha\text{-D-glucoside} + H_2O$$

α-D-glucose methyl-α-D-glucoside

(The hemiacetal carbon of α-D-glucose and the acetal carbon of methyl-α-D-glucoside are shown in color.) The acetals of aldoses are called *glycosides;* those of glucose are specifically known as *glucosides.*

A disaccharide can be formed by the reaction of two monosaccharide molecules. One serves as the hemiacetal; the other, as the alcohol. For example, two glucose molecules could combine to form the disaccharide maltose. If the hemiacetal group of one glucose molecule reacts with the —OH group on carbon 4

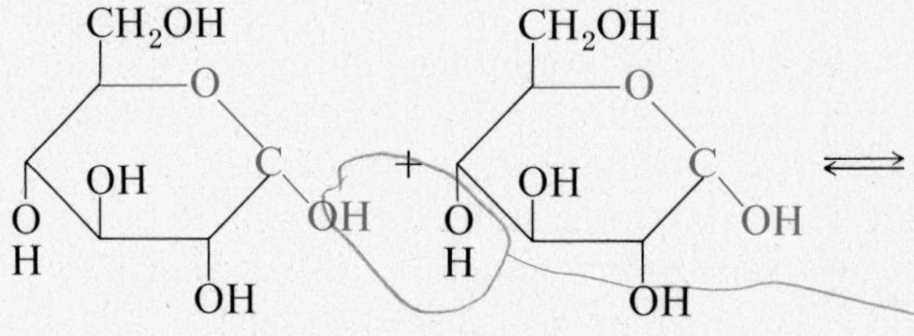

$$\alpha\text{-D-glucose} + \alpha\text{-D-glucose} \rightleftharpoons \alpha\text{-maltose} + H_2O$$

α-D-glucose α-D-glucose

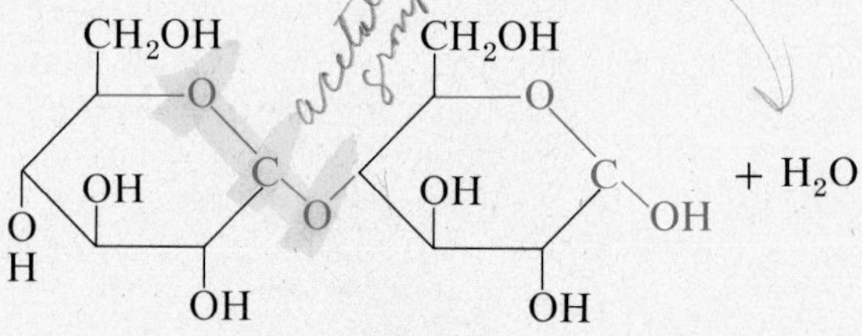

α-maltose

of the second glucose molecule, the disaccharide maltose is produced. The two glucose molecules are joined by an acetal group. Notice that maltose contains both an acetal group and a hemiacetal group. The hemiacetal groups of both glucose and maltose and the acetal group of maltose are shown in color. The presence of a hemiacetal group in maltose means that it is a reducing sugar.

Maltose is not abundant in nature; however, it is easily prepared from starch by the action of either enzymes or dilute solutions of a strong acid. The primary natural source of maltose is sprouting grain seeds. For this reason maltose is often referred to as malt sugar.

exercise 9-30

The acetal grouping in maltose is stable to hydrolysis in basic or neutral solutions. In acidic solutions, maltose is hydrolyzed. Write a balanced equation for the hydrolysis of maltose.

Notice the acetal linkage between the glucose units in maltose has the alpha arrangement between the No. 1 carbon of one glucose molecule and the No. 4 carbon of the second glucose molecule. This is but one of several known ways in which two glucose units combine to form a disaccharide. The disaccharide cellobiose has a 1,4-β acetal bonding arrangement. In gentiobiose and isomaltose the acetal linkage is 1,6-β and 1,6-α, respectively. Figure 9-11 compares the structures of these four disaccharides formed from glucose.

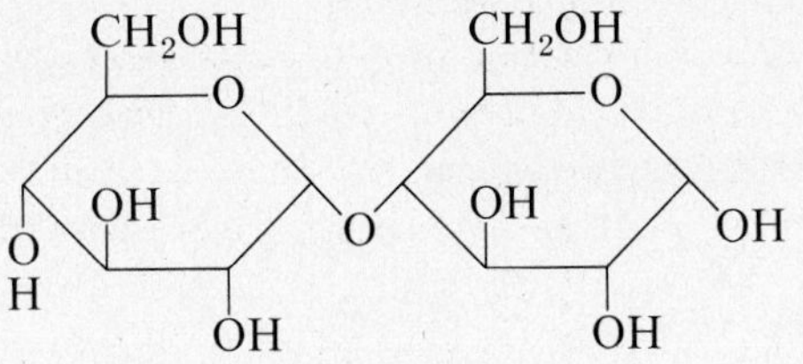

α-maltose

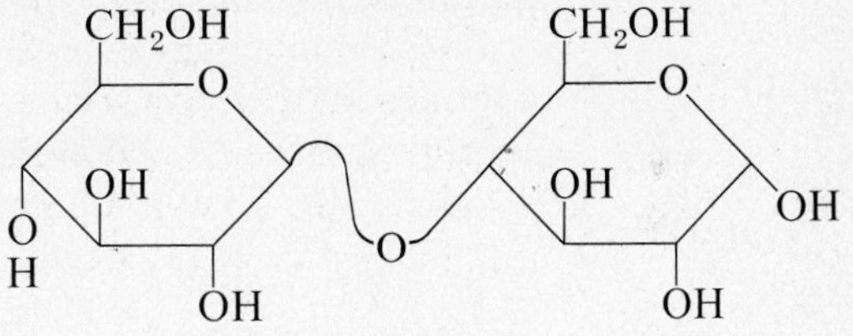

α-cellobiose

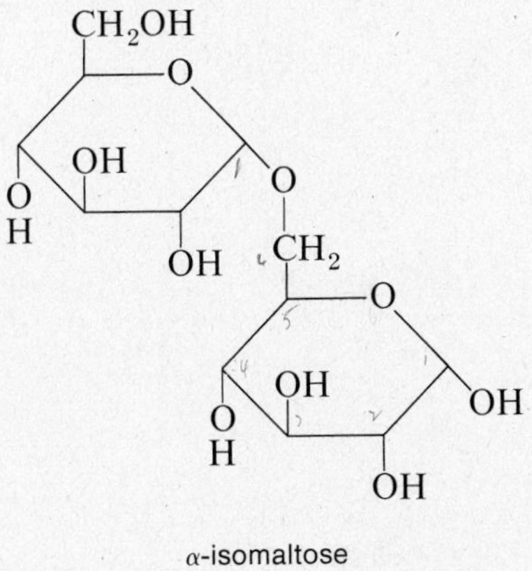

α-isomaltose

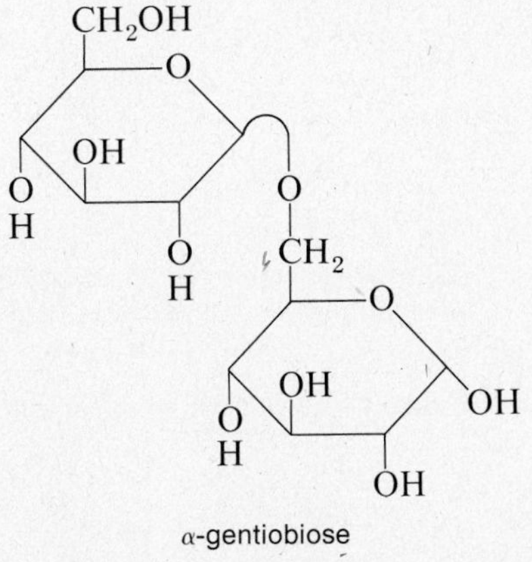

α-gentiobiose

FIGURE 9-11 The structures of four disaccharides formed from glucose.

exercise 9-31

All the disaccharides in Fig. 9-11 are designated as alpha sugars. Explain why. Draw the structure of β-isomaltose.

exercise 9-32

Using structural formulas, write equations for the formation from glucose of the disaccharides shown in Fig. 9-11.

Two additional disaccharides of biological importance are lactose and sucrose. Lactose is found solely in the milk of mammals; hence, its common name is milk sugar. The acetal bonding arrangement is between the beta hemiacetal form of galactose and the —OH group on the No. 4 carbon of glucose.

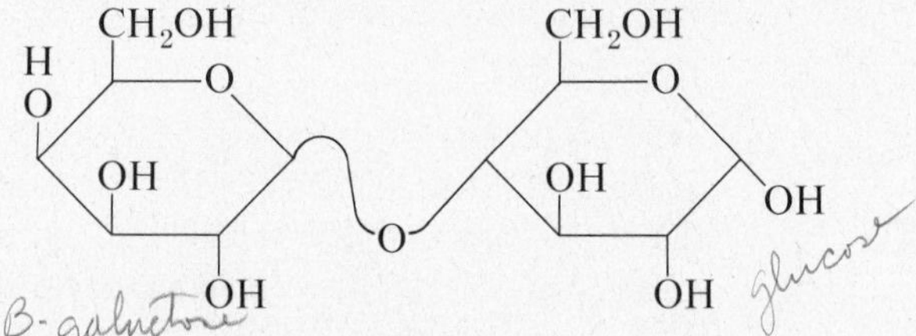

α-lactose

Lactose is a reducing sugar by virtue of the hemiacetal present.

Sucrose, composed of a molecule of glucose and one of fructose, is the most common and familiar disaccharide. Commercially, sucrose is obtained from sugar cane and sugar beets. It is also found in varying concentrations in most fruits, flowers, seeds, and roots. In contrast to most other disaccharides, sucrose is *not* a reducing sugar.

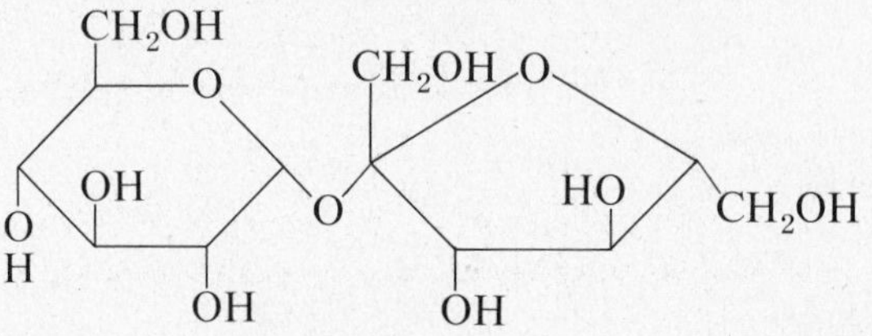

sucrose

exercise 9-33

Carefully examine the structural formula for sucrose and decide if it is derived from α- or β-glucose and from α- or β-fructose.

exercise 9-34

Explain why sucrose is not a reducing sugar.

POLYSACCHARIDES

Polysaccharides are polymers of monosaccharides and are very complex molecules. The great majority of all carbohydrates in nature exist as poly-

High molecular weight compounds (greater than 100,000) are generally insoluble. See solutions and colloids, Chapter 7.

saccharides. These molecules are generally composed of several hundred to several thousand monosaccharide units. As a result, their properties are considerably different than those of monosaccharides and disaccharides. Most are insoluble in water, do not have a sweet taste, and fail to react with mild oxidizing agents.

Although there are naturally occurring polysaccharides of both pentoses and hexoses, the most important are those of glucose. The polysaccharides of glucose can be divided into two basic classes, depending on whether the glucose units are joined by 1,4-α or 1,4-β acetal linkages. Those joined by 1,4-β linkages primarily fulfill the structural needs of plant cells; whereas those in which the glucose molecules are joined by 1,4-α linkages serve nutritional needs in plants and animals.

Cellulose is the most abundant organic compound in the world. It is made up of long chains of glucose joined by 1,4-β acetal linkages.

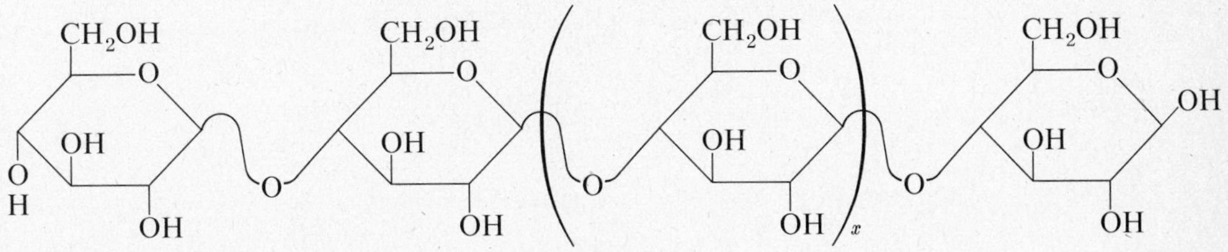

cellulose
(x = 300 to 3000)

The purest natural source of cellulose is cotton, which is greater than 90 percent cellulose. In addition to the use of natural cotton to weave fabrics, large quantities of cotton are use to prepare cellulose derivatives. Cellulose acetate, for example, is used to manufacture film; plastics; yarns and fabrics; and, when dissolved in an organic solvent, fingernail polish. Rayon and cellophane are also produced from cellulose, as are a number of explosives.

exercise 9-35

The complete acid-catalyzed hydrolysis of cellulose yields glucose. Partial hydrolysis yields cellobiose. Write equations for the complete and partial hydrolysis of cellulose.

Starch and glycogen are the glucose polymers in which the glucose molecules are joined by 1,4-α acetal linkages. Starch is a plant polysaccharide, while glycogen is produced by animals. These substances are a form of carbohydrate storage.

Starch is divided into two types, amylose and amylopectin; 10 to 20 percent of starch is amylose. Amylose is thought to consist of long chains of glucose molecules held together by 1,4-α linkages. Amylopectin, the major component of starch, is known to be a branched polysaccharide. Although the glucose units are primarily joined by 1,4-α linkages, in about every 24 to 30 glucose units there is also a 1,6-α linkage. This 1,6-α linkage allows the formation of a branch chain. At each branch point there is a glucose molecule that is bonded to another glucose molecule via the No. 6 carbon in addition to those bonded to it on the No. 1 and No. 4 carbons. A portion of an amylopectin molecule showing the branching is illustrated in Fig. 9-12.

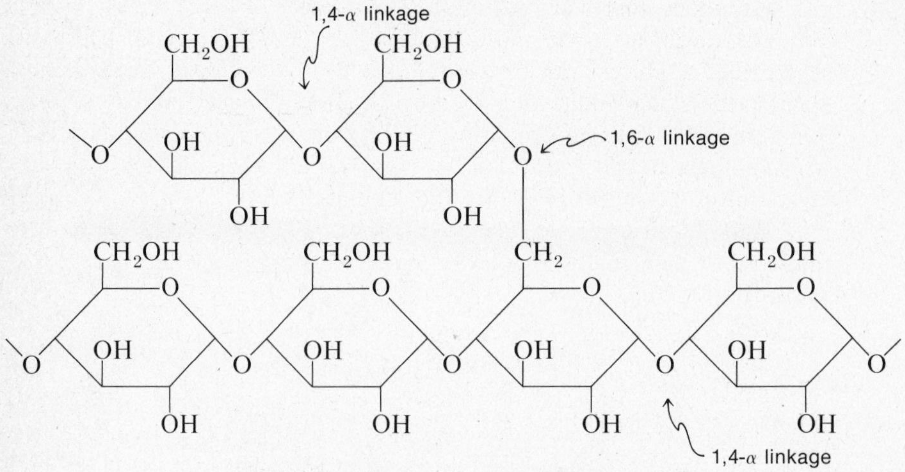

FIGURE 9-12 A portion of an amylopectin molecule showing how branching occurs through 1, 6-α linkages.

Starch is easily hydrolyzed by dilute solutions of strong acids, ultimately leading to the formation of glucose. This breakdown occurs by the random hydrolysis along the chain of the acetal bonds linking the glucose units together. Thus, amylopectin is first broken down into smaller polysaccharide fragments. These smaller fragments are subsequently hydrolyzed into progressively smaller fragments until maltose and finally glucose is produced. The intermediate-sized fragments between maltose and starch are collectively known as *dextrins* (see Fig. 9-13).

A characteristic reaction of starch is with iodine to produce an intense blue color. Dextrins react with iodine to produce various shades and color intensities of red through orange and yellow to colorless solutions. The shade and intensity of color is an indication of the chain length of the fragments produced by starch hydrolysis. These color changes are used to follow the progress of starch hydrolysis reactions by removing small samples of the reaction mixture at various times and treating them with an iodine solution.

exercise 9-36

Mixtures resulting from the partial hydrolysis of starch are found to contain isomaltose as well as glucose, maltose, and dextrins. Explain why the mixture contains isomaltose.

Glycogen is sometimes called animal starch.

Glycogen, found in the liver and muscle tissue of animals, is the animal counterpart of starch. It is a very highly branched polymer of glucose with

FIGURE 9-13 Action of α-amylase on amylopectin. (a) Amylopectin model. (b) Dextrins of medium molecular weight giving violet, purple, or red iodine color produced by splitting of 4 percent of the glucosidic linkages of amylopectin. (c) Possible structures of limit dextrins from amylopectin breakdown; the hepta-, hexa-, and pentasaccharides are probably split, more or less rapidly, into lower-molecular oligosaccharides. (According to P. Bernfeld, *Advances in Enzymol.,* **12,** 379, 1951.)

a structure resembling that of amylopectin. Branching of the chains apparently occurs every 8 to 12 glucose units.

The ability to reversibly make glycogen from glucose is extremely important to the body, which uses glucose as its primary source of energy. We ordinarily take in carbohydrates in large quantities at widely spaced intervals. These carbohydrates are rapidly broken down in the digestive process and the resultant glucose is quickly absorbed. This absorption of large quantities of low molecular weight molecules could cause fatal disturbances to the osmotic pressure relationships in the body. Thus, it becomes a matter of great importance that the organism be able to store large quantities of material required as an energy source in a manner that will have little effect on osmotic pressure. Recall from Chapter 7 that osmotic pressure depends mainly on the number of dissolved particles present and not their kind or size. The reversible formation of the extremely large slightly soluble molecules of glycogen represents a means of storing glucose in a readily accessible form that has little effect on osmotic pressure.

CHAPTER SUMMARY

1. Alcohols contain the —OH group. According to their structure, alcohols are classified as primary, secondary, and tertiary.
2. Alcohols can be prepared by the addition of water to carbon–carbon double bonds. The reaction follows Markovnikov's rule.
3. Alcohols can be dehydrated to alkenes. The elimination of water from an alcohol leads to the most highly substituted double bond.
4. Ethers can be prepared by the intermolecular elimination of water from alcohols.
5. Primary alcohols can be oxidized to aldehydes; secondary alcohols, to ketones; tertiary alcohols do not normally undergo oxidation. Aldehydes can be further oxidized to acids.
6. Fehling's, Benedict's, and Tollen's reagents are useful for determining if a compound is an aldehyde.
7. Aldehydes and ketones react with alcohols to form hemiacetals and hemiketals, respectively. These reactions do not go to completion; an equilibrium mixture is instead obtained.
8. In the presence of an acid catalyst, alcohols react with hemiacetals and hemiketals to form acetals and ketals, respectively. Acetals and ketals are stable in basic solutions but are hydrolyzed in acidic solutions.
9. Aldehydes and ketones can be reduced to alcohols by hydrogen in the presence of a catalyst.
10. Carbohydrates are polyhydroxy aldehydes or polyhydroxy ketones and their derivatives. Carbohydrates are divided into monosaccharides, disaccharides, oligosaccharides, and polysaccharides.

11. Stereoisomerism results from having four different groups attached to a carbon atom. A carbon atom bonded to four different groups is said to be *asymmetric*. Stereoisomers may be enantiomers or diastereomers. Fischer projection formulas are useful for depicting stereoisomers. Additionally, Haworth formulas are used for showing the structures of sugars.
12. Monosaccharides undergo most of the reactions expected for alcohols and aldehydes (ketones). Hemiacetal formation is particularly important. Disaccharides consist of two monosaccharides joined by an acetal bond.
13. Sugars are classified as reducing or nonreducing sugars on the basis of their reactivity toward mild oxidizing agents.
14. Amylose, amylopectin, and glycogen are all polymers of α-D-glucose linked by 1,4 acetal bonds. The major difference among these polymers is the amount of chain branching via 1,6 acetal linkages. Cellulose is also a polymer of glucose. The glucose units in cellulose are joined by 1,4-β acetal linkages. Mammals do not possess the enzymes necessary to catalyze the hydrolysis of cellulose.

QUESTIONS

1. Write structural formulas for the expected products of the indicated reaction. If no reaction is expected, write N.R.

(a) cyclopentene + $H_2O \xrightarrow{H^+}$

(b) $2CH_3OH \xrightarrow{H_2SO_4}$

(c) CH_3–(cyclohexane)–OH $\xrightarrow{K_2Cr_2O_7}$

(d) CH_3–(cyclohexane)–OH $\xrightarrow{H^+}$

(e) CH_3–(cyclohexane)–OH $\xrightarrow[\text{solution}]{\text{Benedict's}}$

(f) $CH_3CH_2C(CH_3)_2$–OH $\xrightarrow{K_2Cr_2O_7}$

(g) methylenecyclohexane ($=CH_2$) + $H_2O \xrightarrow{H^+}$

(h) cyclohexane–CHO $\xrightarrow{K_2Cr_2O_7}$

(i) $CH_3CH_2C(CH_3)_2$–OH $\xrightarrow{H_2SO_4}$

(j) $CH_3CH_2CH_2OH \xrightarrow{K_2Cr_2O_7}$

(k) 1-methylcyclopentanol (CH_3, OH) $\xrightarrow{H_2SO_4}$

(l) cyclohexanone ($=O$) $\xrightarrow[\text{reagent}]{\text{Tollen's}}$

(m) $CH_3CH_2CHO \xrightarrow[\text{reagent}]{\text{Tollen's}}$

(n) 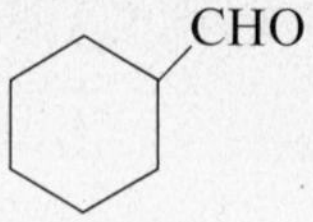 Cyclohexane–CHO $+ H_2 \xrightarrow{\text{Pt}}$

(o) $CH_3CH_2CH{=}CH_2 + H_2O \xrightarrow{H^+}$

(p) 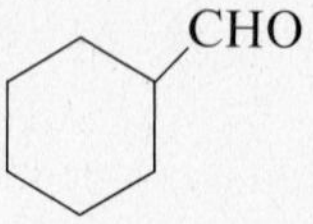 Cyclohexane–CHO $+ CH_3OH \longrightarrow$

(q) 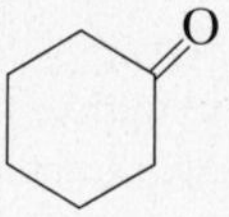 Cyclohexanone $+ H_2 \xrightarrow{\text{Pt}}$

(r) $CH_3CH_2CH_2OH + H_2 \xrightarrow{\text{Pt}}$

(s) $HC(=O)H + CH_3CH_2OH \longrightarrow$

(t) $CH_3CH_2CH{=}CH_2 \xrightarrow[\text{solution}]{\text{Benedict's}}$

(u) $CH_3{-}CH(O{-}CH_3){-}O{-}CH_3 \xrightarrow[H_2O]{H^+}$

(v) $HC(=O)H + 2CH_3CH_2OH \xrightarrow{H^+}$

(w) $CH_3C(=O){-}CH_3 + CH_3OH \longrightarrow$

(x) 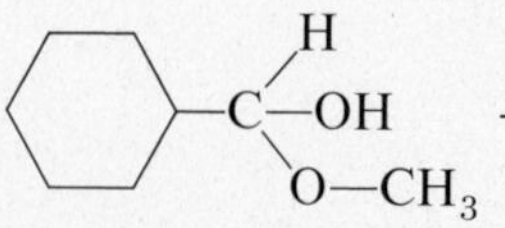 Cyclohexyl–$CH(OH){-}O{-}CH_3 \xrightarrow[\text{reagent}]{\text{Tollen's}}$

2. Give the structure of organic starting material(s) from which each of the following could be prepared in a single step. Also state any other reagents that would be required.

(a) $CH_3CH{=}CH_2$

(b) $CH_3CH_2CH_2CHO$

(c) $CH_3CH_2CH(OH){-}O{-}CH_2CH_3$

(d) $CH_3CH_2CH_2{-}O{-}CH_2CH_2CH_3$

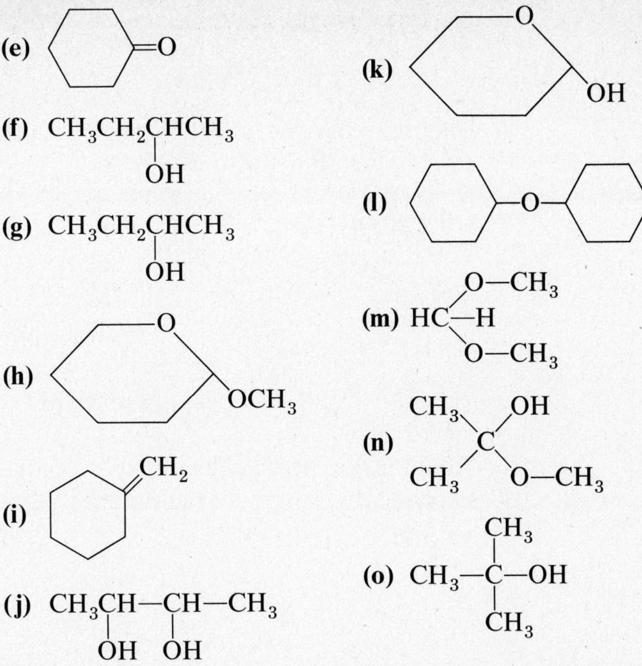

3. Arrange the following alcohols in order of increasing solubility in water.

(a) $CH_3CH_2CH_2CH_2OH$

(b) $CH_3CH(OH)—CH(OH)CH_3$

(c) $CH_2(OH)—CH_2(OH)$

(d) $CH_3CH_2CH_2CH_2CH_2CH_2CH_2CH_2OH$

4. Offer an explanation of why $HOCH_2CH_2OH$ has a much higher boiling point than CH_3CH_2OH.

5. Write Fischer projection formulas for all possible stereoisomers of the following. Where appropriate, identify enantiomeric pairs.

(a) $CH_3CH(OH)CH_2OH$

(b) $CH_3CH_2CH(NH_2)CO_2H$

(c) $CH_3CH(OH)—CH(OH)CHO$

(d) $CH_3CH(OH)—C(=O)—CH(OH)—CH(OH)CHO$

(e) 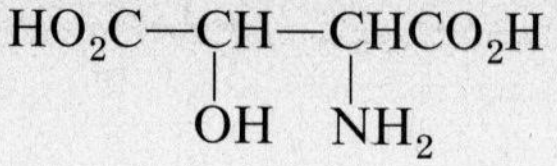 $HO_2C-\underset{\displaystyle OH}{\underset{|}{C}H}-\underset{\displaystyle NH_2}{\underset{|}{C}H}CO_2H$

6. The disaccharaide trehalose can be isolated from mushrooms. The structural formula of trehalose is

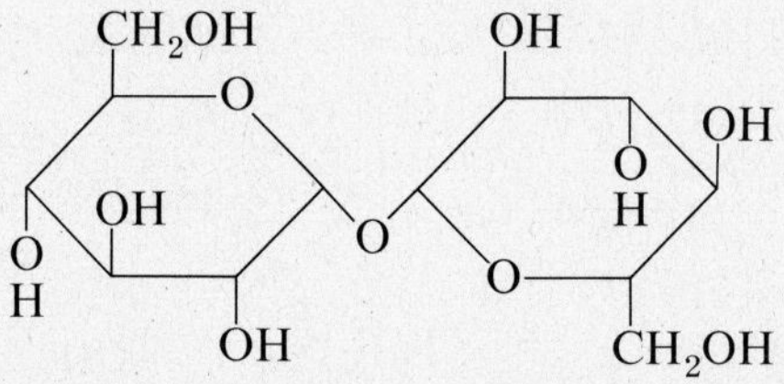

[Structural formula of trehalose: two pyranose rings, each bearing CH_2OH and OH groups, joined through an O bridge]

Would you expect trehalose to be a reducing sugar? Explain.

7. Explain why starch does not reduce Benedict's solution.

10

acid derivatives

INTRODUCTION Chemists study a compound's ability to react with other reagents to form new compounds. The body uses reactions between available materials to synthesize substances required for growth and metabolism. The study of selected reactions has led to a clearer understanding of the body's chemical activities. Among the important types of reactions are those between carboxylic acids and other reagents. By recognizing the presence of certain functional groups in the reactants, you will be able to predict the structural formulas of derivative(s) formed. Several classes of acid derivatives will be introduced in this chapter—acid anhydrides, esters, and triglycerides.

ANHYDRIDES Acid anhydrides make up a biologically important group of compounds. We may represent an anhydride group as

$$-\overset{\overset{\displaystyle O}{\|}}{E}-O-\overset{\overset{\displaystyle O}{\|}}{E}-$$

Add this to the list of functional groups learned in Chapter 4.

where, for our purposes, E may be carbon or phosphorus. The term *anhydride* means *without water*. The anhydride is produced via intermolecular dehydration.

$$\underset{\text{acid}}{-\overset{\overset{\displaystyle O}{\|}}{E}-OH} + \underset{\text{acid}}{HO-\overset{\overset{\displaystyle O}{\|}}{E}-} \rightleftharpoons HOH + \underset{\text{acid anhydride}}{-\overset{\overset{\displaystyle O}{\|}}{E}-O-\overset{\overset{\displaystyle O}{\|}}{E}-}$$

Carboxylic acids have corresponding anhydrides. When E in the general equation above is carbon,

$$\underset{\text{2 moles of acid}}{-\overset{\overset{\displaystyle O}{\|}}{C}-OH + HO-\overset{\overset{\displaystyle O}{\|}}{C}-} \rightleftharpoons HOH + \underset{\text{1 mole of carboxylic acid anhydride}}{-\overset{\overset{\displaystyle O}{\|}}{C}-O-\overset{\overset{\displaystyle O}{\|}}{C}-}$$

Acetic anhydride is formed by the loss of water from 2 moles of acetic acid:

$$\underset{\text{acetic acid}}{CH_3-\overset{\overset{\displaystyle O}{\|}}{C}-OH} + \underset{\text{acetic acid}}{HO-\overset{\overset{\displaystyle O}{\|}}{C}-CH_3} \rightleftharpoons HOH + \underset{\text{acetic anhydride}}{CH_3-\overset{\overset{\displaystyle O}{\|}}{C}-O-\overset{\overset{\displaystyle O}{\|}}{C}-CH_3}$$

For phosphoric acid, E in the general equation must be phosphorus (P). The coupling of two phosphoric acid molecules by the elimination of a water molecule produces a phosphorus-containing anhydride:

$$\mathrm{HO{-}\overset{\displaystyle O}{\overset{\|}{\underset{\displaystyle OH}{\underset{|}{P}}}}{-}OH + HO{-}\overset{\displaystyle O}{\overset{\|}{\underset{\displaystyle OH}{\underset{|}{P}}}}{-}OH \rightleftharpoons HOH + HO{-}\overset{\displaystyle O}{\overset{\|}{\underset{\displaystyle OH}{\underset{|}{P}}}}{-}O{-}\overset{\displaystyle O}{\overset{\|}{\underset{\displaystyle OH}{\underset{|}{P}}}}{-}OH}$$

phosphoric acid — a phosphoric acid anhydride

The coupling process can continue with the incorporation of a third molecule of phosphoric acid into its anhydride form.

$$\mathrm{HO{-}\overset{\displaystyle O}{\overset{\|}{\underset{\displaystyle OH}{\underset{|}{P}}}}{-}O{-}\overset{\displaystyle O}{\overset{\|}{\underset{\displaystyle OH}{\underset{|}{P}}}}{-}OH + HO{-}\overset{\displaystyle O}{\overset{\|}{\underset{\displaystyle OH}{\underset{|}{P}}}}{-}OH \rightleftharpoons HOH + HO{-}\overset{\displaystyle O}{\overset{\|}{\underset{\displaystyle OH}{\underset{|}{P}}}}{-}O{-}\overset{\displaystyle O}{\overset{\|}{\underset{\displaystyle OH}{\underset{|}{P}}}}{-}O{-}\overset{\displaystyle O}{\overset{\|}{\underset{\displaystyle OH}{\underset{|}{P}}}}{-}OH}$$

A third possible combination is the formation of a mixed anhydride by the coupling of two dissimilar acids. A carboxylic acid group and a phosphoric acid group can combine to form a mixed anhydride linkage.

$$\mathrm{R{-}\overset{\displaystyle O}{\overset{\|}{C}}{-}OH + HO{-}\overset{\displaystyle O}{\overset{\|}{\underset{\displaystyle OH}{\underset{|}{P}}}}{-}OH \rightleftharpoons HOH + R{-}\overset{\displaystyle O}{\overset{\|}{C}}{-}O{-}\overset{\displaystyle O}{\overset{\|}{\underset{\displaystyle OH}{\underset{|}{P}}}}{-}OH}$$

a carboxylic acid — phosphoric acid — a mixed carbon–phosphorus anhydride

Energy is stored during the formation of anhydrides (an endergonic reaction). The energy is stored in the

$$\mathrm{{-}\overset{\displaystyle O}{\overset{\|}{E}}{-}O{-}\overset{\displaystyle O}{\overset{\|}{E}}{-}}$$

anhydride bonding arrangement.

The reaction for the formation of an anhydride is reversible. When water is added to an anhydride, the anhydride bond ruptures and the component acids are re-formed. Bond breakage due to the action of water is called *hydrolysis* (Greek: *hydro*, water; *lysis*, to cut).

$$\mathrm{{-}\overset{\displaystyle O}{\overset{\|}{E}}{-}O{-}\overset{\displaystyle O}{\overset{\|}{E}}{-} + HOH \longrightarrow {-}\overset{\displaystyle O}{\overset{\|}{E}}{-}OH + HO{-}\overset{\displaystyle O}{\overset{\|}{E}}{-} + energy}$$

As the equation above indicates, anhydride hydrolysis is an exergonic process. The biological importance of anhydride formation and hydrolysis is that these two reactions provide a means by which an organism can store energy by anhydride formation and then release it upon hydrolysis. This is discussed in Chapter 12.

exercise 10-1

Complete the following table using structural formulas:

ACID	+	ACID	ANHYDRIDE
Butyric		Butyric	____
Phosphoric		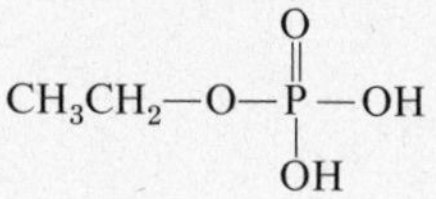 $CH_3CH_2{-}O{-}P(=O)(OH){-}OH$	____
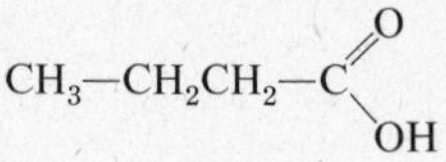 $CH_3{-}CH_2CH_2{-}C(=O)OH$		Acetic	____
$CH_3{-}CH_2CO_2H$		____	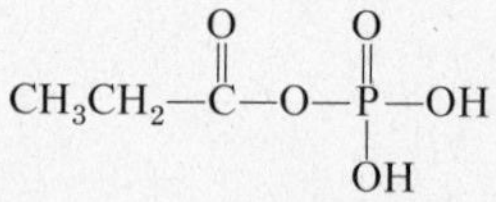 $CH_3CH_2{-}C(=O){-}O{-}P(=O)(OH){-}OH$
____		____	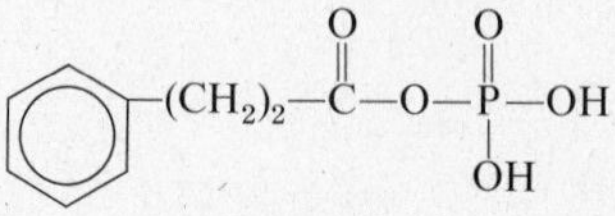 $C_6H_5{-}(CH_2)_2{-}C(=O){-}O{-}P(=O)(OH){-}OH$
1 mole of 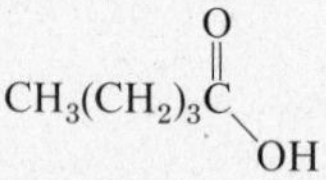 $CH_3(CH_2)_3C(=O)OH$		2 moles of phosphoric acid	____

exercise 10-2

During carbohydrate metabolism, 3-phosphoglyceric acid reacts with $(H_2PO_4)^-$

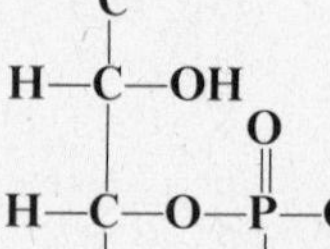

$$\left(HO{-}P(=O)(OH){-}O^{(-)} \right)$$

to form 1,3,-diphosphoglyceric acid, which contains an anhydride group.

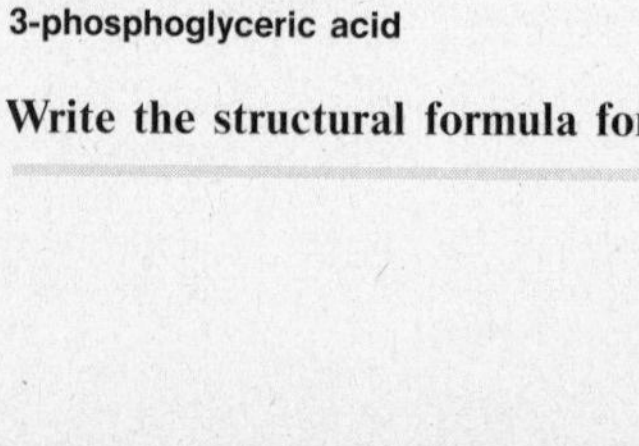

$$HO{-}C(=O){-}CH(OH){-}CH_2{-}O{-}P(=O)(OH){-}OH + H_2PO_4^-$$

3-phosphoglyceric acid

Write the structural formula for 1,3-diphosphoglyceric acid.

ESTERS Many of the pleasant odors of ripening fruits and fragrant blossoms are due to esters, a particular group of chemical compounds. Man-made esters are frequently used as synthetic flavoring agents. Esters are produced by an intermolecular dehydration reaction between an alcohol and an acid. The alcohol and acid used determine the particular ester formed. The reaction may be catalyzed by enzymes or strong acids. In general,

Most often sulfuric acid (H_2SO_4) is used as the acid catalyst.

$$\underset{\text{an alcohol}}{R{-}O{-}H} + \underset{\text{an acid}}{HO{-}\overset{O}{\overset{\|}{C}}{-}R'} \xrightarrow{\text{catalyst}} HOH + \underset{\text{an ester}}{R{-}O{-}\overset{O}{\overset{\|}{C}}{-}R'}$$

Esterification is the formation of an ester by the elimination of the components of water (H— and —OH) from between the reacting alcohol and acid. Elegantly designed experiments indicate that the water formed during esterification arises from the —OH originally present in the acid group and the —H originally present in the alcohol group. The characteristic functional group for an ester is $R{-}O{-}\underset{O}{\underset{\|}{C}}{-}$.

EXAMPLE 10-1

Write the structural formula for the ester formed during the acid-catalyzed reaction between methyl alcohol and acetic acid.

The first step is to write the structural formulas of the reactants:

$$\underset{\substack{\text{methyl alcohol}\\ \text{(methanol)}}}{CH_3{-}O{-}H} + \underset{\text{acetic acid}}{HO{-}\overset{O}{\overset{\|}{C}}{-}CH_3} \xrightarrow{\triangle,\, H^+}$$

Knowing that water is produced by the combination of —OH from the acid and H— from the alcohol, we can next represent this as

$$CH_3{-}O{-}H + H{-}O{-}\overset{O}{\overset{\|}{C}}{-}CH_3 \xrightarrow{\triangle,\, H^+} HOH + \text{ester}$$

Finally, to represent the ester by its structural formula, join the oxygen of the alcohol group to the carbonyl group of the acid:

$$CH_3{-}O{-}H + HO{-}\overset{O}{\overset{\|}{C}}{-}CH_3 \xrightarrow{\triangle,\, \text{trace } H^+} HOH + \underset{\text{methyl acetate}}{CH_3{-}O{-}\overset{O}{\overset{\|}{C}}{-}CH_3}$$

To relate this particular example to the general equation for ester formation, R is $-CH_3$ and R′ is also $-CH_3$. Recall the characteristic $R{-}O{-}\underset{O}{\underset{\|}{C}}{-}$ grouping of an ester.

exercise 10-3

Given below are compounds, some of which contain ester functional groups. Circle the ester functional group where appropriate:

(a) 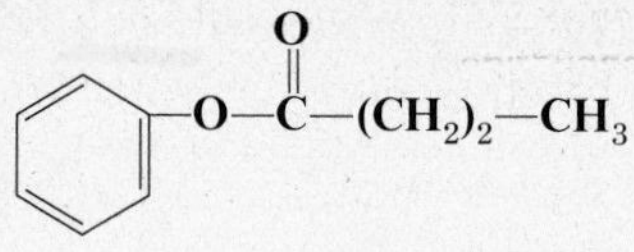 $C_6H_5-O-\overset{\overset{O}{\parallel}}{C}-(CH_2)_2-CH_3$

(b) $CH_3-(CH_2)_4-\overset{OH}{\underset{H}{C}}-O-C_6H_5$

(c) $CH_3-\overset{\overset{O}{\parallel}}{C}-O-\overset{\overset{O}{\parallel}}{C}-CH_2-CH_3$

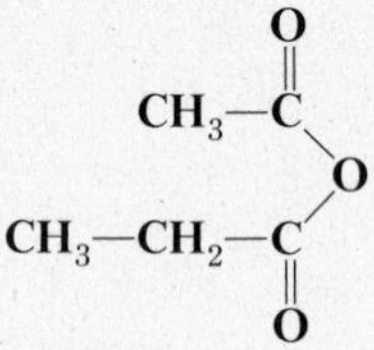

(d) 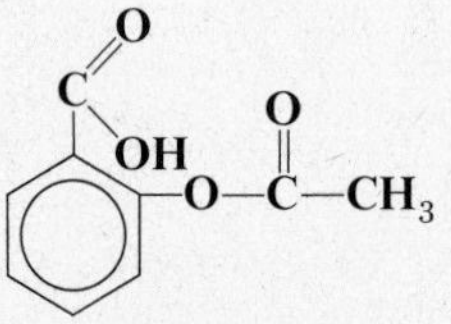 $C_6H_4(COOH)-O-\overset{\overset{O}{\parallel}}{C}-CH_3$

(e) 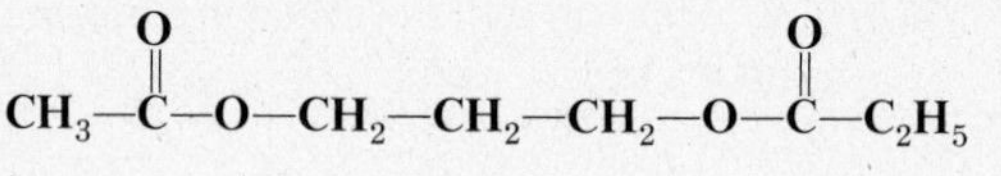 $CH_3-\overset{\overset{O}{\parallel}}{C}-O-CH_2-CH_2-CH_2-O-\overset{\overset{O}{\parallel}}{C}-C_2H_5$

(f) 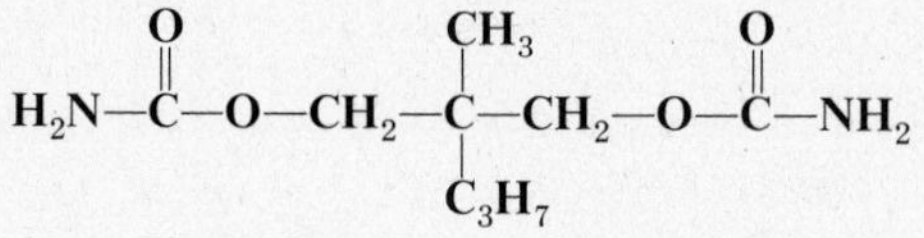 $H_2N-\overset{\overset{O}{\parallel}}{C}-O-CH_2-\overset{CH_3}{\underset{C_3H_7}{C}}-CH_2-O-\overset{\overset{O}{\parallel}}{C}-NH_2$

(g) 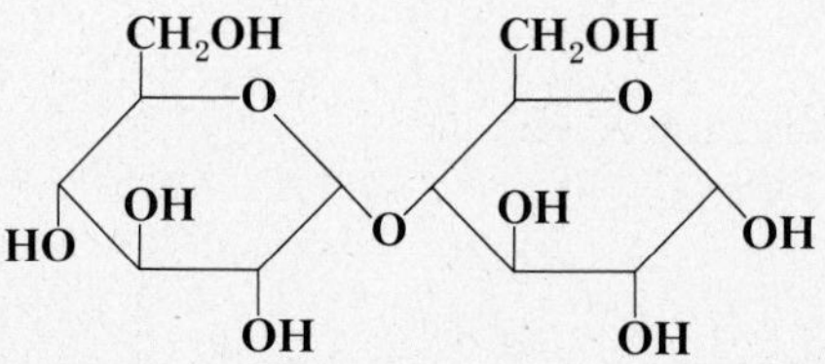 Disaccharide: two pyranose rings, each bearing CH_2OH, ring O, and OH groups (HO, OH, OH on the left ring; OH, OH, OH on the right ring), linked through an O bridge.

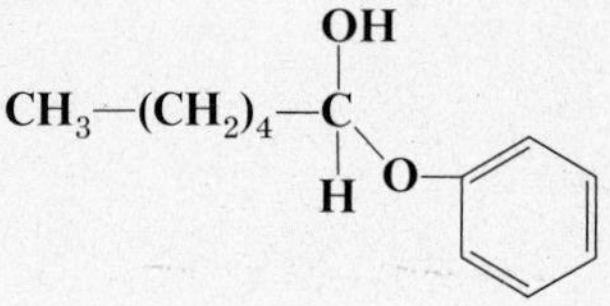

Esters can also be formed by the reaction between an alcohol and an acid anhydride.

$$\mathrm{R{-}O{-}H} + \mathrm{R{-}\overset{\overset{\displaystyle O}{\|}}{E}{-}O{-}\overset{\overset{\displaystyle O}{\|}}{E}{-}R} \xrightarrow{\Delta,\ H^+} \mathrm{R{-}O{-}\overset{\overset{\displaystyle O}{\|}}{E}{-}R} + \mathrm{R{-}\overset{\overset{\displaystyle O}{\|}}{E}{-}OH}$$

an alcohol; an acid anhydride; an ester; an acid

Because an acid anhydride is used, an acid rather than water is the other product.

$$\mathrm{CH_3CH_2CH_2{-}O{-}H} + \mathrm{CH_3{-}\overset{\overset{\displaystyle O}{\|}}{C}{-}O{-}\overset{\overset{\displaystyle O}{\|}}{C}{-}CH_3} \xrightarrow{\Delta,\ H^+}$$

propanol; acetic anhydride

$$\mathrm{CH_3CH_2CH_2{-}O{-}\overset{\overset{\displaystyle O}{\|}}{C}{-}CH_3} + \mathrm{CH_3{-}\overset{\overset{\displaystyle O}{\|}}{C}{-}OH}$$

propyl acetate (an ester); acetic acid

exercise 10-4

Complete the following equations by giving the appropriate structural formulas:

(a) $\mathrm{CH_3OH} + \mathrm{CH_3\overset{\overset{\displaystyle O}{\|}}{C}{-}OH} \xrightarrow{\Delta,\ H^+}$ ____ + ____

(b) $\mathrm{C_6H_5{-}OH} + \mathrm{CH_3CH_2\overset{\overset{\displaystyle O}{\|}}{C}{-}OH} \xrightarrow{\Delta,\ H^+}$ ____ + ____

(c) $\mathrm{C_6H_5{-}\overset{\overset{\displaystyle O}{\|}}{C}{-}OH} + \mathrm{CH_3CH_2OH} \xrightarrow{\Delta,\ H^+}$ ____ + ____

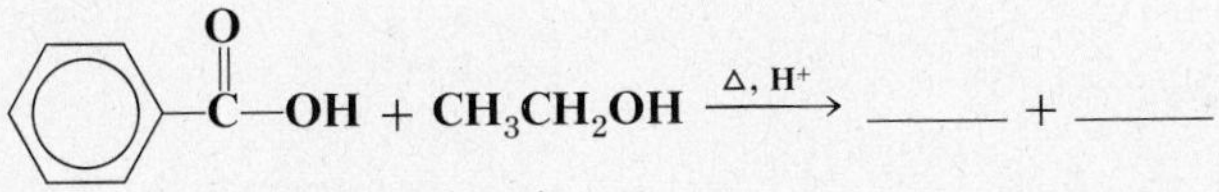

(d) ____ + $\mathrm{H{-}\overset{\overset{\displaystyle O}{\|}}{C}{-}OH} \xrightarrow{\Delta,\ H^+} \mathrm{H_2O} + \mathrm{CH_3(CH_2)_2{-}O{-}\overset{\overset{\displaystyle O}{\|}}{C}{-}H}$

(e) 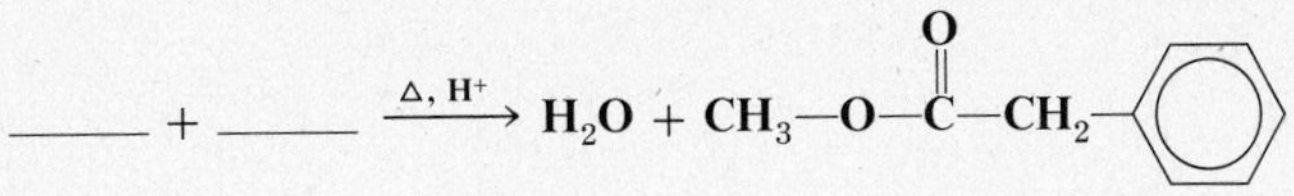 ____ + ____ $\xrightarrow{\Delta,\ H^+} \mathrm{H_2O} + \mathrm{CH_3{-}O{-}\overset{\overset{\displaystyle O}{\|}}{C}{-}CH_2{-}C_6H_5}$

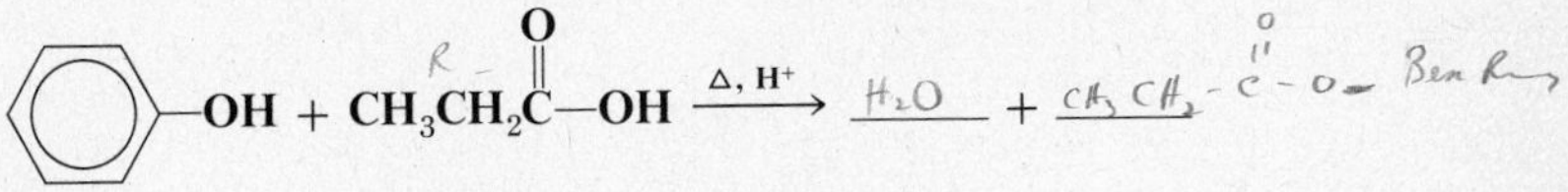

(f) $\mathrm{CH_3(CH_2)_3OH} + \mathrm{CH_3{-}\overset{\overset{\displaystyle O}{\|}}{C}{-}O{-}\overset{\overset{\displaystyle O}{\|}}{C}{-}CH_3} \xrightarrow{\Delta,\ H^+}$ ____ + ____

(g) 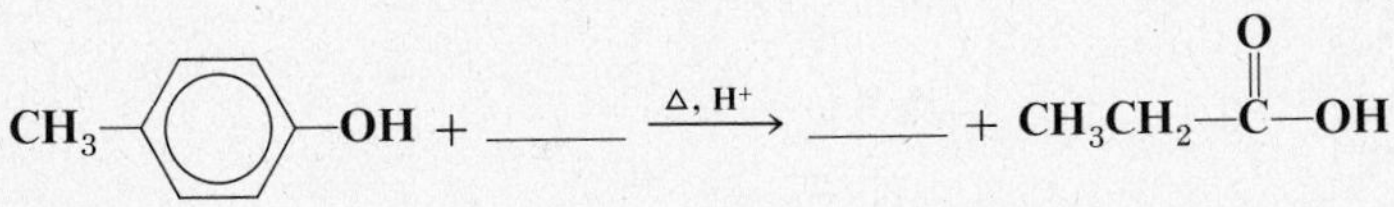 $\mathrm{CH_3{-}C_6H_4{-}OH}$ + ____ $\xrightarrow{\Delta,\ H^+}$ ____ + $\mathrm{CH_3CH_2{-}\overset{\overset{\displaystyle O}{\|}}{C}{-}OH}$

NAMING ESTERS Table 10-1 is a listing of acids and the esters derived from them by reaction with various alcohols.

From the table, it should be noted that the naming of esters is simple: First, name the alcohol component; then change the *-ic* ending of the acid to *-ate* and add it to the alcohol name.

TABLE 10-1
Name relationships of esters to parent alcohols and acids

ALCOHOL	ACID	ESTER
Methyl (methanol)	Formic $\left(H{-}C\begin{matrix}{=}O\\ \diagdown OH\end{matrix}\right)$	Methyl formate $\left(CH_3{-}O{-}\overset{\overset{\displaystyle O}{\|}}{C}{-}H\right)$
Ethyl (ethanol)	Acetic $\left(CH_3{-}C\begin{matrix}{=}O\\ \diagdown OH\end{matrix}\right)$	Ethyl acetate $\left(CH_3{-}CH_2{-}O{-}\overset{\overset{\displaystyle O}{\|}}{C}{-}CH_3\right)$
Ethyl	Propionic $\left(CH_3CH_2C\begin{matrix}{=}O\\ \diagdown OH\end{matrix}\right)$	Ethyl propionate $\left(CH_3{-}CH_2{-}O{-}\overset{\overset{\displaystyle O}{\|}}{C}{-}CH_2{-}CH_3\right)$
Ethyl	Butyric $\left(CH_3(CH_2)_2C\begin{matrix}{=}O\\ \diagdown OH\end{matrix}\right)$	Ethyl butyrate $\left(CH_3{-}CH_2{-}O{-}\overset{\overset{\displaystyle O}{\|}}{C}(CH_2)_2CH_3\right)$
Methyl	Salicylic $\left(C_6H_4(OH){-}\overset{\overset{\displaystyle O}{\|}}{C}{-}OH\right)$	Methyl salicylate $\left(C_6H_4(OH){-}\overset{\overset{\displaystyle O}{\|}}{C}{-}O{-}CH_3\right)$
Propyl (propanol)	Stearic $\left(CH_3(CH_2)_{16}C\begin{matrix}{=}O\\ \diagdown OH\end{matrix}\right)$	Propyl stearate $\left(CH_3(CH_2)_2{-}O{-}\overset{\overset{\displaystyle O}{\|}}{C}{-}(CH_2)_{16}{-}CH_3\right)$
Ethyl	Benzoic $\left(C_6H_5{-}\overset{\overset{\displaystyle O}{\|}}{C}{-}OH\right)$	Ethyl benzoate $\left(C_6H_5{-}\overset{\overset{\displaystyle O}{\|}}{C}{-}O{-}CH_2CH_3\right)$

exercise 10-5

Complete the following table:

	ALCOHOL	ACID	ESTER
NAME AND STRUCTURAL FORMULA	Methyl	Acetic	____
	CH_3OH	CH_3COOH	$CH_3{-}O{-}\overset{O}{\overset{\parallel}{C}}{-}CH_3$
	____	____	____
	CH_3CH_2OH	$H{-}C(=O){-}OH$	____
	____	Stearic (see Table 10-1)	Methyl stearate
	____	____	____
	____	Butyric ($CH_3(CH_2)_2CO_2H$)	____
	$CH_3(CH_2)_2OH$	____	$CH_3(CH_2)_2{-}O{-}\overset{O}{\overset{\parallel}{C}}{-}(CH_2)_2CH_3$

ESTERS OF PHOSPHORIC ACID

Esterification is not limited to carboxylic acids. Phosphoric acid

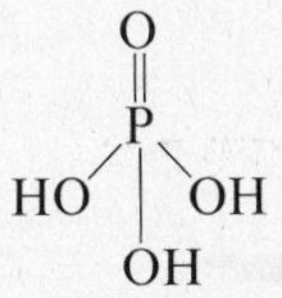

$$O{=}P(OH)_3$$

Other acids, e.g., sulfuric and nitric, can also be esterified.

can be esterified. We can see that phosphoric acid has a central nonmetal atom double-bonded to an oxygen and single-bonded to an —OH group, a feature common to carboxylic acids.

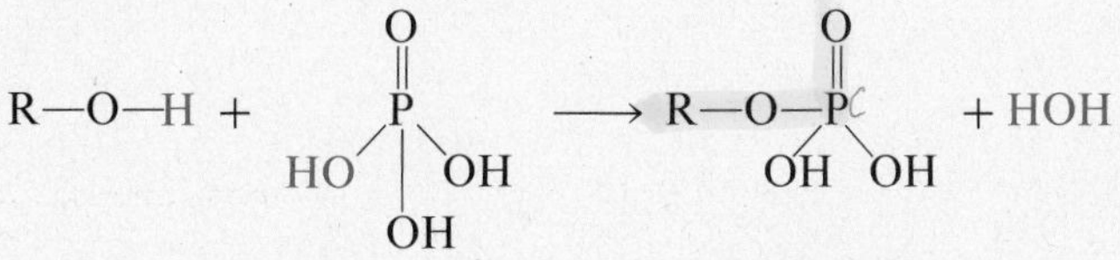

$$R{-}O{-}H + O{=}P(OH)_3 \longrightarrow R{-}O{-}P(=O)(OH)_2 + HOH$$

an alcohol   phosphoric acid   a phosphate ester

The ester still contains a phosphorus atom single bonded to —OH and double-bonded to an oxygen. Thus, phosphate ester molecules can liberate hydrogen ions:

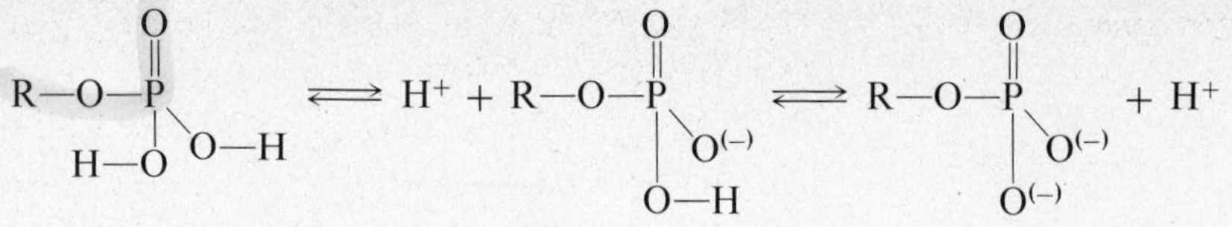

$$R{-}O{-}\overset{\overset{\Large O}{\|}}{\underset{\underset{\Large H-O}{|}}{P}}{-}O{-}H \rightleftharpoons H^+ + R{-}O{-}\overset{\overset{\Large O}{\|}}{\underset{\underset{\Large O-H}{|}}{P}}{-}O^{(-)} \rightleftharpoons R{-}O{-}\overset{\overset{\Large O}{\|}}{\underset{\underset{\Large O^{(-)}}{|}}{P}}{-}O^{(-)} + H^+$$

exercise 10-6

Write the structural formula of the phosphate ester formed by the reaction between ethyl alcohol (ethanol) and phosphoric acid.

ESTERS OF THIOLS

Thiols (thio alcohols) are sulfur-containing compounds analogous to alcohols, with sulfur substituting for oxygen in the alcohol grouping: R—S—H, a thiol; R—O—H, an alcohol. Thiols can react with acids to form the corresponding thioester:

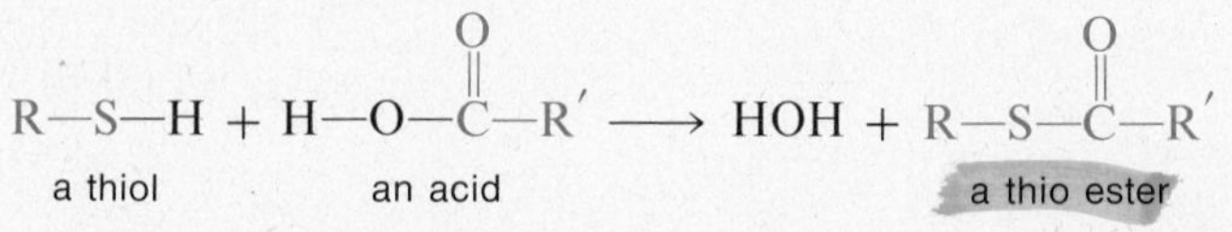

$$\underset{\text{a thiol}}{R{-}S{-}H} + \underset{\text{an acid}}{H{-}O{-}\overset{\overset{\Large O}{\|}}{C}{-}R'} \longrightarrow HOH + \underset{\text{a thio ester}}{R{-}S{-}\overset{\overset{\Large O}{\|}}{C}{-}R'}$$

The prefix thio *means an oxygen atom has been replaced by a sulfur atom.*

Coenzyme A (CoA—S—H) is a metabolically significant thiol that is esterified to produce acetyl coenzyme A, another compound important to body metabolism. The abbreviation CoA represents a structurally complex portion of the coenzyme A molecule. $CH_3{-}\underset{\underset{\Large O}{\|}}{C}{-}$ is known as an acetyl group.

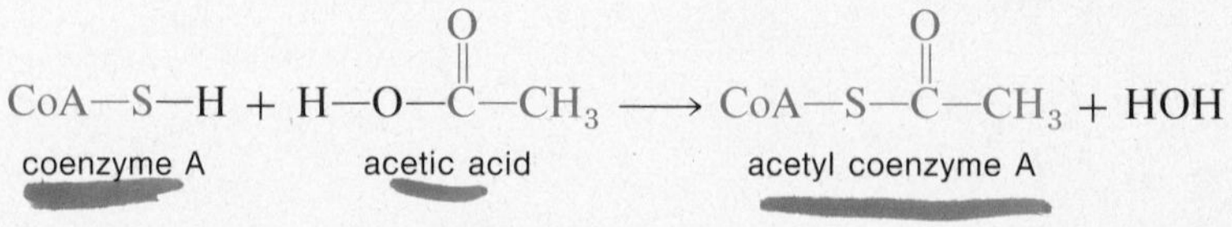

$$\underset{\text{coenzyme A}}{CoA{-}S{-}H} + \underset{\text{acetic acid}}{H{-}O{-}\overset{\overset{\Large O}{\|}}{C}{-}CH_3} \longrightarrow \underset{\text{acetyl coenzyme A}}{CoA{-}S{-}\overset{\overset{\Large O}{\|}}{C}{-}CH_3} + HOH$$

TRIGLYCERIDES

Glycerol is a polyhydroxy alcohol containing three —OH groups per molecule:

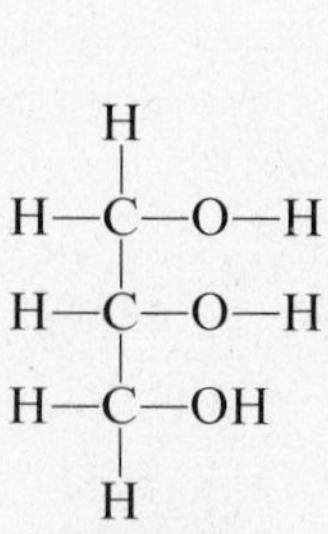

$$\begin{array}{c} H \\ | \\ H{-}C{-}O{-}H \\ | \\ H{-}C{-}O{-}H \\ | \\ H{-}C{-}OH \\ | \\ H \end{array}$$

glycerol

It is an extremely important compound in metabolism because of the esterification reactions it undergoes. Its most important esterification reaction involves combination with those carboxylic acids called *fatty acids*. Fatty acids have the following general properties:

1. They usually contain one carboxylic acid group per molecule, i.e., monocarboxylic, and an alkyl group.
2. The alkyl group is usually an unbranched carbon chain.
3. The fatty acid molecule contains an even number of carbon atoms, usually from 12 to 26 carbons.
4. Due to the long hydrocarbon alkyl chain, fatty acids are water insoluble.
5. The hydrocarbon portion of the chain may be saturated or unsaturated. If unsaturated, it may contain one, two, three, or four carbon-to-carbon double bonds. Those containing two to four double bonds are called *polyunsaturated*.
6. Usually, the greater the degree of unsaturation in a fatty acid, the lower its melting point. Table 10-2 lists several of the more important fatty acids.

Table 10-2
Important fatty acids

NAME	FORMULA	MELTING POINT (°C)
Lauric acid	$CH_3(CH_2)_{10}-C(=O)OH$	44
Myristic acid	$CH_3(CH_2)_{12}-C(=O)OH$	58
Palmitic acid	$CH_3(CH_2)_{14}C(=O)OH$	63
Stearic acid	$CH_3(CH_2)_{16}C(=O)OH$	70
Oleic acid	$CH_3(CH_2)_7CH{=}CH(CH_2)_7C(=O)OH$	4
Linoleic acid*	$CH_3(CH_2)_4CH{=}CHCH_2CH{=}CH(CH_2)_7C(=O)OH$	−5
Linolenic acid*	$CH_3CH_2CH{=}CHCH_2CH{=}CHCH_2CH{=}CH(CH_2)_7C(=O)OH$	−11

*The body cannot synthesize these fatty acids and therefore they must be included in the diet.

Very few free (uncombined) fatty acids are found in a cell. Rather, they are chemically combined by the esterification reaction with glycerol to form a class of compounds called *triglycerides*. If all three fatty acids are the same, the product is a simple triglyceride. More commonly, a variety of both saturated and unsaturated fatty acids react with glycerol to produce a mixed triglyceride. This process can be represented as

$$\underbrace{R_1{-}\overset{\overset{\displaystyle O}{\|}}{C}{-}OH + R_2{-}\overset{\overset{\displaystyle O}{\|}}{C}{-}OH + R_3{-}\overset{\overset{\displaystyle O}{\|}}{C}{-}OH}_{\text{fatty acids}} + \underbrace{\begin{array}{c} H \\ | \\ HO{-}C{-}H \\ | \\ HO{-}C{-}H \\ | \\ HO{-}C{-}H \\ | \\ H \end{array}}_{\text{glycerol}} \xrightarrow{\text{enzyme}} 3HOH + \underbrace{\begin{array}{c} R_3{-}\overset{\overset{\displaystyle O}{\|}}{C}{-}O{-}\overset{\overset{\displaystyle H}{|}}{C}{-}H \\ | \\ R_2{-}\overset{\overset{\displaystyle O}{\|}}{C}{-}O{-}C{-}H \\ | \\ R_1{-}\overset{\overset{\displaystyle O}{\|}}{C}{-}O{-}\underset{\underset{\displaystyle H}{|}}{C}{-}H \end{array}}_{\text{a mixed triglyceride}}$$

Note that 3 moles of water are formed during esterification of 1 mole of glycerol with 3 moles of fatty acids. The triglyceride is a solid or liquid at normal temperatures, depending on the nature of the component fatty acids. Triglycerides containing a large portion of unsaturated fatty acids are normally liquids and are described as oils (not the automobile-lubricating variety). Fat is the collective term for triglycerides containing a high portion of saturated fatty acid components and are usually solids at room temperature. Triglycerides provide the greatest source of chemical energy during body metabolism.

exercise 10-7

Which of the acids listed in Table 10-2 would be expected to be the main fatty acid components of (a) a vegetable oil? (b) An animal fat?

exercise 10-8

(a) Using structural formulas of saturated and unsaturated fatty acids listed in Table 10-2, write a balanced equation representing the formation of a mixed triglyceride.
(b) Are other mixed triglycerides possible from the combination of these same fatty acids?

Lipids are insoluble in water.

Triglycerides make up the majority of a diverse group of compounds called *lipids*. Lipids are described as compounds of biological origin that are soluble in ether or other organic solvents. The diversity of these compounds ranges from salad oils, butter, sex hormones, certain vitamins, and cholesterol to important components of nerve and brain tissues. Waxes and phospholipids are two other subclasses of lipids that will be discussed at this point.

A wax is the ester product of the reaction between a long-chain fatty acid and a monohydroxy alcohol. Thus its general formula is $R{-}\overset{\overset{\displaystyle O}{\|}}{C}{-}O{-}R'$. R is the fatty acid component containing a saturated hydrocarbon side chain ranging from 24 to 36 carbons; R′ is a high molecular weight alcohol containing 16 to 36 carbon atoms. Waxes are insoluble in water and are chemically unreactive. Carnauba wax (used in automobile polishes), earwax, and beeswax are examples of naturally occurring waxes.

Phospholipids (phosphatides) are complex lipids containing phosphate, glycerol, fatty acids, and a nitrogen-containing base. A typical example is phosphatidyl choline, an important component of nerve tissue membranes. Its common name is lecithin:

$$\begin{array}{l} H{-}\overset{\overset{\displaystyle H}{|}}{C}{-}O{-}\overset{\overset{\displaystyle O}{\|}}{C}{-}R_1 \\ \quad | \\ H{-}C{-}O{-}\overset{\overset{\displaystyle O}{\|}}{C}{-}R_2 \\ \quad | \\ H{-}\underset{\underset{\displaystyle H}{|}}{C}{-}O{-}\underset{\underset{\displaystyle O^{(-)}}{|}}{\overset{\overset{\displaystyle O}{\|}}{P}}{-}O{-}CH_2{-}CH_2{-}{}^{\oplus}\underset{\underset{\displaystyle CH_3}{|}}{\overset{\overset{\displaystyle CH_3}{|}}{N}}{-}CH_3 \end{array}$$

phosphatidyl choline (lecithin)

Choline: $-(CH_2)_2{-}N^+{-}(CH_3)_3$

Note that the phosphatide above may be considered the glycerol ester of two fatty acids $\left(R_1{-}C{\genfrac{}{}{0pt}{}{\nearrow O}{\searrow OH}} \text{ and } R_2{-}C{\genfrac{}{}{0pt}{}{\nearrow O}{\searrow OH}}\right)$ plus a derivative of phosphoric acid $\left(HO{-}\underset{\underset{\displaystyle OH}{|}}{\overset{\overset{\displaystyle O}{\|}}{P}}{-}OH\right)$.

ESTER HYDROLYSIS

The acid is a catalyst. Dilute acid is used so there is a large excess of water present, driving the reaction to completion.

Esterification is actually a reversible process. Ester bonds are susceptible to hydrolysis. The presence of an appropriate enzyme or dilute acid catalyzes the hydrolysis reaction. Acid or enzymatic hydrolysis of an ester causes the ester bond to break with the reformation of the original acid and alcohol components. Generalizing the ester hydrolysis,

$$\underset{\text{ester}}{R{-}\overset{\overset{\displaystyle O}{\|}}{C}{-}O{-}R'} + HOH \xrightarrow{H^+ \text{ or enzyme}} \underset{\text{acid}}{R{-}C{\genfrac{}{}{0pt}{}{\nearrow O}{\searrow OH}}} + \underset{\text{alcohol}}{R'{-}OH}$$

The acid or enzymatic hydrolysis of an ester is the reverse of the esterification process. During esterification, we envisioned the —OH from the acid and H from the alcohol joining to form water and the ester forming by the union of the residual alcohol and acid fragments. During hydrolysis,

the ester bond breaks at this point $-\overset{\overset{\displaystyle O}{\|}}{C}\nleftrightarrow O-$ and it may be envisioned

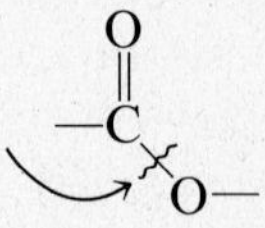

that the OH from water adds to the *carbonyl* carbon and the H from water adds to the oxygen:

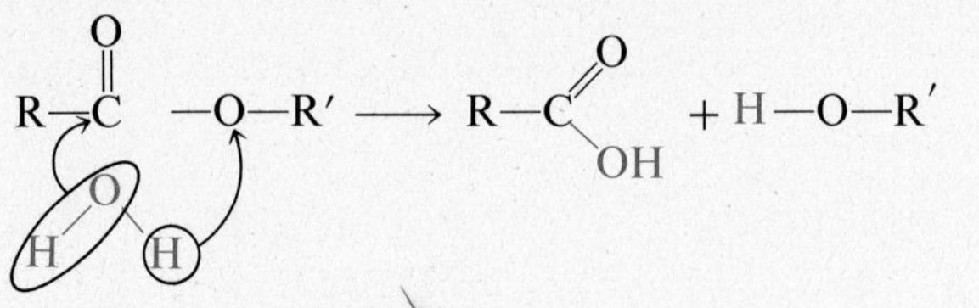

$$R-\overset{\overset{\displaystyle O}{\|}}{C}-O-R' + H-O-H \longrightarrow R-C(=O)OH + H-O-R'$$

Thus, the component acid and alcohol are liberated. Note that water is shown as a reactant in the balanced chemical equation.

EXAMPLE 10-2

What are the structural formulas of the acid and alcohol formed upon acid hydrolysis of $C_6H_5-CH_2-C(=O)-O-CH_2CH_3$? First, set up the question in the form of a chemical equation:

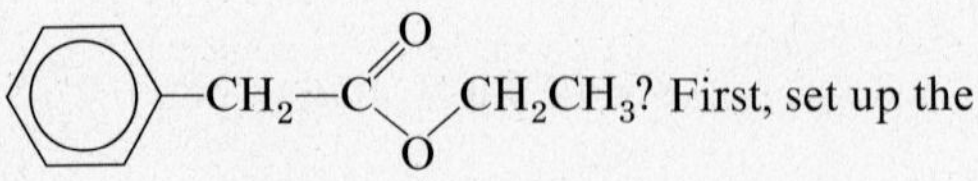

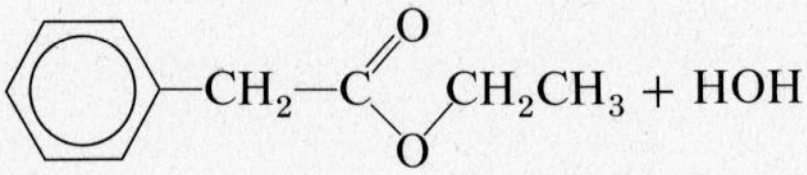

$$C_6H_5-CH_2-C(=O)-O-CH_2CH_3 + HOH \xrightarrow{H^+} \text{acid} + \text{alcohol}$$

Secondly, we know $C_6H_5-CH_2-C(=O)-O-CH_2CH_3$ contains an ester grouping outlined by the dotted lines; $C_6H_5-CH_2-$ is analogous to R in the general equation and $-CH_2-CH_3$ is equivalent to R′.

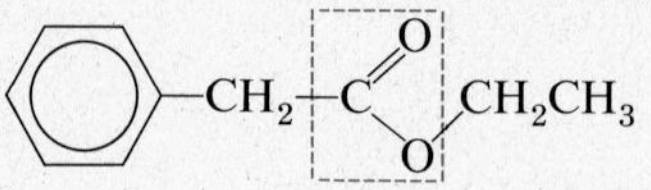

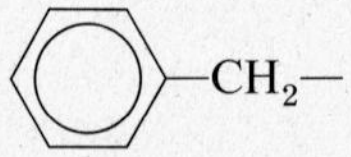

Hydrolysis of the ester linkage occurs by rupture of the carbon-to-oxygen single bond $-C(=O)\!\!\times\!\!O-$.

The carbon–oxygen *double* bond remains intact and will eventually become part of the acid liberated.

Next, imagine that the rupture of the ester bond generates two fragments:

$$C_6H_5-CH_2-C(=O) \quad \text{and} \quad -O-CH_2-CH_3$$

The OH from water adds to the carbonyl carbon; the H from water adds to the $-O-CH_2CH_3$ group.

The net result is formation of the acid $C_6H_5-CH_2-C(=O)OH$ and the alcohol $H-O-CH_2CH_3$.

exercise 10-9

Using structural formulas, write balanced equations representing

(a) The acid hydrolysis of aspirin

$$C_6H_4(COOH)(O-C(=O)-CH_3)$$

(b) The enzymatic hydrolysis of glycerol-3-phosphate

$$HO-CH_2-C(H)(OH)-C(H)(H)-O-P(=O)(OH)OH$$

We have previously seen that esterification with glycerol is a convenient way for the body to store fatty acids; however, esterification is a reversible process. The component fatty acids and glycerol can be obtained by hydrolysis.

Since 1 mole of a triglyceride is formed from the reaction of 3 moles of fatty acids with 1 mole of glycerol, acid or enzymatic hydrolysis of a triglyceride regenerates 3 moles of fatty acid and 1 mole of glycerol. For hydrolysis, 3 moles of water per 1 mole of triglyceride are required:

$$\begin{array}{l} \text{H}-\overset{\text{H}}{\overset{|}{\text{C}}}-\text{O}-\overset{\text{O}}{\overset{\|}{\text{C}}}-\text{R}_1 \quad \text{HOH} \\ \quad | \\ \text{H}-\text{C}-\text{O}-\overset{\text{O}}{\overset{\|}{\text{C}}}-\text{R}_2 + \text{HOH} \xrightarrow[\text{enzyme}]{\text{H}^+ \text{ or}} \\ \quad | \\ \text{H}-\underset{\text{H}}{\underset{|}{\text{C}}}-\text{O}-\overset{\text{O}}{\overset{\|}{\text{C}}}-\text{R}_3 \quad \underbrace{\text{HOH}} \end{array}$$

a mixed triglyceride — 3 moles of water

$$\begin{array}{l} \text{H}-\overset{\text{H}}{\overset{|}{\text{C}}}-\text{OH} \\ \quad | \\ \text{H}-\text{C}-\text{OH} + \underbrace{\text{H}-\text{O}-\overset{\text{O}}{\overset{\|}{\text{C}}}-\text{R}_1 + \text{H}-\text{O}-\overset{\text{O}}{\overset{\|}{\text{C}}}-\text{R}_2 + \text{H}-\text{O}-\overset{\text{O}}{\overset{\|}{\text{C}}}-\text{R}_3} \\ \quad | \\ \text{H}-\underset{\text{H}}{\underset{|}{\text{C}}}-\text{OH} \end{array}$$

glycerol — 3 moles of fatty acids

exercise 10-10

A triglyceride has the following structural formula:

$$\begin{array}{l} \text{H}-\overset{\text{H}}{\overset{|}{\text{C}}}-\text{O}-\overset{\text{O}}{\overset{\|}{\text{C}}}-(\text{CH}_2)_{10}-\text{CH}_3 \\ \quad | \\ \text{H}-\text{C}-\text{O}-\overset{\text{O}}{\overset{\|}{\text{C}}}-(\text{CH}_2)_7\text{CH}=\text{CHCH}_2\text{CH}=\text{CH}(\text{CH}_2)_4\text{CH}_3 \\ \quad | \\ \text{H}-\underset{\text{H}}{\underset{|}{\text{C}}}-\text{O}-\overset{\text{O}}{\overset{\|}{\text{C}}}-(\text{CH}_2)_7-\text{CH}=\text{CH}(\text{CH}_2)_7\text{CH}_3 \end{array}$$

Using structural formulas, write a balanced equation for the acid hydrolysis of this triglyceride.

SAPONIFICATION Ester bonds can also be broken by the action of aqueous solutions of strong bases like sodium hydroxide. This process is called *saponification*.

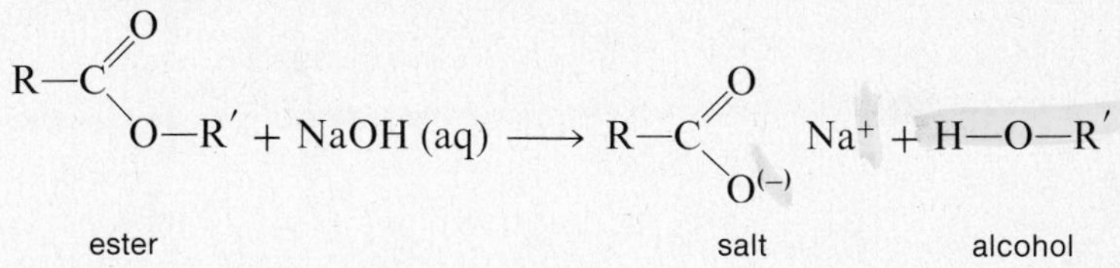

$$R{-}C(=O){-}O{-}R' + NaOH\,(aq) \longrightarrow R{-}C(=O){-}O^{(-)}\;Na^+ + H{-}O{-}R'$$

ester — salt — alcohol

Saponification of an ester thus liberates the original alcohol as in acid or enzymatic hydrolysis. The original acid cannot be isolated, however, because it reacts with excess metal hydroxide present to form the corresponding metal salt.

exercise 10-11

Using structural formulas, complete the following equations:

(a) $NaOH + CH_3CH_2\overset{O}{\overset{\|}{C}}{-}O{-}CH_3 \xrightarrow{H_2O,\ \Delta}$

(b) $CH_3CH_2\overset{O}{\overset{\|}{C}}{-}O{-}CH_3 + HOH \xrightarrow{H^+,\ \Delta}$

(c) 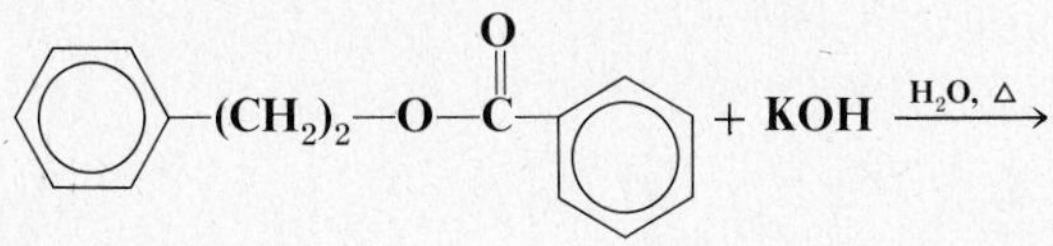 $C_6H_5{-}(CH_2)_2{-}O{-}\overset{O}{\overset{\|}{C}}{-}C_6H_5 + KOH \xrightarrow{H_2O,\ \Delta}$

(d) 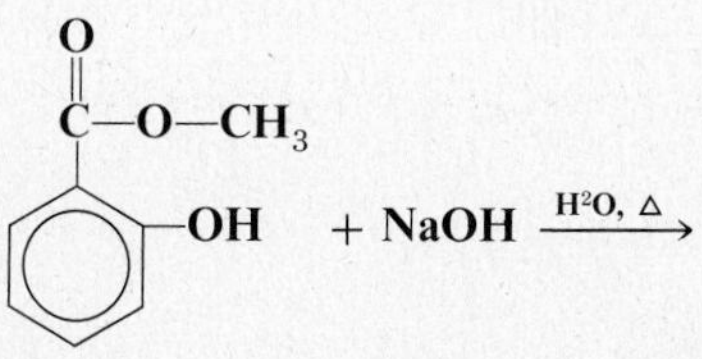 $o\text{-}(CH_3{-}O{-}\overset{O}{\overset{\|}{C}}{-})C_6H_4{-}OH + NaOH \xrightarrow{H^2O,\ \Delta}$

If the saponification of a triglyceride is carried out in an alkali metal hydroxide solution (usually NaOH or KOH), the salts of the component fatty acids are formed. These salts are collectively described as soaps.

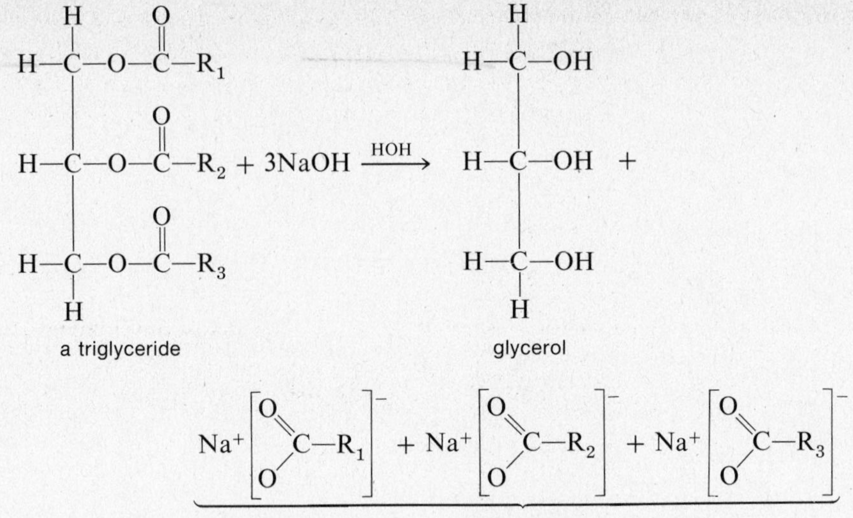

$$\begin{array}{l} \text{H--}\underset{|}{\overset{\text{H}}{\overset{|}{\text{C}}}}\text{--O--}\overset{\text{O}}{\overset{\|}{\text{C}}}\text{--R}_1 \\ \text{H--}\underset{|}{\overset{|}{\text{C}}}\text{--O--}\overset{\text{O}}{\overset{\|}{\text{C}}}\text{--R}_2 + 3\text{NaOH} \xrightarrow{\text{HOH}} \\ \text{H--}\underset{\text{H}}{\underset{|}{\overset{|}{\text{C}}}}\text{--O--}\overset{\text{O}}{\overset{\|}{\text{C}}}\text{--R}_3 \end{array} \quad \begin{array}{l} \text{H--}\overset{\text{H}}{\overset{|}{\text{C}}}\text{--OH} \\ \text{H--}\overset{|}{\text{C}}\text{--OH} + \\ \text{H--}\underset{\text{H}}{\underset{|}{\text{C}}}\text{--OH} \end{array}$$

a triglyceride — glycerol

$$\underbrace{\text{Na}^+\left[\text{O}{=}\text{C}(\text{--R}_1)\text{--O}\right]^- + \text{Na}^+\left[\text{O}{=}\text{C}(\text{--R}_2)\text{--O}\right]^- + \text{Na}^+\left[\text{O}{=}\text{C}(\text{--R}_3)\text{--O}\right]^-}$$

soap; sodium salts of triglyceride component fatty acids

In the past, housewives used fats (mainly triglycerides) saved from cooking plus household lye (NaOH) to make their own supply of soap periodically for home use. Anyone who has used this product can attest to the irritating action toward skin caused by the incomplete removal of the lye from the final product.

exercise 10-12

Write structural formulas for the products formed in the following reactions:

(a) 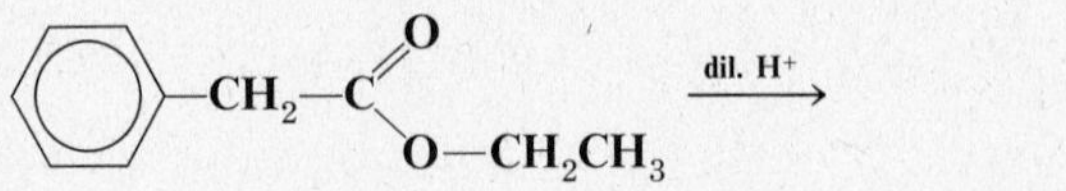 $C_6H_5\text{--CH}_2\text{--C}(=\text{O})\text{--O--CH}_2\text{CH}_3 \xrightarrow{\text{dil. H}^+}$

(b) $C_6H_5\text{--CH}_2\text{--C}(=\text{O})\text{--O--CH}_2\text{CH}_3 + \text{NaOH} \xrightarrow{\text{H}_2\text{O}, \Delta}$

(c)

$$\begin{array}{l} \quad\;\text{CH}_2\text{--O--}\overset{\text{O}}{\overset{\|}{\text{C}}}\text{--(CH}_2)_2\text{CH}_3 \\ \text{H--}\overset{|}{\text{C}}\text{--O--}\overset{\text{O}}{\overset{\|}{\text{C}}}\text{--(CH}_2)_2\text{--C}_6\text{H}_5 \quad + 3\text{NaOH} \xrightarrow{\text{H}_2\text{O}, \Delta} \\ \text{H--}\underset{\text{H}}{\underset{|}{\overset{|}{\text{C}}}}\text{--O--}\overset{\text{O}}{\overset{\|}{\text{C}}}\text{--(CH}_2)_{16}\text{CH}_3 \end{array}$$

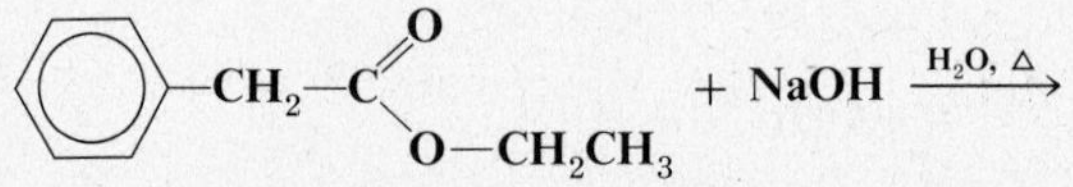

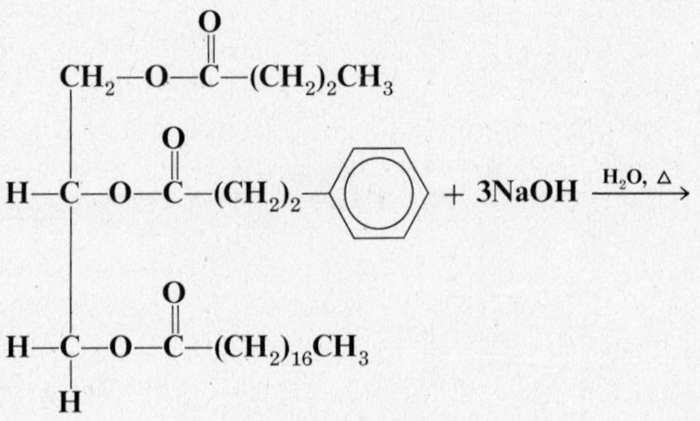

DECARBOXYLATION Carboxylic acids undergo many reactions, one of which is decarboxylation, the loss of carbon dioxide upon heating. The CO_2 loss is a way in which the number of carbon (and oxygen) atoms in a molecule can be decreased. Loss of carbon dioxide occurs via the breakdown of a carboxylic acid group:

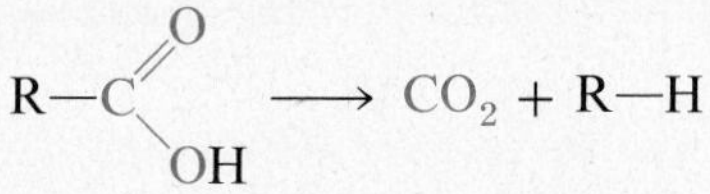

$$R{-}C(=O){-}OH \longrightarrow CO_2 + R{-}H$$

Decarboxylation occurs during many physiologically important processes. In carbohydrate metabolism, oxalosuccinic acid is converted in this way to α-ketoglutaric acid in the presence of the appropriate decarboxylase enzyme.

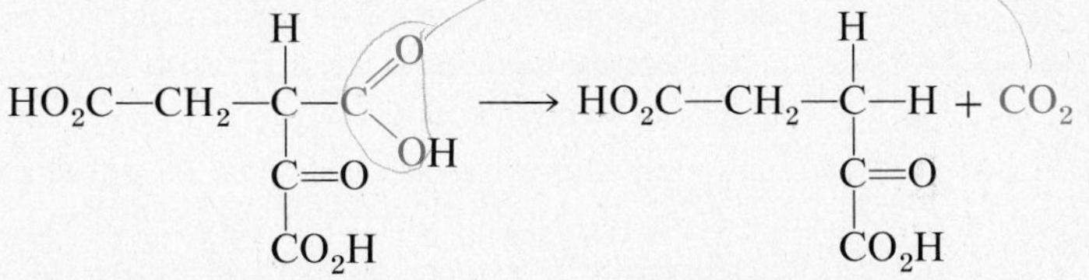

$$HO_2C{-}CH_2{-}CH(C(=O){-}CO_2H){-}C(=O){-}OH \longrightarrow HO_2C{-}CH_2{-}CH(C(=O){-}CO_2H){-}H + CO_2$$

oxalosuccinic acid α-ketoglutaric acid

What do you think an antihistamine does?

The amino acid histidine can be decarboxylated to histamine, which is a powerful allergen that lowers blood pressure by dilating capillaries.

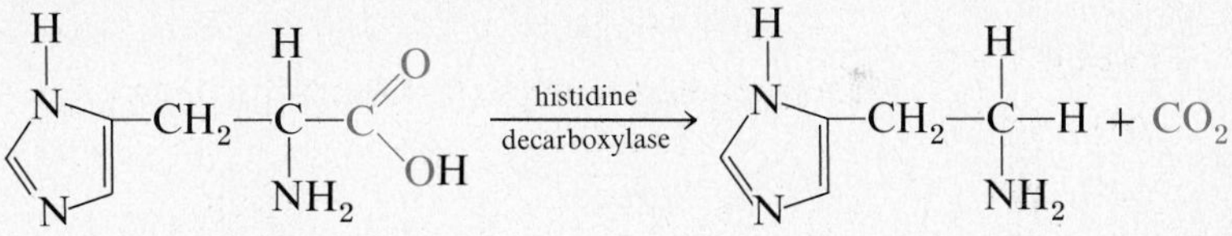

$$\text{(imidazolyl)}{-}CH_2{-}CH(NH_2){-}C(=O){-}OH \xrightarrow[\text{decarboxylase}]{\text{histidine}} \text{(imidazolyl)}{-}CH_2{-}CH(NH_2){-}H + CO_2$$

histidine histamine

exercise 10-13

Cadaverine is produced by the decarboxylation of the amino acid lysine:

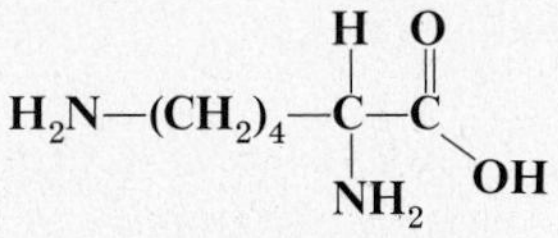

$$H_2N{-}(CH_2)_4{-}CH(NH_2){-}C(=O){-}OH$$

Write the structural formula of cadeverine.

CHAPTER SUMMARY

1. The presence of various functional groups in its molecules dictates the chemical behavior of a substance when it reacts with various reagents.

Functional groups studied in this chapter are

(a) $-C(=O)OH$ Carboxylic acid

(b) $-\overset{\overset{O}{\|}}{C}-O-\overset{\overset{O}{\|}}{C}$ Anhydride

(c) $-\overset{\overset{O}{\|}}{C}-O-R$ Ester

(d) R—S—H Thiol

(e) $-\overset{\overset{O}{\|}}{C}-S-R$ Thioester

Groups b, c, and e are acid derivatives.

2. Anhydride linkages are bonding arrangements in which large amounts of energy can be stored. Hydrolysis of the anhydride bond releases energy and forms an acid.
3. Esters and water are produced by the reaction between an alcohol and an acid. Acids other than carboxylic acids can also be used for ester formation. Esters can also be produced by the reaction between an alcohol and an acid anhydride. Thiols react with acids to produce thioesters.
4. Ester formation is a reversible reaction. Catalyzed by an appropriate enzyme or acid, hydrolysis of an ester produces the parent alcohol and acid. Saponification is ester hydrolysis using alkali metal hydroxide solution to produce alcohol and salts (soaps).
5. Triglycerides are esters formed from glycerol and fatty acids. Hydrolysis of 1 mole of a triglyceride liberates 1 mole of glycerol and 3 moles of fatty acids.
6. Decarboxylation is the loss of carbon dioxide from the carboxylic acid group of a molecule. This is a method by which the number of carbon and oxygen atoms in a molecule is decreased.

QUESTIONS

1. Given below are structural formulas for several compounds. Each compound contains at least one of the functional groups given in the following list. A particular functional group is designated by a numbered arrow. Match the group designated by the arrow with the proper functional group listed below:

(a) primary alcohol	**(g)** acetal
(b) secondary alcohol	**(h)** ester
(c) tertiary alcohol	**(i)** ether
(d) ketone	**(j)** thioester
(e) aldehyde	**(k)** anhydride
(f) hemiacetal	**(l)** carboxylic acid

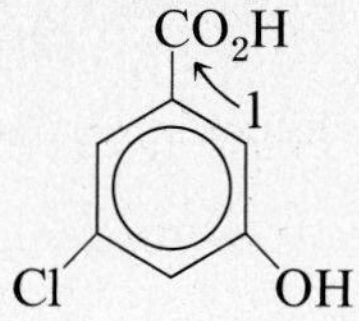

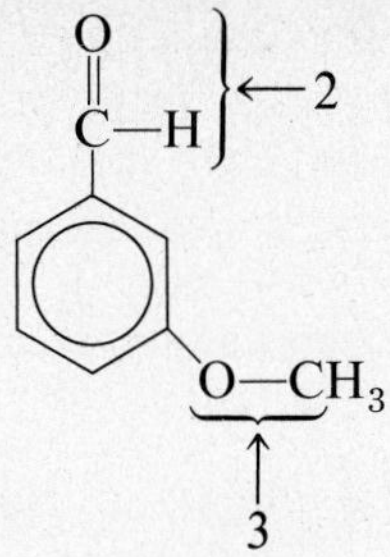

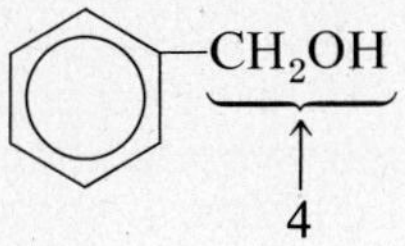

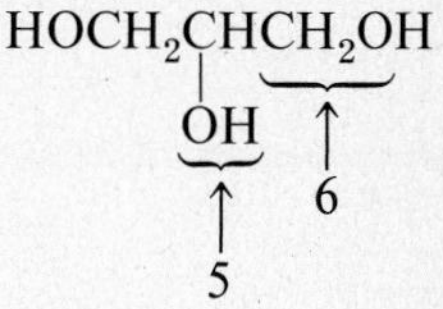

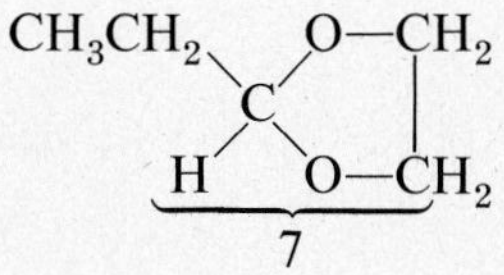

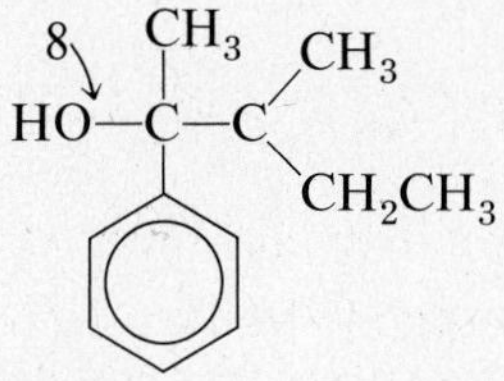

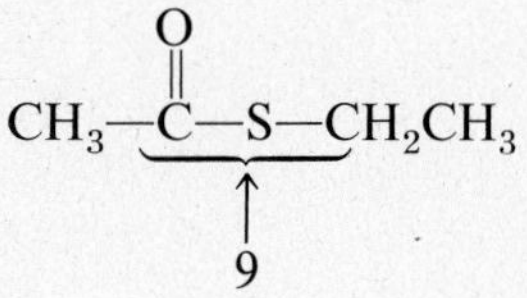

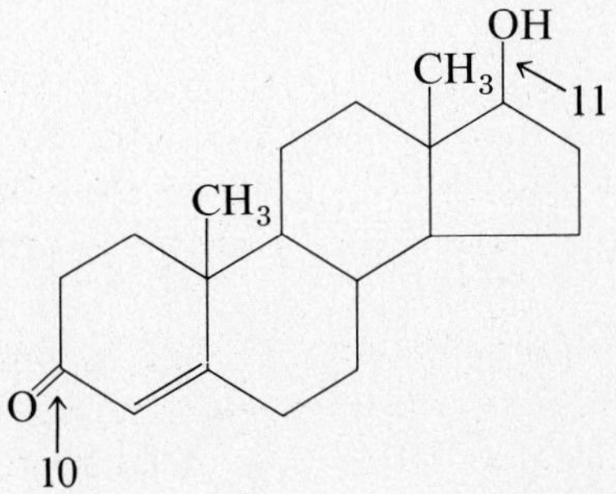

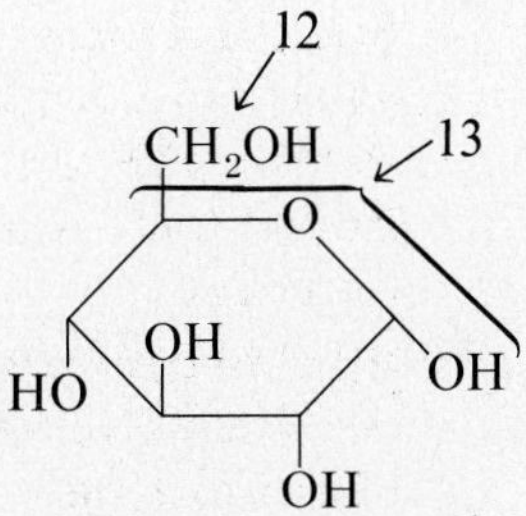

NUMBER	FUNCTIONAL GROUP
1	____
2	____
3	____
4	____
5	____
6	____
7	____
8	____
9	____
10	____
11	____
12	____
13	____

2. Listed below are several equations.
   (a) Complete the equations using organic structural formulas for reactants and/or products.
   (b) Indicate the general type of reaction each equation represents, i.e., esterification, ester hydrolysis, anhydride formation, etc.

(i) $CH_3OH + CH_3(CH_2)_3CO_2H \xrightarrow{\triangle,\ H^+}$ ______ + ______

(ii) $CH_3(CH_2)_2CO_2H + CH_3CO_2H \longrightarrow HOH +$ ______

(iii) 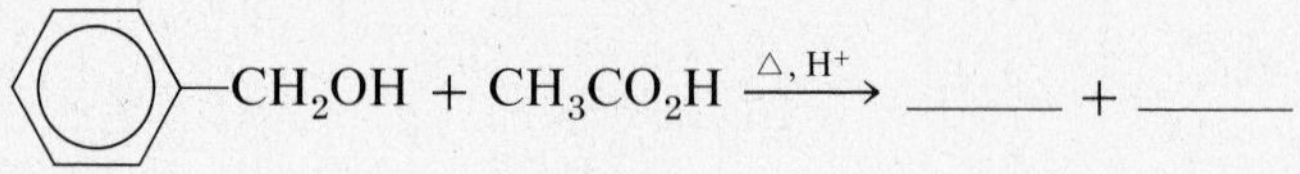 $C_6H_5{-}CH_2OH + CH_3CO_2H \xrightarrow{\triangle,\ H^+}$ ______ + ______

(iv) 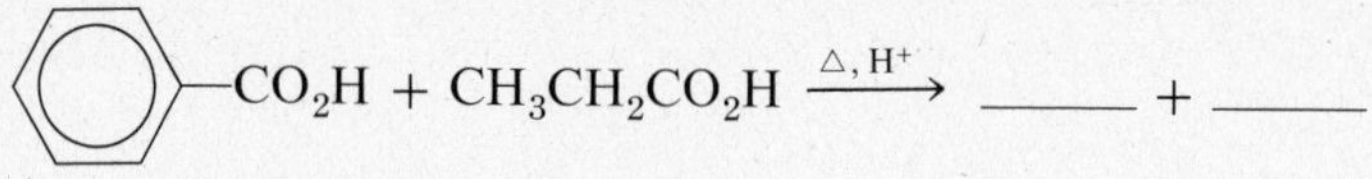 $C_6H_5{-}CO_2H + CH_3CH_2CO_2H \xrightarrow{\triangle,\ H^+}$ ______ + ______

(v) 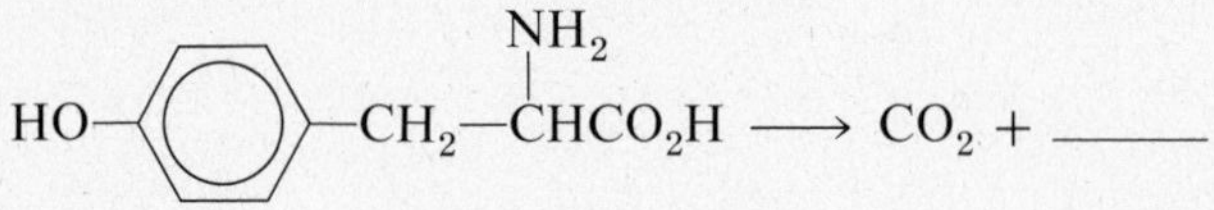 $HO{-}C_6H_4{-}CH_2{-}CH(NH_2)CO_2H \longrightarrow CO_2 +$ ______

(vi) 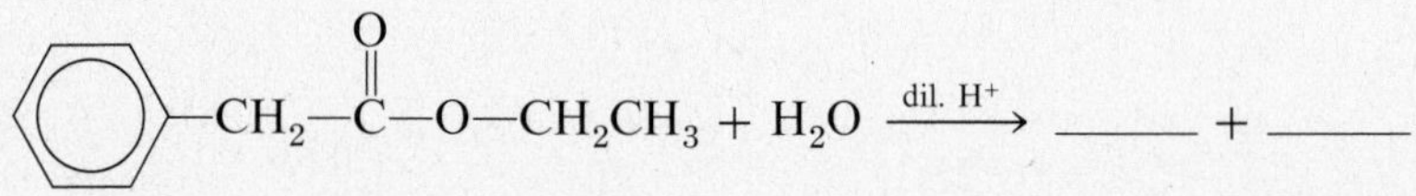 $C_6H_5{-}CH_2{-}\overset{O}{\overset{\|}{C}}{-}O{-}CH_2CH_3 + H_2O \xrightarrow{\text{dil. } H^+}$ ______ + ______

(vii) $CH_3CO_2H + H_3PO_4 \longrightarrow H_2O +$ ______

(viii) glycerol + lauric acid + 2 stearic acid $\longrightarrow 3H_2O +$ ______

(ix)

$$\begin{array}{l} H_2C{-}O{-}\overset{O}{\overset{\|}{C}}{-}(CH_2)_{10}CH_3 \\ \quad | \\ H{-}C{-}O{-}\overset{O}{\overset{\|}{C}}{-}(CH_2)_7{-}CH{=}CH(CH_2)_7CH_3 + H_2 \xrightarrow[\text{pressure}]{\triangle,\ Ni} \text{\_\_\_\_\_\_} \\ \quad | \\ H_2C{-}O{-}\overset{O}{\overset{\|}{C}}{-}(CH_2)_{14}CH_3 \end{array}$$

(x)

$$\begin{array}{l} H_2C{-}O{-}\overset{O}{\overset{\|}{C}}{-}(CH_2)_{10}CH_3 \\ \quad | \\ H{-}C{-}O{-}\overset{O}{\overset{\|}{C}}{-}(CH_2)_7{-}CH{=}CH(CH_2)_7CH_3 + 3HOH \xrightarrow{\triangle,\ H^+} \\ \quad | \\ H_2C{-}O{-}\overset{O}{\overset{\|}{C}}{-}(CH_2)_{14}CH_3 \end{array}$$

(xi)
$$\begin{array}{l} \quad\quad\quad\quad\ \ O \\ \quad\quad\quad\quad\ \ \| \\ H_2C-O-C-(CH_2)_{10}CH_3 \\ \quad | \quad\quad\quad\ O \\ \quad | \quad\quad\quad\ \| \\ H-C-O-C-(CH_2)_7CH{=}CH(CH_2)_7CH_3 + 3NaOH\,(aq) \xrightarrow{H_2O} \\ \quad | \quad\quad\quad\ O \\ \quad | \quad\quad\quad\ \| \\ H_2C-O-C-(CH_2)_{14}CH_3 \end{array}$$

(xii) _____ + _____ $\xrightarrow{\text{dil. } H^+}$ $(CH_3)_2CHCH_2OH$ + $C_6H_5-CO_2H$

(xiii) $CH_3CH_2SH + CH_3(CH_2)_4CO_2H \xrightarrow{\Delta,\, H^+}$ _____ + _____

(xiv) $CH_3-\overset{\overset{\displaystyle O}{\|}}{C}-CO_2H \longrightarrow CO_2 +$ _____

(xv) $HO-\overset{\overset{\displaystyle O}{\|}}{C}-\overset{\overset{\displaystyle O}{\|}}{C}-OH + 2CH_3CH_2OH \xrightarrow{\Delta,\, H^+}$

(xvi) _____ + _____ $\xrightarrow{\text{dil. } H^+}$ $C_6H_4(CO_2H)(OH)$ (salicylic acid: CO_2H and OH ortho on benzene ring) + CH_3OH

(xvii) _____ + _____ $\longrightarrow CH_3CH_2CO_2H + CH_3(CH_2)_4SH$

3. Upon ingestion, aspirin slowly hydrolyzes. Some people cannot take aspirin because it causes irritation of their stomach lining. What products of aspirin hydrolysis could cause this to occur (see Chapter 9)?
4. Salicylic acid is reacted with an aqueous solution of sodium hydroxide. The water is then removed by evaporation leaving a white solid. Using structural formulas, write an equation representing the formation of the white solid. What is the white solid?
5. Four compounds, each having the molecular formula $C_7H_{14}O_2$, were reacted individually with aqueous sodium hydroxide. Each reaction produced ethanol and a different sodium salt. Write structural formulas for each of the four starting compounds.
6. Acid hydrolysis of an ester is a reversible reaction. Base hydrolysis of an ester is nonreversible. Explain.
7. Nitroglycerin is a vasodilator and antispasmodic used to treat angina pectoris, a heart malfunction. Nitroglycerin is an ester formed by the complete esterification of glycerol by its reaction with nitric acid

(HO—NO_2). Using structural formulas, write the reaction representing the formation of nitroglycerin.

8. Explain why waxes are water insoluble.

9. Outline a series of specific reactions that would allow you to prepare ethyl acetate ($CH_3CH_2—O\overset{\overset{\displaystyle O}{\|}}{C}—CH_3$) from ethene. Write structural formulas for all organic reactants and products. Assume that any inorganic reagents you require are available.

11
amine derivatives

INTRODUCTION

The four most abundant elements in living organisms are hydrogen, carbon, oxygen, and nitrogen. Reactions involving compounds of the first three of these elements have been described in the preceding two chapters. This chapter will focus attention on compounds of biological significance that additionally contain nitrogen. We will study the formation and reactions of simple compounds like amides and then progress to more complex materials like proteins. The building up and breaking down of proteins will be described and related to enzyme activity. The chemistry of respiration study will examine the role of hemoglobin, a protein-containing material, in oxygen transport.

AMIDES

Ammonia behaves as a base by accepting a hydrogen ion. It is able to accept a hydrogen ion because it has a pair of unshared electrons.

Amides are related to the products of the reaction between ammonia and a carboxylic acid: Ammonia, acting as a base, can react with an acid to produce an ammonium salt.

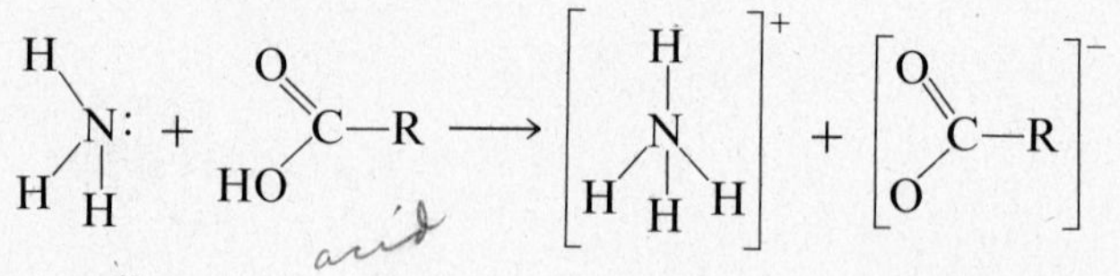

$$NH_3: + HO{-}C({=}O){-}R \longrightarrow [NH_4]^+ + [O{=}C(O){-}R]^-$$

ammonium ion — anion of the carboxylic acid

At elevated temperatures, the salt decomposes by the elimination of water to form an amide:

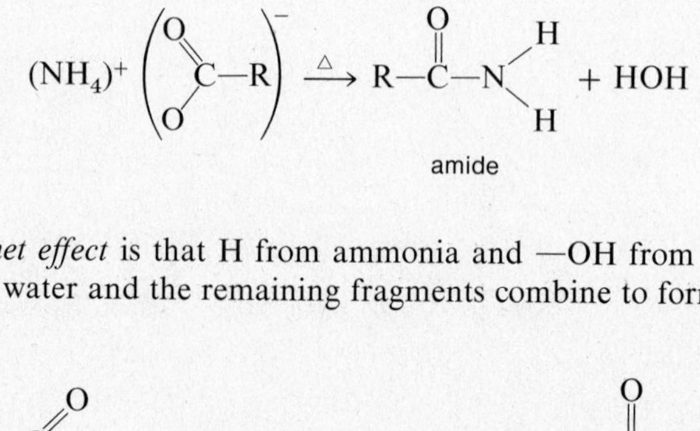

$$(NH_4)^+ \left(O{=}C(O){-}R\right)^- \xrightarrow{\Delta} R{-}\overset{O}{\overset{\|}{C}}{-}NH_2 + HOH$$

amide

Thus, the *net effect* is that H from ammonia and —OH from the acid are split out as water and the remaining fragments combine to form an amide:

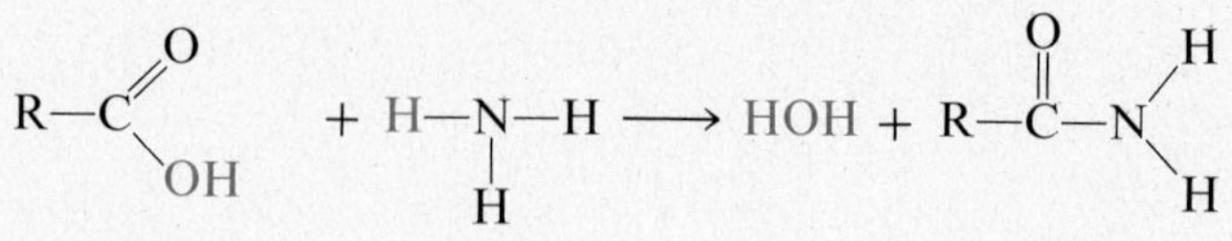

$$R{-}C({=}O){-}OH + H{-}N(H){-}H \longrightarrow HOH + R{-}\overset{O}{\overset{\|}{C}}{-}N(H){-}H$$

carboxylic acid — ammonia — amide

EXAMPLE 11-1

Nicotinic acid (niacin, a vitamin) can react with ammonia to form nicotinamide.

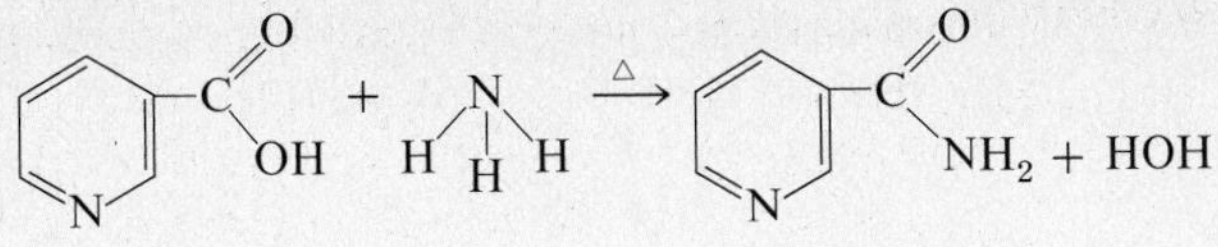

$$\text{C}_5\text{H}_4\text{N}{-}\text{COOH} + \text{NH}_3 \xrightarrow{\Delta} \text{C}_5\text{H}_4\text{N}{-}\text{CONH}_2 + \text{HOH}$$

nicotinic acid — nicotinamide

In Chapter 8 we saw that amines chemically behave similarly to ammonia. Therefore, amines possessing one or two hydrogens bonded at the nitrogen are also able to undergo amide formation:

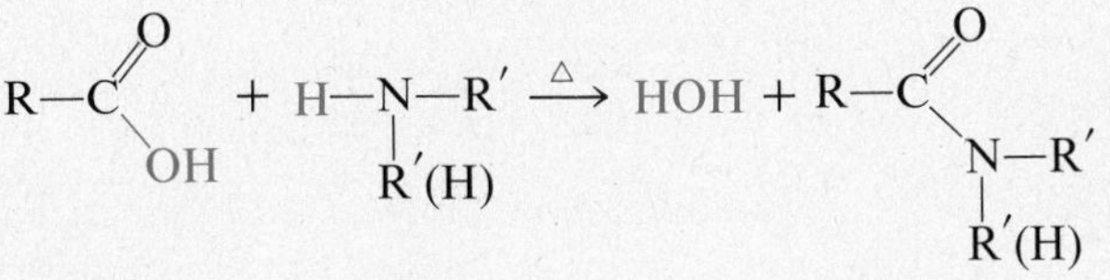

$$\text{R}{-}\text{C}(=\text{O}){-}\text{OH} + \text{H}{-}\text{N}(\text{R}'(\text{H})){-}\text{R}' \xrightarrow{\Delta} \text{HOH} + \text{R}{-}\text{C}(=\text{O}){-}\text{N}(\text{R}'(\text{H})){-}\text{R}'$$

carboxylic acid — amine — amide

EXAMPLE 11-2

Although chemically known since 1938, public awareness of LSD has been limited until recently. LSD is an abbreviation for lysergic acid diethylamide. Lysergic acid reacts with diethyl amine to form LSD.

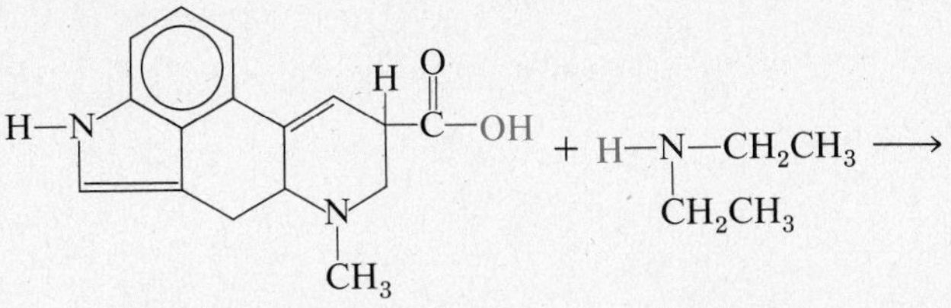

$$\text{lysergic acid (}{-}\text{C}(=\text{O}){-}\text{OH)} + \text{H}{-}\text{N}(\text{CH}_2\text{CH}_3){-}\text{CH}_2\text{CH}_3 \longrightarrow$$

lysergic acid — diethylamine

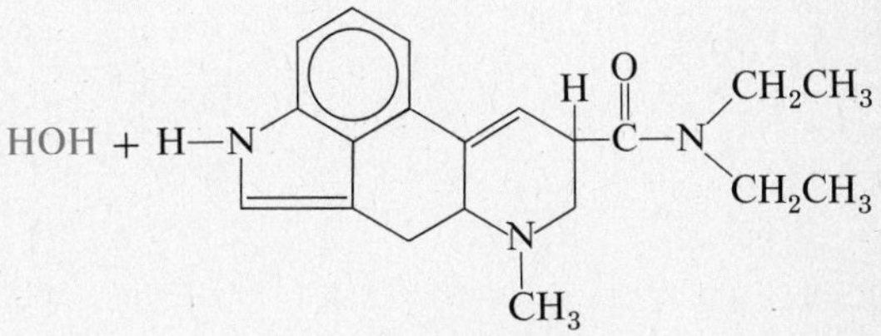

$$\text{HOH} + \text{lysergic acid diethylamide (}{-}\text{C}(=\text{O}){-}\text{N}(\text{CH}_2\text{CH}_3)_2\text{)}$$

lysergic acid diethylamide

The characteristic of an amide is its functional group:

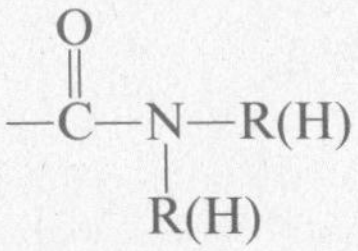

$$-\overset{\displaystyle \text{O}}{\overset{\|}{\text{C}}}-\underset{\displaystyle \text{R(H)}}{\underset{|}{\text{N}}}-\text{R(H)}$$

Both esters and amides are derivatives of carboxylic acids and a comparison of their formation points out this similarity:

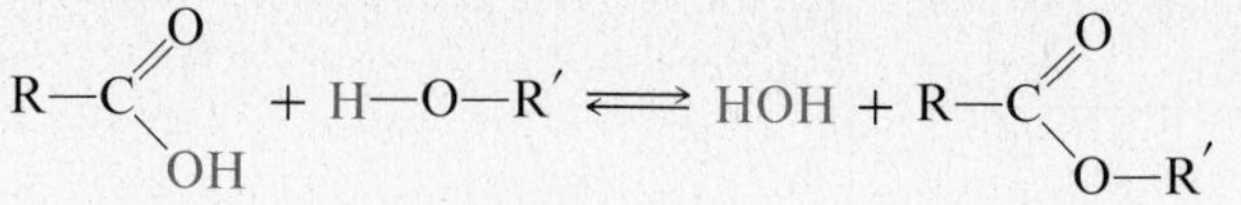

$$\underset{\text{acid}}{R{-}C({=}O)OH} + \underset{\text{alcohol}}{H{-}O{-}R'} \rightleftharpoons HOH + \underset{\text{ester}}{R{-}C({=}O){-}O{-}R'}$$

$$\underset{\text{acid}}{R{-}C({=}O)OH} + \underset{\text{amine}}{H{-}N(R'(H))(R'(H))} \rightleftharpoons HOH + \underset{\text{amide}}{R{-}C({=}O){-}N(R'(H))(R'(H))}$$

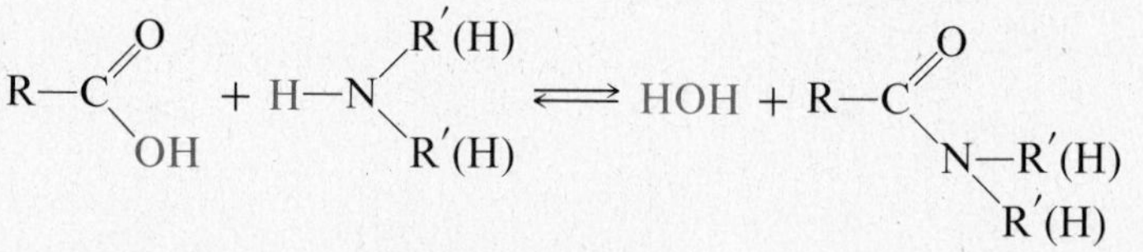

An amine must have a hydrogen directly bonded to the nitrogen atom in order to form an amide. Primary amines ($R{-}NH_2$) and secondary amines (R_2NH) meet this requirement and form amides when reacted with carboxylic acids. Tertiary amines (R_3N) do not have a hydrogen directly bonded to the amine nitrogen. Consequently, tertiary amines do not form amides. The amine nitrogen of all amines has an unshared electron pair, however, that can form a coordinate covalent bond with a hydrogen ion from a carboxylic acid. In this way tertiary amines react with carboxylic acids to produce amine salts analogous to ammonium salts:

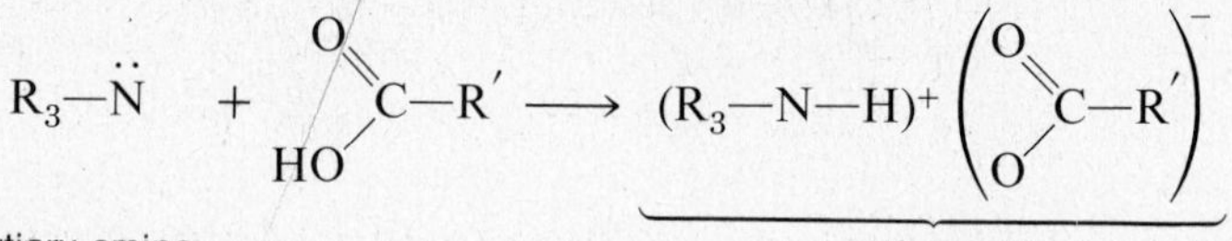

Why must there be a hydrogen atom on the nitrogen atom of the amine group in order to form an amide?

$$\underset{\text{tertiary amine}}{R_3{-}\ddot{N}} + HO{-}C({=}O){-}R' \longrightarrow \underbrace{(R_3{-}N{-}H)^+ \left(O{=}C(O){-}R'\right)^-}_{\text{amine salt}}$$

exercise 11-1

Write structural formulas for the organic product(s) formed in the following reactions:

(a) 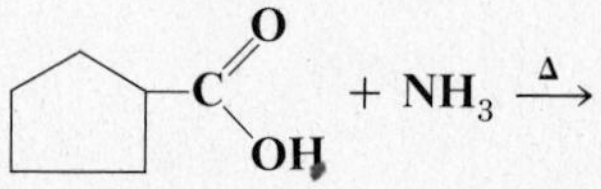 $C_5H_9{-}C({=}O)OH + NH_3 \xrightarrow{\Delta}$ (cyclopentyl carboxylic acid)

(b) $CH_3{-}CH(NH_2){-}C({=}O)OH + H_2N{-}CH_2{-}C({=}O)OH \longrightarrow$

(c) $HO{-}C({=}O){-}(CH_2)_4{-}C({=}O)OH + 2CH_3NH_2 \longrightarrow$

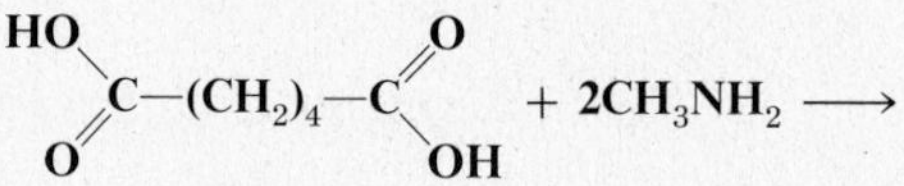

(d) 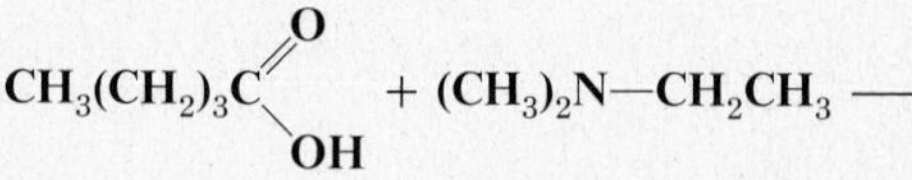 $CH_3(CH_2)_3C({=}O)OH + (CH_3)_2N{-}CH_2CH_3 \longrightarrow$

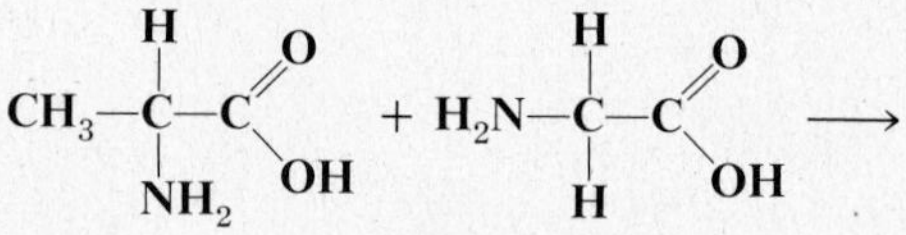

AMINO ACIDS AND PROTEINS What do skin, fingernails, hair, milk, and egg whites have in common? They all contain proteins, complex aggregates of amino acids. In Chapter 8 amino acids were described as important cellular components containing carboxylic acid and amine groups in the same molecule. The most important biochemical function amino acids perform is their reaction with each other to form proteins, materials present in all living cells. The key reaction in the buildup of a protein by the combination of amino acids is the formation of an amide linkage between amino acid units. This linkage is called a *peptide bond.* Peptide bond formation results from the reaction of an amine group on one amino acid molecule with the acid group on another amino acid molecule. Generalizing,

A peptide bond is an amide linkage.

$$R_1\text{-}CH(NH_2)\text{-}COOH + H\text{-}NH\text{-}CH(R_2)\text{-}COOH \longrightarrow HOH + R_1\text{-}CH(NH_2)\text{-}CO\text{-}NH\text{-}CH(R_2)\text{-}COOH$$

amino acid 1 — amino acid 2 — a dipeptide (a peptide bond)

The combination of two amino acids joined by a peptide bond is called a *dipeptide.* Using glycine and alanine, let's write reactions for dipeptide formation between these two amino acids. Being amino acids, they both contain acid and amine groups. This means that two possibilities arise:

1. Glycine can act as the acid and alanine as the amine.

$$H_2N\text{-}CH_2\text{-}COOH + H\text{-}NH\text{-}CH(CH_3)\text{-}COOH \longrightarrow HOH + H_2N\text{-}CH_2\text{-}CO\text{-}NH\text{-}CH(CH_3)\text{-}COOH$$

glycine — alanine — glycylalanine, a dipeptide (peptide bond)

2. Alanine can act as the acid and glycine as the amine.

$$H_2N\text{-}CH(CH_3)\text{-}COOH + H_2N\text{-}CH_2\text{-}COOH \longrightarrow HOH + H_2N\text{-}CH(CH_3)\text{-}CO\text{-}NH\text{-}CH_2\text{-}COOH$$

alanine — glycine — alanylglycine, a dipeptide (peptide bond)

Structural isomers have different properties, e.g., melting points and chemical reactivity.

You might be tempted to say the two dipeptides are the same. Note, however, that in glycylalanine the peptide bond nitrogen is bonded directly to a carbon bearing a hydrogen and a methyl group; in alanylglycine this nitrogen is bonded directly to a carbon bearing two hydrogens. Therefore, these two compounds are structural isomers.

exercise 11-2

Using structural formulas, write equations representing dipeptide formation between valine and leucine (see Table 8-8) for amino acid structures).

Since the dipeptide still contains uncombined amine and carboxyl groups, the process of peptide bond formation can continue. R represents the particular side chain of each alpha amino acid.

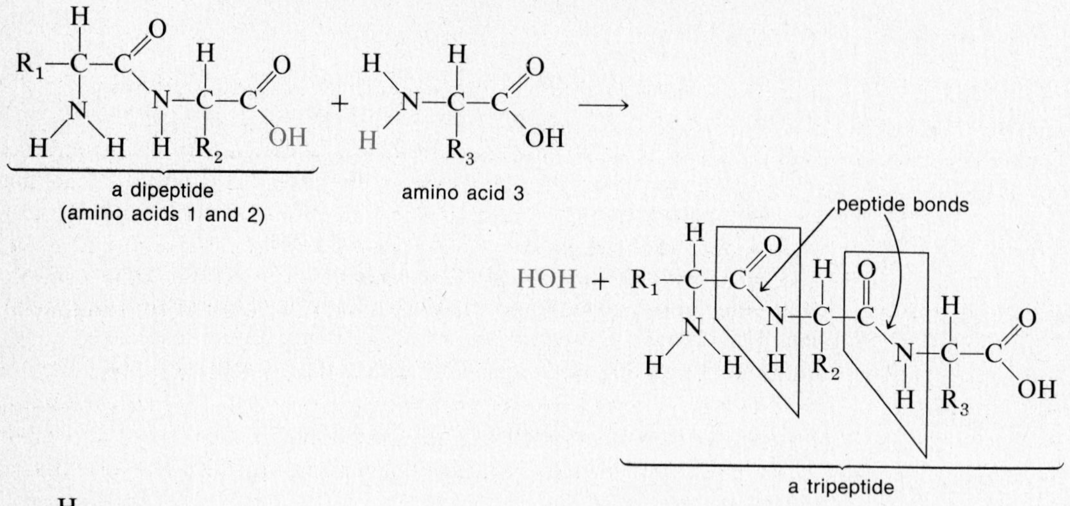

a dipeptide (amino acids 1 and 2) + amino acid 3 ⟶ HOH + a tripeptide

peptide bonds

a tripeptide

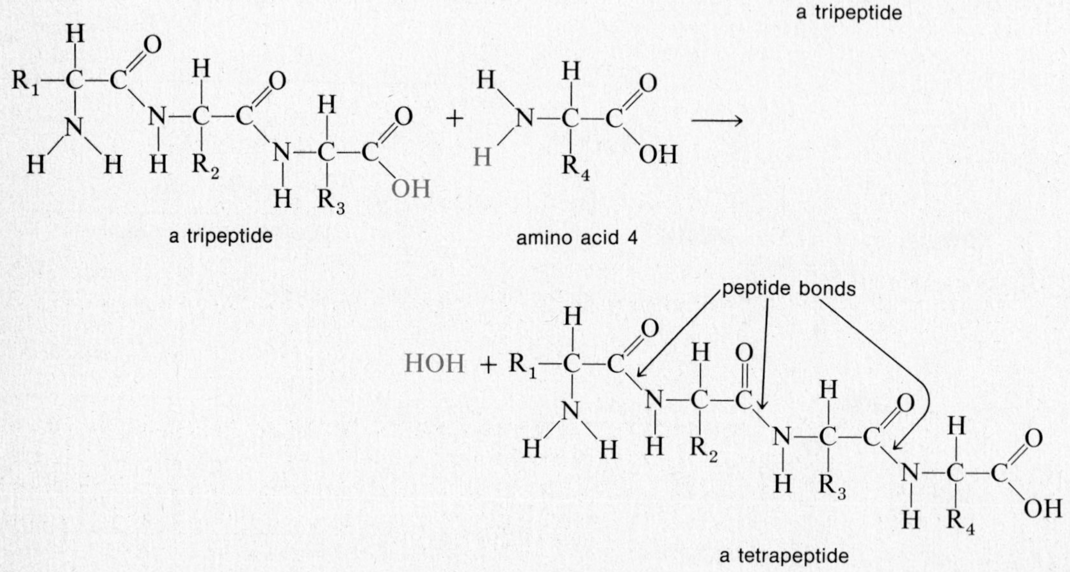

a tripeptide + amino acid 4 ⟶ HOH + a tetrapeptide

peptide bonds

a tetrapeptide

Peptide bond formation can continue by the joining of additional amino acids to the lengthening peptide. Pentapeptides (five amino acids), hexapeptides (six amino acids), etc., can be formed. Combinations of more than 10 amino acids are called *polypeptides*. Large numbers of amino acid units can be incorporated in this way, eventually culminating in the formation of a protein:

$$\text{N-terminal end}\quad H_2N-\underset{R}{\overset{H}{C}}-C(=O)-\left[N(H)-\underset{R}{\overset{H}{C}}-C(=O)\right]_x-N(H)-\underset{R}{\overset{H}{C}}-C(=O)OH\quad\text{C-terminal end}$$

X is a large number, usually greater than 50. Thus, proteins have molecular weights ranging from approximately 6000 amu to over a million amu. It should also be noted that a protein or polypeptide contains an uncombined amine group at one end of the molecule, the so-called N-terminal end, and an uncombined carboxyl group at the other, the C-terminal end. By convention, peptides are named starting from the N-terminal amino acid unit and proceeding by naming all component amino acids, ending with the C-terminal amino acid. Review the first several formulas to see this nomenclature used.

The molecular weights of amino acids average about 100 amu.

Previously we saw that two different amino acids could combine to yield two different dipeptides. How many tripeptides could three different amino acids form? Let's play "molecular scrabble" and consider the possible combinations using letters to represent the amino acids glycine (G), alanine (A) and valine (V):

G–A–V A–V–G V–G–A
G–V–A A–G–V V–A–G

Each combination represents a different tripeptide.

exercise 11-3

Write structural formulas for the tripeptides represented by A–G–V and G–A–V above.

From the material above it should be noted that as the number of different amino acids increases, the number of possible polypeptides sharply increases. The number of theoretically possible combinations is given by the expression $n!$ (read *n factorial*) where *n* is the number of different amino acid units available for combination.

n	n!	n	n!
2	2 × 1 = 2	6	6 × 5 × 4 × 3 × 2 × 1 = 720
3	3 × 2 × 1 = 6	7	7 × 6 × 5 × 4 × 3 × 2 × 1 = 5040
4	4 × 3 × 2 × 1 = 24	8	8 × 7 × 6 × 5 × 4 × 3 × 2 × 1 = 40,320
5	5 × 4 × 3 × 2 × 1 = 120	9	9 × 8 × 7 × 6 × 5 × 4 × 3 × 2 × 1 = 362,880

Twenty different amino acids give 2.4×10^{18} theoretically possible combinations using each amino acid only once.

To say the possible combinations, increases sharply is really understating things. It is easy to understand how, using approximately 20 naturally occurring amino acids as building blocks, nature can fashion a tremendous diversity of proteins. Not all possible combinations are used or are present in the same species. By analogy, the English language is fashioned from just 26 letters, some of whose combinations do not make recognizable English words (ever play Scrabble*?).

PRIMARY STRUCTURAL FEATURES OF PROTEINS

The protein globin of hemoglobin is actually made up of four protein molecules.

Carboxylic acid groups can hydrogen bond with water; hydrocarbons groups cannot.

The peptide bond is the major structural factor in animal tissue. *The sequence of amino acids constitutes the primary structure of a protein or polypeptide.* Sickle cell anemia is an example of what may happen if improper sequencing of a protein's amino acids occurs. Normal human hemoglobin contains the protein globin which consists of two pairs of protein chains: the alpha chains, each containing 141 amino acids, and the beta chains, each containing 146 amino acids. In normal individuals, the first seven amino acids in the beta chain are valine–histidine–leucine–threonine–proline–glutamic acid–glutamic acid–. In an individual suffering from sickle cell anemia, the seventh amino acid is valine instead of glutamic acid. When combined with oxygen, both normal and sickle cell hemoglobin are soluble in plasma. As the oxygenated blood passes from arteries into capillaries, oxygen is released to the surrounding tissues. When this occurs, the solubility of sickle cell hemoglobin is sharply reduced and, within the red blood sickle cell, hemoglobin crystallizes out of solution. This causes the red blood cells to become misshapen. The misshapen cells block capillaries and small blood vessels and reduce normal blood flow. The difference in solubility between normal and sickle cell hemoglobin has been traced to the difference in amino acid sequence. Glutamic acid's side chain is $—CH_2—CH_2—CO_2H$, while that of valine is $—CH—(CH_3)_2$. The presence of the carboxylic acid side-chain group in glutamic acid makes the group much more water (plasma) soluble than valine's hydrocarbon-like side chain. Thus, here on the molecular level is the basis for a disease brought on by changing only 1 out of 146 amino acid units. This is an outstanding example of how chemical structure molecularly determines biological function.

SECONDARY AND TERTIARY STRUCTURAL FEATURES

The primary structure of all proteins is dictated by the number, kind, and order of their component amino acids. Many proteins can be isolated as crystalline solids. Observations like this led to the concept that additional

*Registered trademark of Scrabble, Selchow and Righter Co., Bay Shore, N.Y., for the Production and Marketing Co.

Proteins are not simply long, floppy molecules.

factors play a role in determining the overall structure of a protein molecule. These factors are categorized as contributing to the secondary and tertiary features of a protein's structure. Hydrogen bonding between peptide bond groupings gives secondary structural features to protein molecules. Factors contributing to tertiary features include disulfide bonds, salt bridges and hydrogen bonds between amino acid side-chain groups, and hydrophobic interactions.

Hydrophobic—*water hating.*

SECONDARY STRUCTURE AND HYDROGEN BONDING

The protein or polypeptide chain is primarily linked together by peptide linkages. The $\left[\overset{\delta-}{\underset{|}{\text{—N}}}\text{—}\overset{\delta+}{\text{H}}\right]$ and $\left[\text{—}\overset{\delta+}{\underset{|}{\text{C}}}\text{=}\overset{\delta-}{\text{O}}\right]$ bonds of the peptide linkages are highly polar because of electronegativity differences between the bonded atoms. This creates a condition favorable for hydrogen bond formation. Linus Pauling and R. B. Corey proposed two possible ways by which hydrogen bonding could stabilize the folding of a protein in a particular way. One way is the pleated-sheet configuration, a network of adjacent polypeptide chains held together in a nearly planar arrangement by *intermolecular* hydrogen bonding. The hydrogen bonding occurs between atoms already involved in peptide bonds. Figure 11-1 (a) and (b) shows this arrangement. Silk and muscles are natural proteins whose structural properties seem to be related to this type of arrangement.

Secondary structure of proteins results from hydrogen bonding between amide groups.

The more commonly occurring configuration is the alpha helix arrangement in which hydrogen bonds exist intramolecularly between $\underset{|}{\text{—N}}\text{—H}$ groups and carbonyl–oxygen groups of the peptide linkages down the chain. This intramolecular hydrogen bonding causes the chain to coil into a helix (see Fig. 11-1(c)). Winding a piece of string or thin strip of paper along the length of a pencil will generate a helix. Because of the sequence of amino acids present in a protein, the primary structure dictates the coiling of the protein.

The curliness of hair and the flexibility of skin are attributed to the secondary structure of proteins.

exercise 11-4

Using the structural formulas completed in Exercise 11-3, show how intermolecular hydrogen bonding can occur between two tripeptide molecules.

TERTIARY STRUCTURAL FEATURES

Tertiary structure refers to folding of a polypeptide chain into a unique three-dimensional structure stabilized by covalent forces (disulfide bonds), ionic forces (salt bridges), hydrogen bonds, and London interactions.

Cysteine is an amino acid containing a thiol group (—S—H) that is not involved in peptide bond formation. The presence of cysteine units in a protein provides the possibility for disulfide bond formation. Under mild oxidizing conditions, hydrogen is lost from the thiol and combination of sulfur atoms between cysteine units occurs. In this way, disulfide bond formation can create cross-linking between protein chains or within a protein

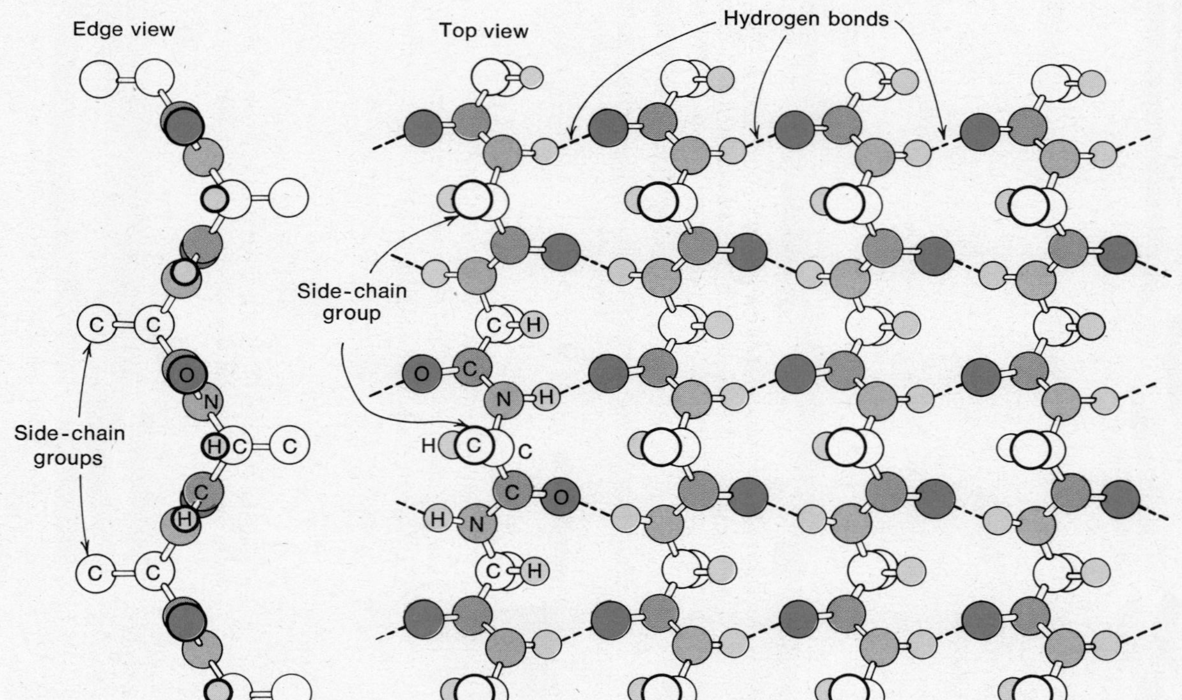

FIGURE 11-1(a) Pleated-sheet structure of extended polypeptide chains. Groups indicated by heavy lines are meant to project forward; those by light lines, backward. Dashed lines indicate hydrogen bonds that hold parallel polypeptide chains to each other and stabilize the configuration. (According to L. Pauling and R. B. Corey.)

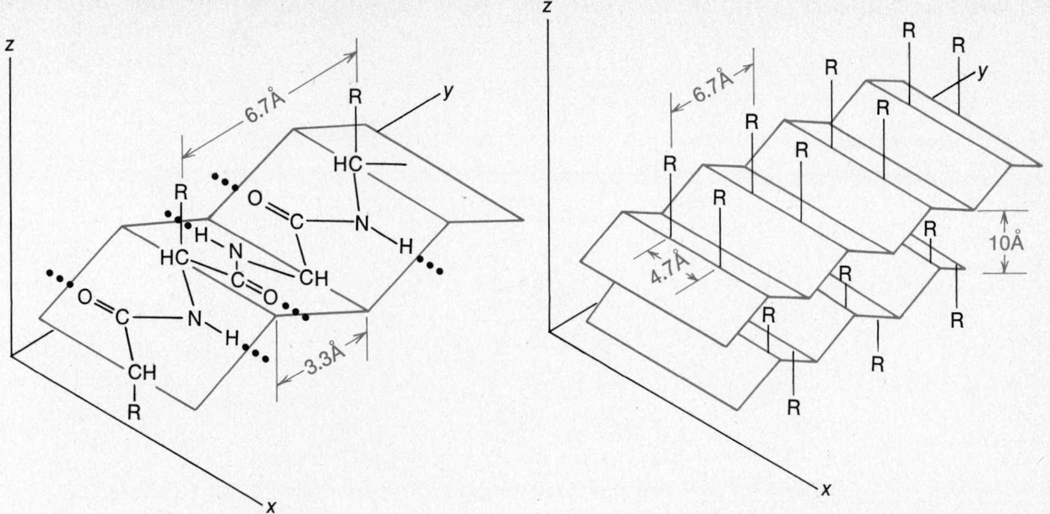

Extended polypeptide chain

Array of chains to give pleated sheets

FIGURE 11-1(b) Schematic diagram of the pleated sheet model of β-keratin. The peptide bonds lie in the plane of the pleated sheet; the side chains lie above or below the sheet and alternate along the chain. The polypeptide chains are held together by hydrogen bonds.

FIGURE 11-1(c) Secondary structure—the peptide chain twists into a helix, in which the C=O from one turn forms a hydrogen bond with the N—H of the turn below.

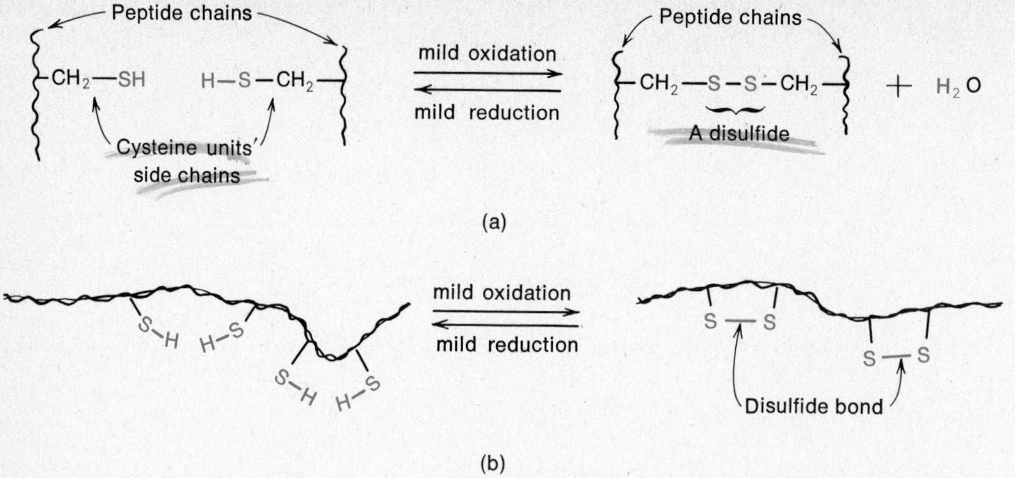

FIGURE 11-2 Disulfide bond formation. (a) Intermolecular. (b) Intramolecular.

Insulin is a protein containing several disulfide linkages between two polypeptide chains.

helix (see Fig. 11-2). Note from Fig. 11-2 that disulfide bond formation is reversible.

The side chains of the acidic and basic amino acids in a protein contain free carboxyl and amine groups. These can undergo Brønsted–Lowry acid–base interactions that produce sites of opposite charges which attract each other. These attractions are termed *salt bridges* and help further stabilize the molecule's tertiary structure (see Fig. 11-3).

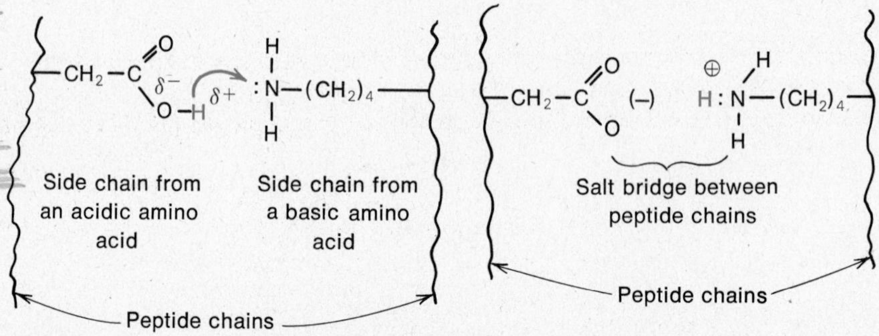

Salt bridges result from the neutralization of the acid groups in the side chains of the acidic amino acids by the amine groups of basic amino acids making up the protein.

FIGURE 11-3 Salt bridge formation.

Several amino acids have side chains capable of hydrogen bonding with side chains of other amino acids in a polypeptide or protein. Several examples are given in Fig. 11-4. Note that this type of hydrogen bonding involves side chains, not peptide bond sites.

This interpretation has been used in describing the behavior of cell membranes.

Cellular activity occurs in an aqueous environment. Many proteins contain amino acids with hydrocarbon-like side chains. These side chains cannot hydrogen bond to water and thus do not promote the protein's water solubility. Instead, the protein molecule folds so that the alkane-like hydrophobic side chains are folded inward. This minimizes hydrophobic group

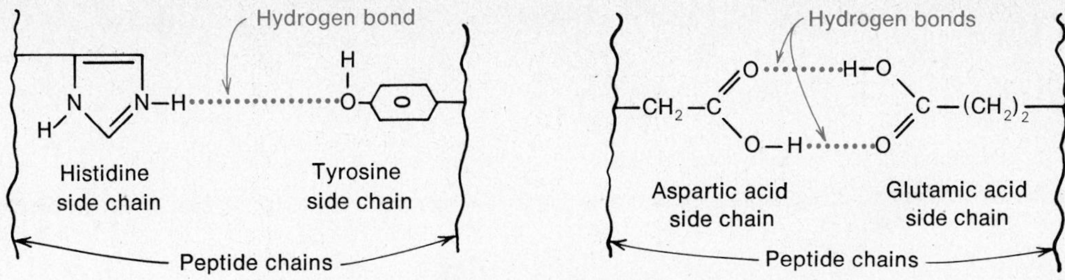

FIGURE 11-4 Hydrogen bonding between amino acid side chains.

contact with water while exposing more polar side chains for interaction with polar water molecules. The interactions occurring between hydrophobic groups are London forces.

London forces are discussed in Chapter 5.

pH AND PROTEIN SOLUBILITY

Protein solubility is pH dependent. An example of this is the souring of milk. Bacteria in the milk produce lactic acid as a metabolic waste product. As lactic acid concentration increases, acidity increases and the pH of milk falls. When the pH reaches 4.6, casein (the major protein in milk) precipitates and forms a curd.

exercise 11-5

Sour milk needed for baking can be made by adding lemon juice or vinegar to regular milk. Explain.

*Be sure you understand why a protein has a **net** plus charge at a pH below the isoelectric point and a **net** minus charge at a pH above the isoelectric point.*

Proteins are least soluble at their *isoelectric point, the pH at which the protein molecule has no* **net** *charge* (see Chapter 8). Oppositely charged groups in a protein molecule cause the molecule to be attracted to neighboring molecules bearing opposite charges. This attraction is strong enough to cause the molecules to coagulate and precipitate from solution. At pH values below its isoelectric point, a protein molecule has a *net* positive charge; pH values greater than the isoelectric point cause protein molecules to have a *net* negative charge. In both cases, similarly charged molecules will repel each other, thus remaining in solution as a colloidal suspension.

PROTEINS AND BUFFER ACTION

Because proteins may contain thousands of component amino acids, there are large numbers of side-chain amine and carboxylic acid groups available for buffering.

The presence of acidic and basic amino acids in a protein provides side chains that can exhibit acid and base behavior toward added reagents. Depending on the pH, the carboxylic acid and amine groups in side chains of component amino acids may be in various forms. Generally, the amine group is in the form $-\overset{\oplus}{N}H_3$ (—N⊕ bonded to three H); the carboxylic acid is in the form $-C(=O)O^{(-)}$. Thus, proteins can function as buffers (see Fig. 11-5). Blood proteins are the main buffers that help to maintain blood pH between 7.3 and 7.5.

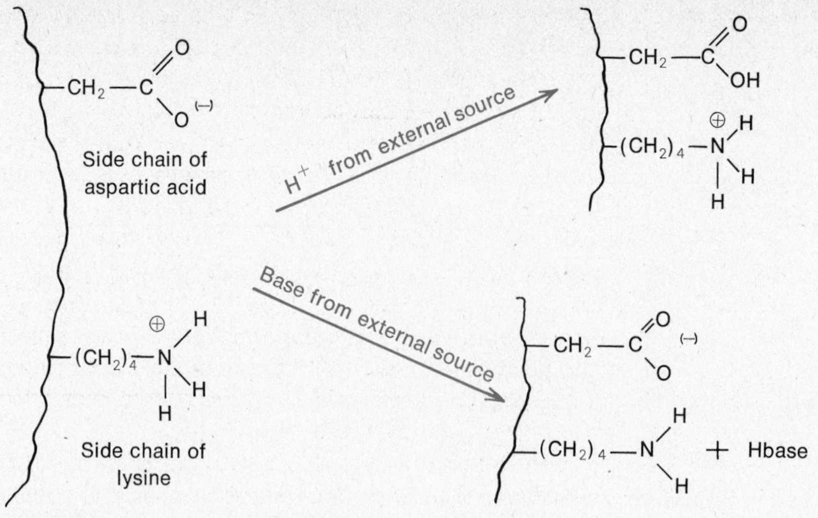

FIGURE 11-5 Protein buffering action.

DENATURATION

Now that we've discussed factors giving a protein a particular structure, let's consider conditions which may cause changes in protein structure. Changes in structure that cause a change in biological activity or function are called *denaturation*. This term applies to alterations of secondary and tertiary structural features. Denaturation can be reversible or irreversible depending on the extent and type of structural change. Denaturation is not usually used to describe changes in a protein's primary structure.

Cauterizing a wound is denaturation.

Common experiences give evidence that heat, sunlight, and/or chemical reagents can denature proteins. Egg whites contain the protein albumin. Frying an egg causes denaturation of the protein and converts the white from a liquid to a solid. Anyone who has suffered a painful sunburn can attest to the action of ultraviolet radiation on protein (skin). A 70 percent ethanol solution serves as a disinfectant because of its denaturing action on proteins in bacteria. Ethanol and various other polar organic solvents are capable of forming hydrogen bonds to proteins, thus interrupting the protein's normal hydrogen bonding pattern.

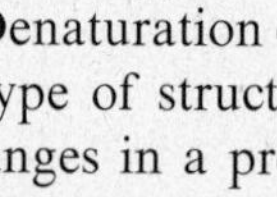

The (—S—H)-containing amino acid cysteine is a component of many proteins. Its ability to form disulfide bonds has been described. Unshared electron pairs on sulfur atoms in disulfide linkages make these atoms excellent electron pair donors for coordinate covalent bond formation. The carboxylate groups $\left[-C\begin{matrix} \diagup\!\!\!\!= O \\ \diagdown O^{(-)} \end{matrix} \right]$ of glutamic and aspartic acids also can form coordinate covalent bonds. Hg^{2+}, Ag^{+}, and Pb^{2+} ions (so-called heavy metal ions) form very strong, coordinate covalent bonds with these groups in a protein. This denatures the protein. Compounds containing these ions, if taken internally, act as poisons by combining with sulfur atoms in disulfide linkages and carboxylate groups. If accidentally swallowed, these heavy metal ions can be kept from general circulation by coagulation in the stomach. Raw egg white or milk are given as antidotes for heavy metal poisoning because they contain proteins that readily combine with the heavy metal ions to form insoluble solids. An emetic is then given that removes the metal ion-containing precipitate before digestive juice action can destroy the protein and reliberate the harmful ions.

Mercury nitrate solution was once used to help stiffen felt used in the manufacture of felt hats. Unwittingly, workers in this area constantly were exposed to absorption of Hg^{2+} ions, which gave these hatters chronic mercury poisoning. This was associated with symptoms of highly eccentric behavior and personality changes, a condition that gave rise to the expression *mad as a hatter*. You may recall reading of such unstable behavior on the part of the Mad Hatter in Lewis Carroll's *Alice in Wonderland*.

Curiously, mercurials and other heavy metal compounds were among the earliest used antiseptics. In low concentration, their compounds are still used. Eyes of newborn babies are sometimes treated with a 1 percent silver nitrate solution to prevent gonorrhea infection. Mercurochrome and the salicylic acid derivative merthiolate are topical, mercury-containing disinfectants.

Tannic acid and picric acid also denature proteins. They are common components of medications used in burn treatment. A physiological danger associated with a burn injury is loss of body fluids and attack by infectious bacteria at the burned region. Upon denaturation, proteins coagulate and form a protective, impervious scab over the burned area.

ANOTHER LOOK AT ENZYMES

Recall from Chapter 6 that enzymes consist of a large protein part, the apoenzyme portion, and a smaller portion, the cofactor. The apoenzyme is nonactive catalytically unless combined with the cofactor. The cofactor usually is the active site in the enzyme. Other regions of the enzyme molecule act as binding sites to hold the substrate in the most favorable position for reaction to occur. This implies that a definite structural arrangement of the enzyme must be present for proper catalytic action.

From an earlier discussion, we have learned that, in general, increasing the temperature causes an increase in reaction rate. Up to a point this is also true for enzymatic reactions. Figure 11-6 is a plot of reaction rate versus temperature for an enzymatic reaction.

Not all enzymes have the same optimum temperature.

From the graph we see that initially a small temperature change produces a large increase in reaction rate. As temperature increases, a point is reached at which the reaction proceeds at a maximum rate. This is called the *optimum* temperature. Most of our enzymes have an optimum temperature of 37 °C.

Beyond the optimum temperature, an increase in temperature actually lowers the reaction rate. This might seem like strange behavior unless we remember that enzymes contain proteins that are denatured by heat. Heating alters the secondary and tertiary structural features of the enzyme. This causes a change in the orientation of the active site in relation to the position

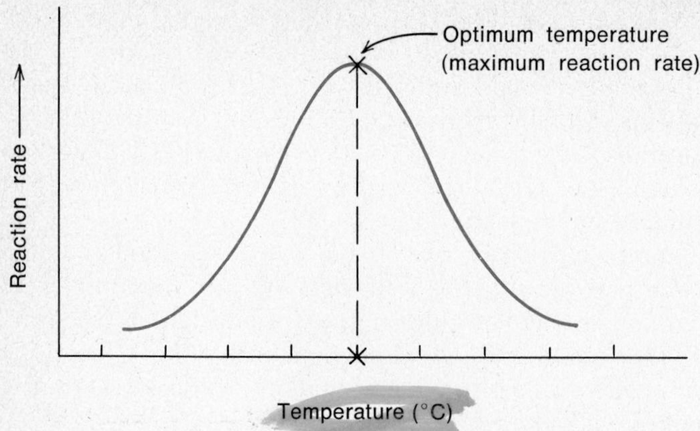

FIGURE 11-6 Effect of temperature changes on the rate of an enzymatic reaction.

of the substrate. The change in orientation prevents the enzyme from catalyzing the reaction and thus reaction rate decreases.

EFFECT OF pH ON ENZYME ACTIVITY

We have seen that pH changes cause changes in secondary and tertiary protein structure. Since enzymes are primarily proteins, we anticipate that pH changes would have an effect on enzyme activity. Experimentally it is found that an enzyme exhibits maximum activity over the narrow pH range in which its molecules possess the proper net charge for substrate binding. This is shown diagrammatically in Fig. 11-7.

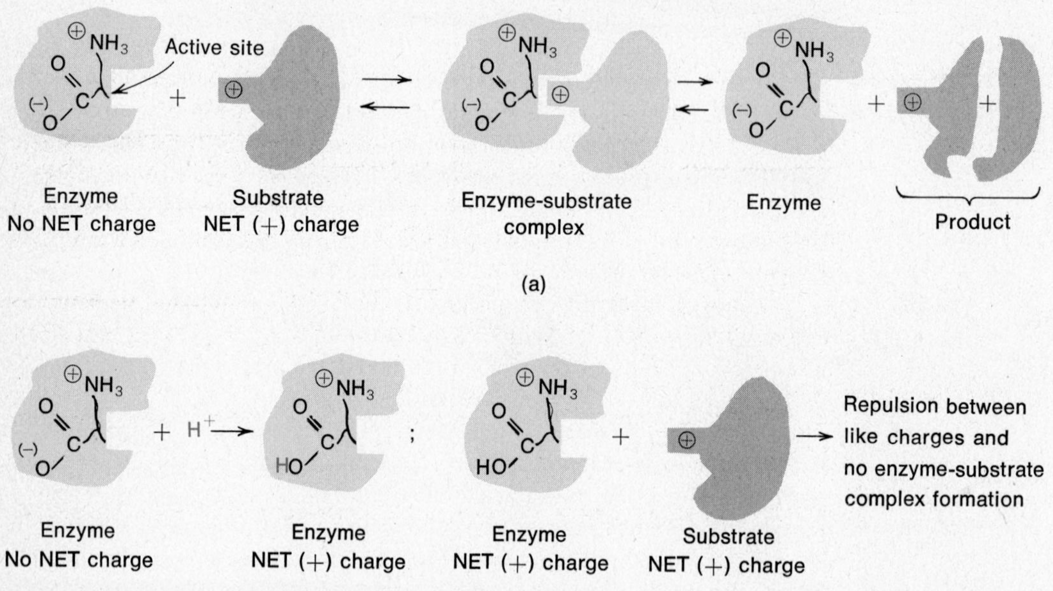

FIGURE 11-7 pH change and enzyme activity. (a) Compatible net charges lead to catalytic enzyme activity. (b) Incompatible net charges due to pH change (addition of H^+) lead to no catalytic activity.

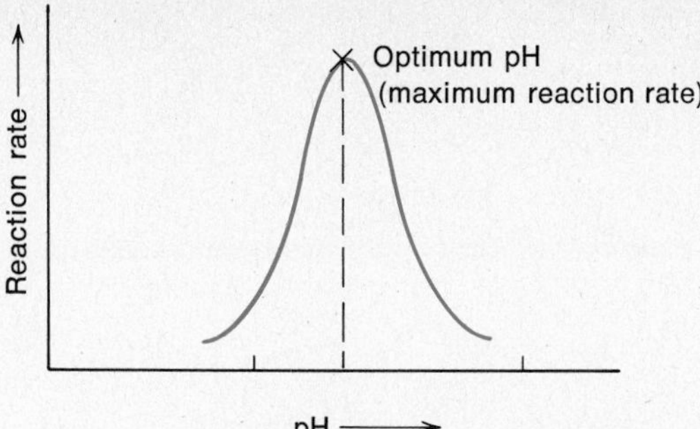

FIGURE 11-8 Enzyme activity as a function of pH.

Most body fluid enzymes have optimum pH values between 6 to 8. Gastric juice enzymes are notable exceptions. Pepsin is most active at pH 1.5–2.5 and rapidly loses its catalytic behavior above pH 6.9 (see Fig. 11-8).

EFFECT OF HEAVY METAL IONS ON ENZYME ACTIVITY

In earlier sections, the inhibiting action that poisons and drugs have on enzyme activity was described. Because enzymes are present in the body in very low concentration, very small doses of poisons can have lethal effects. If taken internally or absorbed through the skin, compounds containing heavy metal ions can have the same effect. By forming coordinate covalent bonds with carboxylate groups, thiol, and disulfide groups present in enzyme proteins, heavy metal ions cause denaturation and coagulation of the protein. This radical change in the shape of the protein renders the enzyme useless for catalysis.

PROTEIN HYDROLYSIS

Protein digestion means the hydrolysis of peptide bonds to liberate the component amino acids.

Amides will undergo hydrolysis. This means that peptide bonds in proteins can be hydrolyzed and the primary structure destroyed. The component amino acids are liberated if acidic or enzymatic hydrolysis is used. During digestion, both acidic and enzymatic hydrolysis occur and the water-soluble amino acids are absorbed into the general circulation for transport to metabolic sites. Figure 11-9(a) and (b) illustrates the complete hydrolysis of a dipeptide and a pentapeptide, respectively.

It is important to note from Fig. 11-9 that a molecule of water is required for the hydrolysis of *each* peptide bond in a polypeptide or protein.

exercise 11-6

Using structural formulas, write a balanced equation for the enzymatic hydrolysis of alanylglycylleucine.

exercise 11-7

Hydrolysis of a hexapeptide molecule would require how many water molecules?

Compare the hydrolysis of a peptide bond to the hydrolysis of an ester.

In vitro, aqueous sodium hydroxide or barium hydroxide are commonly used to hydrolyze proteins. The liberated amino acids are neutralized by the solution of aqueous base so that the component amino acids are converted to their sodium or barium salts. The hydrolysis generally requires

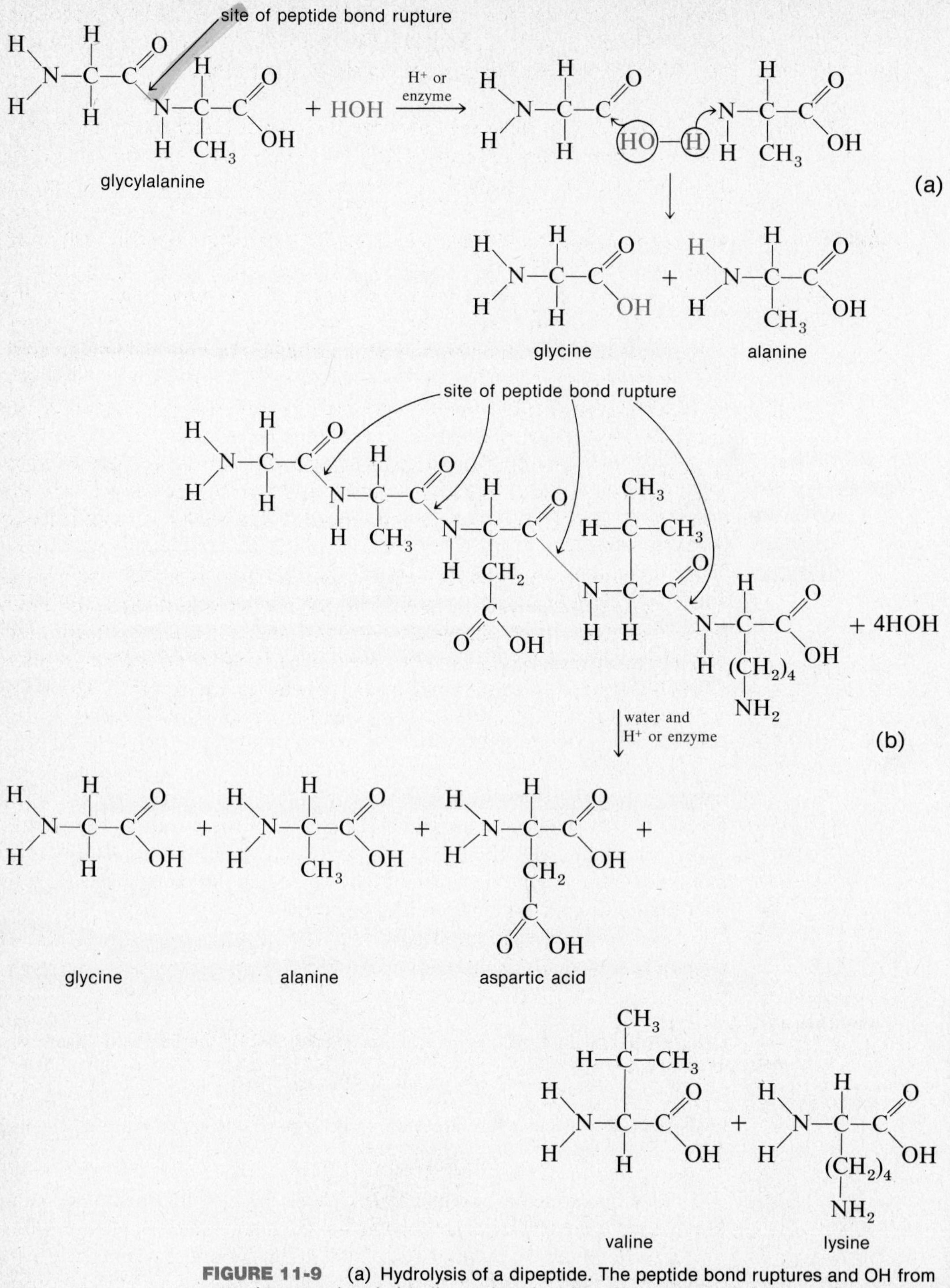

site of peptide bond rupture

glycylalanine + HOH $\xrightarrow{H^+ \text{ or enzyme}}$

glycine + alanine

(a)

site of peptide bond rupture

+ 4HOH

water and H^+ or enzyme

(b)

glycine + alanine + aspartic acid +

valine + lysine

FIGURE 11-9 (a) Hydrolysis of a dipeptide. The peptide bond ruptures and OH from water becomes bonded to the carbonyl carbon; H— from water bonds to the nitrogen formerly in the peptide bond. (b) Hydrolysis of a pentapeptide.

In vitro, outside a living system; in vivo, within a living system.

heating the base–protein mixture at 100°C for 12 to 24 hr for complete hydrolysis of the protein into its amino acid salts. This method is commonly the first step in determining the structure of a protein.

AN EXAMPLE OF PROTEIN STRUCTURE AND FUNCTION—THE CHEMISTRY OF RESPIRATION

While performing life's functions, the body must remove vast amounts of carbon dioxide produced during cellular metabolism. Associated with this is the requirement for tissues to have a continuous supply of oxygen. The solubility of oxygen in blood plasma at body temperature is very low. Plasma can dissolve only about 3 percent of the oxygen required. Obviously, some other method is needed to transport oxygen.

Hemoglobin is the remarkable red blood cell protein that transports about 97 percent of the required oxygen. Each red blood cell contains approximately 280 million hemoglobin molecules. In a normal human adult, each hemoglobin molecule consists of the protein globin that is bonded to heme. Heme consists of an interconnected group of nitrogen-containing ring structures that form coordinate covalent bonds to a central Fe^{2+} ion as shown in Fig. 11-10. The arrangement of alternating single and double bonds causes the molecule to absorb light. Wavelengths corresponding to the blue portion of the visible spectrum are absorbed and red, the complementary color, is seen.

A hemoglobin molecule contains four helical polypeptide chains, each associated with a heme unit. Two of the polypeptide chains are *alpha* chains, each consisting of 141 amino acid groups, and two are *beta* chains, each containing 146 amino acid groups. Because each polypeptide chain contains a heme unit, there are four heme units per hemoglobin molecule. The globin protein portion of the molecule folds and curls about each heme.

Don't attempt to memorize this structure.

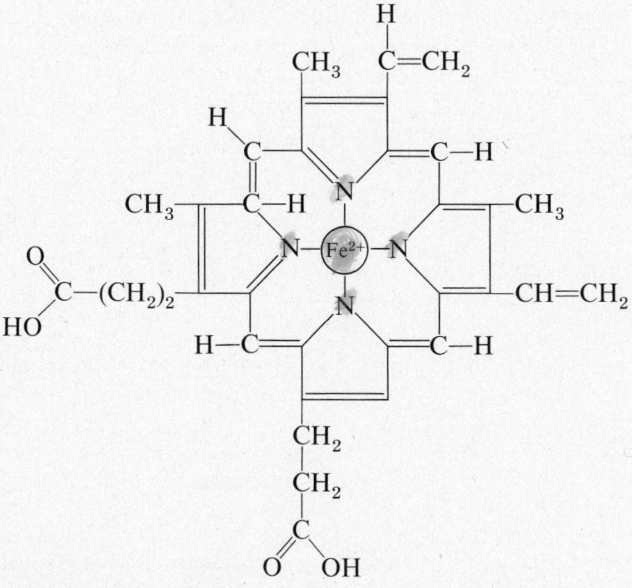

FIGURE 11-10 Heme.

Histidine

is a component amino acid in globin. Its side chain contains the imidazole group

The notation

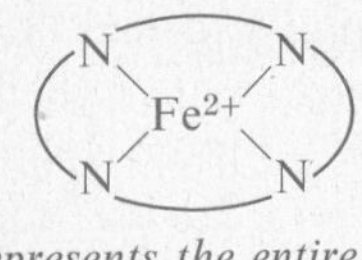

represents the entire heme unit.

This group forms a coordinate covalent bond to the iron ion of each heme group, thereby joining the polypeptide chains to heme. When oxygen combines with hemoglobin to form oxyhemoglobin, oxygen occupies the sixth position about the central iron ion by forming a coordinate covalent bond to the metal ion. This is shown diagrammatically in Fig. 11-11(a) and (b).

exercise 11-8

What is the maximum number of oxygen molecules that can bond to one molecule of hemoglobin?

When the first oxygen molecule is picked up, this changes the protein's tertiary structure. The two beta chains move closer to each other. As a result of this change in the protein structure, other oxygen molecules can bond more readily to the remaining hemes in the hemoglobin molecule. When deoxygenation occurs, the beta chains shift apart. Thus, the action of hemoglobin as an oxygen carrier is more complex than just acting as a storage space in which oxygen can ride along in the red blood cells.

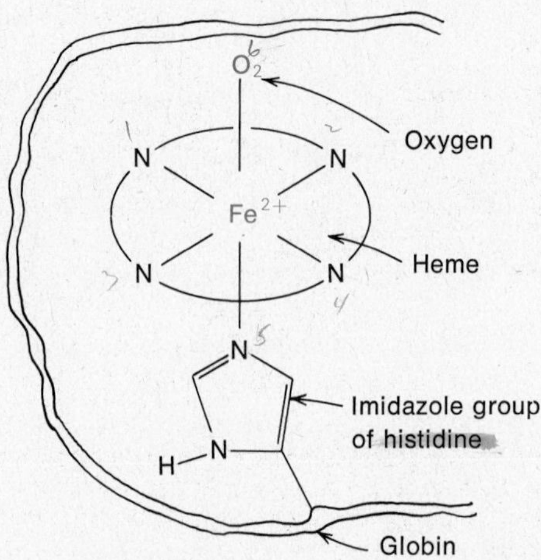

FIGURE 11-11(a) Histidine binding to heme in hemoglobin.

FIGURE 11-11(b) A diagrammatic sketch of a hemoglobin molecule. The alpha chains (light blocks) and beta chains (darker blocks) each enfold a heme group, the site of oxygen (O_2) binding. The lower figure shows two globin chains, each bonded to heme by a histidine group.

Another unusual structural feature of hemoglobin is the maintenance of iron ion to a +2 oxidation state. In aqueous solution, Fe^{2+} is normally oxidized to Fe^{3+} by the action of dissolved oxygen. In hemoglobin, the +2 state is stabilized by bonding to heme and globin. Certain reagents such as nitrite ion (NO_2^-) and bromate ion (BrO_3^-) can cause oxidation of Fe^{2+} to Fe^{3+} in hemoglobin. This results in its conversion to brown methemoglobin, a substance incapable of carrying oxygen.

OXYGEN–HEMOGLOBIN INTERACTIONS

The reaction of oxygen with hemoglobin to form oxyhemoglobin is reversible. In tissue capillaries where oxygen concentration is low, oxyhemoglobin dissociates and releases oxygen. To understand the uptake and release of oxygen more fully, let's first consider how oxygen enters the bloodstream.

In the lungs, inhaled air enters an extensive system of sac-like structures called *alveoli*. Within each alveoli lies a vast network of capillaries. In the alveoli the inhaled air is separated from the capillary blood flow by only the very thin alveolar membrane. The oxygen concentration in venous blood entering the alveolar capillaries is only about 0.0021 mole/liter. The concentration of inhaled oxygen in the alveoli is about 0.0052 mole/liter. Gas will diffuse from a region of higher concentration to a region of lower concentration. Because of the difference in oxygen concentration between alveolar oxygen and alveolar capillary oxygen, oxygen diffuses from the alveoli, crosses the alveolar membrane, and enters the blood plasma of the capillaries. Upon entering a capillary, the oxygen diffuses through the plasma into the red blood cells where it combines with hemoglobin to produce oxyhemoglobin. Because of the flow (influx) of oxygen from the alveoli into the blood, arterial blood leaving the lungs has had its oxygen concentration increased to about 0.0052 mole/liter (see Table 11-1 for the composition of respiratory gases). At this concentration, oxygen readily combines with hemoglobin.

A normal person's blood contains 15 g of hemoglobin in each 100 ml of blood. Each gram of hemoglobin can bind a *maximum* of about 1.34 ml of oxygen and thus the maximum amount of oxygen per 100 ml of blood can be calculated.

$$\frac{\cancel{15\text{ g hemoglobin}}}{100\text{ ml blood}} \times \frac{1.34\text{ ml oxygen}}{\cancel{1\text{ g hemoglobin}}} = \frac{20\text{ ml oxygen}}{100\text{ ml blood}} = 20\text{ volumes percent}$$

The 20 volumes percent represents the maximum amount of oxygen that 100 ml of blood can carry when saturated with oxygen. Under these conditions, the maximum amount of oxygen is combined with hemoglobin as oxyhemoglobin. Under conditions existing in circulating blood, hemoglobin is not completely saturated with oxygen because some of the oxygen dissolves in the plasma rather than uniting with hemoglobin. At an oxygen

TABLE 11-1
Composition of the respiratory gases

	INSPIRED AIR (moles/liter)	ALVEOLAR AIR (moles/liter)	EXPIRED AIR (moles/liter)
N_2	0.031	0.029	0.029
O_2	0.0082	0.0052	0.0061
CO_2	0.00016	0.0021	0.0015
H_2O	0.00026	0.0619	0.0619

concentration of 0.0052 mole/liter, however, 97 percent of the hemoglobin present is combined with oxygen. The hemoglobin saturation is described as being 97 percent.

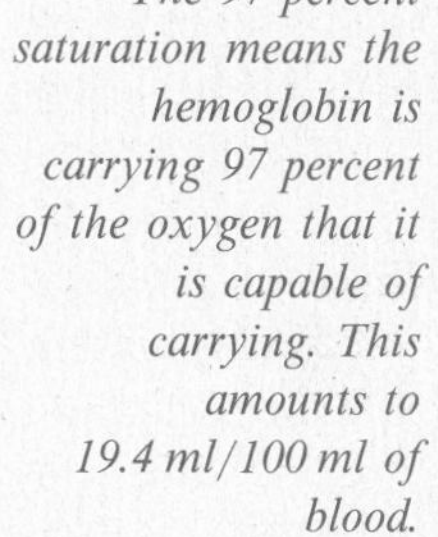
The 97 percent saturation means the hemoglobin is carrying 97 percent of the oxygen that it is capable of carrying. This amounts to 19.4 ml/100 ml of blood.

Figure 11-12 represents the relative amounts of oxyhemoglobin versus oxygen concentration. The left vertical axis refers to the volumes percent of oxygen incorporated as oxyhemoglobin. The right vertical axis relates the percentage of hemoglobin saturated with oxygen.

Figure 11-12 indicates that as oxygen concentration increases from 0.0010 to 0.0036 mole/liter, there is a very rapid uptake of oxygen by hemoglobin. At oxygen concentrations near 0.0052 mole/liter, the curve levels off indicating that the hemoglobin is combining with as much oxygen as possible; i.e., saturation occurs. Beyond that, an increase in oxygen concentration has little effect on the amount of oxyhemoglobin produced.

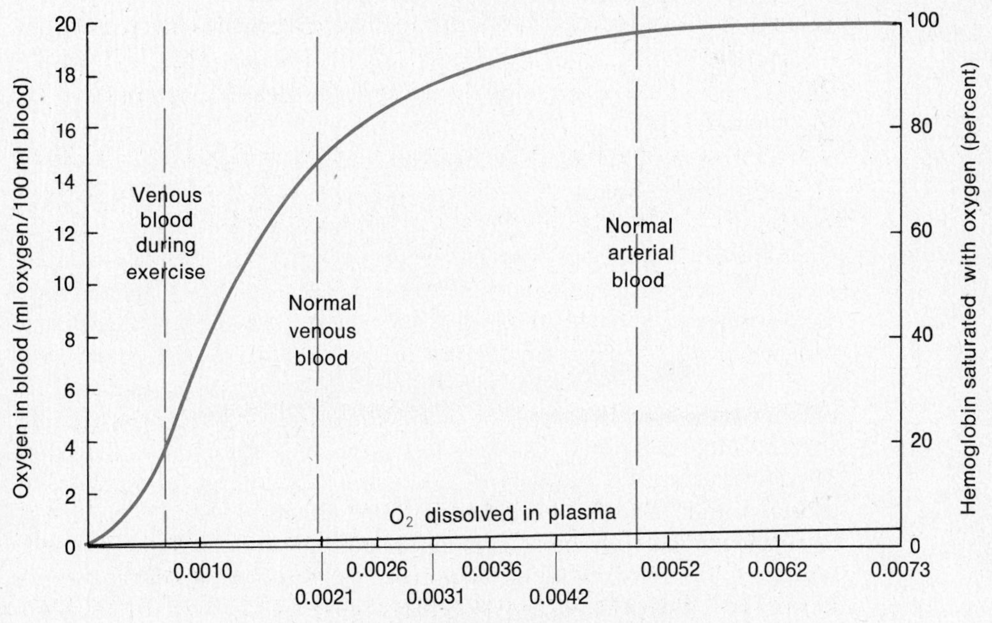

FIGURE 11-12 Oxygen–hemoglobin dissociation curve.

In normal venous blood having an oxygen concentration of 0.0021 mole/liter, note that 75 percent of the hemoglobin is saturated.

How is oxygen released to the tissues for use by cells during metabolism as oxyhemoglobin is transported through arteries into tissue capillaries? In part, it is due to the difference in oxygen concentration that exists between the arterial blood coming into a capillary and that present in the interstitial fluid between a tissue and a capillary. The oxygen concentration in the arterial blood entering a capillary is much greater than that in the interstitial fluid. Oxygen diffuses from the capillary through the interstitial fluid into the tissues. This process is essentially the reverse of that occurring in the lungs. Tissue capillaries usually have an oxygen concentration below 0.0026 mole/liter. The dissociation curve drops off sharply at this concentration, indicating rapid oxyhemoglobin dissociation into oxygen and hemoglobin. The blood's oxygen concentration is reduced from 0.0052 mole/liter of blood found in arterial blood to 0.0021 mole/liter found in venous blood, a loss of about 0.0031 mole/liter of blood.

EFFECTS OF CO_2 ON OXYGEN–HEMOGLOBIN BINDING

Reflecting back to Le Chatelier's principle, we can see that the oxygen–hemoglobin dissociation curve describes a situation in which changes in oxygen concentration are factors creating a stress on the hemoglobin–oxyhemoglobin equilibrium:

$$HHb + O_2 \rightleftharpoons HHbO_2$$

1. At high oxygen concentration, the forward reaction is favored and O_2 uptake occurs.
2. At low oxygen concentration, the reverse reaction is favored and O_2 release occurs.
3. During heavy exercise, O_2 demand is high and rapid oxyhemoglobin dissociation occurs.

The oxyhemoglobin dissociation curve is not linear. This can be interpreted to indicate that changes in oxygen concentration are not the only factors influencing oxyhemoglobin formation and dissociation. Associated with the release of oxygen is the flow of carbon dioxide from tissues into capillaries. Carbon dioxide is a waste product continuously generated by tissues during the breakdown of carbohydrates and fats. Carbon dioxide concentration in arterial blood is less than its concentration in tissues. Therefore, as arterial blood flows into tissue capillaries releasing oxygen, carbon dioxide diffuses from tissues into the blood. Experimentally, it has been shown that oxyhemoglobin dissociation is increased (hemoglobin saturation decreased) by an increase in carbon dioxide concentration in the blood. Figure 11-13 shows the effect that changes in CO_2 concentration has on hemoglobin saturation. As CO_2 concentration increases, the amount of oxygen combined as oxyhemoglobin decreases.

The complete metabolism of one molecule of glucose produces six molecules of CO_2.

Recall from Chapter 8 the dependence of O_2 binding on pH and how CO_2 concentration effects pH.

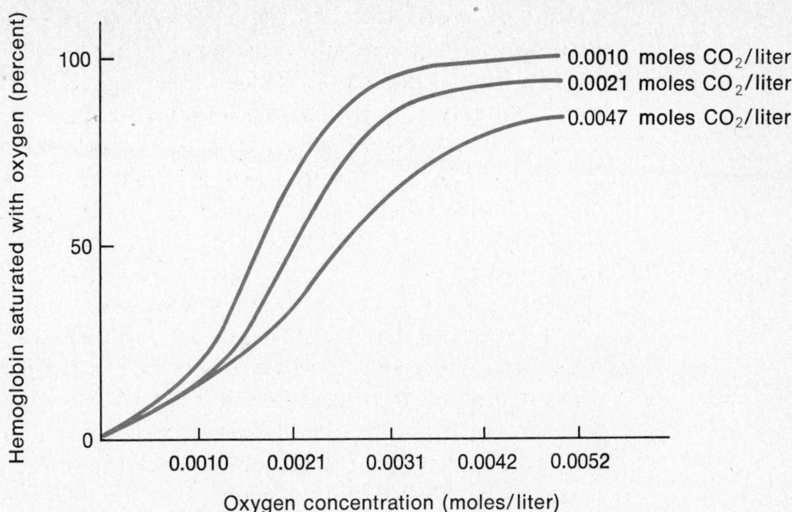

FIGURE 11-13 Effect of CO_2 concentration on oxyhemoglobin dissociation in blood.

REGULATION OF CARBON DIOXIDE CONCENTRATION

Carbon dioxide entering blood plasma from tissues suffers several fates. About 7 percent of it dissolves in the plasma and is transported to the lungs.

Globin contains amino acid groups that have side-chain aliphatic amine groups. A small percentage of the total CO_2 reacts with these aliphatic amine side chains to form carbaminohemoglobin, which dissociates:

$$\mathrm{HHb{-}NH_2 + CO_2 \rightleftharpoons \underset{\text{carbaminohemoglobin}}{HHb{-}\underset{\displaystyle H}{\underset{|}{N}}{-}\overset{\displaystyle O}{\overset{\|}{C}}{-}OH}} \rightleftharpoons \mathrm{H^+ + HHb{-}\underset{\displaystyle H}{\underset{|}{N}}{-}C(=O)O^{(-)}}$$

Carbonic anhydrase, like other enzymes, catalyzes both the forward and reverse reactions.

The majority (about 95 percent) of the total CO_2 in the red blood cells reacts with water to form carbonic acid (H_2CO_3). This reaction is rapidly catalyzed by the specific enzyme carbonic anhydrase. At the normal pH of the blood, carbonic acid molecules dissociate into hydrogen ions and bicarbonate ions. Therefore, the majority of CO_2 in blood exists as bicarbonate ions:

$$H_2O + CO_2 \underset{\longleftarrow}{\xrightarrow[\text{anhydrase}]{\text{carbonic}}} H_2CO_3 \rightleftharpoons H^+ + HCO_3^-$$

When combining with oxygen, hemoglobin releases hydrogen ions.

$$HHb + O_2 \rightleftharpoons HHbO_2 \rightleftharpoons H^+ + HbO_2^-$$

The hydrogen ion is lost from the imidazole group of histidine. Prior to oxygenation, this group exists in the form

hemoglobin—(imidazolium ring: N, $N^{\oplus}$ with H, H)

Upon oxygenation, a hydrogen ion is released from this group.

hemoglobin—(imidazolium ring, $N^{\oplus}$ with H, H) $+ O_2 \rightleftharpoons$ oxyhemoglobin—(imidazole ring, N—H) $+ H^+$

Oxyhemoglobin is a stronger acid than hemoglobin. The weaker the acid, the stronger its conjugate base.

In this way, oxyhemoglobin functions as an acid. Both carbaminohemoglobin and carbonic acid formation result in species whose dissociation releases hydrogen ions. This would tend to lower the pH in the red blood cells. The buffering action of hemoglobin offsets this. By releasing its oxygen, oxyhemoglobin is converted into hemoglobin. In addition, hydrogen ions released by carbonic acid and carbaminohemoglobin are picked up by the imidazole group nitrogen. This is the reverse of the reaction described above.

oxyhemoglobin—(imidazole ring, N—H) $+ H^+ \rightleftharpoons$ hemoglobin—(imidazolium ring, $N^{\oplus}$ with H, H) $+ O_2$

Thus, oxygenation of hemoglobin affects the acidity of the solution. Because the reaction is reversible, the acidity of the solution also affects oxygenation of hemoglobin. Figure 11-14 summarizes these relationships.

When the venous blood reaches the alveolar capillaries, the concentration of carbon dioxide is higher in the blood than in the alveoli. Carbon dioxide diffuses from the red blood cells through the plasma and into the alveoli. From there it is expelled from the lungs. The reduction of carbon dioxide concentration causes the carbaminohemoglobin and carbonic acid equilibria both to shift to the left. Simultaneously, oxygen in the alveoli enters the capillaries. Hemoglobin is oxygenated and the process is repeated as the red blood cells circulate.

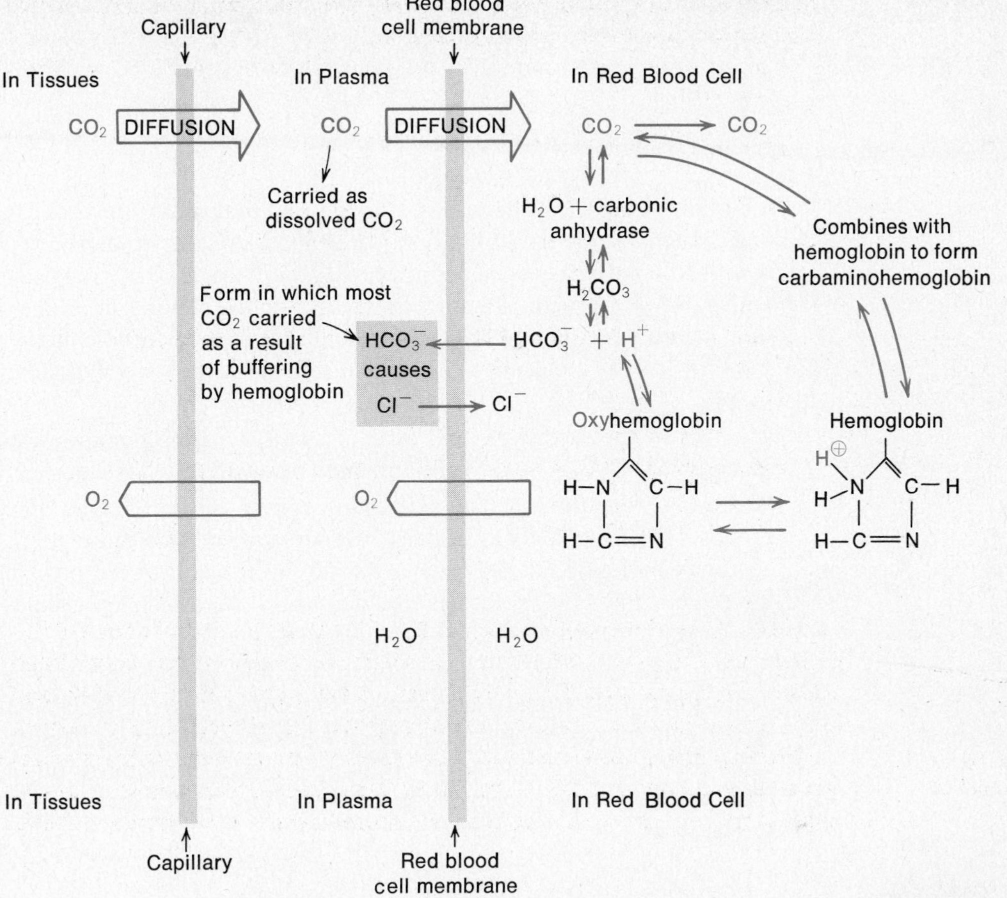

FIGURE 11-14 Overview of changes occurring when CO_2 passes from tissues into plasma and red blood cells. Note release of oxygen due to the change in $[H^+]$.

CHAPTER SUMMARY

1. Functional groups studied in this chapter are

GROUP	TYPE	GROUP	TYPE
(a) $R—NH_2$	Primary amine	(d) $(R_3—NH)^+(RCO_2)^-$	Amine salt
(b) $R_2—NH$	Secondary amine	(e) $R—\underset{\underset{O}{\Vert}}{C}—NR_2(H_2)$	Amide
(c) $R_3—N$	Tertiary amine		

2. Primary and secondary amines react with carboxylic acids to form amides. Enzymatic or acid-catalyzed hydrolysis of amides reforms the parent amine and acid.

3. There are approximately 20 naturally occurring amino acids. Reaction between an amino group of one amino acid molecule with the acid group of another amino acid molecule forms a peptide bond. Many amino acid units can be joined by peptide bonds $\left(\begin{array}{c}\text{O}\\ \|\\ \text{—C—N—}\\ \quad\ \ |\end{array}\right)$ to form a protein. A protein's chemical properties and biological functions are determined by (a) its amino acid sequence (primary structure); (b) hydrogen bonding between peptide bond groups (secondary structure); and (c) tertiary structural features such as disulfide bonds, salt bridges, side-chain hydrogen bonding, and London interactions. These factors cause a protein molecule to be folded and twisted into a particular shape.
4. Changes in the secondary and tertiary structure cause denaturation of a protein. Hydrolysis of peptide bonds results in protein digestion.
5. Because of the presence of side-chain amine and carboxylic acid groups, proteins can act as buffers. At a certain pH (isoelectric point) a protein molecule has no *net* charge and attraction between protein molecules causes the molecules to coagulate and the protein becomes insoluble.
6. Hemoglobin takes up oxygen, binding it to the central Fe^{2+} ion of heme. Loss of H^+ from hemoglobin accompanies oxygen binding. An increase in hydrogen ion concentration (such as when tissues liberate CO_2) causes oxyhemoglobin to release oxygen. Blood buffers offset changes in hydrogen ion concentration during oxygen uptake and release.

QUESTIONS

1. Identify each of the following amines as a primary, secondary, or tertiary amine:

   (a) $CH_3—\underset{\displaystyle H}{\underset{|}{N}}—H$

   (b) 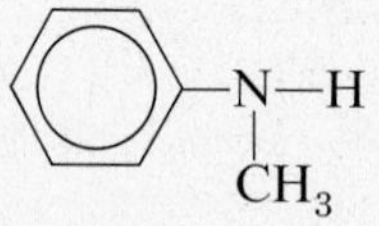 $C_6H_5—\underset{\displaystyle CH_3}{\underset{|}{N}}—H$

   (c) $CH_3CH_2—NH_2$

   (d) $(CH_3)_2N—CH_2CH_3$

   (e) $CH_3\underset{\displaystyle H}{\underset{|}{N}}CH_2CH_3$

   (f) 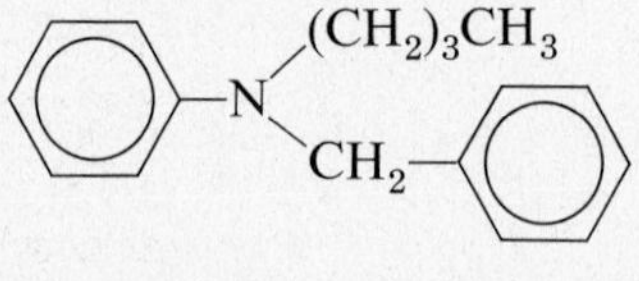 $C_6H_5—N(CH_2)_3CH_3(CH_2—C_6H_5)$

2. For the structural formulas given below, name the functional group indicated by the respective numbered arrow.

(a)

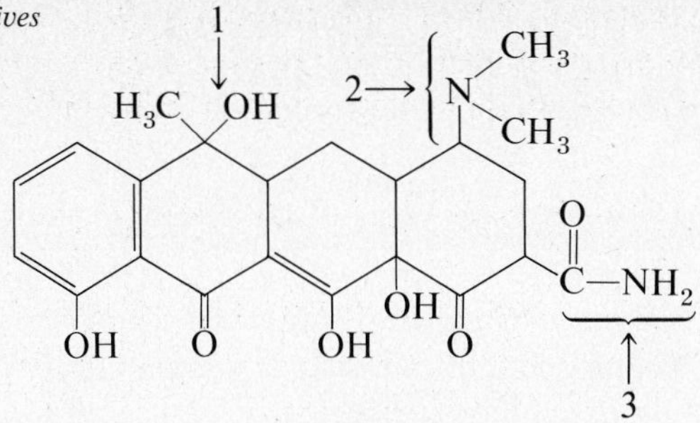

tetracycline
(an antibiotic)

(c)

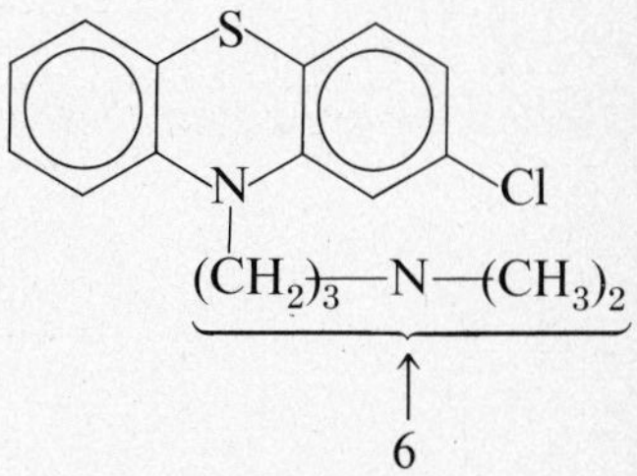

chlorpromazine (Thorazine®*, a tranquilizer)

(b) $H_2N-\overset{O}{\overset{\|}{C}}-O-CH_2-\underset{C_3H_7}{\overset{CH_3}{C}}-CH_2-O-\overset{O}{\overset{\|}{C}}-NH_2$

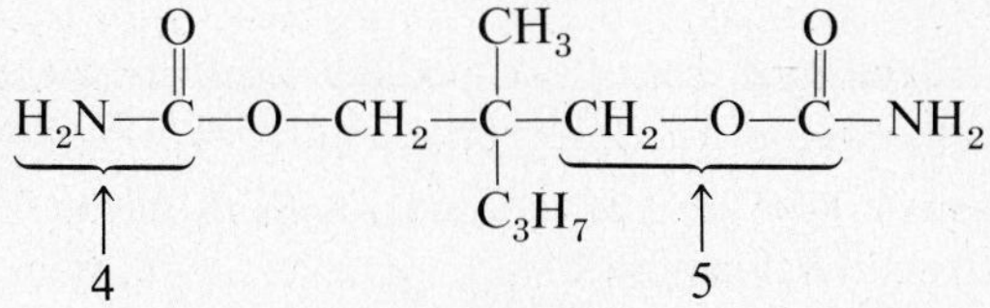

meprobamate (Equanil or Miltown®*, a tranquilizer)

(d)

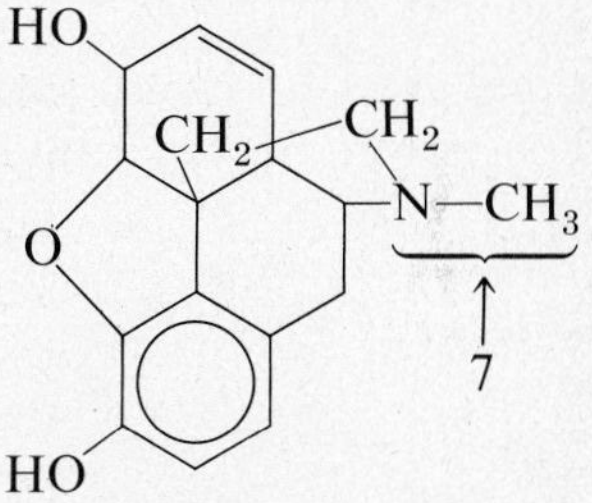

morphine (a narcotic)

3. An aqueous solution of diethylamine $[(CH_3CH_2)_2NH]$ has a pH greater than 7.0. Write an equation that shows how this amine can react with water to produce the indicated effect.
4. Phenacetin

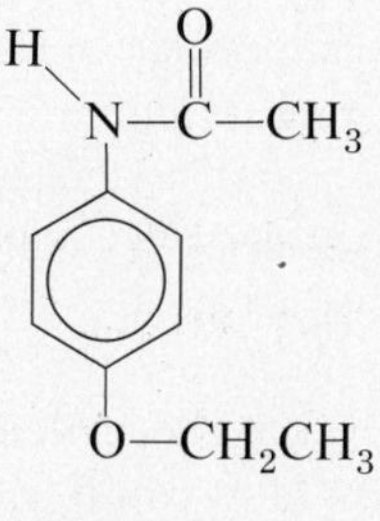

is an analgesic and antipyretic.
This compound hydrolyzes. Using structural formulas, write an equation representing its hydrolysis.
5. Can $CH_3N(CH_2CH_3)_2$ react with CH_3CO_2H to produce an amide? Explain your answer.
6. Can $CH_3CH_2N(CH_3)_2$ react with CH_3CO_2H to produce a salt? Explain your answer.

*® Miltown—registered trademark of Wallace Laboratories. Thorazine—registered trademark of Smith Kline & French Laboratories.

7. Hippuric acid is present in small amounts in normal human urine. The liver is the principal site for hippuric acid formation. Hippuric acid is formed by peptide bond formation between benzoic acid $\left(C_6H_5\text{—}CO_2H\right)$ and glycine. Write the structural formula for hippuric acid (see Chapter 8 for glycine's structural formula).
8. Sphingosine is a component of compounds found in nerve tissues. The structural formula of sphingosine is

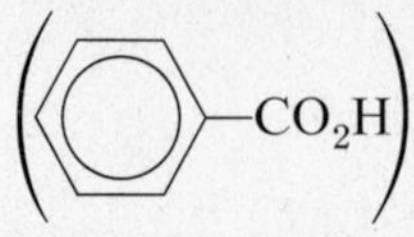

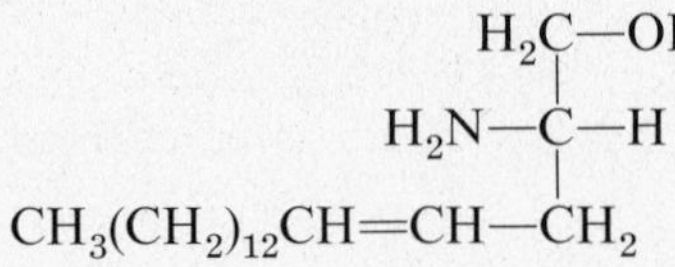

$$\begin{array}{r} H_2C\text{—}OH \\ | \\ H_2N\text{—}C\text{—}H \\ | \\ CH_3(CH_2)_{12}CH{=}CH\text{—}CH_2 \end{array}$$

Write the structural formulas of two possible organic products formed by reacting sphingosine with propionic acid ($CH_3CH_2CO_2H$).
9. Sphingomyelins, a type of sphingolipid found in nerve tissues, have the general structural formula

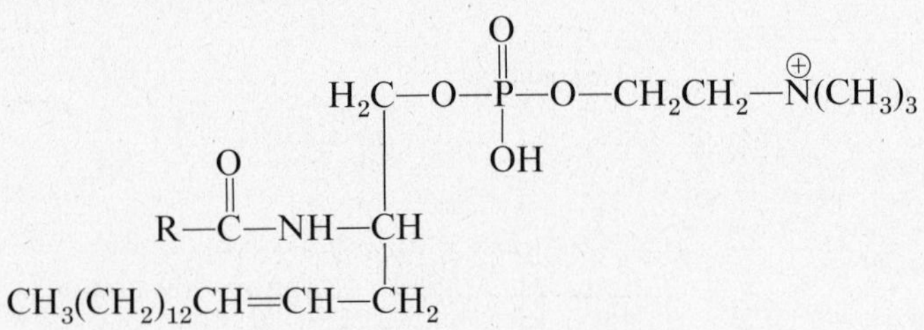

$$\begin{array}{l} \qquad\qquad\qquad\quad O \qquad\qquad\qquad\quad \oplus \\ \qquad\qquad\qquad\quad \| \\ \qquad\qquad H_2C\text{—}O\text{—}P\text{—}O\text{—}CH_2CH_2\text{—}N(CH_3)_3 \\ \quad O \qquad\quad | \qquad\quad | \\ \quad \| \qquad\quad | \qquad\ OH \\ R\text{—}C\text{—}NH\text{—}CH \\ \qquad\qquad\qquad | \\ CH_3(CH_2)_{12}CH{=}CH\text{—}CH_2 \end{array}$$

Using structural formulas, write a balanced equation for the hydrolysis of the sphingolipid shown above.
10. A hexapeptide contains Phe–Asp–Ser–Lys–Glu–Cys. Phenylalanine is the free amino end (N-terminal) amino acid.

Phe = phenylalanine	Lys = lysine
Asp = aspartic acid	Glu = glutamic acid
Ser = serine	Cys = cystine

Use Table 8-8 for the structural formulas of the amino acids.
(a) Write the structural formula of the hexapeptide.
(b) Where could a salt bridge be formed in the molecule?
(c) What part of the molecule would be effected by mild reducing conditions?
(d) How many water molecules are required for the hydrolysis of one molecule of the hexapeptide?
11. An octapeptide yields the following fragments upon hydrolysis: V–A–P, L–T–I, G–V, and P–H–L. V represents valine; A, alanine; P, phenylalanine; L, leucine; T, tyrosine; I, isoleucine; G, glycine; and H, histidine. What is the amino acid sequence in the original octapeptide?
12. Insulin is a protein. Discuss why it cannot be successfully administered orally.

13. Match the statements in Column A with their appropriate choice in Column B. A choice from Column B may be used for more than one answer but each item in Column A has only one correct answer.

Column B is continued on page 312.

COLUMN A	COLUMN B
(a) Product of a dehydration reaction between two alcohol molecules.	(a) A reduction reaction
(b) A compound that reacts with H_2 and Ni under appropriate conditions to form an alkane.	(b) $(CH_3)(CH_3CH_2)N{-}H$
(c) The reaction product of $CH_3{-}CH_2C({=}O)H$ and $HOC(H)(CH_3)_2$	(c) $CH_3CH_2{-}N(CH_3)_2$
(d) Its reaction with an alcohol yields an acetal.	(d) $H_2N{-}C({=}O){-}(CH_2)CH_3$
(e) An amine that won't form an amide.	(e) An oxidation reaction
(f) Its reaction with $KMnO_4$ yields an organic acid.	(f) $C_6H_5{-}CH_2{-}C(H)(NH_2){-}C({=}O)OH$
(g) An amino acid.	(g) decalin ring with CH_3, OH and CH_2CH_3 substituents
(h) A compound capable of forming more than one ester linkage when reacted with an appropriate alcohol.	(h) $CH_3{-}C({=}O){-}O{-}C({=}O){-}C_6H_5$
(i) The product of the reaction of H_2O with [bicyclic ring (octahydronaphthalene) with CH_3 and CH_2CH_3 substituents and a ring double bond]	(i) $CH_3CH_2{-}O{-}C_6H_{11}$ (cyclohexyl)
(j) $HO(O{=})C{-}CH_2{-}CH_2{-}C({=}O)OH \longrightarrow HO(O{=})C{-}CH{=}CH{-}C({=}O)OH + 2H$	(j) $CH_3CH_2{-}C(OH)(H){-}O{-}C(CH_3)(H){-}CH_3$

COLUMN B (*cont.*)

(k) $CH_3CH_2-C(=O)-O-CH(CH_3)-CH_3$

(l) 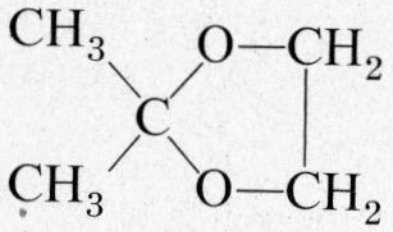

(m) $CH_3CH_2CH_2OH$

(n) $CH_3-CH(OH)-CH_3$

(o) $HO-C(=O)-CH_2CH_2CH_2C(=O)-OH$

(p) $H-CH_2-C(=O)-CH_2-H$

(q) 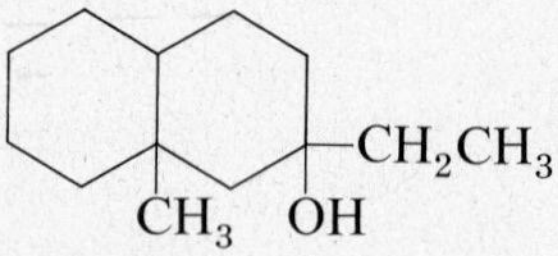

(r) 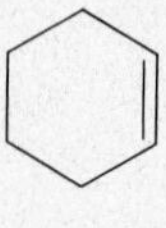

12
energy and biochemical reactions

INTRODUCTION In this chapter, changes of energy and structure that occur during important biochemical processes are described. Many of the substances involved in these changes have complex structures. It is our intent that you *do not* attempt to memorize these structures or reactions. Instead, we want you to concentrate on looking for the functional group(s) in these substances and relate the reactions of these functional groups with the general type of reactions studied in Chapters 8, 9, 10, and 11. Among these reactions are hydrolysis, ester formation, oxidation, reduction, and double bond formation.

THE TRANSFER OF ENERGY For living things to have a continuing existence, a utilizable energy source must be available. Not just any source of energy will do. Only a source that makes its energy available in useful form will suffice. The sun's radiant energy is the source for the sustainment of life as we know it on this planet. Green plants absorb radiant energy and store it in chemical bonds during the process of photosynthesis. The overall equation representing this endergonic process is

$$\text{radiant energy} + 6CO_2 + 6H_2O \xrightarrow{\text{chlorophyll}} \underset{\text{glucose}}{C_6H_{12}O_6} + 6O_2$$

The equation above indicates that in the conversion of simple molecules like water and carbon dioxide, bonds are broken and the atoms are reorganized to synthesize molecules of the more complex substance, glucose. Glucose is a higher energy compound than carbon dioxide or water. The energy absorbed during this reorganization process becomes trapped in the bonding arrangement of atoms in glucose molecules.

The metabolism of glucose to H_2O and CO_2 is called respiration.

oxidation process

Essentially all animals rely on this trapped energy as a utilizable energy source. Green plants photosynthesize glucose; cows eat green plants; in turn, we eat the meat from cows to furnish us with a source of energy (among other things). By converting glucose back to carbon dioxide and water, we extract energy from glucose molecules. This conversion of glucose is a highly exergonic reaction yielding 686 kcal/mole of glucose:

$$C_6H_{12}O_6 + 6O_2 \longrightarrow 6CO_2 + 6H_2O + 686\ \text{kcal/mole glucose}$$

The overall effect is an energy transfer from the sun to green plants to us. Figure 12-1 shows this diagrammatically.

There is a problem with this simplified picture that we have been drawing of these processes. Experimental evidence indicates that photosynthesis occurs as a stepwise process requiring many interrelated steps to convert CO_2 and H_2O to oxygen and glucose. The equation for photosynthesis merely summarizes the results of the overall process. It does **not** imply that carbon dioxide and water molecules bang together and out pops glucose and oxygen. Respiration is also a stepwise oxidation process requiring many carefully controlled and coordinated molecular events for its completion.

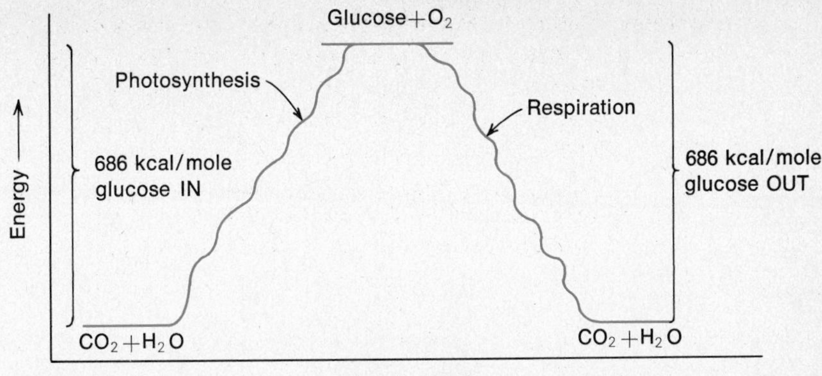

FIGURE 12-1 Relationship of photosynthesis to respiration.

Enzymes control each of the substep reactions during the oxidation. Carbon dioxide and water are only waste products. The actual benefit to the body during respiration is not these waste products but rather the energy produced during their formation. The energy released during respiration is used by the body in many ways: some to maintain a relatively constant body temperature and much is stored for later use by energy-consuming processes, such as muscle action, nerve action, cellular growth, and repair.

In a coupled reaction, an exergonic reaction supplies energy to an endergonic reaction.

In general, the synthesis of cellular components is endergonic. How is energy made available so that these reactions can occur? The key step here is to couple the endergonic reaction with an exergonic reaction. The net energy released during the exergonic reaction supplies the net energy required to carry out the endergonic change. To be coupled successfully, the net energy evolved by the exergonic change must be greater than the energy required for the endergonic reaction. The overall effect is a release of energy. In relation to this, consider the following analogy. A flashlight bulb does not emit light by itself because the metal filament in the bulb cannot give off light by itself; that is, the lighting of the bulb requires energy, an endergonic condition. A flashlight battery acts as a source of energy because of exergonic chemical reactions taking place among the battery's components. If an operating flashlight battery is properly connected to a flashlight bulb, energy released by the battery is sufficient to heat the filament in the bulb to incandescence and light is emitted. The energy released by the net exergonic battery reaction is sufficient to power the endergonic change.

ATP, HYDROLYSIS, AND ENERGY

In a cell, endergonic processes are coupled with exergonic ones so as to allow the endergonic ones to occur. In all living systems, the principal substance used for the exchange of intracellular energy is adenosine triphosphate (ATP), present in cellular fluid at a concentration of 0.5–2.5 mg/ml. Adenosine triphosphate is a substance composed of three main components: (1)

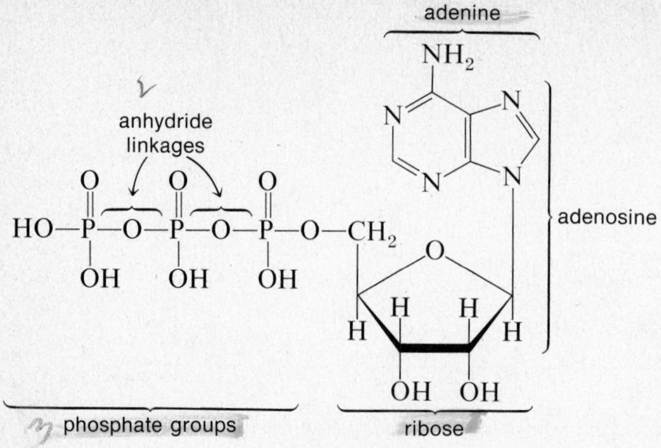

FIGURE 12-2 Adenosine triphosphate, ATP.

adenine; (2) the monosaccharide, ribose; and (3) phosphate groups. The structural formula for ATP is shown in Fig. 12-2.

The phosphate groups are joined by the anhydride linkage, $-\overset{\displaystyle O}{\overset{\|}{P}}-O-\overset{\displaystyle O}{\overset{\|}{P}}-$. Recall from Chapter 10 that anhydride bond formation is endergonic and energy is stored in anhydride bonds. ATP is frequently described as an *energy-rich compound* with bonds called high-energy bonds. *A high-energy bond is one that liberates a large amount of energy when hydrolyzed* (>5 *kcal/mole*). Two such high-energy groups exist in the ATP anhydride linkage. Let's use a block diagram of ATP so that we can focus our attention on its high-energy groups.

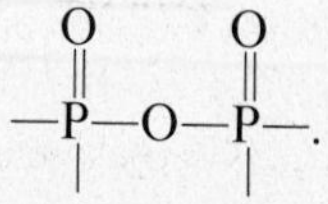

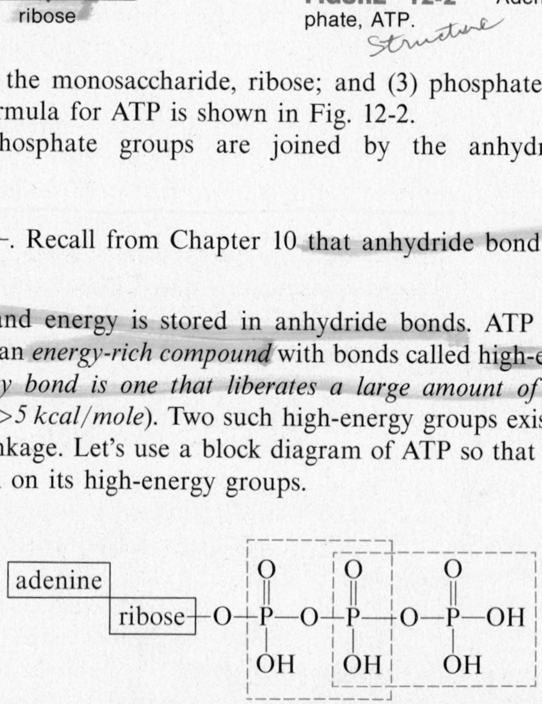

Hydrolysis of phosphate anhydride bonds liberates energy.

Also from Chapter 10, recall that anhydrides can be hydrolyzed. Upon hydrolysis, 1 mole of ATP liberates about 8000 cal of energy available for useful work under the operating conditions (pH, Mg^{2+} concentration, temperature) of a cell.*

*It is extremely difficult to determine the energy of ATP hydrolysis accurately in an intact cell. By taking into account factors such as the pH and temperature at which the cell usually operates, however, chemists have calculated the hydrolysis energy to be normally about 8000 cal released per mole of ATP under these conditions. It is possible that changes in conditions can cause the hydrolysis to release as much as 12,000 cal/mole of ATP.

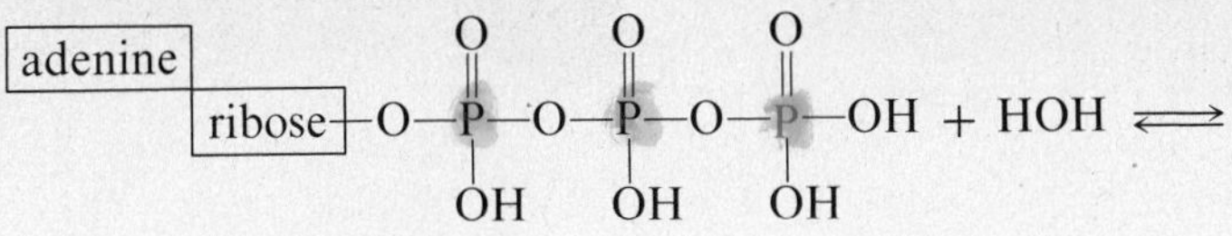

$$\text{adenine-ribose}-O-\overset{O}{\overset{\|}{\underset{OH}{\underset{|}{P}}}}-O-\overset{O}{\overset{\|}{\underset{OH}{\underset{|}{P}}}}-O-\overset{O}{\overset{\|}{\underset{OH}{\underset{|}{P}}}}-OH + HOH \rightleftharpoons$$

adenosine triphosphate (ATP) + water

$$\text{adenine-ribose}-O-\overset{O}{\overset{\|}{\underset{OH}{\underset{|}{P}}}}-O-\overset{O}{\overset{\|}{\underset{OH}{\underset{|}{P}}}}-OH + HO-\overset{O}{\overset{\|}{\underset{OH}{\underset{|}{P}}}}-OH + \text{energy (8000 cal/mole)}$$

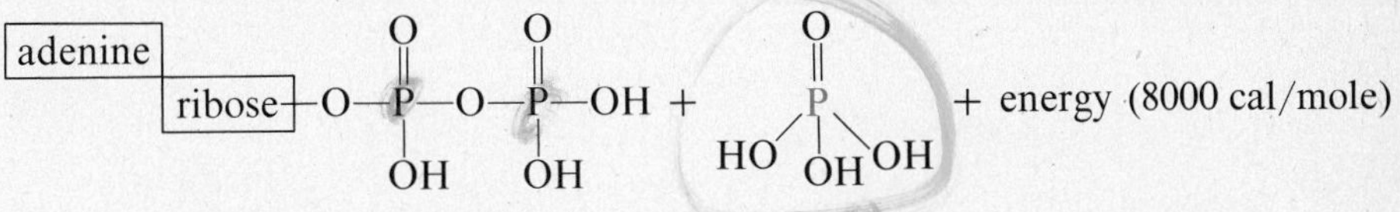

adenosine diphosphate (ADP) + P_i + energy

Shown in the equation above is phosphoric acid (H_3PO_4), which dissociates:

$$H_3PO_4 \rightleftharpoons H^+ + H_2PO_4^-$$
$$H_2PO_4^- \rightleftharpoons H^+ + HPO_4^{2-}$$

Because these equilibria are responsive to changes in hydrogen-ion concentration, the particular form of inorganic phosphate present, i.e., $H_2PO_4^-$ or HPO_4^{2-}, depends on the pH of the cellular environment. P_i is a general notation for both of these types of inorganic phosphate.

exercise 12-1

Which of the three forms, HPO_4^{2-}, $H_2PO_4^-$, or H_3PO_4, would least likely be present at a low pH?

Inorganic phosphate is considered to be all forms of phosphate that are not combined with a carbon compound.

ATP hydrolysis also occurs with the loss of a pyrophosphate group (PP_i)

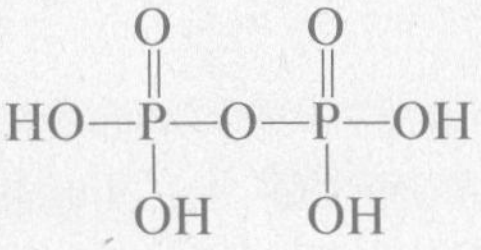

$$HO-\overset{O}{\overset{\|}{\underset{OH}{\underset{|}{P}}}}-O-\overset{O}{\overset{\|}{\underset{OH}{\underset{|}{P}}}}-OH$$

This reaction is also highly exergonic.

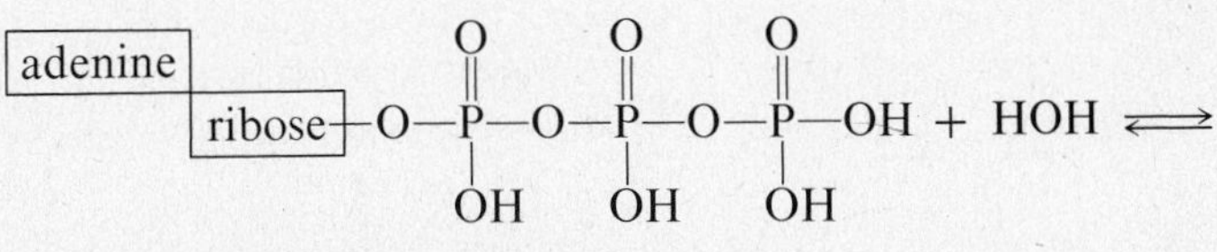

$$\text{adenine-ribose}-O-\overset{O}{\overset{\|}{\underset{OH}{\underset{|}{P}}}}-O-\overset{O}{\overset{\|}{\underset{OH}{\underset{|}{P}}}}-O-\overset{O}{\overset{\|}{\underset{OH}{\underset{|}{P}}}}-OH + HOH \rightleftharpoons$$

adenosine triphosphate (ATP) + water

$$\boxed{\text{adenine}}\boxed{\text{ribose}}-O-\overset{\overset{\large O}{\|}}{\underset{\underset{\large OH}{|}}{P}}-OH + HO-\overset{\overset{\large O}{\|}}{\underset{\underset{\large OH}{|}}{P}}-O-\overset{\overset{\large O}{\|}}{\underset{\underset{\large OH}{|}}{P}}-OH + \text{energy}$$

adenosine monophosphate (AMP) + PP_i + energy

PPi is used to represent pyrophosphate:

$$HO-\overset{\overset{\large O}{\|}}{\underset{\underset{\large O(-)}{|}}{P}}-O-\overset{\overset{\large O}{\|}}{\underset{\underset{\large O(-)}{|}}{P}}-OH$$

ADP could be hydrolyzed to AMP and P_i with the release of about 6500 cal/mole; however, this reaction does not occur frequently biochemically.

$$\boxed{\text{adenine}}\boxed{\text{ribose}}-O-\overset{\overset{\large O}{\|}}{\underset{\underset{\large OH}{|}}{P}}-O-\overset{\overset{\large O}{\|}}{\underset{\underset{\large OH}{|}}{P}}-OH + HOH \rightleftharpoons$$

$$\boxed{\text{adenine}}\boxed{\text{ribose}}-O-\overset{\overset{\large O}{\|}}{\underset{\underset{\large OH}{|}}{P}}-OH + P_i + \text{energy (6500 cal/mole)}$$

adenosine diphosphate (ADP) + water $\rightleftharpoons$
adenosine monophosphate (AMP) + P_i + energy

The remaining phosphate group on AMP may be removed by hydrolysis; however, its linkage to adenosine is not a high-energy anhydride bond and only about 2000 cal/mole are liberated during its hydrolysis.

$$\boxed{\text{adenosine}}-O-\overset{\overset{\large O}{\|}}{\underset{\underset{\large OH}{|}}{P}}-OH + HOH \rightleftharpoons \text{adenosine} + P_i + \text{energy (2000 cal/mole)}$$

AMP + water $\rightleftharpoons$ adenosine + P_i + energy

By exergonically hydrolyzing ATP to either ADP or AMP, a cell liberates energy that can be used to drive an endergonic process (see Fig. 12-3).

ATP SYNTHESIS—THE RESPIRATORY CHAIN OF EVENTS

Energy is released upon ATP hydrolysis. This means that the reverse process is endergonic and energy is stored in ATP during its formation (Fig. 12-4).

$$\text{energy} + ADP + P_i \rightleftharpoons ATP + HOH$$
$$\text{energy} + AMP + PP_i \rightleftharpoons ATP + HOH$$

During catabolism, cells rearrange the bonding of metabolite molecules, breaking them down into simpler substances of lower energy. A net

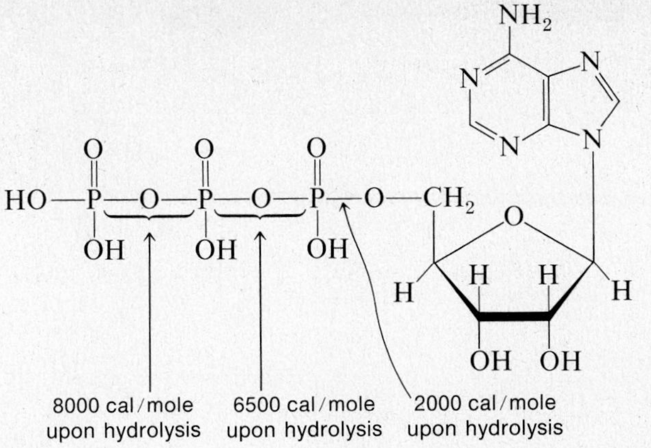

NH2
8000 cal/mole upon hydrolysis
6500 cal/mole upon hydrolysis
2000 cal/mole upon hydrolysis

FIGURE 12-3 Energy obtained by hydrolysis of various bonds in ATP.

ATP+HOH
8000 cal/mole
ADP+Pi
Energy
Reaction path
ATP+HOH
6000 cal/mole
AMP+PPi
Energy
Reaction path

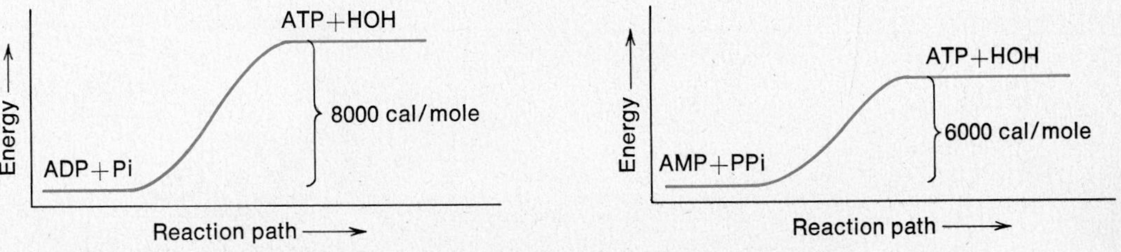

FIGURE 12-4 ATP formation as an endergonic process.

The three names respiratory chain, cytochrome chain, *and* electron transport system *are used for the same sequence of reactions.*

release of energy results and some of this energy is captured via ATP formation. The majority of ATP is produced through the action of the *respiratory chain,* also known as the *cytochrome chain* or the *electron transport system* (ETS). The components of the ETS are found on the inner membrane surface of mitochondria. Mitochondria are rod-shaped structures occurring in most living cells, with their number varying from about 50 to 5000 per cell depending on cell type and function (see Fig. 12-5).

Mitochondria are commonly called the powerhouse *of the cell.*

Even after over 50 years of research, the details about how the electron transport system works have not been completely determined and remains a topic for research today. What has been established is that the enzymes of the ETS function as a sequence in which hydrogens are passed along carriers. Ultimately, the hydrogens combine with oxygen to produce water.

Hydrogen atoms can be imagined to consist of component hydrogen ions and electrons.

$$H\cdot + H\cdot = 2e^- + 2H^+$$

Recall that the loss of hydrogen with its associated electron from a substance is called *oxidation;* gain of hydrogen and its electron is called *reduction.* The formation of water involves the reaction

$\frac{1}{2}O_2$ means $\frac{1}{2}$ mole of diatomic oxygen.

$$2H^+ + 2e^- + \tfrac{1}{2}O_2 \longrightarrow H_2O$$

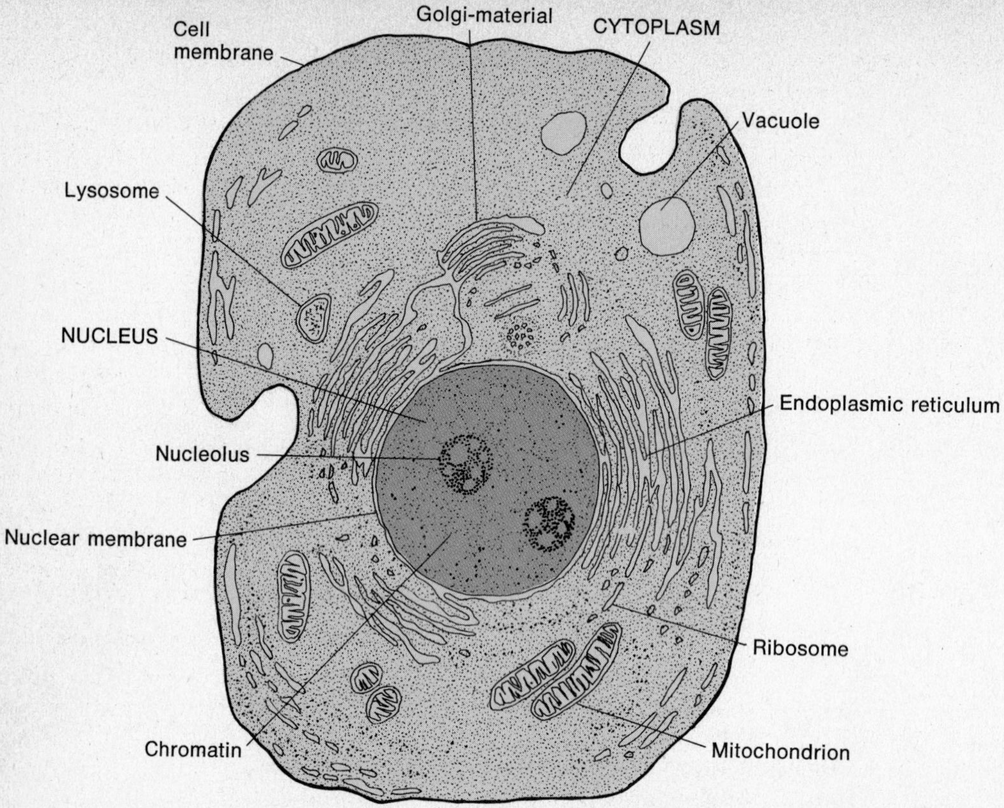

Golgi-material
Cell membrane
CYTOPLASM
Vacuole
Lysosome
NUCLEUS
Endoplasmic reticulum
Nucleolus
Nuclear membrane
Ribosome
Chromatin
Mitochondrion

FIGURE 12-5(a) An idealized cell diagrammatically represented showing all details that can be observed under very high magnification.

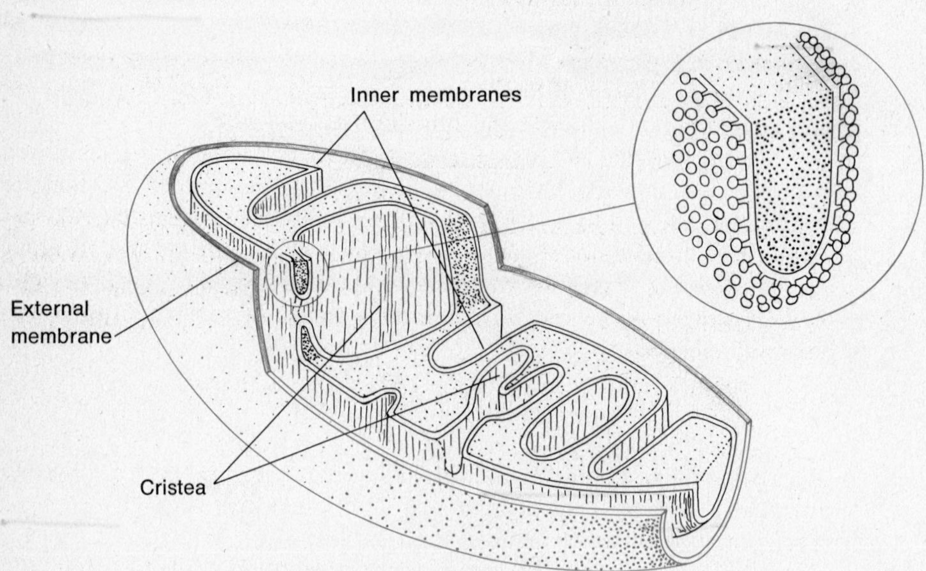

FIGURE 12-5(b) Diagrammatic representation of the structure of a mitochondrion. The respiratory enzymes are part of the inner membrane.

Therefore, a biological requirement exists for hydrogen ion and electron carriers.

Two kinds of carriers are present in the electron transport system: (1) coenzymes, which are electron and hydrogen ion carriers, and (2) cytochromes, which are exclusively electron carriers. The coenzymes presently known to participate in the ETS are (1) nicotinamide adenine dinucleotides (NAD and NADP), (2) flavin nucleotides (FAD and FMN), and (3) coenzyme Q (CoQ).

NICOTINAMIDE NUCLEOTIDE COENZYMES

The structural formula of NAD is given in Fig. 12-6. Despite the formidable-looking formula, we have already discussed most of its components. Think of the coenzyme as consisting of two parts. The first part is adenine bonded to ribose, which is linked to a phosphate as in AMP. The other part of the coenzyme consists of a phosphate-to-ribose-to-nicotinamide sequence. Nicotinamide is a derivative of the vitamin, nicotinic acid (niacin). Because the body cannot produce nicotinamide, it must be included in the diet. The two parts are joined by a pyrophosphate linkage to form a dinucleotide. The coenzyme is bonded to an apoenzyme to form a total enzyme.

Vitamins or their derivatives are often coenzymes.

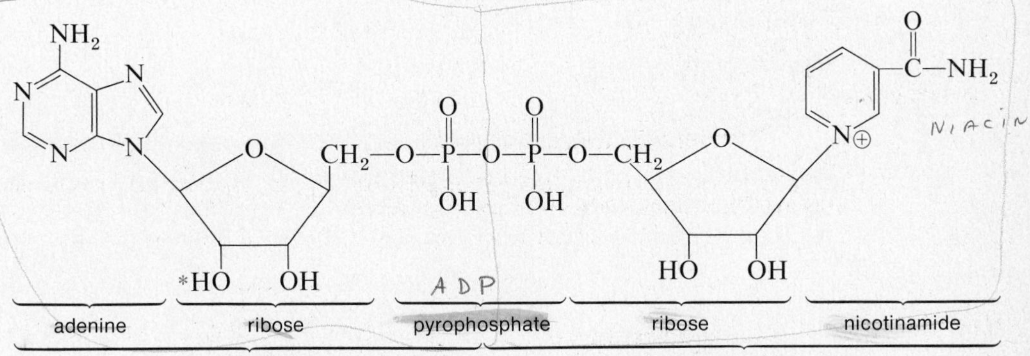

NH_2 N N N N O CH_2—O—P(=O)(OH)—O—P(=O)(OH)—O—CH_2 O N⊕ C(=O)—NH_2

*HO OH HO OH

adenine | ribose | pyrophosphate | ribose | nicotinamide

adenine nucleotide | nicotinamide nucleotide

FIGURE 12-6 Nicotinamide dinucleotide (NAD).

Substitution of a phosphate group for the starred hydrogen in Fig. 12-6 forms a related coenzyme, nicotinamide dinucleotide phosphate (NADP). The acceptance and release of hydrogens occur in the nicotinamide portion of the coenzyme.* R in the formulas at the top of page 322 represents the remaining portions of the coenzyme molecule.

Thus, the coenzyme undergoes two types of reactions:

Reduction of NAD: NAD + $2H\cdot$ $\longrightarrow$ $NADH_2$
($2H^+ + 2e^-$) reduced form

*We have chosen to use the NAD + 2H $\rightleftharpoons$ $NADH_2$ notation. The reaction is more correctly written as $NAD^+ + 2e^- + 2H^+ \rightleftharpoons NADH + H^+$; however, we will use the simpler representation for convenience.

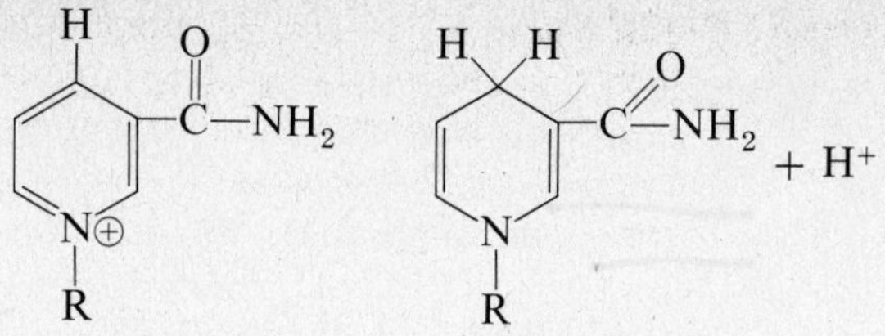

(NAD or NADP) (oxidized form) — NAD⊕ structure: pyridine ring with N⊕–R and –C(=O)–NH$_2$

$NADH_2$ or $NADPH_2$ (reduced form) — pyridine ring with H, H at C4, N–R and –C(=O)–NH$_2$ + H^+

Oxidation of $NADH_2$: $NADH_2 \longrightarrow NAD + 2H\cdot$
oxidized form

Note that these reactions are the reverse of each other.

The hydrogens are acquired from a metabolite MH_2 undergoing oxidation. Since oxidation and reduction occur simultaneously, two half reactions may be written:

Oxidation	$MH_2 \longrightarrow 2H\cdot + M$
Reduction:	$NAD + 2H\cdot \longrightarrow NADH_2$
Net reaction:	$MH_2 + NAD \longrightarrow M + NADH_2$

A curved double arrow representation of these coupled half reactions is commonly used. Note that both arrows go in the *same* direction. The fate of a particular reactant is indicated by following the path of the arrow for that half reaction. For the reactions above, the diagram

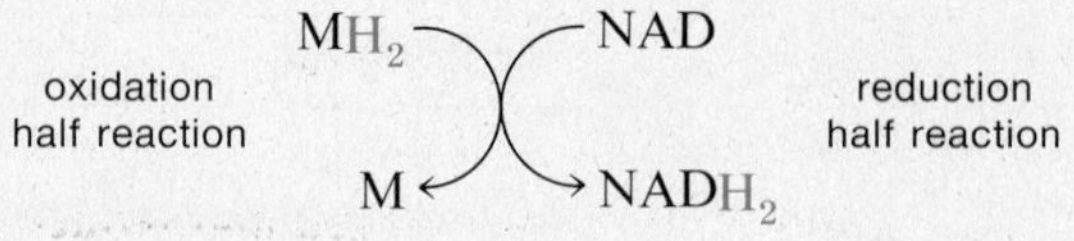

oxidation half reaction: $MH_2 \rightarrow M$ reduction half reaction: $NAD \rightarrow NADH_2$

is interpreted to indicate loss of hydrogen from the metabolite (oxidation) and the simultaneous gain of hydrogen by NAD (reduction).

exercise 12-2

(a) Write individual half reactions for the oxidation of $NADH_2$ by a metabolite, M.

(b) Write the overall reaction using the curved double arrow representation.

NAD and NADP act as coenzymes for dehydrogenases, enzymes that catalyze transfer of hydrogens during oxidation–reduction reactions. NAD and NADP function primarily during reactions involving carbon-to-oxygen double bonds. For example, lactic acid dehydrogenase catalyzes the conversion of lactic acid to pyruvic acid. NAD is required as the coenzyme.

$$CH_3{-}\underset{\underset{H}{|}}{\overset{\overset{OH}{|}}{C}}{-}C\begin{smallmatrix}{/\!/}O\\ \backslash OH\end{smallmatrix} \quad \text{(lactic acid)} \xrightarrow[\text{lactic acid dehydrogenase}]{NAD \rightarrow NADH_2} CH_3{-}\overset{\overset{O}{\|}}{C}{-}C\begin{smallmatrix}{/\!/}O\\ \backslash OH\end{smallmatrix} \quad \text{(pyruvic acid)}$$

lactic acid

lactic acid dehydrogenase

pyruvic acid

exercise 12-3

Describe the reaction above in terms of oxidation and reduction. Indicate which substance has been oxidized and which substance has been reduced and the reasons for your choices.

FLAVIN NUCLEOTIDES

There are two types of flavin nucleotides that take part in the respiratory chain: flavin mononucleotide (FMN) and flavin adenine dinucleotide (FAD). FMN consists of phosphate, ribitol, and flavin bonded together as shown in Fig. 12-7(a). The combination of ribitol and flavin forms riboflavin (vitamin B_2). FAD is the combination of FMN and AMP joined by an anhydride linkage.

exercise 12-4

In Figure 12-7(b), identify the FMN, AMP, and anhydride linkage portions in FAD.

Hydrogens are picked up and released by the flavin portion of the molecule to form oxidized and reduced forms of the coenzymes:

$$\underset{\text{oxidized form}}{FMN} \underset{-2H\cdot}{\overset{+2H\cdot}{\rightleftharpoons}} \underset{\text{reduced form}}{FMNH_2} \qquad \underset{\text{oxidized form}}{FAD} \underset{-2H\cdot}{\overset{+2H\cdot}{\rightleftharpoons}} \underset{\text{reduced form}}{FADH_2}$$

In the electron transport system, FMN removes hydrogens from $NADH_2$. This makes NAD available to oxidize additional metabolites.

$$MH_2 \rightarrow M \qquad NAD \rightarrow NADH_2 \qquad NADH_2 \rightarrow NAD \qquad FMN \rightarrow FMNH_2$$

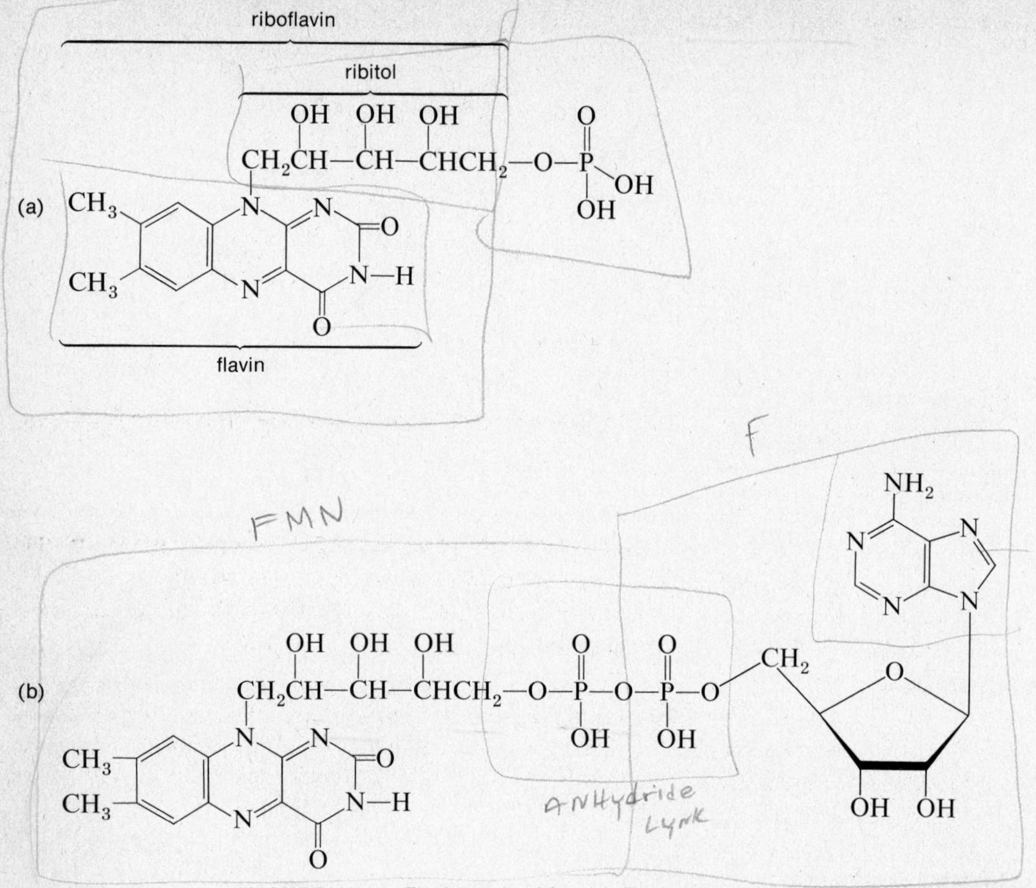

(a) riboflavin

ribitol

flavin

(b)

FIGURE 12-7 Flavin nucleotides. (a) Flavin mononucleotide, FMN. (b) Flavin adenine dinucleotide, FAD.

Contrast the cell's use of FAD as a hydrogen atom acceptor to the cell's use of NAD.

A metabolite may react directly with FAD. The coenzyme is used primarily during reactions in which hydrogens are removed from adjacent carbon atoms during conversion of carbon–carbon single bonds to double bonds.

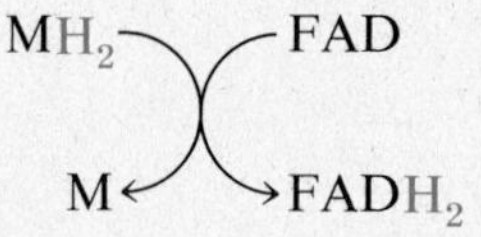

$$MH_2 + FAD \longrightarrow M + FADH_2$$

During carbohydrate and lipid metabolism, the conversion of succinic acid to fumaric acid involves such a change. The reaction is catalyzed by succinic dehydrogenase through the action of its coenzyme, FAD. In this case, enzyme action directs the formation of the *trans* isomer exclusively.

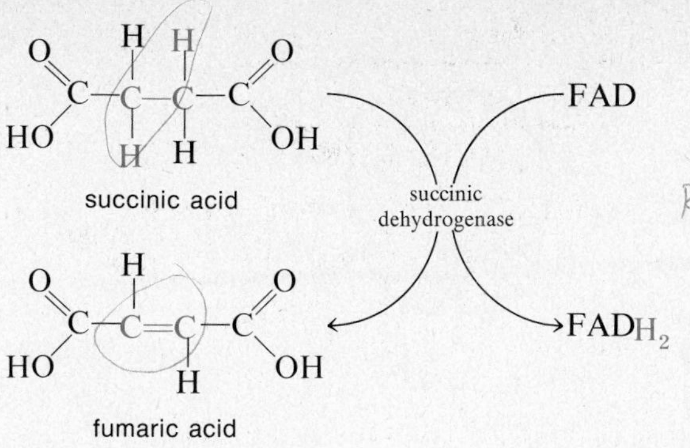

$$\text{HOOC-CH}_2\text{-CH}_2\text{-COOH (succinic acid)} + \text{FAD} \xrightarrow{\text{succinic dehydrogenase}} \text{HOOC-CH=CH-COOH (fumaric acid)} + \text{FADH}_2$$

succinic acid

succinic dehydrogenase

fumaric acid

exercise 12-5

Describe the conversion of succinic acid to fumaric acid in terms of oxidation and reduction.

Both FMN and FAD are coenzymes for dehydrogenases. If the metabolite is FAD dependent, the metabolite loses hydrogens to FAD directly. A metabolite which is NAD dependent initially loses hydrogens to NAD which passes them on to FMN.

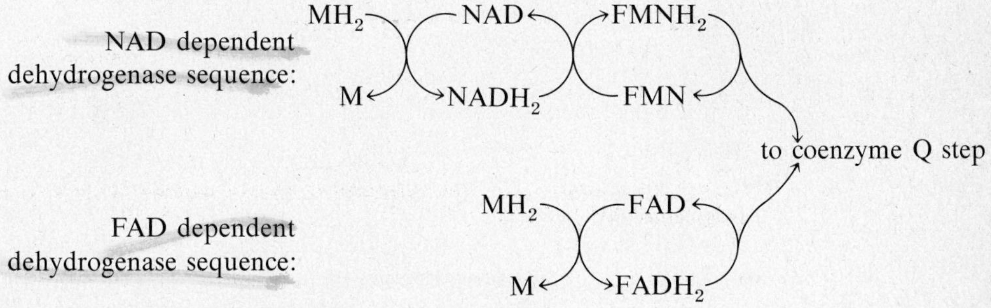

NAD dependent dehydrogenase sequence:

$$MH_2 \rightarrow M;\quad NAD \rightarrow NADH_2;\quad NADH_2 + FMN \rightarrow NAD + FMNH_2 \rightarrow \text{to coenzyme Q step}$$

FAD dependent dehydrogenase sequence:

$$MH_2 \rightarrow M;\quad FAD \rightarrow FADH_2 \rightarrow \text{to coenzyme Q step}$$

In either case, the reduced flavin nucleotide interacts with coenzyme Q, the next carrier in the respiratory chain.

COENZYME Q

Coenzyme Q is a lipid material whose role has only recently been completely established. The coenzyme can exist in oxidized and reduced forms, a role consistent with its activity in an oxidation–reduction sequence.

Reduction: $CoQ + 2H\cdot \longrightarrow CoQH_2$ (reduced form)

Oxidation: $CoQH_2 \longrightarrow 2H\cdot + CoQ$ (oxidized form)

CoQ accepts hydrogens from $FADH_2$ or $FMNH_2$ causing oxidation of the $FADH_2$ and reduction of CoQ.

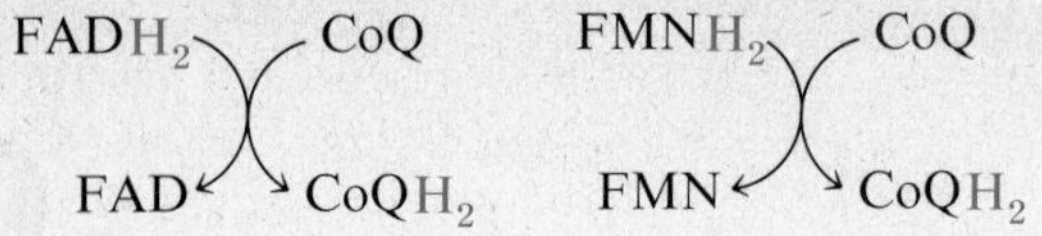

$$FADH_2 + CoQ \rightarrow FAD + CoQH_2 \qquad FMNH_2 + CoQ \rightarrow FMN + CoQH_2$$

exercise 12-6

Complete the following sequence by supplying the appropriate form of the coenzyme in the respective blank:

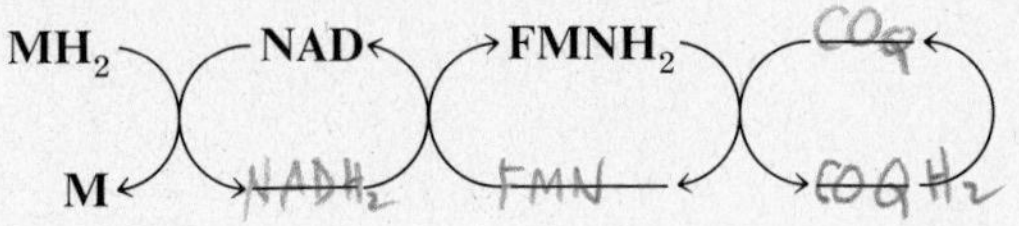

$$MH_2 \rightarrow M \quad NAD \rightarrow \_\_\_ \quad \_\_\_ \rightarrow FMNH_2 \quad \_\_\_ \rightarrow \_\_\_$$

THE CYTOCHROMES

These enzymes were probably the first substances for which experimental evidence indicated involvement in biochemical electron transfer reactions. Sequentially in the ETS these are the cytochromes designated as b, c_1, c, a, and a_3. All cytochromes contain a protein bonded to a coenzyme similar to heme. Although subtle differences exist among the cytochromes due to (1) their protein composition, (2) the bonding between the coenzyme and protein, and (3) variations in side chains attached to the coenzyme, each cytochrome possesses a central iron ion in the coenzyme. Unlike the central iron ion in hemoglobin, *the iron ion in a cytochrome is able to readily undergo reversible oxidation and reduction.*

Oxidation: $\quad Fe^{2+} \longrightarrow e^- + Fe^{3+}$

Reduction: $\quad Fe^{3+} + e^- \longrightarrow Fe^{2+}$

An electron is accepted from one cytochrome molecule and passed on to the next one in the sequence. This is only a one-electron change. Because two electrons are removed from the metabolite MH_2 and passed on from NAD to FMN to CoQ, two cytochrome molecules are required to accept the two electrons from $CoQH_2$.

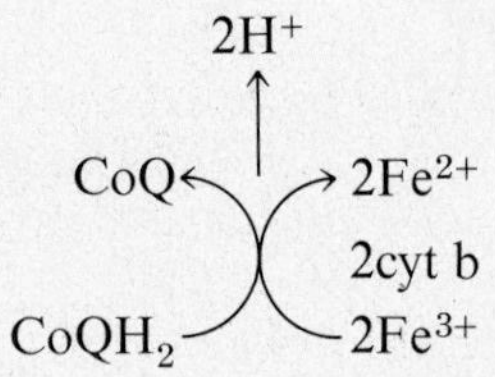

$$CoQH_2 \rightarrow CoQ + 2H^+ \qquad 2cyt\ b:\ 2Fe^{3+} \rightarrow 2Fe^{2+}$$

Oxidation: $\quad CoQH_2 \longrightarrow CoQ + 2H^+ + 2e^-$

Reduction: $\quad 2\text{ cyt b}_{Fe^{3+}} + 2e^- \longrightarrow 2\text{ cyt b}_{Fe^{2+}}$

Net Reaction: $\quad 2\text{ cyt b}_{Fe^{3+}} + CoQH_2 \longrightarrow 2\text{ cyt b}_{Fe^{2+}} + CoQ + 2H^+$

Cytochromes are not hydrogen ion carriers. The hydrogen ions liberated during oxidation of $CoQH_2$ are released into the intracellular environment. They are used later in the final step of the respiratory chain where they combine with electrons and oxygen to form water.

The cytochrome sequence of the electron transport system can be represented as

$2H^+$

$2Fe^{2+}$ — $2Fe^{3+}$ — $2Fe^{2+}$ — $2Fe^{3+}$ — $2Fe^{2+}$ — $\frac{1}{2}O_2$
cyt b — cyt c_1 — cyt c — cyt a — cyt a_3
$2Fe^{3+}$ — $2Fe^{2+}$ — $2Fe^{3+}$ — $2Fe^{2+}$ — $2Fe^{3+}$ — H_2O

cytochrome oxidase

Cytochromes a and a_3 operate as a unit. This unit is the only cytochrome in the ETS that transfers electrons directly to oxygen and is called *cytochrome oxidase*. The vast majority of the oxygen used by cells is involved in this final step of the cytochrome chain.

exercise 12-7

Identify each of the individual reactions below as either an oxidation or reduction reaction:

(a) cyt $c_1Fe^{2+} \longrightarrow$ cyt c_1Fe^{3+}

(b) cyt $bFe^{3+} \longrightarrow$ cyt bFe^{2+}

(c)
cyt cFe^{2+} — cyt aFe^{3+}
cyt cFe^{3+} — cyt aFe^{2+}

(d)
CoQ — $2Fe^{2+}$
cyt b
$CoQH_2$ — $2Fe^{3+}$

(e) $2H^+ + 2e^- + \frac{1}{2}O_2 \longrightarrow H_2O$

The overall sequence for the electron transport system is shown in Fig. 12-8.

OXIDATIVE PHOSPHORYLATION

During the operation of the respiratory chain, metabolites are oxidized, energy is liberated, and water is produced. The driving force for the respiratory chain is the *net* effect of hydrogen reacting with oxygen to produce water. Released energy can be stored in chemical bonds, but bonds have only a certain capacity for picking up energy. If sufficient energy is available, a molecule may acquire an amount of energy it can store. If the amount of energy released is more than the molecule can store, however, the energy

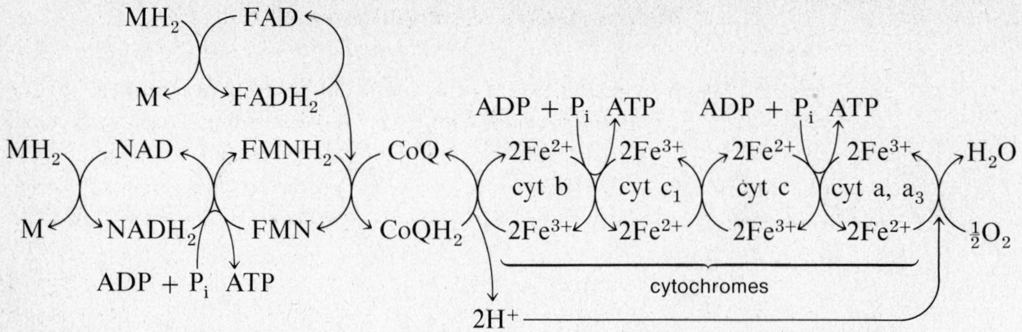

MH_2 FAD
M $FADH_2$
ADP + P_i ATP ADP + P_i ATP
MH_2 NAD $FMNH_2$ CoQ $2Fe^{2+}$ $2Fe^{3+}$ $2Fe^{2+}$ $2Fe^{3+}$ H_2O
cyt b cyt c_1 cyt c cyt a, a_3
M $NADH_2$ FMN $CoQH_2$ $2Fe^{3+}$ $2Fe^{2+}$ $2Fe^{3+}$ $2Fe^{2+}$ $\frac{1}{2}O_2$
ADP + P_i ATP cytochromes
$2H^+$

FIGURE 12-8 The electron transport system (respiratory chain). Note the sites of ATP production.

Notice that oxygen participates only in the final step of the ETS.

not picked up by the molecule is dissipated. Therefore, it is inefficient to release in a single step all the energy extracted from a metabolite during oxidation. A more efficient way is to have the energy released in several steps. A few of the steps can release sufficient energy to form high-energy bonds in the substance picking up the energy. We have already seen that the formation of ATP involves storage of energy in high-energy bonds, about 8000 cal being required to convert a mole of ADP to ATP. The two types of energy-releasing processes are shown diagrammatically in Fig. 12-9.

In the process in Fig. 12-9(a), too much energy is released for molecules to store much of it. The process in Fig. 12-9(b) releases the energy in a series of respiratory chain steps. Since approximately 8 kcal is required per mole of ATP formed from ADP, sufficient energy is released in steps 1, 4, and 6 to generate 3 moles of ATP.

The operation of the respiratory chain efficiently provides the energy for the endergonic conversion of ADP to ATP:

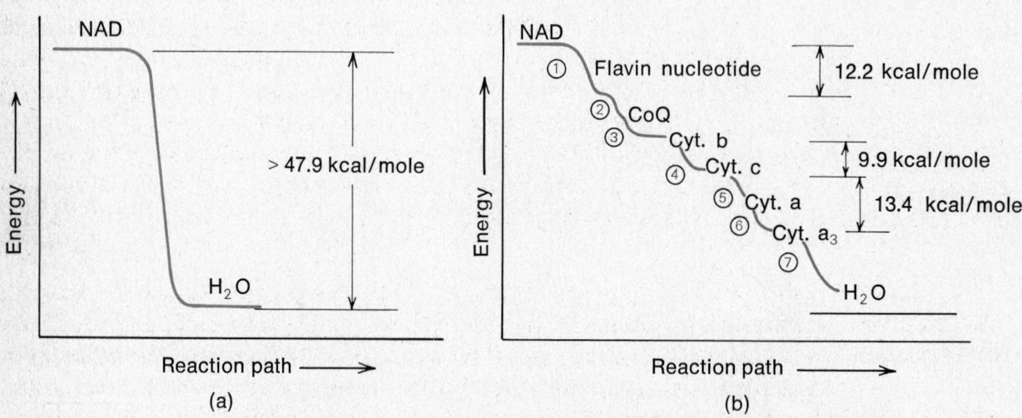

FIGURE 12-9 Energy release during an exergonic process. (a) An inefficient process—too much energy is released in one step. (b) An efficient process—the same amount of energy is released as in (a) but in a series of steps.

$$8000 \text{ cal/mole} + \text{ADP} + \text{P}_i \longrightarrow \text{ATP} + \text{HOH}$$

The coupling of the exergonic respiratory chain processes with the endergonic formation of ATP is called *oxidative phosphorylation*. The way in which this coupling occurs has not been clearly determined. It has been shown experimentally however, that the respiratory chain makes sufficient energy available for oxidative phosphorylation to occur at three places (see Fig. 12-8).

2 ATP/FADH$_2$; 3 ATP/NADH$_2$

Each of the following electron transfers releases sufficient energy to generate ATP: (1) from NAD to CoQ through FMN; (2) from CoQ to cytochrome c_1 through cytochrome b; (3) from cytochrome c to oxygen through cytochrome a, a_3. The passing of 2 moles of electrons from a mole of metabolite through NAD and the remainder of the respiratory chain results in the net production of 3 moles of ATP, one at each of the steps mentioned above. If electrons from the metabolite are passed directly to FAD and then through the rest of the respiratory chain, the initial oxidative phosphorylation step involving NAD is bypassed. In this case, there is a net production of only 2 moles of ATP: one at step 4 and a second at step 6 [see Fig. 12-9(b)].

Catabolism *is the biochemical breakdown of materials;* anabolism *is the synthesis of new materials.*

Thus, during catabolism, energy is extracted from food molecules by chemical action. Energy is then stored in ATP for subsequent release, by hydrolysis, as needed.

TRIGLYCERIDE CATABOLISM—AN OVERVIEW

In the previous sections we have talked in terms of a general metabolite MH_2 undergoing oxidation. We now turn our attention to the catabolism of a particular type of foodstuff, simple lipids (triglycerides). Our discussion will primarily deal with the breakdown of the fatty acids from triglycerides. Hydrogens will be removed from a fatty acid as its catabolism progresses.

When completely catabolized, fatty acids yield carbon dioxide, water, and energy. The oxidation of a fatty acid proceeds through a series of steps in which two-carbon fragments are split off and combine with coenzyme A to produce acetyl coenzyme A. This compound is incorporated into a sequence called the Krebs cycle in which carbon dioxide is given off. Hydrogens from metabolite fragments in the Krebs cycle are passed along the respiratory chain enzymes producing water, energy, and ATP. This series of steps is outlined in Fig. 12-10 on page 330.

LIPID DIGESTION AND ABSORPTION

Bile also serves to neutralize the HCl of the stomach juices.

Lipid digestion begins when the lipids enter the upper region of the small intestine (duodenum). Their presence stimulates the gallbladder to release bile salts into the duodenum. Bile salts emulsify the water-insoluble lipids into small globules (see Chapter 5). The pancreas is also stimulated to release pancreatic juice into the intestine. Pancreatic juice contains pancreatic lipase, an enzyme that catalyzes the hydrolysis of triglycerides into monoglycerides, diglycerides, fatty acids, and glycerol.

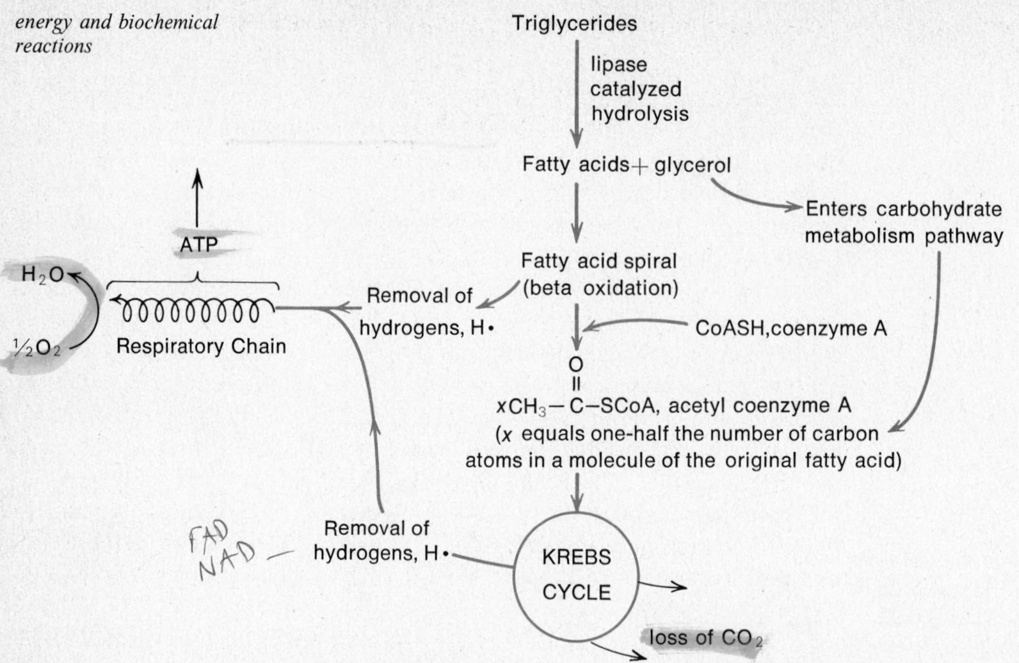

FIGURE 12-10 Metabolic sequence for triglycerides.

The triglycerides combine with the hydrophobic portion of the proteins. The combination occurs via London forces.

Fatty acids are small enough to be absorbed through the intestinal membrane into the lymph circulation.* During absorption into the lymph, fatty acids are combined with glycerol to re-form small droplets of triglycerides. Triglycerides are not water soluble and ordinarily would be insoluble in blood plasma; however, they combine with certain water-soluble blood proteins to form *lipoproteins*. Lipids are transported in this form in the blood. A limited amount of fatty acid catabolism occurs in skeletal muscles and heart muscles. Fatty acid oxidation principally occurs in liver cells because the enzymes necessary for fatty acid oxidation are found in high enough concentration in the mitochondria of these cells.

FATTY ACID SPIRAL—BETA OXIDATION

The fatty acids made available by triglyceride hydrolysis must be activated before oxidation can begin. This should not be unexpected because fatty acids contain long chains of chemically inactive hydrocarbon groups bonded to a carboxylic acid group. A summary of the sequence of steps in the fatty acid spiral is shown in Fig. 12-11. We shall use the oxidation of the 16-carbon acid, palmitic acid, as an example of beta oxidation. The goal is to take

*Lymph is a body fluid similar to plasma in that it contains no red blood cells. Lymph flows through a network of vessels called the *lymphatics*. Although the lymphatics and blood vessels are separate systems, there are certain places in which lymph is discharged into veins. The largest lymph vessel, the thoracic duct, discharges lymph into a vein near the base of the neck.

palmitic acid to the point where it loses two-carbon fragments

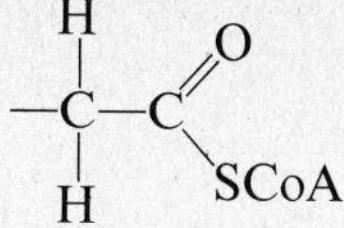

to form acetyl CoA.

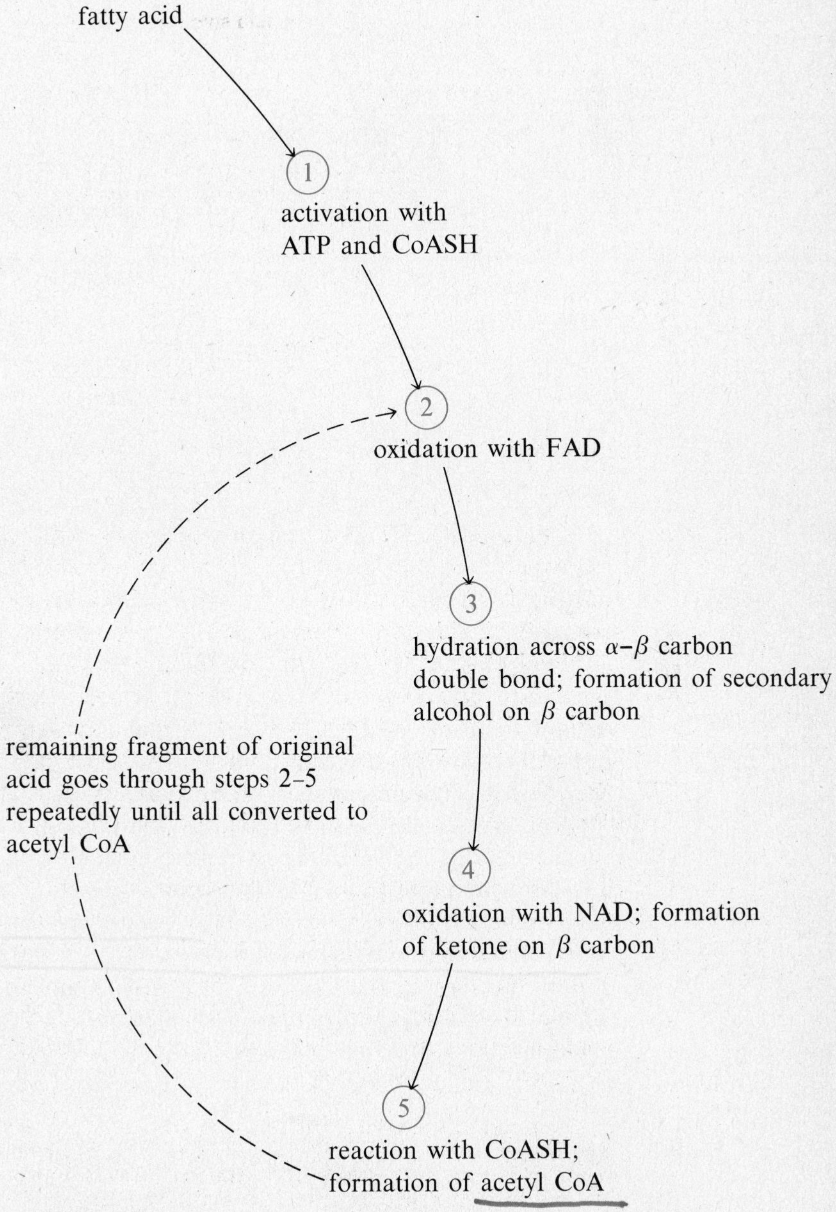

fatty acid

① activation with ATP and CoASH

② oxidation with FAD

③ hydration across α–β carbon double bond; formation of secondary alcohol on β carbon

④ oxidation with NAD; formation of ketone on β carbon

⑤ reaction with CoASH; formation of acetyl CoA

remaining fragment of original acid goes through steps 2–5 repeatedly until all converted to acetyl CoA

FIGURE 12-11 Summary of steps in fatty acid spiral (beta oxidation).

Step 1. Activation requires an input of energy. This involves the two-step coupling of palmitic acid with ATP and coenzyme A. Thiokinase catalyzes the reaction.

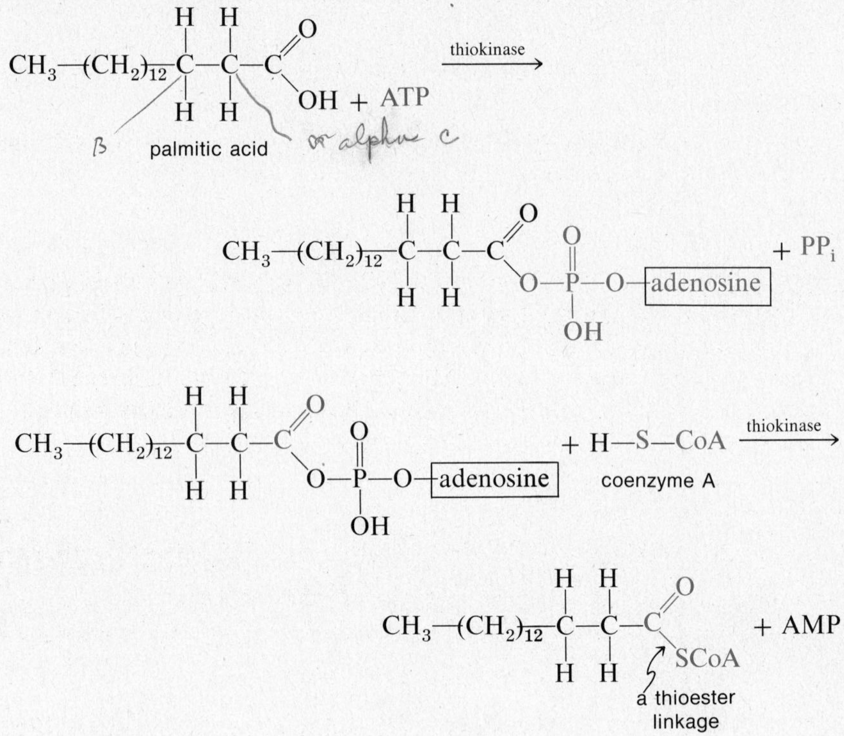

$$CH_3{-}(CH_2)_{12}{-}CH_2{-}CH_2{-}C(=O){-}OH + ATP \xrightarrow{\text{thiokinase}}$$

palmitic acid

$$CH_3{-}(CH_2)_{12}{-}CH_2{-}CH_2{-}C(=O){-}O{-}P(=O)(OH){-}O{-}\boxed{\text{adenosine}} + PP_i$$

$$CH_3{-}(CH_2)_{12}{-}CH_2{-}CH_2{-}C(=O){-}O{-}P(=O)(OH){-}O{-}\boxed{\text{adenosine}} + H{-}S{-}CoA \xrightarrow{\text{thiokinase}}$$

coenzyme A

$$CH_3{-}(CH_2)_{12}{-}CH_2{-}CH_2{-}C(=O){-}SCoA + AMP$$

a thioester linkage

Step 2. Palmitic acid appears to be in the form for which we are aiming. It now ends in $-CH_2-\overset{O}{\overset{\|}{C}}-SCoA$. If this unit were removed, this would leave only the long chemically unreactive hydrocarbon chain. The remaining steps of the spiral prepare part of the hydrocarbon chain so that two carbon fragments can continue to be removed from it. Converting a carbon–carbon single bond of the saturated hydrocarbon side chain into a double bond creates a site that can then undergo further reaction. Step 2 is the removal of hydrogens from adjacent carbons by FAD to create a double bond. (Recall that FAD primarily functions during reactions involving double bonds). The double bond is created between the alpha (α) and beta (β) carbons. The alpha carbon is defined as the one bonded to the carbonyl group; the beta carbon follows the alpha carbon.

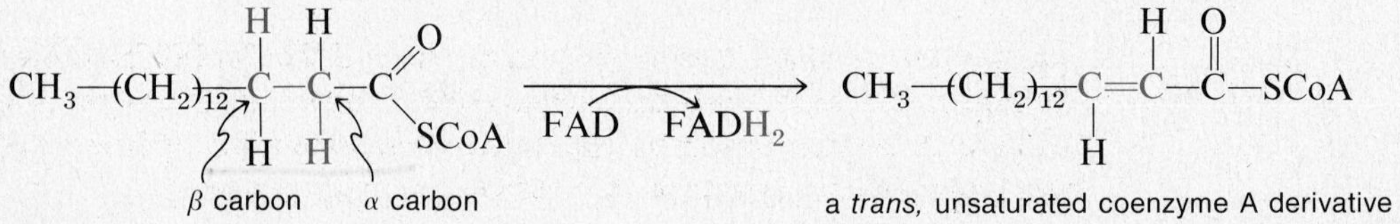

$$CH_3{-}(CH_2)_{12}{-}CH_2{-}CH_2{-}C(=O){-}SCoA \xrightarrow[\text{FAD} \;\; \text{FADH}_2]{} CH_3{-}(CH_2)_{12}{-}CH{=}CH{-}C(=O){-}SCoA$$

β carbon α carbon

a *trans,* unsaturated coenzyme A derivative

Step 3. Recall from Chapter 9 that water can be added across a double bond. This now occurs in the fatty acid spiral, forming a secondary alcohol group at the beta carbon. The addition of water is catalyzed by an enzyme specific for the *trans* form.

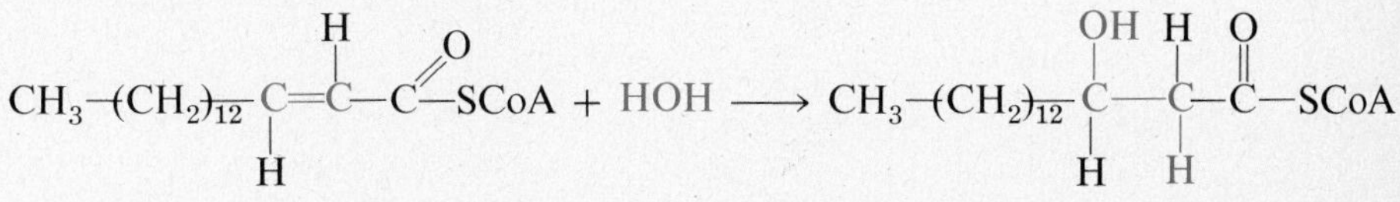

$$CH_3{-}(CH_2)_{12}{-}\overset{\displaystyle H}{\underset{}{C}}{=}\underset{\displaystyle H}{C}{-}\overset{\displaystyle O}{\overset{\|}{C}}{-}SCoA + HOH \longrightarrow CH_3{-}(CH_2)_{12}{-}\overset{\displaystyle OH}{\underset{\displaystyle H}{C}}{-}\overset{\displaystyle H}{\underset{\displaystyle H}{C}}{-}\overset{\displaystyle O}{\overset{\|}{C}}{-}SCoA$$

Step 4. What is really needed is a ketone group at the beta carbon. From Chapter 9 you know that hydrogens can be removed from a secondary alcohol, oxidizing it to a ketone. This now happens in the spiral with NAD acting as the oxidizing agent. (Recall that NAD functions primarily in reactions involving carbon-to-oxygen double bonds.)

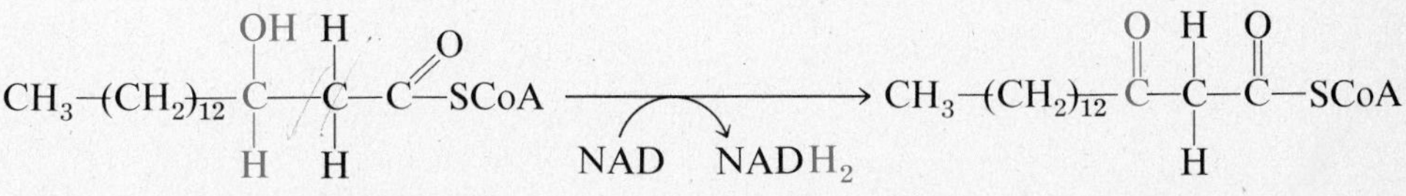

$$CH_3{-}(CH_2)_{12}{-}\overset{\displaystyle OH}{\underset{\displaystyle H}{C}}{-}\overset{\displaystyle H}{\underset{\displaystyle H}{C}}{-}\overset{\displaystyle O}{\overset{\|}{C}}{-}SCoA \xrightarrow[\text{NAD} \quad \text{NADH}_2]{} CH_3{-}(CH_2)_{12}{-}\overset{\displaystyle O}{\overset{\|}{C}}{-}\overset{\displaystyle H}{\underset{\displaystyle H}{C}}{-}\overset{\displaystyle O}{\overset{\|}{C}}{-}SCoA$$

a beta keto derivative of coenzyme A

Steps 2–4 changed the beta carbon from a $—CH_2—$ group to a ketone group. During steps 2 and 4, oxidation occurred at the beta carbon. Consequently, the fatty acid spiral is also called *beta oxidation.* The formation of a ketone group on the beta carbon weakens the alpha-to-beta carbon–carbon bond.

Step 5. The goal of the fatty acid spiral is production of acetyl coenzyme A

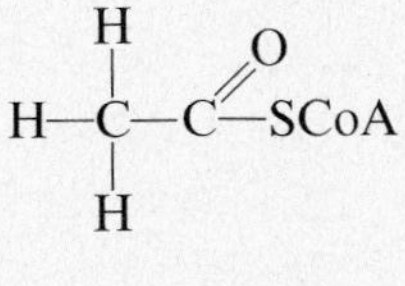

$$H{-}\overset{\displaystyle H}{\underset{\displaystyle H}{C}}{-}\overset{\displaystyle O}{\overset{\|}{C}}{-}SCoA$$

The product of step 4 has a grouping at the end that looks almost like this. In step 5, the alpha-to-beta carbon bond breaks and hydrogen from coenzyme A is transferred to the alpha carbon-containing fragment. The remainder of the coenzyme A bonds to the beta carbon.

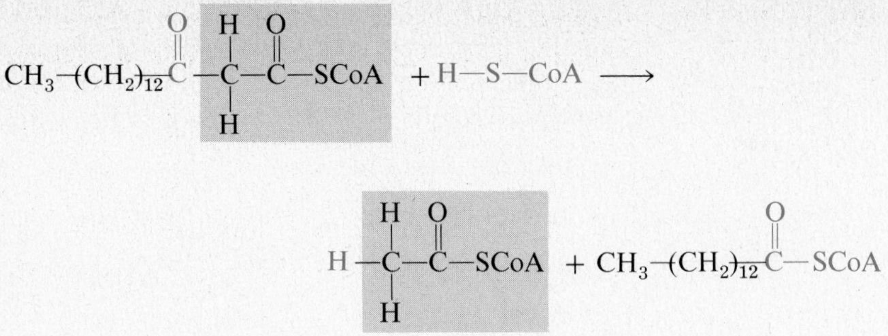

$$CH_3-(CH_2)_{12}-\overset{O}{\overset{\|}{C}}-\underset{H}{\overset{H}{\overset{|}{\underset{|}{C}}}}-\overset{O}{\overset{\|}{C}}-SCoA + H-S-CoA \longrightarrow$$

$$H-\underset{H}{\overset{H}{\overset{|}{\underset{|}{C}}}}-\overset{O}{\overset{\|}{C}}-SCoA + CH_3-(CH_2)_{12}-\overset{O}{\overset{\|}{C}}-SCoA$$

As a result of step 5, three important changes take place:

1. Acetyl coenzyme A is produced.
2. Two carbons (shaded portion above) from the original fatty acid chain are removed from the chain and produce acetyl coenzyme A.
3. The fatty acid, now minus two carbons, is in the form for step 2 to occur. No further activation or consumption of ATP is required.

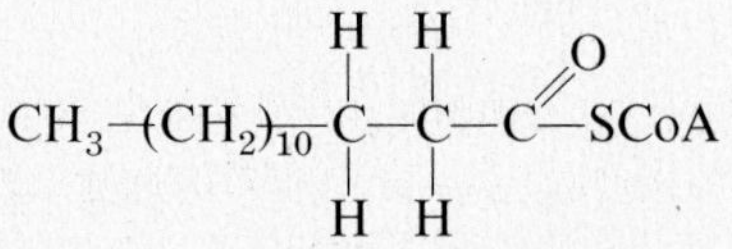

$$CH_3-(CH_2)_{10}-\underset{H}{\overset{H}{\overset{|}{\underset{|}{C}}}}-\underset{H}{\overset{H}{\overset{|}{\underset{|}{C}}}}-\overset{O}{\overset{\|}{C}}-SCoA$$

undergoes steps 2–5.

A diagrammatic summary of the fatty acid spiral for palmitic acid is shown in Fig. 12-12.

exercise 12-8

Starting with the structural formula of lauric acid ($CH_3(CH_2)_{10}CO_2H$), write equations to show the activation of the acid prior to beta oxidation.

exercise 12-9

Using the structural formula for the activated form of lauric acid from Exercise 12-8, write reaction equations showing the steps in beta oxidation during *one* turn of the fatty acid spiral for lauric acid.

After going through six turns of the spiral, palmitic acid is shortened to

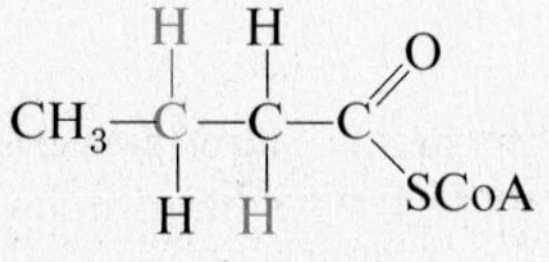

$$CH_3-\underset{H}{\overset{H}{\overset{|}{\underset{|}{C}}}}-\underset{H}{\overset{H}{\overset{|}{\underset{|}{C}}}}-C\begin{matrix}\nearrow O \\ \searrow SCoA\end{matrix}$$

This compound goes through steps 2–5.

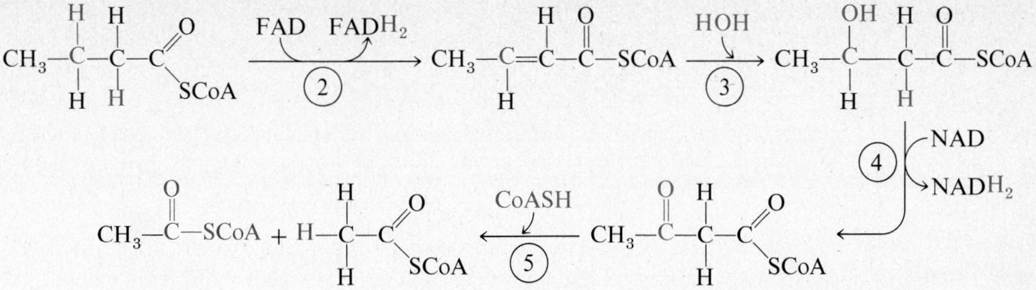

$$CH_3-\underset{H}{\overset{H}{C}}-\underset{H}{\overset{H}{C}}-C(=O)SCoA \xrightarrow[\text{(2)}]{FAD \rightarrow FADH_2} CH_3-\underset{H}{C}=\overset{H}{C}-\overset{O}{\overset{\|}{C}}-SCoA \xrightarrow[\text{(3)}]{HOH} CH_3-\underset{H}{\overset{OH}{C}}-\underset{H}{\overset{H}{C}}-\overset{O}{\overset{\|}{C}}-SCoA$$

$$\xrightarrow[\text{(4)}]{NAD \rightarrow NADH_2} CH_3-\overset{O}{\overset{\|}{C}}-\underset{H}{\overset{H}{C}}-C(=O)SCoA \xrightarrow[\text{(5)}]{CoASH} CH_3-\overset{O}{\overset{\|}{C}}-SCoA + H-\underset{H}{\overset{H}{C}}-C(=O)SCoA$$

Note that in Step 5 the last two-carbon fragment of palmitic acid is converted directly to acetyl coenzyme A. This last two-carbon fragment does not have to go through steps 2–5. The net result is that 2 acetyl coenzyme A's are produced in the final step of the fatty acid spiral. Since two-carbon fragments are removed from palmitic acid to produce acetyl coenzyme A, the overall beta oxidation of this 16-carbon acid yields 8 acetyl coenzyme A's:

$$\text{16-carbon acid} \times \frac{\text{1 acetyl CoA}}{\text{two-carbon fragment from acid}} = \text{8 acetyl coenzyme A's}$$

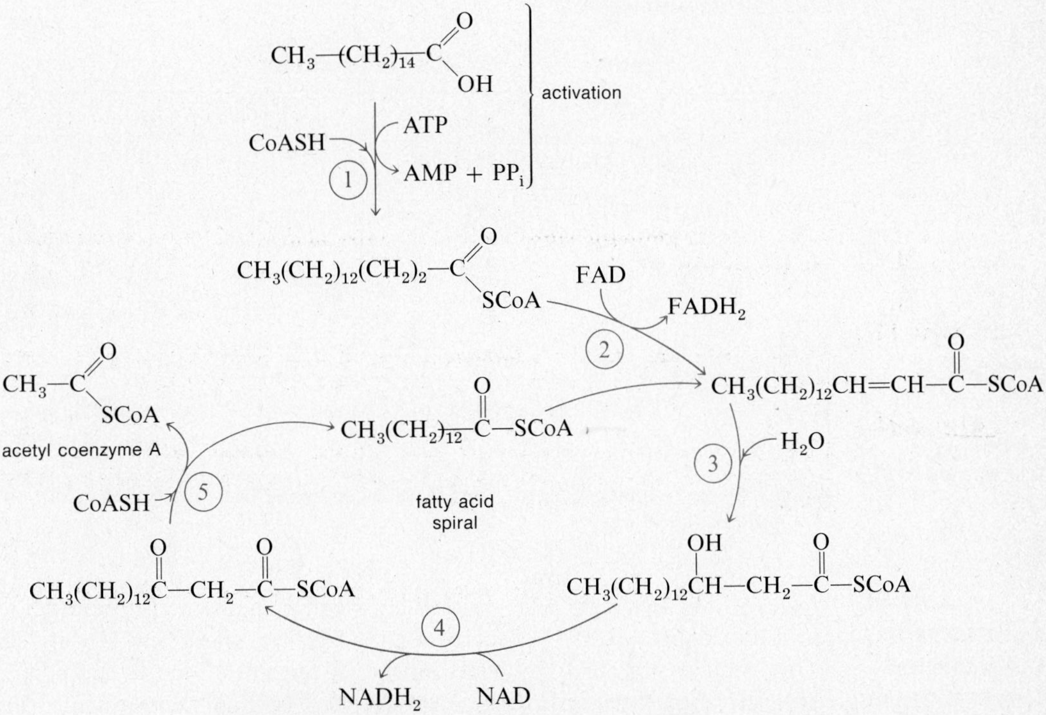

activation:

$$CH_3-(CH_2)_{14}-C(=O)OH \xrightarrow[\text{(1)}]{CoASH;\ ATP \rightarrow AMP + PP_i} CH_3(CH_2)_{12}(CH_2)_2-C(=O)SCoA$$

fatty acid spiral:

$$\xrightarrow[\text{(2)}]{FAD \rightarrow FADH_2} CH_3(CH_2)_{12}CH=CH-\overset{O}{\overset{\|}{C}}-SCoA \xrightarrow[\text{(3)}]{H_2O} CH_3(CH_2)_{12}\overset{OH}{\overset{|}{C}H}-CH_2-\overset{O}{\overset{\|}{C}}-SCoA$$

$$\xrightarrow[\text{(4)}]{NAD \rightarrow NADH_2} CH_3(CH_2)_{12}\overset{O}{\overset{\|}{C}}-CH_2-\overset{O}{\overset{\|}{C}}-SCoA \xrightarrow[\text{(5)}]{CoASH} CH_3-C(=O)SCoA \text{ (acetyl coenzyme A)} + CH_3(CH_2)_{12}-\overset{O}{\overset{\|}{C}}-SCoA$$

FIGURE 12-12 Fatty acid spiral (beta oxidation) of palmitic acid. The term spiral is used because after the initial activation process (step 1), each turn of the spiral acts on a substance having two fewer carbons than its predecessor through the spiral. The numbers refer to processes described individually in the text. A molecule of acetyl coenzyme A is liberated by each turn of the spiral per molecule of fatty acid.

exercise 12-10

How many acetyl CoA's are produced by the fatty acid spiral oxidation of lauric acid?

exercise 12-11

Using structural formulas, write equations for the complete fatty acid spiral oxidation of palmitic acid into acetyl coenzyme units.

FATTY ACID SPIRAL ENERGETICS

Although 8 moles of acetyl CoA are produced per mole of palmitic acid, the last step of beta oxidation does not require FAD or NAD. Thus, only 7 moles of $FADH_2$ and 7 moles of $NADH_2$ are produced per mole of palmitic acid.

The activation step in the oxidation of palmitic acid converts ATP to AMP. Obviously, cells can't keep converting ATP to AMP or ADP without needing to generate new ATP. The initial investment of an ATP to activate the fatty acid spiral is paid back many times.

The FAD and NAD used during the beta oxidation of palmitic acid are part of the respiratory chain; 1 mole of palmitic acid reduces 7 moles of FAD to $FADH_2$; 7 moles of NAD are reduced to $NADH_2$. In the respiratory chain, these reduced coenzymes are oxidized back to FAD and NAD. The hydrogens acquired during beta oxidation are passed along the respiratory chain eventually producing water, energy, and ATP. Each mole of $FADH_2$ generates 2 moles of ATP; each mole of $NADH_2$ yields 3 moles of ATP. The production of ATP by beta oxidation of a mole of palmitic acid can be summarized

Credit	ATP production
$7 \text{ moles } FADH_2 \times \dfrac{2 \text{ moles ATP}}{1 \text{ mole } FADH_2} =$	14 moles
$7 \text{ moles } NADH_2 \times \dfrac{3 \text{ moles ATP}}{1 \text{ mole } NADH_2} =$	21 moles
	35 moles ATP
	−1 mole ATP used to activate the process
Net production:	34 moles ATP/mole palmitic acid

exercise 12-12

What is the net number of moles of ATP produced by the fatty acid spiral oxidation of a mole of lauric acid?

ACETYL COENZYME A CONVERSION—YOUR CELLS RIDE A KREBS CYCLE

We have just seen that ATP production accompanies acetyl coenzyme A formation by beta oxidation of fatty acids. Additional energy can be derived from acetyl coenzyme A by processing it in the mitochondria through a sequence of reactions called the Krebs cycle. The English biochemist Sir Hans Krebs received the 1953 Nobel Prize in medicine and physiology for his work in determining the steps of the cycle. The Krebs cycle and the respiratory chain act cooperatively to produce carbon dioxide and water from an acetyl coenzyme A derivative. The carbon dioxide is released by the decarboxylation of certain substances produced in the Krebs cycle. NAD and FAD in the respiratory chain remove hydrogens from certain substances formed in the Krebs cycle. The hydrogens are passed along the respiratory chain where they eventually combine with oxygen to form water.

A metabolic cycle is a series of reactions having a substance common to its initial and final steps. Oxaloacetic acid begins the cycle as a reactant by combining with acetyl coenzyme A to form citric acid. Citric acid then undergoes a series of several reactions, the last of which forms oxaloacetic acid as a product so that the cycle can continue (see Fig. 12-13). *Any substance that can be metabolized to acetyl coenzyme A can enter the cycle.* During their metabolism, lipids, carbohydrates, and proteins produce acetyl coenzyme A. Thus their breakdown involves the Krebs cycle and respiratory chain. An overview of the Krebs cycle steps is shown in Fig. 12-14. The production of acetyl coenzyme A from carbohydrates and proteins is described in Chapters 13 and 14, respectively.

The Krebs cycle is common to the catabolism of lipids, carbohydrates, and proteins.

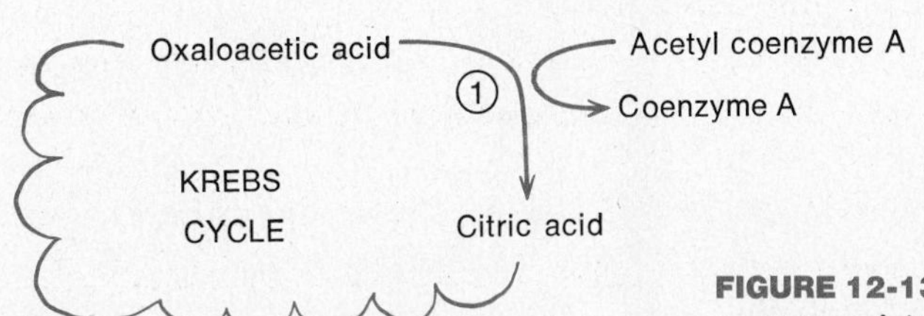

FIGURE 12-13 Relationship of oxaloacetic acid and acetyl coenzyme A to the Krebs cycle.

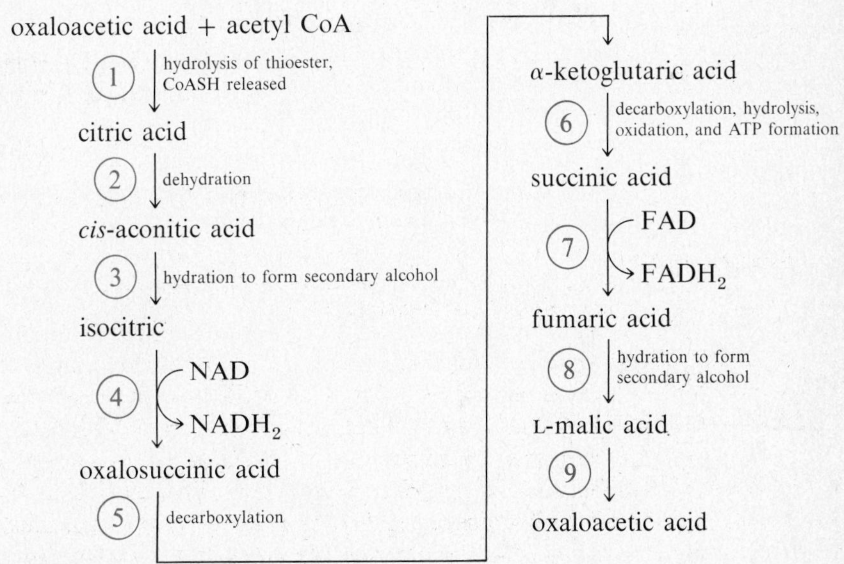

FIGURE 12-14 Overview of steps in the Krebs cycle. Steps 7–9 also occur in the fatty acid spiral.

THE KREBS CYCLE Starting with acetyl coenzyme A (acetyl CoA) and oxaloacetic acid, reactions that lead to the reformation of oxaloacetic acid in the Krebs cycle are

Step 1. Conversion of oxaloacetic acid to citric acid.

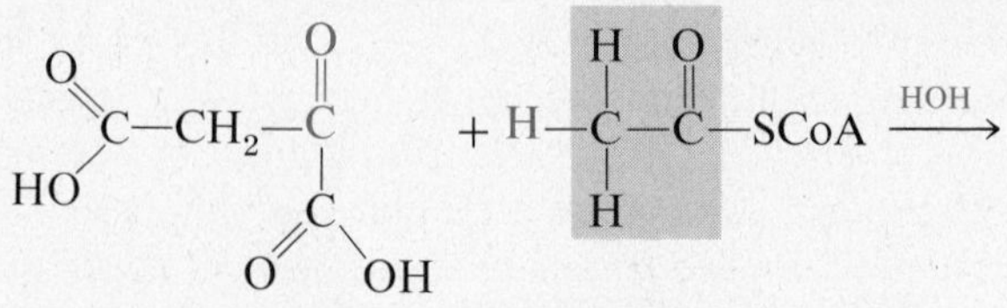

$$HOOC{-}CH_2{-}C({=}O){-}COOH + H{-}CH_2{-}C({=}O){-}SCoA \xrightarrow{HOH}$$

oxaloacetic acid
(a four-carbon acid)

acetyl CoA

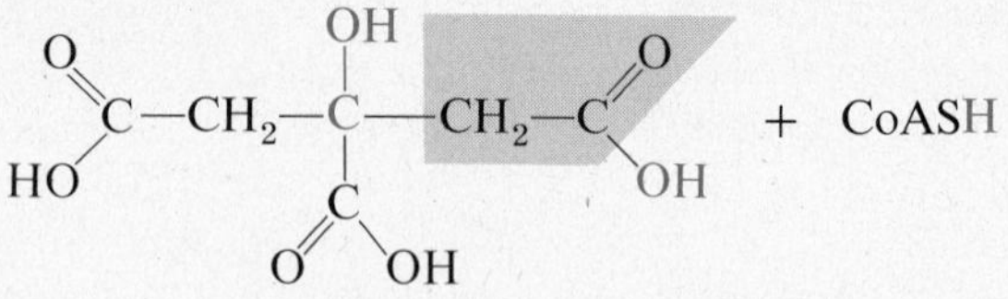

$$HOOC{-}CH_2{-}C(OH)(COOH){-}CH_2{-}COOH + CoASH$$

citric acid
(a six-carbon acid)

coenzyme A

This reaction is complex, involving several steps. The equation above shows the overall changes that occur. The thioester linkage in acetyl CoA is a high-energy bond. When hydrolyzed, it yields coenzyme A and energy. The coenzyme A can be used to initiate the fatty acid spiral. The energy released by the hydrolysis drives the formation of citric acid essentially to completion. Because the subsequent reactions in the Krebs cycle involve citric acid and its derivatives, the name citric acid cycle may also be used for the Krebs cycle. It is also called the tricarboxylic acid cycle. All three names are commonly used and the choice of name is largely one of personal preference. We shall use the term Krebs cycle.

For reversible reactions, the greater the amount of energy released, the more likely the products of the reaction are favored.

exercise 12-13

Look at the structural formula of citric acid. Explain why the term tricarboxylic acid cycle is also appropriate for the Krebs cycle.

Steps 2 and 3. The result of these steps is the formation of a citric acid isomer, isocitric acid. This is accomplished by a dehydration to form a double bond followed by hydration of the double bond.

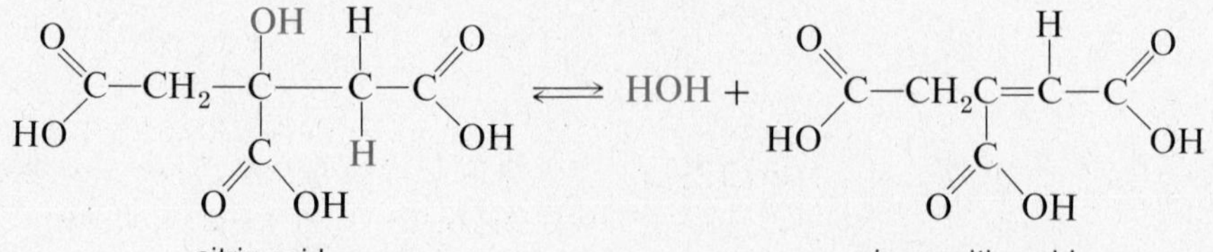

$$HOOC{-}CH_2{-}C(OH)(COOH){-}CH_2{-}COOH \rightleftharpoons HOH + HOOC{-}CH_2{-}C(COOH){=}CH{-}COOH$$

citric acid

cis-aconitic acid

$$\underset{\textit{cis}\text{-aconitic acid}}{\mathrm{HOOC{-}CH_2C(COOH){=}CH{-}COOH}} \rightleftharpoons \underset{\text{isocitric acid}}{\mathrm{HOOC{-}CH_2{-}CH(COOH){-}CH(OH){-}COOH}}$$

Recall from Chapter 9 how a reversible reaction may lead to isomerization.

Your first impression might be nothing was accomplished by taking out water and putting it back. However, the H and OH switched carbons. The alcohol group in citric acid is a tertiary alcohol that cannot be oxidized. The alcohol group in isocitric acid is a secondary alcohol capable of being oxidized.

Step 4. Isocitric acid is oxidized to oxalosuccinic acid. The secondary alcohol group in isocitric acid is oxidized to a ketone by NAD.

$$\underset{\text{isocitric acid}}{\mathrm{HOOC{-}CH_2{-}CH(COOH){-}CH(OH){-}COOH}} \xrightleftharpoons[]{\text{NAD} \rightarrow \text{NADH}_2} \underset{\text{oxalosuccinic acid}}{\mathrm{HOOC{-}CH_2{-}CH(COOH){-}C(=O){-}COOH}}$$

One of the two carbon atoms added to oxaloacetic acid in step 1 is lost in an essentially irreversible reaction.

Step 5. Experimental evidence indicates that oxalosuccinic acid is decarboxylated as it forms. The decarboxylation converts oxalosuccinic acid to α-ketoglutaric acid. By the loss of carbon dioxide, the six-carbon compound is converted to a five-carbon compound. Recall decarboxylation of carboxylic acids from Chapter 10.

$$\underset{\substack{\text{oxalosuccinic acid} \\ \text{(a six-carbon triacid)}}}{\mathrm{HOOC{-}CH_2{-}CH(COOH){-}C(=O){-}COOH}} \longrightarrow CO_2 + \underset{\substack{\alpha\text{-ketoglutaric acid} \\ \text{(a five-carbon diacid)}}}{\mathrm{HOOC{-}CH_2{-}CH_2{-}C(=O){-}COOH}}$$

The second carbon atom added in step 1 is now lost.

Step 6. Essentially complete conversion of α-ketoglutaric acid to succinic acid. This conversion is complex and involves decarboxylation, hydrolysis, *and* oxidation. Only the net reaction is shown. The overall changes release sufficient energy for the *direct* conversion of ADP to ATP.

$$HOOC{-}CH_2{-}CH_2{-}\overset{\displaystyle O}{\overset{\|}{C}}{-}COOH + NAD + ADP + P_i \longrightarrow$$

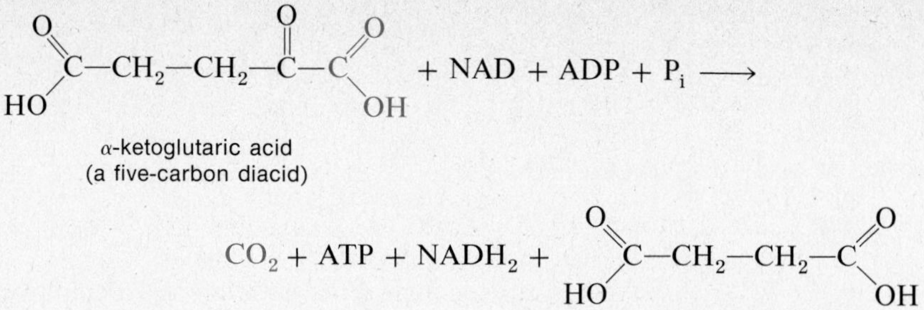

α-ketoglutaric acid
(a five-carbon diacid)

ATP produced in this step is in addition to that produced by the ETS.

$$CO_2 + ATP + NADH_2 + HOOC{-}CH_2{-}CH_2{-}COOH$$

succinic acid
(a four-carbon diacid)

The loss of carbon dioxide shortens the carbon chain to four carbons.

Oxaloacetic acid is needed to start the cycle over again.

Steps 7, 8, and 9. At this point, let's compare the structural formulas of what we have with what we need—oxaloacetic acid:

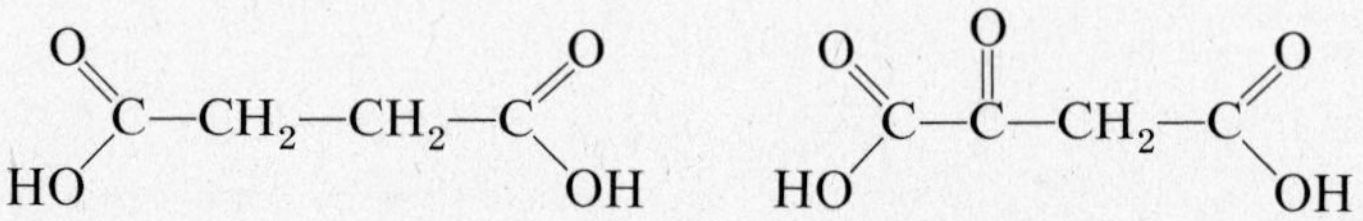

$$HOOC{-}CH_2{-}CH_2{-}COOH \qquad HOOC{-}\overset{\displaystyle O}{\overset{\|}{C}}{-}CH_2{-}COOH$$

succinic acid — oxaloacetic acid

Both contain terminal carboxylic acid groups. Each contain at least one $-CH_2-$ group. The difference is the presence of a ketone group in oxaloacetic acid. We see two adjacent hydrocarbon-like $-CH_2-$ groups in the succinic acid formula. Recall from the beta-oxidation sequence that a $-CH_2-$ group beta to an acid group can be converted to a $-\overset{\displaystyle O}{\overset{\|}{C}}-$ group. The same three-step sequence occurs in the Krebs cycle.

Step 7. Succinic acid is converted to fumaric acid.

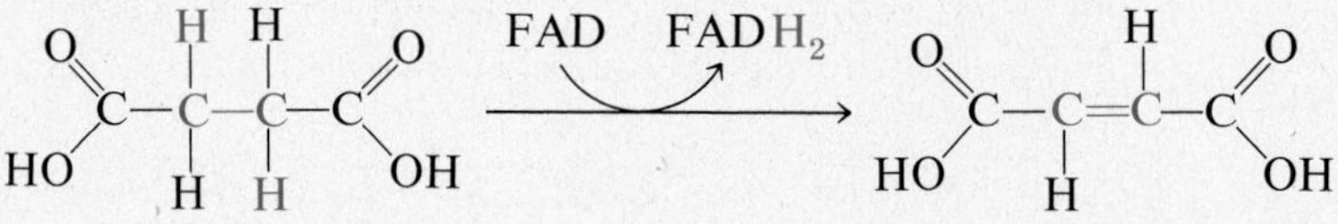

$$HOOC{-}CH_2{-}CH_2{-}COOH \xrightarrow{FAD \ \ FADH_2} HOOC{-}CH{=}CH{-}COOH$$

succinic acid — fumaric acid

Why is fumaric acid called the trans *isomer?*

The reaction is catalyzed by succinic dehydrogenase and only the *trans* isomer, fumaric acid, is formed. Maleic acid, the *cis* isomer, is absent.

Step 8. Addition of water to fumaric acid forms L-malic acid.

$$\mathrm{HOOC{-}CH{=}CH{-}COOH} + \mathrm{HOH} \xrightleftharpoons{\text{fumarase}} \mathrm{HOOC{-}CH(OH){-}CH_2{-}COOH}$$

fumaric acid — L-malic acid

Fumarase is a highly specific enzyme that catalyzes the addition of water across a double bond. (Most enzymes also catalyze their reverse reaction. Fumarase can catalyze the removal of water to form a double bond.) In animal cells, only the L-optical isomer of malic acid is formed. Note that malic acid contains a secondary alcohol group.

Step 9. Formation of oxaloacetic acid from L-malic acid by oxidation of the secondary alcohol group.

$$\mathrm{HOOC{-}CH(OH){-}CH_2{-}COOH} \xrightarrow{\mathrm{NAD} \;\; \mathrm{NADH_2}} \mathrm{HOOC{-}CO{-}CH_2{-}COOH}$$

L-malic acid — oxaloacetic acid

The oxaloacetic acid may now combine with acetyl CoA to produce citric acid and coenzyme A. The citric acid then continues the cycle.

exercise 12-14

Why is the term *beta oxidation* appropriate for steps 7-9 as shown for the Krebs cycle?

ATP PRODUCTION AND THE KREBS CYCLE

A mole of acetyl CoA going through the Krebs cycle produces 3 moles of $NADH_2$, 1 mole of $FADH_2$, and *directly* 1 mole of ATP. $NADH_2$ and $FADH_2$ are part of the respiratory chain and are oxidized back to NAD and FAD. The hydrogens are passed by the respiratory chain to form water eventually by combination with oxygen. ATP's are produced from ADP and P_i with energy made available from the respiratory chain reactions 3 ATP/$NADH_2$ and 2 ATP/$FADH_2$. We can summarize the amount of ATP produced from 1 mole of acetyl CoA being processed through the Krebs cycle (See steps on pp. 337–341).

Step 4. 1 $NADH_2$ × 3 ATP/$NADH_2$	= 3 ATP
Step 6. 1 $NADH_2$ × 3 ATP/$NADH_2$	= 3 ATP
Step 6. 1 mole ATP *directly* from ADP + Pi	= 1 ATP
Step 7. 1 $FADH_2$ × 2 ATP/$FADH_2$	= 2 ATP
Step 9. 1 $NADH_2$ × 3 ATP/$NADH_2$	= 3 ATP
Net result:	12 ATP/acetyl CoA

How many CO_2 are produced by one cycle of the Krebs cycle?

In the preceding section on the fatty acid spiral, we saw that a mole of palmitic acid yields 34 moles of ATP and 8 moles of acetyl CoA upon beta oxidation. After combining with oxaloacetic acid to form citric acid, the acetyl CoA proceeds through the Krebs cycle with a production of 12 ATP/acetyl CoA. Therefore, the overall amount of ATP produced from a mole of palmitic acid catabolized to carbon dioxide and water is

$$\text{1 mole palmitic acid through fatty acid spiral} \longrightarrow \text{34 moles ATP}$$

$$\text{In Krebs cycle: 8 moles acetyl CoA from palmitic acid} \times \frac{\text{12 moles ATP}}{\text{1 mole acetyl CoA}} \longrightarrow \underline{\text{96 moles ATP}}$$

$$\text{Net production: } \frac{\text{130 moles ATP}}{\text{mole palmitic acid}}$$

Note that most of the ATP production occurs via energy released during the Krebs cycle and respiratory chain processes.

exercise 12-15

Calculate the amount of energy, in calories, required to form ATP from ADP and P_i during oxidation of a mole of palmitic acid to carbon dioxide and water.

CHAPTER SUMMARY

1. Photosynthesis is an endergonic process; respiration is exergonic. Both processes occur by stepwise energy changes.
2. When coupled successfully, exergonic reactions supply the energy necessary to drive endergonic reactions. Most biochemical reactions involve molecules that must be initially activated in order to eventually release energy.
3. ATP is the principal substance for cellular energy exchange. ATP, containing adenine, ribose, and three phosphate groups, liberates about 8000 cal/mole when hydrolyzed to ADP + P_i.

$$\text{ATP} + \text{HOH} \rightleftharpoons \text{ADP} + P_i + 8000 \text{ cal}$$

The reaction is reversible. When ATP is formed, the energy is stored in high-energy phosphate anhydride linkages.
4. The respiratory chain (electron transport system) contains hydrogen carriers and electron carriers. Hydrogen ions and electrons are passed along the respiratory chain and they are eventually combined with oxygen to form water. Cytochrome a, a_3 is the only cytochrome that interacts directly with oxygen.
5. Cytochromes act as electron carriers by containing iron ions that can undergo reversible oxidation–reduction: $Fe^{2+} \rightleftharpoons Fe^{3+} + e^-$. The

coenzymes NAD, NADP, FMN, FAD, and CoQ carry hydrogen ions and electrons. NAD primarily functions during reactions involving carbon-to-oxygen double bonds. FAD primarily functions during reactions involving carbon–carbon double bonds. Vitamins are usually coenzyme components. Thus the body's need for vitamins and the role they play in metabolism is seen.

6. ATP is produced by the energy liberated at three sites in the respiratory chain: 3 moles of ATP are produced from 1 mole of $NADH_2$; 2 moles ATP are produced from 1 mole of $FADH_2$.
7. Coupled reactions can be represented by curved double arrows such as in

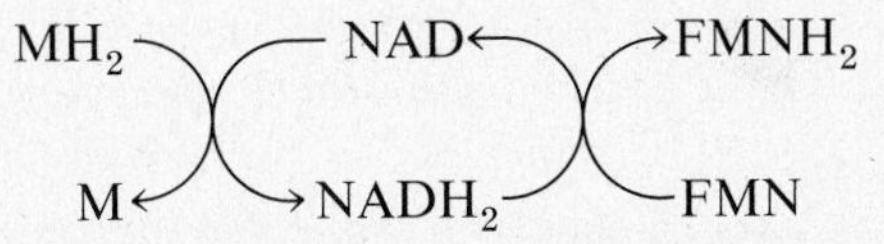

   MH_2 — NAD ← → $FMNH_2$

   M ← → $NADH_2$ — FMN

   The fate of a particular reactant is indicated by the path of the appropriate arrow.
8. The process of oxidation of a metabolite to supply enough energy to form ATP from ADP + P_i is called *oxidative phosphorylation.* The energy liberated during oxidation is released in a series of steps, some of which are energetic enough to cause ATP formation.
9. Digestion of lipids produces fatty acids and glycerol that are transported to the blood via the lymph system. Bile salts emulsify lipids. In the bloodstream, the insoluble lipids are solubilized by combining with proteins to form lipoproteins.
10. In the fatty acid spiral, acetyl CoA (a thioester) is liberated from a fatty acid during beta oxidation. The number of acetyl CoA molecules produced equals half the number of carbon atoms in the fatty acid molecule undergoing oxidation. The acetyl CoA then enters the Krebs cycle. Both processes liberate energy and produce ATP; 12 moles of ATP are produced per acetyl CoA processed in the Krebs cycle.
11. Enzymes catalyze fatty acid spiral and Krebs cycle reactions. Many of the enzymes are specific in their action for catalyzing the formation of a specific geometric or optical isomer.
12. Oxaloacetic acid is a reactant and product in the Krebs cycle. Citric acid and α-ketoglutaric acid formation are essentially irreversible reactions.
13. The same three step sequence of **(a)** double bond formation, **(b)** hydration of a double bond, and **(c)** oxidation of a secondary alcohol is common to the fatty acid spiral and the Krebs cycle.
14. Reactants in the Krebs cycle do not directly interact with oxygen. The use of oxygen in respiration occurs when oxygen reacts with hydrogen ions and electrons passed along the respiratory chain to produce water.

QUESTIONS

1. The cellular concentration of ATP ranges from 0.5 *to* 2.5 mg/ml. Calculate the range of ATP concentration in terms of percent concentration.
2. In AMP, what kind of bond links ribose to phosphate?
3. Cyanide ions (CN^-) form strong coordinate covalent bonds to metal ions. Taken internally, a soluble cyanide compound causes a decrease in respiration and, consequently, death occurs. Explain.
4. In this chapter it is mentioned that only L-malic acid is produced in the Krebs cycle. Using its structural formula, explain why malic acid is optically active.
5. The *trans* isomer, fumaric acid, is produced in the Krebs cycle. Maleic acid is the *cis* isomer. Write its structural formula.
6. In the Krebs cycle, oxalosuccinic acid is converted to α-ketoglutaric acid. This reaction is essentially irreversible.

   $$\text{oxalosuccinic acid} \longrightarrow \alpha\text{-ketoglutaric acid} + \text{carbon dioxide}$$

   What factor causes the forward reaction to be so highly favored?
7. Starting with 1 mole of lauric acid, calculate the total overall ATP production as the acid undergoes fatty acid spiral and Krebs cycle oxidation. How many calories of energy does this ATP production store?
8. Transamination is the transferring of an amine group from one molecule to another. Transamination converts the amino acid, glutamic acid $\left(HO_2CCH_2CH_2\overset{\overset{NH_2}{|}}{C}HCO_2H\right)$ to α-ketoglutaric acid $(HO_2CCH_2CH_2\overset{\overset{O}{\|}}{C}CO_2H)$. Can the body metabolize the α-ketoglutaric acid produced in this way?
9. In the Krebs cycle, malic acid is oxidized to oxaloacetic acid. The forward reaction is highly favored over the reverse reaction. Using Le Chatelier's principle, explain why this is so.
10. Starting with 1 mole of oleic acid $[CH_3(CH_2)_7CH{=}CH(CH_2)_7CO_2H]$, calculate the total overall ATP production as the acid undergoes fatty acid spiral and Krebs cycle oxidation to carbon dioxide and water.
11. How many calories of energy does the ATP production from the catabolism of 1 mole of oleic acid store?

12. The vitamin pantothenic acid is a component of coenzyme A. The structural formula of pantothenic acid is

$$\begin{array}{ccccccc} & & CH_3 & OH & O & & \\ & & | & | & \| & & \\ HO-CH_2 & - & C & -CH- & C & -N- & CH_2CH_2CO_2H \\ & & | & & & | & \\ & & CH_3 & & & H & \end{array}$$

In coenzyme A, pantothenic acid is bonded to a form of ATP by a phosphate ester linkage and is also bonded to β-mercaptoethylamine ($H_2NCH_2CH_2SH$).

(a) What kind of linkage would you predict for the bonding of pantothenic acid to β-mercaptoethylamine?

(b) Write the structural formula for the product formed by the combination of the two substances in part **(a)**.

13
carbohydrate metabolism

INTRODUCTION

In this chapter we will consider the fate of carbohydrates following ingestion. We will see how they are broken down and absorbed, how they react in the cells to produce energy, and how they can be stored in the body. Additionally, we will study their interaction with lipid metabolism and compare the energy available to the body from carbohydrate metabolism to that available from lipids. Finally, the chemical basis and consequences of abnormalities in carbohydrate metabolism, such as galactosemia and diabetes mellitus, will be studied. It is not necessary to memorize the many complex reactions and structures in this chapter.

DIGESTION AND ABSORPTION OF CARBOHYDRATES

When we eat a meal, we take in a vast assortment of compounds. These are usually classified as belonging to one of the three major groups of foodstuffs: proteins, fats, and carbohydrates. Among the compounds in our diet classed as carbohydrates are starch, cellulose, sucrose, glucose, fructose, and lactose, plus minor amounts of maltose, sorbitol, and arabinose. The first three compounds account for the majority of the carbohydrates ingested, probably greater than 95 percent of them.

Prior to absorption into the body, all carbohydrates are broken down to monosaccharides in a process called *digestion*. In order for the nutrients to be absorbed into the bloodstream they must first pass through the mucosal cells of the small intestine. Remember that cell membranes are only permeable to small molecules. Thus, it is necessary that the large food molecules be broken down to their simpler components before absorption can occur. Any foodstuffs that are not broken down are simply excreted. The breakdown of some carbohydrates begins in the mouth, although most occurs in the upper portion of the small intestine known as the duodenum.

Saliva contains the enzyme *salivary α-amylase*, which catalyzes the hydrolysis of starch (both amylose and amylopectin) to dextrins and maltose. The contribution of salivary amylase to the overall digestion of starch is uncertain because of the variability of the length of time of contact between the enzyme and the starch. The pH required for the optimum activity of salivary amylase is between 6 and 8. This means that after the food is swallowed, the enzyme is deactivated by the low pH of the stomach juices. A small amount of starch hydrolysis may occur in the stomach due to the acid-catalyzed hydrolysis of the acetal bonds linking the glucose units together.

Most carbohydrate digestion takes place in the slightly basic media of the small intestine. The pancreas releases *pancreatic amylase*, which catalyzes the hydrolysis of starch to maltose and a mixture of oligosaccharides (sugars containing three to eight glucose units). We can see the specificity of enzymes and the biological importance of the stereochemistry of molecules in the digestive process. Both salivary and pancreatic amylases are enzymes specific for the hydrolysis of 1,4-α acetal linkages between glucose units. Neither of these enzymes is capable of catalyzing hydrolysis of 1,4-β acetal linkages between glucose molecules. Consequently, we cannot digest cellulose.

The undigested materials act as roughage in the digestive tract.

In humans and most other animals, cellulose is of negligible nutritional value. There are certain bacteria and other microorganisms that have enzymes capable of catalyzing the hydrolysis of 1,4-β linkages. Thus, in the overall scheme of nature, a means is provided for the breakdown and utilization of the bulk of dead vegetation. A few mammals (such as cows, horses, and other ruminants) and certain insects (such as termites) harbor in their gastrointestinal tract bacteria whose enzymes will hydrolyze cellulose.

The intestinal digestive juices also contain the enzymes *maltase, lactase,* and *sucrase,* which catalyze the hydrolysis of maltose, lactose, and sucrose, respectively. Also contained in the intestinal juices is *oligo-1,6-glucosidase,* which completes the hydrolysis of the oligosaccharides containing 1,6-α acetal bonds and the isomaltose produced during the breakdown of starch. Thus, the digestion of the main carbohydrates contained in the diet leads to formation of the three monosaccharides: glucose, galactose, and fructose. Glucose is overwhelmingly the predominant monosaccharide.

The rates of absorption have been found to be glucose > fructose > galactose

The mixture of monosaccharides is transported across the intestinal barrier into the blood. Although subjected to intensive study, relatively little is yet known about the manner in which these sugars penetrate the intestinal barrier. Diffusion across membranes due to concentration differences may play a part in sugar transport. It has long been known, however, that transport involves more than simple diffusion. In recognition of this, the term *active transport* is applied to the movement of monosaccharides from the intestines into the blood. This idea of active transport implies the existence of a carrier of some type that combines with the sugar and enables the sugar molecule to move from one side of the membrane to the other. However, no satisfactory explanation of the nature of carrier or its mode of operation has been given to date.

GLUCOSE METABOLISM

Glucose entering the cells of an organism carries with it energy that was stored in its chemical bonds during photosynthesis. This energy is released to the cells during a process called metabolism. The overall reaction for the metabolism of glucose by respiration is

$$C_6H_{12}O_6 + 6O_2 \longrightarrow 6H_2O + 6CO_2 + \text{energy}$$

The reaction is not carried out in a single step but is accomplished by a series of reactions, some of which release a small amount of the total available energy. By release of the energy in small amounts, a large portion of the total energy can be used for endergonic reactions and thus stored by the organism.

The first step in the use of glucose by the body is its conversion to *glucose-6-phosphate*. This step is endergonic since an ATP is used.

$$\text{glucose} + ATP \xrightarrow{\text{hexokinase}} \text{glucose-6-phosphate} + ADP$$

glucose

glucose-6-phosphate

The enzyme *hexokinase* catalyzes the phosphorylation of glucose by ATP to form the ester glucose-6-phosphate.

After the formation of glucose-6-phosphate, there are several paths for its utilization. One path leads to glycogen and is called *glycogenesis*—the formation of glycogen. The other path leads to the breakdown of glucose to pyruvic acid and then eventually to carbon dioxide and water. The conversion of glucose to pyruvic acid is called *glycolysis*. We shall examine glycolysis first.

GLYCOLYSIS

Note that in glycolysis, glucose, a six-carbon material, is broken down to two 3-carbon fragments of pyruvic acid.

The initial steps of glycolysis are the same for most organisms. Some of the later steps may differ, however. Consequently biochemists refer to *fermentation, aerobic glycolysis,* and *anaerobic glycolysis,* depending on the final products produced during glycolysis. Fermentation leads to formation of ethanol; aerobic glycolysis, to carbon dioxide and water; and anaerobic glycolysis, to lactic acid. In each of these processes, pyruvic acid is produced as an intermediate. Aerobic means that oxygen is required, while anaerobic

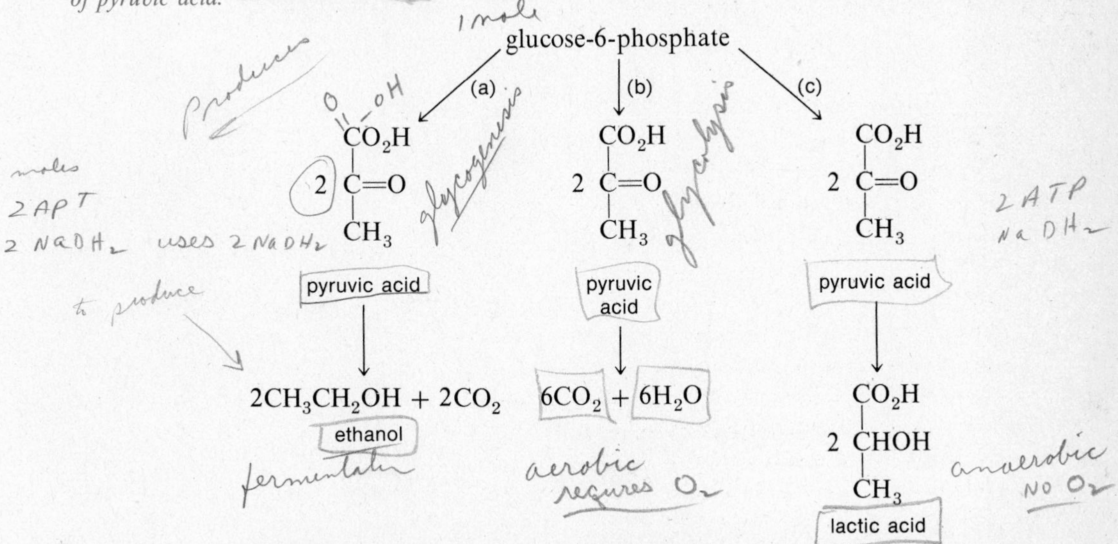

FIGURE 13-1 The three possible fates of glucose during glycolysis. Path (a) is called fermentation, path (b) is aerobic glycolysis, and path (c) is anaerobic glycolysis.

glycolysis occurs in the absence of oxygen. These processes are summarized in Fig. 13-1.

Fermentation is essentially restricted to microorganisms, primarily yeast. Some microorganisms are capable of aerobic glycolysis, while others are restricted to anaerobic glycolysis. Mammals are capable of both anaerobic and aerobic glycolysis, although the latter is the normal reaction pathway.

The overall sequence for the anaerobic conversion of glucose to lactic acid is shown in Fig. 13-2.

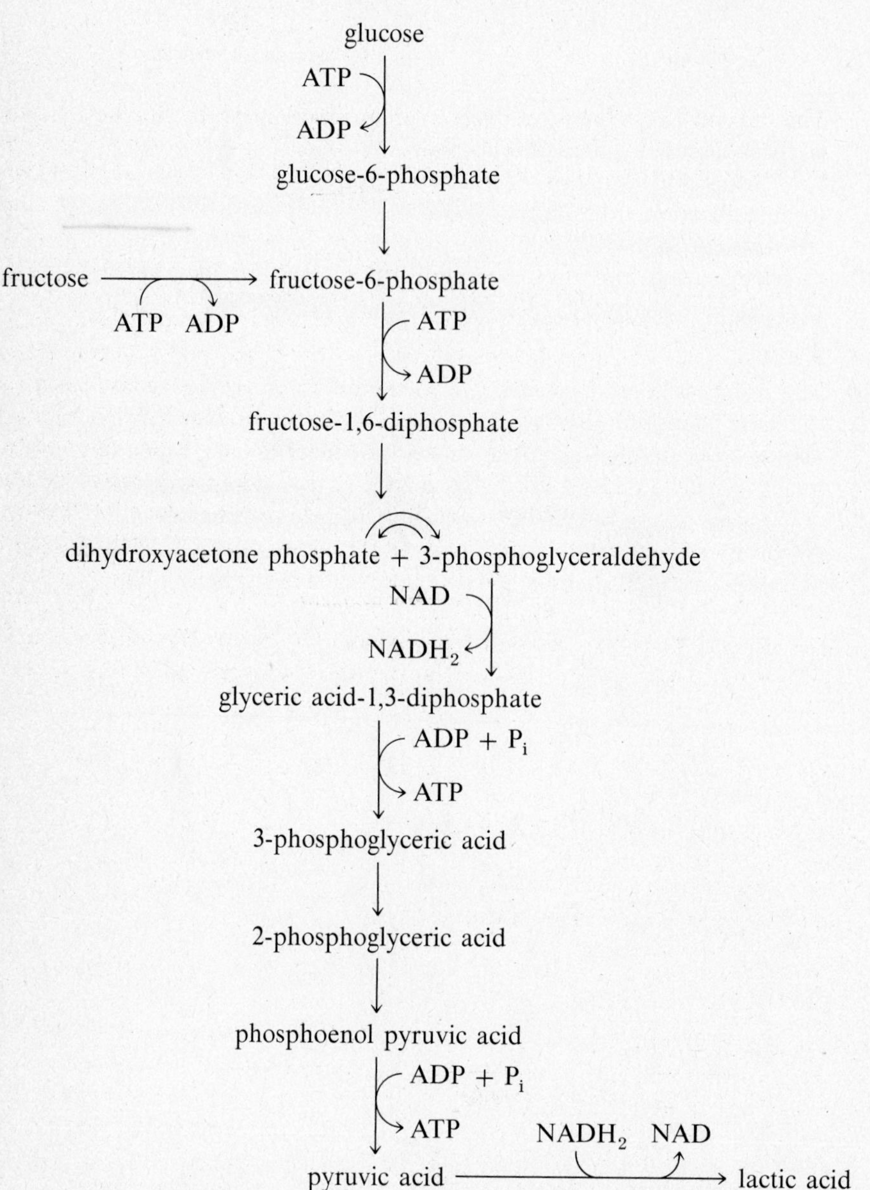

FIGURE 13-2 Summary of glycolysis leading to lactic acid formation.

PYRUVIC ACID AND ATP PRODUCTION The initial step in the breakdown of pyruvic acid is its conversion to acetyl coenzyme A. The reaction involves both decarboxylation and oxidation as well as ester formation. The oxidation is accomplished by NAD.

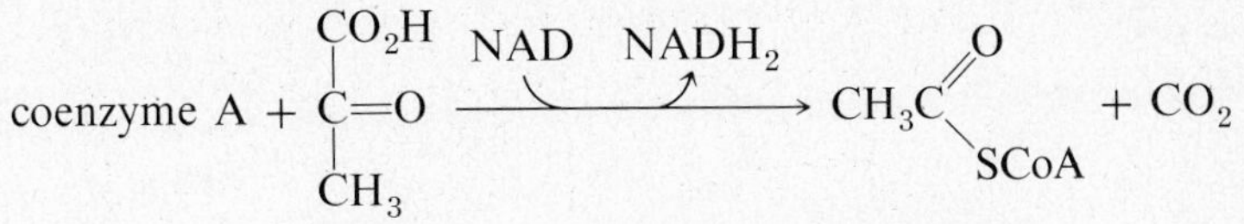

$$\text{coenzyme A} + \underset{\substack{\text{pyruvic}\\ \text{acid}}}{\begin{matrix} CO_2H \\ | \\ C{=}O \\ | \\ CH_3 \end{matrix}} \xrightarrow{\text{NAD} \quad \text{NADH}_2} \underset{\substack{\text{acetyl}\\ \text{coenzyme A}}}{CH_3C\begin{matrix} {}^{\nearrow}O \\ \\ {}_{\searrow}SCoA \end{matrix}} + CO_2$$

exercise 13-1

If the sequence of steps in the conversion of pyruvic acid to acetyl coenzyme A is (a) decarboxylation, (b) oxidation, and (c) esterification, write a balanced equation for each of these steps using structural formulas for pyruvic acid and its derivatives.

Since each glucose molecule yields two pyruvic acid molecules, two acetyl coenzyme A molecules are produced from each glucose. Additionally, two $NADH_2$ molecules are formed from metabolism of each glucose. Upon interaction with the respiratory chain, each of these $NADH_2$ molecules leads to the formation of three additional ATP molecules. Thus, the cell obtains six more ATP molecules from the original glucose molecule. To this point, the cell has a net gain of 14 ATP molecules from a single glucose molecule.

Each of the acetyl coenzyme A molecules produced from glucose are converted to CO_2 and water via the Krebs cycle. For each acetyl coenzyme A reacting via the Krebs cycle, 12 more ATP molecules are produced. This allows the formation of 24 more ATP molecules from the initial glucose molecule, *for a grand total of 38 ATP molecules gained by the cell during the conversion of a single molecule of glucose to carbon dioxide and water.* The accounting for these 38 ATP molecules is summarized in Fig. 13-3.

glucose

↓ glycolysis

2 pyruvic acid + 2ATP + $2NADH_2$

↓

$2CO_2$ + 2 acetyl coenzyme A + $2NADH_2$

↓ Krebs cycle and respiratory chain

$4CO_2 + 6H_2O + 24ATP$

Subtotal: $6CO_2 + 6H_2O + 26ATP + 4NADH_2$

12ATP ← respiratory chain (from $4NADH_2$)

Grand total: $6CO_2 + 6H_2O + 38ATP$

FIGURE 13-3 Summary of ATP production from glucose.

A CLOSER LOOK AT GLYCOLYSIS

Step 1. The first step in glycolysis is the conversion of glucose-6-phosphate to fructose-6-phosphate, catalyzed by the enzyme *phosphohexose isomerase.*

$$\text{glucose-6-phosphate} \xrightleftharpoons{\text{phosphohexose isomerase}} \text{fructose-6-phosphate}$$

glucose-6-phosphate fructose-6-phosphate

As in most biochemical reactions, this isomerism is a reversible reaction. In this particular case, the equilibrium favors glucose-6-phosphate over fructose-6-phosphate. At equilibrium there would be 70 percent glucose-6-phosphate and 30 percent fructose-6-phosphate. As the fructose-6-phosphate is used in subsequent steps, however, Le Chatelier's principle tells us that more glucose-6-phosphate will be isomerized to fructose-6-phosphate in an attempt to maintain equilibrium. Thus, the reaction is eventually driven to completion and the organism can utilize all the available glucose.

Glucose-6-phosphate is not the sole source of fructose-6-phosphate. Any fructose present from a dietary source can be phosphorylated directly by ATP under the action of fructokinase. Thus, a cell can use fructose as an energy source as well as glucose. These relationships are shown in Fig. 13-4.

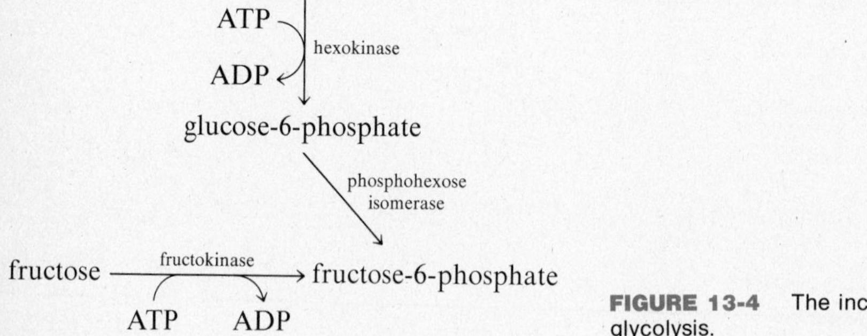

FIGURE 13-4 The incorporation of fructose into glycolysis.

Step 2. The next step in glycolysis is phosphorylation of fructose-6-phosphate to fructose-1,6-diphosphate. This requires ATP and is catalyzed by the enzyme phosphofructokinase.

fructose-6-phosphate $\xrightarrow{\text{phosphofructokinase}}$ fructose-1,6-diphosphate

In this reaction, note the use of a second ATP molecule. To this point in glycolysis, the cell has invested two ATP molecules in each monosaccharide molecule and, as yet, has not received any energy back. Thus, so far the process has been *endergonic* rather than *exergonic.*

Step 3. The fructose-1,6-diphosphate molecule is now split into two three-carbon fragments. To aid in visualizing this reaction more easily, the fructose-1,6-diphophate molecule is written in the open-chain form.

fructose-1,6-diphosphate $\xrightarrow{\text{aldolase}}$ dihydroxyacetone phosphate + 3-phosphoglyceraldehyde

Remember that the goal of glycolysis is production of pyruvic acid, a three-carbon material.

The net effect of this reaction is to break the bond between carbon 3 and carbon 4 of fructose-1,6-diphosphate. Simultaneously, the hydrogen of the —OH on carbon 4 is lost and a carbon–oxygen double bond is formed producing 3-phosphoglyceraldehyde. Carbon 3 of the fructose-1,6-diphosphate gains a hydrogen to yield dihydroxyacetone phosphate.

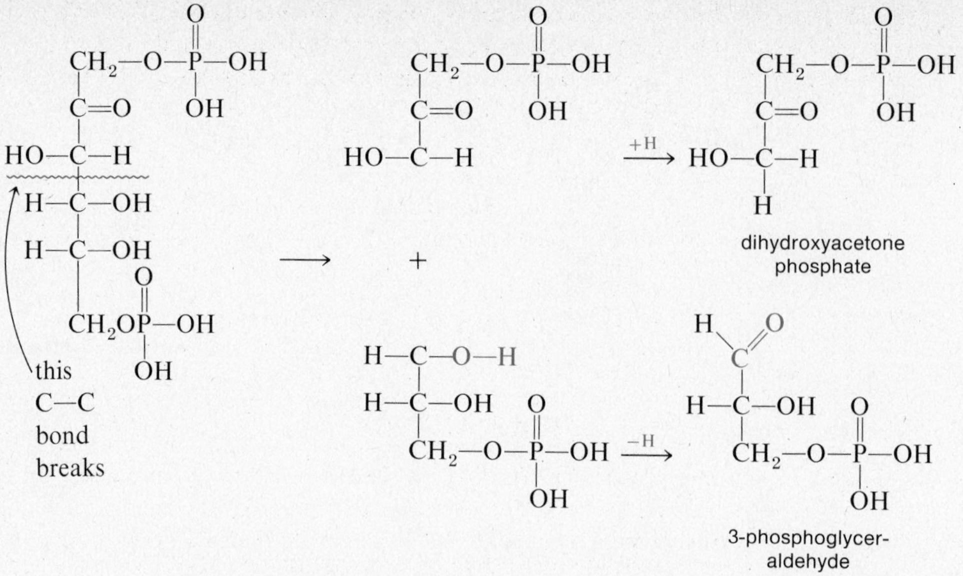

$$
\begin{array}{c}
\text{fructose 1,6-diphosphate: } CH_2{-}O{-}P(=O)(OH){-}OH,\ C{=}O,\ HO{-}C{-}H \ \wr\wr\ H{-}C{-}OH,\ H{-}C{-}OH,\ CH_2OP(=O)(OH){-}OH \\
\longrightarrow \\
CH_2{-}O{-}P(=O)(OH){-}OH,\ C{=}O,\ HO{-}C{-}H \xrightarrow{+H} CH_2{-}O{-}P(=O)(OH){-}OH,\ C{=}O,\ HO{-}C{-}H,\ H \\
+ \\
H{-}C{-}O{-}H,\ H{-}C{-}OH,\ CH_2{-}O{-}P(=O)(OH){-}OH \xrightarrow{-H} H{-}C{=}O,\ H{-}C{-}OH,\ CH_2{-}O{-}P(=O)(OH){-}OH
\end{array}
$$

this C—C bond breaks

dihydroxyacetone phosphate

3-phosphoglyceraldehyde

Step 4. Cells can only utilize the 3-phosphoglyceraldehyde directly. This doesn't mean that the dihydroxyacetone phosphate is wasted. An enzyme capable of isomerizing dihydroxyacetone phosphate to 3-phosphoglyceraldehyde is present in cells.

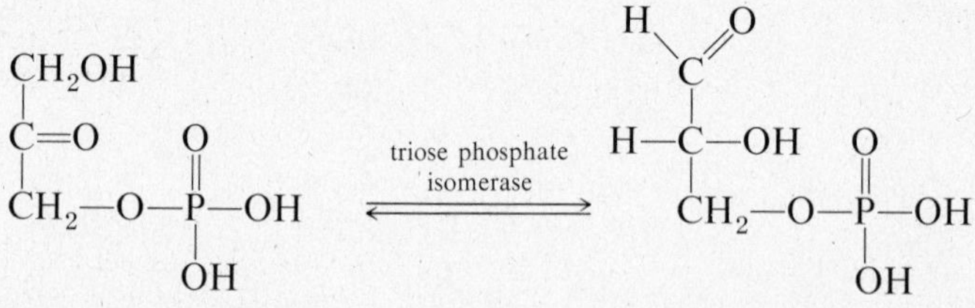

$$
\underset{\text{dihydroxyacetone phosphate}}{CH_2OH{-}C(=O){-}CH_2{-}O{-}P(=O)(OH){-}OH} \underset{\text{isomerase}}{\overset{\text{triose phosphate}}{\rightleftharpoons}} \underset{\text{3-phosphoglyceraldehyde}}{H{-}C(=O){-}CH(OH){-}CH_2{-}O{-}P(=O)(OH){-}OH}
$$

The enzyme effects the movement of a hydrogen atom from carbon 1 to carbon 2. Also, a hydrogen from the —OH group of carbon 1 is moved onto the oxygen of carbon 2. In essence, the primary alcohol of dihydroxyacetone phosphate is oxidized to an aldehyde and the ketone group of carbon 2 is reduced to an alcohol. Again we see a reversible reaction but as 3-phosphoglyceraldehyde is used, dihydroxyacetone phosphate is isomerized in an attempt to maintain equilibrium.

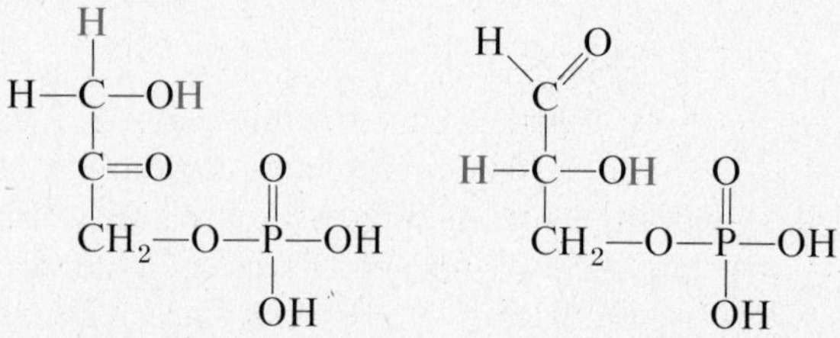

$$
\underset{\text{dihydroxyacetone phosphate}}{H{-}CH(OH){-}C(=O){-}CH_2{-}O{-}P(=O)(OH){-}OH} \qquad \underset{\text{3-phosphoglyceraldehyde}}{H{-}C(=O){-}CH(OH){-}CH_2{-}O{-}P(=O)(OH){-}OH}
$$

Some of the dihydroxyacetone phosphate can be converted to glycerol phosphate.

Eventually the reaction is driven to completion and all the dihydroxyacetone phosphate is converted to 3-phosphoglyceraldehyde. We might also notice that an asymmetric carbon is formed during this isomerization. The specificity of the enzyme is such that only the required D-3-phosphoglyceraldehyde is produced and none of the L isomer.

Step 5. The 3-phosphoglyceraldehyde undergoes oxidation and phosphorylation in the next step. The oxidizing agent for this step is NAD.

$$\underset{\text{3-phosphoglyceraldehyde}}{\text{H(O=)C—CH(OH)—CH}_2\text{—O—P(=O)(OH)—OH}} \xrightarrow[\text{NAD} \;\to\; \text{NADH}_2]{\text{P}_i} \underset{\text{glyceric acid-1,3-diphosphate}}{\text{HO—P(=O)(OH)—O—C(=O)—CH(OH)—CH}_2\text{—O—P(=O)(OH)—OH}}$$

Although the product of the reaction is named as glyceric acid, we see that what is really formed is the mixed anhydride of glyceric acid and phosphoric acid.

exercise 13-2

Identify the anhydride group in glyceric acid-1,3-diphosphate. How should the other phosphate group be classified? Draw the structural formula of glyceric acid.

Step 6. Recall from Chapter 10 that an anhydride linkage is a high-energy group and hydrolysis of the linkage releases energy. The hydrolysis of the mixed anhydride of glyceric acid-1,3 diphosphate releases sufficient energy for the formation of ATP.

$$\underset{\text{glyceric acid-1,3-diphosphate}}{\text{HO—P(=O)(OH)—O—C(=O)—CH(OH)—CH}_2\text{—O—P(=O)(OH)—OH}} \xrightarrow[\text{ADP} + \text{H}_2\text{PO}_4^- \;\to\; \text{ATP}]{} \underset{\text{3-phosphoglyceric acid}}{\text{CO}_2\text{H—CH(OH)—CH}_2\text{—O—P(=O)(OH)—OH}}$$

So far, 2 ATP's are used and 2 ATP's are formed.

Each glucose produces two glyceric acid-1,3-diphosphate molecules. From each glyceric acid-1,3-diphosphate, one ATP is formed. Thus, we see the formation of two ATP molecules from each glucose molecule. *At this point in the breakdown of glucose there has been no net change in the number of*

ATP molecules available to the cell. Remember two ATP molecules have been used in glycolysis and now two ATP molecules have been regenerated. More energy is available from 3-phosphoglyceric acid, as we shall see.

Step 7. Another isomerization occurs next, in which 3-phosphoglyceric acid is converted to 2-phosphoglyceric acid.

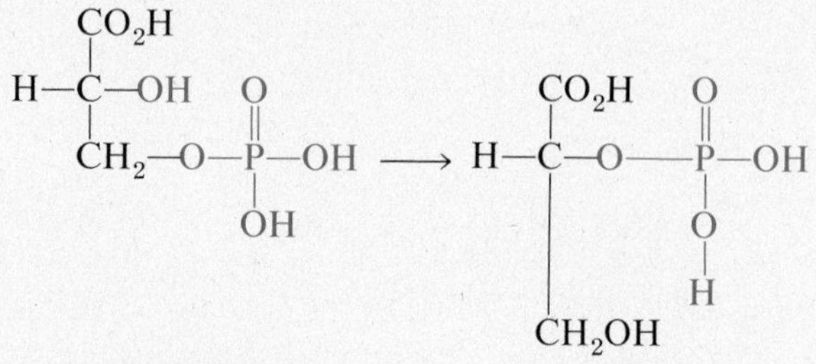

```
    CO2H
     |
 H—C—OH    O                         CO2H    O
     |      ||                          |      ||
    CH2—O—P—OH   ⟶   H—C—O——P—OH
              |                         |      |
             OH                         |      O
                                        |      |
                                        |      H
                                       CH2OH
```

3-phosphoglyceric acid　　　2-phosphoglyceric acid

Step 8. Water is removed from 2-phosphoglyceric acid to produce phosphoenol pyruvic acid. This reaction is the dehydration of an alcohol to form a carbon–carbon double bond.

Phosphoenol pyruvic acid is sometimes abbreviated PEP.

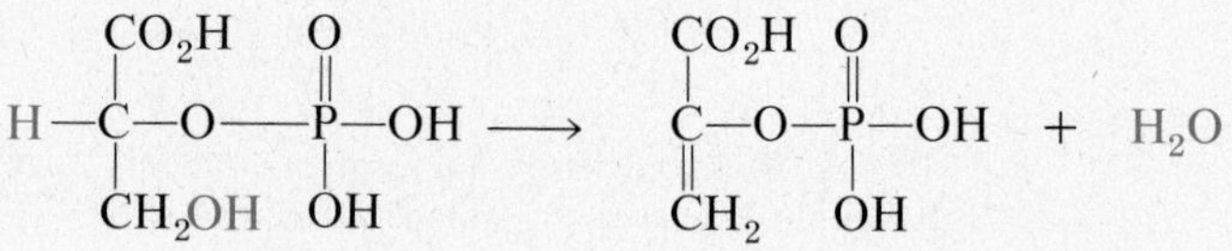

```
     CO2H     O                 CO2H O
      |       ||                  |   ||
  H—C—O——P—OH   ⟶    C—O—P—OH  +  H2O
      |       |                   ||   |
     CH2OH   OH                  CH2  OH
```

2-phosphoglyceric acid　　　phosphoenol pyruvic acid

Step 9. The structure of phosphoenol pyruvic acid is rather interesting if we compare it to the structure of a mixed carbon–phosphate anhydride. Pay particular attention to the colored portion of phosphoenol pyruvic acid in the structure below in comparison to the structure of an anhydride.

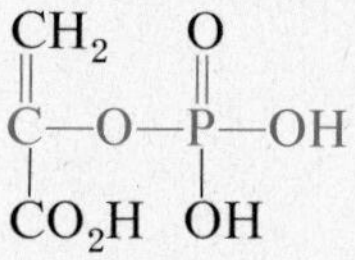

```
   CH2    O             O     O
   ||     ||            ||    ||
   C—O—P—OH            C—O—P—OH
   |      |             |     |
  CO2H   OH             R     O
                              |
                              H
```

phosphoenol pyruvic acid 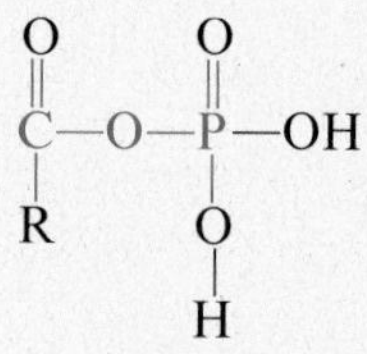　　　a mixed carbon–phosphate anhydride

Except for the fact that phosphoenol pyruvic acid has a carbon-to-carbon double bond instead of a carbon-to-oxygen double bond, the structures look very similar. (The $—CO_2H$ group is merely the R group of the mixed anhydride.) This similarity in structure is apparently sufficient to make the phosphate linkage in phosphoenol pyruvic acid a high-energy one. Chemists find that the hydrolysis of phosphoenol pyruvic acid releases enough energy to allow the cell to produce ATP from ADP and inorganic phosphate.

$$\underset{\text{phosphoenol pyruvic acid}}{\begin{array}{c}CO_2H\ \ O\\ |\qquad \|\\ C{-}O{-}P{-}OH\\ \|\qquad |\\ CH_2\quad OH\end{array}} \xrightarrow[\text{ADP} + P_i \ \to\ \text{ATP}]{} \underset{\text{pyruvic acid}}{\begin{array}{c}CO_2H\\ |\\ C{=}O\\ |\\ CH_3\end{array}}$$

Two pyruvic acids per glucose.

Since two phosphoenol pyruvic acid molecules result from each molecule of glucose undergoing glycolysis, there has been a net gain to the cell of two ATP molecules from each glucose. Thus, the cell has been able to obtain and store for its own purposes some of the energy locked into the glucose molecule during photosynthesis.

Step 10. The fate of the pyruvic acid depends both on the organism and the circumstances in which it finds itself. Remember that NAD is reduced to $NADH_2$ during the oxidation of 3-phosphoglyceric acid to glyceric acid-1,3-diphosphate. Thus, from one molecule of glucose, the cell has had a net gain of two ATP molecules and is left with two $NADH_2$ molecules.

How much energy has been stored in the production of 8 moles of ATP from a mole of glucose?

The cell has only a limited amount of NAD and cannot afford to discard the $NADH_2$ formed. In order to conserve NAD, the $NADH_2$ must be reoxidized by some means. A mammal operating under normal body conditions reoxidizes the $NADH_2$ to NAD via the respiratory chain. Recall from Chapter 12 that three ATP molecules are produced for each $NADH_2$ interacting with the respiratory chain. Thus, six more ATP molecules are available to a mammal from each glucose molecule. *This leads to a net formation of eight ATP molecules* (six from the two $NADH_2$ molecules plus the two produced directly during glycolysis) *from each glucose molecule undergoing glycolysis.*

How much energy has been stored during the anaerobic glycolysis of a mole of glucose to 2 moles of ATP?

You might be wondering what organisms without a respiratory chain do with the $NADH_2$. How do they manage to reoxidize it to NAD? Such organisms reoxidize the $NADH_2$ to NAD by reducing pyruvic acid. Yeasts, for example, decarboxylate the pyruvic acid to acetaldehyde and then reduce the acetaldehyde to ethyl alcohol. We call the breakdown of glucose to ethanol and carbon dioxide *fermentation.*

$$\begin{array}{c}CO_2H\\ |\\ C{=}O\\ |\\ CH_3\end{array} \longrightarrow \underset{\text{carbon dioxide}}{CO_2} + \underset{\text{acetaldehyde}}{CH_3C({=}O)H} \qquad CH_3C({=}O)H \xrightarrow[NADH_2 \ \to\ NAD]{} \underset{\text{ethyl alcohol}}{\begin{array}{c}CH_3CH{-}OH\\ |\\ H\end{array}}$$

Other microorganisms directly reduce the pyruvic acid to lactic acid.

$$\underset{\substack{\text{pyruvic}\\\text{acid}}}{\begin{array}{c}CO_2H\\|\\C{=}O\\|\\CH_3\end{array}} \xrightarrow[\text{NADH}_2 \quad \text{NAD}]{} \underset{\substack{\text{lactic}\\\text{acid}}}{\begin{array}{c}CO_2H\\|\\H{-}C{-}O{-}H\\|\\CH_3\end{array}}$$

The conversion of glucose to lactic acid is often referred to as *anaerobic glycolysis*. Notice that anaerobic glycolysis yields only a net gain of two ATP. The reactions involved in glycolysis are summarized in Fig. 13-5 (page 359).

THE FATE OF PYRUVIC ACID UNDER AEROBIC CONDITIONS

We have discussed the fate of pyruvic acid by anaerobic organisms and have seen that it is converted either to lactic acid or to ethanol. In mammals and other organisms capable of respiration, the pyruvic acid undergoes further reactions to eventually produce CO_2, water, and ATP via the Krebs cycle. During aerobic oxidation, 38 molecules of ATP are produced per molecule of glucose.

ANAEROBIC VERSUS AEROBIC GLYCOLYSIS—ENERGY PRODUCTION

Organisms capable of only anaerobic glycolysis lack the enzymes required for aerobic glycolysis.

Recall from Chapter 12 that 12 ATP/acetyl CoA are produced via the Krebs cycle and *ETS.*

There is a great difference in the energy produced by anaerobic glycolysis compared to that produced by aerobic glycolysis. Organisms capable of anaerobic glycolysis gain 2 ATP molecules per molecule of glucose metabolized; whereas, those capable of aerobic glycolysis obtain 38 ATP molecules per glucose molecule metabolized. Aerobic glycolysis leads to a gain of 19 times as much energy as does anaerobic glycolysis. Without the development of aerobic respiration it is doubtful that organisms incapable of photosynthesis would have evolved much beyond unicellular ones. In other words, without aerobic respiration it is doubtful that the animal kingdom as we know it would exist.

In Chapter 12, it was mentioned that 686 kcal of energy was trapped in each mole of glucose during photosynthesis. The formation of each mole of ATP from ADP stores 8 kcal of energy. Production of 38 moles of ATP from each mole of glucose means that the cell has stored 304 kcal (44 percent) of the energy in a mole of glucose. The remainder of the energy is lost as heat.

One other aspect of glycolysis that should be discussed is the possibility of anaerobic glycolysis in muscle tissue. Under normal conditions a mammal uses aerobic glycolysis. For short periods of time, however, it is possible for muscles to receive energy via anaerobic glycolysis. This situation arises during strenous exercise. During exercise, the demand for oxygen may exceed the body's ability to carry enough of it to the muscles. In order to continue functioning, the muscle cells resort to anaerobic glycolysis, producing lactic acid. If the condition prevails for any extended time, lactic

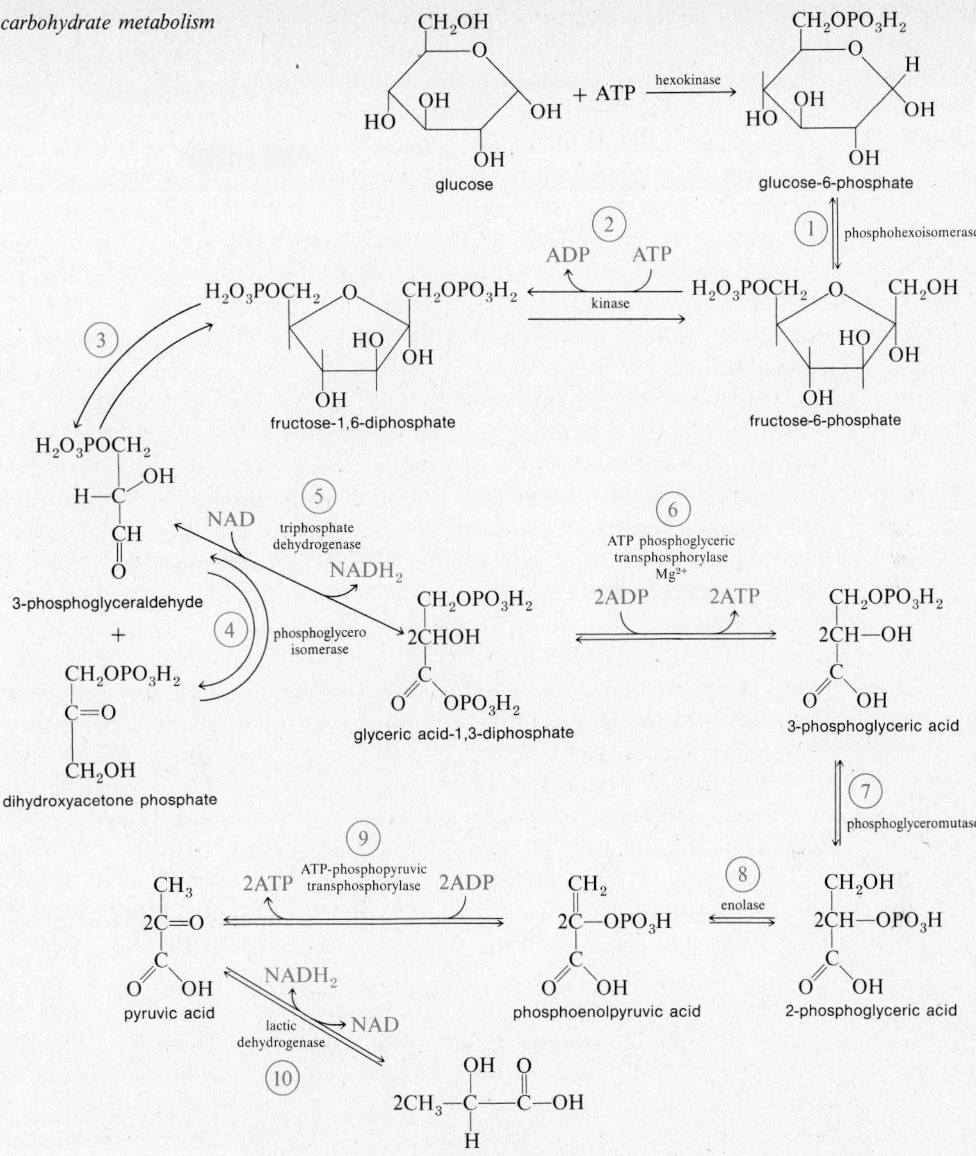

FIGURE 13-5 The anaerobic conversion of glucose to lactic acid.

acid begins to accumulate in blood and muscles. The lactic acid lowers the blood pH and may lead to acidosis and its accompanying consequences. The lactic acid accumulated in muscle irritates the tissues and may lead to considerable soreness. After vigorous exercise ceases and oxygen demand regains balance with oxygen supply, the lactic acid is reoxidized to pyruvic acid and thence to acetyl coenzyme A and is metabolized in the normal fashion.

THE FORMATION OF GLYCOGEN—GLYCOGENESIS AND GLYCONEOGENESIS

Glycogenesis—*making glycogen from glucose;* gluconeogenesis—*making glycogen from lactic acid.*

Earlier it was mentioned that glucose could be stored in the body as glycogen. This storage can occur in both liver and muscles. Glycogen formation is a form of short-term storage of carbohydrates and in normal individuals it occurs when the glucose level in the blood exceeds approximately 100 mg of glucose per 100 ml of blood. The first step in *glycogenesis* is the isomerization of glucose-6-phosphate to glucose-1-phosphate. Following the formation of glucose-1-phosphate, the glucose molecules are linked together by 1,4-α acetal linkages and 1,6-α acetal linkages to form glycogen. The individual steps by which this is accomplished are very complex and will not be studied in this text. We should note, however, that glycogenesis is an energy-consuming process. The amount of glycogen stored is controlled by several hormones.

Glycogen can also be formed from lactic acid. The lactic acid produced in muscles during vigorous exercise must be removed from the muscle tissue. Seldom is this lactic acid immediately oxidized to carbon dioxide and water. The oxidation occurs only if ATP is needed at the time. If the demand for ATP is small, the body reconverts the majority of the lactic acid back to glucose and then to glycogen. The conversion of lactic acid to glycogen is called *gluconeogenesis*. This process occurs primarily in the liver and kidneys. Muscle cells do not contain a sufficient concentration of the enzymes required to catalyze the conversion of lactic acid to glucose. The overall relationships among blood glucose, liver glycogen, muscle glycogen, and lactic acid are shown in Fig. 13-6.

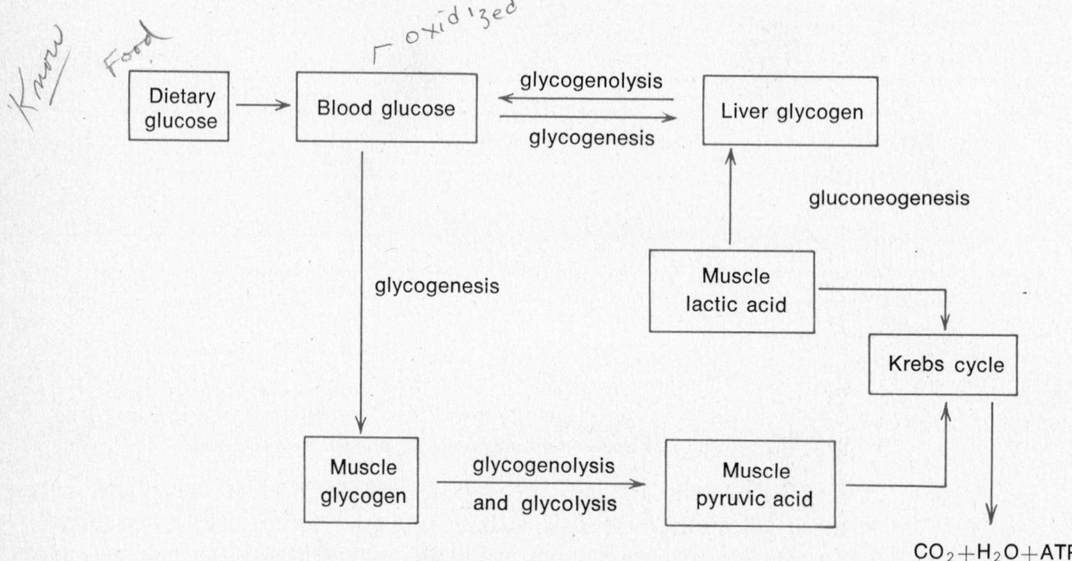

FIGURE 13-6 The pathways for interconversion of glucose, glycogen, and lactic acid.

The relationships between glyconeogenesis and glycolysis are shown in Fig. 13-7.

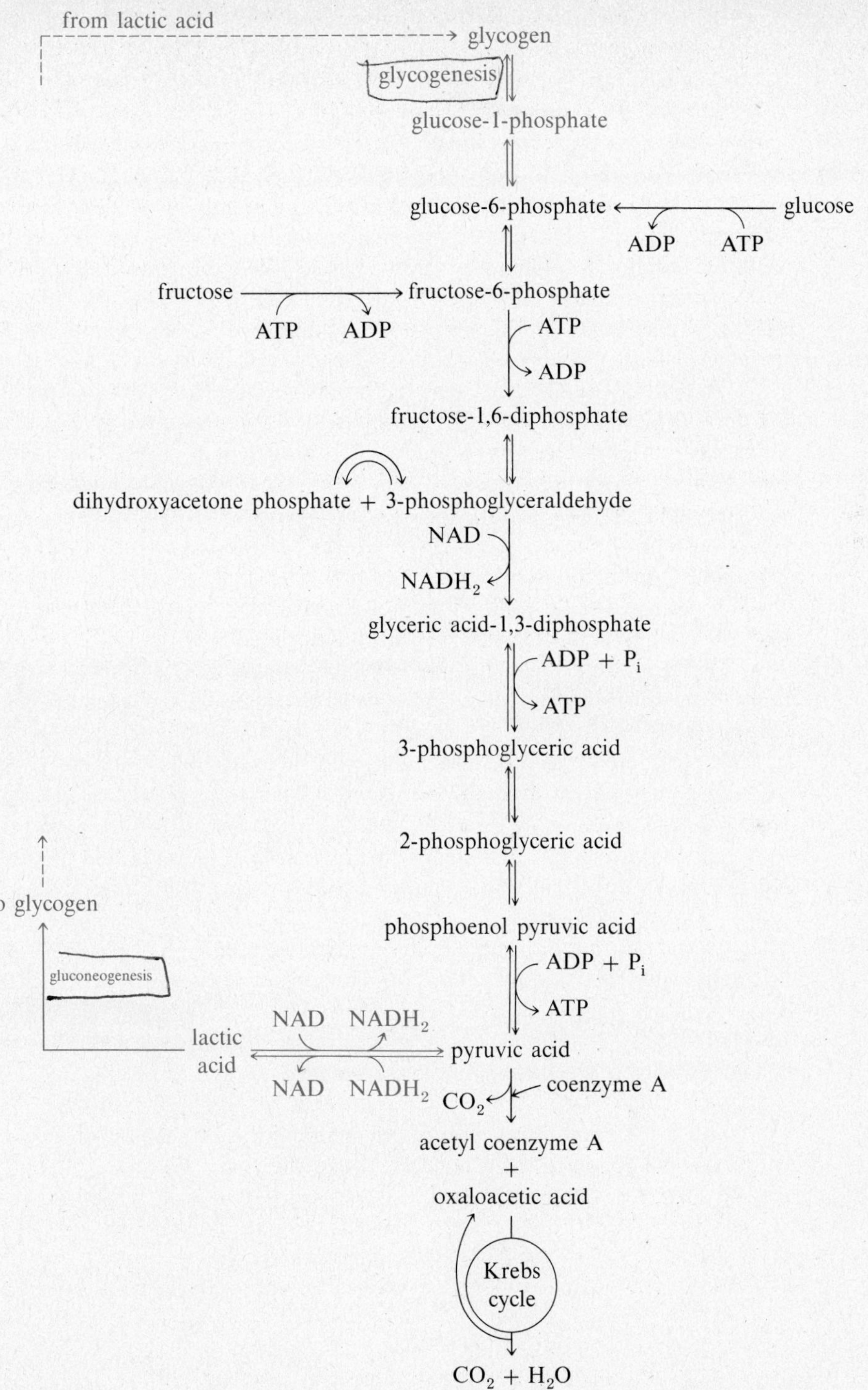

FIGURE 13-7 Metabolic paths for monsaccharides in aerobic organisms.

INTERACTION OF CARBOHYDRATE AND LIPID METABOLISM—LONG-TERM STORAGE

If we eat more than necessary, we gain weight. To say we are gaining weight is merely a slightly less offensive way of saying we are getting fat. This fat is manufactured by the body from excess nutrients included in the diet. The production of fat is *lipogenesis*.

Lipogenesis is the body's means of long-term storage of excess nutrients. The lipids can be made from both fats and the carbohydrates in our diet and to a much lesser extent from dietary protein. Lipids are deposited in regions widely distributed over various parts of the body known as adipose tissue or fat depots. When we speak of lipid storage in the fat depots, it is not meant that the fat is totally inactive. The cells within the adipose tissue are constantly producing new fat molecules and consuming old fat molecules. Thus, there is a continuing turnover of the stored lipids.

The energy reserves present in the fat depots of the normal person are very large compared to other energy reserves within the body. At least 10 percent of body weight is lipid and is sufficient to fulfill the energy requirements of an individual for over a month. In contrast, the body has only enough glycogen to take care of energy requirements for about 24 hr. An advantage to the body of storing energy reserves as fat rather than as glycogen is the high-energy density of fats. There are over twice as many calories per gram of fat as there are per gram of carbohydrate. Fat contains about 9 kcal/g versus 4 kcal/g for carbohydrates.

Correspondingly, when the body metabolizes fats, it obtains a little more than twice as much ATP per gram of fat as it does per gram of carbohydrate metabolized. For example, we saw that a cell makes 38 moles of ATP per mole of glucose metabolized. A mole of glucose weighs 180 g. Therefore, the body produces 38 moles of ATP per 180 g of glucose metabolized or 0.211 mole of ATP per gram of glucose. Recall from Chapter 12 that the body produces 130 moles of ATP per mole of palmitic acid metabolized. This is equivalent to 0.50 mole of ATP per gram of palmitic acid.

FAT SYNTHESIS

The synthesis of fats within the body involves the synthesis of both glycerol and fatty acids. We shall examine the reactions of fatty acid synthesis first. The beginning point is acetyl coenzyme A. Thus, we can see immediately how carbohydrates in the diet can be converted to fat since sugars can be converted to acetyl coenzyme A by glycolysis.

Step 1. The initial reaction is the combination of acetyl coenzyme A with carbon dioxide to yield malonyl coenzyme A.

$$CH_3C(=O)-S-CoA + CO_2 \xrightarrow{ATP \rightarrow ADP} HO_2C-CH_2-C(=O)-S-CoA$$

acetyl coenzyme A → malonyl coenzyme A

Step 2. The second step is the combination of another acetyl coenzyme A with malonyl coenzyme A. This step is accompanied by the loss of carbon dioxide. It is interesting to note that although CO_2 is required to start the process, it is lost almost immediately.

$$\underset{\text{malonyl coenzyme A}}{HO_2C{-}CH_2{-}\overset{O}{\overset{\|}{C}}{-}S{-}CoA} + \underset{\text{acetyl coenzyme A}}{CH_3\overset{O}{\overset{\|}{C}}{-}S{-}CoA} \xrightarrow[]{\text{coenzyme A},\ CO_2} \underset{\text{acetoacetyl coenzyme A}}{CH_3\overset{O}{\overset{\|}{C}}{-}CH_2\overset{O}{\overset{\|}{C}}{-}S{-}CoA}$$

Prior to the discovery of the formation and conversion of malonyl CoA, fatty acid synthesis was considered to be exactly the reverse of beta *oxidation.*

Other than the gain and loss of carbon dioxide, the synthesis of fatty acids is essentially the reverse of beta oxidation of fatty acids studied in Chapter 12.

Step 3. The ketone group in acetoacetyl coenzyme A is reduced to an alcohol. The reducing agent is $NADPH_2$.

$$\underset{\text{acetoacetyl coenzyme A}}{CH_3\overset{O}{\overset{\|}{C}}{-}CH_2\overset{O}{\overset{\|}{C}}{-}S{-}CoA} \xrightarrow{NADPH_2 \quad NADP} \underset{\beta\text{-hydroxybutyryl coenzyme A}}{CH_3{-}\overset{OH}{\overset{|}{C}H}{-}CH_2\overset{O}{\overset{\|}{C}}{-}S{-}CoA}$$

Step 4. Next, a loss of water occurs:

$$\underset{\beta\text{-hydroxybutyryl coenzyme A}}{CH_3\overset{OH}{\overset{|}{C}H}{-}\overset{H}{\overset{|}{C}H}{-}\overset{O}{\overset{\|}{C}}{-}S{-}CoA} \xrightarrow{H_2O} \underset{\text{crotonyl coenzyme A}}{CH_3CH{=}CH{-}\overset{O}{\overset{\|}{C}}{-}S{-}CoA}$$

Step 5. The double bond is now reduced by $NADPH_2$ to yield butyryl coenzyme A.

$$\underset{\text{crotonyl coenzyme A}}{CH_3CH{=}CH{-}\overset{O}{\overset{\|}{C}}{-}S{-}CoA} \xrightarrow{NADPH_2 \quad NADP} \underset{\text{butyryl coenzyme A}}{CH_3\underset{H}{\underset{|}{C}H}{-}\underset{H}{\underset{|}{C}H}{-}\overset{O}{\overset{\|}{C}}{-}S{-}CoA}$$

Step 6. The butyryl coenzyme A may combine with another malonyl coenzyme A with the loss of CO_2. This is followed by repeating steps 3,

4, and 5 to yield the coenzyme A derivative of a six-carbon acid. The process of combination with malonyl coenzyme A and the three subsequent steps are repeated over and over until eventually a fatty acid containing 14 to 20 carbon atoms is formed.

exercise 13-3

Starting with butyryl coenzyme A and using structural formulas, write equations for the formation of

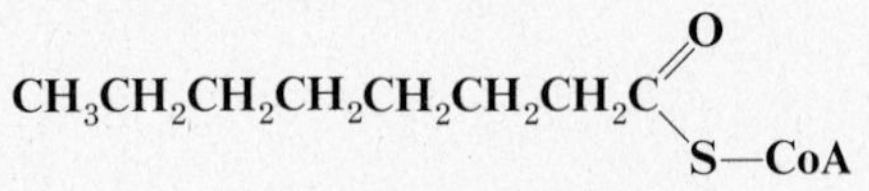

$$CH_3CH_2CH_2CH_2CH_2CH_2CH_2C(=O)\text{—S—CoA}$$

exercise 13-4

Explain why the majority of all naturally occurring fatty acids contains an even number of carbon atoms.

Since glucose is the major source of acetyl coenzyme A for fatty acid synthesis, it is interesting to compare the energy in fatty acids to that in glucose. For the synthesis of palmitic acid from glucose via acetyl CoA one can write the overall equation

$$\underset{\text{glucose}}{4C_6H_{12}O_6} + O_2 \xrightarrow{\text{lipogenesis}} \underset{\text{palmitic acid}}{C_{16}H_{32}O_2} + 8CO_2 + H_2O$$

4 moles glucose $\times \frac{686 \text{ kcal}}{\text{mole glucose}} = 2744 \text{ kcal}$

$\frac{2400 \text{ kcal}}{2744 \text{ kcal}} \times 100 \simeq 88\%$

Four moles of glucose releases 2744 kcal, while one mole of palmitic acid releases 2400 kcal when these amounts of compounds are converted to carbon dioxide and water. We can calculate that about 88 percent of the available energy in glucose is stored in palmitic acid, making the energy efficiency of lipogenesis very high.

The final step of fat synthesis is the combination of the fatty acids with glycerol. The source of glycerol is the $NADH_2$ reduction of dihydroxyacetone phosphate, produced during glycolysis to glycerol phosphate.

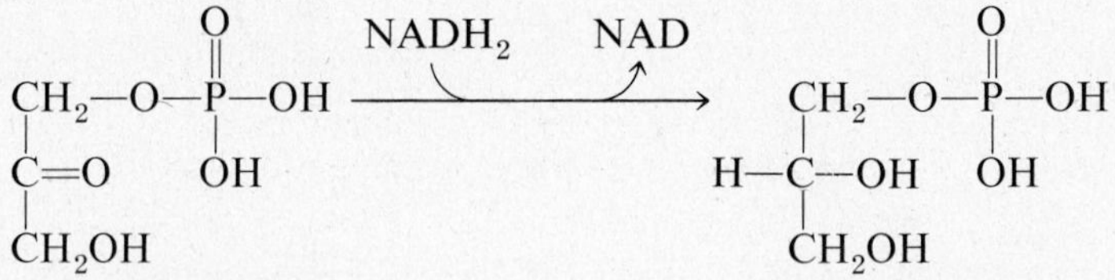

$$\underset{\text{dihydroxyacetone phosphate}}{CH_2\text{—O—P(=O)(OH)—OH},\ C{=}O,\ CH_2OH} \xrightarrow{NADH_2 \to NAD} \underset{\text{glycerol phosphate}}{CH_2\text{—O—P(=O)(OH)—OH},\ H\text{—C—OH},\ CH_2OH}$$

Three moles of fatty acids react with 1 mole of glycerol phosphate to yield triglyceride plus inorganic phosphate.

exercise 13-5

Using structural formulas, write an equation for the formation of a triglyceride from 3 moles of palmitic acid and glycerol phosphate.

The reduction of dihydroxyacetone phosphate to glycerol phosphate is a reversible reaction. In the catabolism of lipids, glycerol is released. This glycerol is phosphorylated by ATP to produce glycerol phosphate.

$$\begin{array}{l} CH_2OH \\ | \\ CHOH \\ | \\ CH_2OH \\ \text{glycerol} \end{array} \xrightarrow{ATP \quad ADP} \begin{array}{l} CH_2{-}O{-}\overset{\overset{\displaystyle O}{\|}}{P}{-}OH \\ | \qquad\qquad\ \ | \\ CHOH \qquad OH \\ | \\ CH_2OH \\ \text{glycerol phosphate} \end{array}$$

The glycerol phosphate is oxidized to dihydroxyacetone by NAD. The dihydroxyacetone enters into the glycolysis scheme just as if its source were glucose (see Fig. 13-8 on page 366).

GALACTOSEMIA Galactosemia was mentioned in Chapter 9 as a disease involving an inborn defect in the metabolism of galactose that allows galactose to accumulate in the body. Now after studying the metabolism of carbohydrates we can understand this disease on a molecular basis. In a normal individual, ingested galactose is phosphorylated by ATP to yield galactose-1-phosphate. This reaction is catalyzed by galactokinase.

$$\text{galactose} \xrightarrow[\text{ATP} \quad \text{ADP}]{\text{galactokinase}} \text{galactose-1-phosphate}$$

The galactose-1-phosphate is next isomerized to glucose-1-phosphate by the enzyme *phosphogalactose uridyl transferase.*

$$\text{galactose-1-phosphate} \xrightarrow[\text{uridyl transferase}]{\text{phosphogalactose}} \text{glucose-1-phosphate}$$

exercise 13-6

Using structural formulas for the monosaccharides, write equations for the two-step conversion of galactose to glucose-1-phosphate.

The glucose-1-phosphate may enter the normal metabolic pathways for glucose. Thus, galactose can be utilized to meet the body's demands for carbohydrates.

A few individuals are born with an inability to synthesize phosphogalactose uridyl transferase. In the absence of this enzyme the dietary

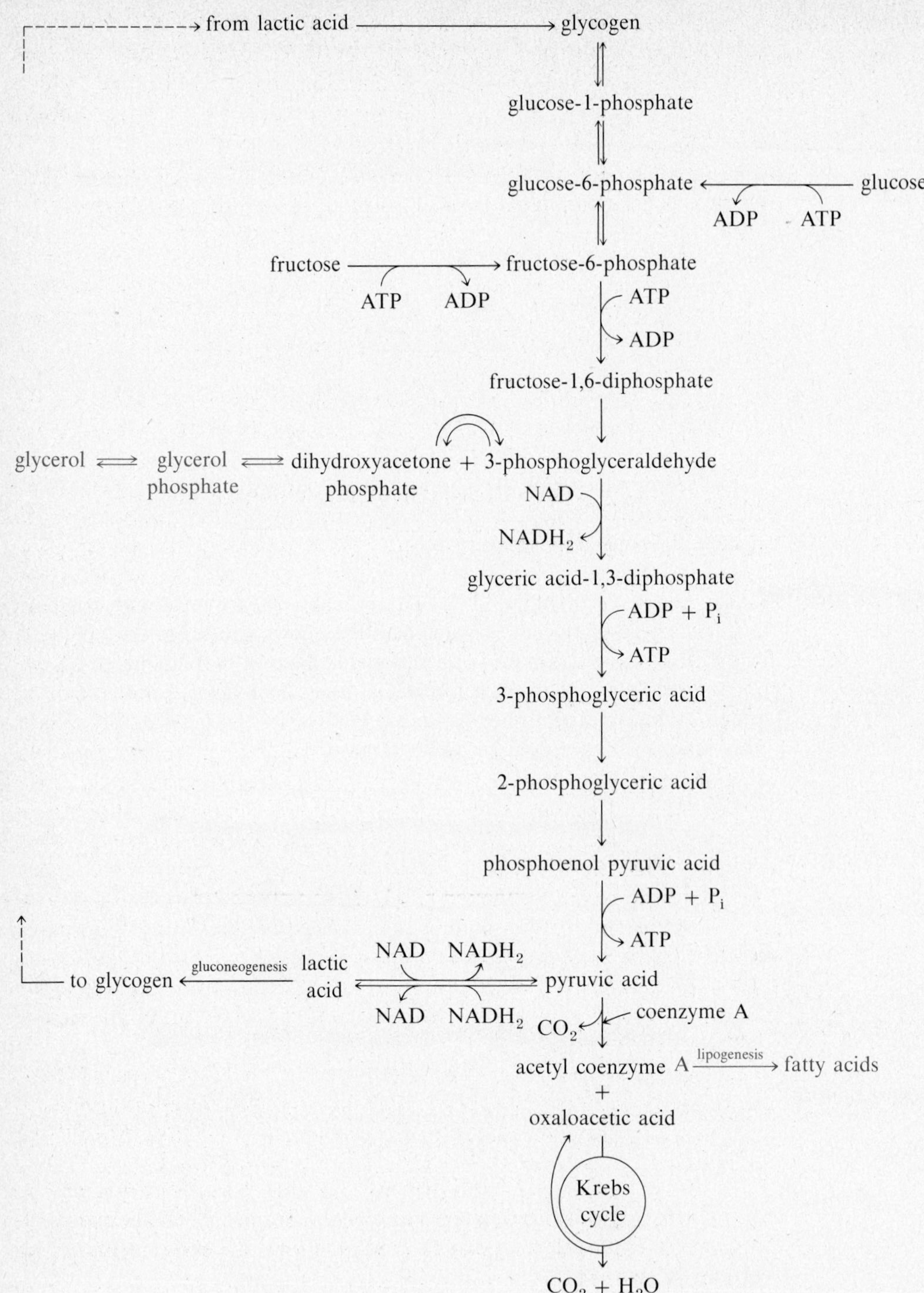

FIGURE 13-8 Interaction of carbohydrate and lipid metabolism.

galactose cannot be converted to glucose-1-phosphate. No alternate metabolic pathway for galactose exists in infants. Consequently the galactose accumulates in the body, primarily in the brain, liver, and eyes. The buildup of galactose in the eyes leads to the formation of *cataracts,* opacity of the optic lens.

Since the body apparently neither requires nor synthesizes large quantities of galactose, the disease can be controlled by exclusion of galactose from the diet of young humans. Later in life an alternate pathway for the metabolism of galactose develops as the concentration of the enzyme required for the alternate pathway (uridine diphosphate galactose pyrophosphorylase) increases. The presence of this enzyme in the liver allows an adult lacking the transferase to metabolize normal amounts of galactose.

DIABETES

Diabetes mellitus is a more widespread abnormality of carbohydrate metabolism than is galactosemia. Diabetes is the result of the inability of the pancreas to provide adequate insulin, which regulates glucose concentration in the blood by promoting glycogenesis in the liver and muscles and by controlling cellular oxidation of glucose by influencing glucose transport across cell membranes.* In the absence of insulin, blood glucose concentration rises and if it exceeds the renal threshold, glucose appears in the urine (see Fig. 13-9). The appearance of glucose in the urine is indicative that other adverse conditions may exist in the body as a consequence of the inability of the person to metabolize glucose properly. Possible conditions are ketosis, acidosis, electrolyte imbalance, and fatty liver. Let's see how these conditions arise.

Presence of glucose in urine is called glucosuria.

Since glucose cannot enter the cells without insulin, it cannot be oxidized to provide energy to the cells. The body responds to the cellular energy demand by resorting to its alternate energy source, the oxidation of fats. The depot fat is mobilized and brought to the liver by the bloodstream (recall from Chapter 12 that the liver is the major site of fat degradation). In the liver, beta oxidation breaks down fatty acids to acetyl coenzyme A. To this point everything appears to be proceeding smoothly, and we would expect that acetyl coenzyme A would be utilized via the Krebs cycle to produce ATP. Unfortunately, operation of the Krebs cycle is impaired in the diabetic. This impairment can be traced directly back to the lack of glucose metabolism.

Referring back to Fig. 12-13 we see that, theoretically, each turn of the Krebs cycle regenerates oxaloacetic acid to begin the cycle over again. This process is not completely efficient, however, and some of the Krebs cycle's intermediates are lost to alternate metabolic pathways.

The loss of some of these intermediates means some oxaloacetic acid is not regenerated to initiate the next turn of the cycle. Thus, to keep the Krebs cycle active and continue to utilize acetyl coenzyme A, there must be a continuing supply of new oxaloacetic acid.

*Discussion of some details of how insulin is thought to regulate blood glucose level is in Chapter 16.

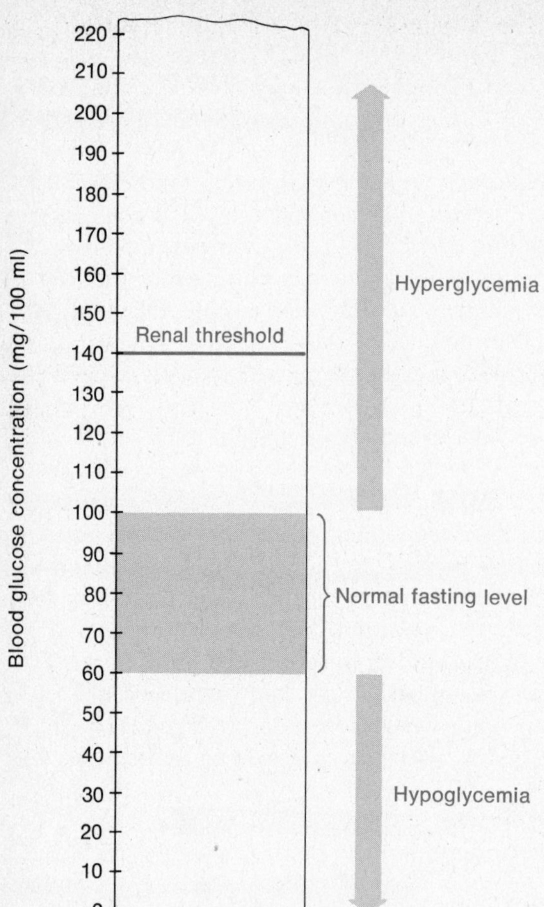

FIGURE 13-9 Summary of blood glucose levels and normal responses. During hypoglycemia, glycogenolysis and glyconeogenesis occur to increase glucose concentration. When glucose concentration is in the normal range, glycogenesis and glycogenolysis occur to maintain concentration. During hyperglycemia, glycogenesis and lipid synthesis occur to decrease glucose concentration. When glucose concentration exceeds the renal threshold, glucose is removed from the blood by the kidneys and appears in the urine.

Carboxylation increases carbon chain length.

There are two major sources of oxaloacetic acid in animal tissue. One of these is the carboxylation (addition of CO_2) of pyruvic acid to produce L-malic acid:

$$\underset{\text{pyruvic acid}}{\begin{array}{c} CO_2H \\ | \\ C{=}O \\ | \\ CH_3 \end{array}} + CO_2 \xrightarrow{NADPH_2 \;\; \curvearrowright \;\; NADP} \underset{\text{L-malic acid}}{\begin{array}{c} CO_2H \\ | \\ HO{-}C{-}H \\ | \\ CH_2 \\ | \\ CO_2H \end{array}}$$

The L-malic acid is then oxidized to oxaloacetic acid by NAD.

exercise 13-7

Using structural formulas, write a balanced equation for the oxidation of L-malic acid to oxaloacetic acid by NAD.

The other major source of oxaloacetic acid is the carboxylation of phosphoenol pyruvic acid:

$$\underset{\text{phosphoenol pyruvic acid}}{\begin{array}{l} CO_2H\ \ O \\ |\qquad\ \ \| \\ C-O-P-OH \\ \|\qquad\ \ | \\ CH_2\quad OH \end{array}} + CO_2 \longrightarrow \underset{\text{oxaloacetic acid}}{\begin{array}{l} CO_2H \\ | \\ C=O \\ | \\ CH_2 \\ | \\ CO_2H \end{array}} + P_i$$

Notice that the source of both pyruvic acid and phosphoenol pyruvic acid is glycolysis (see Fig. 13-10). In the diabetic, glycolysis is impaired,

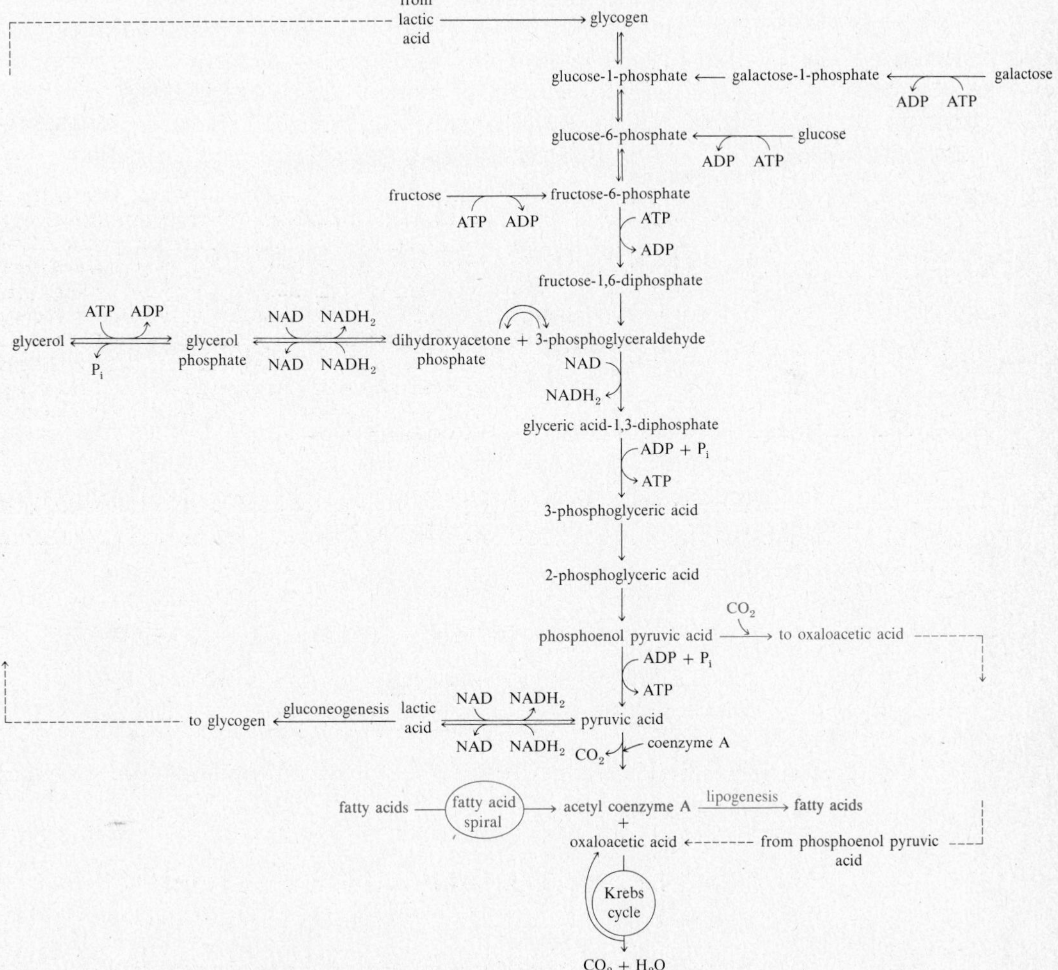

FIGURE 13-10 Overall summary of carbohydrate and lipid metabolism in a normal individual.

Recall from Chapter 6 that concentration changes alter reaction rates.

limiting the amount of pyruvic acid and phosphoenol pyruvic acid available to produce oxaloacetic acid. The resultant decreased amount of oxaloacetic acid leads to a slowdown in the rate of operation of the Krebs cycle. The combination of a decreased rate of acetyl coenzyme A utilization in the Krebs cycle and the increased rate of acetyl coenzyme A formation by beta oxidation of fatty acids in the liver leads to a buildup of acetyl coenzyme A concentration in the body.

In a normal individual, an increased acetyl coenzyme A concentration is a signal for the liver to increase its rate of fatty acid synthesis and simultaneously to decrease the rate of fatty acid breakdown. In the diabetic, however, the energy demands of the body are increasing the rate of fatty acid breakdown. Furthermore, for some very complicated and as yet poorly understood reasons, the liver of a diabetic experiences a decreased ability to synthesize fatty acids. Consequently, in the diabetic, there is an unabated tendency for the buildup of acetyl coenzyme A to continue.

THE FORMATION OF KETONE BODIES IN DIABETES

In an effort to cope with the increased amounts of acetyl coenzyme A, the liver converts acetyl coenzyme A to the so-called *ketone bodies*, which are released into the bloodstream and then eliminated via urine. The ketone bodies are acetoacetic acid, β-hydroxybutyric acid, and acetone.

$$\underset{\text{acetoacetic acid}}{CH_3\overset{\overset{\displaystyle O}{\|}}{C}CH_2CO_2H} \qquad \underset{\beta\text{-hydroxybutyric acid}}{CH_3\underset{\underset{\displaystyle OH}{|}}{C}HCH_2CO_2H} \qquad \underset{\text{acetone}}{CH_3\overset{\overset{\displaystyle O}{\|}}{C}CH_3}$$

Notice that one of these ketone bodies is not a ketone; β-hydroxybutyric acid is included among the ketone bodies since it arises from reduction of acetoacetic acid. In essence, acetoacetic acid arises from the coupling of two acetyl coenzyme A molecules:

$$\underset{\text{acetyl coenzyme A}}{2CH_3{-}C(=O){-}S{-}CoA} \rightleftharpoons \underset{\text{acetoacetyl coenzyme A}}{CH_3{-}\overset{\overset{\displaystyle O}{\|}}{C}{-}CH_2{-}C(=O)SCoA} + CoASH$$

The acetoacetyl coenzyme A undergoes hydrolysis to acetoacetic acid and coenzyme A:

$$\underset{\text{acetoacetyl coenzyme A}}{CH_3{-}\overset{\overset{\displaystyle O}{\|}}{C}{-}CH{-}C(=O)SCoA} + H_2O \rightarrow \underset{\text{acetoacetic acid}}{CH_3{-}\overset{\overset{\displaystyle O}{\|}}{\underset{\uparrow}{C}}{-}CH_2{-}C(=O)OH} + CoASH$$

Acetoacetic acid readily undergoes two reactions in the body. One of these is reduction by $NADH_2$:

$$CH_3\overset{\overset{O}{\|}}{C}CH_2C\begin{matrix}\nearrow O\\ \searrow OH\end{matrix} + NADH_2 \rightleftharpoons CH_3\underset{\underset{OH}{|}}{C}HCH_2C\begin{matrix}\nearrow O\\ \searrow OH\end{matrix} + NAD$$

acetoacetic acid β-hydroxybutyric acid

The other reaction is decarboxylation to acetone:

$$CH_3\overset{\overset{O}{\|}}{C}-CH_2C\begin{matrix}\nearrow O\\ \searrow OH\end{matrix} \longrightarrow CH_3\overset{\overset{O}{\|}}{C}CH_3 + CO_2$$

The formation and release of acetoacetic acid in *small* amounts is a normal continuing process. In a healthy individual the total amounts of ketone bodies is less than 1 mg/100 ml of blood. This is because the ketone bodies can be degraded back to acetyl coenzyme A in muscle tissue and oxidized by the Krebs cycle. In the diabetic, because of the increased rate of acetoacetic acid production and the slowdown of the Krebs cycle, ketone body concentration rises (see Fig. 13-11 on page 372).

If the concentration of ketone bodies in the blood exceeds 1 mg/100 ml, the condition of ketonemia is said to exist. The average daily excretion via urine of ketone bodies in the normal individual is 20 mg. Greater than normal amounts of ketone bodies in the urine is called *ketonuria*. The combination of ketonemia and ketonuria is called *ketosis*. A diabetic individual with severe ketosis may exhibit a blood concentration of ketone bodies as high as 90 mg/100 ml and excrete as much as 5000 mg of ketone bodies per hour!

The odor of acetone can often be detected on the breath of an untreated diabetic.

DIABETIC ACIDOSIS AND ELECTROLYTE IMBALANCE

Two of the ketone bodies are carboxylic acids. As the concentration of these acids in the blood increases, the pH of blood starts to decrease. A continued decrease of the blood pH could lead to *acidosis*. To prevent acidosis, the blood buffers react with the acids to neutralize them. Recall that one of the blood buffers is the H_2CO_3/HCO_3^- buffer system. As acetoacetic acid and β-hydroxybutyric acid enter the bloodstream, they react with bicarbonate ions to produce H_2CO_3, which decomposes to water and carbon dioxide. The CO_2 is exhaled by the lungs.

exercise 13-8

Write balanced equations for the neutralization of acetoacetic acid and β-hydroxybutyric acid by HCO_3^-.

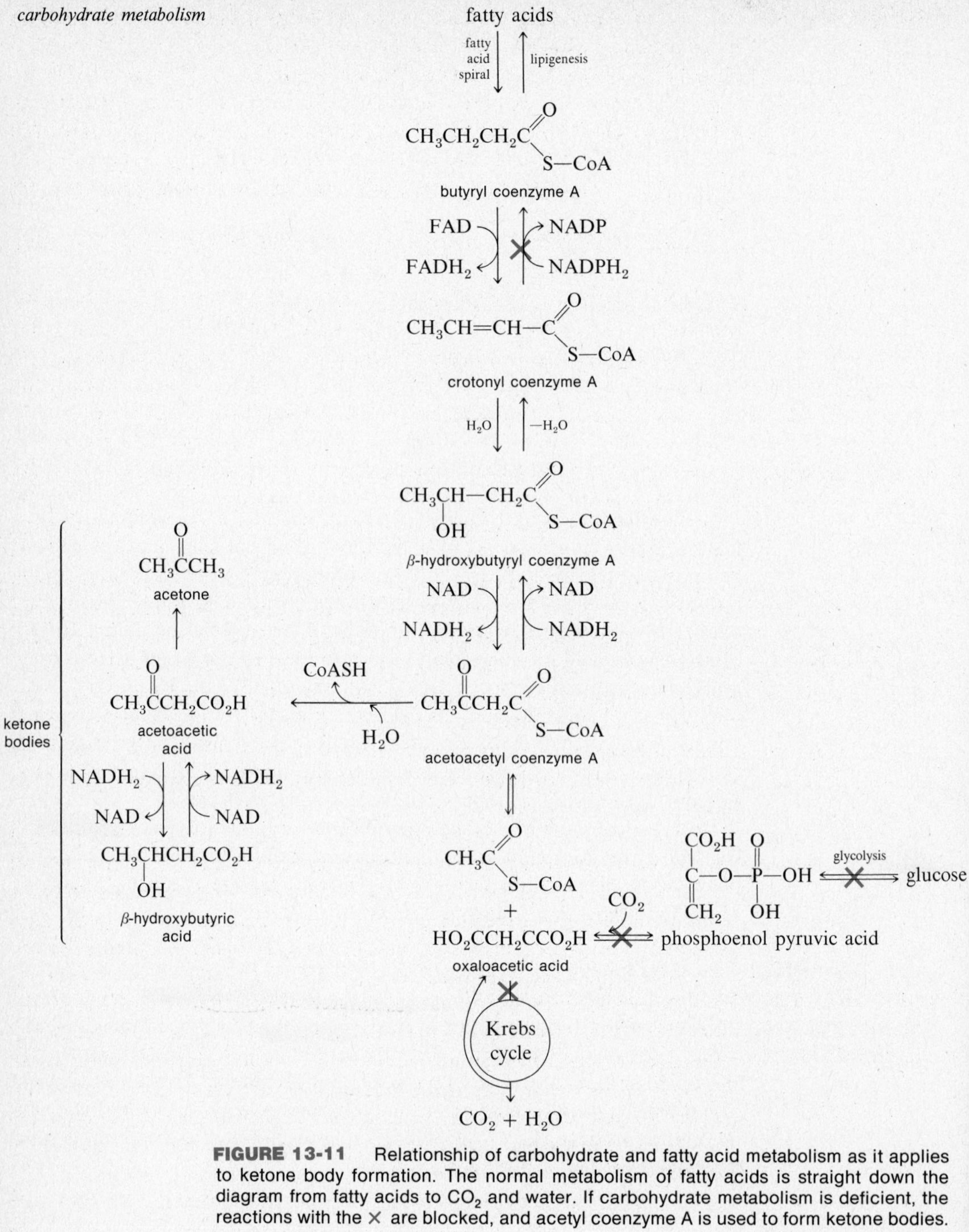

fatty acids

fatty acid spiral ↓ ↑ lipigenesis

$CH_3CH_2CH_2C(=O)S{-}CoA$

butyryl coenzyme A

FAD → $FADH_2$ ↓ ↑ $NADPH_2$ → NADP (×)

$CH_3CH{=}CH{-}C(=O)S{-}CoA$

crotonyl coenzyme A

H_2O ↓ ↑ $-H_2O$

$CH_3CH(OH){-}CH_2C(=O)S{-}CoA$

β-hydroxybutyryl coenzyme A

NAD → $NADH_2$ ↓ ↑ $NADH_2$ → NAD

$CH_3C(=O)CH_2C(=O)S{-}CoA$

acetoacetyl coenzyme A

⇅

$CH_3C(=O)S{-}CoA$

+

$HO_2CCH_2CCO_2H$ ⇌ (×, CO_2) phosphoenol pyruvic acid

oxaloacetic acid

(×) Krebs cycle

$CO_2 + H_2O$

$CO_2H{-}C(=CH_2){-}O{-}P(=O)(OH){-}OH$ ⇌ (glycolysis, ×) glucose

ketone bodies:

$CH_3C(=O)CH_3$

acetone

↑

$CH_3C(=O)CH_2CO_2H$

acetoacetic acid

(← from acetoacetyl coenzyme A: H_2O in, CoASH out)

$NADH_2$ → NAD ↓ ↑ $NADH_2$ ← NAD

$CH_3CH(OH)CH_2CO_2H$

β-hydroxybutyric acid

FIGURE 13-11 Relationship of carbohydrate and fatty acid metabolism as it applies to ketone body formation. The normal metabolism of fatty acids is straight down the diagram from fatty acids to CO_2 and water. If carbohydrate metabolism is deficient, the reactions with the × are blocked, and acetyl coenzyme A is used to form ketone bodies.

Continued release of acetoacetic and β-hydroxybutyric acids into the blood begins to deplete the bicarbonate ion concentration. Depletion of HCO_3^- ion concentration reduces the ability of the blood to neutralize the acids and eventually, as more and more acid is released into the bloodstream, acidosis results. This in turn leads to still further complications. As the pH of the blood goes down, the ability of hemoglobin to transport oxygen decreases and an oxygen deficiency develops. Breathing may become rapid, or even difficult and painful.

As the plasma HCO_3^- concentration continues to fall, chloride ions from plasma enter the red blood cells in exchange for HCO_3^-. This makes more HCO_3^- available in plasma to neutralize the acids and to aid in the effort to return the plasma pH to normal levels. Unfortunately, this chloride/bicarbonate shift is not sufficient to restore plasma pH and instead serves to initiate electrolyte imbalance. Electrolyte imbalance is further enhanced by kidney action. To counteract acidosis, the kidneys produce a more acidic urine. Even in the most acidic urine the kidneys are capable of producing, however, acetoacetic acid and β-hydroxybutyric acid exist as anions. In order to maintain electrical neutrality of the urine, cations must also be excreted. These are primarily sodium ions and as they are excreted in urine, their concentration in plasma is diminished.

The situation in the plasma has now reached the point where HCO_3^-, Cl^-, and Na^+ concentrations have decreased because of the release of ketone bodies into the blood. The decreased concentration of ions in the plasma, if uncompensated, would lead to decreased osmotic pressure of the plasma. To prevent this and to dissolve the greater quantities of materials being excreted, the kidneys increase the urine volume. Thus, these factors further complicate acidosis by inducing a state of dehydration. Hence, metabolic ketosis eventually causes acidosis and dehydration that ultimately may lead to coma and finally death.

The aims of clinical treatment of diabetic acidosis become obvious from the discussion above. In the short-term treatment, acidosis, sodium loss, and water loss must be controlled. This can be accomplished by intravenous injection of an isotonic solution of sodium chloride to replace needed water, sodium ions, and chloride ions. In severe cases, an isotonic solution of sodium bicarbonate or a mixture of NaCl and $NaHCO_3$ may be given intravenously. The addition of bicarbonate serves to give an immediate increase in the blood pH. In less severe cases, sodium lactate or the sodium salt of a Krebs cycle intermediate may be included in the intravenous solution. These materials allow the rate of the Krebs cycle operation to be increased. As the Krebs cycle operation increases, the normal concentration of HCO_3^- is restored by oxidation of acetyl coenzyme A and blood pH is controlled by natural means.

For the long term, treatment must establish normal carbohydrate metabolism to prevent a reoccurring accumulation of ketone bodies. For a diabetic individual, normal carbohydrate metabolism can usually be established by a combination of diet control and repeated insulin injections.

CHAPTER SUMMARY

1. Carbohydrates must be broken down to monosaccharides before digestion. Animals can only hydrolyze polysaccharides of glucose containing 1,4-α linkages. Most carbohydrate digestion occurs in the upper portion of the small intestine.
2. The first step in glucose metabolism is phosphorylation by ATP to glucose-6-phosphate. Glucose-6-phosphate may undergo glycogenesis or glycolysis.
3. Glycolysis is the breakdown of glucose. Fermentation yields ethanol; anaerobic glycolysis yields lactic acid; aerobic glycolysis yields CO_2 and H_2O. All three glycolytic pathways produce pyruvic acid as an intermediate. In aerobic glycolysis the pyruvic acid is converted to acetyl coenzyme A, which is converted to CO_2 and H_2O via the Krebs cycle and electron transport system, respectively.
4. Fermentation and anaerobic glycolysis yield 2 ATP molecules per glucose molecule. Aerobic glycolysis yields 38 ATP molecules per glucose molecule.
5. Lactic acid produced in muscles during anaerobic glycolysis can be converted to glycogen by the liver in a process called *gluconeogenesis*.
6. Fatty acids can be synthesized from acetyl coenzyme A in a process called *lipogenesis*. Lipogenesis allows the body to store excess nutrients. Acetyl coenzyme A is the crossroads between carbohydrate and lipid metabolism.
7. Glycerol from lipids is converted to dihydroxyacetone phosphate and is metabolized.
8. Galactose can be isomerized to glucose and metabolized. A genetic defect leading to the absence of the enzyme required for the isomerization allows galactose to accumulate in the body and cause galactosemia.
9. Inadequate insulin prevents glucose metabolism. The body then metabolizes abnormally large amounts of fats. The abnormal fat metabolism causes formation of excessive amounts of ketone bodies.
10. Two of the ketone bodies are acids and lead to the condition of acidosis. The body's efforts to counteract acidosis eventually leads to electrolyte imbalance and dehydration. If uncontrolled, these conditions may lead to coma or death.

QUESTIONS

1. Write an equation, using structural formulas, for the hydrolysis of isomaltose.
2. Explain why the reaction pyruvic acid $\longrightarrow$ lactic acid is readily

reversible but the reaction pyruvic acid $\longrightarrow$ acetyl coenzyme A is essentially irreversible.

3. At the end of a race, a runner is found to be experiencing mild acidosis. Explain how acidosis might arise in this case.
4. Explain how blood glucose concentration is maintained at a nearly constant level although carbohydrates are ingested only three or four times daily.
5. Explain why exercise leads to a rise in body temperature.
6. Explain the difference between glycogenesis and glyconeogenesis.
7. Glyconeogenesis is an endergonic process. Explain why.
8. In what major way does lipigenesis differ from a simple reversal of the beta oxidation of fatty acids?
9. What would be the structural formula of the product from the acid-catalyzed dehydration of β-hydroxybutyric acid? How does the product of the reaction above compare structurally to the biological dehydration of β-hydroxybutyryl coenzyme A to crotonyl coenzyme A?
10. Explain the difference between hypoglycemia and hyperglycemia.
11. During starvation, ketosis often develops. Offer a simple explanation for this observation.
12. Explain how an increase in urine volume in a diabetic prevents a decrease in the osmotic pressure of plasma.

14
protein metabolism

INTRODUCTION

Generally speaking, the use and treatment of proteins by the body is considerably different than that of carbohydrates and lipids. Proteins have very specific structures serving specific purposes. In addition to the insoluble proteins of epithelial, muscle, connective, glandular, and brain tissues, there are the soluble protein enzymes, plasma proteins, hormones, hemoglobin, antibodies, and nucleoproteins. In contrast to lipids and carbohydrates, the body possesses no storage depots for proteins nor are proteins used as a major energy source.

Excluding water, proteins are the main material of animal tissue and constitute about 75 percent of the dry weight of mammalian tissue. Table 14-1 gives the approximate percent composition of selected mammalian tissue.

TABLE 14-1
Percent composition of selected mammalian tissues

COMPONENT	MUSCLE	WHOLE BLOOD	LIVER	BRAIN	SKIN
Water	72–78	79	60–80	78	66
Protein	18–20	19	15	8	25
Lipid	3	1	3–20	12–15	7
Carbohydrate	0.6	0.1	1–15	0.1	Trace

DIGESTION AND ABSORPTION

Foreign proteins taken into the body may create allergic reactions.

Hydrolysis of peptide bonds in proteins is called *digestion*. The proteins are hydrolyzed to their constituent amino acids in the gastrointestinal tract. Protein digestion is important for two reasons: (1) High molecular weight compounds are converted to low molecular weight compounds that can be transported through cell membranes. (2) Biological specificity is destroyed, preventing antigenic reaction.

Digestion takes place in both the stomach and small intestine under the influence of proteolytic enzymes secreted in gastric and pancreatic juices. Those enzymes contained in gastric juice require a very acidic media for optimum activity. Pepsin, the primary enzyme in the stomach, has its maximum activity at a pH between 1.5 and 2.5. This low pH is attained by the secretion of 0.16 M HCl by the parietal cells of the stomach. Blood plasma is the source of the required hydrogen ions and chloride ions. The concentration of H^+ in the parietal cells is approximately 1 million times that of blood. Formation and maintenance of this large concentration difference requires an expenditure of energy by the body. Calculations suggest the body expends a minimum of 1500 cal to produce 1 liter of this HCl solution. Secretion rates of gastric juices vary from 150 to 500 ml/hr, depending on both appetite and the presence of food in the stomach. Typically, secretion continues throughout and for 10 to 20 min after eating.

exercise 14-1

Assume a person has three meals and two snacks during a typical day. If gastric juice is secreted at a rate of 400 ml/hr for 45 min at breakfast and again at lunch and for 90 min at dinner and at a rate of 250 ml/hr for 30 min after each snack, calculate
(a) The minimum total volume of HCl solution secreted.
(b) The minimum number of moles of HCl secreted.
(c) The minimum number of grams of HCl secreted.
(d) The minimum amount of energy in calories expended for HCl production.

exercise 14-2

Explain the difference between protein denaturation and digestion.

The reasons why HCl doesn't normally destroy the stomach are still being investigated.

Overproduction and secretion of gastric juices leads to the condition of hyperacidity, which causes heartburn and indigestion. Continued hyperacidity for prolonged lengths of time may lead to peptic or duodenal ulcers because of digestion of the stomach or duodenum mucosa.

Pernicious anemia is associated with a condition known as *anacidity* (without acidity) in which virtually no hydrochloric acid is produced. In this disease very little protein digestion can take place in the stomach since pepsin is inactive. (Why?) The result is a nutritional deficiency of amino acids even though the individual is apparently on an adequate diet.

Following protein digestion in the stomach and small intestine, the resultant amino acids are absorbed directly by active transport into the bloodstream through the intestinal mucosa. The rate of absorption differs for each amino acid and is an energy-requiring process.

AMINO ACID POOL AND NITROGEN BALANCE

Since the late 1930's, it has been known that body proteins are constantly undergoing degradation and resynthesis. Thus, a dynamic equilibrium exists for body proteins. The degradation and resynthesis rates vary widely for different proteins. These rates are referred to as *protein turnover rates*. Blood and liver proteins experience a rapid turnover rate with about one-half of their protein undergoing degradation and resynthesis every 6 days. Collagen proteins apparently experience the slowest turnover rates, with a half-life of approximately 3 years (see Table 14-2).

The combination of amino acids from protein degradation and those from dietary sources are referred to as the body's *amino acid pool*. The amino acid pool may be considered as all the free amino acids in the body. They are *not* located in a specific depot. The amino acids in the pool are available to all tissues and circulate via the bloodstream. The normal amino acid concentration is 4 to 8 mg/100 ml of plasma. Whenever amino acid concentration exceeds this range, the amino acids are rapidly removed from circulation by the liver. In the liver, they are degraded by removal of the *amine* group. The non-nitrogen-containing residue is usually converted to acetyl coenzyme A. The amine group may be used for synthesis of other nonprotein nitrogen-containing compounds or converted to urea and excreted via the urine.

Conversion of the amino acid residue to acetyl CoA requires several steps.

TABLE 14-2
Protein types and their sources

TYPE	SOURCE
FIBROUS	
1. Collagens	Connective tissue
2. Elastins	Tendons, arterial walls
3. Keratins	Hair
GLOBULAR	
1. Albumins	Serum
2. Globulins	Serum, muscles
3. Histones	Glandular tissue

A positive nitrogen balance results from the need to synthesize more protein than is degraded.

A normal healthy adult on a correct diet will excrete as much nitrogen as taken in. Such an individual is said to be in nitrogen balance. A growing child or a person recovering from prolonged illness experiences a positive nitrogen balance in which less nitrogen is excreted than is included in the diet. Negative nitrogen balance may result from starvation, wasting diseases, or a lack of sufficient amounts of the essential amino acids in the diet (see Table 8-8).

OVERVIEW OF AMINO ACID METABOLISM

A wide variety of metabolic reactions exists for amino acids. Some of the reactions are specific for certain amino acids, while others are of a general type for all or most amino acids. The variety of reactions that a particular amino acid might undergo is independent of how it entered the amino acid pool. Amino acids from dietary sources are immediately mixed in the blood with those from tissue protein degradation with the result that no metabolic distinction can be made. Furthermore, since there are several possible metabolic fates for most amino acids, it is likely that different molecules of a particular amino acid may simultaneously undergo different metabolic reactions.

The general metabolic pathways for amino acid utilization can be summarized:

1. Synthesis of proteins.
2. Synthesis of nonprotein nitrogen-containing compounds such as heme, nucleic acids, and creatine.
3. Transamination to form required nonessential amino acids.
4. Oxidative deamination involving the removal of the amine group. The amine group is generally converted to urea via ammonia. The remainder of the amino acid molecule is frequently metabolized as an energy source.
5. Undergoing a specific reaction usually related to the presence of a particular element (sulfur) or a particular chemical group (aromatic or heterocyclic rings).

TRANSAMINATION

Transamination is the transfer of an amine group from an amino acid to an α-keto acid. This transfer results in the formation of a different amino acid and a different α-keto acid. The general equation for this reaction may be written:

$$\mathrm{R{-}\underset{\displaystyle NH_2}{\underset{|}{CH}}{-}CO_2H + R'{-}\underset{\displaystyle O}{\underset{\|}{C}}{-}CO_2H \rightarrow R'{-}\underset{\displaystyle NH_2}{\underset{|}{CH}}{-}CO_2H + R{-}\underset{\displaystyle O}{\underset{\|}{C}}{-}CO_2H}$$

This reaction gives the body the ability to synthesize the nonessential amino acids that it requires. With the exception of lysine and threonine, every natural amino acid has been found to undergo transamination reactions. The enzymes required for transamination are known as *transaminases* and require a derivative of the B vitamin pyridoxine.

A specific example of transamination is the reaction between L-glutamic acid and pyruvic acid to produce α-ketoglutaric acid and alanine.

$$\underset{\text{L-glutamic acid}}{\mathrm{HO_2CCH_2CH_2\underset{\displaystyle NH_2}{\underset{|}{C}}HCO_2H}} + \underset{\text{pyruvic acid}}{\mathrm{CH_3\underset{\displaystyle O}{\underset{\|}{C}}CO_2H}} \rightarrow \underset{\text{α-ketoglutaric acid}}{\mathrm{HO_2CCH_2CH_2\underset{\displaystyle O}{\underset{\|}{C}}CO_2H}} + \underset{\text{alanine}}{\mathrm{CH_3\underset{\displaystyle NH_2}{\underset{|}{C}}HCO_2H}}$$

exercise 14-3

Write the structural formulas for the amino acids resulting from transamination between L-glutamic acid and

(a) $\mathrm{HO_2CCH_2\underset{\displaystyle O}{\underset{\|}{C}}CO_2H}$ **oxaloacetic acid**

(b) $\mathrm{HSCH_2\underset{\displaystyle O}{\underset{\|}{C}}CO_2H}$ **β-mercaptopyruvic acid**

(c) $\mathrm{C_6H_5{-}CH_2\underset{\displaystyle O}{\underset{\|}{C}}CO_2H}$ **phenylpyruvic acid**

To aminate is to add an amine group.

Most transamination reactions apparently involve L-glutamic acid as the source of amine groups. This is probably because the Krebs cycle may furnish abundant amounts of α-ketoglutaric acid and the liver can directly aminate α-ketoglutaric acid with ammonia to produce L-glutamic acid.

$$\underset{\text{α-ketoglutaric acid}}{\mathrm{HO_2CCH_2CH_2\underset{\displaystyle O}{\underset{\|}{C}}CO_2H}} + \mathrm{NH_3} \xrightarrow[\text{glutamic acid dehydrogenase}]{\mathrm{NADH_2}\ \ \mathrm{NAD}} \underset{\text{L-glutamic acid}}{\mathrm{HO_2CCH_2CH_2\underset{\displaystyle NH_2}{\underset{|}{C}}HCO_2H}} + \mathrm{H_2O}$$

OXIDATIVE DEAMINATION Oxidative deamination is the removal of the amine group from an amino acid to produce ammonia and an α-keto acid. The overall reaction can be written

$$\underset{\displaystyle NH_2}{\underset{|}{R{-}CH}}{-}CO_2H + \tfrac{1}{2}O_2 \longrightarrow \underset{\displaystyle O}{\underset{\|}{R{-}C}}{-}CO_2H + NH_3$$

Compare oxidative deamination with transamination.

The reaction takes place in two steps. In the first, either NAD or FMN is the oxidizing agent and requires the enzyme amino acid oxidase. The second step is a spontaneous hydrolysis to produce ammonia and the α-keto acid.

$$\text{Step 1.}\ HO_2CCH_2CH_2\underset{\displaystyle NH_2}{\underset{|}{C}H}{-}CO_2H \xrightarrow[\text{amino acid oxidase}]{NAD \quad NADH_2} HO_2CCH_2CH_2\underset{\displaystyle NH}{\underset{\|}{C}}{-}CO_2H$$

$$\text{Step 2.}\ HO_2CCH_2CH_2\underset{\displaystyle NH}{\underset{\|}{C}}{-}CO_2H + H_2O \longrightarrow HO_2CCH_2CH_2\underset{\displaystyle O}{\underset{\|}{C}}{-}CO_2H + NH_3$$

The $NADH_2$ produced in step 1 is reoxidized to NAD via the cytochrome chain. The overall equation for this reoxidation is

$$NADH_2 + \tfrac{1}{2}O_2 \longrightarrow NAD + H_2O$$

Only glutamic acid undergoes oxidative deamination in large amounts.

The reaction sequence involving NAD as shown above is specific for glutamic acid.

Oxidative deamination of other amino acids occurs primarily via transamination with α-ketoglutaric acid to form glutamic acid. This is followed by oxidative deamination of the glutamic acid.

exercise 14-4

Using structural formulas, write equations for the oxidative deamination of (a) serine and (b) phenylalanine.

Minor amounts of amino acids undergo direct oxidative deamination rather than via glutamic acid formation. In this case FMN is required as the oxidizing agent instead of NAD.

DISPOSAL OF AMMONIA Ammonia is a toxic compound and its concentration must not build up in cells. Therefore, if the ammonia produced via deamination of amino acids is not required for anabolic purposes, it must be excreted. The form in which the ammonia is excreted depends in part on where it is produced. Ammonia produced by kidney cells during metabolism may be released directly into the urine. Excess ammonia produced elsewhere in the body is converted

Besides being toxic, ammonia is a base and accumulation of large amounts of it could change the pH of body fluids.

to either urea or glutamine. Most deamination reactions occur in the liver. Any excess ammonia produced in the liver is transformed to urea and excreted via urine. The overall equation for urea formation may be written

$$2NH_3 + CO_2 \longrightarrow \underset{\text{urea}}{H_2N{-}\overset{\displaystyle O}{\overset{\|}{C}}{-}NH_2} + H_2O$$

Urea formation takes place in a series of reactions known as the *ornithine cycle,* shown in Fig. 14-1. Carbamylphosphate reacts with ornithine

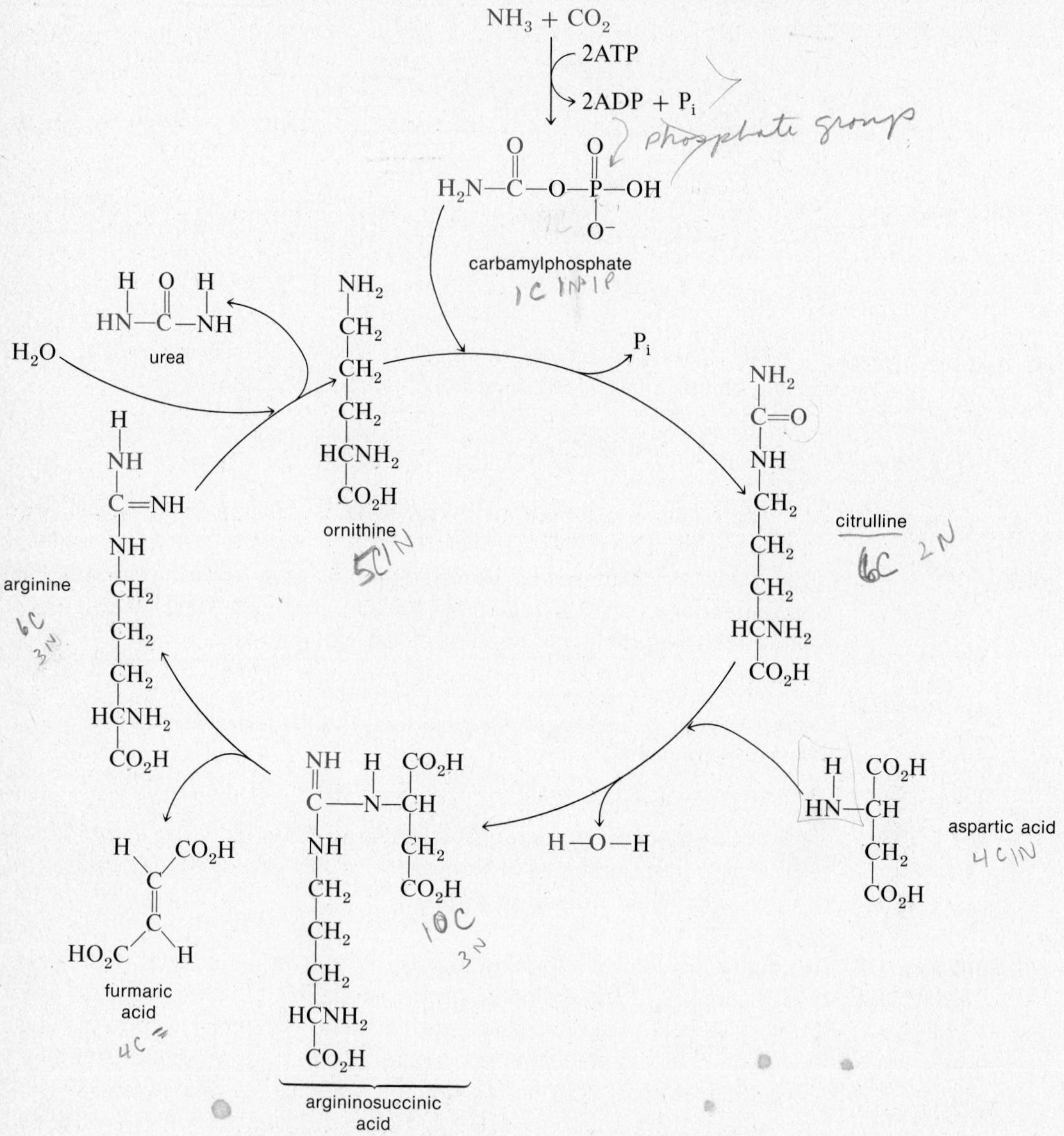

$NH_3 + CO_2$ → (2ATP → 2ADP + P_i) → $H_2N{-}C(=O){-}O{-}P(=O)(O^-){-}OH$ carbamylphosphate

$NH_2CH_2CH_2CH_2CH(NH_2)CO_2H$ ornithine

P_i

$NH_2C(=O)NHCH_2CH_2CH_2CH(NH_2)CO_2H$ citrulline

$HN(H){-}CH(CO_2H)CH_2CO_2H$ aspartic acid

$H{-}O{-}H$

argininosuccinic acid

$HO_2C{-}CH{=}CH{-}CO_2H$ furmaric acid

$HN(H){-}C(=NH){-}NHCH_2CH_2CH_2CH(NH_2)CO_2H$ arginine

H_2O

$HN(H){-}C(=O){-}NH(H)$ urea

FIGURE 14-1 Urea formation via the ornithine cycle.

You should not memorize all the reactions in Fig. 14-1 and 14-2.

to initiate the cycle. It is formed from the combination of ammonia and carbon dioxide with a phosphate group from ATP.

$$NH_3 + CO_2 \xrightarrow[]{2ATP \quad 2ADP + P_i} H_2N-\overset{\overset{O}{\|}}{C}-O-\overset{\overset{O}{\|}}{\underset{\underset{O^{(-)}}{|}}{P}}-OH$$

carbamylphosphate

Notice in Fig. 14-1 that the second ammonia molecule required for urea formation comes from aspartic acid and not from free ammonia. Thus, the ornithine cycle uses ammonia, carbon dioxide, and aspartic acid to produce urea and fumaric acid.

The fumaric acid may be converted to oxaloacetic acid via a portion of the Krebs cycle. Figure 14-2 shows the use and regeneration of aspartic acid in relation to the ornithine cycle. Transamination between oxaloacetic acid and glutamic acid regenerates aspartic acid for use in the ornithine cycle. Since glutamic acid can be synthesized in the liver directly from ammonia and α-ketoglutaric acid, the overall equation for urea formation is

$$2NH_3 + CO_2 \longrightarrow H_2N-\overset{\overset{O}{\|}}{C}-NH_2 + H_2O$$

Although amine groups act as bases, amide groups do not exhibit base behavior.

Ammonia produced by the body elsewhere than in the kidneys or liver is principally used in glutamine synthesis. Glutamine apparently serves two functions: (1) as a means of transporting ammonia in the bloodstream in a nontoxic form and (2) as a temporary store of ammonia in the body. Glutamine is synthesized from glutamic acid and ammonia in a reaction catalyzed by glutamine synthetase and requiring ATP.

$$\underset{\text{glutamic acid}}{HO_2C-CH(NH_2)-CH_2-CH_2-CO_2H} + NH_3 \xrightarrow[ATP \;\rightarrow\; ADP + P_i]{\text{glutamine synthetase}} \underset{\text{glutamine}}{HO_2C-CH(NH_2)-CH_2-CH_2-C(=O)-NH_2} + H_2O$$

exercise 14-5

Classify glutamic acid and glutamine as acidic, basic, or neutral amino acids.

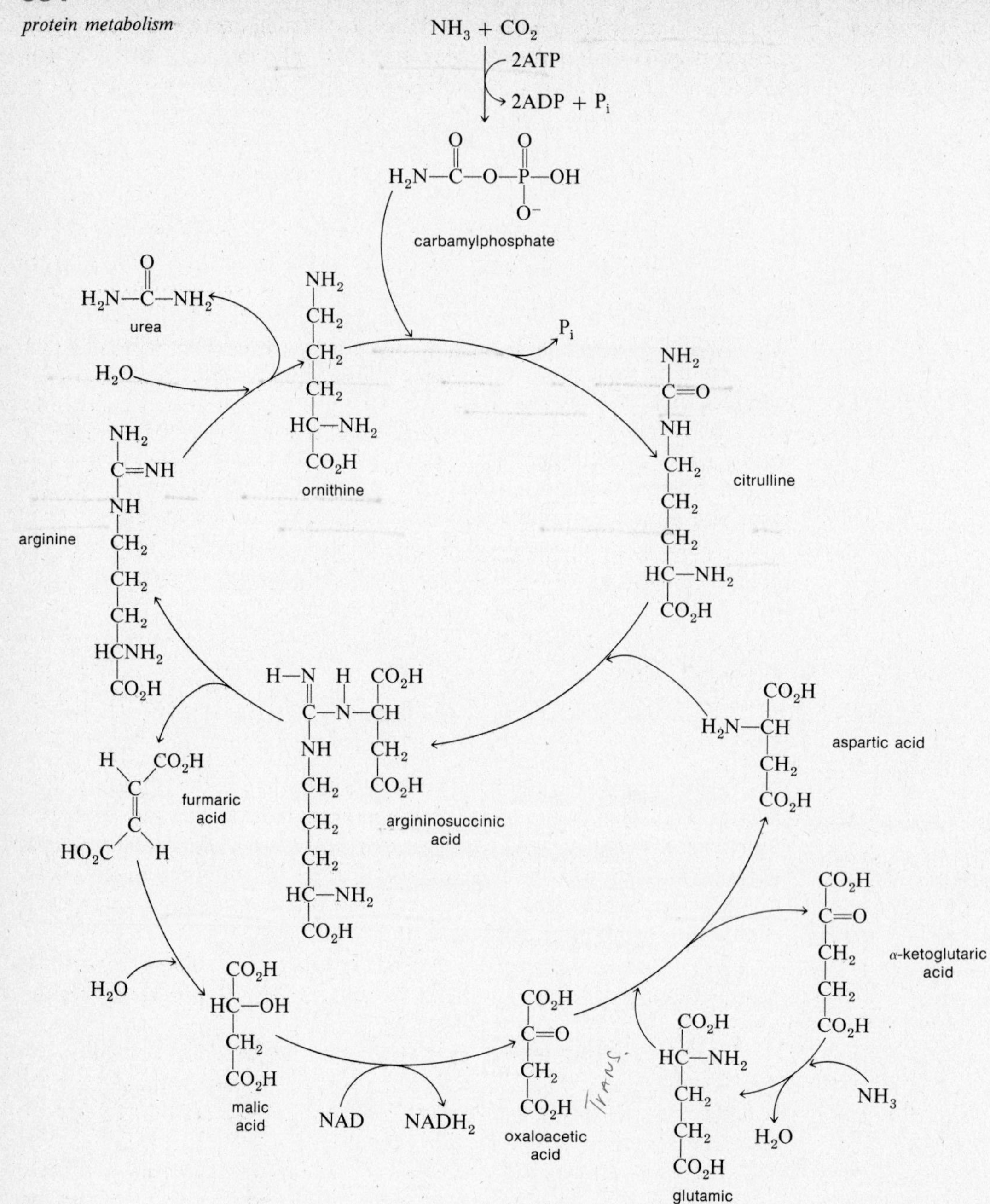

FIGURE 14-2 The use and regeneration of aspartic acid in relation to the ornithine cycle.

The enzyme *glutaminase* is widely distributed in the body and catalyzes the hydrolysis of glutamine to glutamic acid and ammonia. This reaction makes ammonia available for amination of α-keto acids to produce non-essential amino acids.

$$\text{glutamine} + H_2O \xrightarrow{\text{glutaminase}} \text{glutamic acid} + NH_3$$

For a healthy individual on a normal diet, most of the glutamine produced is transported to the liver via the bloodstream. In the liver, glutamine is hydrolyzed and the released ammonia is converted to urea and excreted.

We have examined the general degradation reactions for amino acids. In addition to these general reactions, there are many specific reactions for individual amino acids; however, the details of these are not of concern in this course. Some hereditary diseases result however from the inability of the body to perform a specific reaction of a particular amino acid because of the lack of the required enzyme. Recall that enzymes are primarily proteins. Thus, the absence of an enzyme is in some way related to a malfunction in protein synthesis. We shall examine several hereditary diseases involving amino acid metabolism after our study of protein synthesis.

PROTEIN SYNTHESIS

For many years, the synthesis of proteins was thought to be the reversal of protein hydrolysis. According to this idea certain enzymes catalyzed the formation of a peptide bond between amino acids by removal of a molecule of water.

$$H_2N{-}\underset{\displaystyle R}{\underset{|}{CH}}{-}C\begin{matrix}\diagup\!\!\!\!\diagup O\\ \diagdown OH\end{matrix} + H_2N{-}\underset{\displaystyle R'}{\underset{|}{CH}}{-}C\begin{matrix}\diagup\!\!\!\!\diagup O\\ \diagdown OH\end{matrix} \xrightarrow{\text{enzyme}} H_2N\underset{\displaystyle R}{\underset{|}{CH}}{-}C\begin{matrix}\diagup\!\!\!\!\diagup O\\ \diagdown \end{matrix}\overset{\displaystyle H}{\overset{\diagup}{N}}{-}\underset{\displaystyle R'}{\underset{|}{CH}}{-}C\begin{matrix}\diagup\!\!\!\!\diagup O\\ \diagdown OH\end{matrix} + H_2O$$

However, this proposal could not explain how a cell was able to ensure the proper sequencing of the amino acid residues in the protein being synthesized.

As research on proteins continued, it was discovered that many hereditary diseases were associated with a malfunction of protein synthesis, most notably enzymes and proteinaceous hormones. A partial list of diseases, and the affected protein is given in Table 14-3.

Knowledge that hereditary diseases involve abnormalities in protein synthesis and that inheritance of traits is associated with genes carried on chromosomes in the cell nucleus led workers to seek a relationship between protein synthesis and genes. Eventually as studies on both unicellular and multicellular organisms progressed, the *one gene–one enzyme* theory evolved. According to this theory, genes control metabolism by controlling synthesis of enzymes. A specific gene is present for each enzyme produced by a cell.

TABLE 14-3
A partial list of hereditary diseases in man resulting from a lacking or modified protein

DISEASE	AFFECTED PROTEIN
Acanthocytosis	Lipoproteins
Albinism	Tyrosinase
Alkaptonuria	Homogentisate oxidase
Diabetes mellitus	Insulin
Disaccharide intolerence	Sucrase, maltase, lactase
Dwarfism	Growth hormone
Galactosemia	Galactose-1-phosphate uridylyl transferase
Gout	Hypoxanthine-guanine phosphoribosyltransferase
Hartnup's disease	Tryptophan-2,3-dioxygenase
Hemophilia A	Antihemophilic factor A
Hemophilia B	Antihemophilic factor B
Hyperammonemia	Ornithine transcarbamoylase
Phenylketonuria	Phenylalanine hydroxylase
Tay–Sachs disease	Hexosaminidase A
Wilson's disease	Ceruloplasmin

Consequently, mutation of a single gene results in an abnormality of a single specific enzyme.

Portions of DNA molecules are called genes. *Thus, DNA controls protein synthesis.*

The nuclei of cells have been found to contain *deoxyribonucleic acids* (DNA). Furthermore, the DNA is found only in the chromosomes and the amount of DNA present is proportional to the number of genes. Additionally, the amount of DNA in a cell is a constant, characteristic of each living species. This information, coupled with additional unmentioned evidence, has lead scientists to conclude that *DNA is the genetic material of life* (except for certain viruses) *and possesses the ability to control the formation of proteins, including enzymes.* DNA must also have the ability to replicate (form an exact replica of itself) and also be capable of undergoing mutation. Given the importance of DNA to a cell, let us now inquire into the nature of these molecules.

THE CHEMISTRY AND STRUCTURE OF DNA

DNA molecules are polymers with molecular weights exceeding 1×10^9. Thus, DNA is among the largest molecules known. Complete hydrolysis of DNA results in the isolation of six compounds: adenine, guanine, cytosine, thymine, β-2-deoxyribose, and phosphoric acid.

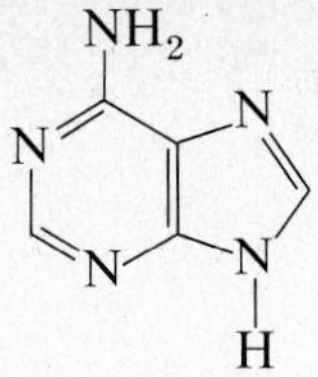

NH$_2$ N N N N H adenine · O H—N N H$_2$N N N H guanine · NH$_2$ N O N H cytosine · O CH$_3$ H—N O N H thymine

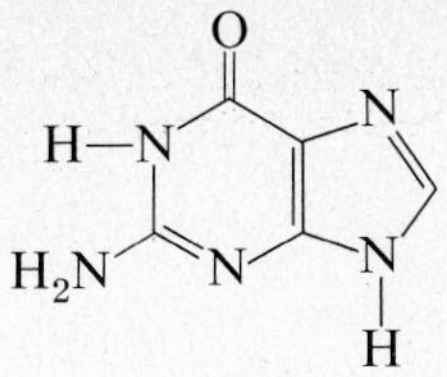

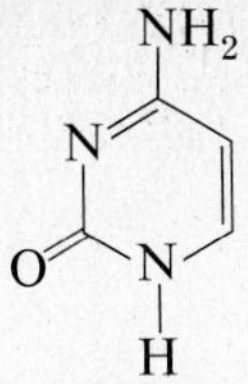

HOCH$_2$ O OH OH

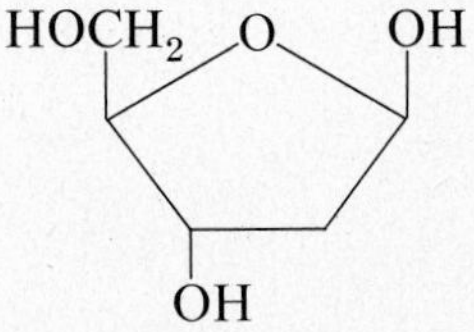

β-D-2-deoxyribose

Heterocyles are ring structures containing other atoms in addition to carbon in the ring.

Adenine, guanine, cytosine, and thymine are heterocyclic bases. They are amines and consequently behave as bases because of the presence of unshared electron pairs on the nitrogen atoms. Deoxyribose is a carbohydrate. In DNA, deoxyribose is bonded to each of the heterocyclic bases via the No. 1 carbon of the sugar. The resultant molecules are known as *nucleosides;* their structural formulas are shown below.

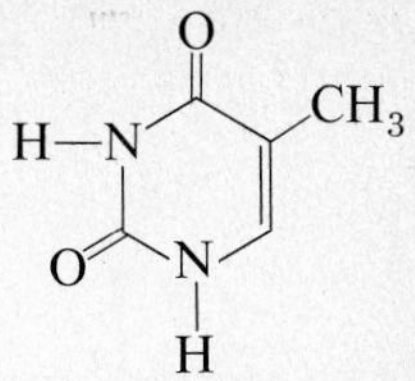

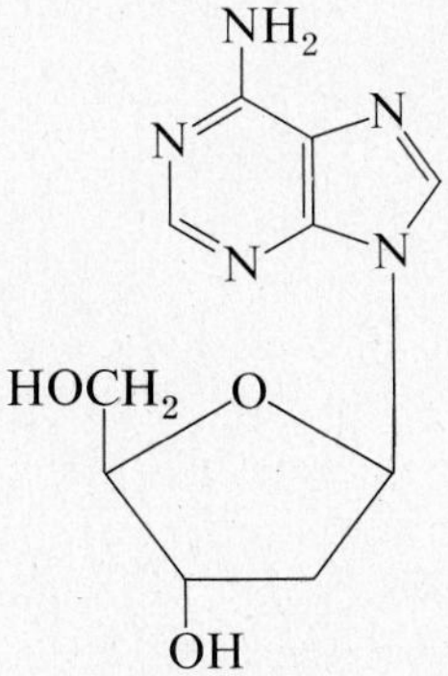

deoxyadenosine 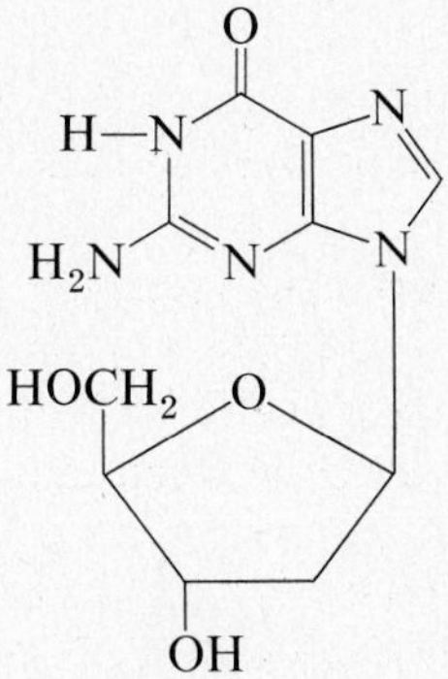 · deoxyguanine 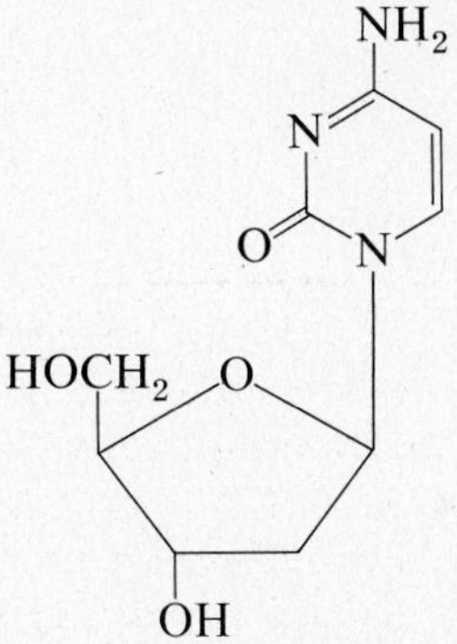 · deoxycytidine 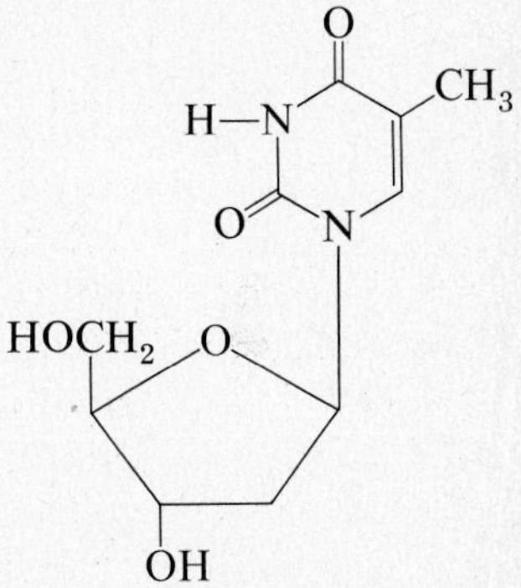 · thymidine

Note: *Although thymidine contains deoxyribose, the prefix—deoxy—is not used.*

The DNA of all living organisms is made up of six different kinds of molecules.

In DNA, these nucleosides are joined together by phosphate ester linkages between the No. 5 carbon of the deoxyribose of one nucleoside and the No. 3 carbon of the deoxyribose of another nucleoside molecule. *The total number of nucleoside molecules and the order in which the four different nucleosides are bonded is what makes one gene different from another gene*. This is true whether we are comparing two different genes in a human cell or comparing a human gene to a plant gene. A portion of a DNA molecule showing the phosphate ester bonds between nucleosides is depicted in Fig. 14-3. Notice in this figure that the alternating deoxyribose–phosphate

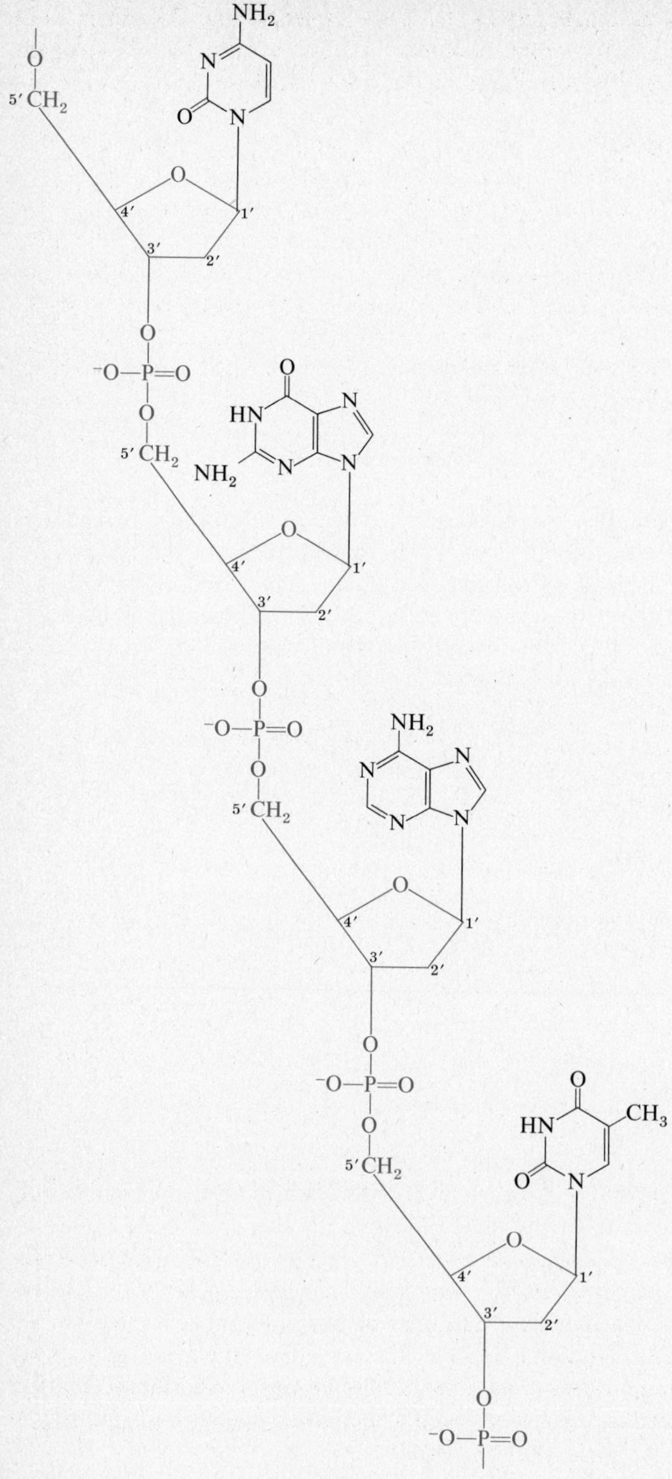

FIGURE 14-3 A portion of a strand of DNA showing the phosphate deoxyribose units forming the backbone.

sequence forms the backbone of the DNA molecule and the heterocyclic bases extend outward from the backbone. (Compare this to the arrangement of the position of the side chains of the amino acid residues in a protein molecule.)

Adenine hydrogen bonds only to thymine; cytosine only to guanine.

Just as proteins do not normally exist as long, floppy molecules, neither does DNA. The DNA molecules have a highly ordered structure that is maintained via hydrogen bonding between the heterocyclic bases. This hydrogen bonding is very specific: Adenine will hydrogen bond to thymine and cytosine will hydrogen bond to guanine. There are two hydrogen bonds formed between each adenine–thymine pair and three between each cytosine–guanine pair. Other hydrogen bonding combinations do not normally occur in DNA. This specific hydrogen bonding between the heterocylic bases is referred to as *base pairing* (see Fig. 14-4).

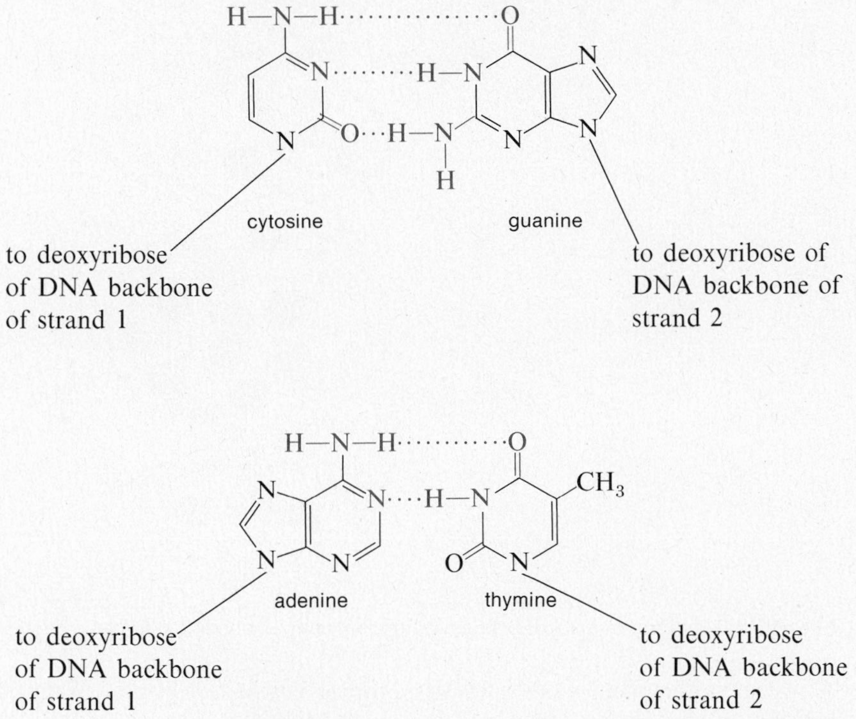

FIGURE 14-4 Hydrogen bonding between heterocyclic bases of the two strands in a DNA molecule.

The hydrogen bonding between complementary base pairs does *not* occur between nucleosides within the same chain but occurs between nucleosides in two separate chains. Thus, the hydrogen bonding or base pairing serves to join two complementary chains. An adenine in the first chain hydrogen bonds to a thymine in the second chain; a guanine in the first chain to a cytosine in the second chain; a thymine in the first to an adenine in the second, and so on (see Fig. 14-4).

The two long chains, joined by thousands of hydrogen bonds, are twisted or intertwined to form a helical coil. This helical coiling of the paired chains produces the well-publicized *double helix* of DNA shown in Fig. 14-5.

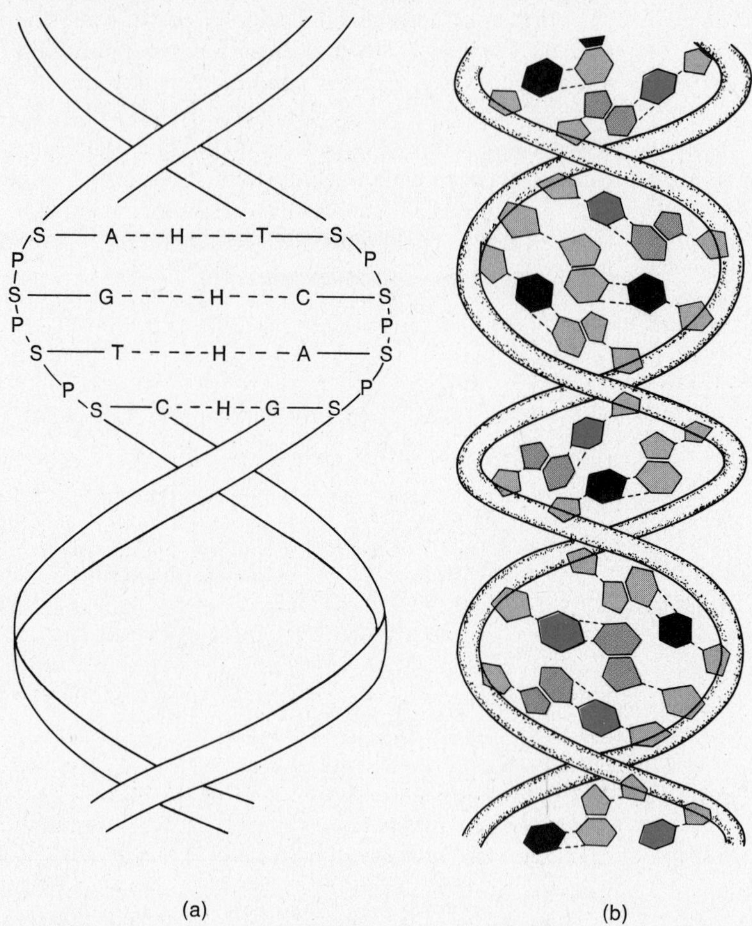

FIGURE 14-5 Double helix of DNA showing the hydrogen bonding between strands.

This structural model for DNA was first proposed by M. Crick and J. Watson in 1954.

MESSENGER RNA

The DNA of the genes in the chromosomes is present in the nucleus of a cell. Protein synthesis occurs on cytoplasmic particles called *ribosomes* that are attached to the endoplasmic reticulum (see Fig. 12-5). If the information contained in the DNA is to control the synthesis of a protein on a ribosome, there must be a carrier of some type that can transmit the information from DNA to the ribosomes. A type of ribonucleic acid (RNA) known as messenger RNA, or *m*RNA, serves as the information carrier. Messenger RNA not only transmits the information required for the synthesis of a specific protein but also determines the rate of protein synthesis by the cell. The lifetime of an individual *m*RNA molecule is relatively short and the concentration of these molecules changes with time. Their concentration at any time determines the number of protein molecules that can be synthesized.

Messenger RNA has many similarities to DNA and yet has its own characteristics. Like DNA, *m*RNA is a polymer made up of heterocyclic bases and a five-carbon sugar and is held together by phosphate ester bonds. The major differences are (1) *m*RNA is a much smaller polymer than DNA, (2) RNA contains uracil instead of thymine, and (3) the sugar is ribose rather than deoxyribose. The structures of uracil and ribose are

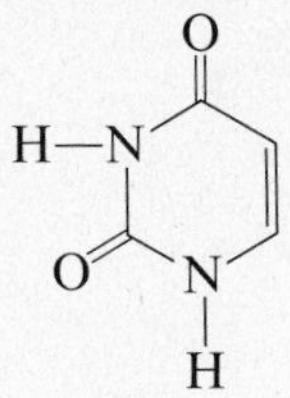

uracil

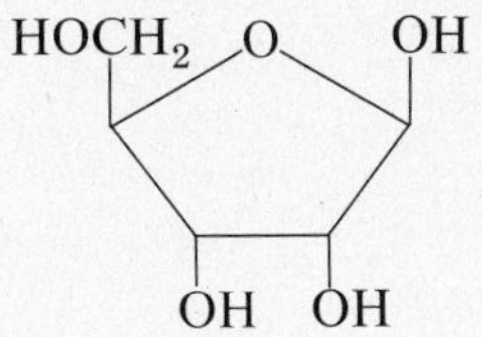

β-D-ribose

exercise 14-6

(a) What is the difference between uracil and thymine?
(b) Uracil will pair with what base?

The structure of a portion of *m*RNA is shown in Fig. 14-6. Another major difference between RNA and DNA is that RNA exists largely as single chains instead of the double chains of DNA.

TRANSCRIPTION

Transcription is the name given to the process by which the information contained in a gene for the synthesis of a specific protein is incorporated into the *m*RNA molecule. A gene is a portion of a DNA molecule. It is thought that there are two types of genes in DNA: structural genes and regulator genes. The structural genes determine the structure of *m*RNA, thus determining the synthesis of a specific protein. The regulatory genes control the rate at which the *m*RNA is produced. The details of how the regulatory genes control the formation of *m*RNA by the structural genes is complex and will not be discussed in this text. We will only briefly examine how *m*RNA is synthesized.

When the conditions in a cell indicate the need for the synthesis of a specific protein, the structural gene portion of the DNA molecule for that particular protein unravels. During this unraveling the helical coils of DNA become uncoiled and the two strands separate. One of the strands serves as a template (pattern) for the formation of *m*RNA. This formation occurs in several steps. The nucleosides are phosphorylated and form nucleotides, also called nucleic acids. The nucleotides hydrogen bond to their complementary base in the DNA molecule. Under the action of an enzyme, the

Note that the DNA is not used up during the formation of m*RNA.*

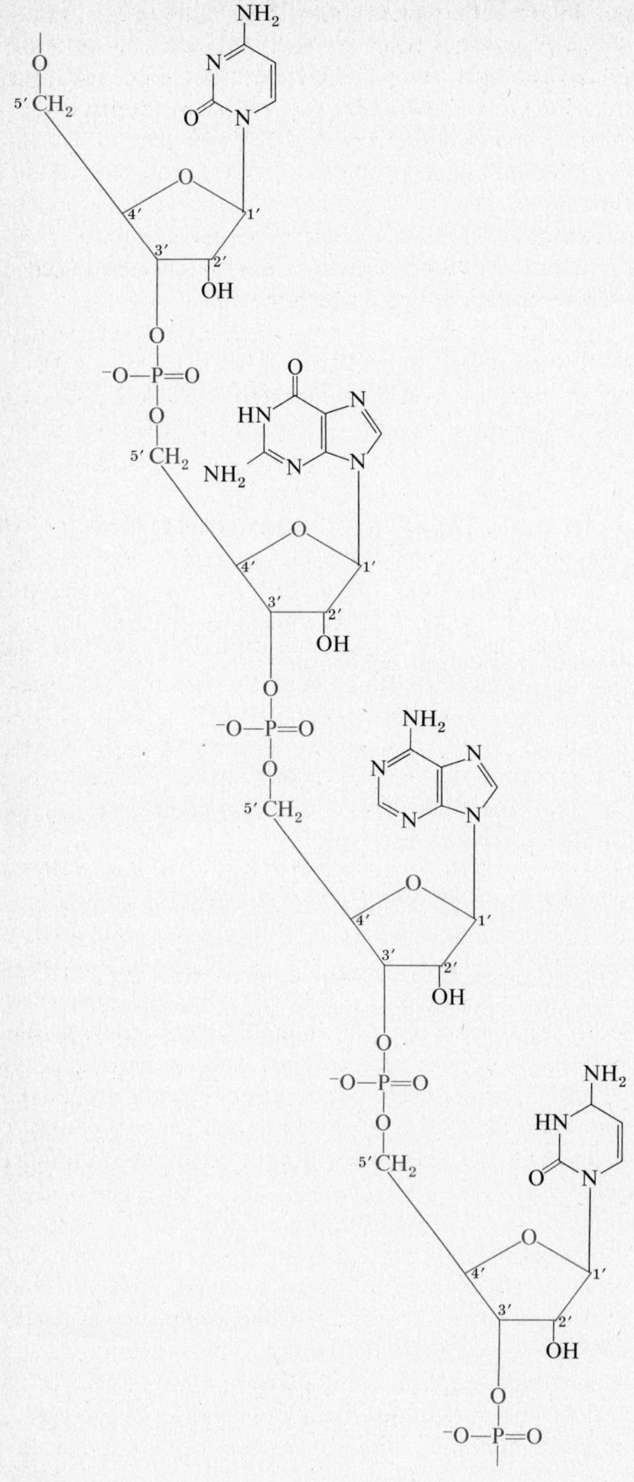

FIGURE 14-6 A portion of a strand of RNA showing the phosphate ribose units forming the backbone.

nucleotides are joined by phosphate ester bonds to form the *m*RNA molecule. When the *m*RNA molecule has been synthesized, it breaks away from the DNA strand and diffuses from the cell nucleus to the ribosomes. After the newly synthesized *m*RNA has diffused away, the two complementary strands of the DNA are rejoined into the normal DNA double helix. This process is depicted in Fig. 14-7 on page 394.

TRANSLATION The order of the nucleosides in the *m*RNA determines the sequence of the amino acids in the protein to be synthesized. The *m*RNA nucleoside order must be translated. This translation dictates both the amino acids and their sequence in the protein to be synthesized. The translation is accomplished by another type of RNA called transfer RNA (*t*RNA). Transfer RNA molecules serve both as amino acid carriers and interpreters of the genetic code. For each amino acid there is at least one specific *t*RNA. The *t*RNA molecules bond to the carboxyl group of their particular amino acid by the formation of an ester linkage.

Another portion of the *t*RNA molecule recognizes a code contained in the *m*RNA for the specific amino acid that the *t*RNA carries. The code consists of *three adjacent* heterocyclic bases in the *m*RNA. The three adjacent bases are known as a *triplet* or a *codon*. For example, it has been determined that the base sequence guanine, guanine, uracil in *m*RNA is specific for the *t*RNA molecule that carries glycine. Thus, when this sequence of bases occurs in a *m*RNA molecule, a glycine is incorporated into the protein molecule being synthesized. The recognition by a *t*RNA molecule for its proper triplet in the *m*RNA occurs by base pairing.

exercise 14-7

In proper sequence, what three heterocyclic bases must be present in the recognition site of the *t*RNA molecule carrying glycine?

Table 14-4 contains the codons for the various amino acids. Notice that there are several codons for most of the amino acids.

Consider the beginning portion of a hypothetical *m*RNA molecule with nucleoside order: UUUCCUACGGUAUUGUCG... . Referring to Table 14-4, we see that the first triplet UUU is the code for phenylalanine; the second triplet CCU is for proline; the third, ACG, for threonine. Continuing in this manner we find the portion of the *m*RNA shown above would dictate the synthesis of the polypeptide phenylalanine–proline–threonine–valine–leucine–serine–... .

exercise 14-8

Write the sequence of amino acids in the polypeptides that would be formed from each of the following *m*RNA molecules.

(a) AUGGACGAGUACAAAGAAAUAACUUGCGGG

(b) UGCACGUCACGGCACUACGGCGUGUCAGUCAAGUACCCGCGA

(c) AGACACUUUAAACCCCCGGGGGGUUUUUCCCCCCUUUCUGCCUA

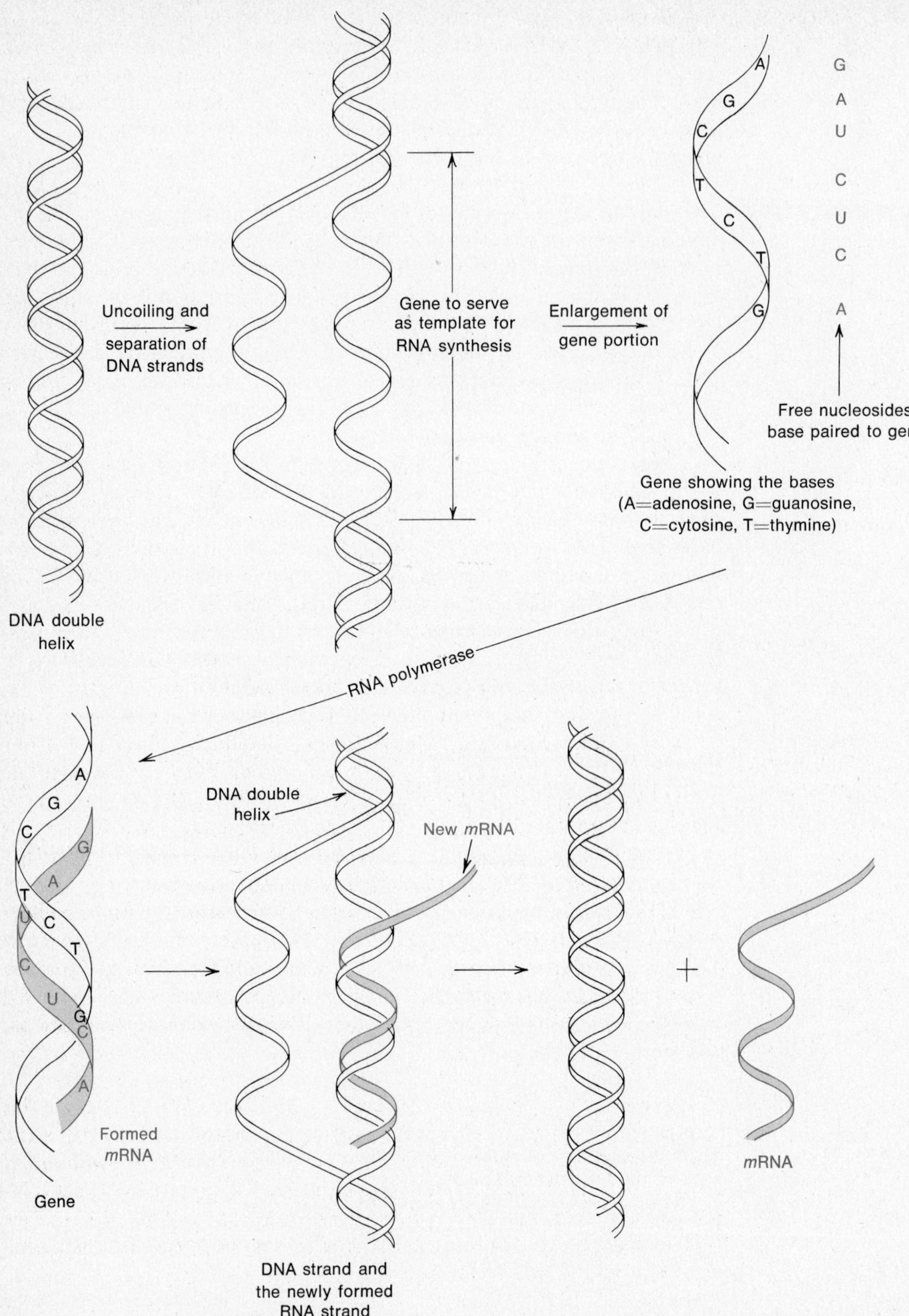

FIGURE 14-7 Schematic representation of *m*RNA formation.

TABLE 14-4
Amino acid codons

AMINO ACID	CODON*
Glycine	GGG, GGA, GGC, GGU
Alanine	GCG, GCA, GCC, GCU
Valine	GUG, GUA, GUC, GUU
Leucine	CUG, CUA, CUC, CUU, UUG, UUA
Isoleucine	AUA, AUC, AUU
Serine	UCG, UCA, UCC, UCU
Threonine	ACG, ACA, ACC, ACU
Cysteine	UGC, UGU
Methionine	AUG
Aspartic acid	GAC, GAU
Glutamic acid	GAG, GAA
Lysine	AAG, AAA
Phenylalanine	UUC, UUU
Tyrosine	UAC, UAU
Arginine	CGG, CGA, CGC, CGU, AGG, AGA
Histidine	CAC, CAU
Tryptophan	UGA, UGG
Proline	CCG, CCA, CCC, CCU

*U, uracil; C, cytosine; A, adenine; G, guanine.

See Figure 12-4 showing the components of a cell.

The translation of the genetic code contained in the *m*RNA and the actual synthesis of the protein molecule takes place on the ribosomes. After its formation in the nucleus, the messenger RNA diffuses to the endoplasmic reticulum where it becomes attached to a ribosome. The attachment occurs in such a manner that the codon for the first *t*RNA carrying the first amino acid is at the ribosomal surface. The *t*RNA carrying the amino acid bonds to the codon. Following bonding of the second *t*RNA to *m*RNA, the amino acid from the first *t*RNA is transferred (under the influence of appropriate enzymes) to the amino group of the second amino acid to form a peptide bond. After this transfer occurs, the first *t*RNA leaves the *m*RNA–ribosome complex. Simultaneously, the ribosome moves to the third codon. Now the process repeats itself: attachment of another *t*RNA, transfer of the growing protein chain, loss of a *t*RNA molecule, and movement of the ribosome (see Fig. 14-8).

MUTATION Mutations in cells can cause a change in the structure of proteins. This change can occur in several ways, such as the substitution of one amino acid for another. An example of this type of change is the substitution of a valine for a glutamic acid in hemoglobin (see Chapter 8). This particular substitution leads to sickling of red blood cells. Another possible change is the omission of an amino acid in a protein; yet another is the substitution of several amino acids in a sequence.

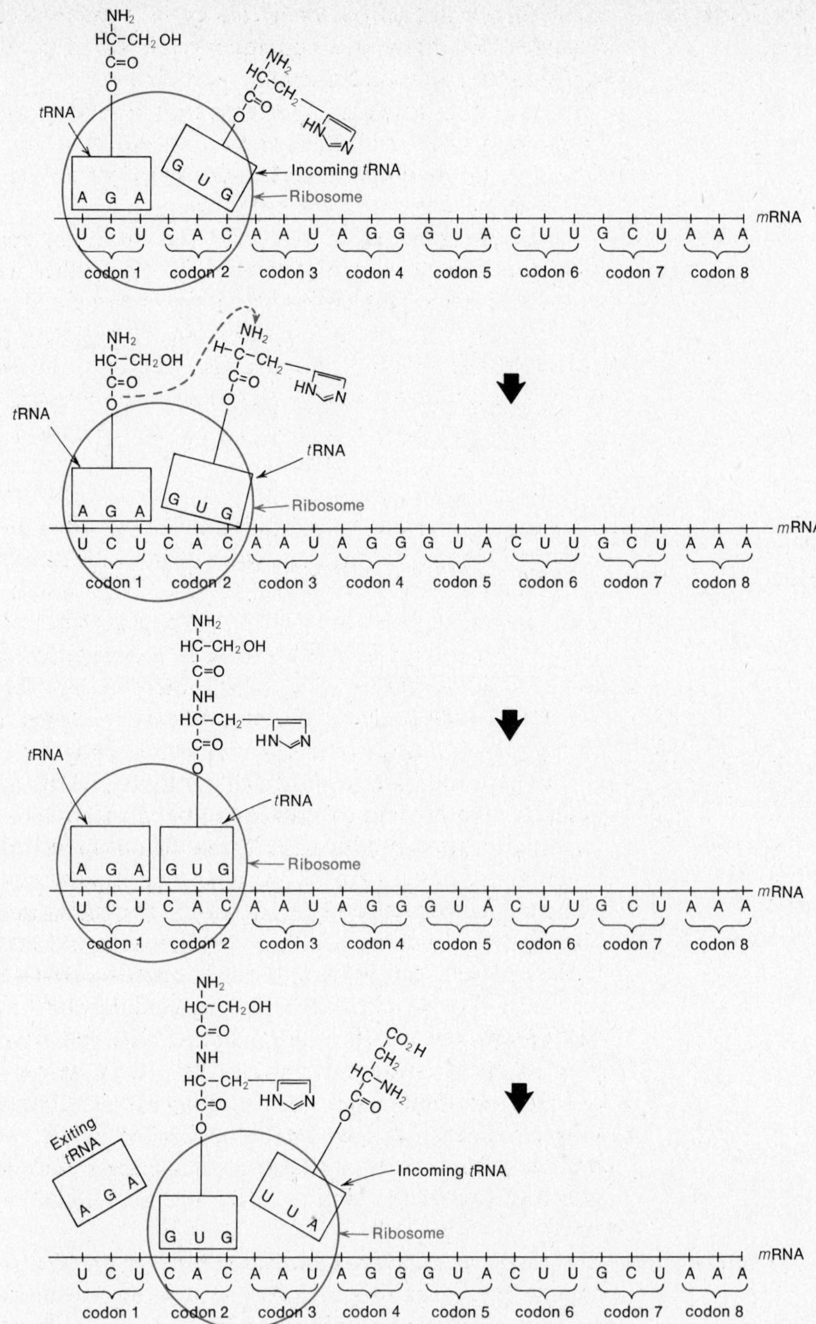

FIGURE 14-8 Protein synthesis by translation of the genetic code.

The opportunities for a change in protein structure are numerous. Changes can result as a consequence of an error in translation or transcription or because of a change in DNA.

Errors in translation or transcription result in a change of an individual or several protein molecules and do not result in inheritable traits; hence, these two errors do not lead to mutations nor do they normally lead to a condition deleterious to the organism.

Mutations result from a change in DNA structure.

A change in DNA structure will cause an abnormality in all the molecules of at least one particular protein synthesized by the cell as long as the cell lives. This lasting variation in protein structure arises because a change in a DNA molecule results in the continous production of altered *m*RNA molecules. Not only will the protein in this cell be abnormal, but this characteristic of producing abnormal proteins will be genetically passed on to new generations of cells during reproduction. Thus, we observe an inheritable alteration; *mutation* has occurred.

To understand this phenomena we must investigate the synthesis of DNA during the processes leading to mitosis. Before doing so, it must be emphasized that mutations are not necessarily bad. Mutation is a part of the evolutionary process and may aid the organism in surviving in a slowly changing or new environment. In the overall scheme of life we must view mutations as nature's experiments. If the mutation has positive survival value, it will be retained and the mutants will flourish, eventually replacing the *normal* individuals. If the mutation offers negative survival value, the mutants will die out. Some mutations apparently have neither a significant negative nor positive survival value and are simply carried along through generations providing differences between individuals of the same species.

Mutations result in the process called evolution.

The process by which DNA is synthesized is called *replication*. In many respects replication of DNA is very similar to the synthesis of *m*RNA. At present our knowledge of the details of the chemical processes involved is very limited. For example, it is not known what initiates the replication process. During the process, the two strands of the double-stranded DNA molecules become unraveled and separated. *Each* strand serves as a template for the synthesis of a new DNA strand. The new strand being synthesized is not identical to the template; rather, it is complementary to the parent strand. The parent strand and the newly synthesized strand join in a double helix coil to form a new DNA molecule. Since each strand of the original DNA molecule serves as a template, two new double-stranded DNA molecules result from the original molecule. This model of DNA replication is depicted in Fig. 14-9.

The enzyme DNA polymerase catalyzes the polymerization of the deoxyribonucleic acids to form the new DNA strands. The triphosphates of the four deoxyribonucleosides found in DNA, Mg^{2+} ions, and the DNA template are all required for the reaction. According to theory, deoxyadenosine triphosphate hydrogen bonds to its complementary base—thymine of the DNA template; thymine triphosphate, to adenine of the DNA; deoxyguanosine triphosphate, to cytosine; and deoxycytidine, to guanine. Follow-

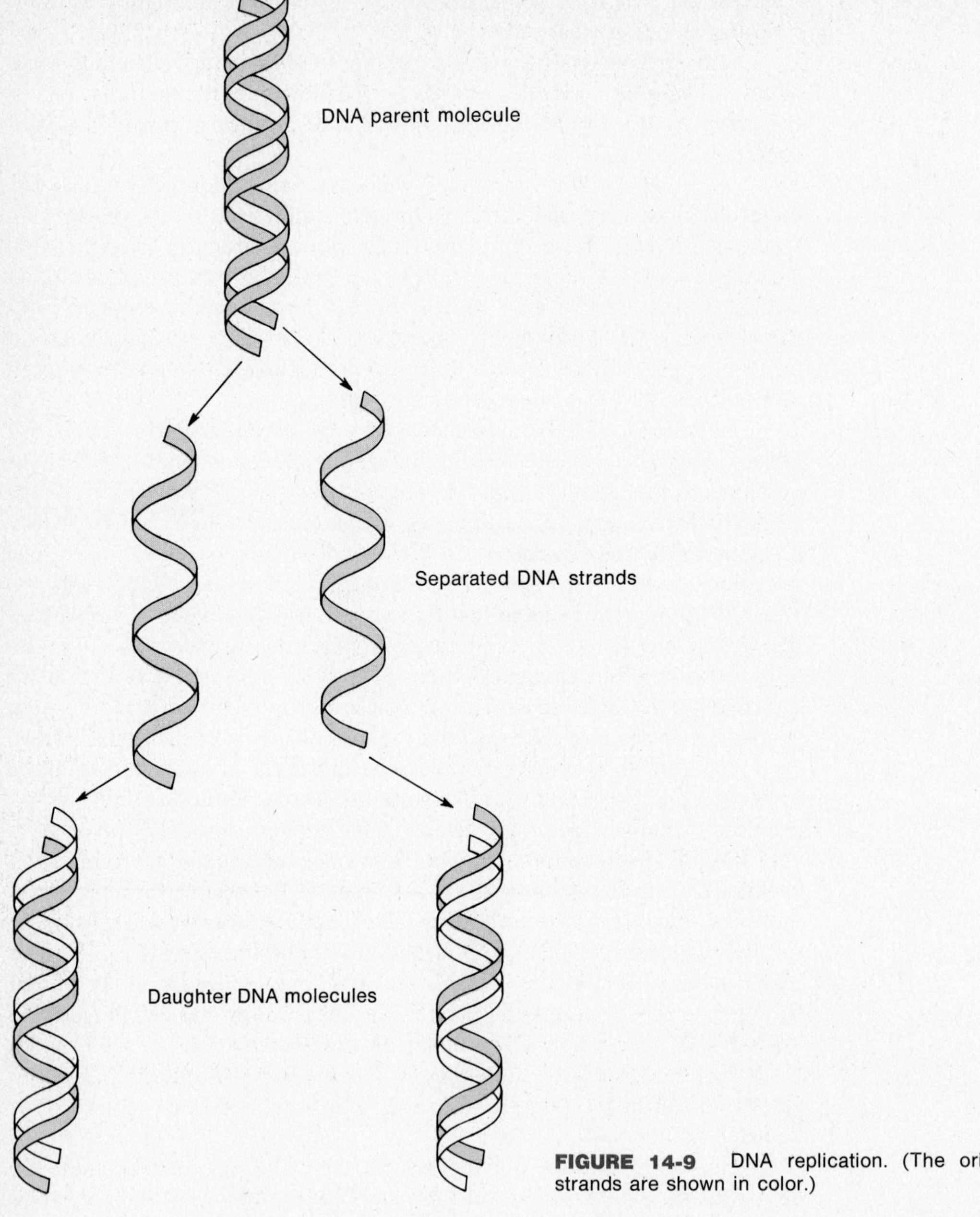

FIGURE 14-9 DNA replication. (The original strands are shown in color.)

ing hydrogen bonding, the deoxyribonucleic acids are joined by 3′,5′-phosphate ester bonds to produce a new DNA strand plus inorganic phosphate (see Fig. 14-10).

Mutations occur whenever the nucleotide sequence in DNA undergoes a change. One way the nucleotide sequence might be changed occurs during

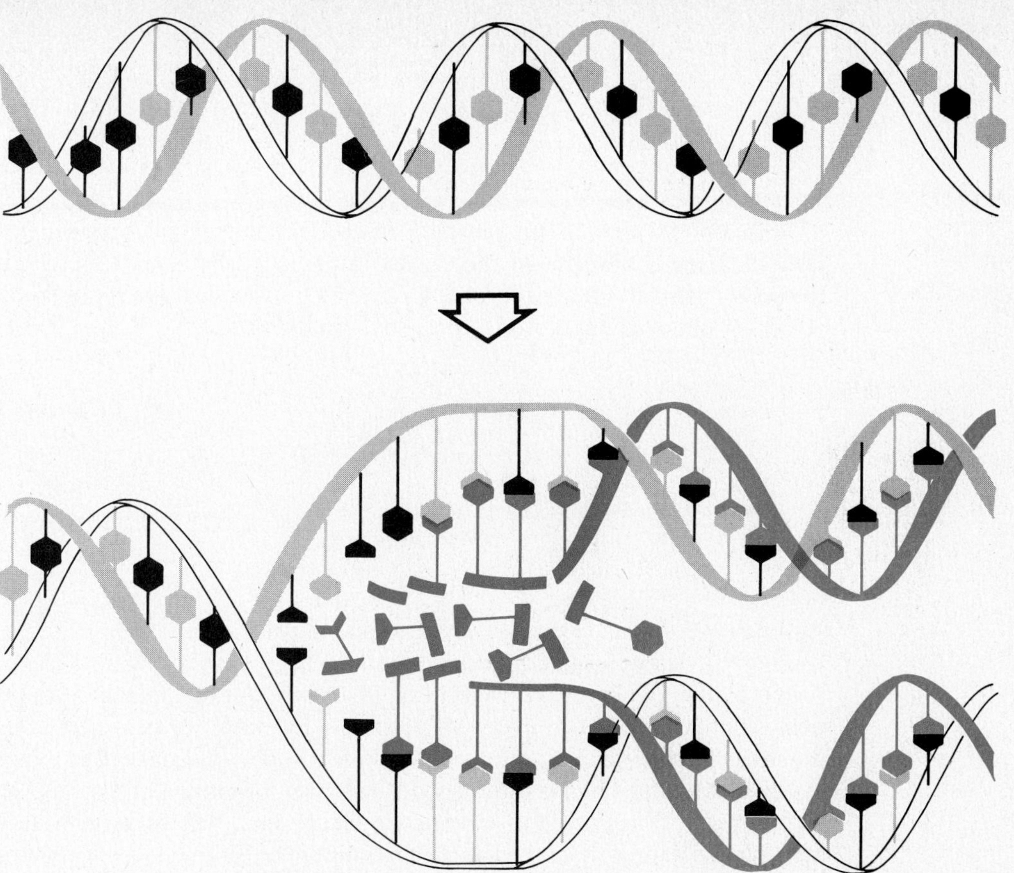

FIGURE 14-10 Schematic representation of DNA replication.

base pairing, resulting in the substitution of one nucleotide by another during replication. High-energy radiation, such as X rays, ultraviolet light, and radioactivity, also cause mutations. Certain chemicals are capable of causing mutations. Particularly effective chemicals are those that react with amino groups. Further, it appears that at least in certain species the cell itself has a built-in mechanism for causing mutations. For example, it has been found that the bacteria *Escherichia coli* contains mutator genes. One of these mutator genes causes the adenine–thymine base pair to be replaced by the cytidine–guanine base pair and vice versa.

HEREDITARY DISEASES

In previous chapters a number of hereditary disorders have been discussed including diabetes, galactosemia, cystinuria, and sickle cell anemia. Other hereditary diseases of sufficient incidence to warrant discussion are phenylketonuria, albinism, and hemophilia.

Phenylketonuria results from a lack of the enzyme phenylalanine hydroxylase, which catalyzes the conversion of phenylalanine to tyrosine. Although phenylalanine is an essential amino acid, the requirement for it is primarily to satisfy the body's need for tyrosine.

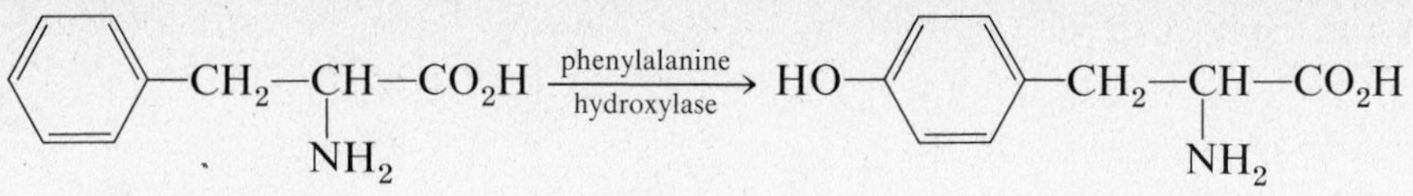

$$C_6H_5{-}CH_2{-}\underset{\underset{NH_2}{|}}{CH}{-}CO_2H \xrightarrow[\text{hydroxylase}]{\text{phenylalanine}} HO{-}C_6H_4{-}CH_2{-}\underset{\underset{NH_2}{|}}{CH}{-}CO_2H$$

phenylalanine tyrosine

In the absence of this enzyme the normal metabolic pathway for phenylalanine is blocked. As the concentration of phenylalanine increases, the body resorts to utilizing it for transamination reactions. Transamination involving phenylalanine produces phenylpyruvic acid.

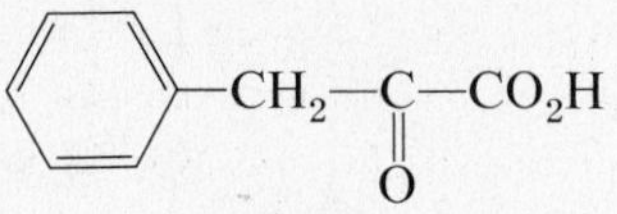

$$C_6H_5{-}CH_2{-}\underset{\underset{O}{\|}}{C}{-}CO_2H$$

phenylpyruvic acid

As much as 1–2 g/day of phenylpyruvic acid may be excreted via the urine. Phenylpyruvic acid is a keto acid, hence the name phenylketonuria (PKU). Accumulation of phenylpyruvic acid rapidly leads to severe mental retardation. Approximately 1 in every 80 people of European descent carry this trait as a recessive gene. Fortunately, a simple blood test for phenylketonuria has been developed and is routinely done in most hospitals 4 or 5 days after birth. Infants displaying this disease are placed on a diet containing very little phenylalanine. This diet must be maintained throughout periods of rapid brain growth and development if these individuals are to be given a chance for a normal adult life.

exercise 14-9

Write an equation using structural formulas for the transamination of phenylalanine with α-ketoglutaric acid.

Albinism is another hereditary disorder associated with abnormal amino acid metabolism. Tyrosine is a precursor of melanin, a dark, polymeric pigment, which is synthesized in melanoblasts located in the basal layer of the skin. Skin color depends on the number and distribution of melanoblasts and melanin concentration. Melanin is also found in the retina, adrenal medulla, and portions of the brain. The absence of this pigment results in albinism characterized by an extremely pale skin color and pink irises in the eyes.

The initial step in the conversion of tyrosine to melanin is the oxidation of tyrosine to dopa, catalyzed by the enzyme *tyrosinase*.

$$\text{HO}-\bigcirc-\text{CH}_2-\underset{\text{NH}_2}{\underset{|}{\text{CH}}}-\text{CO}_2\text{H} \xrightarrow{\text{tyrosinase}} \text{HO}-\underset{\text{HO}}{\bigcirc}-\text{CH}_2-\underset{\text{NH}_2}{\underset{|}{\text{CH}}}-\text{CO}_2\text{H}$$

tyrosine dopa

In albinos the enzyme is missing.

Hemophilia is a sex-linked hereditary disease, occurring primarily in males but transmitted solely through females. The disorder is characterized by a marked increase in the time required for blood clotting. The essential feature of the extremely complex process of clotting is the conversion of fibrinogen to soluble fibrin that reacts to form insoluble fibrin in the presence of Ca^{2+} ions. The conversion of fibrinogen to fibrin is catalyzed by the enzyme thrombin. In order to prevent coagulation in the bloodstream, thrombin itself is not present in blood but is present as inactive prothrombin. Individuals with hemophilia, of which there are three known types, lack the activators necessary for the conversion of prothrombin to thrombin (see Chapter 16).

CHAPTER SUMMARY

1. Proteins serve specific purposes in the body such as structural, enzymatic, and hormonal. They are seldom metabolized as an energy source.
2. Proteins are hydrolyzed to their constituent amino acids during digestion prior to absorption into the body.
3. The amino acid pool is all the free amino acids in the body. These amino acids are not stored in a specific site but are circulated via the bloodstream and are available to all tissues.
4. A nitrogen balance exists when a person excretes as much nitrogen as ingested. Positive nitrogen balance results when less nitrogen is excreted than taken in; negative nitrogen balance results when more nitrogen is excreted than ingested.
5. Primary metabolic paths for amino acids are protein synthesis, heme and nucleic acid synthesis, transamination, and oxidative deamination.
6. Transamination is the transfer of an amino group from an amino acid to an α-keto acid to form a different amino acid and keto acid. This process enables the body to synthesize nonessential amino acids.
7. Oxidative deamination is the removal of an amino group from an amino acid to produce ammonia and a keto acid.
8. Ammonia is usually combined with carbon dioxide to produce urea via the ornithine cycle prior to excretion.
9. DNA is the genetic material of life and possesses the ability to replicate and control the formation of proteins.

10. DNA is a polymer of the sugar deoxyribose and the heterocyclic bases adenine, guanine, cytosine, and thymine. Polymerization occurs via phosphate ester linkages between deoxyribose units.
11. DNA molecules consist of two strands held together by hydrogen bonding between the heterocyclic bases. These two strands are twisted to form a helical coil.
12. Messenger RNA serves as a carrier of the genetic code from DNA to the ribosomes where protein synthesis actually occurs.
13. RNA is similar to DNA *except* that it is a much smaller polymer that contains ribose instead deoxyribose and uracil instead of thymine. It is a single-stranded molecule.
14. Transcription is the process by which the genetic information contained in DNA is incorporated into *m*RNA.
15. Translation is the process by which the order of the heterocyclic bases in *m*RNA determines the order of amino acids in the protein being synthesized. Transfer RNA serves both as an amino acid carrier and the interpreter of the genetic code.
16. A codon is a sequence of three heterocyclic bases in *m*RNA and specifies the amino acid to used in protein synthesis.
17. Mutations are the result of a change in DNA structure.
18. Most hereditary diseases are the result of a missing or abnormal enzyme caused by mutation.

QUESTIONS

1. To counteract heartburn caused by hyperacidity, people take various patent medicines composed primarily of aluminium hydroxide. By a balanced chemical equation, explain how these medications will relieve heartburn.
2. Serine can be synthesized in the body from 3-phosphoglyceric acid by the reaction sequence (a) oxidation, (b) transamination, (c) hydrolysis of the phosphate ester. Using structural formulas, write the reactions leading to serine from 3-phosphoglyceric acid (refer to Chapters 8 and 13, respectively, for the structural formulas of serine and 3-phosphoglyceric acid).
3. Explain what is meant by the term *dynamic state of tissue proteins.*
4. Explain why the amino acid alanine will be metabolized as a carbohydrate after oxidative deamination.
5. Histamine is a powerful vasodilator and in excess can cause vascular collapse. It is liberated by persons suffering traumatic shock and in localized areas of inflammation. Histamine results from the decarboxylation of the amino acid histidine by action of the enzyme histidine

decarboxylase. Write the structure of histamine (see Chapter 8 for the structural formula of histidine).

6. Maple syrup urine disease—named after the characteristic urine odor—is hereditary and arises from a disorder in the metabolism of valine, leucine, and isoleucine. Following transamination of these amino acids, normal metabolism of the resulting keto acids does not occur and the keto acid concentration rises and imparts the odor. Write the structural formulas of the three keto acids.
7. One of the gastrin hormones secreted by humans is the heptadecapeptide with the amino acid sequence Glu–Gly–Pro–Trp–Leu–Glu–Glu–Glu–Glu–Glu–Ala–Tyr–Gly–Trp–Met–Asp–Phe. Write a possible sequence of nucleosides for the *m*RNA that could direct the synthesis of this small protein.
8. From your answer to Question 7, write a possible sequence of nucleosides for both strands of the portion of DNA responsible for the production of gastrin.
9. Explain and differentiate among replication, transcription, and translation.

15
hormones and vitamins

INTRODUCTION

Humoral, from Latin humor, *fluid.*

Endocrine, from Greek endon, *within;* krino, *to separate.*

Hormone, from Greek, hormon, *arouse, excite.*

Discussions in Chapters 13 and 14 have described how the presence or absence of various substances can effect cellular activities. The cascade of effects caused by insulin insufficiency in a diabetic illustrates how the breakdown of one system, insulin production, creates widespread physiological consequences. This serves to remind us that higher organisms depend on the proper integration of many chemical events among a vast number of cells in order to maintain good health. Proper integration requires that cells successfully respond to internal and external changes such as changes in temperature, concentrations of various metabolites, and pH. These changes and the response to them are transmitted through chemical *messengers*. Transmission may occur in two ways: (1) nerve action, organ-to-organ transfer of information or (2) humoral communication, information transfer via body fluids, mainly blood plasma. Endocrine glands are specialized organs that synthesize and secrete hormones directly into the bloodstream. These substances can function as chemical messengers (see Fig. 15-1).

This chapter describes several hormones and vitamins and their effects on various biochemical processes.

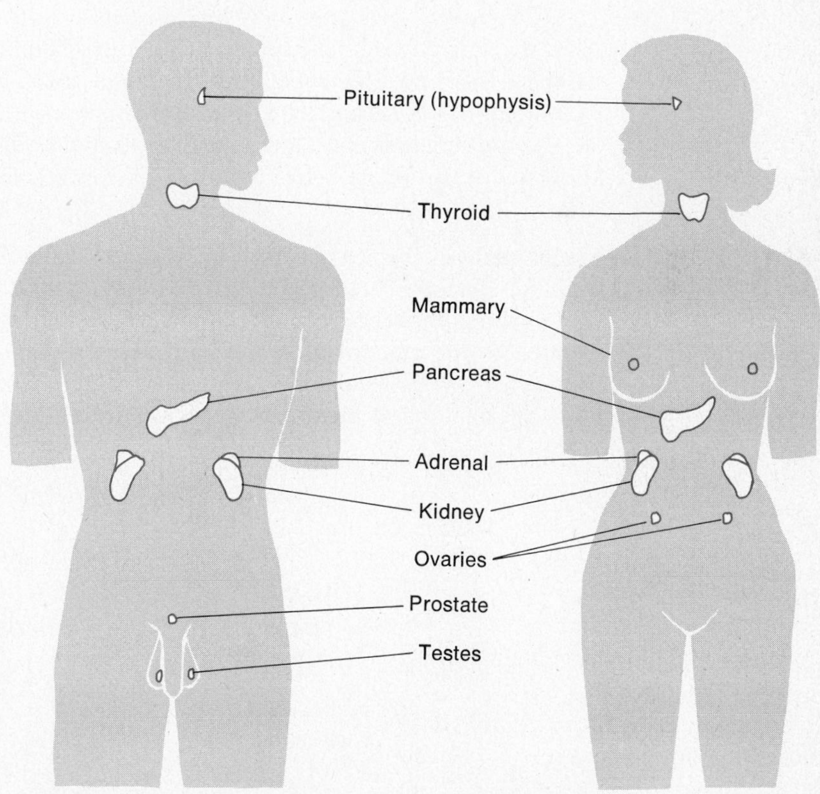

FIGURE 15-1 Locations of the endocrine glands in humans.

HORMONES—SOME GENERAL COMMENTS

Physiologically, hormones have several distinctive properties:

1. They are produced in very small quantities and secreted in nanogram to milligram amounts per day.
2. Although in very low concentration, they are able to bring about profound changes.
3. Normally, secretion rate is dependent on need for the particular hormone.
4. Apparently, a hormone does not directly affect the organ secreting it.
5. Hormones may have an effect on the organism in general or only on specific cells of a given organ called target cells.
6. Hormone effect depends on the amount of hormone present and the responsiveness of the target tissue.
7. Hormones initiate biochemical reactions that may continue long after the hormone concentration is no longer measurable. Many hormones control metabolism by either stimulating or reducing enzyme activity in primary metabolic pathways.

Figure 15-1 shows the location of the endocrine glands in humans. Table 15-1 lists the vertebrate hormones, their classes, targets, and effects.

Chemically, hormones may be classified as (a) proteins and polypeptides, (b) derivatives of the amino acids tryptophan and tyrosine, and (c) steroids—lipids such as cholesterol. In broad terms, hormones are either proteinoid, classes (a) and (b) above, or lipoid, class (c) above. Insulin, vasopressin, and epinephrine (adrenalin) are examples of the former; the sex hormones progesterone, testosterone, and the estrogens are examples of the latter.

HYPOPHYSEAL HORMONES (PITUITARY HORMONES)

The hypophysis, formerly called the pituitary gland, is a small, two-lobed structure located at the base of the brain. Each lobe functions essentially as a separate endocrine gland secreting specific hormones. The hormones secreted by the anterior lobe of the hypophysis gland are all either polypeptides or low molecular weight proteins. Figure 15-2 indicates the secretions of each lobe in mammals.

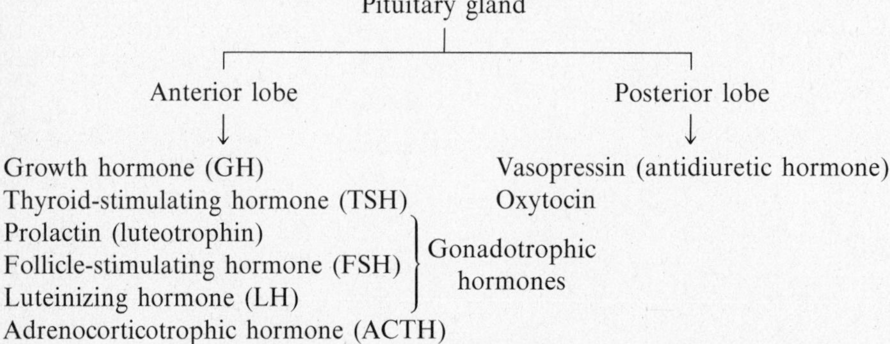

FIGURE 15-2 Hormones secreted by the anterior and posterior lobes of the pituitary gland.

TABLE 15-1
Vertebrate hormones

HORMONE	TARGET	EFFECTS
I. Peptides, protein hormones		
A. Hypophysis—anterior lobe		
Growth hormone (GH, somatotropin)	All tissues	Growth of tissues (easily seen in long bones, metabolism of protein, mobilization of fat)
		Lipolysis
Adrenocorticotrophin (ACTH, corticotrophin)	Adrenal cortex	Synthesis and release of glucocorticoids
	Adipose tissue	Lipolysis
Thyroid-stimulating hormone (TSH, thyrotrophin)	Thyroid gland	Synthesis and secretion of thyroxine and triiodothyronine
Male		
Follicle-stimulating hormone (FSH)	Seminiferous tubules	Production of sperm
Luteinizing hormone, LH	Testes	Synthesis and secretion of androgens
Female		
FSH	Ovary (follicles)	Follicle maturation
LH	Ovary	Final maturation of follicle, estrogen secretion, ovulation, corpus luteum formation, progesterone secretion
Prolactin	Mammary glands (alveolar cell)	Milk production in prepared gland
Melanocyte-stimulating hormones	Melanophores	Pigment dispersal in melanophores (darkening of skin)
B. Hypophysis—posterior lobe		
Oxytocin (let-down) factor, milk-ejection factor)	Uterus	Contraction of smooth muscle, milk ejection
	Mammary glands	
Vasopressin (antidiuretic hormone, ADH)	Kidney	Reabsorption of water
	Arteries	Contraction of smooth muscle
C. Pancreas		
Insulin	All cells	Carbohydrate, fat, and protein metabolism, hypoglycemia
Glucagon	Liver	Hyperglycemia
D. Ovary		
Relaxin	Pelvic ligaments	Separation of pelvic bones
E. Thyroid		
Thyrocalcitonin (calcitonin)	Bones, kidney	Excretion of calcium and phosphorus, inhibited calcium released from bones, decreased blood calcium levels
II. Amino acid derivatives		
A. Thyroid		
Thyroxin Triiodothyronine	Most cells	Increased metabolic rate, growth, and development
B. Adrenal medulla		
Norepinephrine Epinephrine	Most cells	Increased cardiac activity, elevated blood pressure, glycolysis, hyperglycemia
III. Steroids and lipids		
A. Testes		
Androgen (testosterone)	Most cells	Development and maintenance of masculine characteristics and behavior
B. Ovary estrogen	Most cells	Development and maintenance of feminine characteristics and behavior
C. Corpus luteum		
Progesterone	Uterus, mammary glands	Maintenance of uterine endometrium and stimulation of mammary duct formation
D. Adrenal cortex		
Hydrocortisone Cortisone	Most cells	Balanced carbohydrate, protein, and fat metabolism; antiinflammatory action
Aldosterone	Kidney	Reabsorption of Na^+ from urine
E. Prostate, seminal vesicles, brain, nerves		
Prostaglandins	Uterus	Contraction of smooth muscle

GROWTH HORMONE (SOMATOTROPIN)

The removal of the hypophysis from young test animals results in a retardation of their growth. These experiments provided the earliest clues that proper hypophysis functioning is required for normal growth. Growth hormone (GH), also called somatotropin, stimulates skeletal growth and an increase in body weight. Dwarfism results from undersecretion of GH during periods when skeletal growth should normally occur. Oversecretion of growth hormone prior to adulthood results in giantism; oversecretion after full skeletal growth is complete causes acromegaly, a condition in which enlargement of the hands, feet, jaws, and face occurs.

Studies seem to indicate that human growth hormone functions by altering the permeability of cell membranes toward amino acids, thereby affecting the rate of protein synthesis. Somatotropin also effects carbohydrate and lipid metabolism. It decreases the rate at which carbohydrates are utilized by the body, thus causing hyperglycemia and glucosuria (diabetogenic effect). Somatotropin increases the release of lipids from fat depots. This increases fat metabolism; excess acetyl CoA is produced, and ketosis results causing ketonemia and, in severe cases, ketonuria. This is referred to as the ketogenic effect of growth hormone.

THYROID-STIMULATING HORMONE

Thyroid-stimulating hormone (TSH) acts on two sites—fat depots and the thyroid gland. The hormone causes release of fatty acids from fat depots and thus acts as a regulator for fatty acid metabolism. TSH also controls thyroid function by stimulating iodine uptake, formation of iodine-containing hormones such as thyroxine, and release of these hormones into circulation. Using injected radioactive I-131 as a tracer, an animal's rate of iodine uptake can be followed. Figure 15-3 shows a graph typical of iodine uptake in rats.

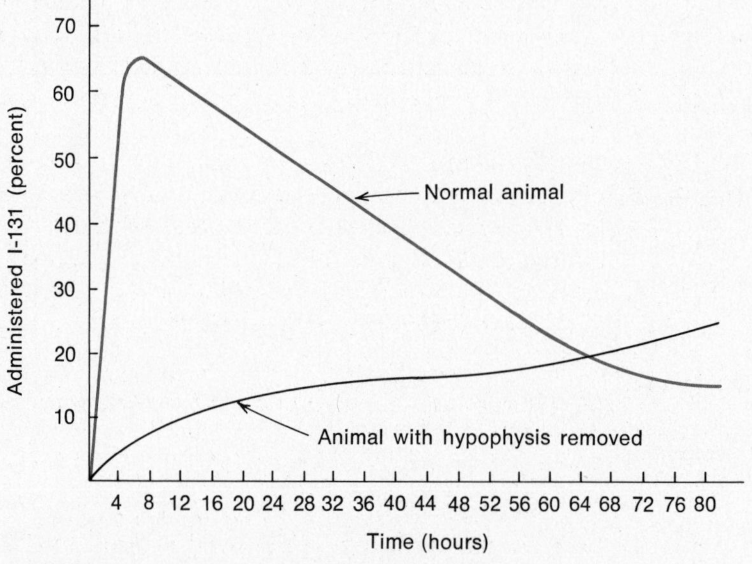

FIGURE 15-3 I-131 uptake by the thyroid gland of rats.

ADRENOCORTICOTROPHIC HORMONE (ACTH) Adrenocorticotrophic hormone (ACTH) acts principally upon the cortex of the adrenal glands. Emotional stress, temperature extremes, and drugs are some of the stimuli that seem to influence the secretion of ACTH and, in turn, effect the secretion of adrenal cortex hormones.

GONADOTROPHIC HORMONES Follicle-stimulating hormone (FSH), luteinizing hormone (LH), and prolactin are collectively called the gonadotrophic hormones because they control the growth and development of the gonads (ovaries in the female and testes in the male). Prior to puberty, only slight amounts of these hormones are secreted. The onset of puberty is marked by an increased secretion of gonadotrophic hormones that leads to the development of secondary sexual characteristics. In males, FSH regulates sperm formation and the production of testosterone, the major male sex hormone. In females, FSH promotes certain changes during the menstrual cycle as well as stimulating production of estrogens. The sex hormones will be more fully discussed later in this chapter.

Luteinizing hormone stimulates specialized cells in the testes of males to produce testosterone. In females, LH stimulates rupture of the ovarian follicle and production of progesterone by the corpus luteum.

Prolactin functions cooperatively with estrogens in promoting development of the mammary glands. Prolactin also is responsible for the onset of milk production in a mother's breasts following the birth of a baby. The hormone also promotes corpus luteum development and, consequently, progesterone production.

Recent research has found that prolactin seems to be essential in causing the induction and growth of mammary cancers in rats and mice. The research data showed that when prolactin secretion was inhibited in rats with mammary cancers, the size of the cancers was reduced. Further research on noncancerous young mice given drugs to suppress prolactin secretion resulted in the development of fewer cases of mammary cancer than in test animals with normal prolactin level. A relationship between prolactin and mammary cancer in humans has not been established.

VASOPRESSIN AND OXYTOCIN Vasopressin and oxytocin are hormones of the posterior lobe of the hypophysis. The chemical structures of both have been found to be similar; they are octapeptides, each containing a disulfide bond. Their symbolized structural formulas are shown in Fig. 15-4. Note that oxytocin and vasopressin differ by only two amino acid residues. Their similarity of structure leads to each exhibiting some activities of the other.

exercise 15-1

Using amino acid structural formulas (Table 8-8), write the structural formula for vasopressin.

The main function of vasopressin is to regulate water loss from the body via the urine. This is discussed in detail in Chapter 16. At elevated levels, vasopressin causes contraction of smooth muscles, especially those

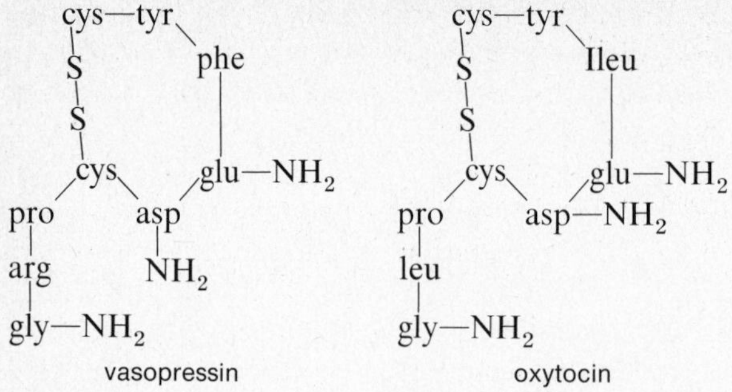

```
cys—tyr                      cys—tyr
 |      \                     |      \
 S       phe                  S       Ileu
 |        |                   |        |
 S        |                   S        |
 |        |                   |        |
 cys      glu—NH2             cys      glu—NH2
/    \   /                   /    \   /
pro    asp                  pro    asp—NH2
|       |                   |
arg     NH2                 leu
|                           |
gly—NH2                     gly—NH2
    vasopressin                 oxytocin
```

FIGURE 15-4 Symbolized structural formulas of vasopressin and oxytocin. Glu-NH_2 and asp-NH_2 represent glutamic acid and aspartic acid residues whose side chain carboxyl groups have been converted to the amide form. In gly-NH_2, the terminal carboxylic acid group has been converted to the amide.

By contraction, the blood vessel's diameter decreases. This causes an increase in blood pressure.

of the blood vessels. This is particularly significant during severe hemorrhaging. A severe hemorrhage stimulates the secretion of large amounts of vasopressin, which causes contraction of coronary and pulmonary arteries, tending to return arterial pressure toward normal level.

Oxytocin, from Greek, rapid birth.

Oxytocin causes powerful contractions of the uterus during childbirth. This hormone is frequently administered by injection to stimulate uterine contraction during childbirth. Studies have shown that a concentration as low as 1 part oxytocin per 2 billion is sufficient to cause contractions of an isolated uterus. Oxytocin also causes milk in the mother's breasts to be released into ducts where it can be obtained by the sucking infant. Thus, oxytocin works in conjunction with prolactin during lactation.

TRYPTOPHAN-RELATED HORMONES

The hormones seratonin and melatonin are synthesized from tryptophan. Seratonin is transformed into melatonin in the pineal gland. Melatonin activates the dispersion of melanin pigments of hair, skin, and the colored portion (irises) of the eyes.

TYROSINE-DERIVED HORMONES

The amino acid tyrosine is the substance from which two important groups of hormones are synthesized—catecholamine-type hormones and thyroid hormones (see Fig. 15-5).

It is known that at least four physiologically active, structurally related, iodine-containing compounds act as thyroid hormones. Under normal conditions, the thyroid gland rapidly takes up iodine for the synthesis of these hormones. The two most important ones are thyroxine and triiodothyronine. These hormones are stored in the thyroid gland by combination with a protein to form thyroglobulin. Hydrolysis of thyroglobulin releases the thyroid hormones. Upon entering the blood, the hormones combine mainly with a plasma protein called thyroxine-binding globulin and are transported in this bound form.

The thyroid hormones regulate both the rate of oxygen consumption and the metabolism in most body tissues. How this is accomplished is still not clear. Experiments on test animals have shown that the number and activity of the mitochondria increase upon administration of thyroxine. It may be that thyroxine stimulates the rate of ATP formation, which in turn can be used to provide the energy necessary for accelerated cellular activity.

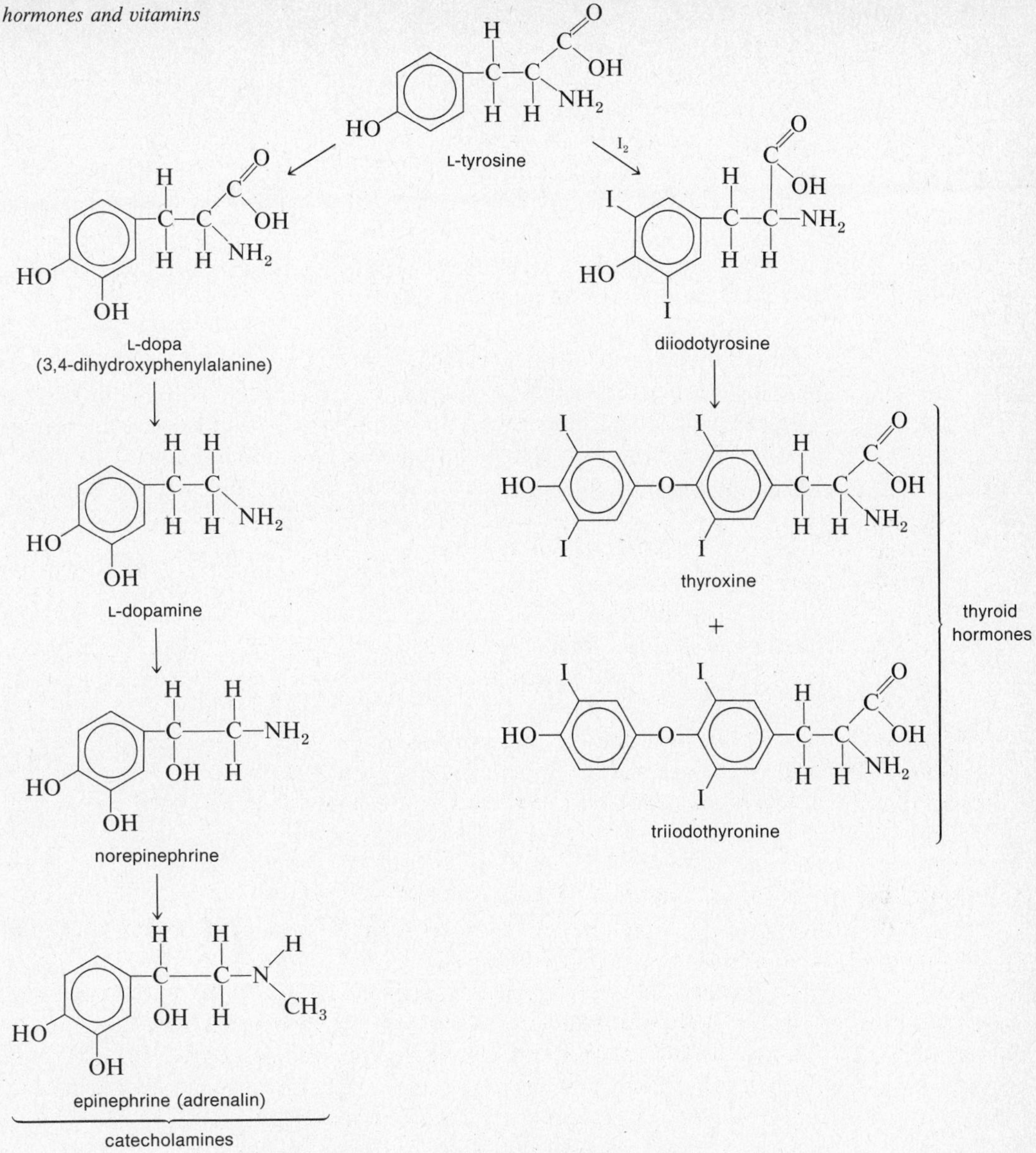

FIGURE 15-5 Hormone derivatives of tyrosine.

Insufficient thyroid secretion causes *hypothyroidism*. In infants, hypothyroidism causes cretinism characterized by retardation of mental, physical, and sexual development. In adults, hypothyroidism leads to myxedema. *Hyperthyroidism* results from excessive stimulation of the thyroid by thyroid-stimulating hormone (TSH). Enlargement of the thyroid occurs and metabolic rate increases significantly (Graves' disease). Exophthalmos (protrusion of the eyeballs) is commonly found as well.

Measurement of protein-bound iodine (PBI) in blood plasma enables diagnosis of hyperthyroidism or hypothyroidism. The normal range of PBI concentration is 4 to 7.5 μg/100 ml plasma. Values greater than 15 μg/100 ml can occur in severe hyperthyroidism. In severe hypothyroidism, PBI concentration falls below 2 μg/100 ml.

Iodized salt is commercially available.

Any enlargement of the thyroid gland is termed a goiter. Simple goiter can be caused by a dietary deficiency of iodine. To form the required amount of thyroxine, about 1 mg/week of ingested iodine is required. Sodium iodide or potassium iodide can be added to table salt (sodium chloride) in about a 1:10,000 ratio to ensure adequate dietary iodine.

Various substances can act as antithyroid agents. Their antithyroid action is usually due to either (1) their ability to prevent iodine incorporation into the tyrosine derivatives or (2) their structural resemblance to the thyroid hormones. Thiourea, thiouracil, and sulfonamides prevent hormone formation by combining with elemental iodine in the thyroid gland. This makes iodine unavailable for hormone formation and thyroid activity decreases.

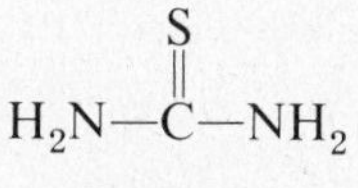

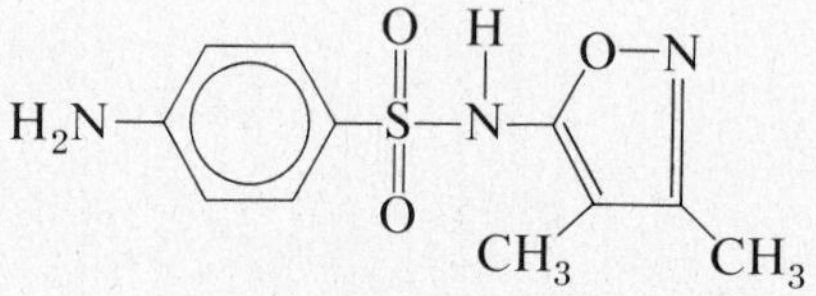

$$H_2N-\overset{\overset{\displaystyle S}{\|}}{C}-NH_2$$

thiourea

thiouracil (HS–C ring structure with N, C–OH, C–H, C–H, N)

a sulfonamide (H_2N–C$_6$H$_4$–SO_2–NH– isoxazole ring with two CH_3 groups)

Compounds such as

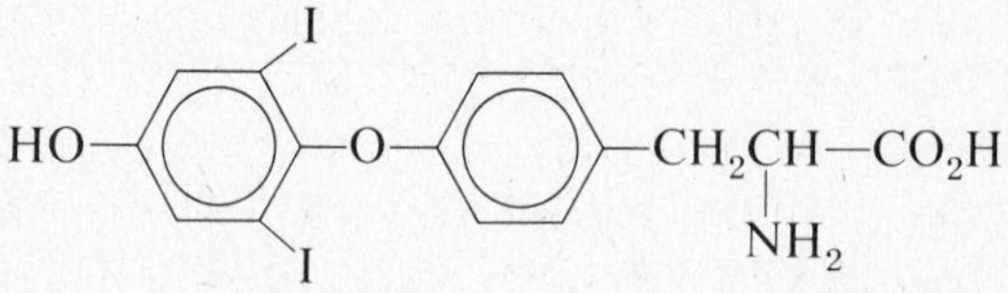

$$HO-C_6H_2I_2-O-C_6H_4-CH_2\underset{\underset{\displaystyle NH_2}{|}}{C}H-CO_2H$$

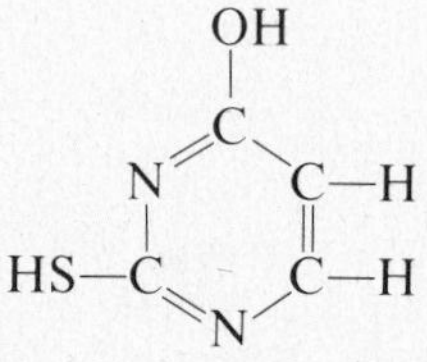

and

$$C_6H_5-CH_2-O-C_6H_2I_2-CO_2H$$

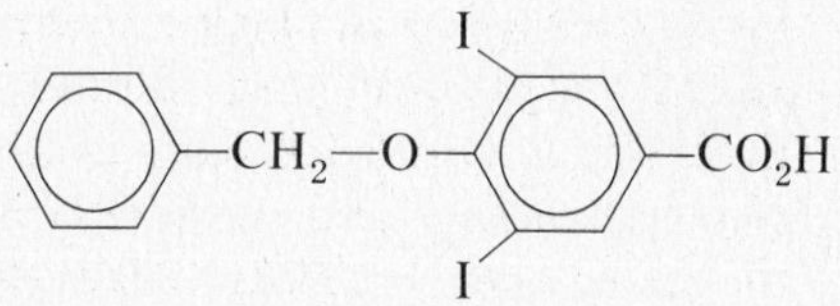

have structures similar to the thyroid hormones and they inhibit thyroid activity. It is thought that this occurs because these substances effectively compete with the natural hormones for the active sites of certain enzymes. By occupying an enzyme's active site, the synthetic substance inactivates the enzyme. This is known as competitive inhibition and reduces thyroid activity.

exercise 15-2

Propylthiouracil is administered to individuals suffering from Graves' disease. Explain how this drug is useful in this situation.

L-Dopa, dopamine, norepinephrine, and epinephrine (adrenalin) are termed catecholamines because of their structural resemblance to catechol

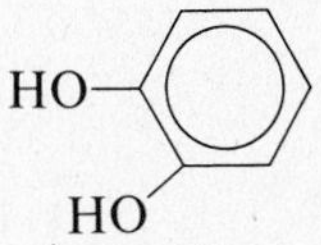

HO
HO

They are produced by the medulla (inner portion) of the adrenal glands. Many individuals suffering from Parkinson's disease are aided by treatment with L-dopa. This material reduces Parkinson symptoms such as tremor and lack of muscle control although exactly how the drug does this is still not known.

Norepinephrine and epinephrine are sometimes called *emergency hormones* because of their effectiveness in rallying the body during times of physical or emotional stress. Epinephrine is in higher concentration and is more effective than norepinephrine. During stress, epinephrine stimulates a rapid increase in blood sugar through liver glycogenolysis. Release of fatty acids from fat depots also occurs. Accompanying these changes are increases in heartbeat, blood flow in limb muscles, and oxygen consumption. Phosphorylase is the enzyme that catalyzes the release of glucose from glycogen. When the body is resting, most of the phosphorylase is inactive so that glucose is not released from glycogen. Epinephrine indirectly activates phosphorylase and glycogenolysis rapidly occurs.

Almost sound like the poetic symptoms for falling in love.

In 1960, E. Sutherland and co-workers investigated promotion of glycogenolysis by epinephrine. They proposed that epinephrine does not directly cause glycogenolysis but activates a chain of events leading to glycogenolysis. Epinephrine, acting as a primary messenger, activates the enzyme adenyl cyclase found in cell membranes. Adenyl cyclase catalyzes the conversion of ATP to cyclic-AMP (*c*-AMP).

HO—P(=O)(OH)—O—P(=O)(OH)—O—P(=O)(OH)—O—$\overset{5'}{C}H_2$ — ribose (4′, 1′, 3′, 2′; HO, OH) — adenine

ribose

ATP

$\longrightarrow$

$HOH + PP_i +$ cyclic-AMP ($^{5'}CH_2$, O, adenine, 3′, O=P, HO, O, OH)

cyclic-AMP
(3′,5′-AMP)

Cyclic AMP acts as *secondary* messenger whose target organ is the liver. The conversion of glycogen to glucose-1-phosphate in the liver is catalyzed by phosphorylase. Phosphorylase may exist in two forms—an inactive form called phosphorylase b and an active form, phosphorylase a. Phosphorylase b is activated to phosphorylase a by the enzyme phosphorylase kinase, which in the presence of Mg^{2+} is activated by *c*-AMP. Phosphorylase a then promotes glycogenolysis. Epinephrine activates glycogenolysis indirectly via messengers. Figure 15-6 shows an overview of the interactions leading to glycogenolysis.

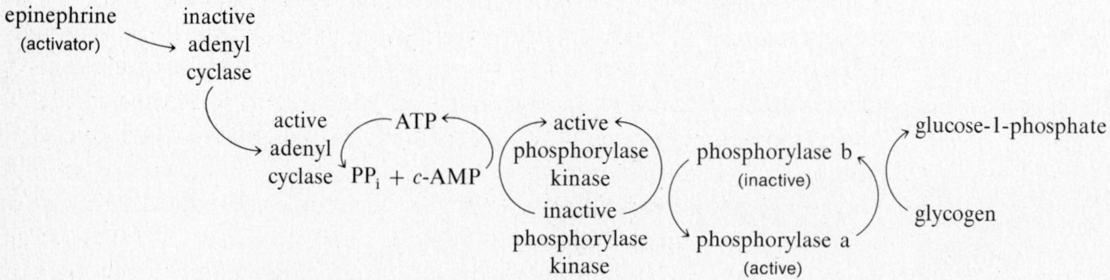

FIGURE 15-6 Epinephrine, the primary messenger, activates formation of a second messenger, *c*-AMP, which activates further events leading up to glycogenolysis.

Norepinephrine's activity toward liver adenyl cyclase is far less than that of epinephrine. The release of fatty acids is also stimulated by norepinephrine.

INSULIN AND GLUCAGON

Insulin is secreted by the beta cells of the islets of Langerhans in the pancreas. As mentioned in Chapter 13, insulin increases the rate of glucose transport across cell membranes especially in muscle and fatty tissues. A person suffering from untreated diabetes mellitus experiences abnormally high glucose concentration in the blood because of failure to transport glucose into cells.

Diagnosis of diabetes mellitus is facilitated by administration of a glucose tolerance test. After fasting, a patient ingests 50 g of glucose in solution. Blood samples are taken periodically and the blood glucose levels are determined. These concentrations are compared to those of a normal individual and a diabetic. Figure 15-7 illustrates typical glucose tolerance test results for a normal individual and a diabetic.

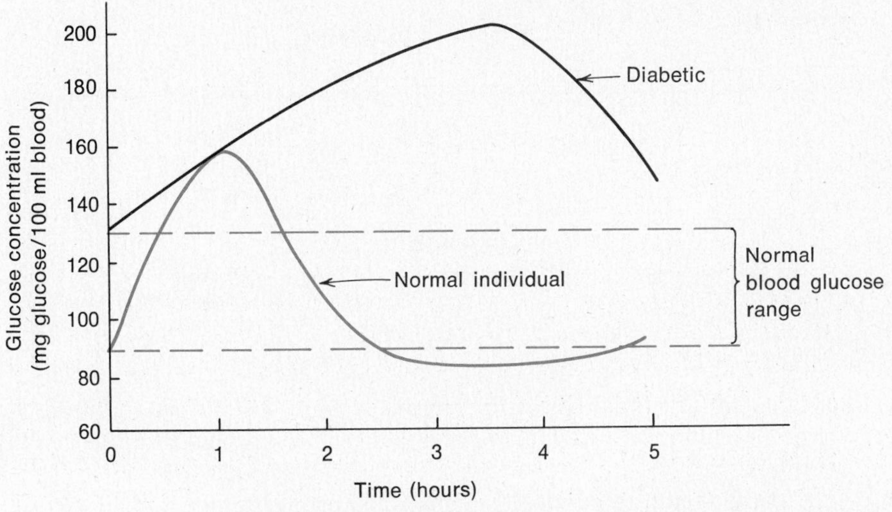

FIGURE 15-7 Glucose tolerance test curves for a normal person and a diabetic.

Following ingestion of the glucose solution, a normal individual's blood glucose concentration quickly rises from its normal fasting value of less than 100 mg of glucose per 100 ml of blood (100 mg%) to about 160 mg%. It then drops to below normal level within 3 hr. In this individual, an increase in insulin secretion follows glucose intake and facilitates rapid transport of glucose into cells for metabolism.

A diabetic usually has a fasting blood glucose level of at least 120 mg%. Upon ingestion of 50 g of glucose, a gradual rise in blood glucose level occurs in the diabetic during the next 2 to 3 hr. The blood sugar level slowly decreases during the third through fifth hours. The slowness of the decline of glucose level in the diabetic following glucose ingestion indicates that an increase in insulin secretion does not occur in the diabetic. Note that the diabetic's glucose level does not fall below the diabetic's normal fasting level during the third to fifth hour following ingestion. This is in contrast to the normal individual's behavior during this period in which glucose level may actually fall below the normal fasting level. In normal individuals, glucose level returns to normal fasting level values after the fifth hour due to glucose release by glycogenolysis.

An excess of insulin (hyperinsulinism) leads to a decrease of blood glucose concentration. Severe shock, convulsions, loss of consciousness, and eventually coma may result as the blood glucose level falls. If glucose is not immediately administered, damage to the central nervous system occurs.

One of your authors has a dog that suffers from glucagon insufficiency.

Glucagon is a low molecular weight protein hormone secreted by the alpha cells of the islets of Langerhans of the pancreas. Glucagon, like epinephrine, causes glycogenolysis by increasing *c*-AMP production, which in turn activates liver phosphorylase.

STEROIDS

Steroids can be considered as derivatives of a perhydrocylopentanophenanthrene-type structure.

The rings of steroids are commonly identified as A, B, C, and D.

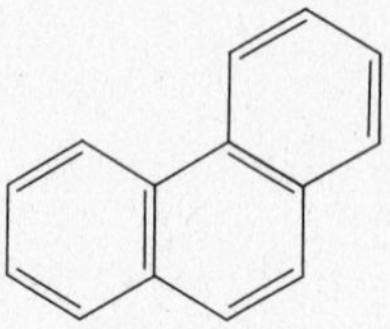

A B C D

phenanthrene

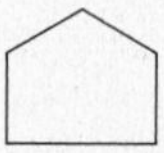

cyclopentane

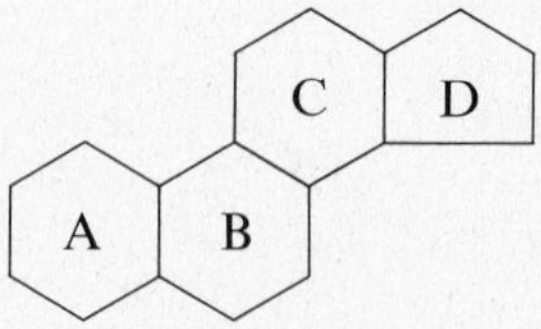

the basic perhydrocyclopentanophenanthrene structure

Systems of four fused rings are synthesized from acetyl CoA and serve as basic structures for the synthesis of cholesterol, cholic acid, and the steroid hormones.

Cholesterol is the most commonly occurring steroid in animals and is the parent compound for the synthesis of the major classes of steroid hormones: estrogens, androgens, progestogens, and the steroid hormones of the adrenal gland. The structural formula of cholesterol shown below indicates its relationship to the basic multi-ring system.

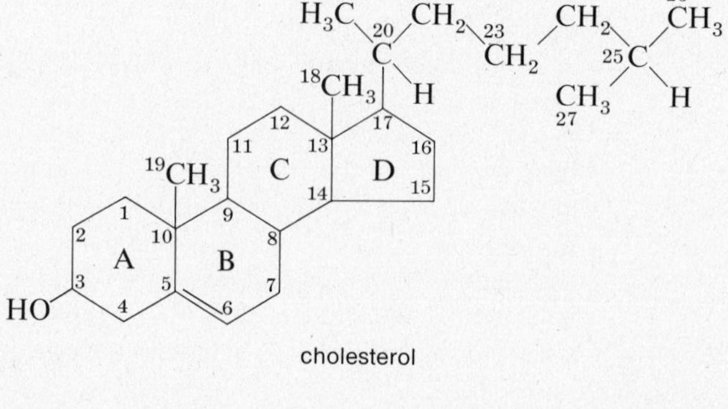

21 22 24 26
H_3C 20 CH_2 23 CH_2 CH_3
C CH_2 25 C
18 CH_3 H CH_3 H
27
12 17
11 13 16
19 CH_3 C D
14 15
1 9
2 10 8
A B
3 5 7
HO 4 6

cholesterol

exercise 15-3

(a) Using the given structural formula for cholesterol, determine its molecular formula.

(b) What feature of the cholesterol molecule serves as the basis for its name ending in -ol?

exercise 15-4

Most of the cholesterol in plasma is present as cholesterol esters. Using $R-C(=O)OH$ **as the general formula for an acid, write the structural formula for a cholesterol ester.**

The structural formulas of some selected steroid hormones are shown below. The term estrogen refers to a group of compounds having similar structures. The two main estrogens are estradiol and estrone.

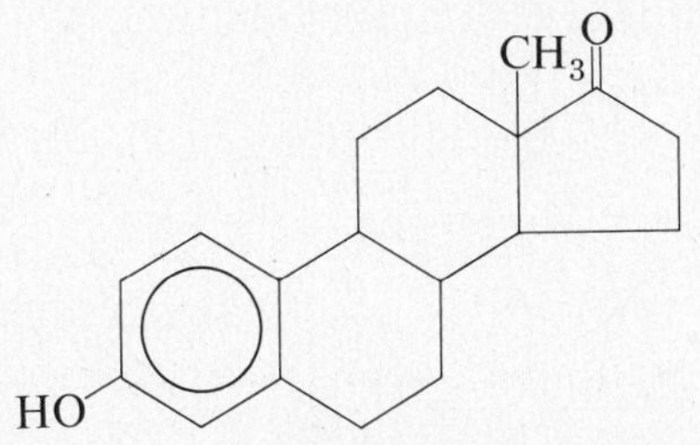

estrone

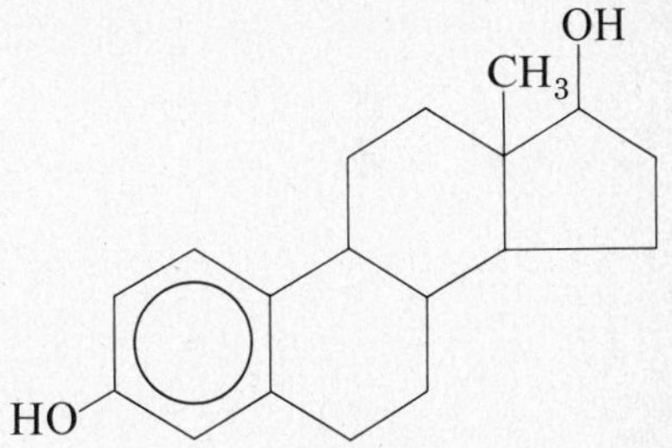

estradiol

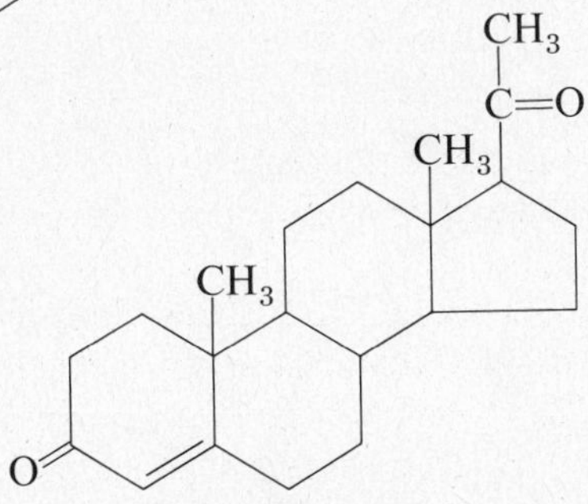

progesterone

female sex hormones

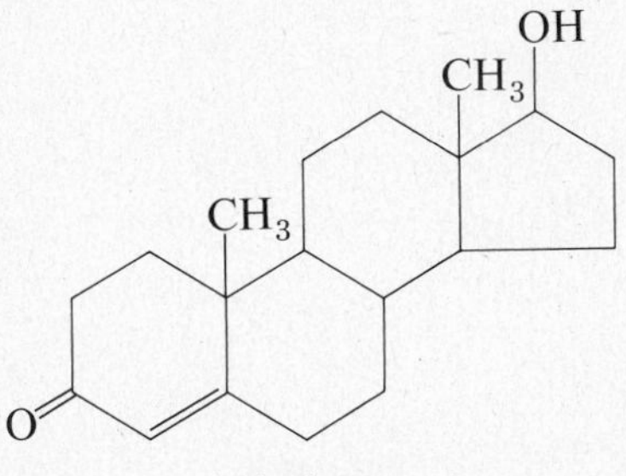

testosterone

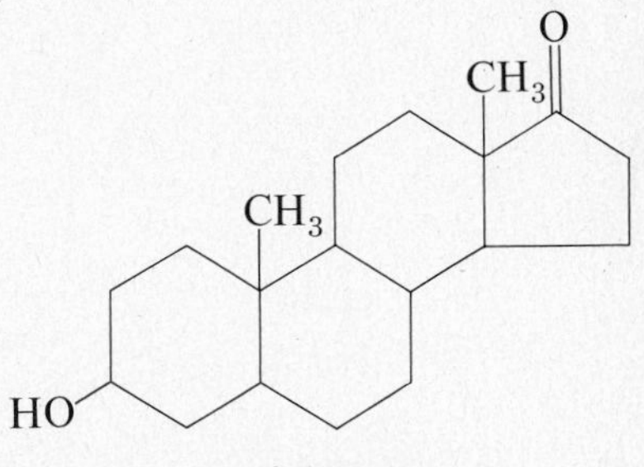

androsterone

male sex hormones

As mentioned previously, the production and activity of the sex hormones is controlled by the action of anterior pituitary hormones FSH and LH.

exercise 15-5

(a) What structural feature of estradiol is the basis for the diol ending in its name?
(b) Is the —OH attached to the D ring in estradiol part of a primary, secondary, or tertiary alcohol group?
(c) What type of reaction converts this group to that found in estrone?

FEMALE SEX HORMONES In human females, changing rates of sex hormone secretion cause a periodic change in the ovaries, uterus, and associated sex organs. This periodic pattern of change is called the menstrual cycle. Beginning at puberty and continuing through menopause, the cycle normally is repeated approximately every 28 days except during pregnancy. Hormonal control of the cycle regulates the release of a mature egg cell and preparation of the uterine lining for the implantation of a fertilized egg cell. This is illustrated in Fig. 15-8.

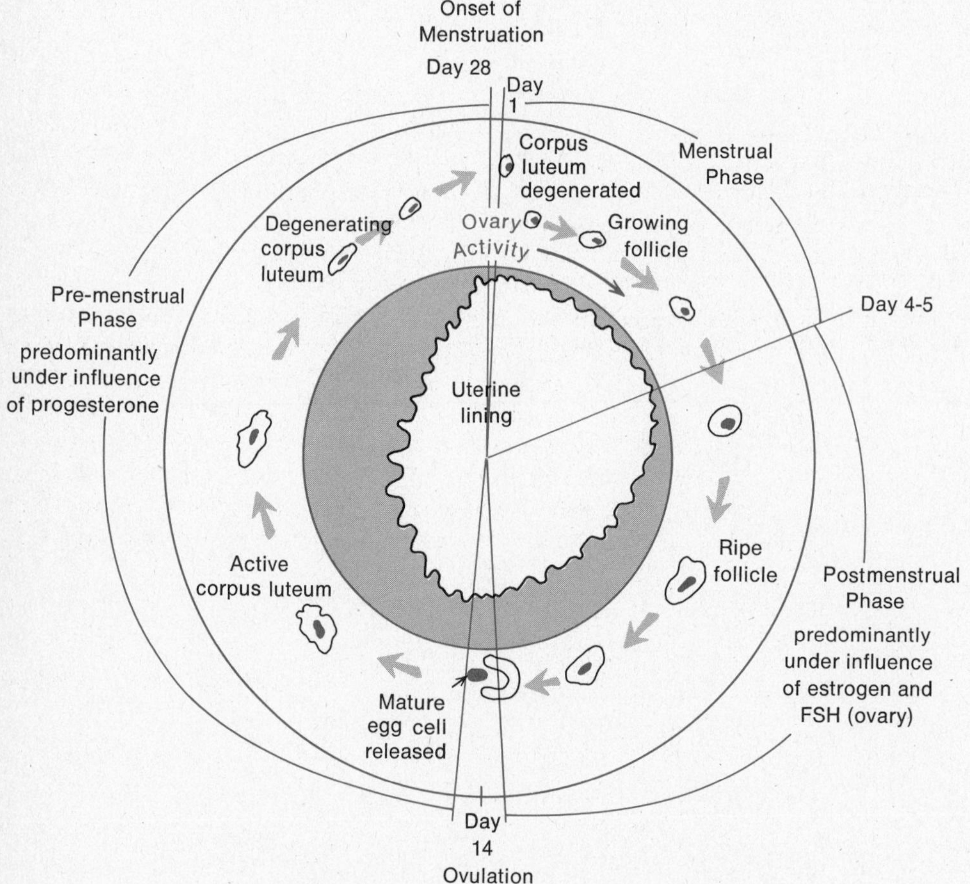

FIGURE 15-8 The human female menstrual cycle.

Following menstrual flow, an ovarian follicle, under the influence of FSH, progressively develops and matures in one of the ovaries. As it matures, the developing follicle secretes increasing quantities of estrogen that aid in the preparation of the uterine lining for arrival of a fertilized egg. Approximately 14 days after the onset of menstrual flow, the influence of luteinizing hormone (LH) causes rupture of the mature follicle, releasing a mature egg cell (ovulation). LH and prolactin promote formation of the corpus luteum within the ruptured follicle. The corpus luteum grows and secretes progesterone, which further promotes uterine lining growth. Progesterone also inhibits development of other ovarian follicles, inhibits uterine contraction, and promotes changes within the breasts.

If the egg is not fertilized, the corpus luteum degenerates causing progesterone secretion to stop and estrogen production to decline. The decrease in concentrations of these hormones stops the thickening of the uterine lining. By about the twenty eighth day after the onset of the previous menstrual flow, the uterine lining begins to slough off accompanied by hemorrhaging and menstruation begins. This marks the beginning of another menstrual cycle. It is believed that the low level of sex hormone concentration prior to menstruation is the cause of symptoms such as irritability and tension.

ORAL CONTRACEPTIVES

For quite some time it has been known that ovulation can be inhibited by injection of sufficient quantities of either estrogens or progesterone. This knowledge initiated a search for a synthetic drug that could prevent ovulation and thus act as a birth control agent. Research indicated that modification of the D-ring substituents in progesterone was the key factor. Incorporation of an —OH group and a —C≡CH group as substituents in the D ring resulted in a compound that could be administered orally and produce the desired contraceptive effect. By the early 1960's, the development and testing had been completed and birth control pills became commercially available. "The Pill" primarily contains a progesterone-like component and a smaller amount of an estradiol-related component. Combination of these prevents ovulation while allowing a normal menstrual cycle in most women. Norethindrone and norethynodrel are progesterone-related compounds commonly used; mestranol is the common estradiol-like component.

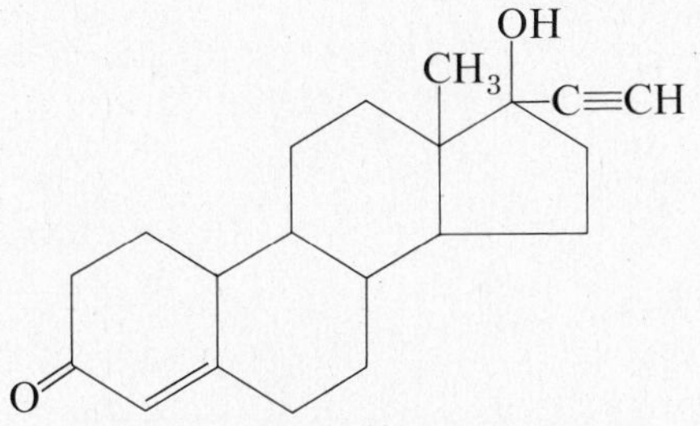

OH
CH_3 —C≡CH
O

norethindrone

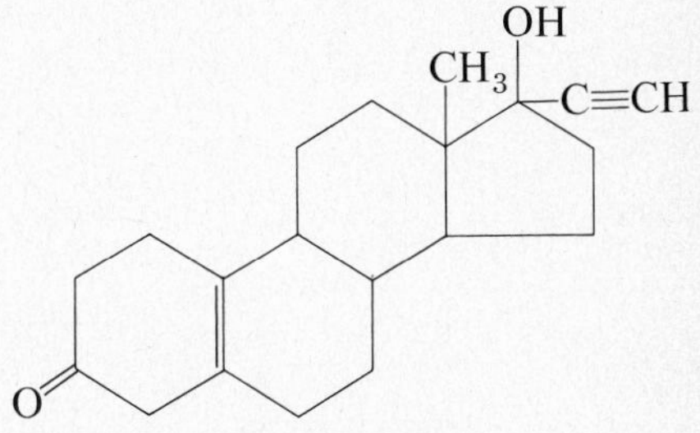

OH
CH_3 —C≡CH
O

norethynodrel

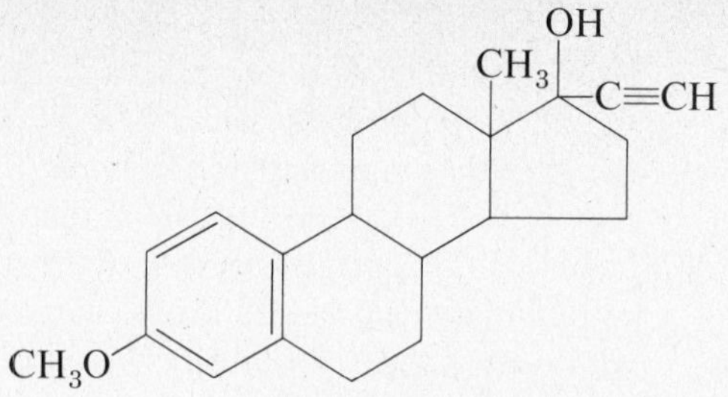

OH
CH_3
C≡CH
CH_3O

mestranol

exercise 15-6

Why is it desirable to have both progesterone-like and estrogen-like components in an oral contraceptive?

MALE SEX HORMONES—ANDROGENS

Male sex hormones are testosterone and androsterone. The concentration of testosterone is much greater than that of androsterone. The androgens have structures resembling those of the estrogens and are produced by specialized cells in the testes. Beginning at about age 10, the testes begin production of testosterone, initially at low levels. By about age 13, testosterone production reaches normal level and is accompanied by the development of male secondary sexual characteristics.

Low concentrations of female hormones occur in normal males; low concentrations of male hormones also are present in normal females. It is the higher concentrations of the testosterone in males and estrogens and progesterone in females that causes development and maintenance of appropriate secondary sexual characteristics.

ADRENAL STEROIDS

The cortex (outer layers) of the adrenal glands produce two major types of hormones—mineralocorticoids and glucocorticoids (see Fig. 15-9). The mineralocorticoids regulate concentrations of mineral ions (mainly sodium and potassium ions). Glucocorticoids principally influence carbohydrate metabolism and are controlled by ACTH secretion.

Adrenal steroids

Mineralocorticoids	Glucocorticoids
Aldosterone	Cortisol
Corticosterone	Corticosterone
Deoxycorticosterone	Cortisone

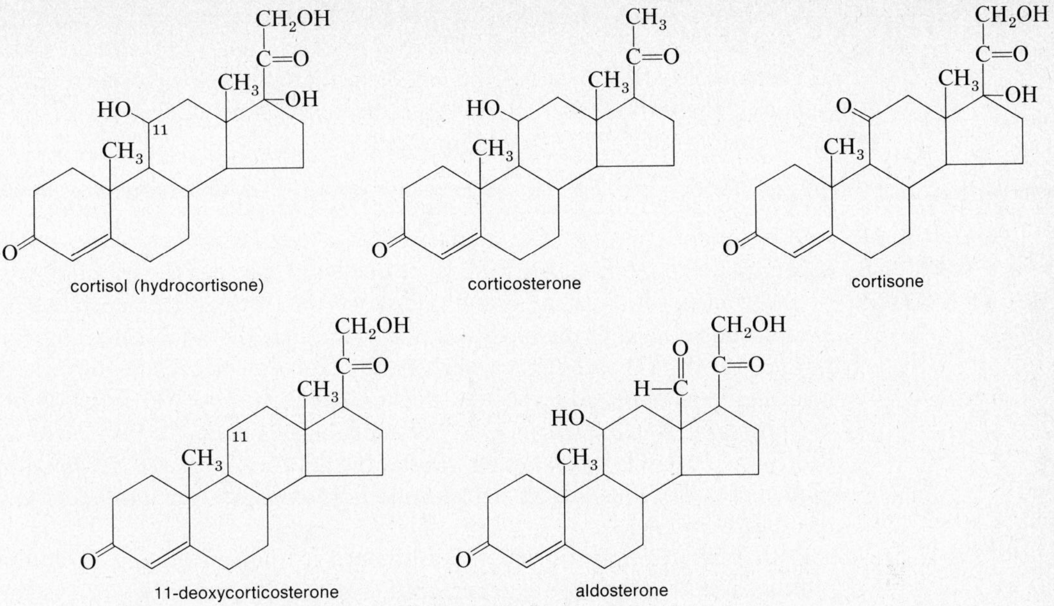

cortisol (hydrocortisone)

corticosterone

cortisone

11-deoxycorticosterone

aldosterone

FIGURE 15-9 Several adrenal corticoids.

MINERALOCORTICOIDS Aldosterone is the major mineralocorticoid hormone with corticosterone and deoxycorticosterone exhibiting minor but nevertheless significant hormone activity. The most important effects of aldosterone are (1) to increase reabsorption of sodium ions by the renal tubules. (2) The simultaneous increase in loss of potassium ions via the urine. (3) Chloride ion reabsorption is enhanced. (4) To control water retention. Lack of aldosterone secretion causes Addison's disease in which sodium ion reabsorption is sharply decreased. Instead, increased amounts of sodium ions, bicarbonate ions, chloride ions, and water are excreted in the urine while potassium ions are retained in blood plasma. Among other effects, these factors lead to muscular weakness, acidosis, and decreased cardiac output. Addison's disease can be successfully treated by administration of supplementary mineralocorticoids, usually in conjunction with glucocorticoids as well.

exercise 15-7

What factor in Addison's disease could cause acidosis to occur?

GLUCOCORTICOIDS Cortisol is the major glucocorticoid hormone. Cortisol promotes gluconeogenesis by the liver, release of free fatty acids from fat depots, and transport of amino acids into liver cells. Because of the enhanced level of amino acids, the liver increases deamination of amino acids, thus promoting conversion of the amino acids to glucose. Cortisol also stimulates the production of key enzymes necessary for glucose formation. Corticosterone and, to a much lesser degree, cortisone produce the same effects.

Cortisol and other glucocorticoids are known to have an antiinflammatory effect that has led to their use in the treatment of conditions such as rheumatoid arthritis and acute inflammation of the glomerlular tubules (glomerulonephritis). Exactly how the glucocorticoids achieve their anti-inflammatory effect is as yet unknown.

FEEDBACK AND CONTROL OF HORMONE CONCENTRATION

ACTH, produced by the anterior pituitary, controls the level of glucocorticoid secretion. This control exemplifies what is called a *feedback relationship* in which a change in concentration of one substance causes an alteration of the concentration of a second substance. For example, a decrease in glucocorticoid level in the blood stimulates an increase in ACTH secretion. In turn, the ACTH secretion causes the adrenal cortex to secrete glucocorticoids. As glucocorticoid levels rise, this inhibits further ACTH secretion by the pituitary. Because an increase in secretion of one substance causes a decrease in secretion of another substance, the term *negative feedback* is used to describe this action. Thus, some hormones can regulate secretion of other hormones.

A type of feedback action is also seen in the functioning of multienzyme systems. Let's use a generalized example of a multienzyme system for descriptive purposes. Suppose the following hypothetical sequence of metabolic reactions occurs, each catalyzed by a specific enzyme:

$$\mathrm{M} \xrightarrow{\text{enzyme M}} \mathrm{N} \xrightarrow{\text{enzyme N}} \mathrm{O} \xrightarrow{\text{enzyme O}} \mathrm{P} \xrightarrow{\text{enzyme P}} \mathrm{Q}$$

The rate at which Q is produced is dependent on the concentration of each substrate M, N, O, and P and the presence and activity of each enzyme. If one of the enzymes is lacking or in insufficient concentration, reaction rates are decreased. Suppose, for example, the concentration of enzyme N is reduced. Under these conditions, product O is produced at a reduced rate. Consequently, the rate of P and Q production is also reduced.

exercise 15-8

Explain in detail why the rate of production of P and Q is reduced by the decrease in the rate of O production.

Q production can also be regulated by a self-limiting type of feedback called end product inhibition (see Fig. 15-10). Initially, enzyme N catalyzes production of O, which in turn ultimately produces Q. If Q has a structure that can bond to enzyme N, the presence of Q can cause the enzyme structure to be modified slightly so that it is no longer active toward catalyzing conversion of N to O.* At very low concentrations of Q, most of enzyme N is in its active form. As the concentration of Q increases, progressively greater amounts of enzyme N are inactivated, slowing the rate of O production. Consequently, Q, the end product of the reaction sequence, inhibits

*This bonding site is frequently called an allosteric site.

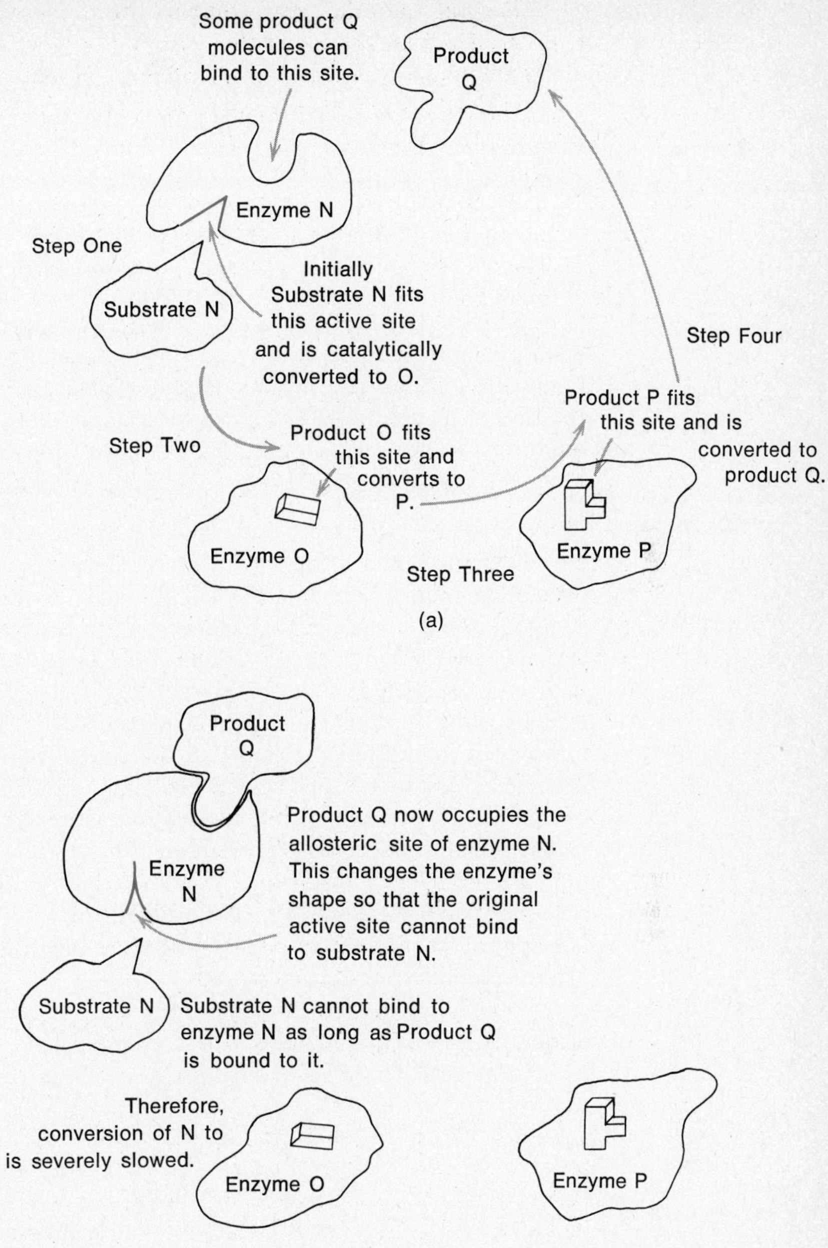

FIGURE 15-10 End-product inhibition. Production of an end product, Q, limits the rates of the reactions which sequentially produce it: N → O → P → Q. By retarding the rate of conversion of N to O, the rate of production of the final product, Q, is also reduced.

its own production. In this way, an organism can prevent overproduction of metabolites.

A specific example of allosteric control occurs in carbohydrate metabolism during glucose-6-phosphate production. In the presence of Mg^{2+} ions, the enzyme hexokinase catalyzes the conversion of glucose to glucose-6-phosphate (G-6-P).

$$\text{ATP} + \text{glucose} \xrightarrow[\text{Mg}^{2+}]{\text{hexokinase}} \text{ADP} + \text{glucose-6-phosphate}$$

When cellular demand for energy is high (such as during exercise), glucose is rapidly metabolized via G-6-P. When an organism is resting, however, a low demand for energy exists and G-6-P concentration builds up in muscle tissue. This has an end product inhibitor effect on muscle hexokinase and the rate of glucose to G-6-P conversion declines. It seems reasonable that organisms conserve glucose during periods of low-energy demand so that it can be available when energy demands increase.

PROSTAGLANDINS

Discovered in the 1930's but not studied until the late 1950's, prostaglandins have been the center of intensive research during the past decade. Although present in minute amounts throughout body tissues, these hormone-like compounds act on localized target sites. Data seem to indicate that prostaglandins are formed in the same cells in which they influence activity. This is in contrast to the action of hormones, which effect target cells a long distance from the point of hormone secretion. Prostaglandins are rapidly inactivated in the bloodstream and the lungs. They have been found to participate in a broad spectrum of activities and are useful in (1) treatment of high blood pressure, peptic ulcers, and asthma; (2) induction of labor; (3) therapeutic abortion; (4) prevention of blood clotting, and (5) controlling inflammation. Their physiological actions seem to depend on the type of organism and target tissue. Prostaglandins apparently function by regulating hormone action and recent research indicates that they influence the activity of the secondary messenger *c*-AMP.

Prostaglandins are 20-carbon fatty acids derived from essential fatty acids such as arachidonic acid. These essential fatty acids are converted to prostanoic acid whose basic structure is common to all prostaglandins.

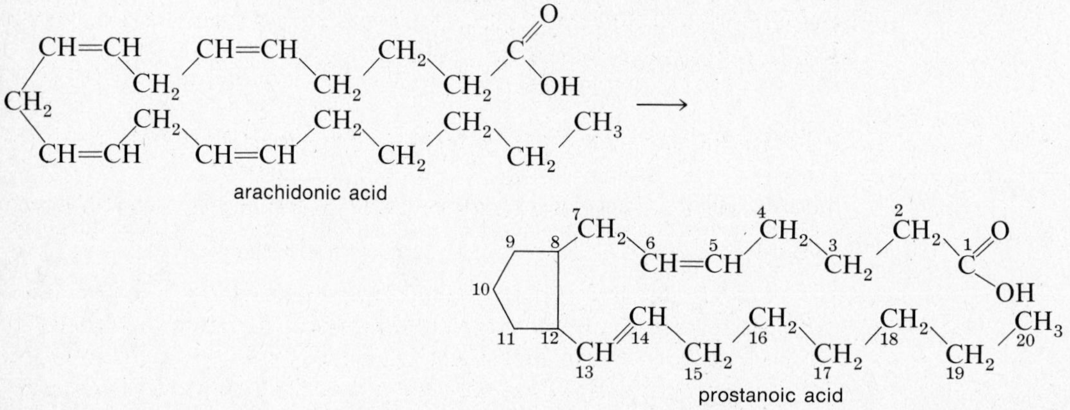

CH=CH CH=CH CH_2 C(=O)OH; CH_2 CH_2 CH_2 CH_2; CH_2 CH_2 CH_2 CH_2 CH_3; CH=CH CH=CH CH_2 CH_2 ⟶

arachidonic acid

9 8 7 CH_2 6 5 CH=CH 4 CH_2 3 CH_2 2 CH_2 1 C(=O)OH; 10; 11 12 CH 13 CH 14 CH_2 15 CH_2 16 CH_2 17 CH_2 18 CH_2 19 CH_3 20

prostanoic acid

Currently, 16 prostaglandins are known to exist in humans and they are arbitrarily classified into four families termed E, F, A, and B. F, A, and B prostaglandins are derived from E prostaglandins. Structural formulas for several prostaglandins (PG) are given below.

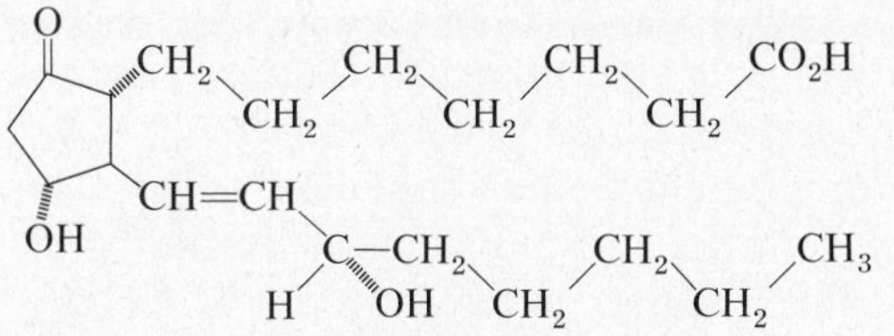

PGE_1

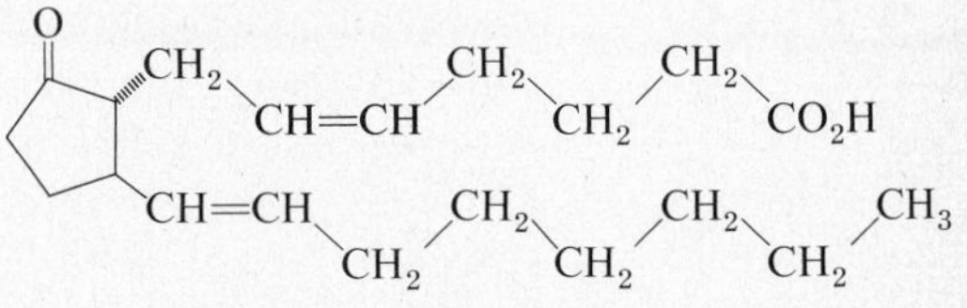

PGE_2

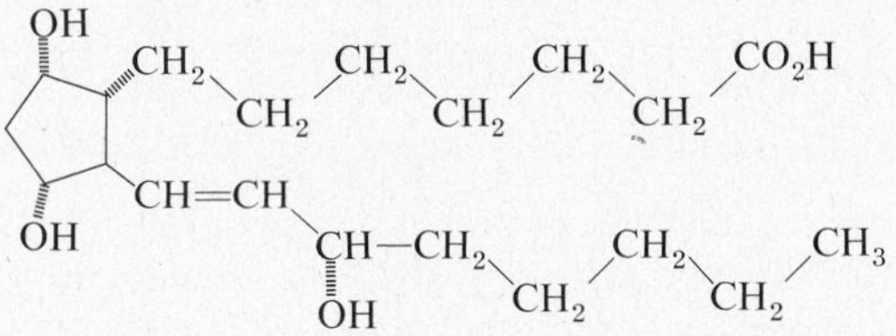

$PGF_{1\alpha}$

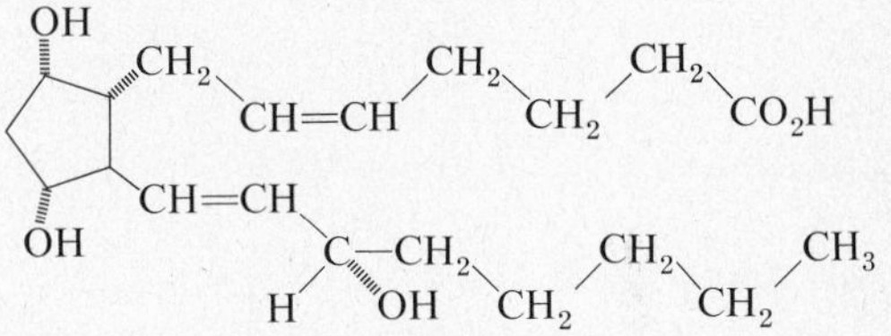

$PGF_{2\alpha}$

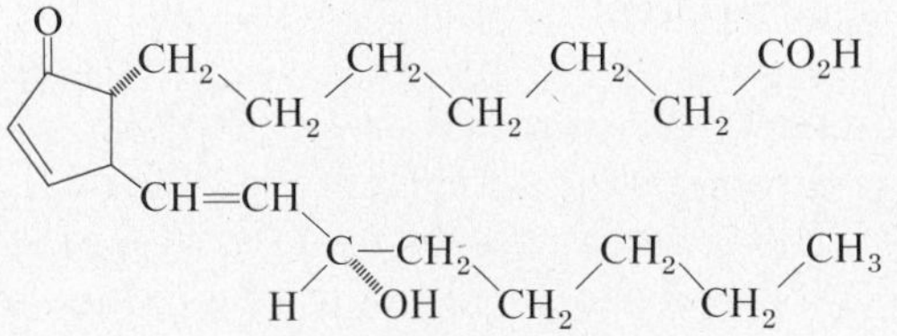

PGA_1

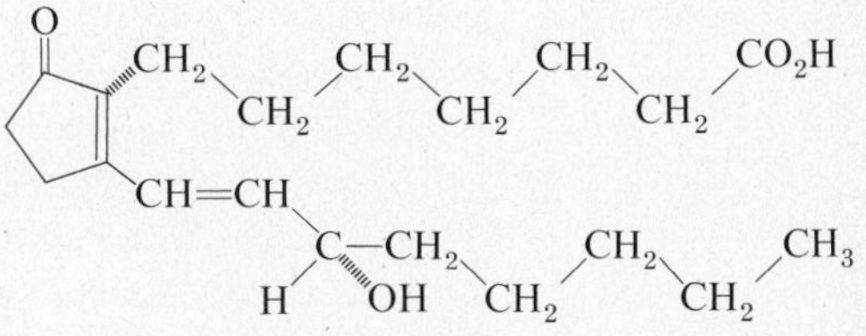

PGB_1

Within a family, members differ by the number of double bonds in the side chains of the five-membered ring. The number of double bonds is indicated by a subscript in the designation, e.g., prostaglandin E_1 (PGE_1), prostaglandin E_2 (PGE_2). The E family contains molecules having a >C=O group at carbon 9 and an —OH group at carbon 11. In the F family, the >C=O group is converted to an —OH. Families A and B both contain oxygen attached directly to the ring only at carbon 9 and have a double bond in the ring portion of the molecule.

exercise 15-9

What general type of reaction is involved in converting PGE_1 to $PGF_{1\alpha}$?

exercise 15-10

What general type of reaction occurs during conversion of PGE_1 to PGA_1?

Prostaglandins $F_{2\alpha}$ and E_2 have been used clinically to induce labor at full-term pregnancy and to speed parturition. They have been found to be more effective than oxytocin and have the advantage that they can be administered orally. They have also been used for therapeutic abortion by inducing labor during early pregnancy. Curiously, infertile men have a subnormal concentration of prostaglandins in their seminal fluid. It could be that their infertility may be due to factors that inhibit prostaglandin production.

Certain prostaglandins can cause tissue inflammation and fever. Recent research findings indicate that aspirin's action in reducing these symptoms could be due to the drug's inhibition of prostaglandin synthesis. E and A varieties inhibit gastric secretions and are used in the treatment of ulcers. Clinically, members of these prostaglandin families have also been demonstrated to reduce blood pressure in hypertensive individuals. By aiding the dilation of bronchial passages, the F variety have found use in relieving respiratory difficulties such as asthma and emphysema.

Much remains to be discovered about prostaglandins, their actions and effects. For example, it is still not known conclusively if excesses or deficiencies of them can result in disease.

AN OVERVIEW OF CERTAIN HORMONE ACTIONS

In this chapter we have seen that hormones play the role of regulators, occasionally in a very dramatic fashion. By their regulatory action, hormones allow the proper integration of systems resulting in homeostasis. In carbohydrate metabolism, glucagon and epinephrine stimulate glycogenolysis, while insulin regulates blood sugar level and glycogenesis. Glucocorticoid and thyroid hormones are also involved in carbohydrate metabolism.

Epinephrine, glucagon, and glucocorticoid hormones stimulate release of fatty acids from fat depots, making fatty acids available as a supplementary energy source. Insulin allows proper carbohydrate metabolism to occur so that two-carbon fragments from fatty acid metabolism can be used for synthesizing cholesterol and triglycerides as well as being further degraded via the Krebs cycle.

Growth hormone, thyroid hormones, and glucocorticoids regulate protein metabolism. Growth hormone increases transport of amino acids into cells, thus enhancing protein synthesis. At greater than normal levels, thyroid hormones overstimulate catabolism and carbohydrates and fats cannot provide adequate amounts of metabolites. Proteins are degraded in an attempt to make up the deficiency. At normal thyroid hormone level, protein synthesis and breakdown are in balance. Glucocorticoids stimulate protein synthesis in the liver but promote catabolism of proteins in muscles and other tissues. When adequate carbohydrate metabolism occurs, amino acids are available for protein synthesis rather than being used extensively for gluconeogenesis. By making carbohydrates available, insulin also aids protein synthesis.

Because of effects such as the secondary messenger action of *c*-AMP, one hypothesis is that hormones function primarily by regulating the rates at which various nutrients, metabolites, and ions are transported across cell membranes. Further research is required before the actual mechanisms for all hormone actions can be described.

VITAMINS Since ancient times, a relationship between proper diet and good health has been recognized. Correct nutrition requires eating more than just carbohydrates, proteins, and lipids. In 1912 the vitamin theory was proposed. According to this theory, the lack of particular nutritional factors in the diet cause certain specific diseases such as scurvy, rickets, and pellagra. Shortly thereafter, specific substances were isolated from foods. When given to test animals suffering from dietary deficiency diseases, these substances cured the condition. These vital food factors were called vitamines, now spelled *vitamins*. The original name was a combination of the words *vital* and *amine* because the first such substance discovered was an amine. It later became evident that not all vitamins are amines and currently a *vitamin* is defined as a naturally occurring organic compound found in foods that is necessary for normal body maintenance and growth. The vitamin may be present in foods as either the vitamin itself or as a compound that can be readily converted into the vitamin. In Chapters 6 and 12 it was mentioned that vitamins are components of enzymes required for metabolism. Thus, it is obvious how a vitamin deficiency can have profound metabolic effects.

Vitamins are classified according to both their functional groups and solubility. Some are soluble in water, while others are water insoluble but soluble in fats (the lipid-soluble vitamins). Figure 15-11 classifies vitamins according to their solubility.

Vitamins

Water soluble	Lipid soluble
Thiamine (vitamin B_1)	Vitamin A
Riboflavin (vitamin B_2)	Vitamin D
Niacin (nicotinic acid)	Vitamin E
Vitamin B_6 (pyroxidal)	Vitamin K
Pantothenic acid	
Biotin	
Folic acid	
Vitamin B_{12}	
Ascorbic acid (vitamin C)	

FIGURE 15-11 Classification of vitamins according to solubility.

Because they are fat soluble, the lipid-soluble vitamins accumulate in body fat. Their availability from this type of storage may prevent deficiencies even when an individual is on a vitamin-deficient diet of a fat-soluble vitamin for some time. In contrast, water-soluble vitamins are not stored and therefore must be ingested daily.

Table 15-2 lists vitamins, their formulas, recommended daily allowance, and conditions caused by their deficiency.

WATER-SOLUBLE VITAMINS Thiamine was the first vitamin isolated. In the body, it is found mainly as thiamine pyrophosphate, which functions as a coenzyme in decarboxylation reactions of α-keto acids. Because of its involvement in metabolism, thiamine

TABLE 15-2
Vitamins important for human well-being

WATER-SOLUBLE VITAMINS

VITAMIN	STRUCTURAL FORMULA	RECOMMENDED (ADULT) DAILY ALLOWANCE	DIETARY SOURCES	DEFICIENCY EFFECT(S)
Thiamine (B_1)	NH_2, CH_2, CH_3, H_3C, N, N, ⊕N, C, C—CH_2CH_2OH, C—S, H	1-2 mg	Whole grains, liver, peas, beans	Beriberi
Riboflavin (B_2)	OH OH OH; CH_2—CH—CH—CH—CH_2OH; H_3C, H_3C, N, N, N, N, O, O	1-2 mg	Wheat germ, liver, green leafy vegetables, milk, eggs	Reduction of rate of energy production from metabolism
Niacin	N, C, O, OH	15-20 mg (estimated)	Wheat germ, meats	Pellagra
Pyridoxal (B_6)	H, C, O, HO, CH_2OH, H_3C, N — pyridoxal; H_2C—NH_2, HO, CH_2OH, H_3C, N — pyridoxamine	Approximately 700 μg (estimated)	Egg yolks, liver, wheat germ, yeast	Convulsions in infants, dermatitis in adults
Pantothenic acid	$HOCH_2$—C(CH_3)(CH_3)—CH(OH)—C(=O)—N(H)—CH_2CH_2C(=O)OH	10 mg	Yeast, liver, eggs	Not known for humans
Biotin	O, HN, NH, S, $(CH_2)_4$—C, O, OH	Approximately 200 μg (estimated)	Yeast, liver, peanuts, chocolate	

Folic acid	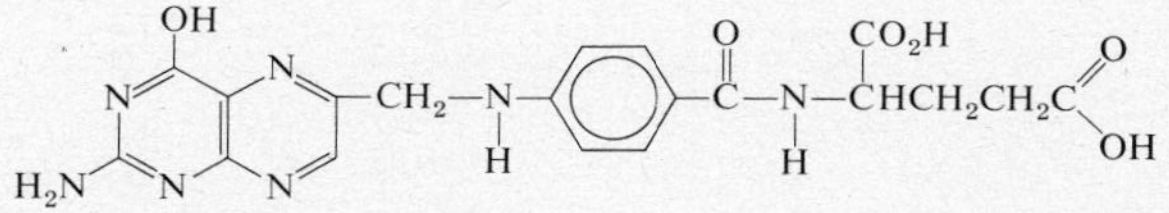		Wheat germ, yeast, spinach	Anemia, growth failure, intestinal tract disorders
Cobalamin (B_{12})	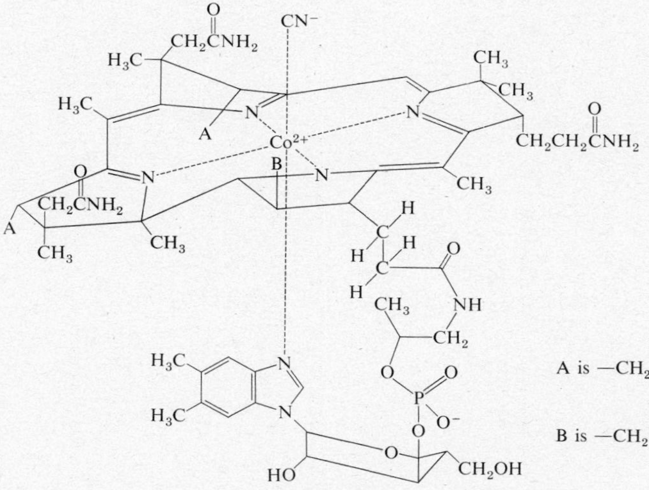 A is $—CH_2CH_2C(=O)NH_2$ B is $—CH_2C(=O)NH_2$	Approximately 1 μg	Liver, meats, eggs	Pernicious anemia
Ascorbic acid (vitamin C)	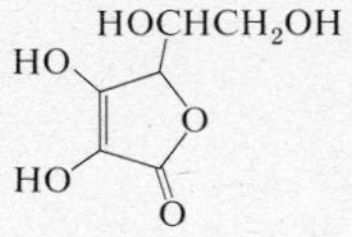	75 mg	Citrus fruits, green vegetables	Scurvy

TABLE 15-2 (continued)

LIPID-SOLUBLE VITAMINS

VITAMIN	STRUCTURAL FORMULA	RECOMMENDED (ADULT) DAILY ALLOWANCE	DIETARY SOURCES	DEFICIENCY EFFECT(S)
Vitamin A	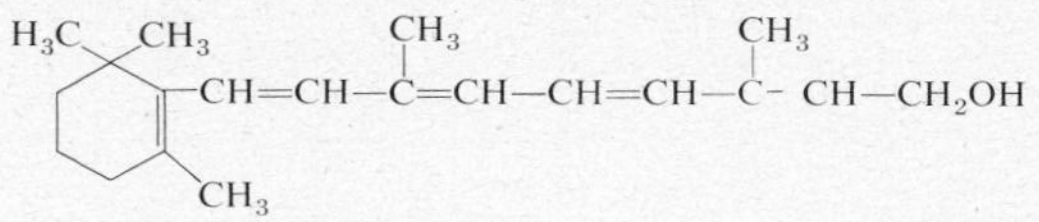	5000 I.U. (excess can cause hypervitaminism) 1 I.U. = 0.3 μg·retinol	Cod liver oil, yellow and green vegetables	Eye disorder (xerophthalmia)
Vitamin D_2	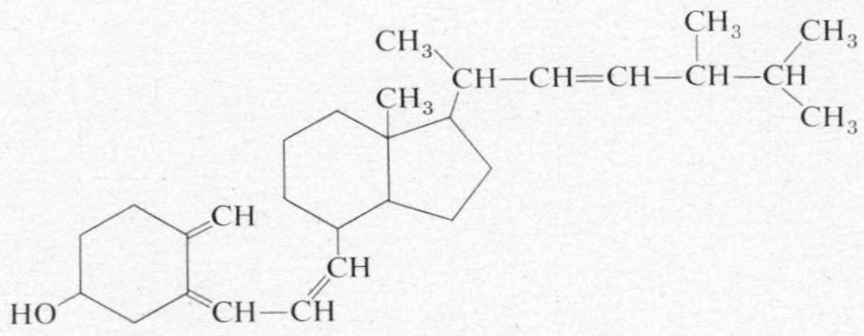 calciferol	400 I.U. (avoid hypervitaminism) 1 I.U. = 0.025 μg cholecalciferol	Fish liver oils	Poor bone formation (hypervitaminism can cause brittle bones)
Vitamin D_3	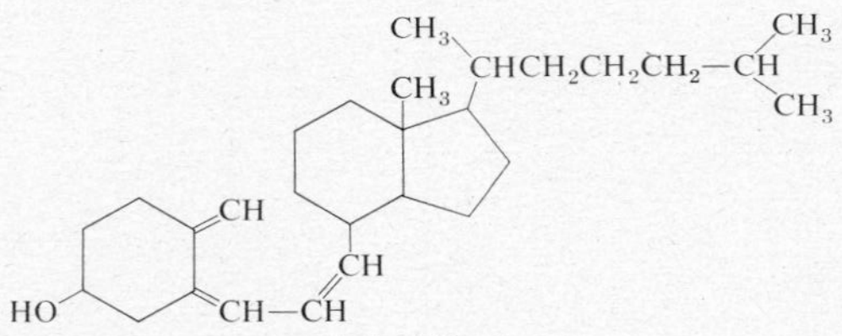 cholecalciferol		Irradiation of compounds in the skin produces the vitamin	
Vitamin E	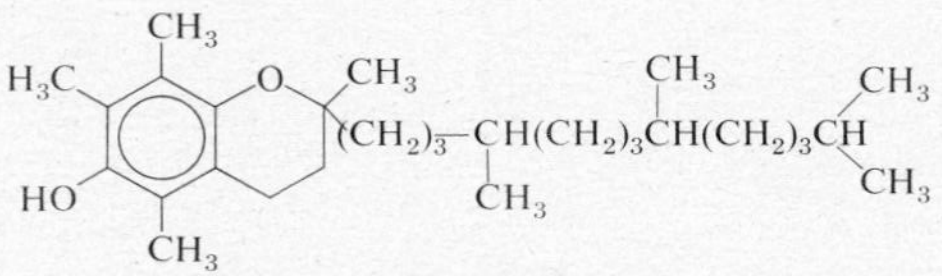 α-tocopherol (other tocopherols also exhibit vitamin E activity)	10–30 mg (estimated)	Wheat germ oil, cottonseed oil, rice, green leafy vegetables	Required by many animals, human requirement and deficiency effects uncertain
Vitamin K	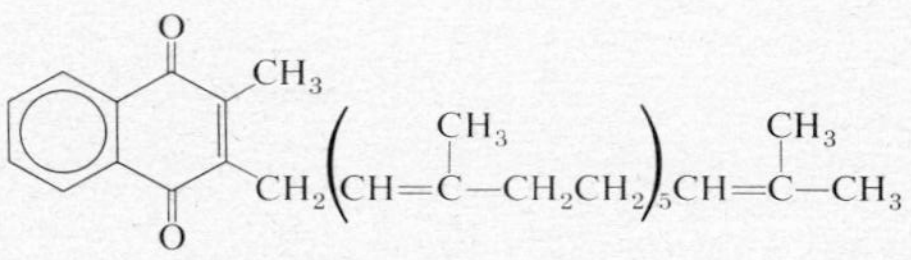 phylloquinone (vitamin K_2) (several forms of vitamin K are known)	Not known	Manufactured by Intestinal bacteria in humans	Impairs blood clotting

is required by all body cells. During thiamine deficiency, glucose utilization by the central nervous system is significantly reduced; this can lead to muscle weakness, cardiac difficulty, and intestinal tract disorders. Beriberi is the term used to describe the composite of these symptoms when caused by thiamine deficiency.

Riboflavin is a component of the coenzymes FAD and FMN, both vital to respiratory chain function. Thus, riboflavin is required by all body cells and a deficiency reduces the rate of energy production during cellular metabolism. The general symptoms are similar to those noted below for niacin deficiency.

Niacin (nicotinic acid) is present in the body as the nicotinamide portion of NAD and NADP. These coenzymes are essential for metabolism and therefore all body cells require niacin. All body functions are reduced during niacin deficiency; dermatitis, diarrhea, and mental disorders can result. These symptoms are characteristic of individuals suffering from pellagra.

Pyridoxine (vitamin B_6) exists in cells mainly as pyridoxal phosphate and pyridoxamine. These three substances as a group are termed vitamin B_6. The vitamin serves as a coenzyme during decarboxylation and transamination reactions. Thus, it is of importance to amino acid and protein synthesis.

Cobalamin (vitamin B_{12}) is a complex molecule containing pyrrole rings (N) bonded to a central cobalt ion. Pernicious anemia results from inadequate absorption of vitamin B_{12}. Gastric juices contain proteins that bind the vitamin and facilitate its absorption. If these proteins (gastric intrinsic factor) are lacking due to defective gastric secretion, pernicious anemia may result even when a sufficient dietary supply of the vitamin is ingested. The manner in which vitamin B_{12} functions is not clearly understood; however, it does seem to have coenzyme action during amino acid metabolism as well as during conversion of ribonucleotides to deoxyribonucleotides.

Ascorbic acid (vitamin C) is essential in human diets although the nature of its function in human metabolism is not exactly known. Most fresh fruits contain ascorbic acid. Cooking foods containing ascorbic acid causes the loss of vitamin activity. Vitamin C is relatively easily oxidized to dehydroascorbic acid. Because it can be reversibly oxidized and reduced, it has been suggested that ascorbic acid can act as a coenzyme during hydrogen transfer; however, no specific coenzyme activity has ever been demonstrated. In 1970, Linus Pauling proposed that massive doses of vitamin C could prevent infection caused by the common cold. This proposal generated considerable controversy among scientists and physicians. Clinical tests reported in the early 1970's seem to indicate that although not able to prevent common cold infections, vitamin C does aid in reducing the length of time cold symptoms persist. It is still not clear how vitamin C does this.

Pauling suggests that 500 to 1000 mg of vitamin C be taken daily.

In Chapter 6, folic acid was mentioned as a vitamin essential for human well-being. Its main role is in the synthesis of thymine and purines, both required for DNA production. Consequently, folic acid is required for gene replication and cellular growth. Growth failure and anemia are characteristic of individuals suffering from folic acid deficiency.

Coenzyme A contains panthothenic acid as a component and the significance of an adequate supply of this vitamin is obvious. Because of the role coenzyme A plays in acetyl CoA formation and the fatty acid spiral, a deficiency of pantothenic acid leads to reduced fat and carbohydrate metabolism. Pantothenic acid–deficient black-haired rats turn gray-haired. No apparent correlation seems to exist between graying hair and pantothenic acid deficiency in humans.

Biotin is another member of the vitamin B complex of vitamins. Its main role is in fatty acid synthesis where it acts as a carrier of carbon dioxide. By accepting CO_2 from biotin, acetyl CoA is converted to malonyl CoA, the first step in fatty acid synthesis (see Chapter 13). It also is involved in the carboxylation of pyruvic acid to malic acid, which is then oxidized to oxaloacetic acid for use in the Krebs cycle.

LIPID-SOLUBLE VITAMINS

β-Carotene is the parent substance for the production of vitamin A. An enzymatically catalyzed cleavage of a β-carotene molecule yields two molecules of vitamin A. All double-bond substituents are in the *trans* form.

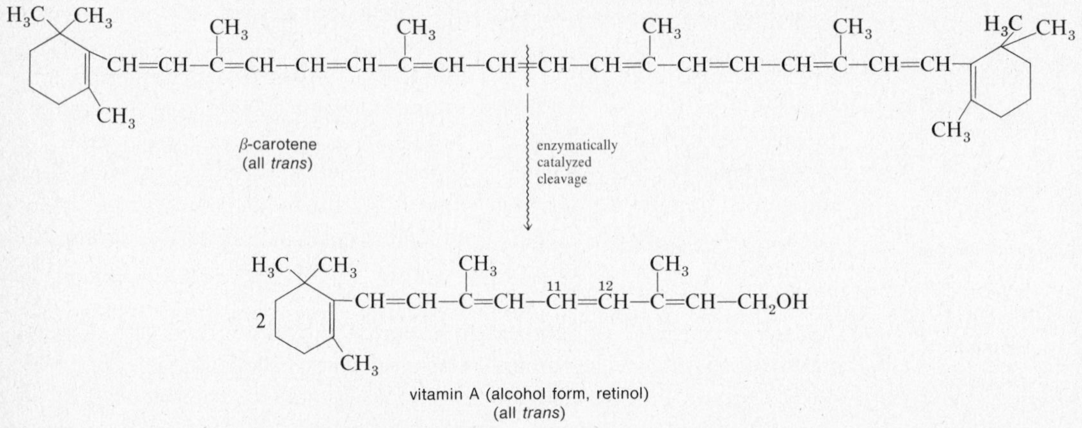

H$_3$C CH$_3$ CH$_3$ CH$_3$ CH$_3$ CH$_3$ H$_3$C CH$_3$

CH=CH—C=CH—CH=CH—C=CH—CH=CH—CH=C—CH=CH—CH=C—CH=CH

CH$_3$ CH$_3$

β-carotene
(all *trans*)

enzymatically
catalyzed
cleavage

H$_3$C CH$_3$ CH$_3$ CH$_3$

2 CH=CH—C=CH—CH=CH—C=CH—CH$_2$OH (11, 12)

CH$_3$

vitamin A (alcohol form, retinol)
(all *trans*)

Because of the system of alternating single and double bonds, both β-carotene and vitamin A absorb light in the visible region of the spectrum and, consequently, are colored. β-Carotene is purple and vitamin A is yellow. The alcohol form of vitamin A can be oxidized to retinene, the aldehyde form. A major function of vitamin A is in vision. Retinene reversibly combines with the protein opsin to form rhodopsin (visual purple), one of the pigments of vision. Usually, the double bonds in the side chain of vitamin

A are in the *trans* form; however, in order to bind to opsin, the double bond between carbons 11 and 12 in retinene is changed to the *cis* form. Light absorbed by rhodopsin changes the *cis* form back to the all *trans* form and rhodopsin dissociates into opsin and all *trans*-retinene. The all *trans*-retinene is reduced by $NADH_2$ to all *trans*-retinol. In the absence of light, the *cis* form of retinene combines with opsin to regenerate rhodopsin so that the visual cycle can continue (see Fig. 15-12 on page 434).

exercise 15-11

Using the structural formula of retinol given, write the structural formula of retinene.

Because of its involvement in vision, a dietary deficiency of vitamin A creates a lack of rhodopsin, which in turn causes a condition known as night blindness. Vitamin A is also necessary for normal growth and proper skin condition.

Vitamin D is sometimes called *the sunshine vitamin.* This is because 7-dehydrocholesterol, a steroid normally found in human skin, is converted to vitamin D_3 when irradiated by ultraviolet light. Ergosterol, found in molds and yeasts, is converted to vitamin D_2 (calciferol) when irradiated by sunlight. The natural vitamin D content of milk and other foods is supplemented by addition of irradiated ergosterol.

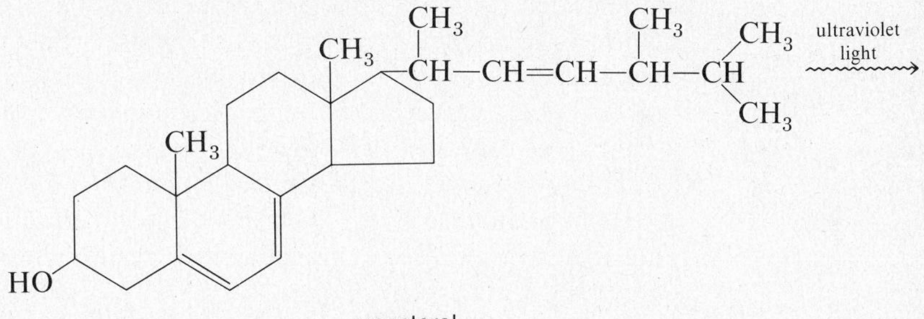

ultraviolet light

ergosterol

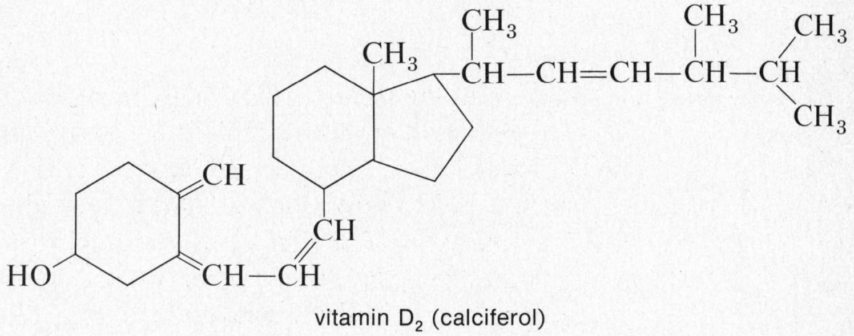

vitamin D_2 (calciferol)

ultraviolet light

7-dehydrocholesterol

vitamin D_3

Vitamin D increases Ca^{2+} and PO_4^{3-} absorption through the intestinal wall, a vital process for bone formation. Lack of dietary vitamin D in infants and children results in abnormal bone formation, a disease called *rickets.* Poor teeth development may also occur. Rickets does not affect adults since their bone formation is completed. Because vitamin D is lipid soluble, it can be stored. This, coupled with excessive intake of the vitamin (hypervitaminosis), can cause bone demineralization (loss of Ca^{2+} from bone) and loss of bone strength, which can lead to multiple fractures even under mild strain. In vitamin D hypervitaminosis, blood serum levels of calcium and phosphate ions increase. This can cause calcium phosphate deposits in soft tissues and in the renal tubules.

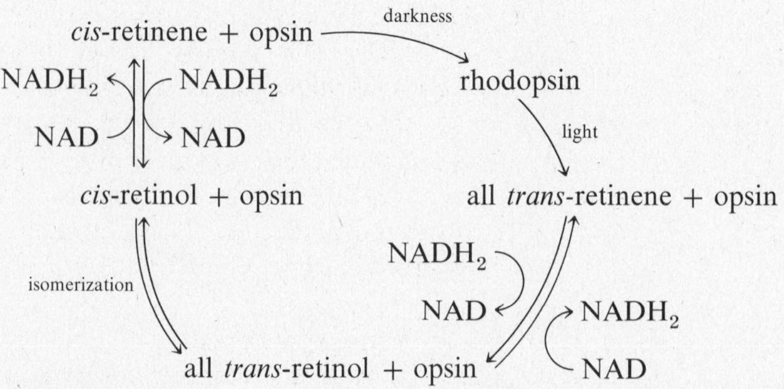

FIGURE 15-12 The visual cycle involving conversion of vitamin A to rhodopsin.

Vitamin E is a collective name given to a group of compounds known as tocopherols. Alpha, beta, and gamma tocopherols are known. Alpha tocopherol exhibits the greatest vitamin E activity. The beta and gamma forms differ in the number and position of methyl groups on the ring system. Tocopherols inhibit oxidation of unsaturated fatty acids and vitamin A. Although there is no sound evidence that vitamin E is essential for humans, known dependency for vitamin E does occur in lower animals. Guinea pigs and rabbits rapidly develop muscular dystrophy when deprived of it. In rats, sterility results from vitamin E deficiency. Administration of vitamin E has not been effective in the treatment of muscular dystrophy or sterility in humans.

Occurring in several forms, vitamin K is required for adequate prothrombin and factor VII formation by the liver. Both of these are necessary for proper clotting of the blood (see Chapter 16). Lack of vitamin K causes clotting time to increase and hemorrhaging occurs. Normally, the widespread occurrence of vitamin K in common foods and its production by intestinal bacteria result in an adequate vitamin level.

CHAPTER SUMMARY

1. Hormones are glandular secretions that directly enter the bloodstream. They are either protenoid or lipoid compounds that help regulate body processes. Some hormones regulate secretion of other hormones by feedback control mechanisms.
2. Many hormones seem to act by altering the cell membrane permeability of target tissues toward certain substances. At least one hormone, epinephrine, has been demonstrated to have its primary message conveyed through the action of a secondary messenger, cyclic AMP.
3. An excess or deficiency of a hormone can cause serious effects. Giantism; acromegaly; retarded physical, mental, and sexual development; and goiter are some of these effects.
4. Systems of four fused rings are the basic structures for cholesterol and the steroid hormones such as estrogen, progesterone, and testosterone. In mature females, sex hormone concentration changes cause the menstrual cycle. Modifications of side chains of the steroid ring system have led to oral contraceptives that prevent ovulation.
5. Prostaglandins resemble hormones in their actions but their effects are very short range. The known prostaglandins in humans are classified into A, B, E, and F families. These compounds have many effects including labor induction, clotting control, and antiinflammation action.
6. Vitamins are required for good health and must be included in the diet. Many vitamins are coenzymes or components of coenzymes necessary for metabolism.

7. Vitamins can be classified as being either water soluble or fat soluble. Those in the latter category are stored in body fat. Accumulation of an excess of a fat-soluble vitamin causes hypervitaminosis and serious side effects. The nonstorage of water-soluble vitamins requires that they be ingested daily.
8. Vitamin deficiency causes many serious effects including scurvy (vitamin C), beriberi (vitamin B_1), pellagra (niacin), anemia (folic acid and vitamin B_{12}), and poor bone and teeth formation (vitamin D). Establishment of a proper diet prevents vitamin deficiency and can cure individuals suffering from the effects of a vitamin deficiency.

QUESTIONS

1. How many water molecules are required to hydrolyze the peptide bonds present in a molecule of oxytocin?
2. Write the symbolic structural arrangement of the organic product formed when vasopressin is subjected to mild reducing conditions.
3. Oxytocin is an octapeptide. Why can't it be successfully administered orally?
4. Cholic acid is the most abundant bile acid in humans.

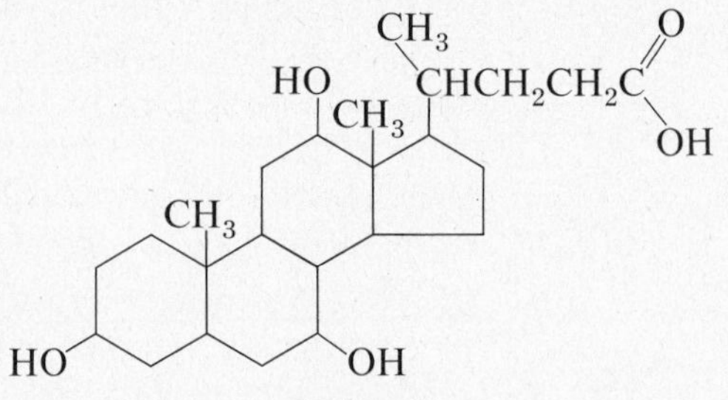

CH_3 O HO $CHCH_2CH_2C$ CH_3 OH CH_3 HO OH

cholic acid

By reacting with glycine, cholic acid is converted to glycocholic acid, which contains an amide linkage. Write the structural formula for glycocholic acid.
5. The sodium salt of glycocholic acid is an important bile salt that aids in the emulsification of fats. Using the structural formula for glycocholic acid from Question 4, write an equation representing the reaction of sodium hydroxide with glycocholic acid to produce the sodium salt.
6. Waxes are naturally occurring esters. Cholesterol palmitate is a wax found in blood plasma. Using structural formulas for the organic species, write an equation representing the formation of cholesterol palmitate.

7. The structural formula for aldosterone given on p. 421 is the aldehyde form. Aldosterone also exists in an internal hemiacetal form. Draw a probable structural formula of the hemiacetal form.
8. Several weeks supply of a fat-soluble vitamin may be administered in a single dose. This is inadvisable for water-soluble vitamins. Explain.
9. Refer to Chapters 12 and 13 and select several reactions in which thiamine is likely to participate. Give a reason for your choice.
10. Upon being administered to humans, pyridoxine is oxidized via pyridoxal to a product that is then excreted. The formula for pyridoxine is

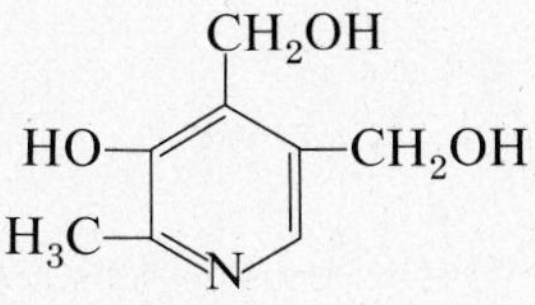

CH_2OH

HO— —CH_2OH

H_3C— N

Write the structural formula for the product of its oxidation via pyridoxal.
11. An individual suffering from pantothenic acid deficiency is likely to be lethargic. Explain why.
12. Retinol is stored in the liver mainly by reacting with long-chain fatty acids to form esters. Using R—CO_2H as a general formula for a long-chain fatty acid, write the structural formula for a retinol ester.
13. Carrots contain β-carotene. Their inclusion in a diet helps prevent night blindness. Explain.

16
body fluids

BODY FLUIDS—AN OVERVIEW Body fluids can be broadly classified into two types: (1) intracellular fluid, that *within* cells; and (2) extracellular fluid, that outside cells. In human adults, intracellular fluid makes up about 50 percent of body weight and extracellular fluid accounts for about 20 percent of body weight. Extracellular fluid is composed mainly (75 percent) of *interstitial* fluid consisting of water and soluble materials. Interstitial fluid bathes the cells and makes the exchange of solutes between the blood and cells occur more readily. The bloodstream accounts for most of the remaining 25 percent of extracellular fluid. Digestive juices, synovial fluid, cerebrospinal fluid, semen, milk, and lymph are other extracellular fluids.

BLOOD A human adult's body contains approximately 6 liters of whole blood, which is a mixture consisting of *plasma* (55–60 percent by volume) and formed elements—red cells (erythrocytes), white cells (leukocytes), and platelets. Plasma can be separated from the formed elements by centrifuging whole blood. Plasma, the liquid portion of circulating or uncoagulated blood, is an aqueous solution containing plasma proteins and various cations and anions. Other solutes such as glucose, uric acid, urea, and polysaccharides are also present. Table 16-1 lists the major organic solutes in plasma and

TABLE 16-1
Some organic solutes in blood plasma

CARBOHYDRATES AND CARBOHYDRATE METABOLITES (fasting levels)		LIPIDS, KETONE BODIES, AND BILE ACIDS (fasting levels)		SELECTED NONPROTEIN NITROGEN-CONTAINING COMPOUNDS (NPN) (fasting levels in venous blood)	
SUBSTANCE	CONCENTRATION (mg/100 ml blood)	SUBSTANCE	CONCENTRATION (mg/100 ml blood)	SUBSTANCE	CONCENTRATION (mg/100 ml blood)
Glucose	60–100	Total lipid	385–675	Total NPN	16–40
Fructose	5–9	Neutral fat	0–260	Urea	10–23
Lactose (during lactation)	0–2	Unesterified		Creatinine	0.8–1.0
Polysaccharides	90–140	Fatty acids	8–31	Creatine	0.2–0.7
Glucosamine	50–90	Phospholipids	110–250	Uric acid	2.5–6
Lactic acid (resting)	5–20	Cholesterol	140–260		
Pyruvic acid	0.4–2	Ketone bodies	0.2–0.9		
Citric acid	1.5–3.2	Cholic acid	0.2–3.0		

their normal concentration ranges. The main plasma proteins are albumin, other globular proteins, and fibrinogen. The main ions present are sodium, potassium, calcium, magnesium, chloride, bicarbonate, monohydrogen and dihydrogen phosphates, and sulfate. If fibrinogen is removed from plasma, the resulting liquid is called *serum*.

Trace amounts of metal ions (Mn^{2+}, Cu^{2+}, Fe^{2+}, Co^{2+}, etc.) are also present.

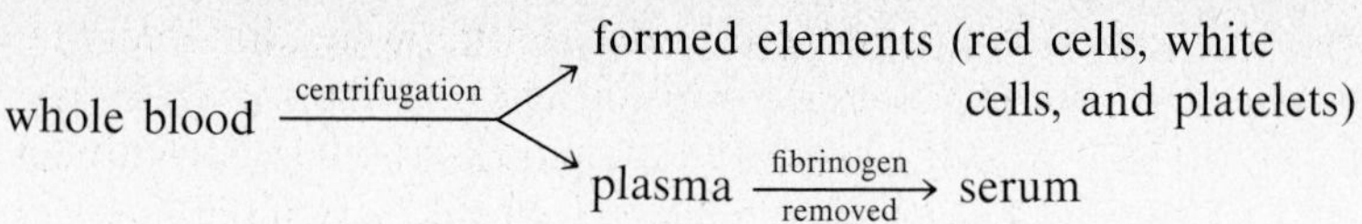

Although many proteins exist in plasma, collectively accounting for about 7 to 8 percent (by weight), we shall limit our discussion to only the three types noted above. Albumin is the most abundant plasma protein. Its main functions are to aid in maintaining osmotic pressure balance between plasma and interstitial fluid and to transport substances that are normally water insoluble such as fatty acids.

$Cu(OH)_2$ and $Fe(OH)_2$ would precipitate.

Three categories of globulins are known: alpha (α), beta (β), and gamma (γ). Alpha and beta globulins combine with metal ions such as Cu^{2+} and Fe^{2+}. Ordinarily, in aqueous solutions having a pH of normal blood (7.4), these metal ions combine with hydroxide ions and precipitate. In the blood, combination of these metal ions with alpha and/or beta globulins produces soluble complexes that allow transport of the ions throughout the body.

Foreign proteins (*antigens*) that enter the blood trigger the production of specific proteins called *antibodies*. Since antibodies in blood are found mainly associated with gamma globulins, these plasma proteins are important in protecting against infectious diseases.

Phospholipids (lecithins), cholesterol, and neutral fats are the most abundant plasma lipids. Practically all lipids in plasma are combined with either alpha or beta globulins. Nonesterified fatty acids are generally associated with albumin. The lipid portion of the lipoprotein generally contains more than one type of lipid. The type of bonding occurring between lipids and proteins is not clearly known but could be due to London forces.

Recall from Chapter 11 that protein molecules fold so as to direct their hydrophobic side chains inward, thus minimizing contact with water. It has been suggested that the solubility of the lipoproteins is due to the orientation of their lipid and protein components. The hydrophobic lipids form a spherical core that can interact with the hydrophobic portion of the protein. The protein covers the lipid by folding to orient its hydrophobic groups toward the lipid and its hydrophilic groups toward the surrounding water.

BLOOD CLOTTING

When blood vessels are cut or broken, bleeding occurs. Obviously bleeding cannot continue very long without severe physiological consequences. The body's major defense against loss of blood is blood clot formation. Clotting requires the participation of many substances and only a brief overview is presented here. Fibrinogen is a soluble plasma protein responsible for clotting. It converts into fibrin, an insoluble, fibrous protein whose strands form a network that traps the formed elements during clotting.

The conversion of fibrinogen to fibrin is catalyzed by the enzyme thrombin. Why, then, doesn't the blood continuously clot? Thrombin exists in the blood in an inactive form, prothrombin, and conversion of inactive

At least 10 factors are known to be involved during activation of prothrombin.

prothrombin to active thrombin requires many factors such as calcium ions and proteins (collectively called thromboplastins). Some of these factors are always present in the blood but others are only released when blood vessels are cut. Hemophilia is a hereditary disease in which an individual (usually male) lacks one or more of the enzymes or enzyme activators necessary for the conversion of prothrombin to thrombin.

$$\underset{\text{(inactive form)}}{\text{prothrombin}} \xrightarrow{\text{Ca}^{2+}\text{, thromboplastins, other factors}} \underset{\text{(active form)}}{\text{thrombin}}$$

$$\underset{\text{(soluble)}}{\text{fibrinogen}} \xrightarrow{\text{thrombin}} \underset{\text{(insoluble)}}{\text{fibrin}}$$

The requirement of calcium ions for clotting is easily demonstrated. The addition of a solution of either sodium oxalate or sodium fluoride to whole blood precipitates calcium oxalate or calcium fluoride. Because calcium ions are removed by this treatment, the blood will not clot.

$$\underset{\text{in blood}}{Ca^{2+}\,(aq)} + \underset{\text{from } Na_2C_2O_4}{C_2O_4^{2-}\,(aq)} \longrightarrow CaC_2O_4\,(s)$$

exercise 16-1

Write a balanced ionic equation for the reaction of aqueous sodium fluoride with calcium ions in the blood to precipitate calcium fluoride.

Sodium citrate solutions are also used to prevent clotting by removing calcium ions. Citrate ions form strong coordinate covalent bonds to calcium ions and prevent calcium from participating in the clotting process.

Heparin is an anticoagulant. Many body tissues contain heparin, especially those bordering blood vessels. By interfering with prothrombin to thrombin conversion, heparin prevents clotting.

Dicumarol and warfarin are anticoagulants administered to prevent blood clot formation in blood vessels (thrombosis).

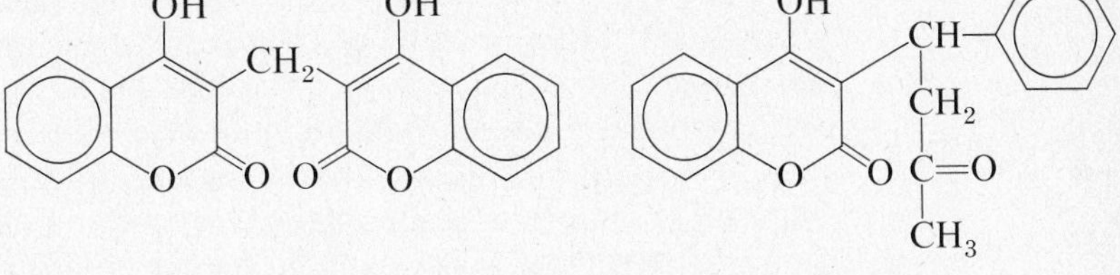

[Structures: dicumarol — two 4-hydroxycoumarin units (OH, O, O) linked by CH_2; warfarin — 4-hydroxycoumarin (OH, O, O) bearing CH(phenyl)–CH_2–C=O–CH_3]

dicumarol

warfarin

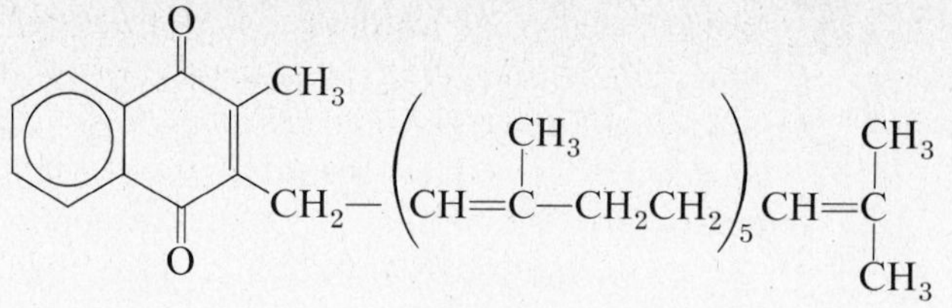

$$\text{(1,4-naphthoquinone ring)}-CH_3;\ -CH_2-\left(CH{=}\underset{\displaystyle|}{\overset{CH_3}{C}}-CH_2CH_2\right)_5 CH{=}C(CH_3)_2$$

vitamin K_2

Normal prothrombin production in the liver depends on adequate dietary intake of vitamin K. The ring structures of dicumarol and warfarin are similar to that in vitamin K. Because of this, they reduce prothrombin production possibly by competing with vitamin K for certain enzyme sites.

BENZIDINE TEST FOR BLOOD The presence of blood may be qualitatively determined using benzidine as the test reagent. The chemical basis for this test is the action of the enzyme peroxidase, present in blood. Peroxidase catalyzes the decomposition of hydrogen peroxide (H_2O_2). If a sample contains blood, addition of benzidine and hydrogen peroxide to the sample results in the formation of an intense blue color. A peroxide decomposition product reacts with benzidine to form the colored substance. Its formation indicates the presence of peroxidase and, therefore, blood.

$$H_2O_2 \xrightarrow{\text{peroxidase}} 2\ \ddot{\underset{..}{O}}{:}H$$

hydroxyl radical (contains an unpaired electron)

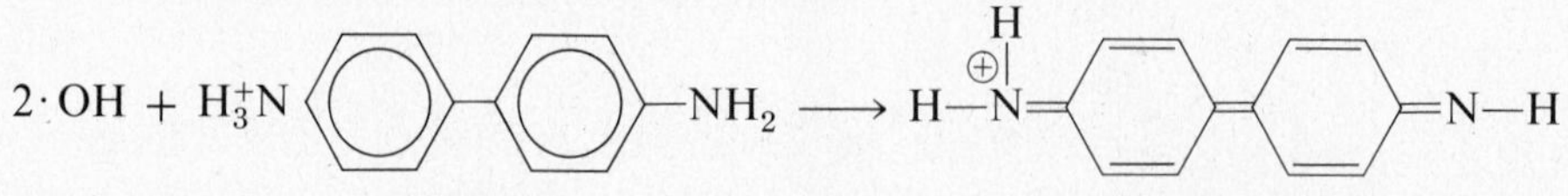

$$2\cdot OH + H_3^+N-C_6H_4-C_6H_4-NH_2 \longrightarrow H-\overset{H}{\underset{\oplus}{N}}{=}C_6H_4{=}C_6H_4{=}N-H$$

benzidine — intensely blue colored substance

URINE Recall from Chapter 7 that the kidneys filter a prodigious amount of blood daily. The kidney tubules dialyze waste products, electrolytes, metabolites, and water from the blood. These substances are either reabsorbed by the blood or released to form urine (diuresis), depending on physiological needs. In this way, the kidneys regulate the composition of body fluids.

The broad fluctuation of urine composition can be seen from specific gravity data. The normal range of urine's specific gravity is 1.008 to 1.030. Variation is due to factors such as fluid intake, muscular activity, digestion, and emotional condition. The normal range of urine production by adults in a 24-hr period is 600 to 2500 ml, averaging about 1500 ml. Clinically, 24- to 48-hr samples are collected for a complete urinalysis so as to minimize variation in composition caused by the factors mentioned above.

Urine is normally a pale amber color due to the presence of pigments that are normal by-products of hemoglobin decomposition. The pigments are collectively called *urobilinogen*. When freshly voided, urine is clear, containing no appreciable sediment. Upon standing, cloudiness may develop due to alkaline conditions caused by bacterial action that precipitates phosphates.

Urine pH can range between the extremes of 4.5 to 8.4 but ordinarily is 5.0 to 7.0. These ranges mainly reflect the ratio of $H_2PO_4^-$ to HPO_4^{2-} ions present, hydrogen ions being exchanged during conversion of these phosphate forms.

$$H_2PO_4^- \rightleftharpoons H^+ + HPO_4^{2-}$$

Diet is normally the chief influence on urine acidity. Meats and other high protein foods yield acids as waste products of their metabolism, causing urine to become more acidic. Most fruits and vegatables contain organic acids such as oxalic and citric acids that are metabolized via the Krebs cycle to bicarbonate ions, causing urine to become less acidic.

BLOOD PLASMA AND URINE RELATIONSHIPS

Plasma and urine are both aqueous solutions. The solutes that are neither retained by plasma nor released into cells are passed into urine by kidney action. Thus, the kidneys (1) eliminate metabolic wastes; (2) control water balance in intracellular and extracellular fluids; (3) control pH by either retaining or releasing hydrogen ions; (4) maintain osmotic pressure of cells by adjusting water content; (5) maintain electrolyte balance by reabsorbing or releasing inorganic ions according to cellular demands; and (5) retain substances such as glucose, proteins, and hormones that are necessary for normal cellular activities. Table 16-2 compares concentrations of solutes common to both urine and plasma.

TABLE 16-2
Comparison of main components of urine and serum

	CONCENTRATION		
COMPONENT	URINE (mg/100 ml)	SERUM (mg/100 ml)	URINE/SERUM RATIO
Sodium	355	335	1.06
Potassium	160	20	8
Calcium	15	10	1.5
Magnesium	20	2.4	8.3
Chloride	500	350	1.4
Inorganic sulfate	8	2.5	4.0
Inorganic phosphorus	160	4	40
Urea	1800	30	60
Creatinine	75	1.5	50
Uric acid	60	3	20

When water retention by the kidneys occurs and normal amounts of water and salts are ingested, the total amount of interstitial fluid increases. As fluid retention continues, abnormally large amounts of fluid accumulate in interstitial regions and swelling occurs. This condition is called *edema*.

Obviously, the body can't allow fluid buildup to continue unchecked. For cells to receive their required amounts of water and solutes, the flow of these materials between intracellular fluid and extracellular fluid must be regulated. The concentration of electrolytes in body fluids is regulated by hormone-controlled kidney action. Hormone action controls the volume of water reabsorbed by the kidneys and the movement of ions across renal membranes. Recall from Chapter 7 that the flow of water and/or solute particles across a membrane affects solute concentration. If solute concentration is too high, normal solute level can be reestablished by either an inflow of water or an outflow of solute through the dialyzing membrane.

Both processes may occur simultaneously.

exercise 16-2

Consider two regions A and B, separated by a dialyzing membrane. If concentration of a solute in region A is too low, describe two ways by which it can be returned to normal level.

Figure 16-1 compares the concentrations of body fluid components in blood plasma, interstitial fluid, and cellular fluid. Except for protein concentrations, notice the similarity between blood plasma and interstitial fluid and their marked dissimilarity with cellular fluid. Extracellular fluid has relatively high concentrations of Na^+, HCO_3^-, and Cl^- ions; compared to intracellular fluid, K^+, SO_4^{2-}, HPO_4^{2-}, and Mg^{2+} are found in relatively low concentrations.

Since sodium ions make up about 90 percent of all positive ions in extracellular fluid, regulation of sodium ion concentration is mandatory and is controlled mainly by aldosterone, which increases the rate of sodium ion reabsorption in the kidney tubules. In the presence of excess aldosterone, few sodium ions are excreted; in its absence, almost no reabsorption of sodium ions occur. Thus, this system determines the amount of sodium ions lost via urine.

exercise 16-3

What effect would reduced aldosterone level have on urinary sodium ion concentration?

Equally important to the regulation of sodium ion level is the regulation of water. Most of the reabsorption of water occurs in the proximal tubules. Final adjustment of the release or reabsorption of water occurs in the distal tubules and is regulated by the action of minute amounts of the antidiuretic hormone, vasopressin. The rate at which vasopressin is released is governed by changes in the blood's osmotic pressure. *Vasopressin promotes*

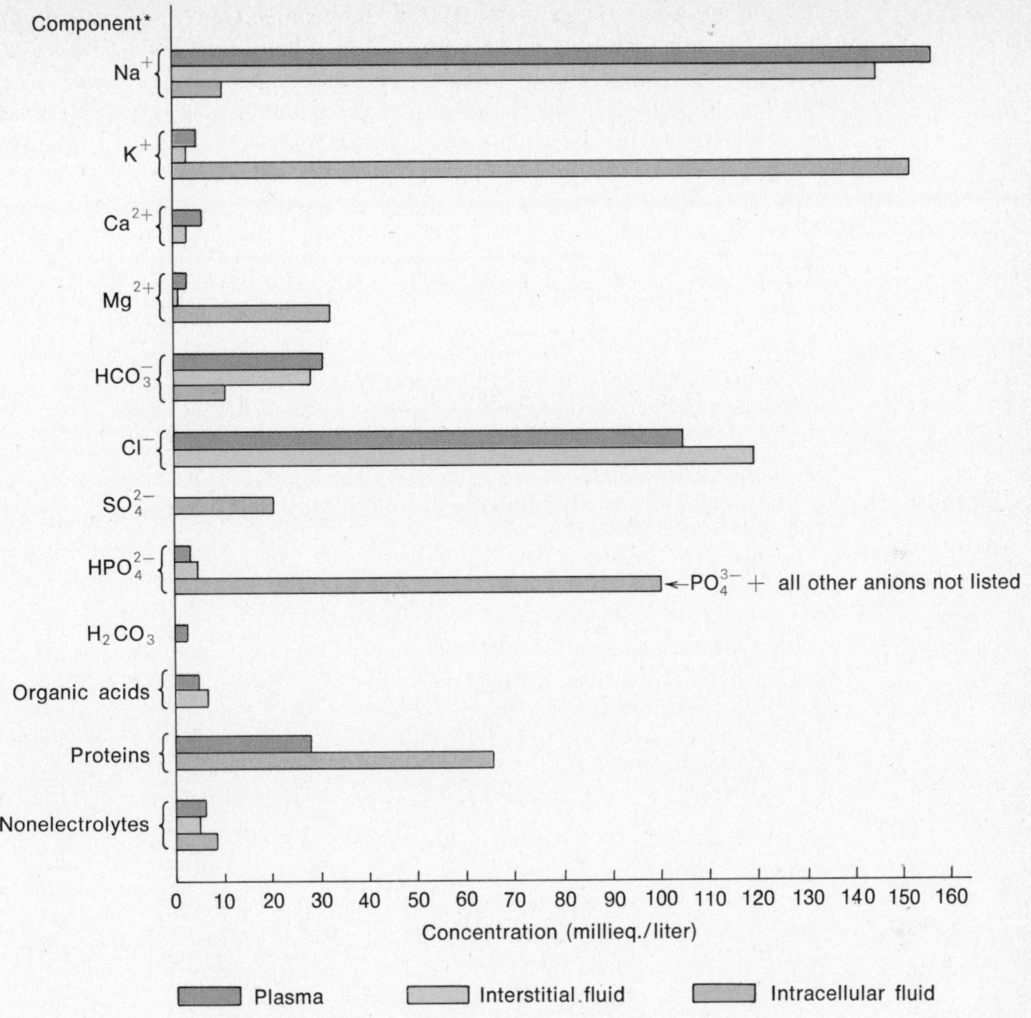

FIGURE 16-1 Comparison of concentrations of components of plasma, interstitial fluid, and intracellular fluid.

increased reabsorption of water from both the distal tubules and collecting ducts of the kidneys. Let's consider some examples of its action.

When blood contains a higher than normal concentration of solutes, it is said to be *hypertonic*. This stimulates release of vasopressin and also stimulates thirst. The increased secretion of vasopressin causes increased reabsorption of water from the distal tubules into the bloodstream. Water intake also aids in returning the osmotic pressure of the blood to normal levels.

In a hypotonic condition, the blood has been diluted by the presence of excess water. To correct this condition, vasopressin secretion is reduced by the posterior pituitary gland's response to a decrease in osmotic pressure. Reduction of vasopressin levels decreases water reabsorption from the distal tubules and collecting ducts and causes an increase in the volume of urine excreted. Consequently, solute concentration in the blood increases and brings the osmotic pressure back to normal level.

Why is osmotic pressure decreased?

exercise 16-4

Diabetes insipidus is a rare condition where the daily volume of urine voided is four to eight times normal level. Could high or low levels of vasopressin secretion be responsible for this?

exercise 16-5

The normal osmolarity of blood plasma is 5454 milliosmols/liter of water. In a hypotonic condition, is the osmolarity greater or less than this value?

exercise 16-6

Alcohol inhibits vasopressin secretion. On a physiological basis, comment on the advisability of rest room facilities in establishments allowing on-premises consumption of alcoholic beverages.

ELECTROLYTE BALANCE

Solute concentration of plasma is in equilibrium with interstitial fluid that, in turn, is in equilibrium with intracellular fluid solutes. Therefore, interstitial fluid acts as an intermediary between the other two fluids. The principal ions in extracellular fluid are sodium and chloride. Potassium, phosphates, and ionized proteins are the chief intracellular ions (see Fig. 16-2). Sodium ions are actively transported from cells into interstitial fluid and plasma; potassium ions are actively transported into cells.* Cellular and capillary membranes are impermeable to large molecules like proteins and such molecules are not exchanged directly between fluids.

The height of the columns in Fig. 16-2 is related to the solute concentration of the various fluids and thus is proportional to a fluid's osmotic pressure. Intracellular fluid contains the greatest concentration of solutes and therefore is expected to have the highest osmotic pressure of the three, i.e., is hypertonic to plasma and interstitial fluid. Cells would thus tend to gain water by osmosis from the surrounding fluids. Interstitial fluid has the lowest solute concentration and water will tend to move from it to plasma or cells.

As noted in the previous section, kidney function regulates electrolyte levels by either reabsorbing electrolytes into the bloodstream or releasing them into urine. This regulation is termed *electrolyte balance*. Additionally,

*Solutes move into and out of cells by two processes—dialysis and active transport. Dialysis refers to the movement of solutes across a semipermeable membrane from a region of *higher* concentration to a region of *lower* concentration. In active transport, solutes may move from a region of *lower* solute concentration to a region of *higher* solute concentration. In order to do so, an input of energy is required; i.e., active transport is endergonic. Energy is supplied by ATP hydrolysis. The exact mechanism by which active transport occurs remains to be determined.

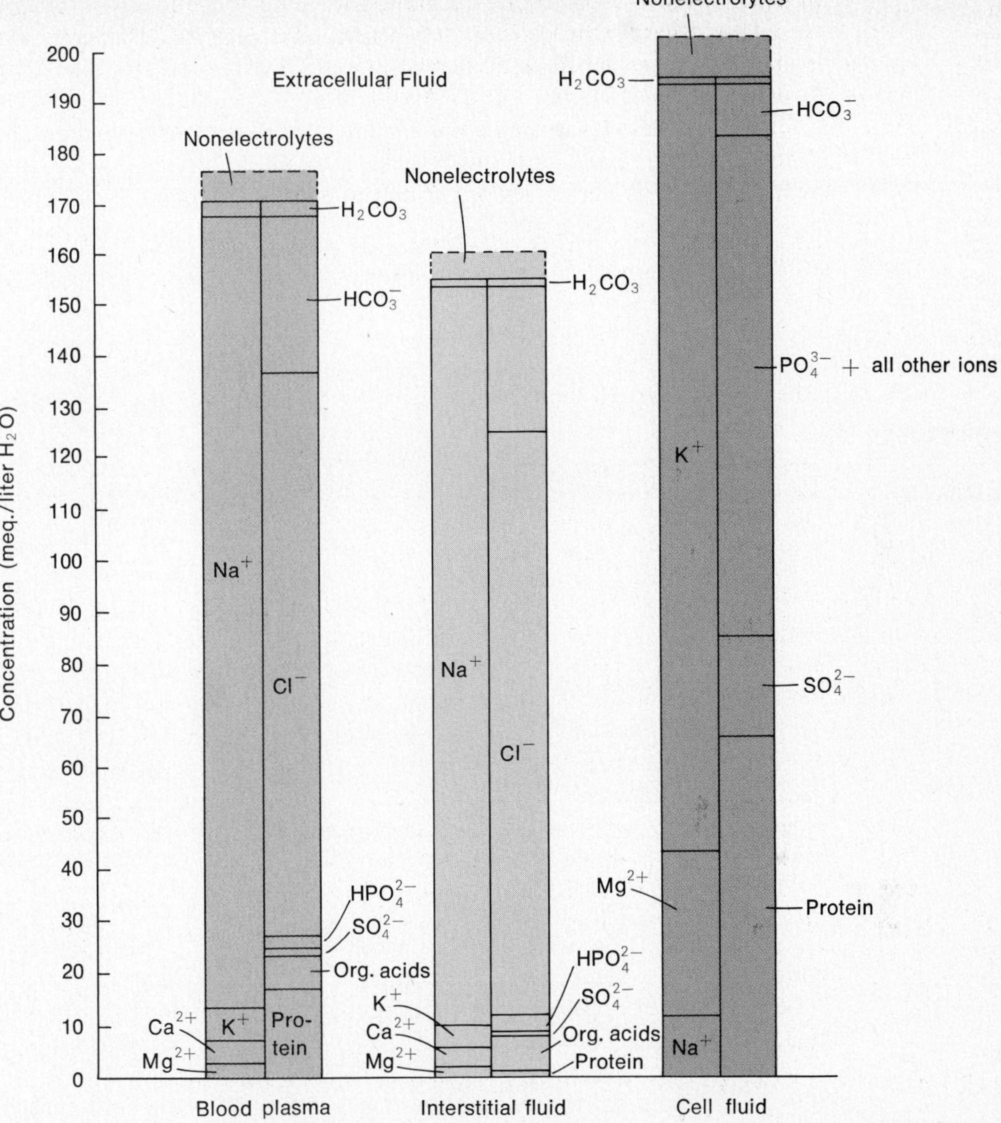

FIGURE 16-2 Comparisons of electrolyte concentrations in plasma, interstitial fluid, and intracellular fluid. Adapted from J. L. Gamble, *Chemical Anatomy, Physiology and Extracellular Fluid,* 6th ed., Harvard University Press, Cambridge, Massachusetts, 1954.

as noted in Chapter 8, the kidneys aid in maintaining acid–base balance of blood.

In the cells lining the kidney tubules and collecting ducts, carbonic anhydrase catalyzes the reaction of carbon dioxide with water to produce carbonic acid. Dissociation of carbonic acid yields bicarbonate ions and hydrogen ions. The hydrogen ions are released into the urine. Normally, urine pH is 5.0 to 7.0; however, release of hydrogen ions can continue until the tubular fluid pH falls to about 4.5, the lower limit of the tubule lining's ability to release hydrogen ions. A pH of 4.5 represents an increase in hydrogen ion concentration of about 800 times that present at a pH of 7.4, the normal pH of blood.

As acidosis develops, the blood's hydrogen ion concentration increases. Additional bicarbonate ions must be made available to buffer the additional hydrogen ions. A mechanism for urine's role in making more bicarbonate ions available has been proposed by R. F. Pitts and is illustrated in Fig. 16-3.

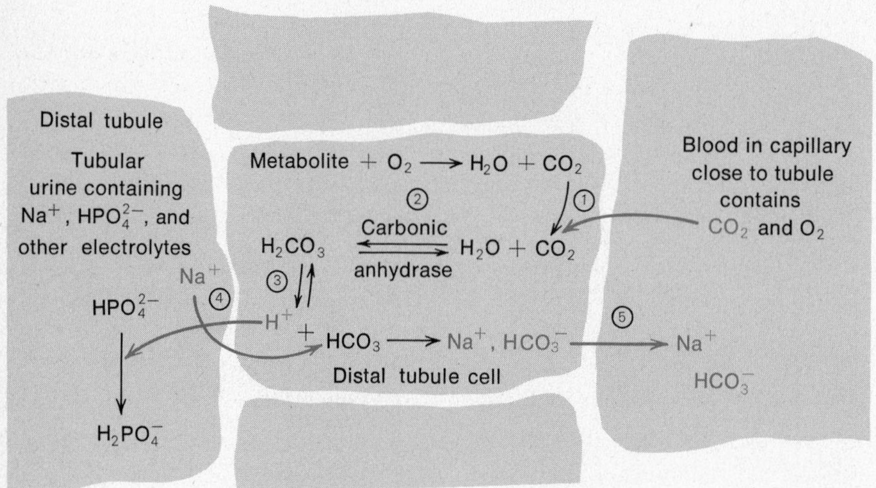

FIGURE 16-3 Cation exchange mechanism for urine acidification according to R. F. Pitts (*American Journal of Medicine,* Vol. 9, 356, 1950). Encircled numbers relate to the text discussion.

Carbon dioxide is available to a cell of the distal tubule from two sources: (1) as a waste product of the cell's own metabolism and (2) by diffusion into the cell from an adjacent capillary (Step 1). Referring to Figure 16-3, the following sequence occurs:

Step 2. Carbonic anhydrase catalyzes the reaction of water and carbon dioxide to produce carbonic acid (H_2CO_3).

Step 3. Carbonic acid dissociates into hydrogen ions and bicarbonate ions. Hydrogen ions from this source then pass through the cell membrane into the distal tubule urine.

Step 4. In order to maintain electrical neutrality, sodium ions from the urine are reabsorbed into the cell and then into the capillary.

Step 5. For each hydrogen ion released, a bicarbonate ion is also formed. These bicarbonate ions, along with reabsorbed sodium ions, diffuse into the bloodstream and the influx of the bicarbonate ions replaces bicarbonate ions used during buffering.

The net effect of excreting hydrogen ions via the urine is to make more bicarbonate ions available for buffering.

exercise 16-7

Which of the steps above would most likely require aldosterone?

exercise 16-8

According to the sequence given above, an increase in blood's hydrogen ion concentration has what effect on urine's pH? Explain.

During alkalosis, a greater than normal concentration of bicarbonate ions are present in blood. The excess bicarbonate ions are then dialyzed into the urine where they are excreted along with sodium ions and other positive ions. Loss of bicarbonate ions via the urine decreases blood bicarbonate concentration to normal level and normal blood pH results.

The four major disturbances of plasma and urine acid–base balance are metabolic acidosis, respiratory acidosis, metabolic alkalosis, and respiratory alkalosis (see Chapter 8). The following equilibria are pertinent:

$$CO_2 + H_2O \underset{\text{anhydrase}}{\overset{\text{carbonic}}{\rightleftharpoons}} H_2CO_3 \rightleftharpoons H^+ (aq) + HCO_3^- (aq)$$

Table 16-3 summarizes the net effects of these conditions on urine pH, plasma bicarbonate ion concentration, and plasma carbonic acid concentration.

TABLE 16-3
Effects of pH change of plasma and urine

DISTURBANCE	URINE pH	PLASMA $[HCO_3^-]$ (meq/liter)	PLASMA $[H_2CO_3]$ (meq/liter)
Normal level	5.5–7.0	25	1.25
1. Respiratory acidosis	↓	↑	↑
2. Respiratory alkalosis	↑	↓	↓
3. Metabolic acidosis	↓	↓	↓
4. Metabolic alkalosis	↑	↑	↑

Respiratory acidosis results from hypoventilation. Decreased breathing causes CO_2 concentration to increase. This causes the equilibria above to shift to the right and additional hydrogen ions are excreted via the urine. Hyperventilation causes the opposite effect. The overproduction of acidic metabolic waste products causes metabolic acidosis. These substances release hydrogen ions that combine with plasma bicarbonate to produce carbonic acid. The carbonic acid is then decomposed to carbon dioxide and water. Hydrogen ions move from distal tubule cells into urine and bicarbonate ions move from the cells into plasma to compensate for reduced bicarbonate levels in plasma. Metabolic alkalosis causes opposite effects.

CEREBROSPINAL FLUID

Cerebrospinal fluid (CSF) is found in cavities within and around the brain and spinal column. Most of the fluid originates in the choroid plexus of the brain and is continuously secreted. A normal adult's brain contains about 125 ml of CSF, which is completely renewed within 3- to 4-hr intervals. Although the composition of CSF and plasma are similar, certain differences exist. Potassium ion concentration is lower and chloride concentration is higher than in plasma. Calcium concentration in CSF does not normally fluctuate despite changes in plasma calcium concentration. These data suggest that CSF is not simply composed of solutes that dialyze from plasma. Cerebrospinal fluid acts as a cushion protecting the brain.

Sodium and potassium ions, glucose, and urea normally pass readily among plasma, interstitial fluid, and cells; however, they do not *readily* diffuse between plasma and cerebrospinal fluid. This reduced permeability is attributed to the presence of physiological barriers called the blood–cerebrospinal fluid and blood–brain barriers. The cause of these barriers is as yet undetermined. One suggestion is that the brain's capillary membranes are less permeable than those elsewhere in the body.

MILK

Milk is produced by the mammary glands of mammals for the nourishment of their young. During pregnancy, the high level of estrogens causes growth and branching of the ducts within the mammary glands. By the end of pregnancy, the mother's breasts are fully developed for nursing. Prolactin initiates milk production and oxytocin ("letdown factor") stimulates release of milk from the breasts to the sucking infant.

Milk contains triglycerides, lactose, calcium, vitamin A, phosphorus, and the proteins casein and lactoglobulin. Because of its composition, milk is often called *nature's perfect food;* however, it is deficient in iron, copper, most of the B vitamins (except riboflavin and pantothenic acid), vitamin C, and vitamin D. Pasteurization further reduces the vitamin C content. Human milk contains more lactose and less protein than cow's milk (see Table 16-4).

TABLE 16-4
Comparison of human milk and cow's milk

COMPONENT	HUMAN (percent)	COW (percent)
Water	87.5	87
Carbohydrates	7.0–7.5	4.5–5.0
Proteins	1.0–1.5	3.0–4.0
Lipids	3.0–4.0	3.5–5.0
Minerals	0.2	0.75

Lactose is found exclusively in mammalian milk. Because galactose is not present in significant amounts in other body tissues or fluids, it must be formed from blood glucose in the mammary glands. The use of carbon-14 radioisotope tracer experiments on test animals has demonstrated this to be true. Glucose containing radioactive carbon-14 was injected into test mammals. Lactose samples withdrawn later had equal amounts of C-14 carbon in their glucose and galactose components. The net overall production of lactose by mammary glands can be summarized:

$$\text{D-glucose in mammary blood vessels} \xrightarrow{\text{isomerizes}} \text{D-galactose}$$

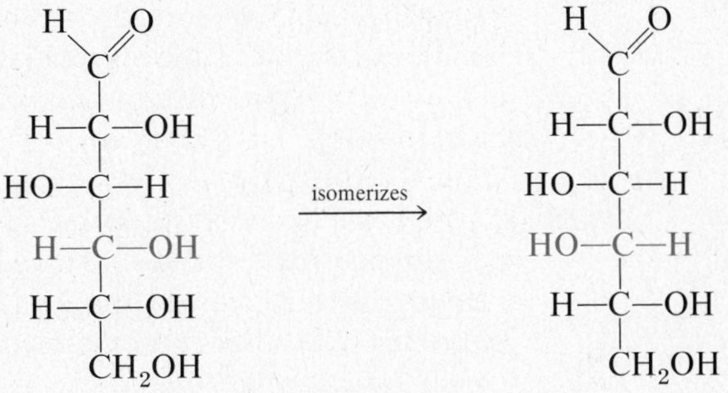

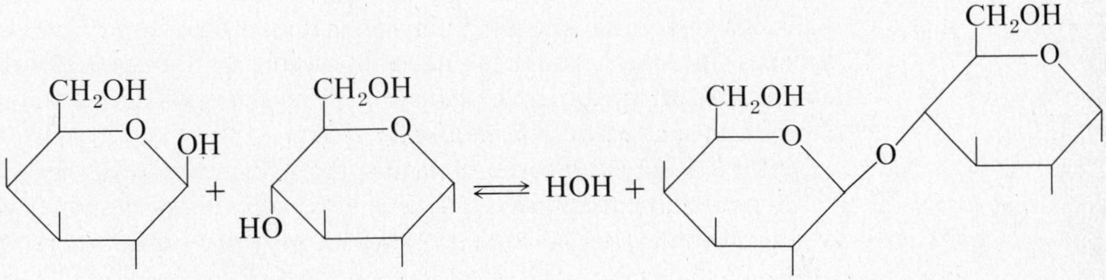

$$\begin{array}{c} \text{H}\quad\text{O} \\ \diagdown\;/\!\!/ \\ \text{C} \\ | \\ \text{H—C—OH} \\ | \\ \text{HO—C—H} \\ | \\ \text{H—C—OH} \\ | \\ \text{H—C—OH} \\ | \\ \text{CH}_2\text{OH} \end{array} \xrightarrow{\text{isomerizes}} \begin{array}{c} \text{H}\quad\text{O} \\ \diagdown\;/\!\!/ \\ \text{C} \\ | \\ \text{H—C—OH} \\ | \\ \text{HO—C—H} \\ | \\ \text{HO—C—H} \\ | \\ \text{H—C—OH} \\ | \\ \text{CH}_2\text{OH} \end{array}$$

$$\beta\text{-D-galactose} + \alpha\text{-D-glucose} \rightleftharpoons \text{HOH} + \beta\text{-lactose}$$

β-D-galactose + α-D-glucose

β-lactose

The proteins casein and lactoglobulin are not found in other tissues and are not related to the plasma proteins. Casein contains residues of the amino acid serine. The —OH side chain groups on the serine residues are esterified by phosphates. Thus, much of the phosphorus content of milk is associated with casein.

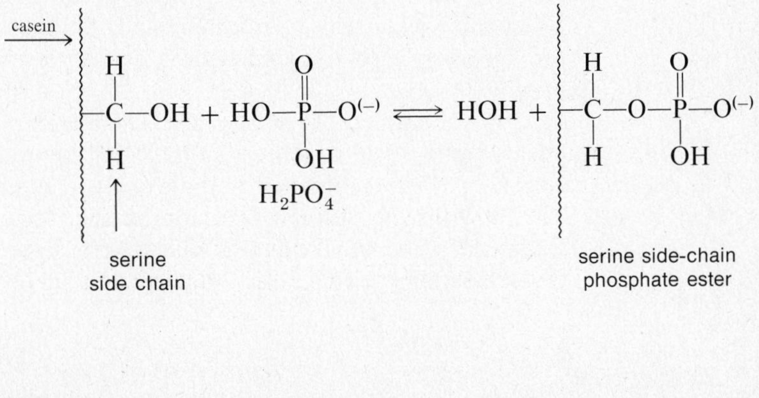

$$\text{casein} \rightarrow \; \begin{array}{c} \text{H} \\ | \\ \text{—C—OH} \\ | \\ \text{H} \end{array} + \begin{array}{c} \text{O} \\ \| \\ \text{HO—P—O}^{(-)} \\ | \\ \text{OH} \\ \text{H}_2\text{PO}_4^- \end{array} \rightleftharpoons \text{HOH} + \begin{array}{c} \text{H}\qquad\quad\text{O} \\ |\qquad\quad\;\; \| \\ \text{—C—O—P—O}^{(-)} \\ |\qquad\quad\;\; | \\ \text{H}\qquad\quad\text{OH} \end{array}$$

serine side chain

serine side-chain phosphate ester

Casein has a net negative charge and Ca^{2+} ions are associated with it to form calcium caseinate, a salt partly responsible for milk's white color. Casein also serves as an emulsifying agent to keep milk butterfat in suspension.

Colostrum is the fluid secreted by the breasts for the first few days following parturition. It is considerably different than ordinary milk. Colostrum contains less fat and carbohydrate and more minerals and proteins, especially the immunoglobulins. These contain the antibodies found in the mother's blood. Therefore, the antibodies required for immunity are transferred in colostrum to the nursing infant. Colostrum contains a protein that inhibits the protein–digestive enzyme trypsin thus preventing digestion of the antibodies.

LYMPH

Lymph is the fluid obtained from the lymphatic ducts, a network of vessels that ultimately empty into the bloodstream at the thoracic duct. Thus, the lymphatic system provides a method by which substances can be carried from interstitial spaces into the blood. Solute concentrations in lymph and interstial fluid are in equilibrium so that lymph composition closely resembles that of the interstitial fluid in the region of the body from which lymph flows.

High molecular weight materials such as proteins and lipids cannot easily pass through venous capillaries. Because of their unique construction, however, lymph capillaries can take in such substances. These materials can then be transported by the lymphatic system to the bloodstream. Protein concentration in lymph varies depending on its point of origin in the body.

Villi are finger-like projections found on the intestinal walls.

Fats are absorbed through the intestinal villi into the lymphatic system and are then transported to the bloodstream by way of the thoracic duct.

BILE AND BILE SALTS

Bile is secreted by the liver as a yellow fluid, pH 7.0–8.5, at a rate of between 500 to 700 ml daily. Its major constituents are cholesterol; bile acids, such as cholic and deoxycholic acids; bile salts; and bilirubin, a yellow bile pigment produced from the breakdown of heme. As bile accumulates, it is stored in the gallbladder where it is concentrated by water reabsorption and also acidified. During storage, bilirubin is oxidized to biliverdin, a green bile pigment.

Glycocholic and taurocholic acids are cholesterol derivatives. The combination of glycine and cholic acid produces glycocholic acid; taurine plus cholic acid yields taurocholic acid. Both of these reactions proceed through cholyl CoA, the coenzyme A derivative of cholic acid, as shown on page 453.

Explain why these acids exist as anions at the normal bile pH.

At the normal pH of bile, glycocholic and taurocholic acids occur as anions. Cations such as sodium and calcium are present to maintain electrical neutrality. The combination of these anions and cations produce bile salts. The bile salts facilitate fat digestion by emulsifying fats into micelles (see Chapter 5). The small droplets have a very large surface area exposed to the surrounding water. This enhances the action of pancreatic lipase in

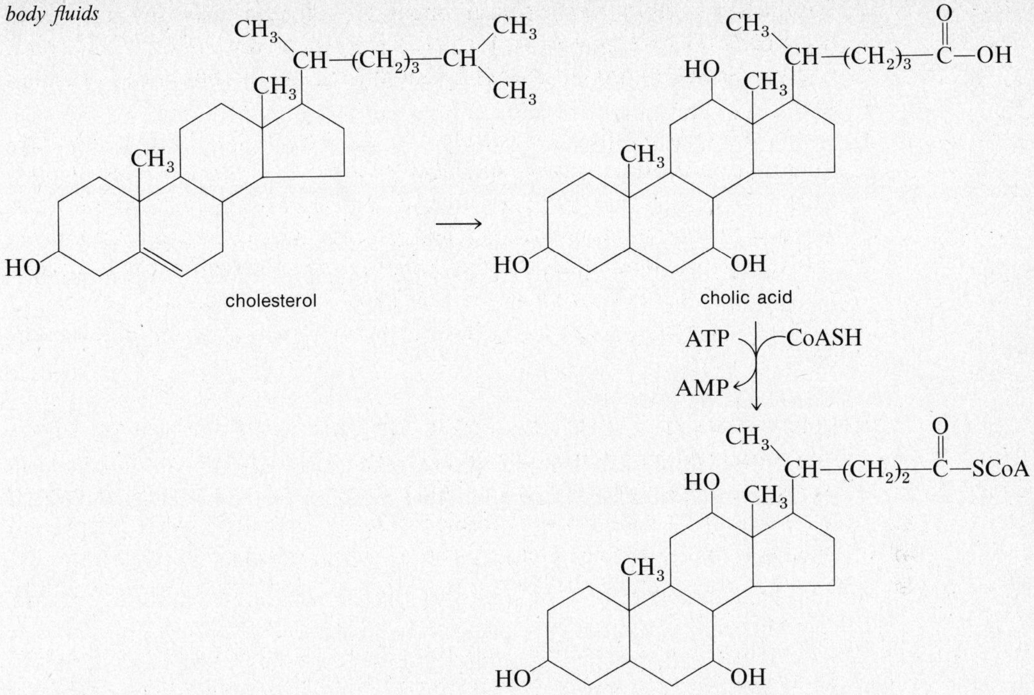

cholylCoA + $H_2N-CH_2-CO_2H$ ⟶
glycine

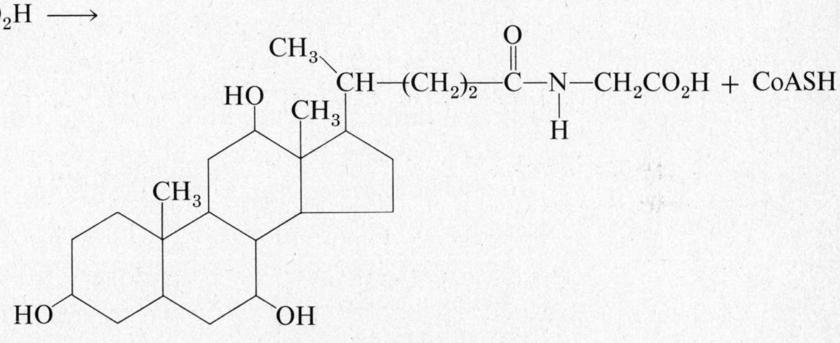

glycocholic acid + coenzyme A

cholylCoA + $H_2NCH_2CH_2SO_3H$ ⟶
taurine

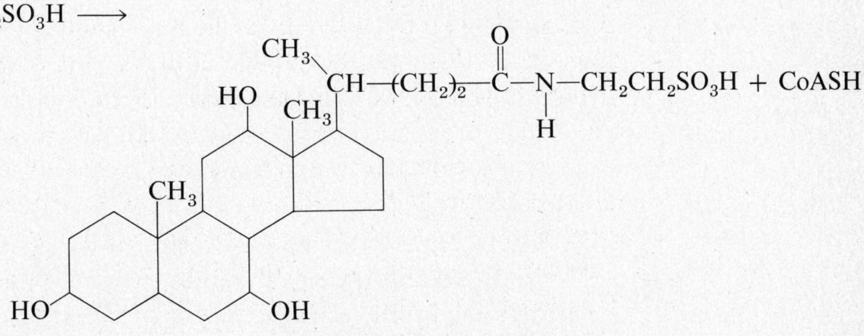

taurocholic acid + coenzyme A

catalyzing the hydrolysis of fats. Free (unesterified) fatty acids and monoglycerides are also emulsified by bile salt action.

A certain amount of bile acids must be present in relation to the amount of cholesterol to keep cholesterol in solution. Failure of the gallbladder to acidify alkaline bile from the liver, excessive absorption of bile salts and fatty acids from the liver, or excessive secretion of cholesterol in the bile can cause the bile acid/cholesterol ratio to fall below a critical level. When this occurs, precipitation of cholesterolic mixtures occurs and forms gallstones. Most commonly, gallstones consist of calcium–bilirubin and cholesterol.

CHAPTER SUMMARY

1. Body fluids are located within cells (intracellular) or outside cells (extracellular). Extracellular fluid is mainly either interstitial fluid or blood.
2. Whole blood is a mixture of formed elements (white cells, red cells, and platelets) and plasma that can be separated by centrifugation. Plasma contains ions—principally sodium, chloride and bicarbonate ions—and proteins—mainly albumin, globulins, and fibrinogen. Albumin aids in fatty acid transport; α and β globulins transport metal ions; γ globulins are associated with antibodies. Fibrinogen is required for blood clotting.
3. Clotting is a complex process. Factors such as calcium ions and thromboplastins activate the conversion of inactive prothrombin to active thrombin. Thrombin activates conversion of soluble fibrinogen into a network of insoluble fibrin strands. The formed elements become enmeshed among the fibrin strands.
4. Heparin, dicumarol, and warfarin are anticoagulants. Removal of fibrinogen from plasma converts plasma into serum, a nonclotting fluid. When added to plasma, fluoride and oxalate salts remove calcium ions by precipitating as insoluble calcium salts. Removal of calcium ions prevents prothrombin activation and clotting does not occur.
5. Urine production results from kidney action. It fluctuates within the normal range (adult) of 600 to 2500 ml/24 hr. Urine's normal range of specific gravity is 1.008 to 1.030 and of pH is 4.5 to 8.4.
6. Aldosterone regulates urinary sodium ion concentration by regulating sodium ion reabsorption in the kidney tubules. Vasopressin controls reabsorption of water from the kidneys' distal tubules and collecting ducts.
7. Electrolyte balance is the adjustment of electrolyte levels between body fluids. Reabsorption of solutes into the bloodstream or their release into the urine are the major mechanisms for electrolyte balance. Sodium, bicarbonate, and chloride ions are the principal ions in plasma

and interstitial fluid. Potassium, phosphate, and ionized proteins are the chief intracellular ions. Hydrogen ions are released into urine by cells lining the distal tubules. This is accompanied by absorption of bicarbonate ions and sodium ions into the blood.

8. Plasma and urine acid-base balance is disrupted mainly by (a) metabolic acidosis due to overproduction of acidic metabolic waste products, (b) respiratory acidosis from hypoventilation, (c) metabolic alkalosis, and (d) respiratory alkalosis from hyperventilation.
9. Cerebrospinal fluid (CSF) is present within and around the cavities of the brain and spinal column. Electrolytes do not *readily* diffuse between blood and CSF. This reduced permeability is termed the *blood–brain barrier.*
10. Milk is produced uniquely by the mammary glands of mammals. It is an aqueous mixture principally containing triglycerides, calcium ions, lactose, phosphorus, vitamin A, casein, and lactoglobulin. Composition varies among species. In mammary glands, glucose is converted to galactose and these two monosaccharides combine to form lactose. Colostrum, a fluid secreted by the breasts during the first several days after parturition, differs from milk in that it contains more immunoglobulins and minerals but less fat and carbohydrates than milk.
11. Lymph is fluid obtained from lymphatic vessels. Lipids and proteins pass through specialized capillaries into lymph and are then transported back to the bloodstream via the thoracic duct.
12. Bile is produced by the liver and stored in the gallbladder. Bile's major components are cholesterol, cholic acid, deoxycholic acid, taurocholic acid, salts of these acids, and bilirubin. Bile salts emulsify fats. Cholesterol and calcium–bilirubin precipitate from bile under alkaline conditions to form gallstones.

QUESTIONS

1. When potassium ions leave cells, hydrogen ions and sodium ions enter. What effect does this have on plasma pH?
2. As sodium ions enter plasma, what effect does this have on the water balance in cells?
3. Shortly after a blood clot forms, it contracts and a colorless fluid flows from around the clot. What is this fluid?
4. Bile aids in fat absorption. Prior to surgery, vitamin K commonly is injected into patients with liver disease or obstructed bile ducts. Explain why vitamin K is administered.
5. A 24-hr urine sample was collected from a patient, Henrietta Hypochondriac. The total sample volume was 1200 ml. The laboratory reported the following data:

SAMPLE	VOLUME	MASS	TEMPERATURE
Water	10.00 ml	9.9604 g	25°C
Urine	10.00 ml	10.083 g	25°C

(a) Calculate the specific gravity of the urine sample.
(b) Is this specific gravity value within the normal range?

6. The posterior hypophysis responds to changes in osmotic pressure. Vasopressin controls water reabsorption from distal tubules and collecting ducts. Explain what is meant by the statement that a feedback mechanism exists between vasopressin secretion and posterior hypophysis action.
7. Explain the differences among whole blood, plasma, and serum. Are there any similarities?
8. Blood accounts for about 7 percent of your total body weight. The density of blood is 1.06 g/ml.
   (a) Calculate the number of liters of blood your body contains.
   (b) If approximately 20 drops equal 1 ml, calculate the number of drops of blood your body contains.
9. When left standing, urine samples change in composition due to bacterial action. One such action is the formation of ammonia from urea. What effect will ammonia formation have on the pH of a urine sample?
10. Why would a person who drinks seawater suffer from cellular dehydration?

17

drugs—chemical agents in disease treatment

INTRODUCTION

Drugs are chemical compounds that interact with various compounds in the body. These interactions may be harmful or helpful. Generally speaking, helpful drugs are those whose interactions are useful in the treatment of diseases. Their possible courses of action are numerous and include relief of symptoms, prevention of bacterial growth, and ability to stimulate the body to produce hormones and enzymes.

There are four main sources of drugs: plants, animals, minerals, and the chemistry laboratory. Regardless of the source, chemistry and chemists are involved in the isolation or preparation and purification of drugs. Examples of drugs obtained from these four sources are given in Table 17-1.

There are many ways in which drugs may be classified. Those in Table 17-1 are classified according to their origin and chemical structure. A more common classification of drugs is based upon their action. Table 17-2 lists ten divisions of drugs according to their action and examples of each. In this chapter we will briefly examine each of these divisions and the mode of drug action.

LOCAL ANESTHETICS

Local anesthetics are drugs capable of blocking nerve impulse conduction along sensory and motor nerve fibers. The sensory fibers are more susceptible to the action of these compounds than are the motor fibers. Hence, it is possible to cause a loss of sensation without producing paralysis by carefully controlling the concentration of the drug. All local anesthetics are very toxic and it is necessary to confine the compound to a small region.

The structural formulas of procaine, ethylaminobenzoate, cyclomethycaine, and cocaine are:

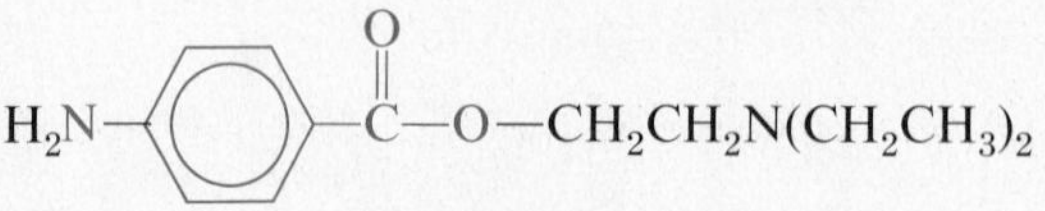

$$H_2N-C_6H_4-\overset{O}{\overset{\|}{C}}-O-CH_2CH_2N(CH_2CH_3)_2$$

procaine (Novocain)

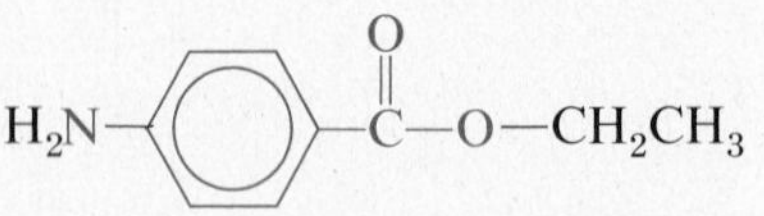

$$H_2N-C_6H_4-\overset{O}{\overset{\|}{C}}-O-CH_2CH_3$$

ethylaminobenzoate (Benzocaine)

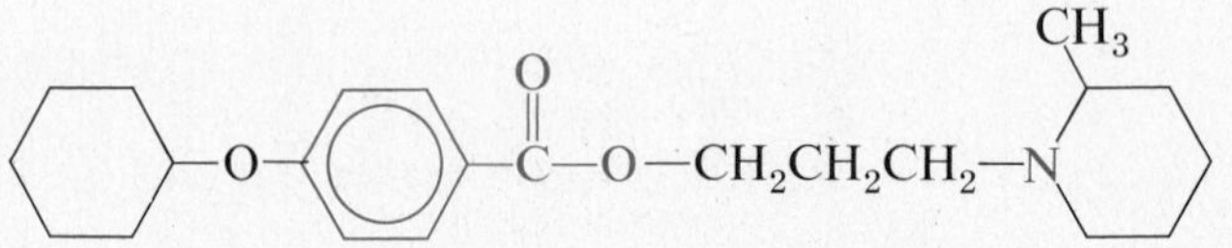

$$C_6H_{11}-O-C_6H_4-\overset{O}{\overset{\|}{C}}-O-CH_2CH_2CH_2-N(C_5H_9CH_3)$$

cyclomethycaine (Surfacaine)

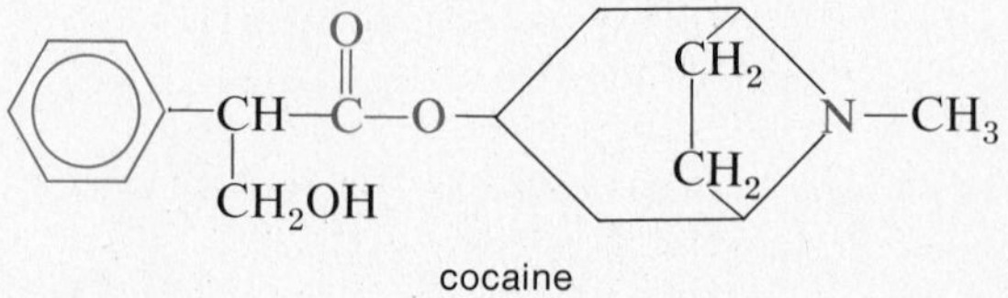

$$C_6H_5-\underset{CH_2OH}{\underset{|}{CH}}-\overset{O}{\overset{\|}{C}}-O-C_6H_{10}(CH_2)(CH_2)N-CH_3$$

cocaine

All of these compounds contain nitrogen and are administered as the salts of a strong acid to improve their solubility. Cyclomethycaine is prepared as the sulfate salt, the others as the hydrochloride. We can see that these

TABLE 17-1
Origin of Selected Drugs

- I. Plants
  - A. Alkaloids
    1. Caffeine
    2. Atropine
    3. Morphine
  - B. Glycosides
    1. Digitoxin
  - C. Gums
  - D. Resins
- II. Animals
  - A. Steroids
    1. Cortisone
    2. Progesterone
    3. Estrone
  - B. Proteins
    - 2. Epinephrine
    - 2. Insulin
    - 3. ACTH
- III. Minerals
  - A. Elements
    1. Iodine
    2. Sulfur
  - B. Salts
    1. Epsom salt ($MgSO_4 \cdot 7H_2O$)
    2. $CaCO_3$
    3. $Ca_3(PO_4)_2$
    4. NaCl
- IV. Laboratory
  - A. Sulfa
    1. Sulfamerazine
    2. Sulfisomidine (Elkosin)
    3. Sulfisoxazole (Gantrisin)
  - B. Barbiturates
    1. Pentobarbital (Nembutal)
    2. Phenobarbital (Luminal)
    3. Secobarbital (Seconal)
  - C. Carbamates
    1. Meprobamate (Equanil, Miltown)
  - D. Hydantoins
    1. Diphenylhydantoin (Dilantin)
    2. Methylphenylethylhydantoin (Mesantoin)
    3. Ethotoin (Peganone)

TABLE 17-2
Drug Classifications

I. Local Anesthetics
   A. Procaine (Novocain)
   B. Ethylaminobenzoate (Benzocaine)
   C. Cyclomethycaine sulfate (Surfacaine)
II. Antihistamines
   A. Diphenhydramine (Benadryl)
   B. Thonzylamine (Anahist)
   C. Dimenhydrinate (Dramamine)
III. Anti-infectives
   A. Antibacterials
      1. Antibiotics
         (a) Penicillins
         (b) Tetracyclines
         (c) Streptomycins
         (d) Neomycin
         (e) Chloramphenicol
   B. Antifungals
      1. Griseofulvin
      2. Undecylenic acid
   C. Anthelmintics
   D. Antiprotozoals
IV. Autonomic
      1. Epinephrine
      2. Ephedrine
      3. Amphetamines
      4. Atropine
V. Cardiovascular
   A. Digoxin (Lanoxin)
   B. Acetyldigitoxin (Acylanid)
   C. Procainamide hydrochloride (Pronestyl hydrochloride)
VI. Central Nervous System Depressants
   A. Analgesics
      1. Morphine
      2. Codeine
      3. Meperidine (Demerol)
      4. Acetylsalicylic acid (aspirin)
      5. Propoxyphene hydrochloride (Darvon)
   B. General Anesthetics
      1. Diethyl ether
      2. Nitrous oxide
      3. Cyclopropane
      4. Thiopental sodium (Pentothal sodium)
   C. Anticonvulsants
      1. Ethotoin (Peganone)
      2. Methylphenylethylhydantoin (Mesantoin)
      3. Diphenylhydantoin (Dilantin)

TABLE 17-2 (continued)

- D. Hypnotics and Sedatives
  1. Barbital (Veronal)
  2. Phenobarbital (Luminal)
  3. Pentobarbital (Nembutal)
  4. Chloral hydrate
- E. Tranquilizers
  1. Reserpine
  2. Chlorpromazine (Thorazine)
  3. Prochlorperazine (Compazine)
  4. Meprobamate (Miltown)
  5. Chlordiazepoxide (Librium)
  6. Hydroxyzine Pamoate (Vistaril)

VII. Central Nervous System Stimulants
- A. Caffeine
- B. Amphetamine
- C. Imipramine (Tofranil)

VIII. Enzymes
1. Chymotrypsin
2. Fibrinolysin
3. Penicillinase (Neutrapen)

IX. Gastrointestinal Agents
1. Aluminum hydroxide gel
2. Sodium bicarbonate
3. Sodium citrate
4. Magnesium trisilicate
5. Calcium carbonate
6. Methylcellulose
7. Magnesium hydroxide (milk of magnesia)
8. Mineral Oil
9. Phenolphthalein
10. Bisacodyl (Dulcolax)

X. Hormones and Hormone Derivatives
1. Cortisone acetate
2. Prednisone
3. Insulin
4. Tolbutamide (Orinase)
5. Norethindrone
6. Progesterone
7. Relaxin
8. Estrone

compounds are chemically related in other ways; all are esters and contain a substituted benzene ring. Cocaine is a naturally occurring compound and is isolated from coca leaves. The other three substances have been synthesized as local anesthetics and are benzoic acid derivatives. This in itself is not sufficient to cause them to behave as anesthetics. In fact, it is interesting to note that both procaine and ethylaminobenzoate contain the amino acid, *para*-aminobenzoic acid. Recall from Chapter 5 the importance of this amino acid in the vitamin tetrahydrofolic acid.

Cocaine is primarily used for surface anesthesia of the eyes, nose, and throat. The usual concentration is 1 to 5% of the hydrochloride; it is administered by swabbing the mucous membranes. Commonly, epinephrine is added to the solution to slow the rate of absorption. Cocaine is an addicting drug; hence its use is regulated by narcotic laws.

Cyclomethycaine is usually prepared as an ointment, cream, or jelly in concentrations of 0.5 to 1%. Its most common use is the relief of pain from skin lesions and abrasions, although it is also used on rectal and vaginal ulcerations or fissures. Ethylaminobenzoate is less soluble than cyclomethylcaine, which means it is absorbed more slowly. Hence, it is useful for burn treatment as well as the purposes for which cyclomethycaine is used. Depending upon its use, the concentration may vary from 1 to 5%.

Procaine is both less toxic and less potent than cocaine. Undoubtedly, it is the single most widely used local anesthetic. Because of its lower toxicity it is one of the safest local anesthetics for infiltration. Generally, 1 to 2% solutions of procaine containing 0.002% epinephrine are used for injection. Absorption of procaine from the mucous membranes is very limited; consequently, it is not very useful for topical administration.

ANTIHISTAMINES

Antihistamines are thought to act by preventing the action of histamine. Their greatest effect is on nasal allergies, particularly hay fever. The structural formulas of the three antihistamines in Table 17-2 are given below.

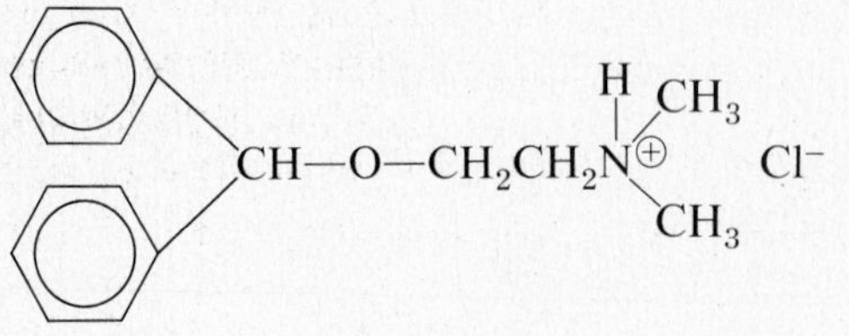

diphenhydramine
(Benadryl)

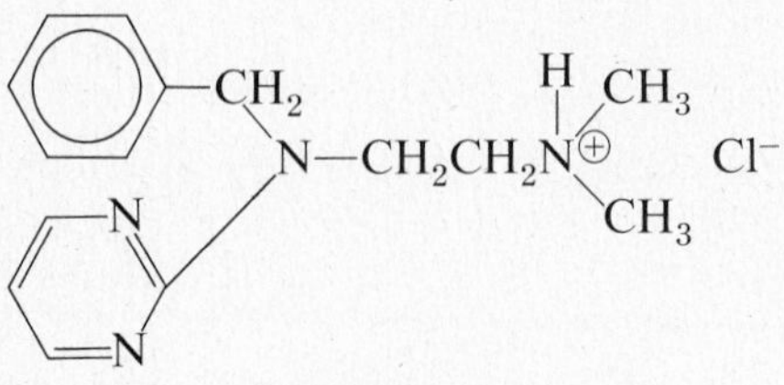

throzylamine
(Anahist)

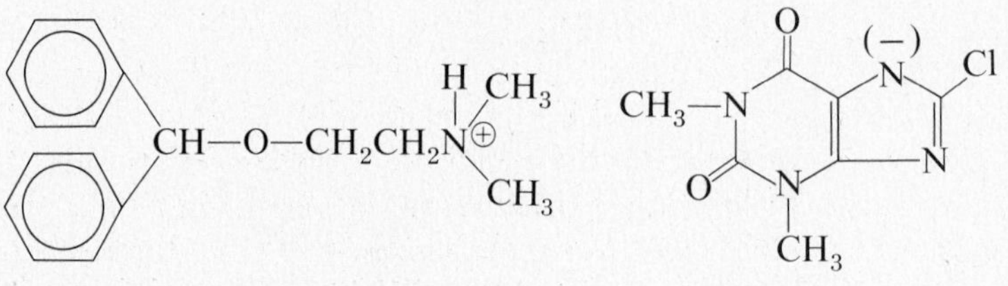

dimenhydrinate
(Dramamine)

Notice that all three compounds contain the grouping

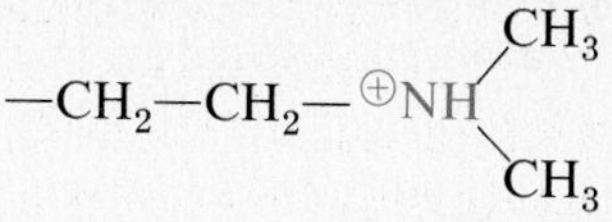

$$-CH_2-CH_2-{}^{\oplus}NH(CH_3)_2$$

resulting from the addition of H^+ to the teritiary amine group. It is thought that this group is responsible for the antihistaminic action of these compounds. Histamine contains the grouping $-CH_2CH_2NH_2$. The antihistamines presumably also bond to the histamine receptor sites. By bonding to these receptor sites more strongly than does histamine they exclude histamine, thereby preventing its normal physiologic action.

Histamine is a vasoconstrictor.

A side effect of the antihistamines is drowsiness, sometimes accompanied by dryness of the mouth and throat. A rather unusual side effect of many antihistamines is their ability to alleviate the discomfort of motion sickness. The one most commonly used for motion sickness is dimenhydrinate (Dramamine). This compound is basically the same as diphenhydramine (Benadryl). The difference is in the acid used to neutralize the tertiary amine group of the antihistamine. This change in the acid used changes the anion in the resulting salt. In Benadryl the anion is chloride, whereas in Dramamine the anion is the more complex group:

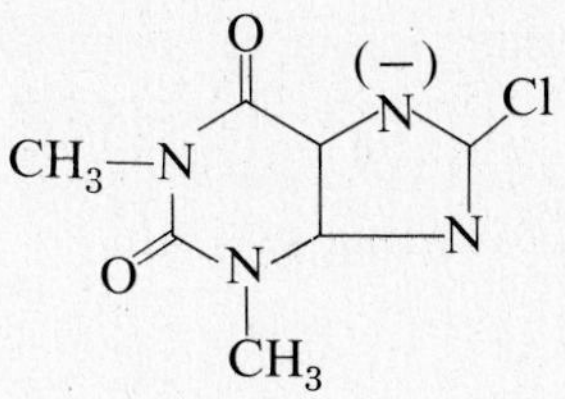

O, (−), N, Cl, CH_3-N, N, O, N, CH_3

exercise 17-1

Relate the action of antihistamines to that of enzyme inhibitors.

ANTI-INFECTIVES

Anti-infectives are compounds used for the treatment or prevention of microbial and parasitic infections. They are divided according to whether their action is primarily effective against bacteria, fungi, protozoa, or helminths.

The most familiar of anti-infectives are antibacterial agents. These compounds may be bactericidal, which means they kill the bacteria, or they may be bacteriostatic, meaning they prevent further bacterial growth. When a bacteriostatic agent is removed, the bacteria can again grow. The most widely used antibacterials are antibiotics. Antibiotics are substances produced by microorganisms which kill or inhibit the growth of other microorganisms. The oldest known antibiotic is penicillin, discovered in 1928 by Alexander Fleming. He extracted it from a mold after noticing the mold prevented the growth of a staphylococcus culture. In 1940 the first pure penicillin was isolated in England at Oxford University. As the need for treating those wounded in World War II grew, manufacture of penicillin began in the United States in 1941. The structural formulas of several classes of antibiotics are shown in Fig. 17-1.

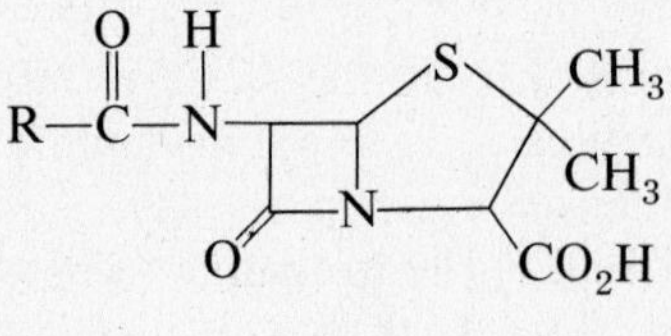

penicillin

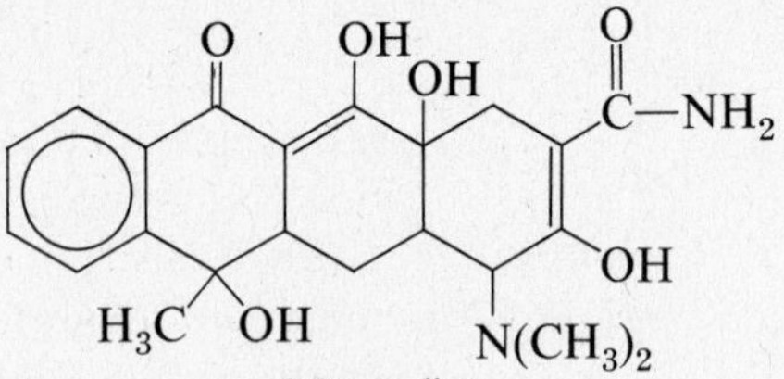

tetracycline

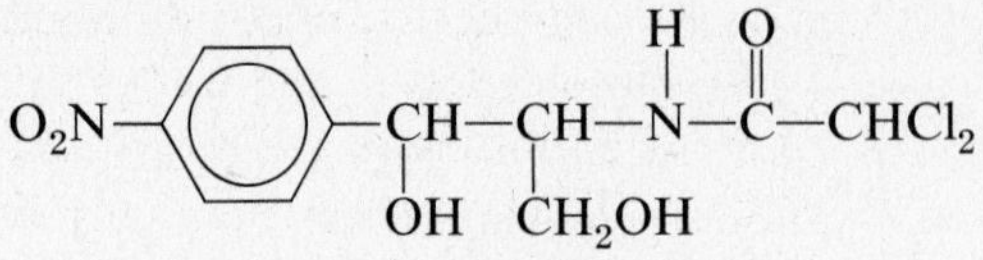

chloramphenicol

FIGURE 17-1 Structural formulas of three common antibiotics.

Antibiotics function in several ways. Their usefulness is dependent upon their ability to interfere with bacterial metabolic reactions which are absent in humans. The interference may be inhibition of a particular enzyme or alteration of cell membrane permeability.

The penicillins remain the most widely used antibiotics. The difference between the various penicillins is the identity of the R group attached to the basic ring structure.

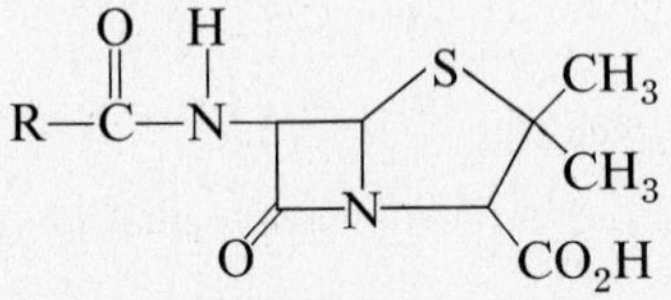

penicillin G, R = $C_6H_5CH_2$—

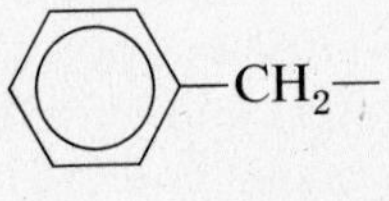

penicillin O, R = $CH_2{=}CHCH_2CH(SH)$—

penicillin V, R = C_6H_5—O—CH_2—

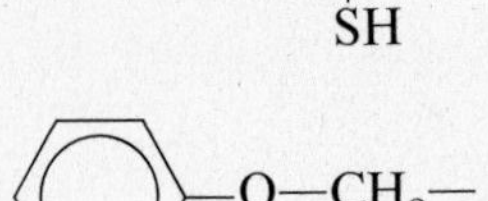

methicillin (Staphcillin), R = 2,6-$(OCH_3)_2C_6H_3$—

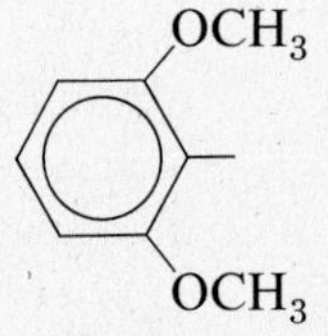

phenethicillin (Darcil), R = C_6H_5—O—CH(CH_3)—

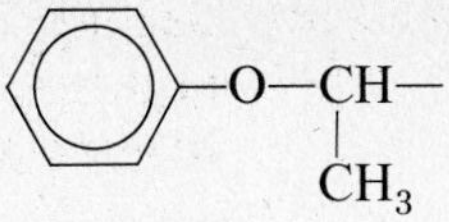

oxacillin (Prostaphlin) R = (5-methyl-3-phenyl-4-isoxazolyl; N–O ring with CH_3)

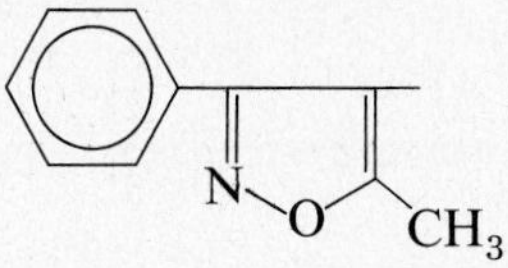

The mold produces the basic penicillin molecule, and the various R groups are attached in the laboratory by chemical synthesis. These R groups introduce specific actions by the penicillin. For example, penicillin V and phenethicillin are stable to gastric juices and can be administered orally, whereas penicillin G and penicillin O are partially inactivated by gastric fluids and are usually administered parenterally. Methicillin and phenethicillin appear to be particularly effective against staphylococci.

Notice that all the penicillins are acids. They are generally neutralized and administered as either the sodium or potassium salts. This conversion increases their water solubility and decreases the likelihood of tissue irritation.

exercise 17-2

Explain why the sodium or potassium salts of penicillin are less likely to cause tissue irritation than the free penicillins.

The tetracyclines are a family of closely related compounds containing four fused six-membered carbon rings. The structural formulas of several are given in Fig. 17-2.

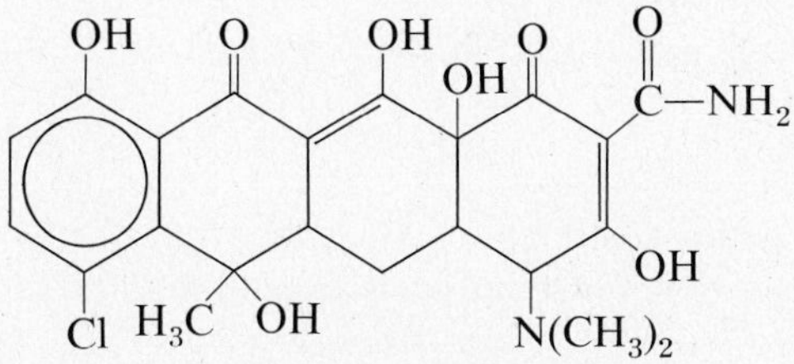

OH O OH O OH O C—NH_2 OH Cl H_3C OH $N(CH_3)_2$

chlortetracycline
(Aureomycin)

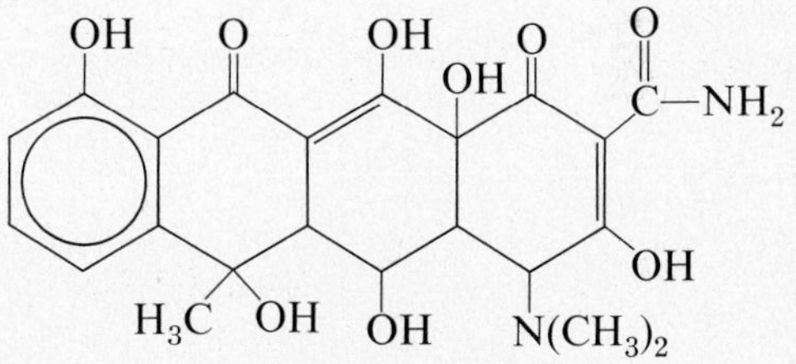

OH O OH O OH O C—NH_2 OH H_3C OH OH $N(CH_3)_2$

oxytetracycline
(Terramycin)

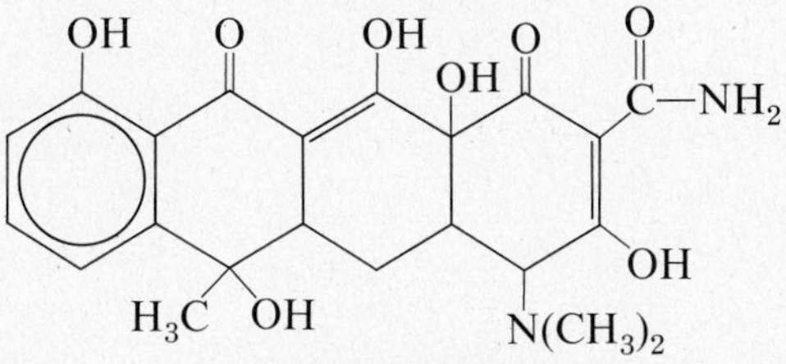

OH O OH O OH O C—NH_2 OH H_3C OH $N(CH_3)_2$

tetracycline
(Achromycin)

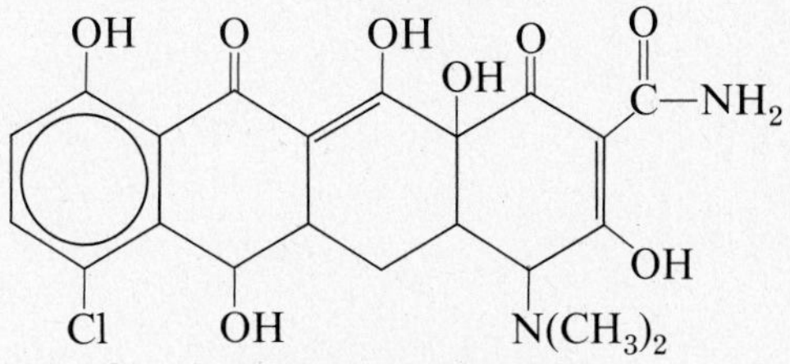

OH O OH O OH O C—NH_2 OH Cl OH $N(CH_3)_2$

demethylchlortetracycline
(Declomycin)

FIGURE 17-2 Structural formulas of selected tetracyclines.

Gram-positive and gram-negative refer to differences in the behavior of bacterial species to dyes used in staining techniques.

All of the tetracyclines are produced by various streptomyces. The tetracyclines are so-called broad spectrum antibiotics, meaning they are effective against a wide variety of microorganisms, including both gram-positive and gram-negative bacteria. They are usually prepared and administered as the hydrochloride salts.

Chloramphenicol is also produced by a streptomyces; however, its chemical structure differs considerably from the tetracyclines. It is a broad-spectrum antibiotic and is now produced synthetically.

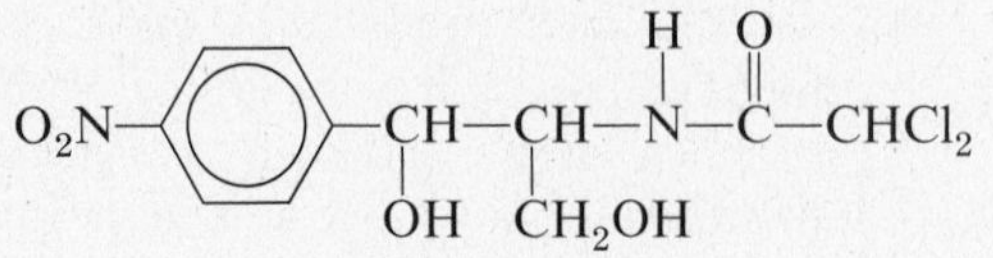

$$O_2N-C_6H_4-\underset{\underset{OH}{|}}{CH}-\underset{\underset{CH_2OH}{|}}{CH}-\overset{\overset{H}{|}}{N}-\overset{\overset{O}{\|}}{C}-CHCl_2$$

chloramphenicol

The use of chloramphenicol is somewhat restricted because of its serious and occasionally fatal side effects. Its primary use is against typhoid fever and in treatment of bacterial infections which are resistant to other antibiotics.

Sulfa drugs, while not antibiotics, are potent antibacterial compounds. Their structures and mode of action were discussed in Chapter 6.

Fungal infections are classified as superficial or deep. Superficial fungal infections are normally treated with various topical medications dispensed in the form of lotions, ointments, and powders. Fatty acids are widely used in the treatment of ringworm of the skin (athlete's foot). Usually a salt of the acid is used. For example, the best-selling medication for athlete's foot contains a mixture of undecylenic acid and zinc undecylenate.

exercise 17-3

The structural formula of undecylenic acid is $CH_2{=}CH(CH_2)_8CO_2H$. Write the formula for zinc undecylenate.

Exfoliation is the sloughing off of dead skin.

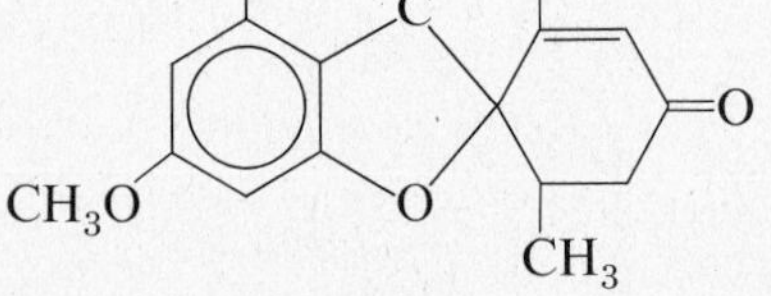

Griseofulvin is used for systemic treatment of superficial fungal infections. Administration of the drug is oral. Griseofulvin is an antibiotic having no antibacterial action. Because of possible serious side effects, griseofulvin is not prescribed unless topicial treatment fails. The substance is fungistatic rather than fungicidal. Its effectiveness results from arresting the growth of the fungus in order to permit the removal of the organism by exfoliation of the infected skin.

OCH₃ O OCH₃ C =O CH₃O O CH₃

griseofulvin

AUTONOMIC DRUGS

Autonomic drugs are those that either mimic or oppose the peripheral effects of nerve impulses from the autonomic (involuntary) nervous system. The autonomic drugs can be subdivided into four classes:

1. Adrenergic drugs
2. Cholinergic drugs
3. Adrenergic blocking drugs
4. Cholinergic blocking drugs

We will examine only the adrenergic agents and the cholinergic blocking drugs.

The adrenergic agents are largely aromatic compounds related to epinephrine. The characteristic structural formula of this class of compounds is:

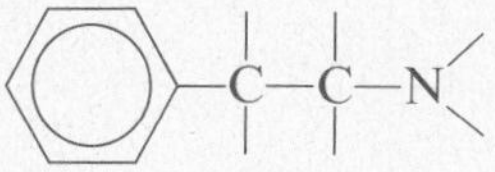

$$C_6H_5-\overset{|}{\underset{|}{C}}-\overset{|}{\underset{|}{C}}-N\langle$$

The differences between the various compounds are the groups attached either to the aromatic ring or to the side chain. The drugs are believed to act at the ganglia and effector cells rather than upon the nerve endings. The structural formulas of epinephrine, ephedrine, and several amphetamines are shown in Fig. 17-3.

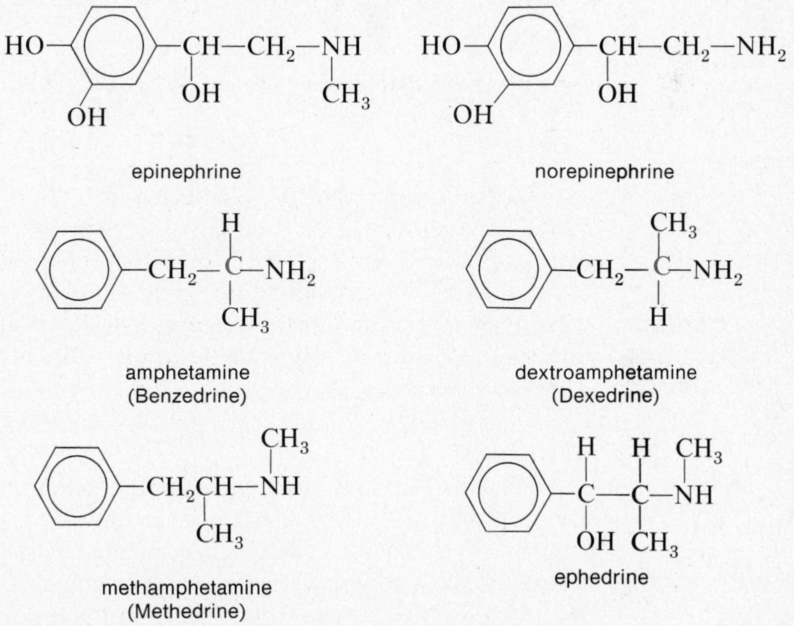

HO–(ring, OH)–CH(OH)–CH_2–NH(CH_3)

epinephrine

HO–(ring, OH)–CH(OH)–CH_2–NH_2

norepinephrine

(ring)–CH_2–C(H)(CH_3)–NH_2

amphetamine
(Benzedrine)

(ring)–CH_2–C(CH_3)(H)–NH_2

dextroamphetamine
(Dexedrine)

(ring)–CH_2CH(CH_3)–NH(CH_3)

methamphetamine
(Methedrine)

(ring)–C(H)(OH)–C(H)(CH_3)–NH(CH_3)

ephedrine

FIGURE 17-3 Structural formulas of selected adrenergic agents.

We can see from Fig. 17-3 how closely related the compounds are. Epinephrine and norepinephrine are substances normally produced by the body and were discussed in Chapter 15 on hormones. The main action of epinephrine is as a stimulant, acting primarily on the heart, blood vessels, and smooth muscles. It raises blood pressure by causing constriction of blood vessels.

Amphetamines, while not addicting, do seem to be habit-forming.

The remaining compounds in Fig. 17-3 are collectively known as amphetamines. In addition to displaying the properties of epinephrine they also stimulate the central nervous system. Notice that Benzedrine and Dexedrine are mirror images. Dexedrine is approximately four times as physiologically active as Benzedrine, once again pointing out how slight changes in chemical structure can cause dramatic biological effects. At one time, amphetamines were prescribed as aids in weight reduction; now, their use in this capacity has been greatly reduced since any weight loss was usually temporary.

Atropine is a cholinergic blocking (anticholinergic) agent. Its action is to block the effect of acetylcholine, thereby blocking nerve transmission. Under the influence of atropine, the heartbeat becomes rapid, smooth muscle tissue relaxes, and the eyes become dilated.

Atropine was originally isolated from *Atropa belladonna* or "deadly nightshade," a member of the potato family. The name was given the plant because of its poisonous characteristics and named after the oldest of the Greek fates—Atropos—who supposedly cut the thread of life. Belladonna means "beautiful lady" and became part of the name because women of ancient Rome put belladonna preparations in their eyes causing them to appear larger and more lustrous.

Chemically, atropine is classified as an alkaloid and has the structural formula:

$$\text{(N-CH}_3\text{ bicyclic)}-O-\overset{\overset{\displaystyle O}{\|}}{C}-\underset{\underset{\displaystyle H}{|}}{\overset{\overset{\displaystyle CH_2OH}{|}}{C}}-C_6H_5$$

atropine

CARDIOVASCULAR DRUGS

Among the oldest and most important drugs available to the medical profession are those of the digitalis family. They have been in use since the 1200's. The digitalis compounds are isolated from the leaves of *Digitalis purpurea* and *Digitalis lanta,* purple foxglove and white foxglove, respectively.

Digitalis is a mixture of compounds resulting from the combination of various steroids and sugars. The structural formulas of digoxin, acetyldigitoxin, procainamide hydrochloride, and reserpine are shown in Fig. 17-4. Acetyldigitoxin is a derivative of digoxin resulting from esterification of an OH group on the sugar portion of digoxin by acetic acid.

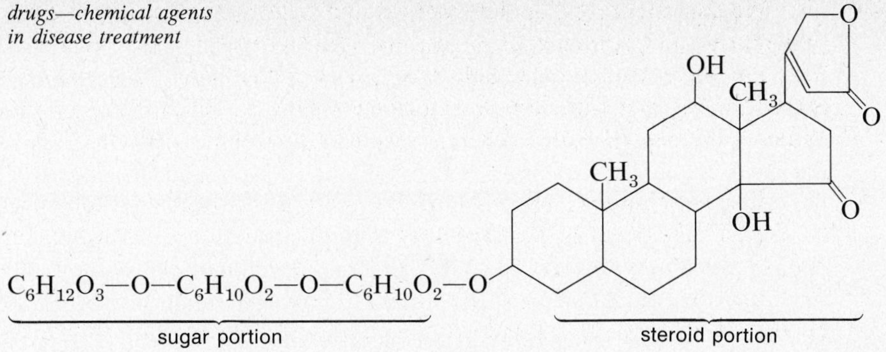

digoxin (Lanoxin)

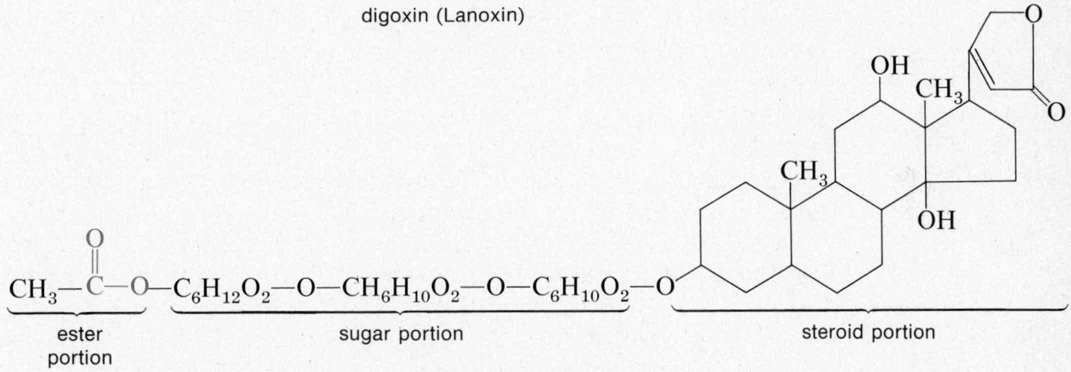

acetyldigitoxin (Acylanid)

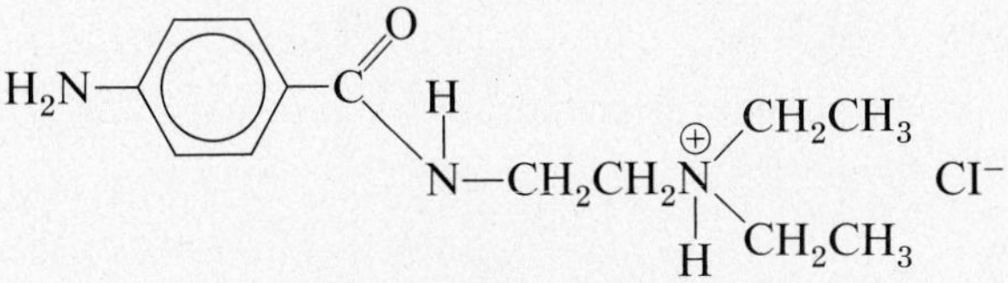

procainamide hydrochloride
(Pronestyl hydrochloride)

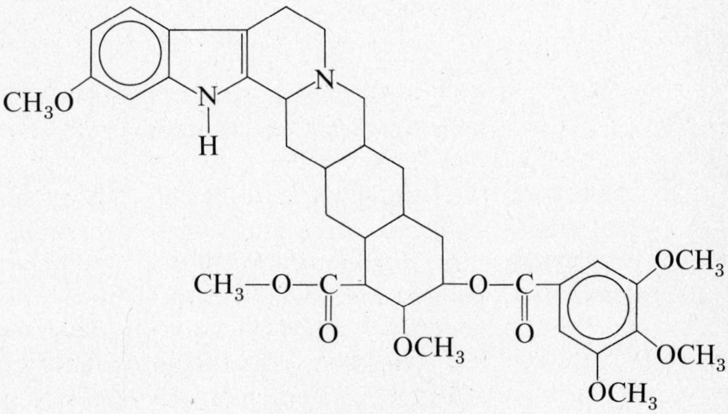

reserpine (Serpasil)

FIGURE 17-4 Structural formulas of selected cardiovascular drugs.

Digoxin (Lanoxin) and acetyldigitoxin (Acylanid) stimulate and improve the quality of heart action. They increase the contraction of a diseased heart thereby causing the ventricles to empty more completely. Procainamide hydrochloride is a heart depressant which slows the rate of heartbeat and makes it more regular. Reserpine is a alkaloid prepared from the root of the shrub *Rauwolfia serpentina*. It lowers blood pressure and is used to treat hypertension. Digoxin, acetyldigitoxin, and reserpine exert their action by decreasing peripheral vasoconstriction. Additionally, reserpine is thought to depress the hypothalamus, thus also affecting the autonomic system.

Let's compare the structural formula of procainamide hydrochloride to that of the local anesthetic, procaine hydrochloride (see Fig. 17-5). We

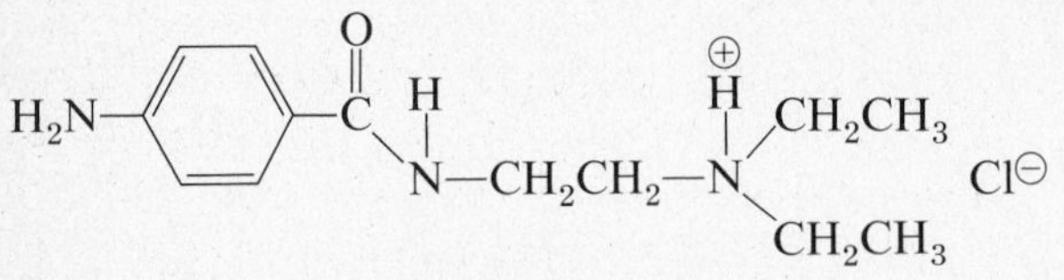

$$H_2N-C_6H_4-C(=O)-NH-CH_2CH_2-N^{\oplus}H(CH_2CH_3)_2 \quad Cl^{\ominus}$$

procainamide hydrochloride

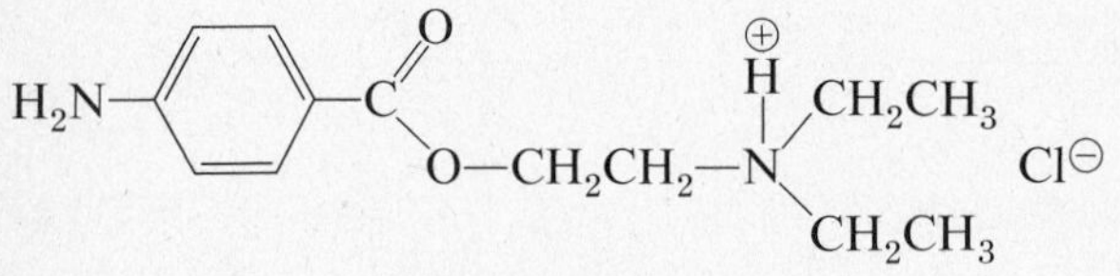

$$H_2N-C_6H_4-C(=O)-O-CH_2CH_2-N^{\oplus}H(CH_2CH_3)_2 \quad Cl^{\ominus}$$

procaine hydrochloride

FIGURE 17-5 Comparison of structural formulas of procainamide hydrochloride and procaine hydrochloride.

see once again how a small change in chemical structure can alter the physiological action of a compound.

exercise 17-4

(a) Describe in detail the difference in chemical structure between procaine hydrochloride and procainamide hydrochloride. (b) Write equations using structural formulas for the acidic hydrolysis of procaine hydrochloride and procainamide hydrochloride.

Procainamide is only about one-third as toxic as procaine; it does not have significant effects on the central nervous system.

CENTRAL NERVOUS SYSTEM DEPRESSANTS

The activity of nerve centers and conducting pathways can be depressed by drugs. Those drugs whose depressant action affects either the brain or spinal cord are classified as central nervous system depressants. The compounds are subclassified according to their action as analgesics, general anesthetics, anticonvulsants, hypnotics, sedatives, and tranquilizers.

Analgesics are substances which act to relieve pain without inducing a loss of consciousness. An ideal analgesic would relieve pain without producing any other change in the central nervous system. But such an analgesic has not yet been discovered; all analgesics produce varying degrees of change in behavior and consciousness. Analgesics are divided according to whether they are addicting or nonaddicting. Generally, the more effective the drug is in ameliorating pain, the more addicting it is likely to be. Nonaddicting analgesics raise the pain threshold by depressing the thalmus, thereby slowing the flow of impulses to the cerebral cortex. The addicting drugs not only raise the pain threshold but also change the person's reaction to pain by suppressing the anxiety associated with pain. Current evidence indicates anxiety suppression and addiction are interlinked.

Addicting analgesics are called narcotics and are either (a) naturally occurring compounds generally derived from opium alkaloids or (b) synthetic compounds with chemical structures related to those of the opium alkaloids. Opium is the dried juice of unripe seed capsules of a poppy native to Asia. Its effect on man has been known for over 2000 years. Opium contains over twenty compounds, three of which occur in high concentration and are used medically. These three are morphine, codeine, and papaverine. Codeine differs from morphine only in that one alcohol group of morphine has been converted to an ether. Papaverine, with a considerably different chemical structure, has little or no effect on the central nervous system but does produce relaxation of smooth muscle tissue.

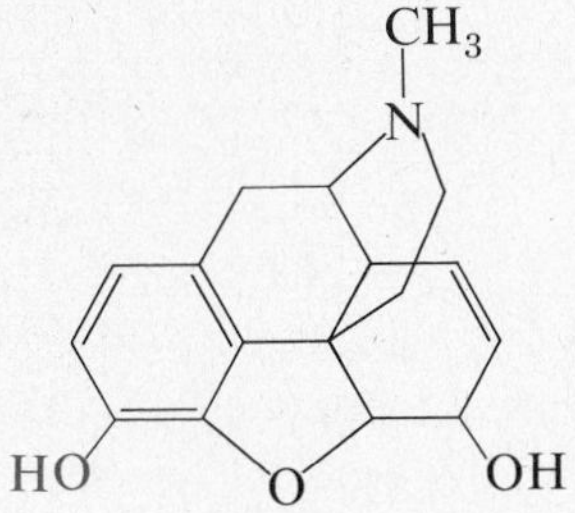

CH_3 N HO O OH

morphine

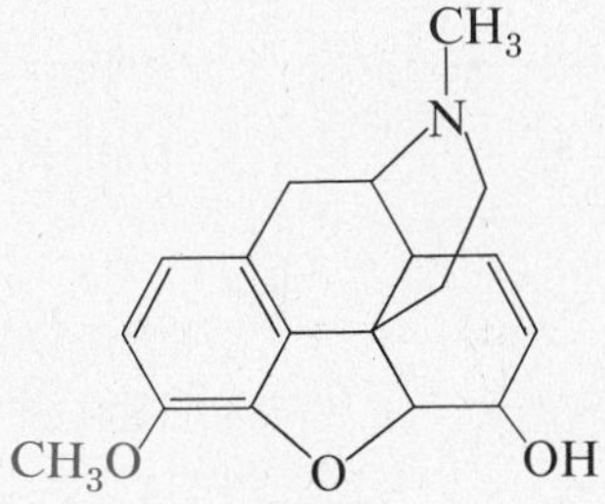

CH_3 N CH_3O O OH

codeine

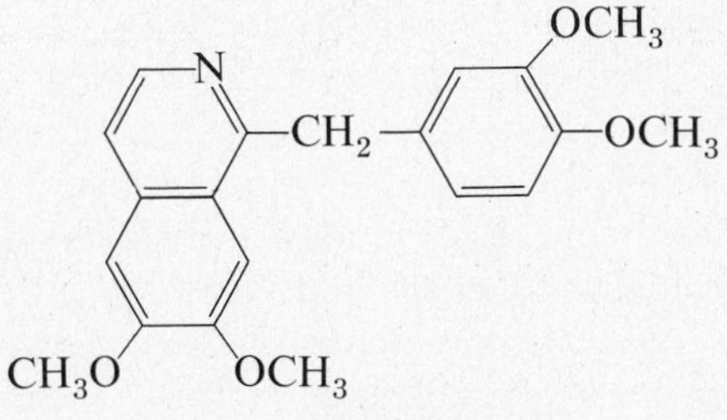

N CH_2 OCH_3 OCH_3 CH_3O OCH_3

papaverine

Morphine was isolated from opium in 1805 by the German druggist and chemist Frederich Sertürner. Throughout the 1800's opium, codeine, and morphine were used in patent medicine cough syrups. They are still used in cough syrups today but are controlled by narcotic laws, and most preparations containing them are legally available only by prescription. Codeine is less potent than morphine and is also less addicting. Accordingly, codeine, in amounts less than 2.2 mg per ml, is exempt from narcotic laws.

Morphine will react with acetic anhydride to produce the diacetyl derivative. The diacetyl of morphine is known as heroin. Interestingly, in the 1890's heroin was proposed as a treatment for morphine addiction. It was quickly discovered, however, that heroin treatment of a morphine addict did not produce the desired effect. Heroin produces addiction faster than morphine and its addiction is harder to cure.

exercise 17-5

Using the structural formulas of morphine and of acetic anhydride (see Chapter 10), write a balanced equation showing the formation and structural formula of heroin.

Meperidine (Demerol) is a synthetic compound used as a substitute for morphine. It is considerably less effective than morphine and is addicting. The advantages of meperidine are its lesser tendency to produce nausea, sleep, and respiratory depression as compared to morphine.

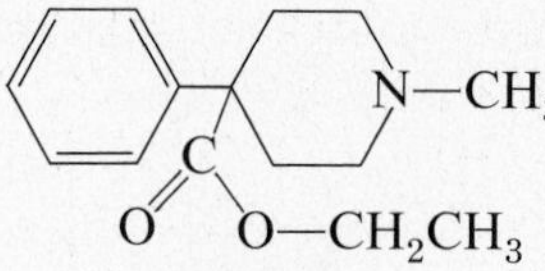

N—CH_3
C
O   O—CH_2CH_3

meperidine
(Demerol)

There are many nonaddicting analgesics. The oldest and most widely used are the salicylates. Acetylsalicylic acid (aspirin) is probably the most common drug used throughout the world. In addition to their analgesic effect, the salicylates exhibit an antipyretic effect. The antipyretic effect is apparently produced by increasing the elimination of heat rather than decreasing the metabolic rate.

Propoxyphene hydrochloride, sold as Darvon, is a synthetic compound used when, for various reasons, it is undesirable to administer either aspirin or codeine. It does not exhibit antipyretic action and most studies indicate it is no more effective than aspirin. Its advantage is the minimizing of the side effects of nausea and vomiting.

General anesthetics produce a reversible loss of consciousness and sensation accompanied by relaxation of muscles. Very little is known about the chemistry of the action of these drugs on the central nervous system. One suggestion has been that they inhibit the enzymes involved in oxidation, thus slowing the metabolic rate.

Substances used as anesthetics include diethyl ether, nitrous oxide (N_2O), cyclopropane, halogenated hydrocarbons, and thiopental sodium (Sodium Pentothal). Thiopental is given intravenously; the others are administered by inhalation. Access to oxygen must always be provided with anesthetics administered by inhalation. This can be done by merely dropping the liquid anesthetic on a gauze covering the patient's nose or mouth. However, this method presents difficulty in controlling the anesthetic state and allows the vapor to escape into the room, creating a safety hazard. The safety problem arises because almost all the gaseous or volatile liquid anesthetics form explosive mixtures with air. The more common and safer method of inhalation anesthesia is the use of an anesthetic machine which does not allow the gaseous mixture to escape into the room. This method requires a means of removing carbon dioxide and water from the mixture since the patient will be rebreathing the oxygen–anesthestic mixture.

Included in the drugs affecting the central nervous system are those which will control various kinds of convulsive seizures. The most common type of seizure is the epileptic seizure. There are over one million epileptic persons in the United States. This affliction is believed to be a symptom of a brain disease and not the disease itself. Distinctive changes in the electrical activity of the cerebral cortex are associated with the disorder and can be detected by an electroencephalogram (EEG). Anticonvulsive drugs' mode of action is largely unknown. They suppress the seizure but may or may not alter the abnormal brain waves. The drugs frequently used in treatment are closely related and known as hydantoins. Three such drugs are shown in Fig. 17-6.

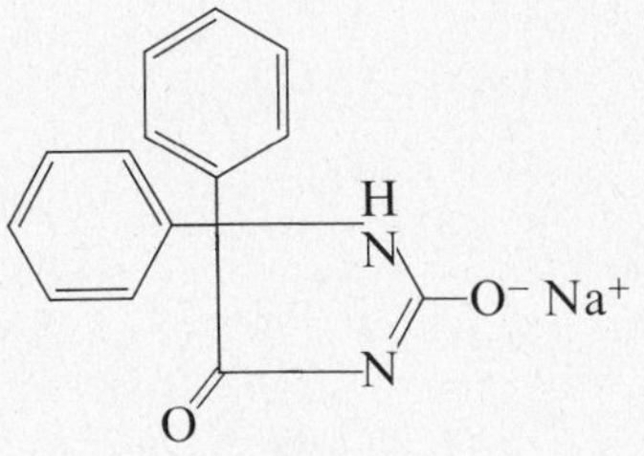

H N N O O^- Na^+

diphenylhydantoin sodium
(Dilantin)

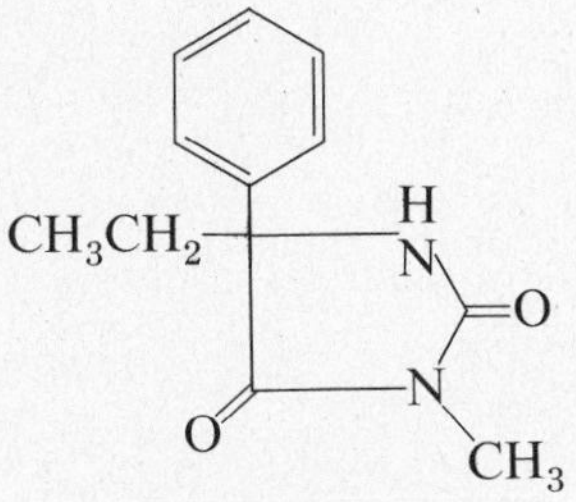

CH_3CH_2 H N O N O CH_3

methylphenylethylhydantoin
(Mesantoin)

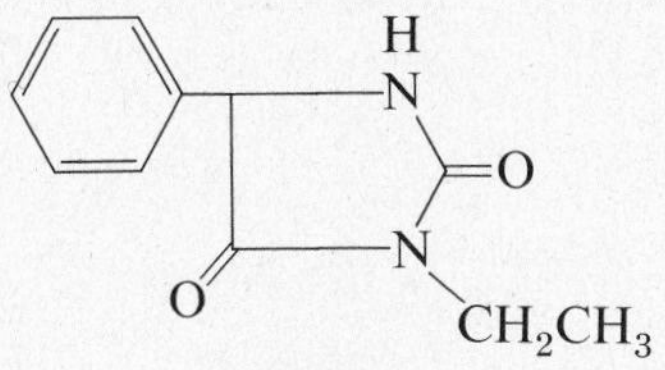

H N O N O CH_2CH_3

ethotoin
(Peganone)

FIGURE 17-6 Drugs useful in the control of epileptic seizures.

Hypnotics and sedatives are a class of drugs used primarily for the relief of anxiety. They are, however, capable of producing depression and in large dosage cause anesthesia or even coma and death. The difference between a hypnotic and a sedative is the degree of action. Hypnotics are normally administered around bedtime and produce sleep very soon after administration. A sedative is given several times throughout the day and produces a calming, relaxing effect. A particular drug may serve both purposes; it may be given in large doses as a hypnotic or in smaller doses as a sedative. The term soporific is synonymous with hypnotic. Hypnotics and sedatives are divided into two classes: barbiturates and nonbarbiturates.

The barbiturates are derivatives of barbituric acid. Barbiturates are among the oldest synthetic drugs and are widely used. Over a million pounds of barbiturates are produced annually in the U.S.A. They are synthesized from urea and substituted malonic acids which react to form cyclic amides.

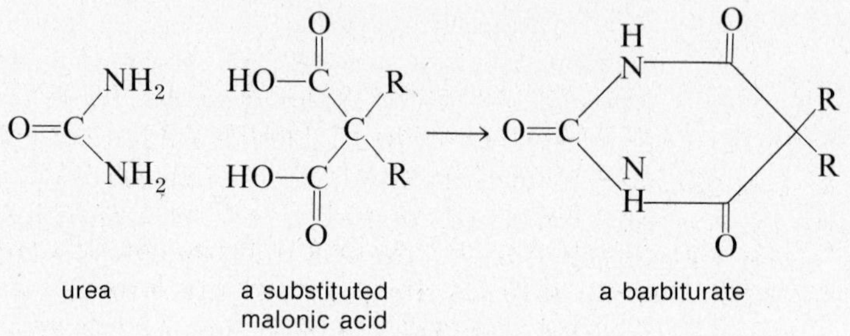

urea + a substituted malonic acid ⟶ a barbiturate

Several common barbiturates are shown in Fig. 17-7.

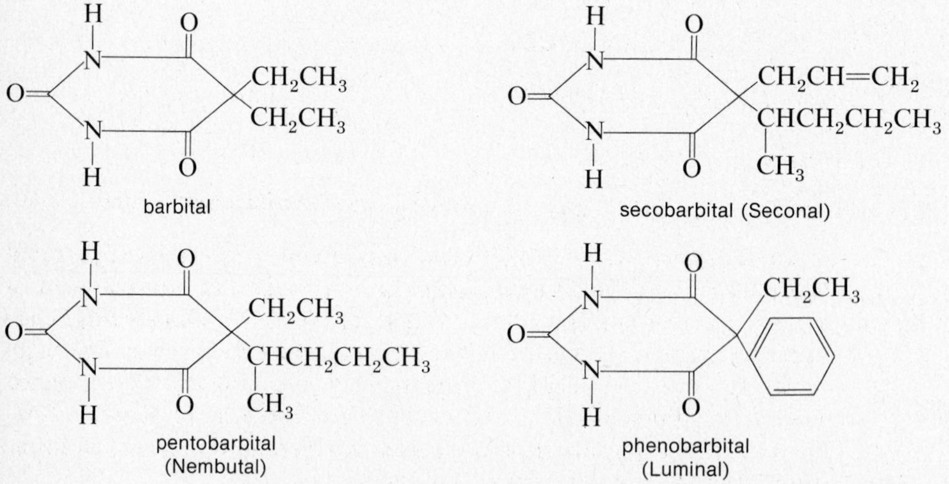

barbital ($R = CH_2CH_3$, CH_2CH_3)

secobarbital (Seconal) ($R = CH_2CH{=}CH_2$, $CH(CH_3)CH_2CH_2CH_3$)

pentobarbital (Nembutal) ($R = CH_2CH_3$, $CH(CH_3)CH_2CH_2CH_3$)

phenobarbital (Luminal) ($R = CH_2CH_3$, phenyl)

FIGURE 17-7 Structural formulas of medically useful barbiturates.

exercise 17-6

Write the structural formulas of the specific substituted malonic acids used in the synthesis of each of the barbiturates shown in Fig. 17-7.

The barbiturates generally depress the central nervous system, thereby inducing sedation. However, they are not useful as analgesics or anesthetics.

They appear to act selectively on the brain stem by diminishing the flow of nerve impulses to the cerebral cortex. Because of their mode of action, barbiturates are also useful as anticonvulsants.

The barbiturates are habit-forming. For this reason there is a continuing search for nonaddicting, nonbarbiturate hypnotics and sedatives. A number of nonbarbiturate hypnotics are very old drugs whose use declined when the barbiturates were first discovered. The use of older compounds has been revived in recent years. One of these older drugs is chloral hydrate.

The term synergistic *is used to describe the enhancement of two drugs' actions by each other. The combination of chloral hydrate and ethyl alcohol is said to produce a synergistic effect.*

Chloral hydrate ($CCl_3C(OH)(H)\text{—}OH$) acts within 10–15 minutes to produce sleep very similar to natural sleep and lasting 5 or more hours. It has little or no analgesic, anticonvulsant, or anesthetic effect, and the patient can be awakened easily. It is usually administered as a capsule or dissolved in a pleasant-tasting solution. It should never be given with or shortly after the ingestion of alcohol. The combination of chloral hydrate with alcohol is referred to as a "Mickey Finn" or knockout drops.

Among the newer nonbarbiturates are ethchlorvynol and ethinamate, whose structural formulas are shown in Fig. 17-8.

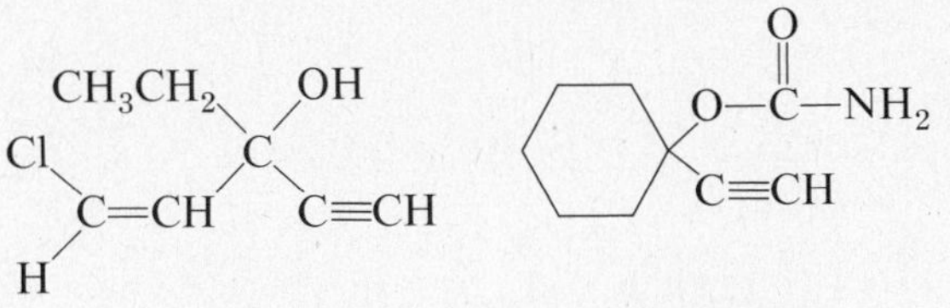

CH_3CH_2 OH
Cl C
C=CH C≡CH
H

ethchlorvynol
(Placidyl)

O
O—C—NH_2
C≡CH

ethinamate
(Valmid)

FIGURE 17-8 Nonbarbiturate hypnotics.

Tranquilizers are divided into two classes, minor and major. The minor tranquilizers are used primarily for their calming effect in treatment of anxiety and tension. They differ from sedatives in that a dosage sufficient to exert a calming influence does not normally induce drowsiness. The major tranquilizers will relieve psychotic symptoms whereas the minor ones cannot. They are used clinically to treat acute and chronic psychoses. However, they do not cure mental illness; in some way they protect the patient from the stimuli causing excessive emotional behavior.

The chemical action of both minor and major tranquilizers is largely unknown. Considering the large variation in the chemical structures of various tranquilizers it is unlikely that they all exert their influence by the same mode of action. Some compounds used for their tranquilizing effect are also used clinically for other purposes. One example of such a compound is reserpine, discussed earlier as a drug used in the treatment of cardiovascular disorders. Figure 17-9 depicts the chemical structures of several tranquilizers.

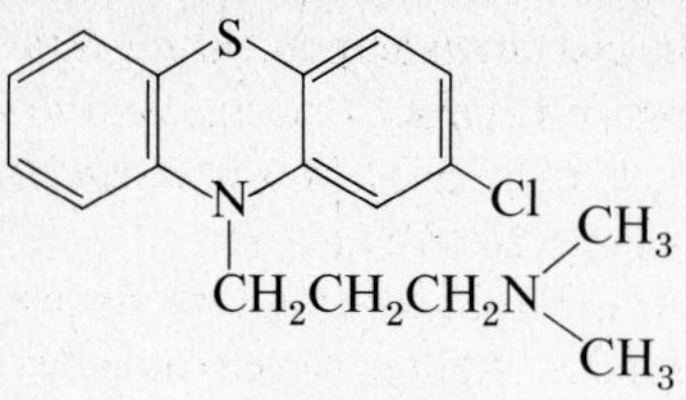

chlorpromazine
(Thorazine)

(a)

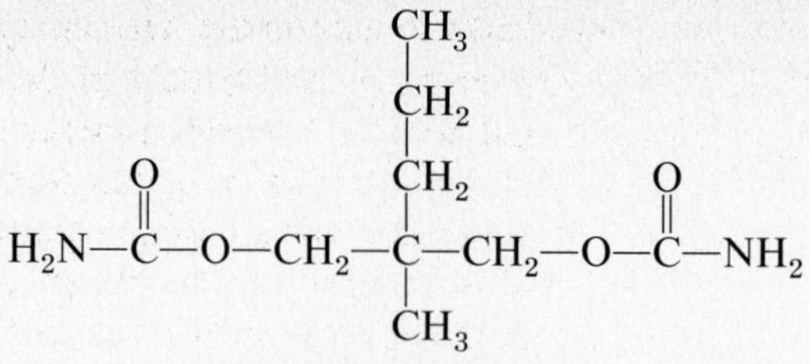

meprobamate
(Equanil, Miltown)

(c)

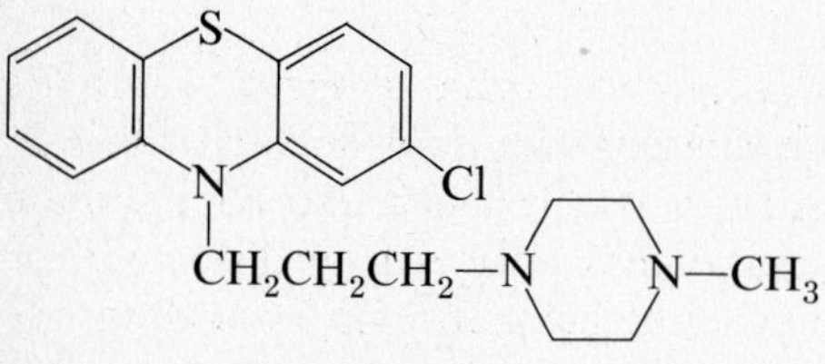

prochlorperazine
(Compazine)

(b)

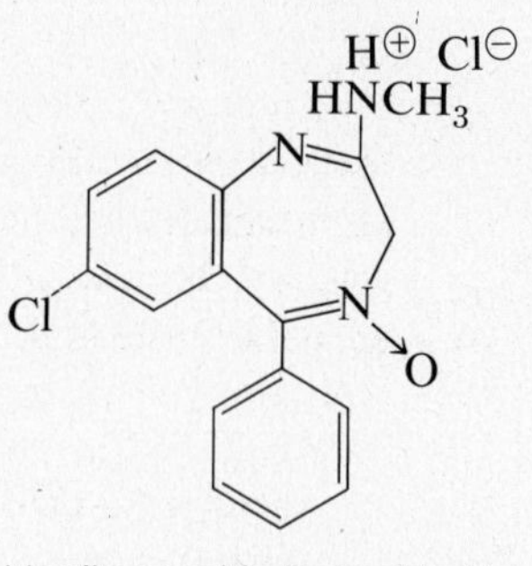

chlordiazepoxide hydrochloride
(Librium)

(d)

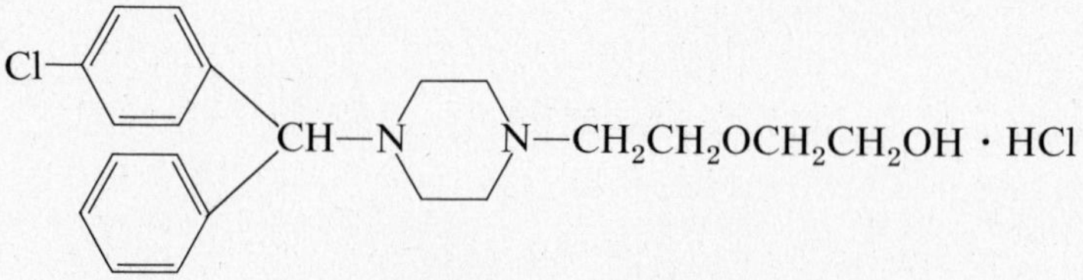

hydroxyzine hydrochloride
(Atarax Hydrochloride, Vistaril Parenteral)

(e)

FIGURE 17-9 Structural formulas of selected tranquilizers. (a) and (b) are major tranquilizers; (c), (d), and (e) are minor tranquilizers.

CENTRAL NERVOUS SYSTEM STIMULANTS

Drugs employed to stimulate the central nervous system may be classified as analeptics, direct stimulants, and indirect stimulants. Analeptics are those used primarily to counteract drug-induced respiratory depression. These compounds stimulate the entire nervous system and may cause muscular twitching or even, in cases of overdosage, convulsions.

Direct stimulants exert their principal effect on the central nervous system rather than on the autonomic nervous system. However, some, such as the amphetamines, have a pronounced effect on both the central nervous system and the autonomic nervous system. Amphetamines suppress appetite and have been used widely in the treatment and management of obesity. The relationship between appetite suppression and central nervous system stimulus has not been explained adequately. It is thought by some researchers to result from modifying the behavior of the hypothalamus.

In contrast to the direct stimulants, the indirect stimulants apparently do not cause direct excitation of neural cells. At one time it was thought these drugs behaved as monoamine oxidase inhibitors, but it has since been shown that many indirect stimulants do not affect the monoamine oxidase activity. The basic mode of action of these compounds remains unknown. The major use of indirect stimulants is in the clinical treatment of emotional depression.

The most commonly used central nervous system stimulant is almost exclusively administered outside the clinical setting. This self-administered stimulant is caffeine, found in coffee, tea, and cola beverages. Caffeine is one of a family of alkaloids known as xanthines. The basic xanthine structure and the structural formula of caffeine are shown in Fig. 17-10. Caffeine

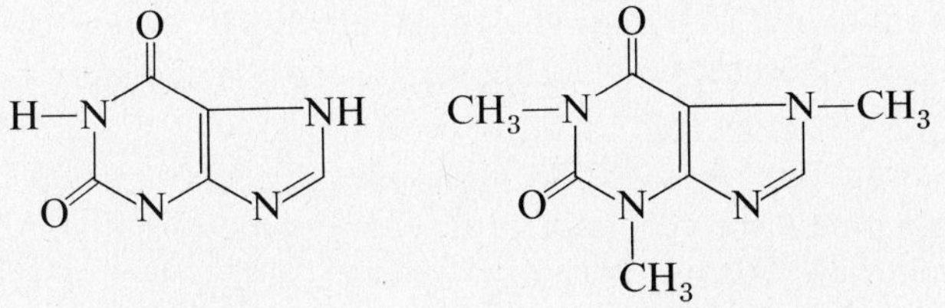

xanthine

caffeine (1,3,7,-trimethylxanthine)

FIGURE 17-10 Structural formulas of xanthine and caffeine.

stimulates not only the central nervous system, but also has a stimulatory effect on the kidneys, heart, skeletal muscles, and smooth muscles. Although caffeine is apparently a mild and safe drug, some recent evidence suggests it may cause chromosome damage, and it has been suggested that people in their childbearing ages should avoid large quantities of caffeine. Interestingly, caffeine has for many years been incorporated into aspirin-based preparations for fever reduction. Evidence gathered in the past several years indicates caffeine actually decreases the effectiveness of aspirin in controlling fever.

OTHER DRUG CLASSES

In addition to the drug classes already discussed, Table 17-2 also lists enzymes, gastrointestinal agents, and hormones. These types of compounds have been discussed throughout the text and further discussion of them is not included in this chapter. Those interested in pursuing further the aforementioned classes of drugs or the several drug classes omitted from Table 17-2 are encouraged to consult a pharmacology text.

CHAPTER SUMMARY

1. All drugs are chemical compounds and their effects are the result of their chemical interactions with enzymes, hormones, and other body chemicals.
2. Drugs are classified according to source, chemical structure, or clinical effect. Until a greater understanding of drug action is achieved, classification based on clinical applications is presently the most useful.
3. Local anesthetics are normally administered in concentrations which will block nerve impulses along sensory nerve fibers but not along the motor nerve fibers. This causes a loss of sensation but not paralysis.
4. Antihistamines block the action of histamine and are useful in treatment of allergies, particularly nasal allergies. A useful side effect is their ability to relieve motion sickness symptoms.
5. Anti-infectives prevent the growth of microorganisms. The majority of anti-infectives are antibacterial agents. Antibiotics are anti-infectives which are produced by microorganisms.
6. Autonomic drugs affect the autonomic (involuntary) nervous system. These drugs are subdivided into four classes: (a) adrenergic drugs, (b) cholinergic drugs, (c) adrenergic blocking drugs, and (d) cholinergic blocking drugs.
7. Cardiovascular drugs are those compounds affecting heart action. Digitalis and reserpine are compounds produced by plants and have been used to treat heart disease for many years.
8. Central nervous system depressants exert a depressant effect on the brain or spinal cord. Subclasses of these drugs are: (a) analgesics, (b) general anesthetics, (c) anticonvulsants, (d) hypnotics, (e) sedatives, and (f) tranquilizers.

QUESTIONS

1. Write a balanced equation for the preparation of the potassium salt of penicillin G starting with pure penicillin G.
2. Identify the asymmetric carbon atom(s) in Benzedrine and Dexedrine and show by appropriate drawings that these two compounds are mirror images.
3. Codeine preparations in concentrations less than 2.2 mg/ml are exempt from narcotic laws. Calculate the maximum percent concentration that a codeine solution may have to be exempt from these laws.
4. Using structural formulas, write the chemical equation for the preparation of benzocaine from *para*-aminobenzoic acid and ethanol. Identify the functional groups present in benzocaine. Write balanced equa-

tions for the formation of benzocaine hydrochloride and cyclomethylcaine sulfate from benzocaine and cyclomethylcaine, respectively.

5. Using structural formulas, write an equation for the acid hydrolysis of phenethicillin.

appendix one mathematics review

This section is a review of common mathematical operations used in text material exercises.

FRACTIONS A fraction is a ratio of two numbers, the numerator and denominator, e.g.,

$$\frac{120}{2} \begin{array}{l} \longleftarrow \text{numerator} \\ \longleftarrow \text{denominator} \end{array}$$

The value of a fraction is determined by dividing the denominator into the numerator. The answer is called the *quotient*. The meaning of the fraction above can be expressed

$$\begin{array}{r} \text{numerator} \longrightarrow \\ \text{denominator} \longrightarrow \end{array} \frac{120}{2} = 120 \div 2 = 2\overline{)\,120} = 60 \longleftarrow \text{quotient}$$

When the numerator is greater than the denominator, e.g.,

$$\tfrac{100}{50} = 2,$$

the quotient is greater than 1. When the numerator equals the denominator, the quotient is 1:

$$\tfrac{4}{4} = 1 \qquad \tfrac{1200}{1200} = 1$$

When the numerator is less than the denominator, the quotient is less than 1:

$$\tfrac{2}{4} = 0.5 \qquad \tfrac{15}{60} = 0.25$$

Fractions in which the denominator is greater than the numerator can be converted into decimal values by performing the division indicated; $\tfrac{15}{60}$ is such a fraction. Converting this to decimal form is done by

$$\begin{array}{r} 0.25 \\ 60\overline{)\,15.00} \\ \underline{12\;0} \\ 300 \\ \underline{300} \end{array} = 0.25$$

Similarly:

$$\tfrac{208}{520} = \begin{array}{r} 0.40 \\ 520\overline{)\,208.00} \\ \underline{208\;0} \\ 0 \\ \underline{0} \end{array} = 0.40$$

exercise A1-1

Perform the following operations:

(a) $75 \div 25 =$ **(d)** $\dfrac{240}{10} =$

(b) $12 \div 60 =$ **(e)** $\dfrac{25}{75} =$

(c) $\dfrac{180}{60} =$ **(f)** $\dfrac{20}{100} =$

COMPLEX FRACTIONS

A complex fraction is a combination of several simpler fractions. An example of a complex fraction is

$$\frac{6 \times 4 \times 8}{3 \times 12 \times 5}$$

This expression is the product of three simpler fractions:

$$\frac{6}{3} \times \frac{4}{12} \times \frac{8}{5} = \frac{6 \times 4 \times 8}{3 \times 12 \times 5}$$

In the general case,

$$a \times \frac{b}{c} \times \frac{d}{e} \times \frac{\cdots}{\cdots} = \frac{a(b)(d)(\cdots)}{c(e)(\cdots)}$$

Thus, rather than performing the operations of each of the simpler fractions, the numerator terms are multiplied to give a single product; the denominator terms are multiplied to give a single term; the denominator is then divided into the numerator. From the example above,

$$\frac{6}{3} \times \frac{4}{12} \times \frac{8}{5} = \frac{6 \times 4 \times 8}{3 \times 12 \times 5} = \frac{192}{180} = 1.07$$

EXAMPLE A1-1

$$6 \times \frac{28}{16} \times \frac{4}{50} = \frac{6 \times 28 \times 4}{16 \times 15} = \frac{672}{800} = 0.84$$

exercise A1-2

Solve each of the following:

(a) $120 \times \dfrac{36}{4} =$ **(d)** $\dfrac{2}{5} \times \dfrac{19}{9} =$ **(g)** $\dfrac{48 \times 3}{6 \times 40 \times 2} =$

(b) $750 \times \dfrac{10}{25} =$ **(e)** $14 \times \dfrac{3}{12} \times \dfrac{4}{12} =$

(c) $\dfrac{840}{16} \times 3 =$ **(f)** $\dfrac{75 \times 3}{15 \times 2 \times 25} =$

EXPONENTS AND EXPONENTIAL NOTATION

Exponents and exponential notation are condensed ways to express large and small numbers. An exponent is a superscript that follows another number. The exponent indicates how many times the preceding number is to be multiplied by itself.

10^2 (read *ten squared*) $= 10 \times 10 = 100$

10^4 (read *ten to the fourth power*) $= 10 \times 10 \times 10 \times 10 = 10{,}000$

A number expressed in standard exponential form has only one number to the left of the decimal point. Exponential form is also called scientific notation. Since exponential notation depends on understanding values of powers of ten, the following table summarizes these relations:

NUMBER	EXPONENTIALLY EXPRESSED	
1,000,000	1.0×10^6	
10,000	1.0×10^4	↓
1000	1.0×10^3	decreasing
100	1.0×10^2	magnitude
10	1.0×10^1	
1	1.0×10^0	
0.1	1.0×10^{-1}	
0.01	1.0×10^{-2}	
0.001	1.0×10^{-3}	
0.0001	1.0×10^{-4}	
0.0000001	1.0×10^{-7}	↓

Note that for positive exponents a *higher-numbered exponent* indicates an expression of *larger value;* for *negative exponents,* as the *exponent increases,* the expression's *numerical value decreases.* For example, $1 \times 10^4 > 1 \times 10^3$ but $1 \times 10^{-4} < 1 \times 10^{-3}$ ($>$ represents *greater than;* $<$ represents *less than*).

Numbers expressed exponentially can be considered as the product of two numbers, one of which is a power of ten. The other number is written to the left of the power of ten term. When written in proper exponential form, this number has only one figure to the left of the decimal point.

$$300 = 3 \times 10 \times 10 = 3.0 \times 10^2$$

$$4000 = 4 \times 10 \times 10 \times 10 = 4.0 \times 10^3$$

$$800{,}000 = 8 \times 10 \times 10 \times 10 \times 10 \times 10 = 8.0 \times 10^5$$

300 is converted to exponential form by moving the decimal point two places to the left; 3.0.0. The original number is 300, not 3.0. To make the exponential form value equal to the original value, 3.0 is multiplied by 10 raised to the correct exponent. The exponent is determined by counting the number

of decimal places that the decimal point was moved. If the decimal was moved to the left, the exponent is a positive number; if the decimal was moved to the right, the exponent is a negative number:

$$4500. = 4.5\,0\,0. = 4.5 \times 10^3$$
$$71{,}000. = 7.1 \times 10^4$$
$$275. = 2.75 \times 10^2$$
$$0.041 = 0.0\,4.1 = 4.1 \times 10^{-2}$$
$$0.00067 = 6.7 \times 10^{-4}$$
$$0.82 = 8.2 \times 10^{-1}$$

exercise A1-3

Express the following in correct exponential form: (a) 2000; (b) 0.002; (c) 0.5; (d) 50; (e) 6,023,000,000; (f) 0.000089; (g) 6; (h) 275,000

There are times when a number expressed in standard exponential form has to be converted to another exponential form. This is done by moving the decimal point and correspondingly changing the exponent.

The number 2000 can be expressed several ways:

$$2000 = 2000. \times 10^0$$
$$200.0 \times 10^1$$
$$20.0 \times 10^2$$
$$2.0 \times 10^3$$

as this decreases, the power of 10 increases

For numbers less than 1, a similar technique is used, but with negative exponents the larger the exponent, the smaller its numerical value; i.e., $10^{-5} < 10^{-4}$ and $10^{-3} < 10^{-2}$. The number 0.0052 can be expressed

$$0.0052 = 0.52 \times 10^{-2}$$
$$= 5.2 \times 10^{-3}$$
$$= 52 \times 10^{-4}$$

as this increases this decreases

$$6.4 \times 10^{-5} = 64 \times 10^{-?}$$

What is the exponent required? Changing 6.4 to 64 corresponds to multiplying by 10; thus, the exponent of 10 must be decreased by 1. Therefore, 10^{-5} is reduced to 10^{-6} and $6.4 \times 10^{-5} = 64 \times 10^{-6}$. Note in the previous example that $5.2 \times 10^{-3} = 52 \times 10^{-4}$.

exercise A1-4

Express the following in standard exponential form (one number to the left of the decimal point): (a) 2700; (b) 0.00067; (c) 84×10^6; (d) 84×10^{-6}; (e) 416×10^5; (f) 57.3×10^{-4}; (g) 0.375×10^5; (h) 0.014

MULTIPLICATION USING EXPONENTIAL NOTATION

When multiplying exponential numbers, the numerical portions are multiplied and the exponents are added algebraically.

$$100 \times 10{,}000 = 1{,}000{,}000$$

$$\underset{\text{numerical portion}}{(1.0} \times 10^2) \times \underset{\text{numerical portion}}{(1.0} \times 10^4) = 1.0 \times 10^{(2+4)} = 1.0 \times 10^6$$

$$0.001 \times 0.01 = 0.00001$$

$$(1.0 \times 10^{-3}) \times (1.0 \times 10^{-2}) = 1.0 \times 10^{(-3)+(-2)} = 1.0 \times 10^{-5}$$

$$(1.0 \times 10^4) \times (6 \times 10^3) = (1.0 \times 6) \times 10^{(4+3)} = 6.0 \times 10^7$$

$$(3.2 \times 10^3) \times (2.2 \times 10^2) = (3.2 \times 2.2) \times 10^{(3+2)} = 7.04 \times 10^5$$

$$(4.0 \times 10^{-7}) \times (1.5 \times 10^{-4}) = (4.0 \times 1.5) \times 10^{(-7)+(-4)} = 6.0 \times 10^{-11}$$

$$(6.02 \times 10^{23}) \times (1.4 \times 10^{-18}) = (6.02 \times 1.4) \times 10^{23+(-18)} = 8.4 \times 10^5$$

$$(6.0 \times 10^{23}) \times (5.0 \times 10^{18}) = (6.0 \times 5.0) \times 10^{(23+18)} = 30 \times 10^{41} = 3.0 \times 10^{42}$$

Note the conversion of the answer to standard exponential form in the last example.

exercise A1-5

Perform each of the indicated operations. Express the final answer in standard exponential form.

(a) $(4 \times 10^3) \times (2 \times 10^2) =$
(b) $(2.3 \times 10^5) \times (4 \times 10^6) =$
(c) $(4 \times 10^7) \times (7 \times 10^9) =$
(d) $(1.6 \times 10^{-4}) \times (5 \times 10^{-6}) =$
(e) $(9 \times 10^{-2}) \times (5 \times 10^{-9}) =$
(f) $(7 \times 10^5) \times (8 \times 10^{-2}) =$
(g) $(3 \times 10^{-8}) \times (2.5 \times 10^4) =$
(h) $(24 \times 10^{-9}) \times (5 \times 10^4) =$
(i) $(72 \times 10^2) \times (0.052) =$
(j) $(16 \times 10^3) \times (5 \times 10^{-3}) =$

DIVISION USING EXPONENTIAL NOTATION

In dividing numbers expressed exponentially, the numerical portions are divided and the exponent of the divisor is algebraically subtracted from the exponent of the numerator.

$$10{,}000 \div 100 = \frac{10{,}000}{100} = 100$$

$$1.0 \times 10^4 \div 1.0 \times 10^2 = \frac{1.0 \times 10^4}{1.0 \times 10^2} = 1.0 \times 10^{(4-2)} = 1.0 \times 10^2$$

$0.001 \div 0.01 =$

$$1.0 \times 10^{-3} \div 1.0 \times 10^{-2} = \frac{1.0 \times 10^{-3}}{1.0 \times 10^{-2}} = 1.0 \times 10^{(-3)-(-2)}$$

$$= 1.0 \times 10^{-1}$$

$$\frac{8 \times 10^5}{4 \times 10^2} = \frac{8}{4} \times \frac{10^5}{10^2} = 2 \times 10^{5-2} = 2 \times 10^3$$

$$\frac{12 \times 10^4}{3 \times 10^{-1}} = \frac{12}{3} \times \frac{10^4}{10^{-1}} = 4 \times 10^{4-(-1)} = 4 \times 10^5$$

$$\frac{27 \times 10^{-16}}{9 \times 10^{-4}} = \frac{27}{9} \times \frac{10^{-16}}{10^{-4}} = 3 \times 10^{-16-(-4)} = 3 \times 10^{-12}$$

exercise A1-6

Do each of the following. Express the final answer in standard exponential form.

(a) $\frac{9 \times 10^4}{3 \times 10^2} =$

(b) $\frac{8 \times 10^6}{2 \times 10^3} =$

(c) $\frac{54 \times 10^{15}}{9 \times 10^7} =$

(d) $\frac{81 \times 10^{-4}}{27 \times 10^{-2}} =$

(e) $\frac{9.3 \times 10^{-8}}{3.0 \times 10^{-3}} =$

(f) $\frac{8 \times 10^{-6}}{2 \times 10^{-3}} =$

(g) $\frac{12 \times 10^{-12}}{4 \times 10^{-7}} =$

(h) $\frac{27 \times 10^4}{9 \times 10^{-6}} =$

(i) $\frac{60 \times 10^8}{5 \times 10^{-15}} =$

(j) $\frac{9 \times 10^9}{0.6 \times 10^{18}} =$

(k) $\frac{15 \times 10^{-5}}{5 \times 10^{-8}} =$

(l) $\frac{24 \times 10^5}{6 \times 10^{23}} =$

(m) $\frac{72 \times 10^5}{0.08} =$

(n) $\frac{18 \times 10^4}{3 \times 10^{-4}} =$

Complex fractions involving numbers expressed exponentially are solved by following the rules for multiplication and division of exponential notation.

EXAMPLE A1-2

$$\frac{(3 \times 10^5)(12 \times 10^4)}{6 \times 10^3} = \frac{36 \times 10^9}{6 \times 10^3} = 6 \times 10^6$$

EXAMPLE A1-3

$$3 \times 10^{-7} \times \frac{4 \times 10^{-6}}{2 \times 10^{-10}} \times \frac{4 \times 10^4}{6 \times 10^5} = \frac{(3 \times 10^{-7})(4 \times 10^{-6})(4 \times 10^4)}{(2 \times 10^{-10})(6 \times 10^5)}$$

$$= \frac{48 \times 10^{-9}}{12 \times 10^{-5}} = 4 \times 10^{-4}$$

exercise A1-7

Do each of the following. Express the final answer in standard exponential form.

(a) $16 \times 10^8 \times \dfrac{7 \times 10^5}{5 \times 10^2} =$

(b) $9 \times 10^7 \times \dfrac{8 \times 10^{-4}}{6 \times 10^2} =$

(c) $25 \times 10^{13} \times \dfrac{6 \times 10^6}{15 \times 10^{-15}} =$

(d) $\dfrac{(5 \times 10^{23})(8 \times 10^{15})}{(4 \times 10^{11})(6 \times 10^{20})} =$

(e) $\dfrac{(7 \times 10^{-9})(15 \times 10^{-12})(6 \times 10^{-8})}{(3 \times 10^{-4})(7 \times 10^{-15})(3 \times 10^{-6})} =$

(f) $7.4 \times 10^8 \times \dfrac{16 \times 10^{18}}{5 \times 10^{-6}} \times \dfrac{13 \times 10^6}{24 \times 10^{-9}} =$

(g) $4 \times 10^{-18} \times \dfrac{6 \times 10^{23}}{15 \times 10^{19}} \times \dfrac{7 \times 10^{-12}}{8 \times 10^7} =$

DIMENSIONAL ANALYSIS (FACTOR-LABEL METHOD)

The key point to recognize in using dimensional analysis is that units can be mathematically manipulated like numbers. They can be multiplied, divided, and canceled.

$$30 \text{ days} = ? \text{ sec}$$

$$30 \cancel{\text{days}} \times \frac{24 \cancel{\text{hr}}}{1 \cancel{\text{day}}} \times \frac{60 \cancel{\text{min}}}{1 \cancel{\text{hr}}} \times \frac{60 \text{ sec}}{1 \cancel{\text{min}}} = 2{,}592{,}000 \text{ sec} = 2.592 \times 10^6 \text{ sec}$$

In using dimensional analysis to set up and solve a problem, make sure you first associate the units given in the problem with certain physical quantities such as meters for length, cubic centimeters or milliliters for volume, and grams for mass. In solving problems, start with the unit you have and then, by using conversion factors, convert the original unit to the unit associated with the quantity of the desired answer.

A conversion factor is a statement of an equality between two quantities. For example, 1 ft = 12 in. and 1 kg = 1000 g are such equalities. Because it is an equality, a conversion factor can be expressed as a ratio. Thus, 1 ft = 12 in. is 1 ft/12 in. or 12 in./1 ft; 1 kg = 1000 g is 1 kg/1000 g or 1000 g/1 kg.

In solving problems, conversion factors are arranged so that all units cancel except those in the final answer. A unit in the denominator of one term will cancel the same unit in the numerator of another term.

EXAMPLE A1-4

How many inches are in 4 ft?

GIVEN	WANT TO KNOW	KNOW
4 ft	How many inches are in 4 ft?	$1\text{ ft} = 12\text{ in}; \frac{1\text{ ft}}{12\text{ in.}}; \frac{12\text{ in.}}{1\text{ ft}}$

Arrange the conversion factor so that feet cancel and inches is the unit remaining:

$$4\,\cancel{\text{ft}} = \frac{12\text{ in.}}{1\,\cancel{\text{ft}}} = 48\text{ in.}$$

EXAMPLE A1-5:

Fifty-five kg is the value commonly used in physiology for the weight (mass) of an average adult human female, how many pounds are equivalent to this value?

GIVEN	WANT TO KNOW
55 kg 1 kg = 2.2 lb	How many pounds does this represent?

$$55\,\cancel{\text{kg}} \times \frac{2.2\text{ lb}}{1\,\cancel{\text{kg}}} = 121\text{ lb}$$

The answer is in the units desired. Note that the conversion factor 1 kg = 2.2 lb is arranged as 2.2 lb/1 kg so that the units, kilogram, cancel. If the factor had been arranged 1 kg/2.2 lb, the units would not cancel and the problem would be answered incorrectly.

$$55\text{ kg} \times \frac{1\text{kg}}{2.2\text{ lb}} = \frac{55\text{ kg}^2}{2.2\text{ lb}}$$

It may be necessary to use several conversion factors in order to obtain the desired units. The problem above could have also been solved using the conversion factors 1 kg = 1000 g; 1 lb = 454 g.

$$55\,\cancel{\text{kg}} \times \frac{1000\,\cancel{\text{g}}}{1\,\cancel{\text{kg}}} \times \frac{1\text{ lb}}{454\,\cancel{\text{g}}} = \frac{55 \times 1000 \times 1\text{ lb}}{1 \times 454} = 121\text{ lb}$$

EXAMPLE A1-6

How many seconds are in 15 days; i.e., 15 days = __?__ sec?

GIVEN	WANT TO KNOW	KNOW
15 days	How many seconds are in 15 days?	1 day = 24 hr; $\frac{24\text{ hr}}{1\text{ day}}$; $\frac{1\text{ day}}{24\text{ hr}}$
		60 min = 1 hr; $\frac{60\text{ min}}{1\text{ hr}}$; $\frac{1\text{ hr}}{60\text{ min}}$
		60 sec = 1 min; $\frac{60\text{ sec}}{1\text{ min}}$; $\frac{1\text{ min}}{60\text{ sec}}$

$$15\,\cancel{\text{days}} \times \frac{24\,\cancel{\text{hr}}}{1\,\cancel{\text{day}}} \times \frac{60\,\cancel{\text{min}}}{1\,\cancel{\text{hr}}} \times \frac{60\text{ sec}}{1\,\cancel{\text{min}}} = 1{,}296{,}000\text{ sec} = 1.296 \times 10^6\text{ sec}$$

exercise A1-8

Given the following conversion factors, calculate the values indicated:
15 hargens = 3 grizzles; 5 sperkles = 1 pferd; 10 pferds = 1 grizzle

(a) 6 grizzles = _____ pferds;
(b) 6 grizzles = _____ hargens;
(c) 6 grizzles = _____ sperkles;
(d) 15 pferds = _____ hargens
(e) 30 hargens = _____ sperkles

EXAMPLE A1-7

A baby aspirin tablet contains 1.25 grains of aspirin. If two tablets are given every 4 hr, how many grams of aspirin have been administered after 12 hr? (15 grains = 1 g.)

GIVEN	WANTED
2 tablets per 4 hr = $\frac{2\text{ tablets}}{4\text{ hr}}$	Number of grams of aspirin after 12 hr
1 tablet per 1.25 grains = $\frac{1\text{ tablet}}{1.25\text{ grains}}$ or $\frac{1.25\text{ grains}}{1\text{ tablet}}$	
15 grains per 1 g = $\frac{15\text{ grains}}{1\text{ g}}$	
12-hr span	

The question is asking to express the amount of aspirin given in grams from its original expression in grains.

As a test of our method, let's first use only units and see if they can be arranged so that after canceling units properly the only unit left is gram(s).

$$\cancel{\text{hour}} \times \frac{\text{tablets}}{\cancel{\text{hour}}}\text{: This gives us tablets.}$$

To cancel tablets,

$$\cancel{\text{hour}} \times \frac{\cancel{\text{tablets}}}{\cancel{\text{hour}}} \times \frac{\text{grains}}{\cancel{\text{tablet}}}\text{: This gives us grains.}$$

To cancel grains,

$$\cancel{\text{hour}} \times \frac{\cancel{\text{tablets}}}{\cancel{\text{hour}}} \times \frac{\cancel{\text{grains}}}{\cancel{\text{tablet}}} \times \frac{\cancel{\text{gram}}}{\cancel{\text{grains}}} = \text{grams}$$

Now that we have confirmed that this sequence of conversion gives us the desired unit, let's put in the numbers associated with the values to calculate the final answer:

$$12\ \cancel{\text{hr}} \times \frac{2\ \cancel{\text{tablets}}}{4\ \cancel{\text{hr}}} \times \frac{1.25\ \cancel{\text{grains}}}{1\ \cancel{\text{tablet}}} \times \frac{1\ \text{g}}{15\ \cancel{\text{grains}}} = \frac{12 \times 2 \times 1.25 \times 1}{4 \times 1 \times 15}\ \text{g} = \frac{30}{60}\ \text{g} = 0.50\ \text{g}$$

exercise A1-9

Calculate each of the following:

(a) The bacterium *Beggiatoa mirabilis* has a width of 45 μ (microns); (1 $\mu = 1 \times 10^{-6}$ m). If arranged in a row so that they were just touching, how many bacteria would be required to form a row reaching from the earth to the sun, a distance of 93 million mi? (1 km = 1000 m; 1 mi = 1.6 km.)

(b) In 1971, approximately 37 million tons of aspirin were produced in the United States. An individual tablet contains 5 grains of aspirin. Calculate the number of aspirin tablets that could have been made from the aspirin produced in 1971.

(c) The greatest recorded human weight is 486 kg. How many pounds is this?

(d) The shortest recorded adult male measured 26.5 in. tall. How many meters is this? (1 meter = 1.1 yard.)

appendix two
quantum mechanical model of the atom

INTRODUCTION

As described in Chapter 2, Niels Bohr proposed that the energy of an electron in an atom is quantized; i.e., an electron has a definite, specific amount of energy. Depending on their energy, electrons in atoms occupy regions of energy called *energy levels* (shells). The lowest energy level is the *K* shell. Electrons of higher energy occupy higher energy levels such as *L, M, N,* etc. The change of an electron's energy is also quantized. An electron can gain or lose only a specific amount of energy in order to reach another energy level. Calculations of electron transitions based upon these postulates agreed with experimental data for the hydrogen atom. The mathematical relationships developed by Bohr to explain electron transitions generated the concept that in a hydrogen atom the electron circled the nucleus at certain fixed distances, much like a miniature solar system.

During the 1920's, experimental data indicated that energy levels are subdivided into sublevels. In searching for a new set of mathematical expressions that would permit calculating the energies of electron transitions between sublevels, physicists deduced that an electron's position and energy could not be *simultaneously* determined. The idea is based upon considering the electron as a particle. Alternatively, Louis de Broglie proposed that electrons exhibit wave behavior and studies using this approach would be fruitful. Using these concepts, physicists derived equations related to wave behavior (wave equations) that permitted calculation of electron transition energy in agreement with experimentally observed values. This new wave theory of atomic structure is called *quantum mechanics.**

ORBITALS, ELECTRONS, AND QUANTUM NUMBERS

Central to this theory is the concept that the exact path along which an electron moves cannot be determined. Instead, for an electron having a particular energy, the wave equations allow calculation of the *probability* of the electron's location in relation to the nucleus. Considered over a period of time, this probability gives a close approximation of an electron's behavior. The probability is related to a pattern called an *electron cloud.* The volume and shape of the electron cloud about a nucleus where an electron having a *particular allowed energy* is most likely to be found is dependent on an electron's energy.

An electron possessing a given amount of energy occupies a given orbital in an atom. Because there are many allowed energies an electron can have, an orbital is present for each allowed energy. The energy of an electron in an atom is described by a set of four integers that are solutions to wave equations. These four values are called *quantum numbers* and are symbolized by n, l, m_l, and m_s. Collectively, these values relate to the orbital in which an electron is located.

In Chapter 2, the various energy levels are designated by numbers 1, 2, . . . in order of increasing energy. Each of these numbers is called the

*The principal originators of quantum mechanics were Louis de Broglie, Werner Heisenberg, Erwin Schrödinger, and Paul Dirac, each of whom received a Nobel Prize for his contributions to this field.

principal quantum number n for that energy level. Each energy level actually contains sublevels. The number of sublevels in an energy level is equal to the principal quantum number of that level.

ENERGY LEVEL'S PRINCIPAL QUANTUM NUMBER, n	NUMBER OF SUBLEVELS WITHIN THE ENERGY LEVEL
1	1
2	2
3	3
.	.
.	.
.	.

Sublevels differ in energy and other characteristics; this will be discussed shortly. Four types of sublevels—*s, p, d,* and *f*—are sufficient in describing all the elements currently known. Other sublevels in the fifth and higher energy levels are theoretically possible. Sublevels are designated by a second quantum number, l. The general shape of the electron cloud for the electron is determined by l. The following assignments are made:

l VALUE	SUBLEVELS
0	s
1	p
2	d
3	f

The relationship between n and l is $l = 0, 1, 2, 3, \ldots, (n - 1)$. An electron occupying the s orbital of the first energy level is designated a 1s electron.

exercise A2-1

Given are n and l values for several electrons. What information about each of the electrons described below is conveyed by the pair of quantum numbers?

	n	l		n	l
(a)	**1**	**0;**	**(d)**	**4**	**2**
(b)	**2**	**0;**	**(e)**	**3**	**0**
(c)	**3**	**1;**	**(f)**	**4**	**3**

The energy of the lower-lying sublevels is shown in Fig. A2-1.

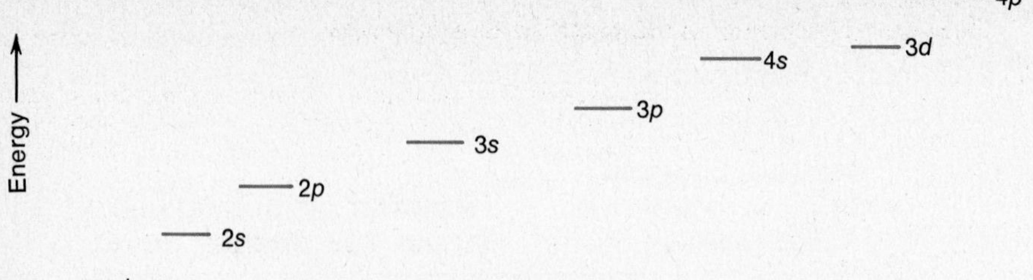

FIGURE A2-1 Relative energies of lower-lying sublevels.

ENERGY AND ELECTRON DISTRIBUTION

The number of orbitals each type of sublevel contains varies depending on the sublevel. An *s* sublevel, regardless of its energy level, contains only one orbital; *p* sublevels have three orbitals; *d* sublevels have five orbitals, and *f* sublevels have seven orbitals. The total number of orbitals in a given energy level equals n^2.

Energy level's principal quantum number, *h*	Number of sublevels in energy level	Types of sublevels $l = 0, 1, 2 \ldots, (n - 1)$		Total Number of orbitals, n^2, in energy level
1	1	$l = 0$; *s*		$1 = 1^2$
2	2	$l = 0$; *s*	1	$4 = 2^2$
		$l = 1$; *p*	3	
3	3	$l = 0$; *s*	1	$9 = 3^2$
		$l = 1$; *p*	3	
		$l = 2$; *d*	5	
4	4	$l = 0$; *s*	1	$16 = 4^2$
		$l = 1$; *p*	3	
		$l = 2$; *d*	5	
		$l = 3$; *f*	7	

exercise A2-2

How many orbitals would (a) the fifth energy level have? (b) the sixth energy level?

Each orbital can be occupied by a *maximum* of two electrons. Since n^2 gives the total number of orbitals in a given energy level, $2n^2$ equals the *maximum* number of electrons that a given energy level can accommodate.

$$\frac{2\ \text{electrons}}{\cancel{\text{orbital}}} \times \frac{n^2\ \cancel{\text{orbitals}}}{\text{energy level}} = \frac{2n^2\ \text{electrons}}{\text{energy level}}$$

This relationship is summarized in Tables A2-1 and A2-2.

TABLE A2-1
Electron distribution in the first four energy levels

n	TYPE OF SUBLEVEL	NUMBER OF ORBITALS IN SUBLEVEL	MAXIMUM NUMBER OF ELECTRONS IN SUBLEVEL	MAXIMUM NUMBER OF ELECTRONS IN ENERGY LEVEL
1	1*s*	1	2	2
2	2*s*	1	2	8
	2*p*	3	6	
3	3*s*	1	2	18
	3*p*	3	6	
	3*d*	5	10	
4	4*s*	1	2	32
	4*p*	3	6	
	4*d*	5	10	
	4*f*	7	14	

TABLE A2-2
Energy levels in atoms

n, principal quantum number of level	1	2	3	4	5	6	7
Maximum number of electrons (theoretical)	2	8	18	32	50	72	98
Number actually found in nature	2	8	18	32	32	10	2

The electron distribution of an atom is such that electrons occupy orbitals of the lowest energy sublevel available. If this sublevel is filled (already occupied by the maximum number of electrons allowed), additional electrons then occupy sublevels of higher energy. Figure A2-1 allows us to predict the order of filling of subshells for atoms of an element. This order is summarized in Table A2-3 for the first 20 elements. Note that the energy of the 4*s* sublevel is slightly lower than that of the 3*d*. Therefore, the 4*s* is filled before the 3*d* in potassium and calcium. The order of filling for a particular atom is called the atom's *electron configuration.* The superscript number refers to the number of electrons present in the sublevel.

exercise A2-3

Predict the electron configurations of scandium, element 21; zinc, element 30; krypton, element 36.

The l quantum number determines the general shape of the electron cloud. The third quantum number, m_l, relates to the orientation of the electron cloud in space. m_l and l are related: $m_l = l, l - 1, l - 2, \ldots, 0, -1, -2, \ldots, -l$.

TABLE A2-3
Electron configuration of elements 1 to 20

ELEMENT	ATOMIC NUMBER	ELECTRON CONFIGURATION
Hydrogen	1	$1s^1$
Helium	2	$1s^2$
Lithium	3	$1s^2 2s^1$ (recall only two electrons may occupy an *s* subshell)
Beryllium	4	$1s^2 2s^2$
Boron	5	$1s^2 2s^2 2p^1$
Carbon	6	$1s^2 2s^2 2p^2$
Nitrogen	7	$1s^2 2s^2 2p^3$
Oxygen	8	$1s^2 2s^2 2p^4$
Fluorine	9	$1s^2 2s^2 2p^5$
Neon	10	$1s^2 2s^2 2p^6$ (*p* sublevels can hold a maximum of six electrons)
Sodium	11	$1s^2 2s^2 2p^6 3s^1$
Magnesium	12	$1s^2 2s^2 2p^2 3s^2$
Aluminum	13	$1s^2 2s^2 2P^6 3s^2 3P^1$
Silicon	14	$1s^2 2s^2 2P^6 3s^2 3p^2$
Phosphorus	15	$1s^2 2s^2 2p^6 3s^2 3p^3$
Sulfur	16	$1s^2 2s^2 2p^6 3s^2 3p^4$
Chlorine	17	$1s^2 2s^2 2p^6 3s^2 3p^5$
Argon	18	$1s^2 2s^2 2p^6 3s^2 3p^6$
Potassium	19	$1s^2 2s^2 2p^6 3s^2 3p^6 4s^1$
Calcium	20	$1s^2 2s^2 2p^6 3s^2 3p^6 4s^2$

l	0	1	2	3
SUBLEVEL	*s*	*p*	*d*	*f*
NUMBER OF ORBITALS	1	3	5	7
ORBITAL m_l ASSIGNMENT	1	1 0 −1	2 1 0 −1 −2	3 2 1 0 −1 −2 −3

THE PAULI EXCLUSION PRINCIPLE AND HUND'S RULE The fourth quantum number, m_s, is related to the spin of an electron. An electron can be imagined to be spinning on its axis. Two directions of spin are possible—clockwise and counterclockwise. To occupy an orbital simultaneously, two electrons must have *opposite* spins—one clockwise and the other counterclockwise. The m_s values are $+\frac{1}{2}$ and $-\frac{1}{2}$, respectively. We will use the notation of an upward arrow, ↑, to denote $+\frac{1}{2}$; a downward arrow, ↓, denotes $-\frac{1}{2}$.

Taken collectively, a set of the four quantum numbers uniquely defines an electron in any atom. In an atom, each electron has its own set of quantum numbers; i.e., no two electrons in an atom can have the *same set* of quantum numbers. This is called the *Pauli exclusion principle* in honor of the physicist Wolfgang Pauli who first proposed it. Table A2-4 summarizes the quantum number relationships. Several examples are given below.

TABLE A2-4
Allowed values of quantum number sets for electrons—first three energy levels of atoms

n	l	m_l	m_s
1	0	0	$+\frac{1}{2}$ or $-\frac{1}{2}$
2	0	0	$+\frac{1}{2}$ or $-\frac{1}{2}$
2	1	1	$+\frac{1}{2}$ or $-\frac{1}{2}$
2	1	0	$+\frac{1}{2}$ or $-\frac{1}{2}$
2	1	-1	$+\frac{1}{2}$ or $-\frac{1}{2}$
3	0	0	$+\frac{1}{2}$ or $-\frac{1}{2}$
3	1	1	$+\frac{1}{2}$ or $-\frac{1}{2}$
3	1	0	$+\frac{1}{2}$ or $-\frac{1}{2}$
3	1	-1	$+\frac{1}{2}$ or $-\frac{1}{2}$
3	2	2	$+\frac{1}{2}$ or $-\frac{1}{2}$
3	2	1	$+\frac{1}{2}$ or $-\frac{1}{2}$
3	2	0	$+\frac{1}{2}$ or $-\frac{1}{2}$
3	2	-1	$+\frac{1}{2}$ or $-\frac{1}{2}$
3	2	-2	$+\frac{1}{2}$ or $-\frac{1}{2}$

EXAMPLE A2-1

In its lowest energy condition, the electron in a hydrogen atom is in the orbital of an s sublevel of the first shell, i.e., $1s^1$. The principal quantum number n equals 1 and because this is an s sublevel orbital, l equals zero and m_l is zero;

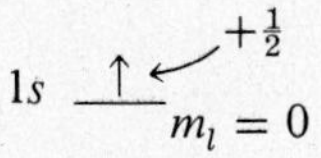

$$1s \ \underline{\uparrow} \leftarrow +\tfrac{1}{2} \qquad m_l = 0$$

m_s is $+\frac{1}{2}$. Therefore the set $1, 0, 0, +\frac{1}{2}$ is assigned to this electron.

EXAMPLE A2-2

A helium atom has an electron configuration of $1s^2$. Each electron has a unique set of quantum numbers: The first is $1, 0, 0, +\frac{1}{2}$; the other is $1, 0, 0, -\frac{1}{2}$;

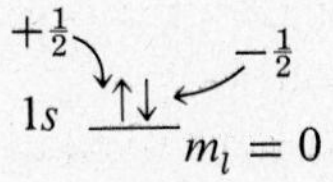

$$+\tfrac{1}{2} \rightarrow \ 1s \ \underline{\uparrow\downarrow} \leftarrow -\tfrac{1}{2} \qquad m_l = 0$$

EXAMPLE A2-3

The electron configuration for the lithium atom is $1s^22s^1$. The sets of quantum numbers would be $1, 0, 0, +\frac{1}{2}$; $1, 0, 0, -\frac{1}{2}$; $2, 0, 0, +\frac{1}{2}$.

$$2s\ \underline{\uparrow}\ m_l = 0$$

$$1s\ \underline{\uparrow\downarrow}\ m_l = 0$$

EXAMPLE A2-4

A boron atom's electron configuration is $1s^22s^22p^1$. For the electron in the p sublevel, the correct interpretation for its set of quantum numbers is $n = 2$; p sublevels have an l value of 1; there are three orbitals in a p sublevel—the first has an m_l value of 1; the first electron in an orbital is assigned an m_s value of $+\frac{1}{2}$. Therefore, the $2p^1$ electron's set of quantum numbers is $2, 1, 1, +\frac{1}{2}$.

$$2p\ \underset{1}{\underline{\uparrow}}\ \underset{0}{\underline{\quad}}\ \underset{-1}{\underline{\quad}} = m_l$$

$$2s\ \underset{0}{\underline{\uparrow\downarrow}} = m_l$$

$$1s\ \underset{0}{\underline{\uparrow\downarrow}} = m_l$$

exercise A2-4

Give the set of quantum numbers for (a) each electron in a beryllium atom; (b) the twelfth electron in a magnesium atom.

One further restriction is placed on electron occupancy of an orbital. In an uncombined atom, orbitals of the same sublevel are of equal energy, e.g., the three orbitals of a p sublevel. *Each* orbital will acquire an electron before any of these orbitals acquires a second electron. This generalization is commonly called *Hund's rule*. In accordance with the Pauli exclusion principle, the second electron must have a spin opposite that of the first electron. The following examples illustrate Hund's rule and the buildup of electron occupancy of available orbitals.

EXAMPLE A2-5

Nitrogen atoms have a $1s^22s^22p^3$ configuration.

$$2p\ \underset{1}{\underline{\uparrow}}\ \underset{0}{\underline{\uparrow}}\ \underset{-1}{\underline{\uparrow}} = m_l$$

$$2s\ \underline{\uparrow\downarrow}\ m_l = 0$$

$$1s\ \underline{\uparrow\downarrow}\ m_l = 0$$

Note that each of the p orbitals is occupied by an electron.

EXAMPLE A2-6

Fluorine atoms have a $1s^22s^22p^5$ configuration.

$$2p \; \underset{1}{\underline{\uparrow\downarrow}} \; \underset{0}{\underline{\uparrow\downarrow}} \; \underset{-1}{\underline{\uparrow}} = m_l$$

$$2s \; \underline{\uparrow\downarrow} \; m_l = 0$$

$$1s \; \underline{\uparrow\downarrow} \; m_l = 0$$

exercise A2-5

Write the electron distribution by orbitals for atoms of the following elements: (a) beryllium, (b) magnesium, (c) oxygen, (d) phosphorus, (e) sulfur.

GEOMETRIC REPRESENTATION OF ELECTRON CLOUDS

It was previously mentioned that the value of l determines the general shape of the electron cloud. The value of m_l determines the orientation of the electron cloud in space. The charge cloud for any s electron, regardless of its primary quantum number, is spherically symmetrical like a ball [see Fig. A2-2(a)]. Recall that an electron cloud is a probability pattern of an electron's existence about a nucleus encompassing about 95 percent of the probable volume where an electron may occur. A spherically symmetrical electron cloud indicates that the probability of finding the electron about the nucleus is independent of direction. As the energy level increases, the s sublevel increases in volume, i.e., $3s$ larger than $2s$, which is larger than $1s$ (see Fig. A2-2).

Orbitals in a p sublevel are lobe-shaped (see Fig. A2-3). This means that the p orbitals are probability patterns of electron distribution that are direction dependent. The three sets of orbitals in a p sublevel are oriented at right angles to each other (like the arrangement between the walls and ceiling at the corner of a room). If three axes x, y, and z are used as reference, the p orbital whose electron probability is greatest along the x-axis is called a p_x orbital (see Fig. A2-3). The p_y orbital is concentrated along the y-axis and the p_z orbital is along the z-axis. As n increases, the size of the p orbitals increases also.

Electron cloud diagrams may also be drawn for d and f orbitals. The orientation and geometry of these orbitals is complex and will not be considered here.

SIGMA AND PI BONDS

A covalent bond results from the sharing of two electrons between two bonded atoms. The electrons forming the covalent bond are originally present in orbitals of their respective atoms. The overlap of an orbital containing an electron from one atom with an orbital containing an electron from a second atom forms a covalent bond. Two types of overlap involving p electrons are possible—end-to-end and sideways. The end-to-end overlap of two orbitals produces a *sigma* (σ) *bond* in which electron density is symmetrical along the bond axis (see Fig. A2-4). The sideways overlap of two orbitals forms a pi (π) bond in which the electron density is concentrated above and below the bond axis.

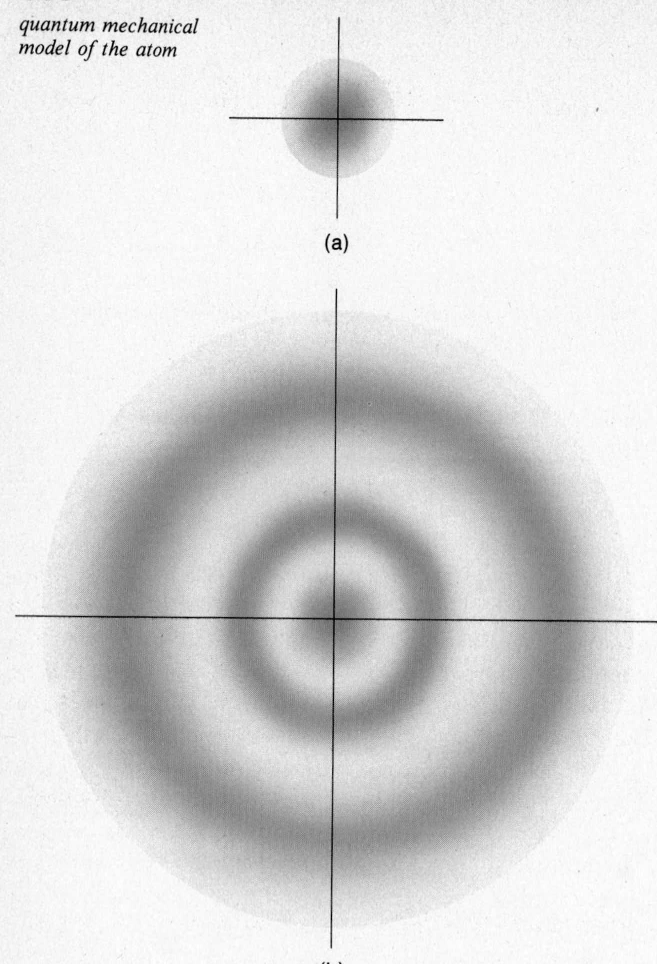

FIGURE A2-2 (a) Charge cloud for a 1s orbital. (b) Charge cloud for a 3s orbital.

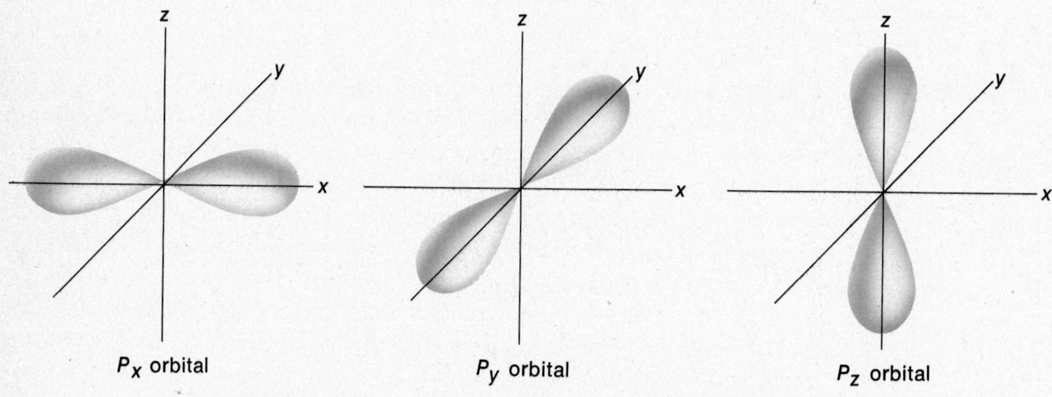

FIGURE A2-3 Electron clouds of P_x, P_y, and P_z orbitals.

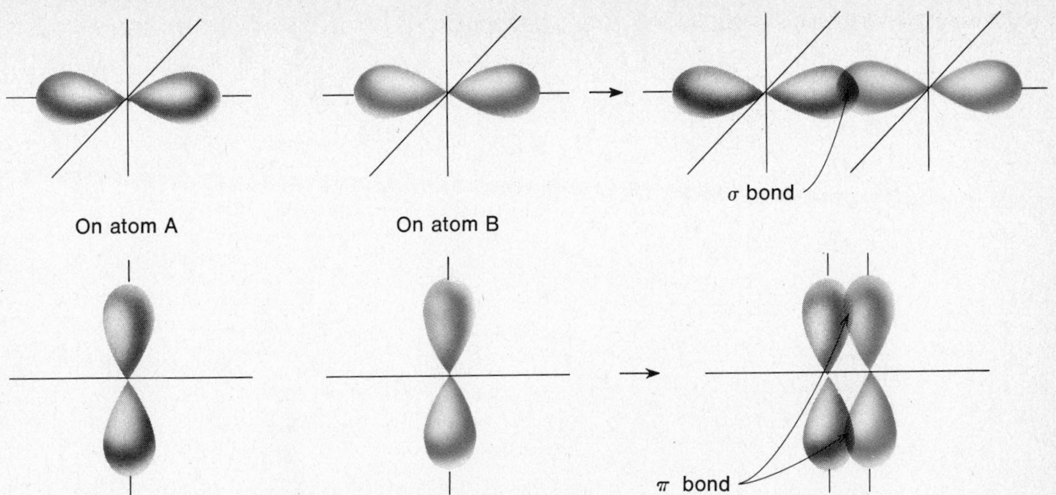

FIGURE A2-4 Sigma (σ) and pi (π) bonds.

Because the orbitals used for overlap are now part of a bond in a molecule, the terms *sigma* and *pi* refer to molecular orbitals that are occupied by electrons of bonded atoms. Benzene (C_6H_6) contains six carbon atoms bonded to each other and individually to six hydrogen atoms. Each carbon uses three electrons individually to form sigma bonds with two other carbon atoms and a hydrogen atom.

$$6\ \cancel{\text{carbon}} \times \frac{3\ \text{electrons}}{\cancel{\text{carbon}}} = 18\ \text{electrons used for sigma covalent bonding}$$

Each carbon has 4 outer electrons available, however.

$$6\ \cancel{\text{carbon atoms}} \times \frac{4\ \text{electrons}}{\cancel{\text{carbon}}} = 24\ \text{electrons}$$

So far we have accounted for 18. The remaining 6 are present in a π electron cloud formed by the edgewise overlap of p_z orbitals from each carbon atom. The electrons are not localized on a particular carbon atom but belong to the entire ring system (see Fig. A2-5). This is symbolically represented by [benzene ring symbol].

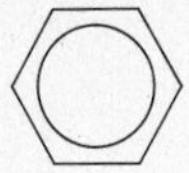

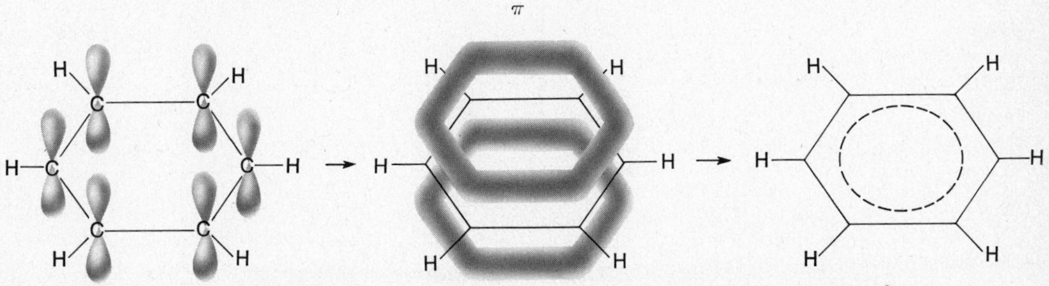

FIGURE A2-5 A representation of the π electron cloud in benzene.

appendix three
answers to in-chapter exercises

CHAPTER ONE

1-1 length; volume; mass; length; mass; length; length; volume

1-2 (a) 0.100 liter; (b) 0.010 km; (c) 10 mg; (d) 2250 ml; (e) 80 mg

1-3 960 drelbes

1-4 (a) 0.000000042 kg; (b) 30 ml; 30,000 μl; (c) 0.000005 km

1-5 95°F

1-6 0.82 g/ml

1-7 (a) 132 g; (b) 13,530 g; (c) 5.0 cc; (d) 0.02 liter

1-8 (a) yes; (b) no; (c) no

1-9 no

1-10 1.26×10^5 cal

1-11 2.6×10^{19} eV

CHAPTER TWO

2-1 (a) $^{65}_{30}Zn$, $^{66}_{31}Ga$; (b) $^{65}_{30}Zn$, $^{66}_{30}Zn$; (c) $^{66}_{30}Zn$, $^{66}_{31}Ga$; (d) $^{65}_{30}Zn$, $^{66}_{30}Zn$

2-2 $^{32}_{15}P$, $^{33}_{15}P$; $^{32}_{16}S$, $^{33}_{16}S$
Isotopes have same number of protons.

2-3 (a) $^{4}_{2}He$; (b) $^{234}_{91}Pa$; (c) $^{210}_{84}Po$

2-4 60 hr

2-5 12.5 mg

2-6 (a) Na and K; N and P; Ne and Ar; (b) Ne and Ar

2-7 *K L M* 2, 8, 1 metal; *K L* 2, 5 nonmetal; *K L* 2, 1 metal; *K L M N* 2, 8, 8, 1 metal; *K L M* 2, 8, 6 nonmetal

2-8 Li, O, Mg, Cl

2-9 $_3Li$: K2, L1; $_{20}Ca$: K2, L8, M8, N2; $_{13}Al$: K2,L8,M3
Li^+: K2; Ca^{2+}: K2, L8, M8; Al^{3+}: K2, L8

2-10 Oxygen: atom *K*2, *L*6; ion, O^{2-} *K*2, *L*8
Sulfur: atom *K*2, *L*8, *M*6; ion, S^{2-} *K*2, *L*8, *M*8

2-11 (a) 35 electrons, 35 protons; (b) 36 electrons, 35 protons

2-12 $X = \frac{1}{2} \times 12 = 6$ amu; $Y = 144$ amu; $Z = 180$ amu

2-13 10.8 amu

CHAPTER THREE

3-1 (a) Potential energy is increased; i.e., energy is stored. (b) Formation of starch represents the storage of energy during photosynthesis whereas the consumption of starch by animals releases this potential energy as chemical energy and heat.

3-2 (a) The symbol C represents the carbon nucleus and the two electrons in the *K* shell; Ne represents the neon nucleus and the two electrons in the *K* shell, and; Cl represents the chlorine nucleus, the two electrons in the *K* shell, and the eight electrons in the *L* shell. (b) Because both oxygen and sulfur have six electrons in the outermost occupied shell.

3-3 Al^{3+}, F^-, Ca^{2+}, Cs^+, I^- and Se^{2-}

3-4 NaBr, Li_2O, MgS, MgF_2, and K_2S

3-5 $MnCl_2$, ZnO, Cu_2O, CuO, and Fe_2S_3

3-6 sodium bromide, lithium oxide, magnesium sulfide, magnesium fluoride, potassium sulfide, manganese(II) chloride, zinc(II) oxide, copper(I) oxide, copper(II) oxide and iron(III) sulfide

3-7
(a) $Mg + I_2 \longrightarrow MgI_2$
(b) $4Fe + 3O_2 \longrightarrow 2Fe_2O_3$
(c) $4Li + O_2 \longrightarrow 2Li_2O$
(d) $2Be + O_2 \longrightarrow 2BeO$
(e) $2K + Br_2 \longrightarrow 2KBr$

3-8
(a) $2Na + 2H_2O \longrightarrow 2NaOH + H_2$
(b) $2CuS + 3O_2 \longrightarrow 2CuO + 2SO_2$
(c) $N_2 + 3H_2 \longrightarrow 2NH_3$
(d) $6CO_2 + 6H_2O \longrightarrow C_6H_{12}O_6 + 6O_2$
(e) $C_{16}H_{32}O_2 + 23O_2 \longrightarrow 16CO_2 + 16H_2O$

3-9 (a) oxidation; (b) reduction; (c) reduction; (d) oxidation

3-10
(a) redox reaction; Mg is reducing agent; O_2 is the oxidizing agent.
$2Mg \longrightarrow 2Mg^{2+} + 4e^-$ oxidation half reaction
$4e^- + O_2 \longrightarrow 2O^{2-}$ reduction half reaction
(b) not a redox reaction
(c) redox reaction; Al is reducing agent; Fe_2O_3 is oxidizing agent.
$2Al \longrightarrow 2Al^{3+} + 6e^-$ oxidation half reaction
$6e^- + Fe_2O_3 \longrightarrow Fe + 3O^{2-}$ reduction half reaction
(d) redox reaction; HCl is oxidizing agent; Al is the reducing agent;
$2Al \longrightarrow 2Al^{3+} + 6e^-$ oxidation half reaction
$6e^- + 6HCl \longrightarrow 6Cl^- + 3H_2$ reduction half reaction
(e) redox reaction; Zn is reducing agent; $CuCl_2$ is oxidizing agent.
$Zn \longrightarrow Zn^{2+} + 2e^-$ oxidation half reaction
$2e^- + CuCl_2 \longrightarrow Cu + 2Cl^-$ reduction half reaction

3-11 (a) F_2 $:\ddot{F}:\ddot{F}:$

(b) NH_3 $H:\ddot{N}:H$ with H (structure: H above N, H to right, H below)

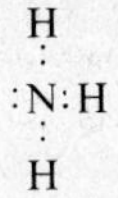

(c) PCl_3 $:\ddot{Cl}:\ddot{P}:\ddot{Cl}:$ with $:\ddot{Cl}:$ below P

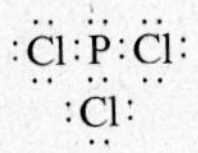

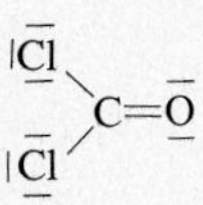

(d) CCl_4 $:\ddot{Cl}:C:\ddot{Cl}:$ with $:\ddot{Cl}:$ above and below C

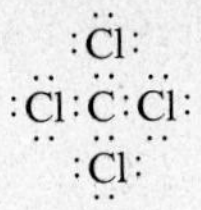

(e) SiF_4 $:\ddot{F}:Si:\ddot{F}:$ with $:\ddot{F}:$ above and below Si

3-12 (a) 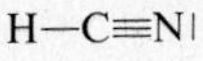 H—C≡N|

(b) $\overline{Cl}$ and $\overline{Cl}$ bonded to C=$\overline{O}$

(c) $\overline{S}$=C=$\overline{S}$

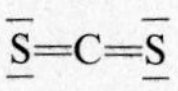

(d) 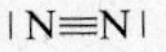 |N≡N|

3-13 (a) Na_2CO_3; (b) $Al_2(SO_4)_3$; (c) NH_4NO_3; (d) $Ca(H_2PO_4)_2$; (e) $Mg(OH)_2$

3-14 (a) 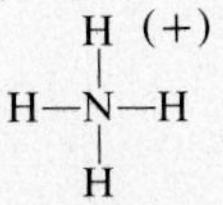 H—N—H with H above and H below N, (+)

(b) 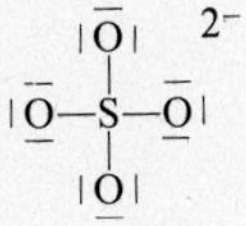 |$\overline{O}$—S—$\overline{O}$| with |$\overline{O}$| above and below S, 2−

(c) 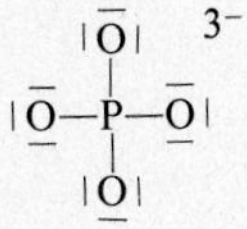 |$\overline{O}$—P—$\overline{O}$| with |$\overline{O}$| above and below P, 3−

(d) 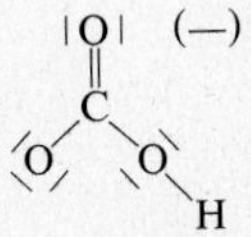 |O| (−) double-bonded to C; C bonded to $\overline{O}$ and to $\overline{O}$—H

(e) 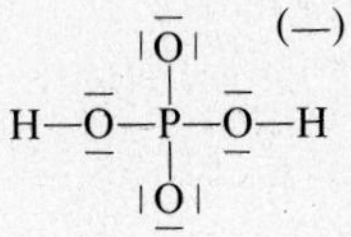 H—$\overline{O}$—P—$\overline{O}$—H with |$\overline{O}$| above and below P, (−)

3-15 (a) nonpolar (c) $\overrightarrow{H—O}$ (e) nonpolar

(b) $\overrightarrow{C—Cl}$ (d) nonpolar (f) $\overrightarrow{C—O}$

3-16 (a) V-shaped (b) triangular planar (c) tetrahedral
(d) tetrahedral (e) linear

3-17 (a) no dipole because is linear
(b) no dipole because is tetrahedral
(c) polar molecule
(d) polar molecule
(e) polar molecule
(f) polar molecule

3-18 (a) CO (b) NH_3 (c) CH_2O (d) H_2S (e) HCN

3-19 (a) 58.44 (b) 141.98 (c) 96 (d) 180.06 (e) 16488.93

CHAPTER FOUR

4-1 methane, CH_4; propane, $CH_3CH_2CH_3$; butane, $CH_3CH_2CH_2CH_3$. Octane of gasoline is based upon the combustion properties of a hydrocarbon containing eight carbon atoms.

4-2 (a) butane, hexane (b) pentane, hexane (c) hexane, octane (d) hexane, octane (e) butane, hexane.

4-3

```
          H                H  H                      H  H  H
          |                |  |                      |  |  |
methyl, H-C-;     ethyl, H-C--C-;        propyl, H-C--C--C-;
          |                |  |                      |  |  |
          H                H  H                      H  H  H

              H
              |
            H-C
              | \
              H  \
isopropyl,        C-H;
              H  / \
              | /
            H-C
              |
              H
```

4-4

```
                                         |
                                    H  H-C-H
                                    |    |
           H                 --->  H-C-----C-H
           |                /        |    |
      H  H-C-H             /         H  H-C-H
      |    |              /                |
   H-C-----C-H                             H
      |    |              \
      H  H-C-H             \               H
           |                \              |
           H                 --->   H  H-C-H
                                    |    |
                                  H-C-----C-
                                    |    |
                                    H  H-C-H
                                         |
                                         H
```

4-5

```
                    H      H       H       H
                    |      |       |       |
(a) 5, methyl    H-C------C-------C-------C-H
                    |      |       |       |
                    H    H-C-H     H     H-C-H
                           |               |
                           H               H

                                  H
                                  |
                            H   H-C-H  H
                            |     |    |
(b) 3, methyl, methyl,   H-C-----C-----C-H
                            |     |    |
                            H   H-C-H  H
                                  |
                                  H

                                  H
                                  |
                            H   H-C-H  H  H  H
                            |     |    |  |  |
(c) 5, methyl, methyl,   H-C-----C-----C--C--C-H
                            |     |    |  |  |
                            H   H-C-H  H  H  H
                                  |
                                  H
```

(d) 10, isopropyl, methyl, ethyl,

4-6 From 4-2, (a) 2,3-dimethylbutane; (b) 2-methylpentane (c) 3,4-dimethylhexane; (d) 2,4-dimethylhexane; (e) 2,3-dimethylbutane From 4-5, (a) 2-methylpentane; (b) 2,2-dimethylpropane; (c) 2,2-dimethylpentane

4-7 (a) propene (b) 2-methyl-1-propene (c) 1-butene (d) 2,6-dimethyl-2-heptene (e) 2,4-dimethyl-1-pentene

4-8 Use model kit.

4-9 (a) no, 2-methyl-1-propene (b) yes, *cis*-2-pentene

and *trans*-2pentene

(c) yes,

trans-3-methyl-3-hexene, *cis*-3-methyl-3-hexene

(d) no,

```
    H  H
    |  |
H—C—C
    |  | \         H
    H  H  \       /
    H  H    C=C
    |  |   /     \  H  H
H—C—C          |  |
    |  |            C—C—H
    H  H            |  |
                    H  H
```

3-ethyl-3-hexene

(e) yes,

```
                            H
                            |
                            C—H
               H           /|
                \         / H
    H        H    C=C
    |        |   /     \   H      H
H—C——C            \  |      |
    |        |          C——C—H
    H   H—C—H           |      |
             |      H—C—H  H
             H          |
                        H
```

cis-2,3,5-trimethyl-3-hexene,

```
             H
             |
    H   H—C—H       H
    |        |          |
H—C——C        H—C—H
    |        |    \    /
    H        H     C=C   H      H
                  /      \  |      |
                 H         C——C—H
                            |      |
                        H—C—H  H
                            |
                            H
```

trans-2,3,5-trimethyl-3-hexene

4-10 (a) C_8H_{16} (b) C_9H_{18} (c) $C_{13}H_{24}$ (d) $C_{17}H_{28}$

4-11 (a) C_3H_4 (b) C_4H_6 (c) $C_{10}H_{14}$ (d) $C_{14}H_{22}$ (e) $C_{40}H_{56}$

4-12 (a) $2C_4H_{10} + 13O_2 \longrightarrow 8CO_2 + 10H_2O$
(b), (c), (d) $C_6H_{12} + 9O_2 \longrightarrow 6CO_2 + 6H_2O$
(e) $2C_6H_6 + 15O_2 \longrightarrow 12CO_2 + 6H_2O$

4-13 (a) amine; (b) acid; (c) aromatic, amine; (d) aromatic; (e) ether; (f) amine, acid—called an amino acid; (g) alcohol (steroid); (h) alcohol; (i) alcohol; (j) aromatic, ester, alcohol

4-14 Ethyl alcohol and water can hydrogen bond with each other.

4-15

```
                                        H
                                        |
                                  H—N—CH3
              δ−                 .·  ··
        ··     ↑               ··
        N—H    |      CH3—N—H
       /  \    |            |
H—C—H  H   |            H
    |          δ+
    H
```

CHAPTER FIVE

5-1 a, b, d, e, f

5-2 Alcohol groups because they are more polar.

5-3 Greater surface area, exposing more ions to the solvent.

5-4 $Ca(CH_3CH_2CH_2CH_2CH_2CH_2CH_2CH_2CH_2CH_2CH_2CH_2CH_2CH_2CH_2CH_2CH_2CO_2)_2$

5-5 (a) Dissolve 25 g boric acid in sufficient water to make 500 ml of solution.
(b) Dissolve 9 g NaCl in enough water to make 1000 ml of solution.
(c) Dissolve 2 g $AgNO_3$ in enough water to make 2 liters of solution.
(d) Dissolve 0.075 g benadryl in enough water to make 30 ml of solution.

5-6 (a) 600 ml (b) 400 ml (c) 3000 ml (d) 40 ml

5-7 (a) 37.5 g (b) 3.15 g (c) 0.0375 g (d) 0.5 g

5-8 (a) 0.3710 g (b) 7.42 g (c) 159.97 g (d) 0.797 lb (e) 1.599 tons

5-9 (a) 215.65 (b) 342.14 (c) 46.02 (d) 180.09 (e) 176.06

5-10 (a) 1 (b) 0.5 (c) 0.02 (d) 13.33 (e) .278

5-11 20.4 g

5-12 (a) 7.31 g (b) 15.23 g (c) 4.2 g (d) 128.4 g (e) 25.65

5-13 (a) 0.3 liter (b) 1.5 (c) 35.1 (d) 0.06 (e) 2500 ml

5-14 (a) 2 (b) 1.5 (c) 0.5 (d) 3.0 (e) 0.5

5-15 (a) 0.1 (b) 0.03 (c) 0.024 (d) 0.024 (e) 0.01

5-16 (a) 500 (b) 1440 (c) 10 (d) 50 (e) 5

5-17 (a) 40 ml (b) 0.03 (c) 50 ml (d) 240 ml

5-18 (a) 300 ml; (b) 5.4 percent; (c) 100 ml of 20 percent glucose diluted to 400 ml with distilled water.

5-19 (a) 0.015; (b) 0.16

CHAPTER SIX

6-1 $(C_6H_{10}O_5)_{500} + 3000O_2 \longrightarrow 3000CO_2 + 2500H_2O$
$(C_6H_{10}O_5)_{1000} + 60{,}000O_2 \longrightarrow 60{,}000CO_2 + 50{,}000H_2O$
$(C_6H_{10}O_5)_{100{,}000} + 600{,}000O_2 \longrightarrow 600{,}000CO_2 + 500{,}000H_2O$

6-2 As heat is given off by the reaction, it increases the kinetic energy of the molecules thereby increasing the number possessing the required activation energy. Placing the reaction vessel in ice water removes heat, lowering the kinetic energy of the molecules thereby slowing down the reaction rate.

6-3 (a) $r/2$ (b) $4r$ (c) r

6-4 65 – 95 mg/100 ml

6-5 Acids and alcohols

6-6 Evaporation of the alcohol requires energy. The energy is absorbed from the child's skin surface, lowering its temperature.

CHAPTER SEVEN

7-1 Condition 2: solution B is hypotonic to solution A.
Condition 3: solution B is hypertonic to solution A.

7-2 Water will leave cells because 9 percent NaCl is hypertonic to the normal 0.9 percent solution in red cells.

7-3 To be isotonic with normal red blood cells

7-4 (a) 2; (b) 1; (c) 3; (d) 2

7-5 (a) 2 osmol; (b) 1.5 osmol; (c) 0.15 osmol; (d) 2 osmol; (e) 0.20 osmol

7-6 (a) 4 *M*; (b) 2 *M*; (c) 1.33 *M*

7-7 (a) Equilibrium shifts to right, forward reaction favored. (b) Same as (a). (c) Equilibrium shifts to left, reverse reaction favored.

7-8 Container contents no longer at equilibrium when cap removed.

$$H_2CO_3 \underset{\text{high press.}}{\rightleftharpoons} CO_2 + H_2O$$

When high pressure is removed, forward reaction is favored and carbon dioxide escapes.

7-9 CO combines with Hb and prevents HbO_2 formation.

7-10 5 moles/liter

7-11 (a) $\dfrac{100 \text{ moles/liter}}{10 \text{ moles/liter}}$; (b) $\dfrac{70 \text{ moles/liter}}{7 \text{ moles/liter}}$; (c) $\dfrac{0.10 \text{ mole/liter}}{0.01 \text{ mole/liter}}$; any others where the ratio is 10/1

7-12 The reaction is not at equilibrium.

7-13 $K = \dfrac{[\text{methyl salicylate}] \times [H_2O]}{[\text{salicylic acid}] \times [\text{methanol}]}$

7-14 (a) O_2 gas would otherwise escape and equilibrium couldn't be achieved.
(b) $K = \dfrac{[H_2O]^2 \times [O_2]}{[H_2O_2]^2}$

7-15 (a) glucose-6-phosphate. (b) 0.03 mole/liter

7-16 The very small equilibrium constant indicates that the formation of nitric oxide is not favored at 37°C.

CHAPTER EIGHT

8-1 A $[OH^-] = 1 \times 10^{-10}\ M$; acidic
B $[H^+] = 1 \times 10^{-5}\ M$; acidic
C $[OH^-] = 1 \times 10^{-14}\ M$; acidic
D $[H^+] = 1 \times 10^{-8}\ M$; neither

8-2 Equilibrium shifts to left and reverse reaction favored.

$$H_2O \rightleftharpoons H^+ + OH^-$$

8-3 $[H^+] = 1 \times 10^{-4}$; pH = 4, acidic; $[H^+] = 1 \times 10^{-8}$, pH = 8, basic; $[H^+] = 1 \times 10^{-3}$, pH = 3, acidic; $[H^+] = 1 \times 10^{-13}$, pH = 13, basic; $[H^+] = 1 \times 10^{-1}$, pH = 1, acidic

8-4 (A) 1–2, highly acidic; (B) 6–7 slightly acidic: (C) 0–1, highly acidic; (D) 12–13, highly alkaline

8-5 1×10^{-4} to 1×10^{-8}

8-6 (a) KCl, potassium chloride; (b) barium hydroxide, $Ba(OH)_2$, and hydrochloric acid, HCl: $2HCl + Ba(OH)_2 \longrightarrow BaCl_2 + 2H_2O$

8-7 $Mg(OH)_2 + 2HCl \longrightarrow MgCl_2 + 2H_2O$

8-8 0.0185 *M*

8-9 Water serves as both a hydrogen ion donor (an acid) and a hydrogen ion acceptor (a base)

8-10 $$H_2O + CH_3{-}CH(OH){-}C(=O){-}OH \rightleftharpoons H_3O^+ + CH_3{-}CH(OH){-}C(=O){-}O^{(-)}$$

acid — conjugate base

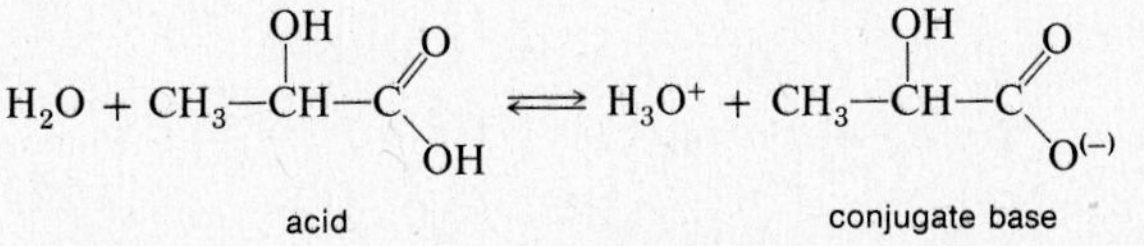

8-11 (a) Water is an acid (H^+ donor), carbonate ion a base (H^+ acceptor).
(b) Hydroxide is the conjugate base of water; bicarbonate is the conjugate acid of carbonate.

8-12 carbonic: $K = \dfrac{[H^+] \times [HCar^-]}{[HCar]}$

salicylic: $K = \dfrac{[H^+] \times [Sal^-]}{[HSal]}$

pyruvic: $K = \dfrac{[H^+] \times [Py^-]}{[HPy]}$

8-13 acetic < lactic < hippuric < acetylsalicylic < salicylic < pyruvic

8-14 increasing conjugate base strength: $Py^- < Sal^- < ASA^- < Ip^- < Lac^- < OAc^-$ (Weak acids have strong conjugate bases.)

8-15 between 2 and 3 ($4.2 \times 10^{-3}\ M = [H^+]$)

8-16 increase $[H^+]$

8-17 4.5×10^{-7} *M*

8-18 Ammonia, acting as a base, accepts hydrogen ion from hydrocholoric acid.

8-19 A base: base + $[(CH_3NH_3]^+Cl^- \longrightarrow$ HBase + CH_3NH_2Cl

8-20 Stomach acid (primarily HCl) furnishes hydrogen ions and chloride ions to convert chlorpromazine to its soluble salt, chlorpromazine hydrochloride.

8-21 $H_3N^{(+)}-C_6H_4-C(=O)-O-(CH_2)_2-N(CH_2CH_3)_2$ or $H_2N-C_6H_4-C(=O)-O-(CH_2)_2-\overset{(+)}{N}H(CH_2CH_3)_2$

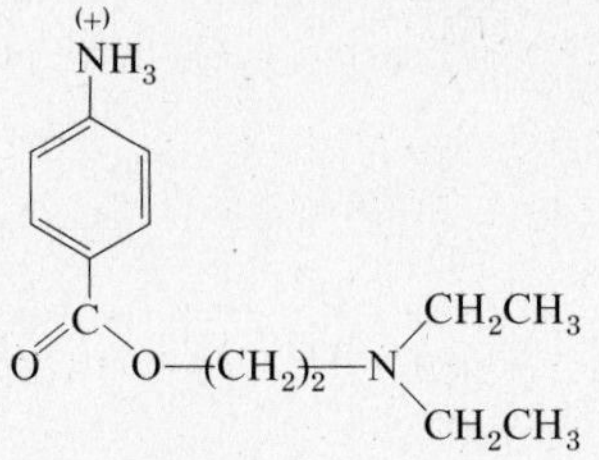

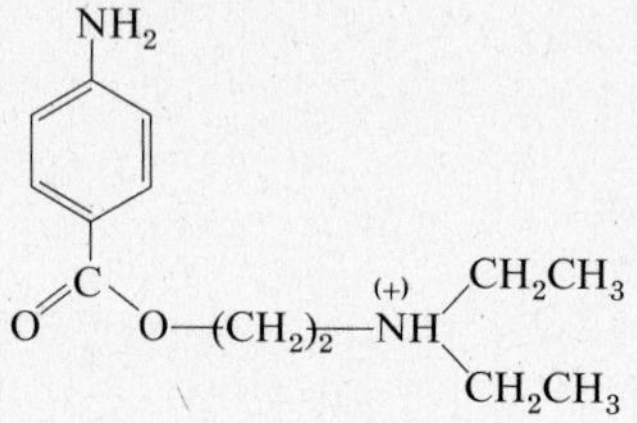

8-22 glycine: $^{(-)}O-C(=O)-CH_2-\overset{(+)}{N}H_3$; valine: $^{(-)}O-C(=O)-CH(\overset{(+)}{N}H_3)-CH(CH_3)_2$;

tyrosine: $^{(-)}O-C(=O)-CH(\overset{(+)}{N}H_3)-CH_2-C_6H_4-OH$; proline: $^{(-)}O-C(=O)-CH$ in ring with CH_2, CH_2, H_2C, $H_2\overset{(+)}{N}$;

arginine: $^{(-)}O-C(=O)-CH(\overset{(+)}{N}H_3)-(CH_2)_3-NH-C(=NH)-NH_2$ or $^{(-)}O-C(=O)-CH(NH_2)-(CH_2)_3-NH-C(=NH)-\overset{(+)}{N}H_3$;

glutamic acid: $^{(-)}O-C(=O)-CH(NH_3)-CH_2CH_2C(=O)OH$

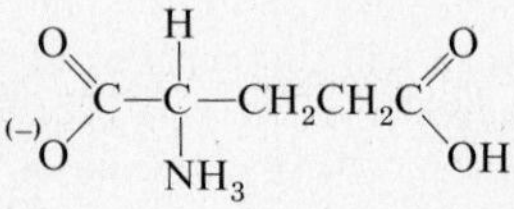

8-23

	Zwitterion	Acid Added	Base Added
Leucine	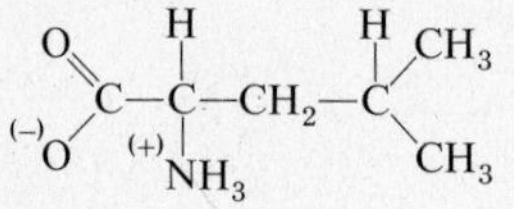 $^{(-)}O-C(=O)-CH(\overset{(+)}{N}H_3)-CH_2-CH(CH_3)_2$	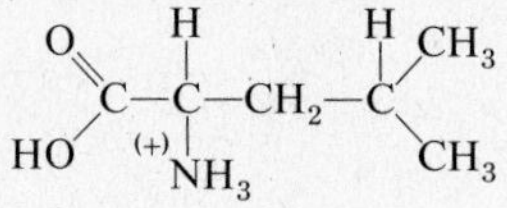 $HO-C(=O)-CH(\overset{(+)}{N}H_3)-CH_2-CH(CH_3)_2$	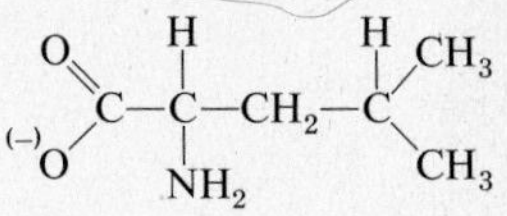 $^{(-)}O-C(=O)-CH(NH_2)-CH_2-CH(CH_3)_2$
Valine	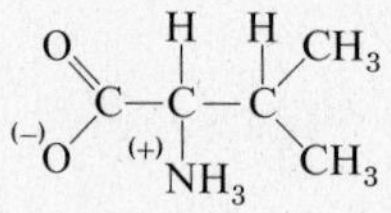 $^{(-)}O-C(=O)-CH(\overset{(+)}{N}H_3)-CH(CH_3)_2$	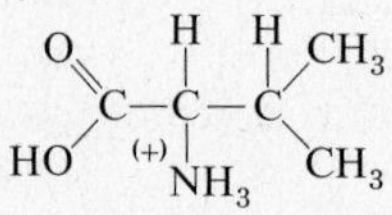 $HO-C(=O)-CH(\overset{(+)}{N}H_3)-CH(CH_3)_2$	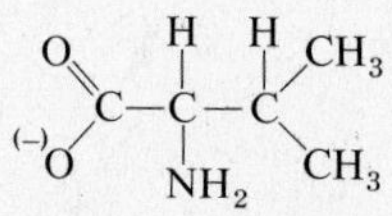 $^{(-)}O-C(=O)-CH(NH_2)-CH(CH_3)_2$
Serine	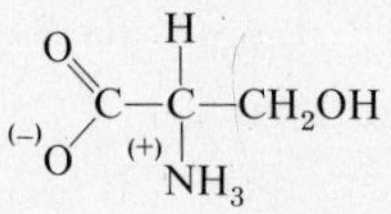 $^{(-)}O-C(=O)-CH(\overset{(+)}{N}H_3)-CH_2OH$	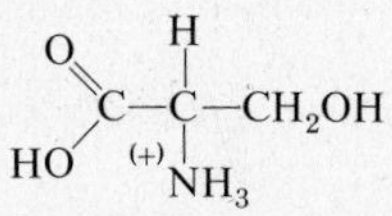 $HO-C(=O)-CH(\overset{(+)}{N}H_3)-CH_2OH$	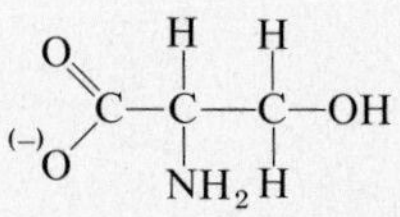 $^{(-)}O-C(=O)-CH(NH_2)-CH_2-OH$

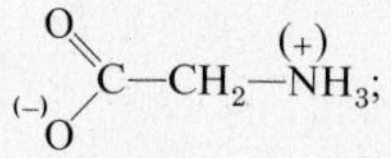

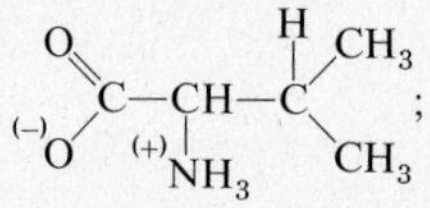

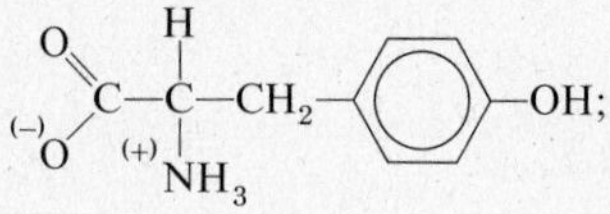

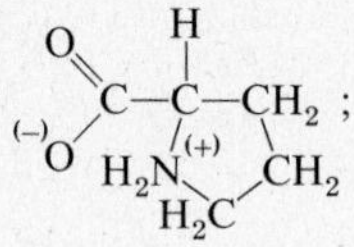

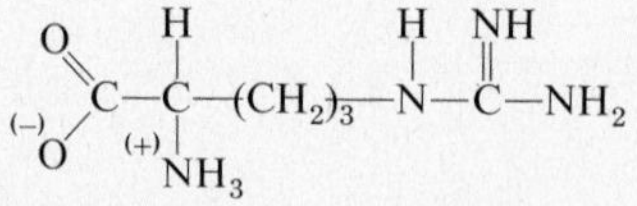

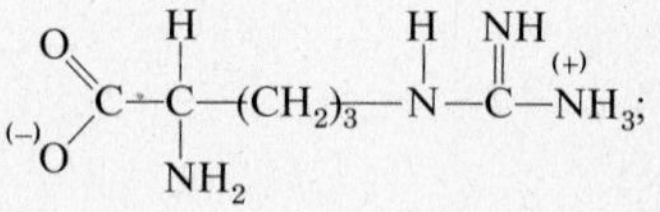

8-24 Breathing into a bag allows $[CO_2]$ to build up in the bag and increased $[CO_2]$ can be inhaled. The increased $[CO_2]$ shifts the equilibria in Eq. 8-2 to the right, thus releasing H^+ to aid in offsetting alkalosis.

CHAPTER NINE

9-1 (a) primary (b) primary (c) secondary (d) secondary (e) tertiary (f) primary

9-2 $CH_3CH_2CH_2CH_2CH_2OH$, primary; $CH_3CH_2CH_2\underset{\displaystyle OH}{\underset{|}{C}}HCH_3$, secondary;

$CH_3CH_2\underset{\displaystyle OH}{\underset{|}{C}}HCH_2CH_3$, secondary; $CH_3\overset{\displaystyle CH_3}{\overset{|}{C}}HCH_2CH_2OH$, primary

$CH_3\overset{\displaystyle CH_3}{\overset{|}{C}}H\underset{\displaystyle OH}{\underset{|}{C}}HCH_3$, secondary; $CH_3\overset{\displaystyle CH_3}{\overset{|}{\underset{\displaystyle OH}{\underset{|}{C}}}}-CH_2CH_3$, tertiary;

$HOCH_2\overset{\displaystyle CH_3}{\overset{|}{C}}HCH_2CH_3$, primary; $CH_3-\overset{\displaystyle CH_3}{\overset{|}{\underset{\displaystyle CH_3}{\underset{|}{C}}}}-CH_2OH$, primary

9-3
(a) 2-methyl-1-propanol (b) cyclopentanol
(c) 1-methylcyclohexanol (d) 2,3-dimethyl-1-pentanol
(e) 2-methyl-1-pentanol (f) 2-ethyl-1-butanol
(g) 2-methylcyclohexanol (h) 4-methylphenol

9-4
(a) $CH_3\underset{\displaystyle OH}{\underset{|}{C}}HCH_3$

(b) $CH_3-\overset{\displaystyle OH}{\overset{|}{\underset{\displaystyle CH_3}{\underset{|}{C}}}}-CH_2-CH_2-CH_2-CH_2-CH_3$

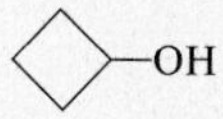

(c) —OH

(d) OH, CH_3

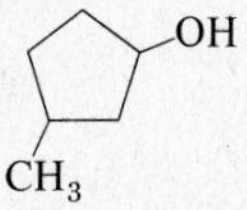

(e) $HOCH_2CH_2CH_2OH$

9-5
(a) $CH_3CH_2\underset{\displaystyle OH}{\underset{|}{C}}H-CH_3$

(b) $CH_3-\overset{\displaystyle CH_3}{\overset{|}{\underset{\displaystyle CH_3}{\underset{|}{C}}}}-OH$

(c) $CH_3-\overset{\displaystyle CH_3}{\overset{|}{\underset{\displaystyle OH}{\underset{|}{C}}}}-CH_2CH_3$

(d) $CH_3CH_2\overset{\displaystyle CH_3}{\overset{|}{\underset{\displaystyle OH}{\underset{|}{C}}}}-CH_2-\overset{\displaystyle CH_3}{\overset{|}{C}}HCH_3$

(e) 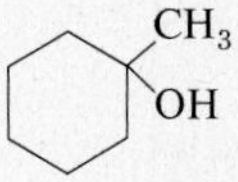 CH_3, OH

(f) 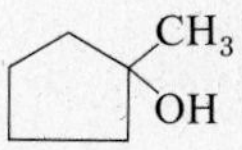 CH_3, OH

(g) $CH_3CH_2\underset{\displaystyle OH}{\underset{|}{C}}HCH_3$

(h) $HO_2CCH_2\underset{\displaystyle OH}{\underset{|}{C}}HCO_2H$

9-6 Both carbon atoms of the double bond in the alkene have the same degree of substitution.

9-7 (a) $CH_3CH{=}CH_2$ (c) cyclohexene (e) 1-methylcyclohexene (CH_3 on the ring)

(b) $CH_3{-}\overset{CH_3}{\overset{|}{C}}{=}CH_2$ (d) 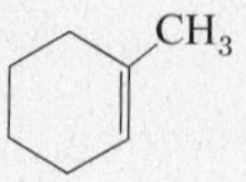 3-methylcyclohexene (CH_3 on the ring) (f) $CH_3\overset{CH_3}{\overset{|}{C}}HCH_2CH{=}C(CH_3)_2$

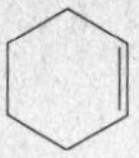

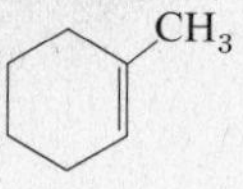

9-8 Addition of water to either 1-butene or 2-butene leads to formation of 2-butanol as the major product.

9-9 (a) CH_3OCH_3 (c) cyclopentyl–O–cyclopentyl

(b) $CH_3\overset{CH_3}{\overset{|}{C}}H{-}O{-}\overset{CH_3}{\overset{|}{C}}HCH_3$ (d) 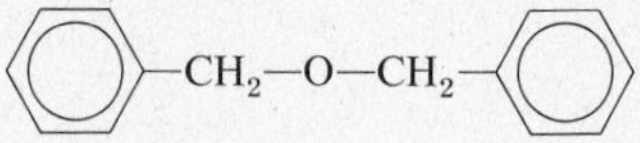 $C_6H_5{-}CH_2{-}O{-}CH_2{-}C_6H_5$

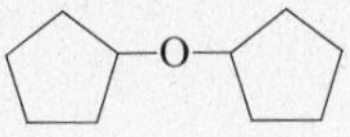

9-10 Tetrahydrofuran (five-membered ring containing O)

9-11 No hydrogen on the carbon next to the one bearing the —OH group; therefore an olefin cannot be formed.

9-12 (a) CH_3CHO (c) $CH_3\overset{CH_3}{\overset{|}{C}}H{-}\underset{O}{\underset{\|}{C}}{-}CH_3$ (e) $OHCCH_2\underset{O}{\underset{\|}{C}}CH_3$

(b) $CH_3\overset{O}{\overset{\|}{C}}CH_3$ (d) $OHC{-}CH_2CHO$ (f) N.R.

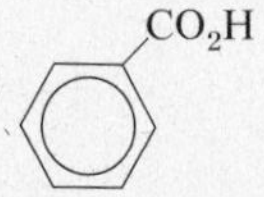

9-13 (a) CH_3CO_2H (e) $HO_2CCH_2CH_2CO_2H$

(d) $C_6H_5CO_2H$ (benzene ring bearing CO_2H) (f) $CH_3CH_2\underset{O}{\underset{\|}{C}}CH_2CO_2H$

9-14 $HCHO + 2Ag(NH_3)_2^+ + 2OH^- \longrightarrow HCO_2^-NH_4^+ + 2Ag + 3NH_3 + H_2O$

9-15 $CH_3\underset{OH}{\underset{|}{C}}H{-}\underset{O}{\underset{\|}{C}}CH_3 + 2Cu^{2+} \xrightarrow{NaOH} CH_3\underset{O}{\underset{\|}{C}}{-}\underset{O}{\underset{\|}{C}}CH_3 + Cu_2O$

9-16 Because it is the reverse of the oxidation of an alcohol.

9-17 (a) $CH_3CH_2CH_2OH$ (c) $CH_3\underset{OH}{\underset{|}{C}}HCH_2CH_2CH_2OH$ (e) cyclohexanol (OH on the ring)

(b) $CH_3\underset{OH}{\underset{|}{C}}HCH_3$ (d) $CH_3CH_2CH_2CH_2OH$

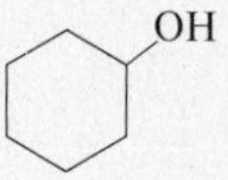

9-18

(a) $CH_3CH(OCH_3)(OH)$

(b) $CH_3CH_2CH(OCH_3)(OH)$

(c) $CH_3CH(OCH_2CH_3)(OH)$

(d) $CH_3C(HO)(OCH_3)CH_3$

(e) $CH_3CH(OH)\text{—}O\text{—}CH_2CH_2\text{—}O\text{—}CH(OH)\text{—}CH_3$

9-19

(a)
```
      CH3
      CH—O
     /     \    OH
CH2        C
     \     /    CH2CH3
      CH2
```

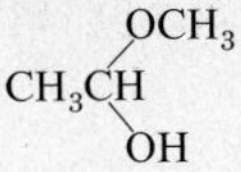

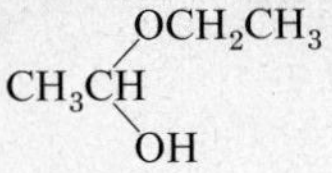

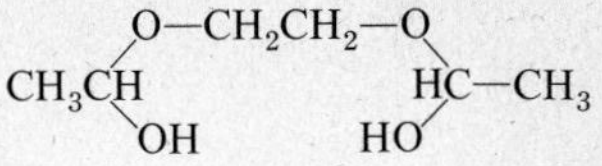

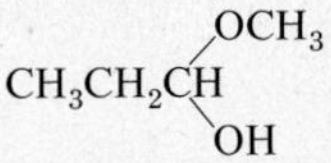

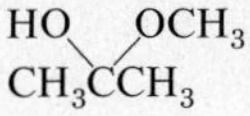

(b)
```
  CH2—CH2    OH
 /        \ /
CH2        C
 \        / \
  CH—O       H
  |
  CH3
```

(c)
```
  CH2—O      CH3
 /      \   /
CH2      C
 \      /   \
  CH2—CH2    OH
```

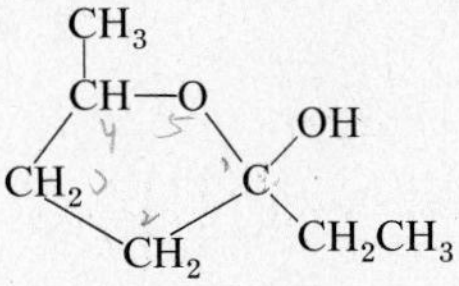

(d)
```
      O       OH
     / \     /
CH2      C
  \     /  \
  CH2—CH2    H
```

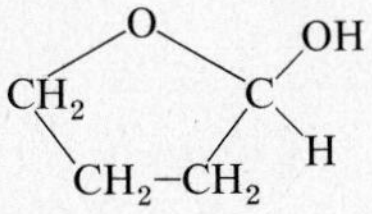

(e)
```
      OH
      |
  CH2—CH     OH
 /       \  /
CH2        C
 \       /  \
  CH2—O      H
```

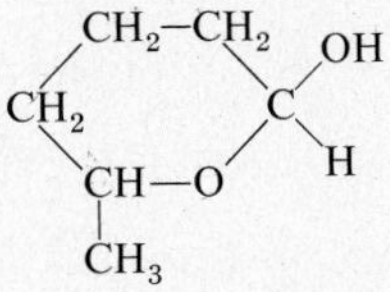

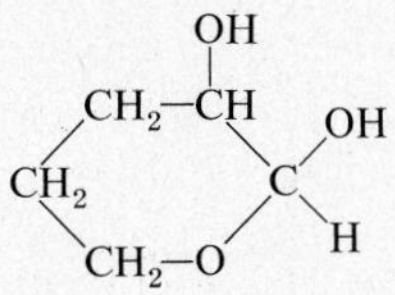

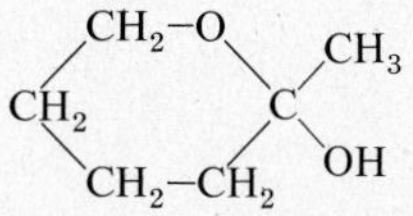

9-20

(a)
```
   H  H   H  H
   |  |   |  |
H—C—C*—C—C—H
   |  |   |  |
   H  O   H  H
      |
      H
```

(b)
```
   H  H  H
   |  |  |
H—C—C—C—H
   |  |  |
   H  O  H
      |
      H
```

(c)
```
   H     H  H
   |     |  |
H—C—C—C—C—H
   |  ‖  |  |
   H  O  H  H
```

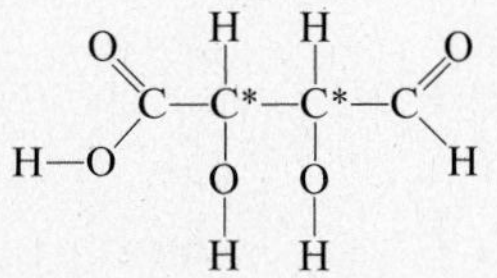

(d)
```
   H     H   H
   |     |   |
H—C—C—C*—C—H
   |  ‖  |   |
   H  O  O   H
         |
         H
```

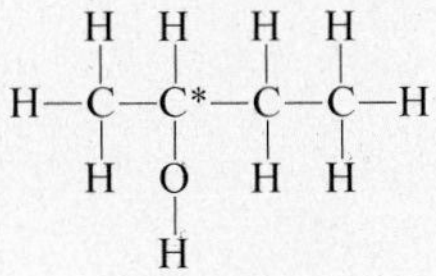

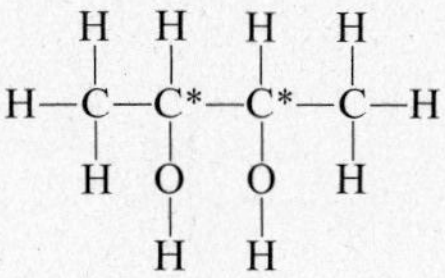

(e)
```
   H
   |
H—C         H
   |  \    /   H  H
   H   C=C     |  |
      /    \  —C—C—H
     H         |  |
               H  H
```

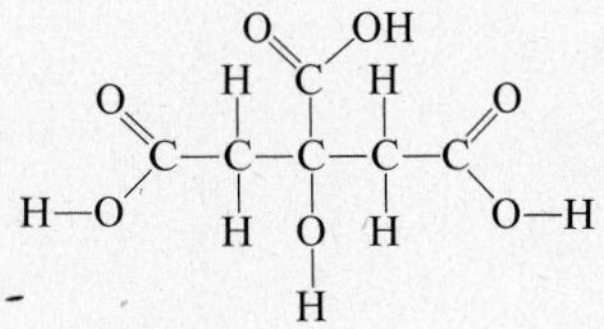

(f)
```
               O  OH
            H   \\C/   H
  O         |    |     |       O
   \\C—C—C—C—C//
   /        |    |     |       \
H—O         H    O     H        O—H
                 |
                 H
```

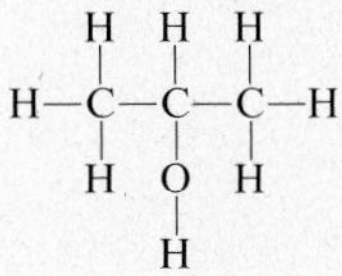

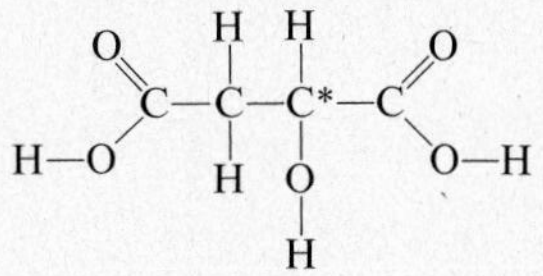

(g)
```
           H  H
  O        |  |        O
   \\C—C—C*—C//
   /       |  |        \
H—O        H  O         O—H
              |
              H
```

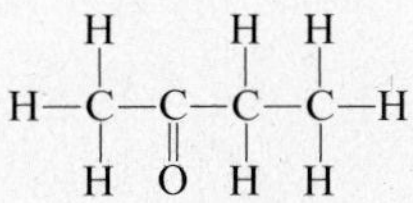

(h)
```
           H    H
  O        |    |       O
   \\C—C*—C*—C//
   /       |    |       \
H—O        O    O        H
           |    |
           H    H
```

(i)
```
   H  H    H   H
   |  |    |   |
H—C—C*—C*—C—H
   |  |    |   |
   H  O    O   H
      |    |
      H    H
```

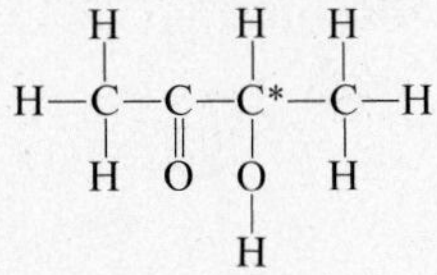

(j)
```
   H  H   H  H   H
   |  |   |  |   |
H—C—C*—C—C*—C—H
   |  |   |  |   |
   H  O   O  O   H
      |   |  |
      H   H  H
```

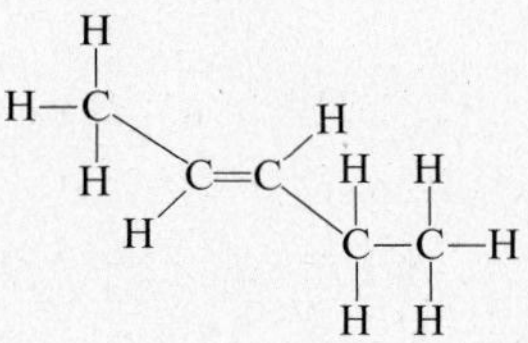

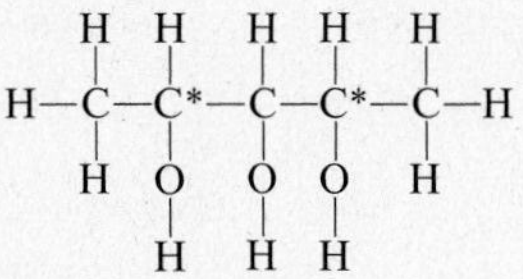

9-21

(a)

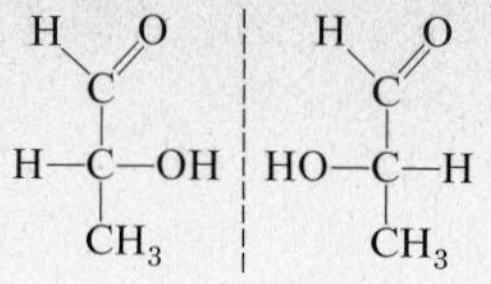

(b)

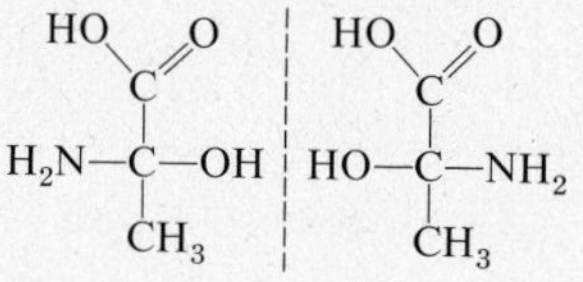

(c)

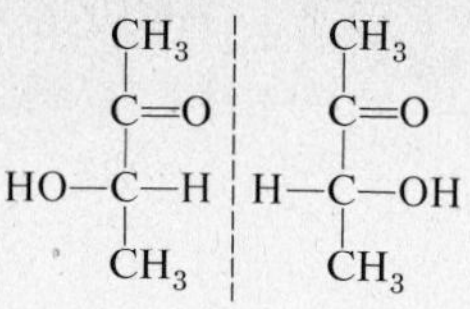

(d)

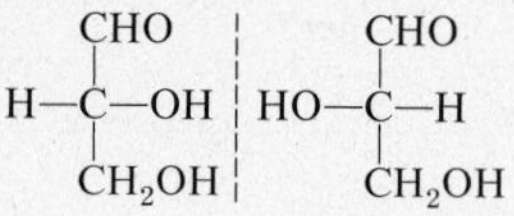

9-22

Column A	Column B
1	4
2	1
3	2
4	5
5	3

9-23

(a)

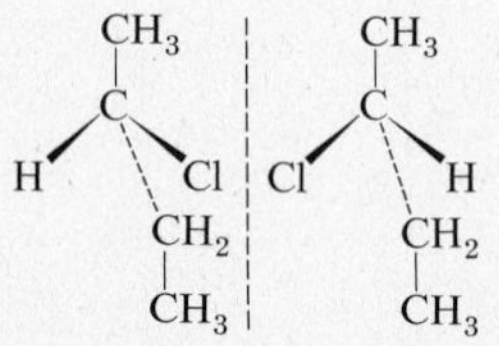

(b)

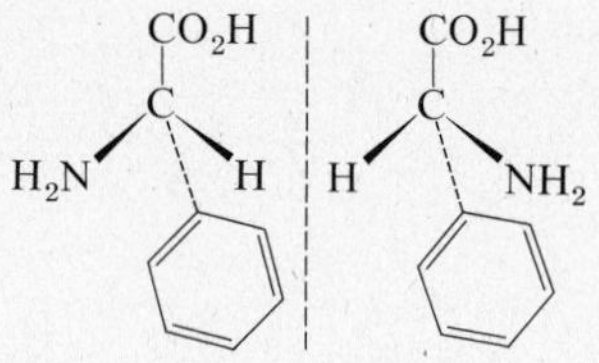

(c)

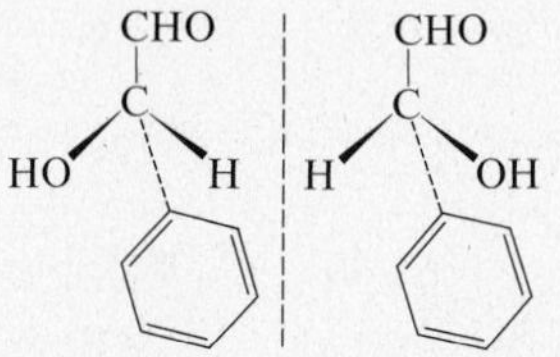

(d)

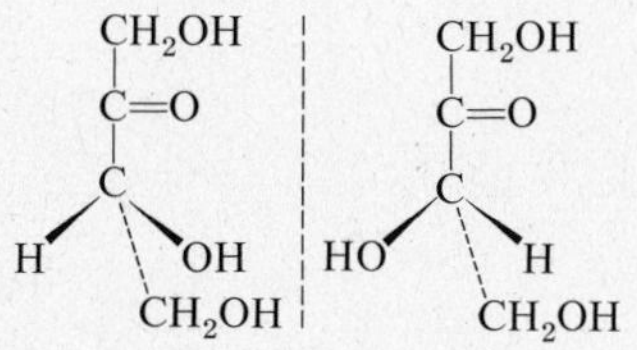

9-24 (a) 2,1 (b) 4,2 (c) 2,1 (d) 4,2 (e) 16,8

9-25 (a) D (b) D (c) L (d) L (e) L (f) D

9-26 D-Lactic acid has the OH written to the right in the Fischer formula and is related to D-glyceraldehyde. L-Lactic acid is related to L-glyceraldehyde.

9-27

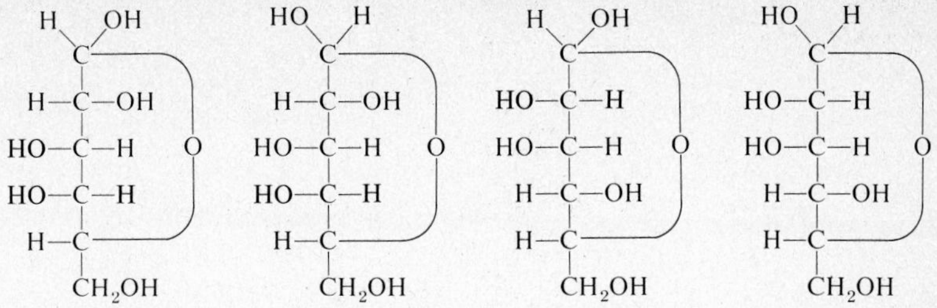

α-D-galactose β-D-galactose α-D-mannose β-D-mannose

9-28

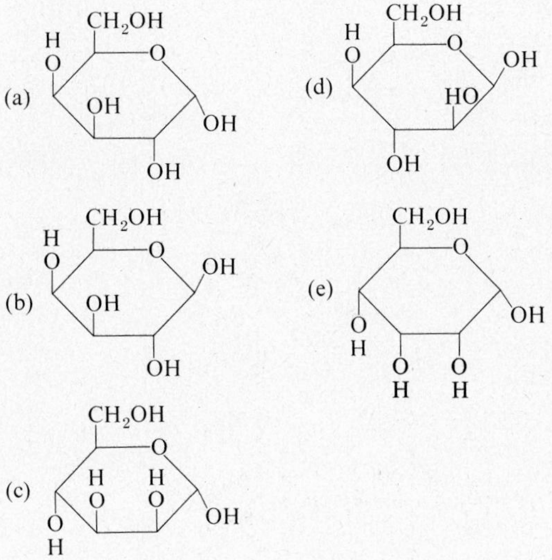

(a) (b) (c) (d) (e)

9-29

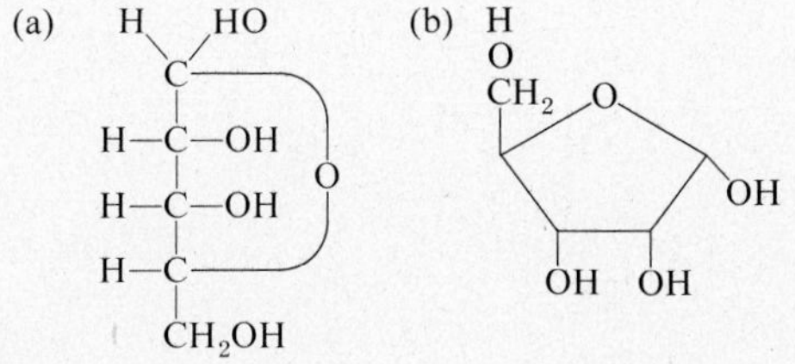

(a) (b)

9-30

$+ H_2O \xrightarrow{H^+} 2$

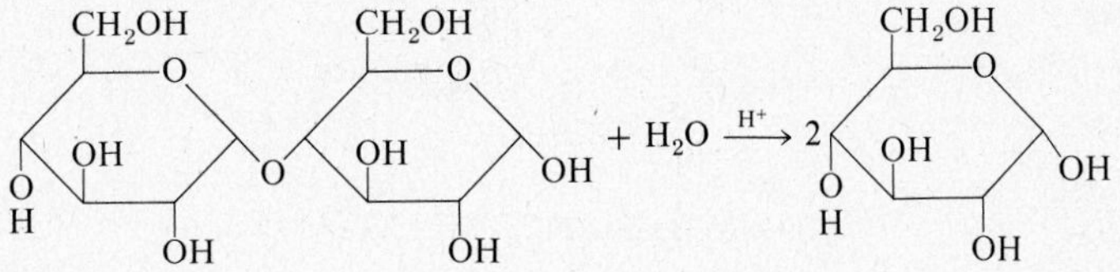

9-31 They are designated the alpha forms of the sugars because the hemiacetal group is written downward in the Haworth formula.

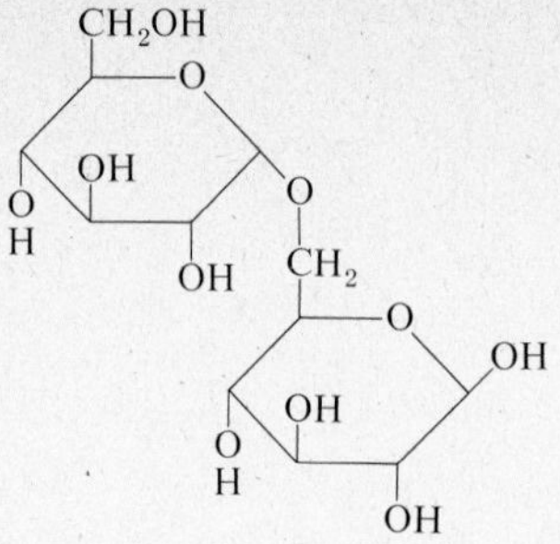

9-32

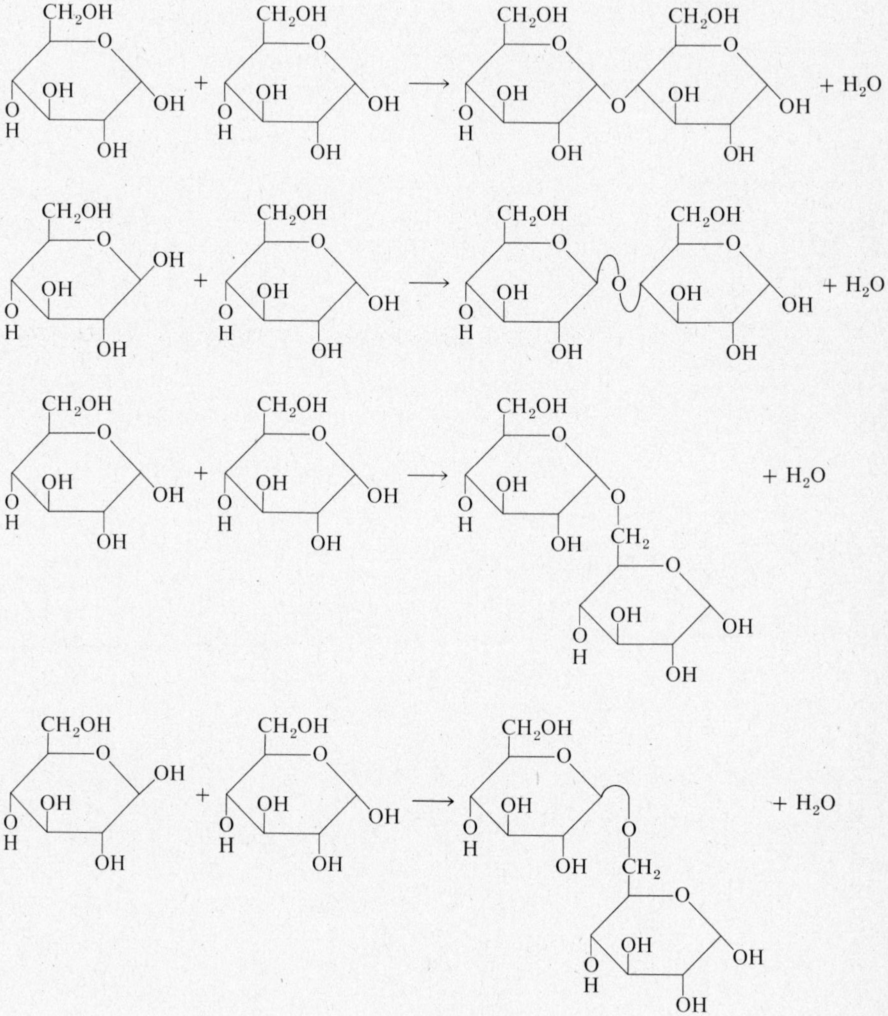

$+ H_2O$

$+ H_2O$

$+ H_2O$

$+ H_2O$

9-33 α-glucose and β-fructose

9-34 There is neither a hemiacetal nor a hemiketal group present.

9-35

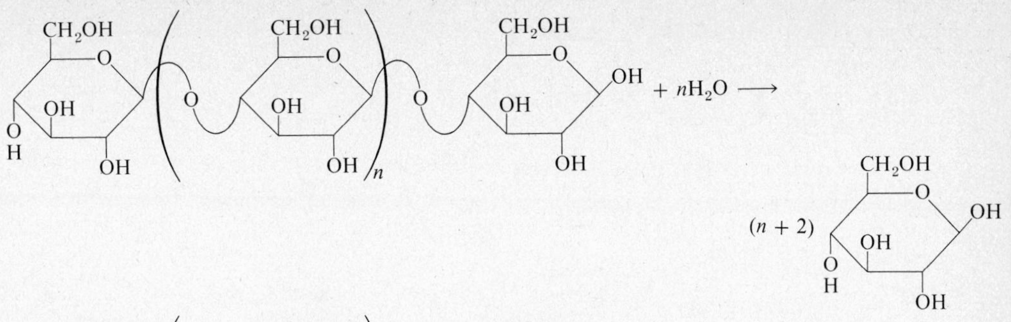

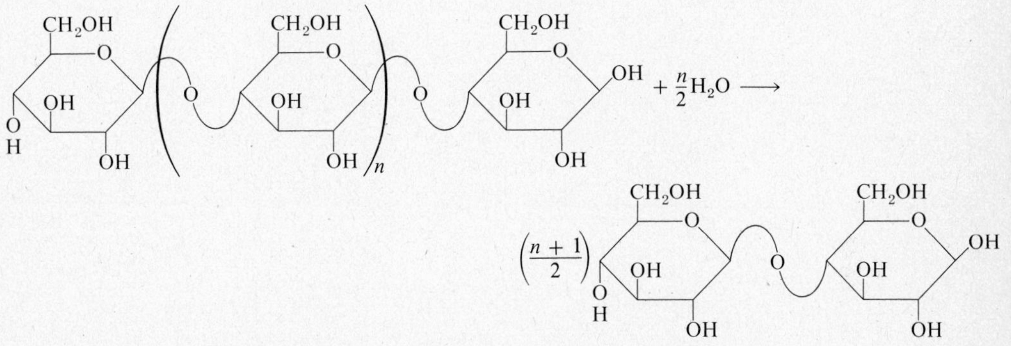

$$\text{(glucose)}-\left(\text{O-glucose}\right)_n-\text{O-glucose} + nH_2O \longrightarrow (n+2)\ \text{glucose}$$

$$\text{(glucose)}-\left(\text{O-glucose}\right)_n-\text{O-glucose} + \frac{n}{2}H_2O \longrightarrow \left(\frac{n+1}{2}\right)\ \text{glucose-O-glucose}$$

9-36 Because starch contains some α-1,6 linkages between glucose units.

CHAPTER TEN

10-1

Acid	Acid	Anhydride
Butyric	Butyric	$CH_3(CH_2)_2-\overset{O}{\overset{\parallel}{C}}-O-\overset{O}{\overset{\parallel}{C}}-(CH_2)_2CH_3$
Phosphoric	$CH_3CH_2-O-P(=O)(OH)(OH)$	$(HO)(HO)P(=O)-O-P(=O)(OH)-O-CH_2CH_3$
$CH_3CH_2CH_2CO_2H$	Acetic	$CH_3CH_2CH_2\overset{O}{\overset{\parallel}{C}}-O-\overset{O}{\overset{\parallel}{C}}-CH_3$
$CH_3CH_2CO_2H$	$HO-P(=O)(OH)(OH)$	$CH_3CH_2-\overset{O}{\overset{\parallel}{C}}-O-P(=O)(OH)-OH$
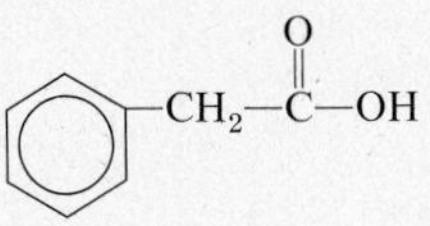 $C_6H_5-CH_2-\overset{O}{\overset{\parallel}{C}}-OH$	$HO-P(=O)(OH)(OH)$	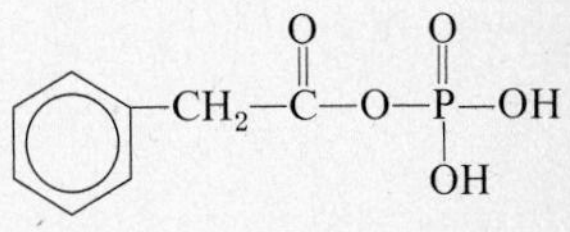 $C_6H_5-CH_2-\overset{O}{\overset{\parallel}{C}}-O-P(=O)(OH)-OH$
$CH_3(CH_2)_3\overset{O}{\overset{\parallel}{C}}-OH$	2 phosphoric	$CH_3(CH_2)_3\overset{O}{\overset{\parallel}{C}}-O-P(=O)(OH)-O-P(=O)(OH)-OH$

10-2

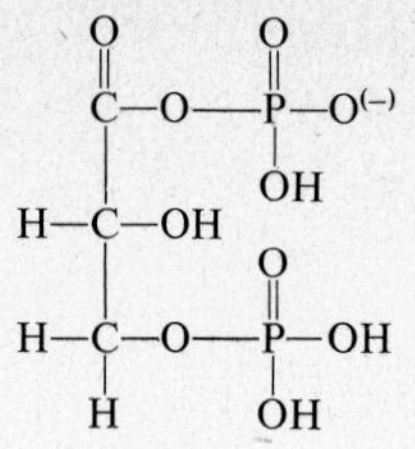

$$\begin{array}{l} \mathrm{O} \qquad\quad \mathrm{O} \\ \| \qquad\quad\ \| \\ \mathrm{C{-}O{-}P{-}O^{(-)}} \\ | \qquad\qquad | \\ | \qquad\quad\ \mathrm{OH} \\ \mathrm{H{-}C{-}OH} \\ | \qquad\qquad \mathrm{O} \\ | \qquad\qquad \| \\ \mathrm{H{-}C{-}O{-}P{-}OH} \\ | \qquad\qquad | \\ \mathrm{H} \qquad\quad \mathrm{OH} \end{array}$$

10-3 (a) $C_6H_5{-}O{-}\overset{O}{\overset{\|}{C}}{-}(CH_2)_2CH_3$; (b) no ester group; (c) no ester group;

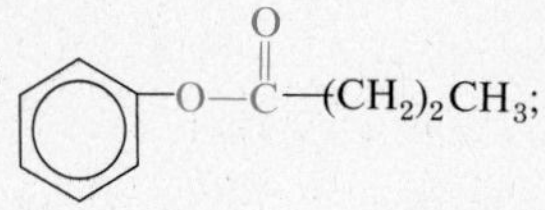

(d) $o\text{-}(HO{-}\overset{O}{\overset{\|}{C}})C_6H_4{-}O{-}\overset{O}{\overset{\|}{C}}{-}CH_3$; (e) $CH_3{-}\overset{O}{\overset{\|}{C}}{-}O{-}(CH_2)_3{-}O{-}\overset{O}{\overset{\|}{C}}{-}C_2H_5$;

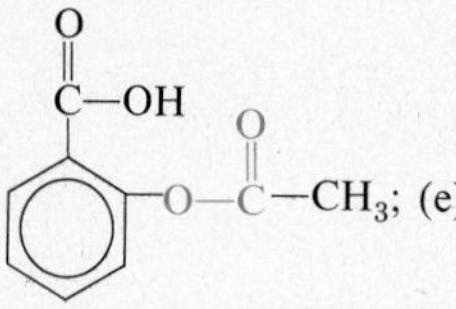

(f) $H_2N{-}\overset{O}{\overset{\|}{C}}{-}O{-}CH_2{-}\underset{C_3H_7}{\underset{|}{\overset{CH_3}{\overset{|}{C}}}}{-}CH_2{-}O{-}\overset{O}{\overset{\|}{C}}{-}NH_2$; (g) no ester group

10-4 (a) $H_2O + CH_3{-}O{-}\overset{O}{\overset{\|}{C}}{-}CH_3$

(b) $H_2O + C_6H_5{-}O{-}\overset{O}{\overset{\|}{C}}{-}CH_2CH_3$

(c) $H_2O + C_6H_5{-}\overset{O}{\overset{\|}{C}}{-}O{-}CH_2CH_3$

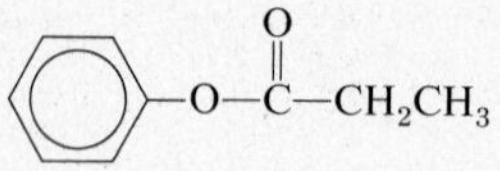

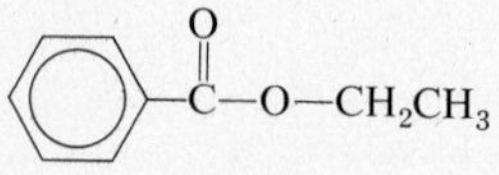

(d) $CH_3(CH_2)_2OH$

(e) $CH_3OH + C_6H_5{-}CH_2\overset{O}{\overset{\|}{C}}{-}OH$

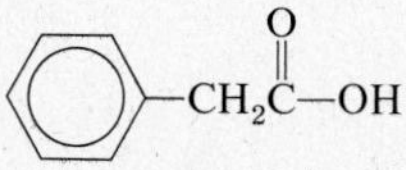

(f) $CH_3\overset{O}{\overset{\|}{C}}{-}OH + CH_3(CH_2)_3{-}O{-}\overset{O}{\overset{\|}{C}}{-}CH_3$

(g) $CH_3CH_2{-}\overset{O}{\overset{\|}{C}}{-}O{-}\overset{O}{\overset{\|}{C}}{-}CH_2CH_3$,

$CH_3{-}C_6H_4{-}O{-}\overset{O}{\overset{\|}{C}}{-}CH_2CH_3$

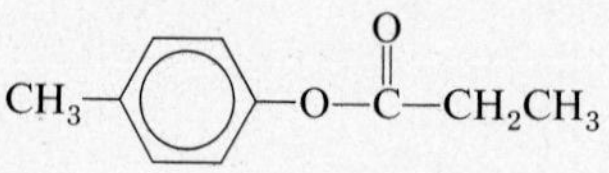

10-5

Alcohol	Acid	Ester
Methyl CH_3OH	Acetic CH_3CO_2H	Methyl acetate $CH_3O\overset{O}{\overset{\parallel}{C}}CH_3$
Ethyl CH_3CH_2OH	Formic $H{-}CO_2H$	Ethyl formate $CH_3CH_2O{-}\overset{O}{\overset{\parallel}{C}}{-}H$
Methyl CH_3OH	Stearic $CH_3(CH_2)_{16}CO_2H$	Methyl stearate $CH_3{-}O{-}\overset{O}{\overset{\parallel}{C}}{-}(CH_2)_{16}{-}CH_3$
Propyl $CH_3(CH_2)_2OH$	Butyric $CH_3(CH_2)_3CO_2H$	Propyl butyrate $CH_3(CH_2)_2{-}O{-}\overset{O}{\overset{\parallel}{C}}{-}(CH_2)_3CH_3$

10-6 $CH_3CH_2{-}O{-}\overset{O}{\overset{\parallel}{P}}{-}OH$, with OH on P ($CH_3CH_2{-}O{-}P(=O)(OH){-}OH$)

10-7 (a) oleic, linoleic, linolenic; (b) stearic, palmitic

10-8 (a)

$$R_1{-}\overset{O}{\overset{\parallel}{C}}{-}OH + R_2{-}\overset{O}{\overset{\parallel}{C}}{-}OH + R_3{-}\overset{O}{\overset{\parallel}{C}}{-}OH + H{-}C{-}OH\ (H_2C{-}OH,\ H_2C{-}OH) \longrightarrow$$

R_1: a saturated fatty acid, e.g., myristic or lauric
R_2: an unsaturated fatty acid, e.g., oleic or linoleic
R_3: a saturated fatty acid, e.g., palmitic or stearic
$H_2C{-}OH$ / $H{-}C{-}OH$ / $H_2C{-}OH$: glycerol

$$3H_2O + \begin{array}{l} H_2C{-}O{-}\overset{O}{\overset{\parallel}{C}}{-}R_1 \\ HC{-}O{-}\overset{O}{\overset{\parallel}{C}}{-}R_2 \\ H_2C{-}O{-}\overset{O}{\overset{\parallel}{C}}{-}R_3 \end{array}$$

(b) Yes, for example,

$$\begin{array}{l} H_2C{-}O{-}\overset{O}{\overset{\parallel}{C}}{-}R_2 \\ HC{-}O{-}\overset{O}{\overset{\parallel}{C}}{-}R_1 \\ H_2C{-}O{-}\overset{O}{\overset{\parallel}{C}}{-}R_3 \end{array} \quad \text{or} \quad \begin{array}{l} H_2C{-}O{-}\overset{O}{\overset{\parallel}{C}}{-}R_2 \\ HC{-}O{-}\overset{O}{\overset{\parallel}{C}}{-}R_3 \\ H_2C{-}O{-}\overset{O}{\overset{\parallel}{C}}{-}R_1 \end{array}$$

10-9 (a)

$$C_6H_4(\overset{O}{\overset{\parallel}{C}}{-}OH)(O{-}\overset{O}{\overset{\parallel}{C}}{-}CH_3) + HOH \xrightarrow{\triangle,\ H^+} C_6H_4(\overset{O}{\overset{\parallel}{C}}{-}OH)(OH) + CH_3\overset{O}{\overset{\parallel}{C}}{-}OH$$

(b) $HO{-}CH_2{-}\underset{\large OH}{\underset{|}{C}}HCH_2{-}O{-}\underset{\large OH}{\underset{|}{\overset{\large O}{\overset{\|}{P}}}}{-}OH + HOH \xrightarrow{\text{enzyme}}$

$$HO{-}CH_2\underset{\large OH}{\underset{|}{C}}HCH_2OH + HO{-}\underset{\large OH}{\underset{|}{\overset{\large O}{\overset{\|}{P}}}}{-}OH$$

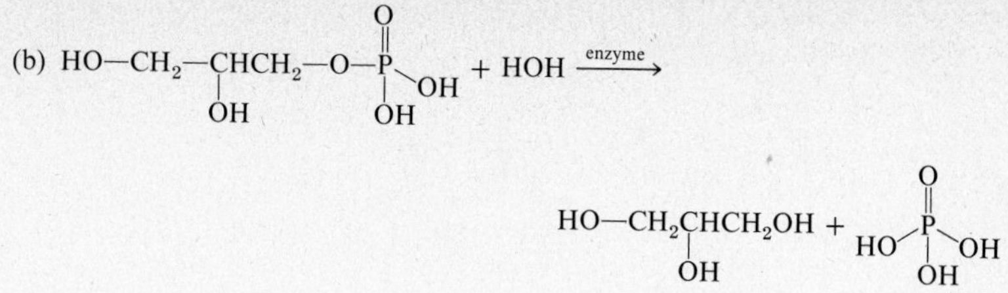

10-10

$$\begin{array}{l} H_2C{-}O{-}\overset{O}{\overset{\|}{C}}{-}(CH_2)_{10}CH_3 \\ \;| \\ HC{-}O{-}\overset{O}{\overset{\|}{C}}{-}(CH_2)_7CH{=}CHCH_2CH{=}CH(CH_2)_4CH_3 + 3H_2O \xrightarrow{\Delta,\ H^+} \\ \;| \\ H_2C{-}O{-}\overset{O}{\overset{\|}{C}}{-}(CH_2)_7CH{=}CH(CH_2)_7CH_3 \end{array}$$

$$CH_3(CH_2)_7CH{=}CH(CH_2)_7CO_2H +$$
$$CH_3(CH_2)_4CH{=}CHCH_2CH{=}CH(CH_2)_7CO_2H +$$
$$CH_3(CH_2)_{10}CO_2H + \begin{array}{l} H_2C{-}OH \\ \;\;| \\ H{-}C{-}OH \\ \;\;| \\ H_2C{-}OH \end{array}$$

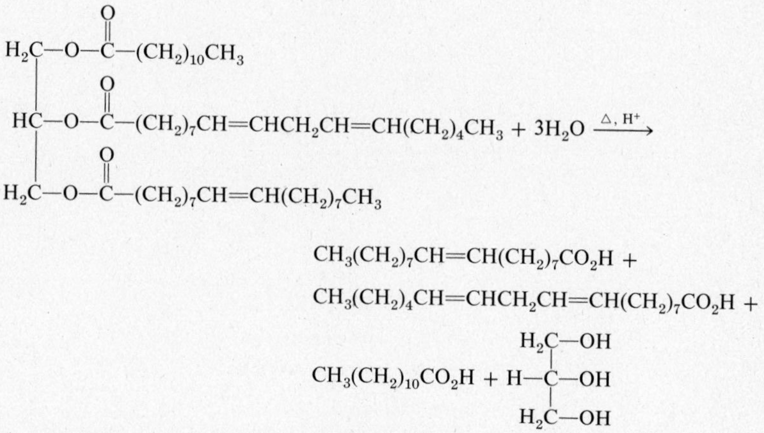

10-11 (a) $CH_3OH + Na^{(+)} \left[CH_3CH_2\overset{O}{\overset{\|}{C}}{-}O\right]^{(-)}$

(b) $CH_3CH_2CO_2H + CH_3OH$

(c) $C_6H_5{-}(CH_2)_2OH + K^{(+)} \left[C_6H_5{-}\overset{O}{\overset{\|}{C}}{-}O\right]^{(-)}$

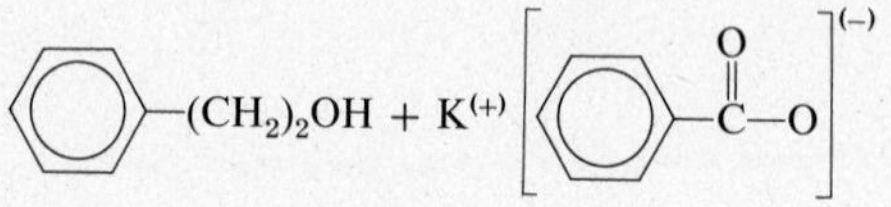

(d) $CH_3OH + Na^{(+)} \left[\text{2-HO-}C_6H_4{-}\overset{O}{\overset{\|}{C}}{-}O\right]^{(-)}$

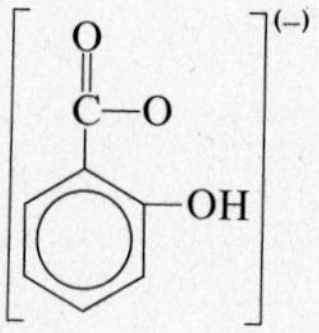

10-12 (a) $CH_3CH_2OH + C_6H_5{-}CH_2CO_2H$

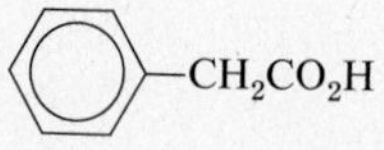

(b) $CH_3CH_2OH + Na^{(+)} \left[C_6H_5{-}CH_2{-}\overset{O}{\overset{\|}{C}}{-}O\right]^{(-)}$

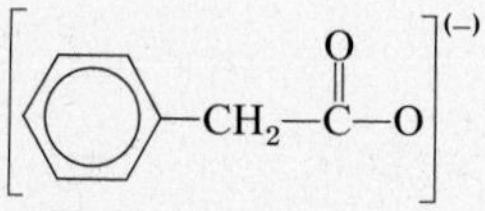

(c) $HOCH_2CH(OH)CH_2OH + Na^{(+)}\left[CH_3(CH_2)_2\overset{O}{\overset{\|}{C}}{-}O\right]^{(-)} +$

$Na^{(+)}\left[C_6H_5{-}(CH_2)_2\overset{O}{\overset{\|}{C}}{-}O\right]^{(-)} + Na^{(+)}\left[CH_3(CH_2)_{16}\overset{O}{\overset{\|}{C}}{-}O\right]^{(-)}$

10-13 $H_2N{-}(CH_2)_4{-}CH_2{-}NH_2$

CHAPTER ELEVEN

11-1 (a) $[NH_4]^{(+)}\left[C_5H_9{-}\overset{O}{\overset{\|}{C}}{-}O\right]^{(-)}$ (cyclopentyl)

(b) $CH_3{-}CH(NH_2){-}\overset{O}{\overset{\|}{C}}{-}NH{-}CH_2CO_2H$

(c) $CH_3{-}NH{-}\overset{O}{\overset{\|}{C}}{-}(CH_2)_4{-}\overset{O}{\overset{\|}{C}}{-}NH{-}CH_3$

(d) $\left[(CH_3)_2NH{-}CH_3\right]^{(+)}\left[CH_3(CH_2)_3\overset{O}{\overset{\|}{C}}{-}O\right]^{(-)}$

11-2 $(CH_3)_2CH{-}CH(NH_2){-}\overset{O}{\overset{\|}{C}}{-}OH$ (valine) $+ (CH_3)_2CHCH_2{-}CH(NH_2){-}\overset{O}{\overset{\|}{C}}{-}OH$ (leucine) $\longrightarrow$

$H_2O + (CH_3)_2CH{-}CH(NH_2){-}\overset{O}{\overset{\|}{C}}{-}NH{-}CH(CO_2H){-}CH_2CH(CH_3){-}CH_3$

valylleucine
(leucylvaline also possible)

11-3 AGV: $CH_3{-}CH(NH_2){-}\overset{O}{\overset{\|}{C}}{-}NH{-}CH_2{-}\overset{O}{\overset{\|}{C}}{-}NH{-}CH(CH(CH_3)_2){-}CO_2H$

GAV: $H_2N{-}CH_2{-}\overset{O}{\overset{\|}{C}}{-}NH{-}CH(CH_3){-}\overset{O}{\overset{\|}{C}}{-}NH{-}CH(CH(CH_3)_2){-}\overset{O}{\overset{\|}{C}}{-}OH$

11-4 (see above) ------ represent hydrogen bonds

11-5 Milk contains the protein casein. Adding lemon juice or vinegar (both are acids) lowers the milk pH to the isoelectric point of casein and causes it to precipitate.

11-6

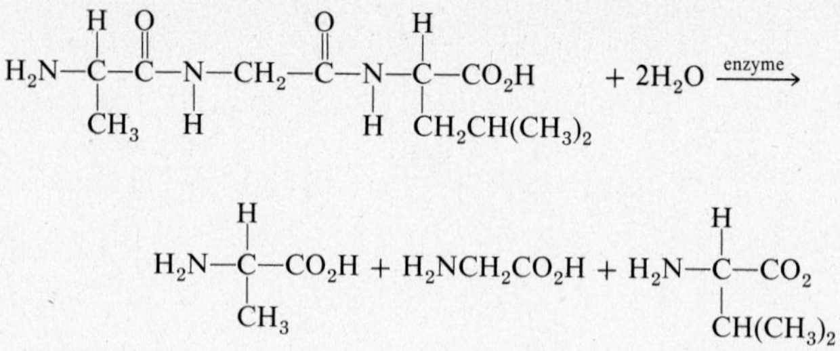

$$H_2N-\underset{CH_3}{\overset{H}{C}}-\overset{O}{\overset{\|}{C}}-\underset{H}{N}-CH_2-\overset{O}{\overset{\|}{C}}-\underset{H}{N}-\underset{CH_2CH(CH_3)_2}{\overset{H}{C}}-CO_2H + 2H_2O \xrightarrow{\text{enzyme}}$$

$$H_2N-\underset{CH_3}{\overset{H}{C}}-CO_2H + H_2NCH_2CO_2H + H_2N-\underset{CH(CH_3)_2}{\overset{H}{C}}-CO_2$$

11-7 five

11-8 four

CHAPTER TWELVE

12-1 Low pH, high $[H^+]$; therefore HPO_4^{2-} is least likely to be present

12-2 (a) $NADH_2 \longrightarrow 2H + NAD$

$M + 2H \longrightarrow MH_2$

(b) M → MH_2 coupled with $NADH_2$ → NAD

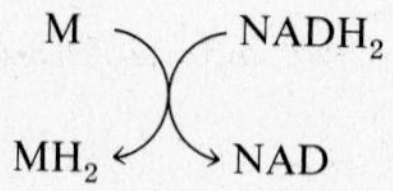

12-3 NAD has become reduced to $NADH_2$ by gaining hydrogen; lactic acid loses hydrogen being oxidized to pyruvic acid.

12-4

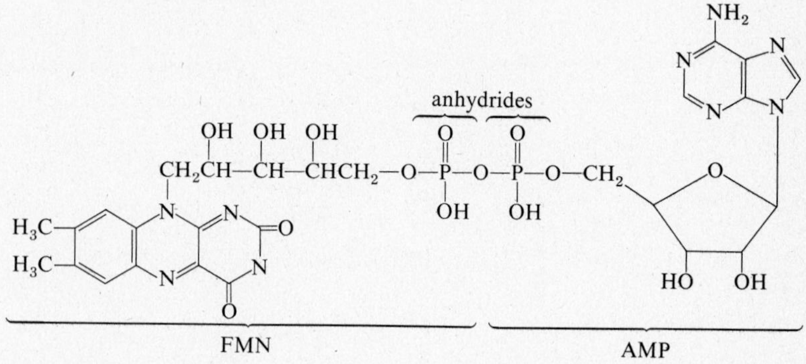

FMN–AMP structure: $CH_2CH(OH)-CH(OH)-CH(OH)CH_2-O-P(=O)(OH)-O-P(=O)(OH)-O-CH_2$ (ribose, adenine); flavin ring with H_3C, H_3C; labels: anhydrides, FMN, AMP, NH_2, HO, OH

12-5 Succinic acid is oxidized (loss of hydrogen) to fumaric acid; FAD is reduced (gain of hydrogen) to $FADH_2$.

12-6 MH_2 → M; NAD → $NADH_2$; $FMNH_2$ → FMN; CoQ → $CoQH_2$ (coupled reactions)

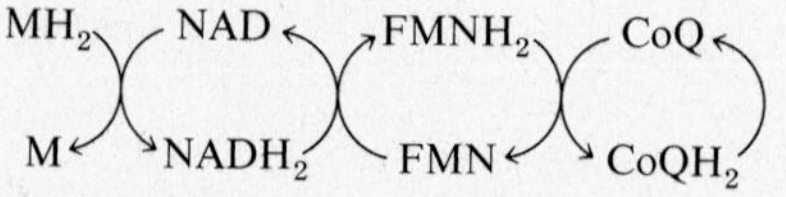

12-7 (a) oxidation; (b) reduction; (c) cyt c oxidation, cyt a reduction; (d) $CoQH_2$ oxidation, cyt b reduction; (e) Oxygen is reduced.

12-8

$$CH_3(CH_2)_{10}CO_2H + ATP \xrightarrow{\text{thiokinase}} CH_3(CH_2)_{10}\overset{O}{\overset{\|}{C}}{-}O{-}\underset{OH}{\underset{|}{\overset{O}{\overset{\|}{P}}}}{-}O{-}\text{adenosine} + PP_i$$

$$\xrightarrow{CoASH} CH_3(CH_2)_{10}{-}\overset{O}{\overset{\|}{C}}{-}SCoA + AMP$$

"active" lauric acid

12-9

$$CH_3(CH_2)_8(CH_2)_2\overset{O}{\overset{\|}{C}}{-}SCoA \xrightarrow{FAD \quad FADH_2} CH_3(CH_2)_8CH{=}CH{-}\overset{O}{\overset{\|}{C}}{-}SCoA$$

$$CH_3(CH_2)_8CH{=}CH{-}\overset{O}{\overset{\|}{C}}{-}SCoA + H_2O \longrightarrow CH_3(CH_2)_8\overset{OH}{\overset{|}{C}}H{-}CH_2{-}\overset{O}{\overset{\|}{C}}{-}SCoA$$

$$CH_3(CH_2)_8\overset{OH}{\overset{|}{C}}H{-}CH_2{-}\overset{O}{\overset{\|}{C}}{-}SCoA \xrightarrow{NAD \quad NADH_2} CH_3(CH_2)_8\overset{O}{\overset{\|}{C}}{-}CH_2{-}\overset{O}{\overset{\|}{C}}{-}SCoA$$

$$CH_3(CH_2)_8\overset{O}{\overset{\|}{C}}{-}CH_2{-}\overset{O}{\overset{\|}{C}}{-}SCoA + CoASH \longrightarrow \underset{\text{acetyl CoA}}{CH_3{-}\overset{O}{\overset{\|}{C}}{-}SCoA} + CH_3(CH_2)_8\overset{O}{\overset{\|}{C}}{-}SCoA$$

12-10 6 acetyl coenzyme A's

12-11 See text; after activation, steps 2 through 5 repeated six times shorten the chain to

$$CH_3{-}CH_2{-}CH_2{-}\overset{O}{\overset{\|}{C}}{-}SCoA.$$

See text for sequence from there.

12-12 24 moles ATP/mole lauric acid

12-13 Citric acid contains three carboxylic acid groups.

12-14 Hydrogens are removed (oxidation) from the beta carbon that had been a secondary alcohol and is then oxidized to a ketone.

12-15 $130 \text{ moles ATP} \times \dfrac{8000 \text{ cal}}{\text{mole ATP}} = 104{,}000 \text{ cal} = 104 \text{ kcal}$

CHAPTER THIRTEEN

13-1

$$\underset{CH_3}{\underset{|}{\overset{CO_2H}{\overset{|}{C}}}}{=}O \longrightarrow \underset{CH_3}{\underset{|}{H{-}C{=}O}} + CO_2$$

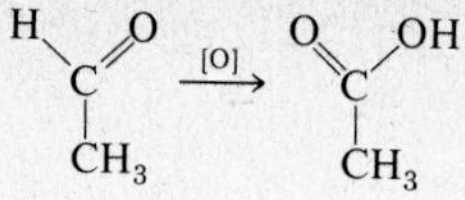

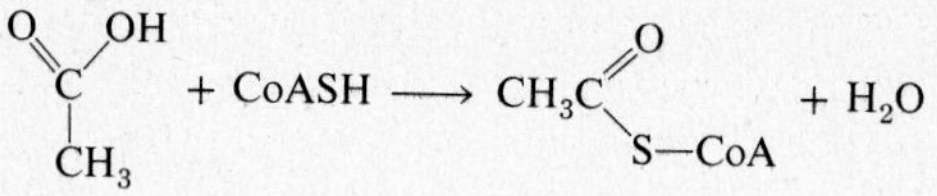

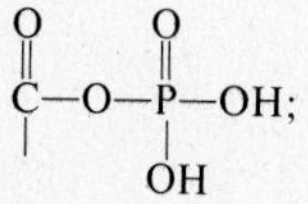

$$CH_3CHO \xrightarrow{[O]} CH_3COOH$$

$$CH_3COOH + CoASH \longrightarrow CH_3C(=O){-}S{-}CoA + H_2O$$

$$-C(=O){-}O{-}P(=O)(OH){-}OH;\quad \text{phosphate ester;}\quad HO{-}C(=O){-}CH(OH){-}CH_2OH$$

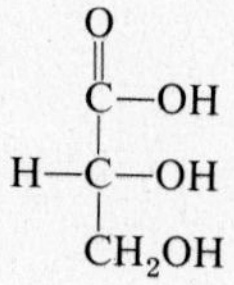

13-3

$$CH_3CH_2CH_2C(=O)SCoA + HO_2C{-}CH_2{-}C(=O){-}SCoA \longrightarrow CH_3CH_2CH_2C(=O){-}CH_2{-}C(=O)SCoA + CoASH + CO_2$$

$$CH_3CH_2CH_2C(=O){-}CH_2{-}C(=O)SCoA \xrightarrow{NADPH_2 \to NADP} CH_3CH_2CH_2CH(OH)CH_2C(=O)SCoA$$

$$CH_3CH_2CH_2CH(OH)CH_2C(=O)SCoA \longrightarrow CH_3CH_2CH_2CH{=}CHC(=O)SCoA + H_2O$$

$$CH_3CH_2CH_2CH{=}CHC(=O)SCoA \xrightarrow{NADPH_2 \to NADP} CH_3CH_2CH_2CH_2CH_2C(=O)SCoA$$

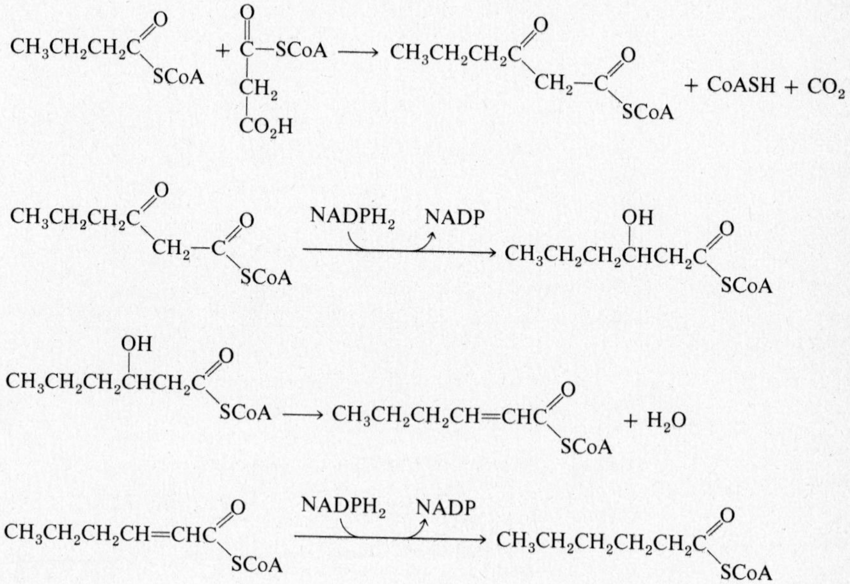

Then repeat steps starting with malonyl CoA step; then $NADPH_2 \longrightarrow NADP$; dehydration; $NADPH_2 \longrightarrow NADP$

13-4 Because they are built up two carbon atoms at a time from acetyl CoA.

13-5

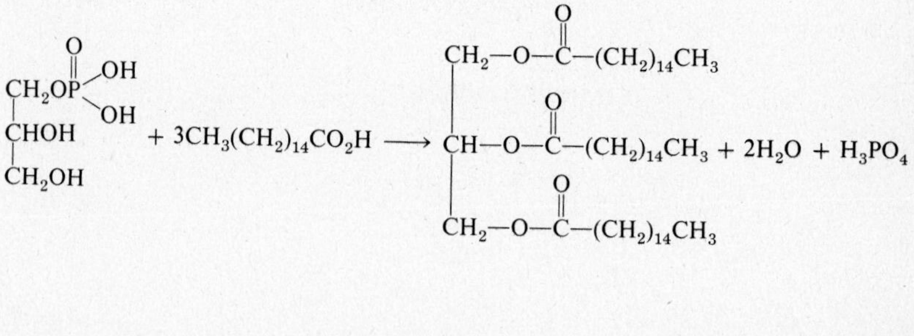

$$CH_2OP(=O)(OH)_2{-}CHOH{-}CH_2OH + 3CH_3(CH_2)_{14}CO_2H \longrightarrow \begin{array}{l} CH_2{-}O{-}C(=O){-}(CH_2)_{14}CH_3 \\ CH{-}O{-}C(=O){-}(CH_2)_{14}CH_3 \\ CH_2{-}O{-}C(=O){-}(CH_2)_{14}CH_3 \end{array} + 2H_2O + H_3PO_4$$

13-6

```
                                                          O
                                                          ‖
   H    OH                              H            O—P—OH
    \  /                                 \         /    |
     C——————┐                             C——————┐      OH
     |      |                             |      |
 H—C—OH     |     ATP         ADP      H—C—OH    |
     |      |        \_______↗            |      |
HO—C—H      O   ——————————————→      HO—C—H      O
     |      |                             |      |
HO—C—H      |                        HO—C—H      |
     |      |                             |      |
 H—C————————┘                          H—C———————┘
     |                                    |
    CH2OH                                CH2OH
```

```
                  O                                     O
                  ‖                                     ‖
   H         O—P—OH                     H          O—P—OH
    \       /    |                       \        /    |
     C——————┐    OH                       C——————┐     OH
     |      |                             |      |
 H—C—OH     |                          H—C—OH    |
     |      |                             |      |
HO—C—H      O        ——→             HO—C—H      O
     |      |                             |      |
HO—C—H      |                          H—C—OH    |
     |      |                             |      |
 H—C————————┘                          H—C———————┘
     |                                    |
    CH2OH                                CH2OH
```

13-7

```
      CO2H                             CO2H
       |                                |
   O—C—H     NAD        NADH2          C=O
       |        \_______↗               |
      CH2   ——————————————————→        CH2
       |                                |
      CO2H                             CO2H
```

13-8

$$\mathrm{CH_3\overset{O}{\overset{\|}{C}}CH_2CO_2H + HCO_3^- \longrightarrow CH_3\overset{O}{\overset{\|}{C}}CH_2CO_2^- + H_2O + CO_2}$$

$$\mathrm{CH_3\underset{OH}{\underset{|}{C}}HCH_2CO_2H + HCO_3^- \longrightarrow CH_3\underset{OH}{\underset{|}{C}}HCH_2CO_2^- + H_2O + CO_2}$$

CHAPTER FOURTEEN

14-1 (a) 1450 ml (b) 0.232
(c) 8.468 g (d) 2175 cal

14-2 Denaturation alters the secondary and tertiary structure of proteins. Digestion hydrolyzes the amide bonds, destroying the primary structure to produce amino acids.

14-3 (a) $\mathrm{HO_2CCH_2\underset{NH_2}{\underset{|}{C}}HCO_2H}$ (c) $\mathrm{C_6H_5{-}CH_2{-}\underset{NH_2}{\underset{|}{C}}H{-}CO_2H}$

(b) $\mathrm{HSCH_2\underset{NH_2}{\underset{|}{C}}HCO_2H}$

14-4 (a) $HOCH_2\underset{\substack{|\\NH_2}}{C}HCO_2H + HO_2CCH_2CH_2\underset{\substack{\|\\O}}{C}{-}CO_2H \longrightarrow$

$$HOCH_2\underset{\substack{\|\\O}}{C}{-}CO_2H + HO_2CCH_2CH_2\underset{\substack{|\\NH_2}}{C}HCO_2H$$

$$HO_2CCH_2CH_2\underset{\substack{|\\NH_2}}{C}HCO_2H \xrightarrow{NAD \quad NADH_2} HO_2CCH_2CH_2\underset{\substack{\|\\NH}}{C}{-}CO_2H$$

$$HO_2CCH_2CH_2\underset{\substack{\|\\NH}}{C}{-}CO_2H + H_2O \longrightarrow HO_2CCH_2CH_2\underset{\substack{\|\\O}}{C}CO_2H + NH_3$$

(b) $C_6H_5{-}CH_2\underset{\substack{|\\NH_2}}{C}HCO_2H + HO_2CCH_2CH_2\underset{\substack{\|\\O}}{C}CO_2H \longrightarrow$

$$C_6H_5{-}CH_2\underset{\substack{\|\\O}}{C}CO_2H + HO_2CCH_2CH_2\underset{\substack{|\\NH_2}}{C}HCO_2H$$

$$HO_2CCH_2CH_2\underset{\substack{|\\NH_2}}{C}HCO_2H \xrightarrow{NAD \quad NADH_2} HO_2CCH_2CH_2\underset{\substack{\|\\NH}}{C}CO_2H$$

$$HO_2CCH_2CH_2\underset{\substack{\|\\NH}}{C}CO_2H + H_2O \longrightarrow HO_2CCH_2CH_2\underset{\substack{\|\\O}}{C}CO_2H + NH_3$$

14-5 Glutamic acid is an acidic amino acid.
Glutamine is a neutral amino acid.

14-6 (a) Tymine has a methyl group while uracil does not.
(b) adenine

14-7 Cytosine, cytosine, adenine for the example given in the text. Other possible answers from information in Table 14-4 are cytosine, cytosine, cytosine; cytosine, cytosine, uracil; cytosine, cytosine, guanine.

14-8 (a) Methionine – aspartic acid – glutamic acid – tyrosine – lysine – glycine – isoleucine – threonine – cysteine – glycine

(b) Cysteine – threonine – serine – arginine – histidine – tryptophan – glycine – valine – serine – valine – lysine – tyrosine – proline – arginine

(c) Arginine – histidine – phenylalanine – lysine – proline – proline – glycine – glycine – phenylalanine – phenylalanine – proline – proline – phenylalanine – cysteine – leucine

14-9 $C_6H_5{-}CH_2\underset{\substack{|\\NH_2}}{C}HCO_2H + HO_2CCH_2CH_2\underset{\substack{\|\\O}}{C}CO_2H \longrightarrow$

$$C_6H_5{-}CH_2\underset{\substack{\|\\O}}{C}CO_2H + HO_2CCH_2CH_2\underset{\substack{|\\NH_2}}{C}HCO_2H$$

CHAPTER FIFTEEN

15-1 [structural formula: peptide chain with NH_2, CH_2CHC, $C{=}O$, $H{-}N$, $CHCH_2CH_2C{-}NH_2$, S—S bridge, $CH_2{-}C{-}N{-}C{-}CHCH_2C{-}NH_2$, H_2C, $CH_2{-}CH_2$, $CHC{-}N{-}CH{-}C{-}N{-}CH_2{-}C{-}NH_2$, $(CH_2)_3$, NH, C=NH, NH_2, OH]

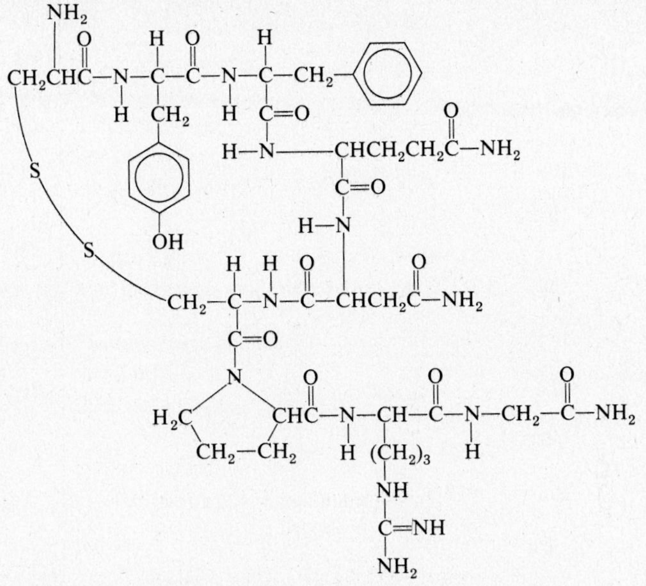

15-2 The drug could compete with the thyroid hormones for occupancy of enzymes' active sites. This would diminish thyroid activity.

15-3 (a) $C_{27}H_{45}O$; (b) the alcohol group at position 3

15-4 [structural formula: steroid ester with H_3C groups, side chain $CH{-}CH_2{-}CH_2{-}CH_2{-}CH(CH_3)_2$, and $R{-}C(=O){-}O{-}$ on ring A]

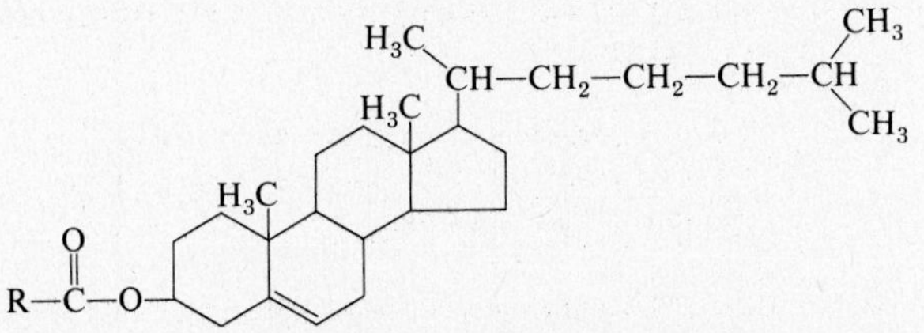

15-5 (a) diol = two alcohol groups; one on ring A, the other on ring D. (b) secondary. (c) Oxidation of a secondary alcohol yields a ketone.

15-6 To prevent ovulation while allowing regular menstruation.

15-7 Loss of bicarbonate ions reduces buffer capacity of the blood toward hydrogen ions.

15-8 $O \longrightarrow P \longrightarrow Q$ represents a chain reaction in which the predecessor must be formed before the next step in the chain can occur.Decreasing O production retards P production, which depends on the availability of O. Q depends on production of P and is thus indirectly dependent on O production.

15-9 Reduction of a ketone group on the ring to a secondary alcohol group.

15-10 Dehydration reaction (removal of H and OH) creating a double bond in the ring.

15-11 $(CH_3)_2C$ (ring) $-CH{=}CH{-}C(CH_3){=}CH{-}CH{=}CH{-}C(CH_3){=}CH{-}CHO$

H₃C CH₃ / CH₃ / CH₃ / O
—CH=CH—C=CH—CH=CH—C=CH—C—H

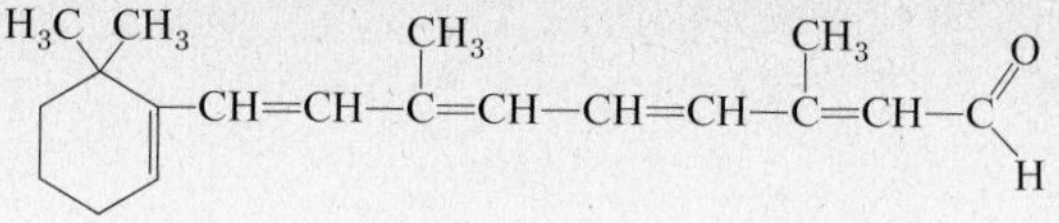

CHAPTER SIXTEEN

16-1 $Ca^{2+}(aq) + 2F^{-(aq)} \longrightarrow CaF_2$

in blood from NaF

16-2 (a) Water can migrate from region A. (b) Solute can migrate into region A.

16-3 (a) Raise urinary sodium ion concentration; (b) Increase aldosterone release.

16-4 Low vasopressin levels

16-5 Less than this value

16-6 Vasopressin inhibition would cause an increase in urinary volume and output. Thus, rest rooms are needed.

16-7 Step 4. Aldosterone regulates sodium reabsorption.

16-8 An increase in blood's hydrogen ion concentration causes an increase in urinary hydrogen ion concentration and therefore decreases urine's pH. (Hydrogen ion concentration is inversely related to pH.)

CHAPTER SEVENTEEN

17-1 Antihistamines mimic the structure of histamine. They bond to histamine receptor sites more strongly than histamine, thus acting as competitive inhibitors.

17-2 Free penicillins contain an acid group which can liberate hydrogen ions as skin irritants. The penicillin salts are formed by the neutralization of these acids by NaOH or KOH.

17-3 $Zn^{(2+)}[CH_2 = CH(CH_2)_8CO_2^{(-)}]_2$

17-4 (a) Both contain a tertiary N⊕; procaine hydrochloride contains an ester group, the other an amide.

(b) 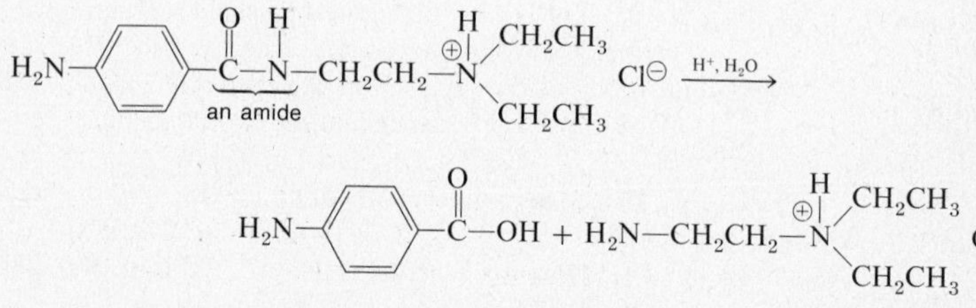 $H_2N{-}C_6H_4{-}\underbrace{C(=O){-}NH}_{\text{an amide}}{-}CH_2CH_2{-}N^{\oplus}H(CH_2CH_3)_2\ Cl^{\ominus} \xrightarrow{H^+,\ H_2O}$

$H_2N{-}C_6H_4{-}C(=O){-}OH + H_2N{-}CH_2CH_2{-}N^{\oplus}H(CH_2CH_3)_2\ Cl^{\ominus}$

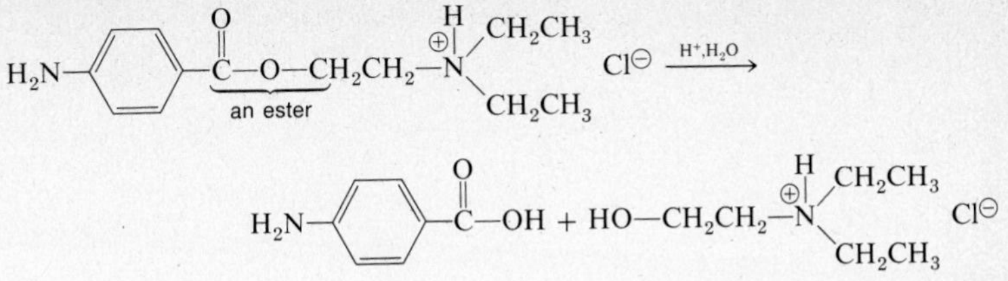

$$H_2N-C_6H_4-\overset{O}{\overset{\|}{C}}-O-CH_2CH_2-\overset{H}{\overset{|}{N^{\oplus}}}(CH_2CH_3)_2\ Cl^{\ominus} \xrightarrow{H^+,H_2O}$$

an ester

$$H_2N-C_6H_4-\overset{O}{\overset{\|}{C}}-OH + HO-CH_2CH_2-\overset{H}{\overset{|}{N^{\oplus}}}(CH_2CH_3)_2\ Cl^{\ominus}$$

17-5 (morphine structure with HO and OH, N–CH_3) + $CH_3-C(=O)-O-C(=O)-H_3C$ → (diacetylated structure with two $O-C(=O)-CH_3$ groups, N–CH_3)

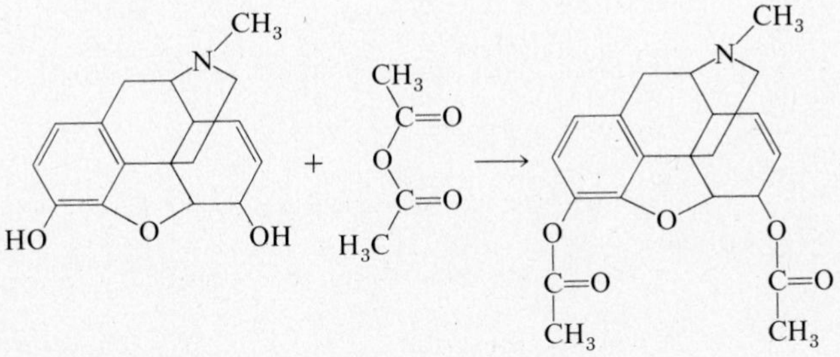

17-6

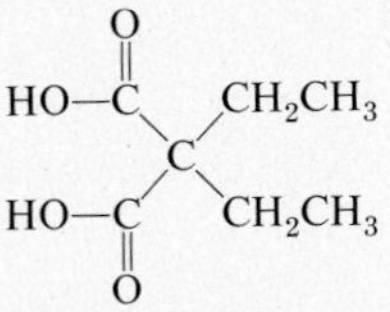

$$(HO-\overset{O}{\overset{\|}{C}})_2C(CH_2CH_3)_2$$

for barbital

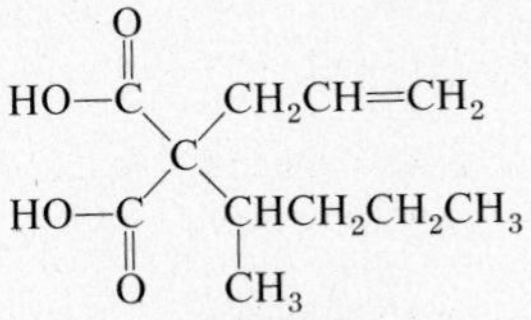

$$(HO-\overset{O}{\overset{\|}{C}})_2C(CH_2CH{=}CH_2)(\underset{CH_3}{\underset{|}{C}}HCH_2CH_2CH_3)$$

for secobarbital

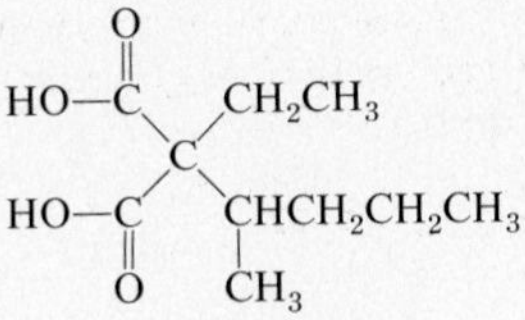

$$(HO-\overset{O}{\overset{\|}{C}})_2C(CH_2CH_3)(\underset{CH_3}{\underset{|}{C}}HCH_2CH_2CH_3)$$

for pentobarbital

$$(HO-\overset{O}{\overset{\|}{C}})_2C(CH_2CH_3)(C_6H_5)$$

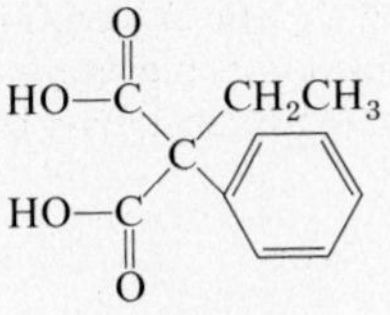

for phenobarbital

APPENDIX ONE

A1-1 (a) 3; (b) 0.2; (c) 3; (d) 24; (e) 0.33; (f) 0.2

A1-2 (a) 1080; (b) 300; (c) 157.5; (d) 0.84; (e) 1.16; (f) 0.3; (g) 0.3

A1-3 (a) 2×10^3; (b) 2×10^{-3}; (c) 5×10^{-1}; (d) 5×10^1; (e) 6.023×10^9; (f) 8.9×10^{-5}; (g) 6×10^0; (h) 2.75×10^5

A1-4 (a) 2.7×10^3; (b) 6.7×10^{-4}; (c) 8.4×10^7; (d) 8.4×10^{-5}; (c) 4.16×10^7; (f) 5.73×10^{-3}; (g) 3.75×10^4; (h) 1.4×10^{-2}

A1-5 (a) 8×10^5; (b) 9.2×10^{11}; (c) 2.8×10^{17}; (d) 8×10^{-10}; (e) 4.5×10^{-10}; (f) 5.6×10^4; (g) 7.5×10^{-4}; (h) 1.2×10^{-3}; (i) 3.7×10^2; (j) 8×10^1

A1-6 (a) 3×10^2; (b) 4×10^3; (c) 6×10^8; (d) 3×10^{-2}; (e) 3.1×10^{-5}; (f) 4×10^{-3}; (g) 3×10^{-5}; (h) 3×10^{10}; (i) 1.2×10^{24}; (j) 1.5×10^{-8}; (k) 3×10^3; (l) 4×10^{-18}; (m) 9×10^7; (n) 6×10^8

A1-7 (a) 2.24×10^{12}; (b) 1.2×10^{2}; (c) 1×10^{35}; (d) 1.6×10^{7}; (e) 1×10^{-3}; (f) 1.3×10^{48}; (g) 1.4×10^{-33}

A1-8 (a) 60 pferds; (b) 30 hargens; (c) 300 sperkles; (d) 7.5 hargens; (e) 300 sperkles

A1-9 (a) 3×10^{15} bacteria. (b) 1.01×10^{14} tablets. (c) 1.07×10^{3} lb; (d) 0.67 m.

APPENDIX TWO

A2-1 (a) in $1s$ orbital. (b) in $2s$ orbital. (c) in $3p$ orbital. (d) in $4d$ orbital. (e) in $3s$ orbital. (f) in $4f$ orbital.

A2-2 (a) 25; (b) 36

A2-3 Sc: $1s^22s^22p^63s^23p^64s^23d^1$, Zn: $1s^22s^22p^63s^23p^64s^23d^{10}$, Kr: $1s^22s^22p^63s^23p^64s^23d^{10}4p^6$

A2-4 (a) 1, 0, 0, $+\frac{1}{2}$; 1, 0, 0, $-\frac{1}{2}$; 2, 0, 0, $+\frac{1}{2}$; 2, 0, 0, $-\frac{1}{2}$; (b) 3, 0, 0, $-\frac{1}{2}$

A2-5 (a) $1s^22s^2$. (b) $1s^22s^22p^63s^2$. (c) $1s^22s^22p^4$. (d) $1s^22s^22p^63s^23p^3$. (e) $1s^22s^22p^63s^23p^4$.

index

a

b

c

d

e

f

g

h

i

k

l

m

n

o

p

q

r

s

t

u

V

W

Z